Research on Extraction Technologies, Analysis Methods and Functional Evaluation of Food Active Components

Research on Extraction Technologies, Analysis Methods and Functional Evaluation of Food Active Components

Editors

Yongliang Zhuang
Hu Hou
Zhipeng Yu
Shengbao Cai

Basel • Beijing • Wuhan • Barcelona • Belgrade • Novi Sad • Cluj • Manchester

Editors
Yongliang Zhuang
Kunming University of Science and Technology
Kunming
China

Hu Hou
Ocean University of China
Qingdao
China

Zhipeng Yu
Hainan University
Hainan
China

Shengbao Cai
Kunming University of Science and Technology
Kunming
China

Editorial Office
MDPI AG
Grosspeteranlage 5
4052 Basel, Switzerland

This is a reprint of articles from the Topic published online in the open access journals *Antioxidants* (ISSN 2076-3921), *Foods* (ISSN 2304-8158), *Molecules* (ISSN 1420-3049), *Nutrients* (ISSN 2072-6643), and *Processes* (ISSN 2227-9717) (available at: https://www.mdpi.com/topics/R2O674M3MT).

For citation purposes, cite each article independently as indicated on the article page online and as indicated below:

Lastname, A.A.; Lastname, B.B. Article Title. *Journal Name* **Year**, *Volume Number*, Page Range.

ISBN 978-3-7258-2047-4 (Hbk)
ISBN 978-3-7258-2048-1 (PDF)
doi.org/10.3390/books978-3-7258-2048-1

Contents

foods

MDPI

Review

Improving the Sustainability of Processing By-Products: Extraction and Recent Biological Activities of Collagen Peptides

Shumin Xu [1], Yuping Zhao [1], Wenshan Song [2], Chengpeng Zhang [1], Qiuting Wang [1], Ruimin Li [1], Yanyan Shen [1], Shunmin Gong [1], Mingbo Li [1] and Leilei Sun [1,*]

1 College of Life Science, Yantai University, No. 30, Qing Quan Road, Yantai 264005, China; lmm199791@163.com (S.X.); water15689@163.com (Y.Z.); cpzhang1997@163.com (C.Z.); wqt1981979471@163.com (Q.W.); 17861135192@163.com (R.L.); yyshenqx1999@163.com (Y.S.); gong1394@foxmail.com (S.G.); sllshd1991@163.com (M.L.)

2 Marine Biomedical Research Institute of Qingdao, No. 23, Hong Kong East Road, Qingdao 266073, China; songwsh@ouc.edu.cn

* Correspondence: llsun@ytu.edu.cn; Tel.: +86-0535-6902638

Abstract: Society and consumers are increasingly concerned about food safety and the sustainability of food production systems. A significant amount of by-products and discards are generated during the processing of aquatic animals, which still needs to be fully utilized by the food industry. The management and sustainable use of these resources are essential to avoiding environmental pollution and resource waste. These by-products are rich in biologically active proteins, which can be converted into peptides by enzymatic hydrolysis or fermentation treatment. Therefore, exploring the extraction of collagen peptides from these by-products using an enzymatic hydrolysis technology has attracted a wide range of attention from numerous researchers. Collagen peptides have been found to possess multiple biological activities, including antioxidant, anticancer, antitumor, hypotensive, hypoglycemic, and anti-inflammatory properties. These properties can enhance the physiological functions of organisms and make collagen peptides useful as ingredients in food, pharmaceuticals, or cosmetics. This paper reviews the general methods for extracting collagen peptides from various processing by-products of aquatic animals, including fish skin, scales, bones, and offal. It also summarizes the functional activities of collagen peptides as well as their applications.

Keywords: by-products of aquatic animals; collagen peptides; bioactivity; industrial application

Citation: Xu, S.; Zhao, Y.; Song, W.; Zhang, C.; Wang, Q.; Li, R.; Shen, Y.; Gong, S.; Li, M.; Sun, L. Improving the Sustainability of Processing By-Products: Extraction and Recent Biological Activities of Collagen Peptides. *Foods* **2023**, *12*, 1965. https://doi.org/10.3390/foods12101965

Academic Editor: Adrián J. Hernández

Received: 12 April 2023
Revised: 2 May 2023
Accepted: 9 May 2023
Published: 12 May 2023

1. Introduction

The demand for fish and its products grows as the world population increases and human consumption behavior changes [1]. These processing by-products also put a lot of pressure on the environment if they are discarded and buried. However, these by-products are rich in protein, mainly collagen, and can be well processed into valuable products. The Food and Agriculture Organization of the United Nations (FAO) has recently published the 2022 edition of The State of World Fisheries and Aquaculture. According to the report, the total production of fisheries and aquaculture reached a new peak of 214 million tons in 2020, with 178 million tons of aquatic animals being produced. Fish by-products make up around 70% of processed fish products and typically comprise of fish heads (9–12% of total fish weight), offal (12–18%), skin (1–3%), bones (9–15%), and scales (approximately 5%). The Marine Raw Materials Organization reports that some fish by-products can be utilized to produce fish oil and fishmeal (Figure 1), but others are discarded as waste, resulting in economic losses and environmental pollution. However, these by-products are rich in protein, primarily collagen, and can be well processed into valuable products.

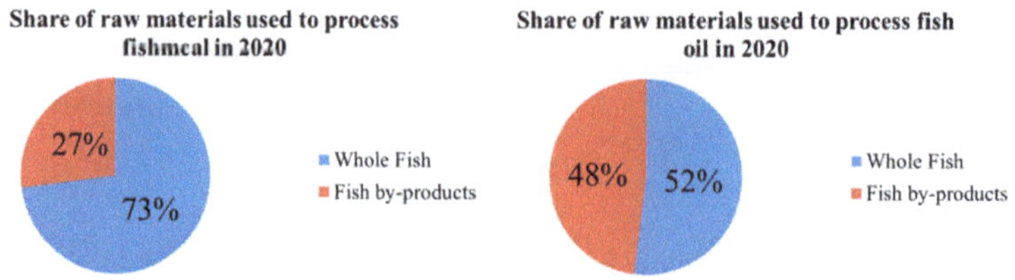

Figure 1. Share of raw materials used to process fishmeal and fish oil in 2020.

Collagen is a major extracellular protein, accounting for about 30% of total protein, and is mainly found in various fibrous tissues such as skin, bone, blood vessels, tendons, muscles, cornea, and ligaments [2,3]. To date, 28 types of collagens have been identified, with type I collagen being the most abundant animal protein (90%) [4]. A repetitive amino acid sequence at the molecular level characterizes collagen. The sequence consists mainly of $(Gly\text{-}X\text{-}Y)_n$, where glycine (Gly) makes up about 1/3 of the collagen peptide chain, X,Y are arbitrary amino acids other than glycine. Typically, the classical composition is proline (Pro) at the X-position and 4-hydroxyproline (Hyp) at the Y-position [3,5]. Collagen has unique physical properties, including homogeneity, tensile strength, flexibility, various bioactive functions, and controlled enzymatic properties [6]. Collagen plays an important role in various biological processes, such as promoting cell proliferation, differentiation, and migration [7]. It also regulates and mediates coagulation while facilitating wound tissue repair and regeneration [8,9]. Degradation of collagen can lead to various diseases and syndromes, such as osteoarthritis and joint pain, hair abnormalities, skin laxity, and muscle weakness.

Many studies have found that collagen has a certain nutritional value. However, realizing its efficacy and value is difficult because of its considerable molecular weight and strong bonding, which cannot be quickly digested, absorbed, and utilized by humans or animals. The collagen peptide is the product of enzymatic digestion of collagen or gelatin and has a noticeable improvement in digestion and absorption, nutrition, and functional effects compared with collagen [10,11]. The market for collagen peptides has seen significant growth both domestically and internationally. The bioactive functions of collagen peptides are increasingly widely recognized, including antioxidative [12–15], anti-fatigue [16–18], hypoglycemic activity [19,20], hypolipidemic activity [21,22], antimicrobial attributes [23,24], and other activities [25,26]. The collagen peptides market is expected to grow at a Compound Annual Growth Rate (CAGR) of 5.42% to exceed USD 22,622 million by 2027 [27]. The utilization of collagen peptides is rising as dietary supplements in food and beverages and as bioactive ingredients in cosmetics [28,29], nutraceuticals [30], and pharmaceuticals [31,32].

The objectives of this review are (1) to examine current approaches on the preparation of collagen peptides from food processing by-products, focusing on their advantages and disadvantages, (2) to summarize the biological activity of collagen peptides, and (3) to discuss the applications and future challenges related to collagen peptides in the food and cosmetic industries.

2. Processing Approaches for the Generation of Collagen Peptides

There are three primary ways to prepare collagen peptides (Figure 2). The first method involves obtaining bioactive peptides with various functions by enzymatically digesting and hydrolyzing collagen (Figure 3) [15]. The second method requires extracellular enzymes produced by microbial fermentation to prepare collagen peptides [23,33]. Finally, the third chemical hydrolysis method is to prepare collagen peptide by acid-base catalytic decomposition of peptide bond, which is a method to improve the functional properties of protein.

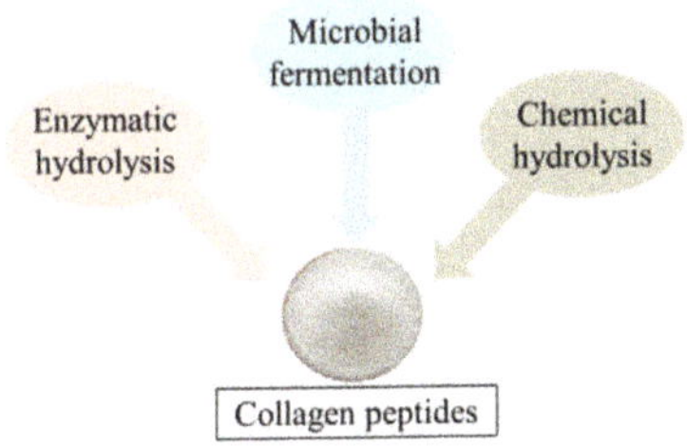

Figure 2. Diagram of the different methods used for the release of collagen peptides.

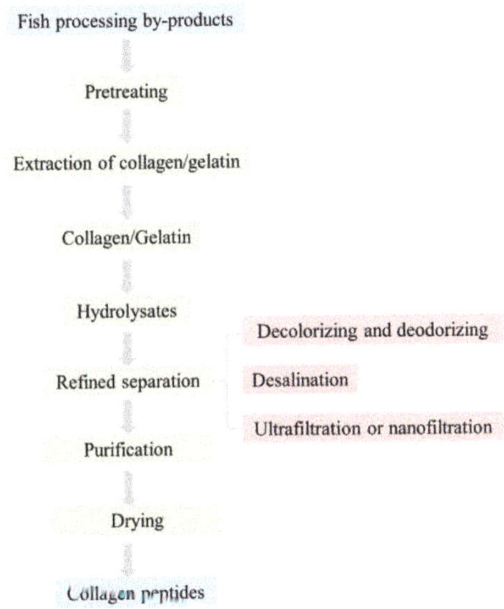

Figure 3. Schematic diagram for preparation of collagen peptides.

In the long chain of amino acids of proteins, peptide fragments are not biologically active. Only after enzymatic digestion can many biologically active peptides be released and function. Proteolysis is the most commonly used method for the preparation of bioactive peptides. Enzymatic digestion refers to using specific enzymes to carry out the corresponding chemical reaction to eliminate some particular structures. Enzymes are commonly classified as either exonucleases or endonucleases. Exonucleases, such as flavourzyme [34], aminopeptidase, and carboxypeptidase, target the peptide bonds at the end of the peptide chain. On the other hand, endonucleases like alkaline proteases [14,15,35], trypsin [31], and papain [36,37], act on specific regions within the peptide chain. Different proteases have different cleavage sites and degrees of hydrolysis (Table 1) [19]. More studies have been reported on the preparation of collagen peptides using ultrasound [37–39] assisted enzymatic digestion, resulting in improved yields and activity. It was found that microwave assisted or steam jet treatment under appropriate conditions could improve the yield and activity of bovine bone collagen peptides. Zhang et al. [40] used steam explosion to treat bovine bone and found that the protein recovery was improved, and the molecular weight of the collagen peptides was significantly reduced. The extracted collagen peptides still had a good calcium binding capacity (44.7 µg/mg) and osteoblast proliferative activity (126.7%) without changing the amino acid composition. Perhaps this method could also be applied to fish bones, fish heads, and other by-products to improve protein recovery.

Table 1. The active position of some common proteases.

Protease	Cutting Site
Chymotrypsin	Tyr-, Trp-, Phe-, Leu-
Alcalase	Hydrolysis of the carboxyl terminal peptide bond of aromatic amino acid residues or hydrophobic amino acid residues
Pepsin	Hydrolysis of Phe residue, Leu residue carboxyl terminal peptide bond
Trypsin	Hydrolyzing the carboxyl terminal peptide bond of Arg residue and Lys residue
Neutral protease	Hydrolysis of the C-terminal peptide bonds of aromatic amino acid residues such as Tyr, Try and Phe
Flavor protease	Mainly exonucleases that catalyse the hydrolysis of hydrophobic amino acids at the ends of brain chains
Papain	Broad hydrolysis of peptide bonds with preference for hydrophobic side chain amino acids

Microbial fermentation is a process that uses a variety of enzymes synthesized and secreted during the growth of microorganisms to degrade proteins and produce specific small-molecule active peptides [41]. The degree of enzymatic digestion depends on the protein source, strain, fermentation temperature, fermentation time, and pH [42]. Compared to enzymatic hydrolysis, some enzyme-producing strains may be toxic to the organism, making purifying microbial fermentation products more difficult. Although this method is cost-effective and productive for protein preparation, it is yet to be widely used in large-scale production due to technical limitations [41].

Acid extraction mainly uses low ionic concentrations of acidic conditions to impregnate the raw material, thus destroying the intermolecular salt bonds and Schiff bases while causing fiber swelling and dissolution. The primary solvents used are hydrochloric or sulfurous acid, phosphoric acid, sulfuric acid, acetic acid, citric acid, and formic acid [43]. The standard treatment agents used for alkali collagen extraction are lime, sodium hydroxide, sodium carbonate, etc. Since the alkali extraction is easy to cause peptide bond hydrolysis, the molecular weight of the obtained hydrolysis product is relatively low. However, the violent reaction makes it easy to cause environmental pollution and destroy the product's nutrients. Although the acid-base method is simple, the reaction environment is more extreme, which is not conducive to maintaining biological activity. The acidic or alkaline waste solution produced after extraction must be treated appropriately. Otherwise, some harmful by-products will be made and will cause environmental pollution [43,44].

As can be seen from Table 2, the most commonly used method for the preparation of collagen peptides is enzymatic hydrolysis, including single enzyme hydrolysis and multiple enzyme complex hydrolysis.

Table 2. Source, preparation methods, and biological activity of collagen peptides.

Source	Preparation Method	Biological Activity	References
Tuna skin	Enzymatic hydrolysis (Alcalase, Neutrase and Savinase)	Antioxidant activity	[45]
Asian sea bass skin	Enzymatic hydrolysis (Papain)	Wound-healing and antioxidant activities	[46]
Sturgeon fish skin	Enzymatic hydrolysis (Trypsin, Alcalase, Neutrase, Flavourzyme, Pepsin, Papain)	Antimicrobial activity	[24]
Swim bladders of giant croaker	Enzymatic hydrolysis (Neutral protease)	Antioxidant activity	[47]
Redlip croaker scales	Enzymatic hydrolysis (Neutral protease)	Antioxidant activity	[48]
Walleye pollock skin	Enzymatic hydrolysis (Flavourzyme and Alcalase)	Attenuated obesity and modulated gut microbiota	[34]
Yellowfin tuna skin	Enzymatic hydrolysis (Alcalase)	Antioxidant activity	[49]
Silver carp skin	Enzymatic hydrolysis (Alcalase)	Antiplatelet activity	[50]

Table 2. *Cont.*

Source	Preparation Method	Biological Activity	References
Silver carp skin	Enzymatic hydrolysis (Alkaline protease, Trypsin, Neutral protease and a mixture of alkaline proteaseand trypsin)	Anti-photoaging activity	[51]
Asian bullfrog skin	Ultrasound-assisted enzymatic Hydrolysis (papain)	Antioxidant activity	[37]
Mixed by-products from various fish species	Enzymatic hydrolysis (Alcalasa)	Antioxidant and antimicrobial activities	[52]
Squid skin	Enzymatic hydrolysis (Papain, Alkaline protease, Neutral protease, Acid protease and Compound protease)	Cryoprotective effect	[53]
Skipjack tuna skins	Enzymatic hydrolysis (Trypsin, Neutrase, Papain, Pepsin, and Alcalase)	Antioxidant activity	[54]
Tilapia skin	Enzymatic hydrolysis (Alcalase, Protamex, Flavourzyme, Neutrase, Papain, Bromelain, and Trypsin)	DPP-IV inhibitory activity	[55]
Tilapia skin	Enzymatic hydrolysis (Alcalase)	ACE-inhibitory activity	[56]
Fresh water fish heads	Microbial fermentation (*E. faecium* NCIM5335, *P. acidilactici* FD3, *P. acidilactici* NCIM5368)	Antioxidant activity	[57]
Salmon by-products with skin	Bacterial extracellular protease	Antioxidant and anti-freezing activities	[33]
Bigeye tuna skin	Subcritical water hydrolysis	Antioxidant activity	[58]
Bigeye tuna skin	Catalyst-assisted subcritical water hydrolysis	Antioxidant and antimicrobial activities	[59]

In addition, fish-derived collagen peptides often suffer from a heavy fishy taste. The methods of deodorization can be classified as sensory masking, physical, chemical, biological, or a combination of several methods [60]. Han Na Cho et al. [61] used solid-phase microextraction (SPME) and solvent-assisted flavor evaporation (SAFE) combined with gas chromatography-mass spectrometry (GC-MS-O) to identify odour compounds from marine collagen peptides and gelatin, exploring effective removal methods. The current study showed that *Lactobacillus plantarum* fermentation was the most effective in removing the odour from the collagen peptides.

3. Biological Characteristics of Collagen Peptide

Collagen peptides are a class of small molecule amino acid sequences with various biological functions. Hydrolyzing the protein with proteinases disrupts its primary structure, even though the active peptide does not exhibit biological activity in the intact protein structure. The organism benefits from the release of the free peptide for its health. Bioactive peptides have superior absorption mechanisms and functional properties that cannot be compared to amino acids [62]. Collagen peptides have been found to exhibit various biological activities, such as hypotensive and blood sugar-regulating effects, antioxidant properties, anti-inflammatory and immunomodulatory actions, anti-thrombotic activity, calcium absorption promotion, antibacterial effects, and skin-condition improving properties [56].

3.1. Anti-Atherosclerotic Activity

Atherosclerosis, a lipid-driven chronic inflammatory disease that forms plaques in medium and large arteries, is a significant cause of ischemic heart disease and stroke.

It occurs mainly in the bends or branches of these arteries, leading to the blockage or rupture of the vessels. As such, it poses one of the greatest threats to human health through cardiovascular disease [63]. A recent study published in a sub-publication of The Lancet assessed the prevalence of carotid atherosclerosis and risk factors in 21 national and regional populations worldwide: nearly 2 billion people will have carotid atherosclerosis in 2021. Atherosclerosis has seriously threatened the human quality of life, health, and longevity [64]. Therefore, effective treatment of atherosclerotic plaques is imperative to slow the disease and save patients' lives.

More information is needed on the anti-atherosclerosis collagen peptides from a by-product of fish processing. It was found that collagen peptides extracted from the skin of Atlantic salmon inhibited lipid plaque formation in the aortic arch and intimal thickening in the thoracic aorta, showing excellent anti-atherosclerotic effects. Although a mixture of collagen peptides extracted from salmon skin, with a molecular weight ranging from 2 to 30 KDa and composed mainly of glutamine (37%) and myostatin (39%), did not significantly improve carotid atherosclerosis or cardiac structure in hypertensive patients during the short-term intervention, it was found to impact systolic cardiac function positively [35].

Therefore, anti-atherosclerosis is one area of research exploration through collagen peptides prepared from aquatic animals processing by-products. It is challenging to find other peptides that reduce the risk of atherosclerosis through bioinformatics.

3.2. *Anticoagulant Activity*

Excessive blood clotting, known as pathological thrombosis, can lead to serious health issues such as pulmonary infarction, cerebral thrombosis, retinal arteriovenous obstruction, myocardial infarction, and embolisms of the extremities. These conditions have become a leading cause of morbidity and mortality in China. While blood clotting is a necessary physiological defense process of the body, it must be regulated to prevent these harmful outcomes [65]. Surveys show that in recent years, 1–3 out of every 1000 people suffer from thrombosis-like diseases, which seriously endanger human health. Commonly used anticoagulant substances in clinical practice, such as heparin, oral coumarin, streptokinase, and urokinase, can cause side effects like thrombocytopenia and subcutaneous purpuric bleeding [66]. Peptides and some small molecule peptides have been studied for their anticoagulant activity, such as Hirulog-1 [67], Argatroban [68], etc., but few peptides have been studied for their anticoagulant activity in fish collagen peptides.

Anticoagulant activity has been found in oligopeptides isolated from the skin of silver carp [52,69], Yellowfin sole (*Limanda aspera*) frame [70], and mackerel skin [71]. All isolated peptides were tested in vitro and showed the ability to prolong activated partial thromboplastin in a dose-dependent manner. The three oligopeptides of silver carp skin, which are not inactivated by gastrointestinal digestion and are absorbed by Caco-2, inhibited adenosine diphosphate and thrombin-induced platelet aggregation in a dose-dependent manner, however had no effect on platelet release. Although weaker than Argatroban, it also significantly prolonged APTT, TT and PT. In addition, the 60 mg/mL tripeptide showed significantly higher ($p < 0.05$) platelet aggregation inhibitory activity (~30%), similar to the synthetic antithrombotic compounds like aspirin and indomethacin [69].

Peptides with an anticoagulant activity have molecular weights below 3.5 KDa, and most are below 2.5 KDa. Therefore, it seems useful to obtain low molecular weight peptides with a good anticoagulant activity by ultrafiltration, microfiltration, or nanofiltration [72]. Although the anticoagulant activity of collagen peptides, a by-product of fish processing, has not been studied much, those peptides that have been isolated have demonstrated satisfactory results. Therefore, further studies are necessary to find more anticoagulant peptides from fish and to assess whether they show the same side effects as warfarin. In conclusion, based on adequate studies, collagen peptides could be a good natural source of alternatives to anticoagulant drugs.

3.3. Antidiabetic Activity

Diabetes mellitus (DM) is a severe chronic metabolic disease characterized by persistent elevation of collective blood glucose levels due to defective insulin secretion and resistance [20]. DM is mainly marked by high blood glucose, with common symptoms such as excessive drinking, polyuria, polyphagia, and lethargy. Long-term persistent high blood glucose will lead to chronic complications, including heart, brain, liver, kidney, spleen, retina, nervous system, skeletal muscle and blood vessels, and other tissues [73]. The dramatic increase in the incidence of diabetes has made it an important global issue. Studies have shown that DM is characterized by long duration, difficulty to cure, and high disability rate, and most patients require long-term drug therapy. There are a wide variety of drugs for the clinical treatment of diabetes, including insulin promoters, insulin sensitizers, biguanides, alpha-glucosidase inhibitors, etc. Clinical diabetes medication has various types, but they may have issues with their effectiveness, how they work, and their adverse side effects. Therefore, searching for new, safe, natural hypoglycemic agents to replace synthetic hypoglycemic agents has become a growing need [74]. The most described antidiabetic peptides are derived from milk and soy protein; however, several recent studies have shown that collagen peptides can be effective in alleviating syndromes associated with diabetes [75]. For example, Zhang et al. [76] reported that a protein hydrolysate (TCP) isolated from the skin of Nile tilapia (*Oreochromis niloticus*) improved the symptoms of "excessive drinking and eating" in diabetic mice, and that a high dose of TCP produced better results than metformin. Similarly, Harnedy et al. [77] reported that protein hydrolysates isolated from Atlantic salmon (*Salmo salar*) skin gelatin and excipients could exhibit in vitro anti-diabetic activity by inducing the release of glucagon-like peptide 1 (GLP-1). Some of the targets of diabetes therapy are glucosidase and alpha-amylase, and therefore hypoglycemic activity is expressed in vitro as the inhibition of α-glucosidase or α-amylase. Most of the antidiabetic peptides have molecular weights below 1 KDa. Collagen peptides produced by collagenase hydrolysis, as assessed by the inhibition of α-amylase, showed antidiabetic activity of 80.45% for CP-5 compared to 75.81% and 71.17% for CP-25 and CP-50, respectively, at a concentration of 1 mg/mL [78]. In a double-blind randomized trial evaluating the effectiveness of collagen peptide (CPT) as a nutritional supplement in patients with T2DM, it was shown that taking 5 g of CPT over the three-month study period significantly reduced fasting blood glucose and glycated hemoglobin in the subjects [79]. The above studies clearly reveal the potential application of fish collagen peptides in the treatment of diabetes. Nevertheless, more clinical trials are still needed to use collagen peptides more rationally in future biomedicine.

3.4. Anti-Inflammatory Activity

Inflammation is an immune stress response, a complex and highly regulated defense response of the immune system to pathogenic microbial infections, physical or chemical stimuli, etc. It manifests primarily as redness, swelling, heat, and pain, killing pathogens, limiting illness, and repairing the damage [72]. However, the targeted destruction and assisted repair stage needs to be corrected. In that case, inflammation can cause continuous tissue damage by leukocytes, lymphocytes, and collagen, which can be harmful to the organism and, in severe cases, life-threatening. Anti-inflammatory drugs including steroids, non-steroidal anti-inflammatory drugs, and monoclonal antibodies can be used to combat the inflammatory response [80]. However, many drug treatments are not sufficient to control chronic inflammatory responses and long-term treatment may lead to adverse events [81]. Therefore, the development of natural products with anti-inflammatory properties is essential as for alternative and safe dietary supplements.

A number of collagen peptides derived from fish have now been shown to have anti-inflammatory activity. Most of the anti-inflammatory peptides are derived from algae and molluscs, but they have also been identified in hydrolysates from tuna cooking juice by-products [82], Salmo salar skin [81], and unicorn leatherjacket (*Aluterus monoceros*) [78]. The results showed that Salmo salar skin collagen hydrolysate had anti-inflammatory effects on LPS-induced RAW264.7 cells, among which the QA peptide which showed strong anti-inflammatory activity with an IC_{50} value of 849.3 μM against the LPS-induced NO secretion in macrophages. Similarly, Kumar et al. [78] conducted in vitro anti-inflammatory assays on human red blood cell (HRBC) membranes and found a slight protective effect on HPBC with increasing collagen peptide concentrations, with a maximum membrane protection of 4.01%, 3.66% and 3.27% for CP-5, CP-25, and CP-50 expression at 0.2 mg/mL concentrations, respectively.

3.5. Anti-Cancer Activity

With increasing human life expectancy, lifestyle changes, and environmental pollution, cancer has surpassed cardiovascular disease as the second leading cause of human death in incidence and mortality [83]. The American Association for Cancer Research (AACR) recently released the 2022 Progress Against Cancer report at the end of July this year. The report shows that by 2040, the total number of cancer patients worldwide will reach about 28 million, and about 16.2 million patients will die. The FDA approved eight new anti-cancer therapies and expanded the approval of 10 existing ones to treat additional types of cancers between August 2021 and July 2022 [84]. However, the side effects of the drug can cause damage to several areas, including the gastrointestinal tract, skin, and nervous system. Dietary interventions have a unique advantage over anti-cancer drugs in treating cancer or alleviating the suffering caused by cancer because of their low side effects and low cost. Therefore, the development of functional foods with an anti-cancer activity is imperative. Developing healthy food with anti-allergic effects is socially necessary to compensate for the shortage of traditional drug therapy.

In the last few years collagen peptides have also been shown to exhibit anti-cancer activity, with several in vitro studies showing potential in different cancer types [85]. Peptides isolated from the skin of Unicorn leatherjacket fish (*Aluterus monoceros*) showed inhibitory effects on COLO320 cells in vitro [78]. Baehaki et al. [86] hydrolyzed Milk Fish (*Chanos chanos*) skins by collagenase extracted from Bacillus licheniformis F11.4 w and showed the highest activity against HeLa cells at 60 min and the greatest activity against HCT-166 cells at 30 min. The by-product of Barred mackerel was hydrolyzed with actinidin followed by alcalase to produce a 3~10 KDa peptide. At a concentration of 1 mg/mL, the antioxidant and anti-cancer activities were 67.75% and 96.93%, respectively [87]. Yaghoubzadeh et al. [88] separated bioactive peptides by membrane ultrafiltration using alcalase (HA) and flavourzyme (HF) and hydrolyzing rainbow trout skin. The results showed that collagen peptides with a molecular weight less than 3 KDa had the highest inhibitory concentration. The above studies clearly demonstrate that collagen peptides exhibit significant anti-cancer activity in in vitro assays. However, more in-depth studies are needed to investigate the relationship between the chemical structure of the peptides and their anti-cancer activity through reliable in vivo studies to elucidate the health benefits of collagen peptides.

3.6. Antihypertensive Activity

Hypertension refers to a clinical syndrome characterized by increased arterial blood pressure (systolic and diastolic) in the body circulation (systolic blood pressure ≥140 mm Hg and diastolic blood pressure ≥90 mm Hg), which may increase the incidence and severity of cardiovascular disease, peripheral arterial disease, heart failure, and stroke with impaired kidney function [89,90]. More than one sixth of the blood pressure-lowering drugs used worldwide are angiotensin I-converting enzyme (ACE) inhibitors, such as captopril, because ACE is the key enzyme in the process of raising blood pressure in humans [91]. Drugs used to control high blood pressure can cause side effects; for this reason, much research has focused on finding alternatives to control or prevent this disease through diet [92]. There is an urgent need for new ACE inhibitors to treat hypertension, and inhibitors from natural sources are promising medicinal and edible supplements due to their efficacy and safety. In the last few years, several researchers have been involved in the development of collagen peptides as ACE inhibitors. Lee et al. [93] found that collagen peptides prepared by trypsin hydrolysis of Chum salmon (*Oncorhynchus keta*) skin exhibited the strongest ACE inhibitory activity. The IC_{50} for ACE inhibitory activity was 18.7 μM for the small molecule heptapeptide Gly-Leu-Pro-Leu-Asn-Leu-Pro with a molecular weight of 770 Da. In vivo experiments demonstrated that Gly-Leu-Pro was comparable to captopril in terms of its antihypertensive effect within 6 h of administration. Since the 1990s, several ACE-inhibiting peptides have been studied in different natural sources such as bovine whey, mushrooms, and walnuts, and most of the anti-hypertensive peptides were found to have molecular weights below 1.5 KDa [72]. Thus, fish processing by-products appear to have great potential as a source of antihypertensive peptides, and notably, collagen peptides deserve higher attention due to their ACE inhibition activity.

3.7. Antimicrobial Activity

Excessive use of antibiotics has created a problem by increasing the prevalence of drug-resistant bacteria, a significant public health concern. To address this issue, researchers are exploring biological antimicrobial agents that have broad activity and can be used in food or supplements due to the strict regulations on chemical preservatives [94]. As collagen peptides exhibit hydrophobicity, they can bind to negatively charged surfaces due to electrostatic interactions, penetrate and disrupt membrane structures, leading to the death of bacteria, fungi, parasites, and viruses [95]. Unlike the single-target bactericidal principle of conventional antibiotics, the antimicrobial properties of the peptides are demonstrated with the ability to destroy pathogens by disrupting multiple targets, which can significantly reduce the emergence of resistant bacteria. Their resistance of a wide range of bacteria makes them one of the best alternatives to combination antibiotics [96,97].

Several studies have shown that enzymatic hydrolysis of fish processing by-products can produce peptides with antimicrobial activity. For instance, antibacterial peptide fractions isolated from Atlantic mackerel (*Scomber scombrus*) by-products were reported to inhibit the growth of Gram-positive and Gram-negative bacteria [98]. Maryam Atef et al. [24] also demonstrated that collagen hydrolysates produced by neutrase and papain showed the most effective inhibitory activity against Gram-negative bacteria (*Salmonella strains*), providing enhanced potential for products against food-borne pathogens. Wald et al. [99] used pepsin to prepare antimicrobial hydrolysates from rainbow trout by-products and found that the degree of hydrolysis (DH%) had a considerable effect on the antimicrobial activity, with minimum inhibitory concentrations of 2 mg/mL and 5 mg/mL at 30% DH against the fish farming bacteria *Flavobacterium psychrophilum* and *Renibacterium salmoninarum*. The antimicrobial peptides mainly prolong the lag period of bacterial growth. Furthermore, the wide range of effects suggests that collagen peptides prepared from hydrolyzed fish processing by-products are highly applicable as preservatives in food processing or in fish farming to protect fish health.

3.8. Antioxidant Activity

Oxidative stress occurs when there is an imbalance between the production of reactive oxygen species (ROS) and the natural antioxidant capacity of cells. This disruption can lead to abnormal cellular and molecular function caused by a combination of ROS and reactive nitrogen radicals. Excessive production of reactive oxygen species (ROS) in the body can be caused by aging, radiation, alcohol, or drugs. This leads to intracellular proteins, lipids, and DNA damage and triggers apoptosis [100]. Over time, this may result in degenerative diseases associated with aging, like diabetes, cancer, cataractogeneses, Parkinson's and Alzheimer's disease. In addition, ROS-induced lipid peroxidation is one of the significant causes of lipid degradation in food and cosmetic matrices [101,102]. The scavenging of ROS is a crucial way to lower health risks associated with oxidation products (such as MDA, carbonylated proteins, etc.). Antioxidants are used to inhibit these free radical-producing compounds. Various antioxidants, such as thiols or vitamin C, vitamin E, uric acid, and lipoic acid, can be used to terminate these oxidation reactions. Butylated hydroxytoluene (BHT), butylated hydroxyanisole (BHA), tert-butylhydroquinone (TBHQ), and propyl hemilactate (PG) are the most commonly used synthetic antioxidants in the food and pharmaceutical industries. Although these synthetic antioxidants have shown greater antioxidant activity than natural antioxidants such as alpha-tocopherol and ascorbic acid, their use has begun to be limited due to the DNA damage and toxicity induced by these compounds [103]. Therefore, searching for antioxidants that significantly inhibit oxidative reactions has become an increasing research focus.

Natural antioxidant peptides are being extracted from different animal or plant species. Several marine species such as mussels, fish, and algae are sources of antioxidant peptides produced by enzymatic hydrolysis [104]. Collagen peptides with antioxidant activity have been developed from a variety of fish processing by-products, such as grass carp scales [14], squid (*Dosidicus gigas*) by-products [105], salmon scales [106], hammerhead shark skin [107], redlip croaker (*Pseudosciaena polyactis*) scales [50], skipjack tuna (*Katsuwonus pelamis*) [108], and lamuru (*Caranx ignobilis*) fishbone [109]. The two most common assays for antioxidant activity of peptides are the scavenging of DPPH radicals and the scavenging of ABTS radicals [72]. These studies found that collagen peptides have better DPPH and ABTS free radical scavenging ability. These studies also found a strong correlation between the amino acid composition and the molecular weight of the peptides and the level of antioxidant activity. In comparison, Wang et al. [50] found that collagen hydrolysate (RSCH) from red fish scales prepared using neutral protease showed the highest hydrolysis (21.36 $\pm$ 1.18%) and 2,2-diphenyl-1-picrylhydrazyl (DPPH·) radical scavenging activity (30.97 $\pm$ 1.56%) among the six hydrolysis products. Subsequently, the amino acid sequences GPEGPMGLE, EGPFGPEG, and GFIGPTE were identified as the most active against DPPH radicals (EC_{50} 0.59, 0.37 and 0.45 mg/mL), hydroxyl radicals (EC_{50} 0.45, 0.33 and 0.32 mg/mL) and superoxide anion radicals (EC_{50} 0.62, 0.47 and 0.74 mg/mL) showed the strongest scavenging activity.

3.9. Skin Improvement

As one of the principal organs of the body and the main barrier, the skin plays a vital role in protecting tissue structures, regulating temperature, and protecting the body from external stimuli such as ultraviolet (UV) light, heat, toxins, and infections [11,110]. Over time, the skin undergoes a natural aging process, from becoming dry and lacking in elasticity, to developing many wrinkles. The mechanism of skin aging is relatively complex, divided into external factors such as UV radiation and pollution and internal factors such as cell metabolism and hormones that lead to the skin's loss of elastin and collagen [11,111]. Therefore, the use of molecules that can act on these characteristics may be an effective way to improve skin conditions. With the world's population aging and increasing concerns about skin health, there is a growing interest in active substances that can improve skin. Collagen peptides have attracted much attention due to their excellent biocompatibility, easy digestibility, and transdermal absorption properties. Studies have shown that collagen peptides extracted from fish skin and fish bone effectively improve skin wrinkles and fine lines and brighten skin tone. Much literature confirms the collagen peptides' physiological activity for anti-aging and skin photoaging prevention. Clinical trials have shown an overall significant increase in skin elasticity in subjects taking the test product. Histological analysis of skin biopsies showed positive changes in skin structure in the test product group, with reduced daylight elasticity gain and improved collagen fibrous tissue [30]. In his study, Wang et al. [112] found that CHs were prepared from Nile tilapia skin by simulating gastrointestinal digestion, and that visual appearance, tissue structure, and matrix homeostasis were significantly improved in mice skin by feeding them to 9-month-old mice for 180 days. In addition, oxidative stress was reduced by increasing the activity of antioxidant enzymes, which was important for the improvement of the skin.

Collagen hydrolysates/peptides derived from fish processing by-products appear to be an efficient and promising functional ingredient of food and cosmetics for the prevention of skin ageing, as it may promote skin improvement by promoting the proliferation of fibroblasts, which are the main cells in the proliferation phase, and the secretion of growth factors to increase skin collagen levels. The use of these discarded fish parts to obtain collagen hydrolysates may contribute to a sustainable circular economy and improve the skin health of consumers. Aquatic animal processing by-products, especially fish skin, have great potential as an essential source of skin-improving active substances and should be widely applied.

3.10. Attenuated Obesity

According to the World Health Organization (WHO), which published the European Regional Obesity Report 2022, overweight and obesity have reached "epidemic" proportions in Europe, affecting nearly 59% of adults and one-third of children in Europe [113]. Being overweight and obese poses health risks beyond personal appearance, including disability and premature death. They are the fourth most common risk factors in Europe after hypertension, diet-related risks, and non-communicable diseases like those caused by the smoking of tobacco. These conditions can lead to type II diabetes, heart disease, musculoskeletal complications, and at least 13 types of cancer, according to recent estimates [34]. The pathogenesis of obesity and metabolic disorders is related to the intestinal microbiota, a complex and dynamic ecosystem whose composition affects nutrient acquisition, energy stabilization, and fat storage [34,114]. Studies have shown that *Bifidobacterium* spp. and *Lactobacillus gasseri* can effectively reduce body fat, improving lipid status and glucose homeostasis [115]. Collagen peptides have also been found to have a beneficial effect on obesity. Chotphruethipong et al. [116] found that the hydrolysis of collagen in the skin of defatted bass (*Lates calcarifer*) inhibited the activity of obesity-related enzymes, such as α-amylase and α-glucosidase. Effective anti-obesity effects of the hydrolysate CP, obtained by enzymatic hydrolysis using walleye cod skin in mice fed a high-fat diet, modulates the overall composition of the intestinal microbiota [34]. Currently, anti-obesity drugs are limited by their high cost and the numerous adverse effects associated with their administration. These results suggest that collagen peptides could be further developed as potential adjuvant therapeutic agents for obesity and related metabolic diseases.

Currently, researchers are primarily focused on preparing collagen peptides with high purity and specific peptides with specific bioactivity through isolation and purification processes. However, these methods are not practical for industrial applications due to their high cost. Additionally, there is a lack of research on the interactions between collagen peptides and other nutrients present in food.

4. Industrial Application

The share of collagen peptides in the international market has increased rapidly in recent years. It has formed several application areas, including beauty, joint and bone health, sports health, and topical products. The unique functions of collagen peptides, such as promoting skin [11] and bone repair [117], antioxidants [102], and others, determine their diverse applications.

It is also applied in cosmetics to improve skin moisture and elasticity through transdermal absorption, etc. Collagen peptides have gained attention for their antioxidant activity and skin-improving properties. They can effectively scavenge reactive oxygen radicals, which slows down the aging process of the skin [118,119]. As a result, they are commonly used in cosmetic formulations such as lotions, creams, and masks.

Collagen peptides can be used as nutritional supplements or food fortification to regulate human body functions through oral intake [120]. Kumar et al. [121] mixed MCP (0–10%) into biscuit powder and found that the addition of MCP significantly reduced the water holding capacity ($p < 0.05$) but had no significant effect on the oil absorption capacity. In addition, the protein content and antioxidant capacity of the biscuits were significantly increased ($p < 0.05$), and the calorie content was slightly decreased ($p < 0.05$). Therefore, the function of collagen peptides can be used as a nutritional health food option by combining them into other appropriate food systems.

However, the application of collagen peptides, some of which are only in vitro and cell studies, with no clinical trials as a guide, is still a field with unknowns and its protective mechanism has not been fully elucidated. At the same time, due to the influence of molecular weight, amino acids, and its order on the activity of bioactive peptides, collagen peptides still face great challenges in large-scale screening, purification, and production applications. With the deepening of the biological activity of collagen peptides and the gradual maturity of production technology, its safety should also be better guaranteed.

5. Conclusions

As discussed in this paper, collagen peptides of aquatic by-products obtained through enzymatic techniques have been shown to have high biological activity and functional properties, which have been widely used in pharmaceuticals, healthcare products, foodstuffs, daily-use chemicals, and other fields. However, there are still problems, such as low levels of high-value transformation, single development, immature scale, and intelligent production technology. Some new extraction processes (ultrasonic, microwave, supercritical extraction, high voltage pulsed electric field technology, etc.) can be applied to improve the extraction rate and industrialization of collagen peptides in combination with each other. In addition, combinations of two or more enzymes during enzymatic digestion have been shown to produce low molecular weight peptides that readily penetrate the intestinal tract, resulting in beneficial health properties. Future research should also focus on exploring protein-rich by-products that are currently underutilized and neglected, as well as their contaminants.

By-products of aquatic animals have high research value in the field of natural bioactive molecules. Fish collagen peptides have been shown to have biological activities such as antioxidant, antibacterial, antitumor, immunomodulatory, and hypoglycemic, and they are widely used in medicinal applications. Although collagen peptides have efficacy in preventing and treating various diseases, further research and clinical trials are still needed to improve the utilization value.

Author Contributions: All authors participated in the preparation of this review article. S.X., Y.Z. and W.S. explored the conceptualization of the viewpoints and conducted the literature search, paper writing and revision. C.Z., Q.W. and R.L. performed the data collection, analysis, and interpretation, respectively. Y.S. and S.G. took part in the discussion and paper revision. M.L. gave the supervision. L.S. contributed to the design of this review article and funding acquisition. All authors have read and agreed to the published version of the manuscript.

Funding: This review was funded by National Natural Science Foundation of China (No. 42106111); and Natural Science Foundation of Shandong Province (No. ZR2021QD030).

Data Availability Statement: No new data were created or analyzed in this study. Data sharing is not applicable to this article.

Acknowledgments: Special thanks to Yantai University for providing us with research condition. In addition, we would like to thank the professors and postgraduate students of the team of processing and comprehensive utilization of aquatic products at the School of Life Sciences, Yantai University, Yantai, China.

Conflicts of Interest: The authors declare no conflict of interest.

Abbreviations

FAO	Food and Agriculture Organization of the United Nations
CAGR	Compound Annual Growth Rate
USD	United States Dollar
APTT	Activated partial thromboplastin time
TT	Thrombin time
PT	Prothrombin time
DM	Diabetes mellitus
GLP-1	Glucagon-like peptide 1
T2DM	Diabetes mellitus type 2
LPS	Lipopolysaccharide
HRBC	Human red blood cell
AACR	American Association for Cancer Research
FDA	Food and Drug Administration
ACE	Angiotensin Converting Enzyme
IC_{50}	Half-maximal inhibitory concentration
DH	Degree of hydrolysis
ROS	Reactive oxygen species
BHT	Butylated hydroxytoluene
BHA	Butylated hydroxyanisole
TBHQ	Tert-butylhydroquinone
PG	Propyl gallate
DPPH	1,1-Diphenyl-2-picrylhydrazyl
ABTS	2,2′-amino-di(2-ethyl-benzothiazoline sulphonic acid-6)ammonium salt
UV	Ultraviolet
WHO	World Health Organization

References

1. Ramakrishnan, S.R.; Jeong, C.R.; Park, J.W.; Cho, S.S.; Kim, S.J. A review on the processing of functional proteins or peptides derived from fish by-products and their industrial applications. *Heliyon* **2023**, *9*, e14188. [CrossRef] [PubMed]
2. Ahmed, M.; Verma, A.K.; Patel, R. Collagen extraction and recent biological activities of collagen peptides derived from sea-food waste: A review. *Sustain. Chem. Pharm.* **2020**, *18*, 100315. [CrossRef]
3. Leon-Lopez, A.; Morales-Penaloza, A.; Martinez-Juarez, V.M.; Vargas-Torres, A.; Zeugolis, D.I.; Aguirre-Alvarez, G. Hydrolyzed Collagen-Sources and Applications. *Molecules* **2019**, *24*, 4031. [CrossRef] [PubMed]
4. Espinales, C.; Romero-Peña, M.; Calderón, G.; Vergara, K.; Cáceres, P.J.; Castillo, P. Collagen, protein hydrolysates and chitin from by-products of fish and shellfish: An overview. *Heliyon* **2023**, *9*, e14937. [CrossRef] [PubMed]
5. Campos, L.D.; de Almeida Santos, V., Jr.; Pimentel, J.D.; Fernandes Carregã, G.L.; Betim Cazarin, C.B. Collagen supplementation in skin and orthopedic diseases: A review of the literature. *Heliyon* **2023**, *9*, e14961. [CrossRef]
6. Tang, C.; Zhou, K.; Zhu, Y.; Zhang, W.; Xie, Y.; Wang, Z.; Zhou, H.; Yang, T.; Zhang, Q.; Xu, B. Collagen and its derivatives: From structure and properties to their applications in food industry. *Food Hydrocoll.* **2022**, *131*, 107748. [CrossRef]
7. Song, X.; Li, Z.; Li, Y.; Hou, H. Typical structure, biocompatibility, and cell proliferation bioactivity of collagen from Tilapia and Pacific cod. *Colloids Surf. B Biointerfaces* **2022**, *210*, 112238. [CrossRef]
8. Guo, J.; Liu, Q.; Cheng, J.; Fu, X.; Zhang, Y.; Yang, H. Hemostatic cotton based on biocompatible poly(vinyl alcohol)/soluble starch-fish skin collagen composites. *Mater. Today Commun.* **2022**, *32*, 103812. [CrossRef]
9. Wang, T.; Yang, L.; Wang, G.; Han, L.; Chen, K.; Liu, P.; Xu, S.; Li, D.; Xie, Z.; Mo, X.; et al. Biocompatibility, hemostatic properties, and wound healing evaluation of tilapia skin collagen sponges. *J. Bioact. Compat. Polym.* **2020**, *36*, 44–58. [CrossRef]
10. Manzoor, M.; Singh, J.; Gani, A. Exploration of bioactive peptides from various origin as promising nutraceutical treasures: In vitro, in silico and in vivo studies. *Food Chem.* **2022**, 373, 131395. [CrossRef]
11. Zhao, X.; Zhang, X.; Liu, D. Collagen peptides and the related synthetic peptides: A review on improving skin health. *J. Funct. Foods* **2021**, *86*, 104680. [CrossRef]
12. Zhang, L.; Zhao, G.-X.; Zhao, Y.-Q.; Qiu, Y.-T.; Chi, C.-F.; Wang, B. Identification and Active Evaluation of Antioxidant Peptides from Protein Hydrolysates of Skipjack Tuna (*Katsuwonus pelamis*) Head. *Antioxidants* **2019**, *8*, 318. [CrossRef] [PubMed]
13. Vijayan, D.K.; Sreerekha, P.R.; Dara, P.K.; Ganesan, B.; Mathew, S.; Anandan, R.; Ravisankar, C.N. Antioxidant defense of fish collagen peptides attenuates oxidative stress in gastric mucosa of experimentally ulcer-induced rats. *Cell Stress Chaperones* **2022**, *27*, 45–54. [CrossRef]
14. Hu, Z.; Sha, X.; Zhang, L.; Huang, S.; Tu, Z. Effect of grass carp scale collagen peptide FTGML on cAMP-PI3K/Akt and MAPK signaling pathways in B16F10 melanoma cells and correlation between anti-melanin and antioxidant properties. *Foods* **2022**, *11*, 391. [CrossRef]

15. Gonzalez-Serrano, D.J.; Hadidi, M.; Varcheh, M.; Jelyani, A.Z.; Moreno, A.; Lorenzo, J.M. Bioactive peptide fractions from collagen hydrolysate of common carp fish by-product: Antioxidant and functional properties. *Antioxidants* **2022**, *11*, 509. [CrossRef] [PubMed]
16. Feng, R.; Zou, X.; Wang, K.; Liu, H.; Hong, H.; Luo, Y.; Tan, Y. Antifatigue and microbiome reshaping effects of yak bone collagen peptides on Balb/c mice. *Food Biosci.* **2023**, *52*, 102447. [CrossRef]
17. Xu, J.; Li, Y.; Regenstein, J.; Su, X. In vitro and in vivo anti-oxidation and anti-fatigue effect of monkfish liver hydrolysate. *Food Biosci.* **2017**, *18*, 9–14. [CrossRef]
18. Ding, J.-F.; Li, Y.-Y.; Xu, J.-J.; Su, X.-R.; Gao, X.; Yue, F.-P. Study on effect of jellyfish collagen hydrolysate on anti-fatigue and anti-oxidation. *Food Hydrocoll.* **2011**, *25*, 1350–1353. [CrossRef]
19. He, L.; Wang, X.; Wang, Y.; Luo, J.; Zhao, Y.; Han, G.; Han, L.; Yu, Q. Production and identification of dipeptidyl peptidase IV (DPP-IV) inhibitory peptides from discarded cowhide collagen. *Food Chem.* **2023**, *405*, 134793. [CrossRef]
20. Zhang, Y.; Chen, R.; Chen, X.; Zeng, Z.; Ma, H.; Chen, S. Dipeptidyl peptidase IV-Inhibitory peptides derived from silver carp (*Hypophthalmichthys molitrix* Val.) proteins. *J. Agric. Food Chem.* **2016**, *64*, 831–839. [CrossRef]
21. Tometsuka, C.; Funato, N.; Mizuno, K.; Taga, Y. Long-term intake of ginger protease-degraded collagen hydrolysate reduces blood lipid levels and adipocyte size in mice. *Curr. Res. Food Sci.* **2021**, *4*, 175–181. [CrossRef] [PubMed]
22. Affane, F.; Louala, S.; el Imane Harrat, N.; Bensalah, F.; Chekkal, H.; Allaoui, A.; Lamri-Senhadji, M. Hypolipidemic, antioxidant and antiatherogenic property of sardine by-products proteins in high-fat diet induced obese rats. *Life Sci.* **2018**, *199*, 16–22. [CrossRef] [PubMed]
23. Rashid, N.Y.A.; Manan, M.A.; Pa'Ee, K.F.; Saari, N.; Wong, F.W.F. Evaluation of antioxidant and antibacterial activities of fish protein hydrolysate produced from Malaysian fish sausage (*Keropok Lekor*) by-products by indigenous Lactobacillus casei fermentation. *J. Clean. Prod.* **2022**, *347*, 131303. [CrossRef]
24. Atef, M.; Chait, Y.A.; Ojagh, S.M.; Latifi, A.M.; Esmaeili, M.; Hammami, R.; Udenigwe, C.C. Anti-Salmonella activity and peptidomic profiling of peptide fractions produced from sturgeon fish skin collagen (*Huso huso*) using commercial enzymes. *Nutrients* **2021**, *13*, 2657. [CrossRef] [PubMed]
25. Chen, Y.P.; Liang, C.H.; Wu, H.T.; Pang, H.Y.; Chen, C.; Wang, G.H.; Chan, L.P. Antioxidant and anti-inflammatory capacities of collagen peptides from milkfish (*Chanos chanos*) scales. *J. Food Sci. Technol.* **2018**, *55*, 2310–2317. [CrossRef] [PubMed]
26. Felician, F.F.; Yu, R.H.; Li, M.Z.; Li, C.J.; Chen, H.Q.; Jiang, Y.; Tang, T.; Qi, W.Y.; Xu, H.M. The wound healing potential of collagen peptides derived from the jellyfish *Rhopilema esculentum*. *Chin. J. Traumatol.* **2019**, *22*, 12–20. [CrossRef]
27. Global Market Insights. *Fish Protein Hydrolysate Market Size, by Technology by Form, by Source, by Application, Swine, Calves, Aquaculture, Pet Food, Food, Cosmetics, Agriculture, Industry Analysis Report, Regional Outlook, COVID-19 Impact Analysis, Application Development Potential, Price-Trends, Competitive Market Share & Forecast. 2023–2032*; Global Market Insights: Selbyville, DE, USA, 2023.
28. Han, S.-B.; Won, B.; Yang, S.-C.; Kim, D.-H. Asterias pectinifera derived collagen peptide-encapsulating elastic nanoliposomes for the cosmetic application. *J. Ind. Eng. Chem.* **2021**, *98*, 289–297. [CrossRef]
29. Aguirre-Cruz, G.; Leon-Lopez, A.; Cruz-Gomez, V.; Jimenez-Alvarado, R.; Aguirre-Alvarez, G. Collagen Hydrolysates for Skin Protection: Oral Administration and Topical Formulation. *Antioxidants* **2020**, *9*, 181. [CrossRef]
30. Czajka, A.; Kania, E.M.; Genovese, L.; Corbo, A.; Merone, G.; Luci, C.; Sibilla, S. Daily oral supplementation with collagen peptides combined with vitamins and other bioactive compounds improves skin elasticity and has a beneficial effect on joint and general wellbeing. *Nutr. Res.* **2018**, *57*, 97–108. [CrossRef]
31. Yang, T.; Zhang, K.; Li, B.; Hou, H. Effects of oral administration of peptides with low molecular weight from Alaska Pollock (*Theragra chalcogramma*) on cutaneous wound healing. *J. Funct. Foods* **2018**, *48*, 682–691. [CrossRef]
32. Praet, S.; Alzyadat, T.; Purdam, C.; Welvaert, M.; Gaida, J.; Chetty, G.; Manzanero, S.; Vlahovich, N.; Hughes, D.; Waddington, G. Oral supplementation of specific collagen peptides accelerates improvement in Achilles tendon pain and function in combination with a tailored exercise program. *J. Bodyw. Mov. Ther.* **2018**, *22*, 862–863. [CrossRef]
33. Wu, R.; Wu, C.; Liu, D.; Yang, X.; Huang, J.; Zhang, J.; Liao, B.; He, H. Antioxidant and anti-freezing peptides from salmon collagen hydrolysate prepared by bacterial extracellular protease. *Food Chem.* **2018**, *248*, 346–352. [CrossRef] [PubMed]
34. Wang, S.; Lv, Z.; Zhao, W.; Wang, L.; He, N. Collagen peptide from Walleye pollock skin attenuated obesity and modulated gut microbiota in high-fat diet-fed mice. *J. Funct. Foods* **2020**, *74*, 104194. [CrossRef]
35. Liu, H.; Yang, Y.; Liu, Y.; Cui, L.; Fu, L.; Li, B. Various bioactive peptides in collagen hydrolysate from salmo salar skin and the combined inhibitory effects on atherosclerosis in vitro and in vivo. *Food Res. Int.* **2022**, *157*, 111281. [CrossRef] [PubMed]
36. Li, W.; Kobayashi, T.; Meng, D.; Miyamoto, N.; Tsutsumi, N.; Ura, K.; Takagi, Y. Free radical scavenging activity of type II collagen peptides and chondroitin sulfate oligosaccharides from by-products of mottled skate processing. *Food Biosci.* **2021**, *41*, 100991. [CrossRef]
37. Indriani, S.; Sae-Leaw, T.; Benjakul, S.; Quan, T.H.; Karnjanapratum, S.; Nalinanon, S. Impact of different ultrasound-assisted processes for preparation of collagen hydrolysates from Asian bullfrog skin on characteristics and antioxidative properties. *Ultrason. Sonochem.* **2022**, *89*, 106163. [CrossRef]
38. Lu, W.-C.; Chiu, C.-S.; Chan, Y.-J.; Mulio, A.T.; Li, P.-H. Characterization and biological properties of marine by-product collagen through ultrasound-assisted extraction. *Aquac. Rep.* **2023**, *29*, 101514. [CrossRef]

39. Yang, L.; Guo, Z.; Wei, J.; Han, L.; Yu, Q.-L.; Chen, H.; Chen, Y.; Zhang, W. Extraction of low molecular weight peptides from bovine bone using ultrasound-assisted double enzyme hydrolysis: Impact on the antioxidant activities of the extracted peptides. *LWT* **2021**, *146*, 111470. [CrossRef]
40. Zhang, H.; Liu, H.; Qi, L.; Xv, X.; Li, X.; Guo, Y.; Jia, W.; Zhang, C.; Richel, A. Application of steam explosion treatment on the collagen peptides extraction from cattle bone. *Innov. Food Sci. Emerg. Technol.* **2023**, *85*, 103336. [CrossRef]
41. Cruz-Casas, D.E.; Aguilar, C.N.; Ascacio-Valdés, J.A.; Rodríguez-Herrera, R.; Chávez-González, M.L.; Flores-Gallegos, A.C. Enzymatic hydrolysis and microbial fermentation: The most favorable biotechnological methods for the release of bioactive peptides. *Food Chem.* **2021**, *3*, 100047. [CrossRef]
42. Bechaux, J.; Gatellier, P.; Le Page, J.-F.; Drillet, Y.; Sante-Lhoutellier, V. A comprehensive review of bioactive peptides obtained from animal byproducts and their applications. *Food Funct.* **2019**, *10*, 6244–6266. [CrossRef]
43. Wang, X.; Yu, H.; Xing, R.; Li, P. Characterization, preparation, and purification of marine bioactive peptides. *Biomed. Res. Int.* **2017**, *2017*, 9746720. [CrossRef] [PubMed]
44. Huang, C.Y.; Kuo, J.M.; Wu, S.J.; Tsai, H.T. Isolation and characterization of fish scale collagen from tilapia (*Oreochromis* sp.) by a novel extrusion-hydro-extraction process. *Food Chem.* **2016**, *190*, 997–1006. [CrossRef]
45. Devita, L.; Lioe, H.N.; Nurilmala, M.; Suhartono, M.T. The bioactivity prediction of peptides from tuna skin collagen using integrated method combining in vitro and in silico. *Foods* **2021**, *10*, 2739. [CrossRef] [PubMed]
46. Chotphruethipong, L.; Sukketsiri, W.; Aluko, R.E.; Sae-Leaw, T.; Benjakul, S. Effect of hydrolyzed collagen from defatted Asian sea bass (*Lates calcarifer*) skin on fibroblast proliferation, migration and antioxidant activities. *J. Food Sci. Technol.* **2021**, *58*, 541–551. [CrossRef]
47. Zheng, J.; Tian, X.; Xu, B.; Yuan, F.; Gong, J.; Yang, Z. Collagen peptides from swim bladders of giant croaker (*Nibea japonica*) and their protective effects against H_2O_2-induced oxidative damage toward human umbilical vein endothelial cells. *Mar. Drugs* **2020**, *18*, 430. [CrossRef]
48. Wang, W.-Y.; Zhao, Y.-Q.; Zhao, G.-X.; Chi, C.-F.; Wang, B. Antioxidant peptides from collagen hydrolysate of redlip croaker (*Pseudosciaena polyactis*) scales: Preparation, characterization, and cytoprotective effects on H_2O_2-damaged HepG2 Cells. *Mar. Drugs* **2020**, *18*, 156. [CrossRef]
49. Nurilmala, M.; Hizbullah, H.H.; Karnia, E.; Kusumaningtyas, E.; Ochiai, Y. Characterization and antioxidant activity of collagen, gelatin, and the derived peptides from yellowfin tuna (*Thunnus albacares*) skin. *Mar. Drugs* **2020**, *18*, 98. [CrossRef]
50. Cui, L.; Li, B. Enrichment of antiplatelet peptides and removal of fishy odor from silver carp skin collagen hydrolysates by macroporous resins: pH value of loading sample affects the peptides separation. *Food Chem.* **2023**, *411*, 135481. [CrossRef] [PubMed]
51. Huang, J.-J.; Li, H.-L.; Xiong, G.-Q.; Cai, J.; Liao, T.; Zu, X.-Y. Extraction, identification and anti-photoaging activity evaluation of collagen peptides from silver carp (*Hypophthalmichthys molitrix*) skin. *LWT* **2023**, *173*, 114384. [CrossRef]
52. Zamorano-Apodaca, J.C.; Garcia-Sifuentes, C.O.; Carvajal-Millan, E.; Vallejo-Galland, B.; Scheuren-Acevedo, S.M.; Lugo-Sanchez, M.E. Biological and functional properties of peptide fractions obtained from collagen hydrolysate derived from mixed by-products of different fish species. *Food Chem.* **2020**, *331*, 127350. [CrossRef] [PubMed]
53. Cao, S.; Cai, J.; Wang, X.; Zhou, K.; Liu, L.; He, L.; Qi, X.; Yang, H. Cryoprotective effect of collagen hydrolysates from squid skin on frozen shrimp and characterizations of its antifreeze peptides. *LWT* **2023**, *174*, 114443. [CrossRef]
54. Zhang, S.-Y.; Zhao, Y.-Q.; Wang, Y.-M.; Yang, X.-R.; Chi, C.-F.; Wang, B. Gelatins and antioxidant peptides from Skipjack tuna (*Katsuwonus pelamis*) skins: Purification, characterization, and cytoprotection on ultraviolet-A injured human skin fibroblasts. *Food Biosci.* **2022**, *50*, 102138. [CrossRef]
55. Xu, Q.; Zheng, L.; Huang, M.; Zhao, M. Exploring structural features of potent dipeptidyl peptidase IV (DPP-IV) inhibitory peptides derived from tilapia (*Oreochromis niloticus*) skin gelatin by an integrated approach of multivariate analysis and Gly-Pro-based peptide library. *Food Chem.* **2022**, *397*, 133821. [CrossRef] [PubMed]
56. Sun, S.; Gao, Y.; Chen, J.; Liu, R. Identification and release kinetics of peptides from tilapia skin collagen during alcalase hydrolysis. *Food Chem.* **2022**, *378*, 132089. [CrossRef]
57. Murthy, P.S.; Rai, A.; Bhaskar, N. Fermentative recovery of lipids and proteins from freshwater fish head waste with reference to antimicrobial and antioxidant properties of protein hydrolysate. *J. Food Sci. Technol.* **2014**, *51*, 1884–1892.
58. Ahmed, R.; Chun, B.-S. Subcritical water hydrolysis for the production of bioactive peptides from tuna skin collagen. *J. Supercrit. Fluids* **2018**, *141*, 88–96. [CrossRef]
59. Haq, M.; Ho, T.C.; Ahmed, R.; Getachew, A.T.; Cho, Y.-J.; Park, J.-S.; Chun, B.-S. Biofunctional properties of bacterial collagenolytic protease-extracted collagen hydrolysates obtained using catalysts-assisted subcritical water hydrolysis. *J. Ind. Eng. Chem.* **2020**, *81*, 332–339. [CrossRef]
60. Wang, X.; Le, B.; Zhang, N.; Bak, K.H.; Zhang, Y.; Fu, Y. Off-flavour compounds in collagen peptides from fish: Formation, detection and removal. *Int. J. Food Sci. Technol.* **2023**, *58*, 1543–1563. [CrossRef]
61. Na Cho, H.; Cho, D.W.; Hurk, B.S.; Bae, H.A.; Kim, D.-E.; Baek, H.H. Characterization of off-odor compounds of collagen peptides from Tilapia Scale using GC-MS-olfactometry. *Food Sci. Biotechnol.* **2015**, *24*, 403–410. [CrossRef]
62. Fu, Y.; Therkildsen, M.; Aluko, R.E.; Lametsch, R. Exploration of collagen recovered from animal by-products as a precursor of bioactive peptides: Successes and challenges. *Crit. Rev. Food Sci. Nutr.* **2019**, *59*, 2011–2027. [CrossRef] [PubMed]
63. Wolf, D.; Ley, K. Immunity and Inflammation in Atherosclerosis. *Circ. Res.* **2019**, *124*, 315–327. [CrossRef] [PubMed]

64. Gutierrez, J.; Turan, T.N.; Hoh, B.L.; Chimowitz, M.I. Intracranial atherosclerotic stenosis: Risk factors, diagnosis, and treatment. *Lancet Neurol.* **2022**, *21*, 355–368. [CrossRef] [PubMed]
65. Indumathi, P.; Mehta, A. A novel anticoagulant peptide from the Nori hydrolysate. *J. Funct. Foods* **2016**, *20*, 606–617. [CrossRef]
66. Kong, Y.; Huo, J.-L.; Xu, W.; Xiong, J.; Li, Y.-M.; Wu, W.-T. A novel anti-platelet aggregation tripeptide from Agkistrodon acutus venom: Isolation and characterization. *Toxicon* **2009**, *54*, 103–109. [CrossRef]
67. Bourdon, P.; Jablonski, J.-A.; Chao, B.H.; Maraganore, J.M. Structure-function relationships of hirulog peptide interactions with thrombin. *FEBS Lett.* **1991**, *294*, 163–166. [CrossRef]
68. Banner, D.W.; Hadváry, P. Crystallographic analysis at 3.0-A resolution of the binding to human thrombin of four active site-directed inhibitors. *J. Biol. Chem.* **1991**, *266*, 20085–20093. [CrossRef]
69. Li, Y.; Wang, B.; Li, B. The in vitro bioavailability of anti-platelet peptides in collagen hydrolysate from silver carp (*Hypophthalmichthys molitrix*) skin. *J. Food Biochem.* **2020**, *44*, e13226. [CrossRef]
70. Rajapakse, N.; Jung, W.K.; Mendis, E.; Moon, S.H.; Kim, S.K. A novel anticoagulant purified from fish protein hydrolysate inhibits factor XIIa and platelet aggregation. *Life Sci.* **2005**, *76*, 2607–2619. [CrossRef]
71. Khiari, Z.; Rico, D.; Martin-Diana, A.B.; Barry-Ryan, C. Structure elucidation of ACE-inhibitory and antithrombotic peptides isolated from mackerel skin gelatine hydrolysates. *J. Sci. Food Agric.* **2014**, *94*, 1663–1671. [CrossRef]
72. Cunha, S.A.; Pintado, M.E. Bioactive peptides derived from marine sources: Biological and functional properties. *Trends Food Sci. Technol.* **2022**, *119*, 348–370. [CrossRef]
73. Yoon, B.H.; Ang, S.M.; Alabd, A.; Furlong, K.; Yeo, C.J.; Lavu, H.; Winter, J.M. Pancreatic cancer-associated diabetes is clinically distinguishable from conventional diabetes. *J. Surg. Res.* **2021**, *261*, 215–225. [CrossRef] [PubMed]
74. Figueira-Goncalves, J.M.; Golpe, R. Impact of oral antidiabetics agents in the prevention of COPD exacerbations. *Arch. Bronconeumol.* **2022**. [CrossRef] [PubMed]
75. Vázquez, J.A.; Fraguas, J.; Mirón, J.; Valcárcel, J.; Pérez-Martín, R.I.; Antelo, L.T. Valorisation of fish discards assisted by enzymatic hydrolysis and microbial bioconversion: Lab and pilot plant studies and preliminary sustainability evaluation. *J. Clean. Prod.* **2020**, *246*, 119027. [CrossRef]
76. Zhang, R.; Chen, J.; Jiang, X.; Yin, L.; Zhang, X. Antioxidant and hypoglycaemic effects of tilapia skin collagen peptide in mice. *Int. J. Food Sci. Technol.* **2016**, *51*, 2157–2163. [CrossRef]
77. Harnedy, P.A.; Parthsarathy, V.; McLaughlin, C.M.; O'Keeffe, M.B.; Allsopp, P.J.; McSorley, E.M.; O'Harte, F.P.; FitzGerald, R.J. Atlantic salmon (*Salmo salar*) co-product-derived protein hydrolysates: A source of antidiabetic peptides. *Food Res. Int.* **2018**, *106*, 598–606. [CrossRef]
78. Kumar, L.V.; Shakila, R.J.; Jeyasekaran, G. In vitro anti-cancer, anti-diabetic, anti-inflammation and wound healing properties of collagen peptides derived from Unicorn Leatherjacket (*Aluterus monoceros*) at different hydrolysis. *Turk. J. Fish. Aquat. Sci.* **2019**, *19*, 551–560.
79. Devasia, S.; Kumar, S.; Ps, S.; Inoue, N.; Sugihara, F.; Koizumi, S. A double blind, randomised, four arm clinical study to evaluate the safety, efficacy and tolerability of collagen peptide as a nutraceutical therapy in the management of type II diabetes mellitus. *J. Diabetes Metab.* **2020**, *10*, 1–7.
80. Gilroy, D.; De Maeyer, R. New insights into the resolution of inflammation. *Semin. Immunol.* **2015**, *27*, 161–168. [CrossRef]
81. Liu, H.; Li, B. Separation and identification of collagen peptides derived from enzymatic hydrolysate of Salmo salar skin and their anti-inflammatory activity in lipopolysaccharide (LPS)-induced RAW264.7 inflammatory model. *J. Food Biochem.* **2022**, *46*, e14122.
82. Cheng, M.-L.; Wang, H.-C.; Hsu, K.-C.; Hwang, J.-S. Anti-inflammatory peptides from enzymatic hydrolysates of tuna cooking juice. *Food Agric. Immunol.* **2015**, *26*, 770–781. [CrossRef]
83. GBD 2015 Mortality and Causes of Death Collaborators. Causes of Death, Global, regional, and national life expectancy, all-cause mortality, and cause-specific mortality for 249 causes of death, 1980–2015: A systematic analysis for the Global Burden of Disease Study 2015. *Lancet* **2016**, *388*, 1459–1544. [CrossRef] [PubMed]
84. AACR. *AACR Cancer Progress Report 2022*; AACR: Orlando, FL, USA, 2022; pp. 3–6.
85. Hernandez-Ledesma, B.; Hsieh, C.C. Chemopreventive role of food-derived proteins and peptides: A review. *Crit. Rev. Food Sci. Nutr.* **2017**, *57*, 2358–2376. [CrossRef] [PubMed]
86. Baehaki, A.; Suhartono, M.T.; Sukarno; Syah, D.; Setyahadi, S. Collagen peptides from fish skin with angiotensin I-Converting enzyme (ACE) inhibitor and cancer antiproliferative activity. *Res. J. Pharm. Biol. Chem. Sci.* **2016**, *7*, 1994–2000.
87. Mirzapour-Kouhdasht, A.; Moosavi-Nasab, M.; Krishnaswamy, K.; Khalesi, M. Optimization of gelatin production from Barred mackerel by-products: Characterization and hydrolysis using native and commercial proteases. *Food Hydrocoll.* **2020**, *108*, 105970. [CrossRef]
88. Yaghoubzadeh, Z.; Peyravii Ghadikolaii, F.; Kaboosi, H.; Safari, R.; Fattahi, E. Antioxidant activity and anticancer effect of bioactive peptides from rainbow trout (*Oncorhynchus mykiss*) skin hydrolysate. *Int. J. Pept. Res. Ther.* **2019**, *26*, 625–632. [CrossRef]
89. Okagu, I.U.; Ezeorba, T.P.C.; Aham, E.C.; Aguchem, R.N.; Nechi, R.N. Recent findings on the cellular and molecular mechanisms of action of novel food-derived antihypertensive peptides. *Food Chem.* **2022**, *4*, 100078. [CrossRef]
90. Ahtesh, F.B.; Stojanovska, L.; Mishra, V.; Donkor, O.; Feehan, J.; Bosevski, M.; Mathai, M.; Apostolopoulos, V. Identification and Effects of Skim Milk-Derived Bioactive Antihypertensive Peptides. *Biologics* **2021**, *2*, 1–14. [CrossRef]

91. Zheng, Y.; Wang, X.; Guo, M.; Yan, X.; Zhuang, Y.; Sun, Y.; Li, J. Two Novel Antihypertensive Peptides Identified in Millet Bran Glutelin-2 Hydrolysates: Purification, In Silico Characterization, Molecular Docking with ACE and Stability in Various Food Processing Conditions. *Foods* **2022**, *11*, 1355. [CrossRef]
92. Morales-Camacho, J.I.; Espinosa-Hernandez, E.; Rosas-Cardenas, F.F.; Semeria-Maitret, T.; Luna-Suarez, S. Insertions of antihypertensive peptides and their applications in pharmacy and functional foods. *Appl. Microbiol. Biotechnol.* **2019**, *103*, 2493–2505. [CrossRef]
93. Lee, J.K.; Jeon, J.-K.; Byun, H.-G. Antihypertensive effect of novel angiotensin I converting enzyme inhibitory peptide from chum salmon (*Oncorhynchus keta*) skin in spontaneously hypertensive rats. *J. Funct. Foods* **2014**, *7*, 381–389. [CrossRef]
94. de Souza, G.S.; de Jesus Sonego, L.; Santos Mundim, A.C.; de Miranda Moraes, J.; Sales-Campos, H.; Lorenzon, E.N. Antimicrobial-wound healing peptides: Dual-function molecules for the treatment of skin injuries. *Peptides* **2022**, *148*, 170707. [CrossRef]
95. Wei, D.; Zhang, X. Biosynthesis, bioactivity, biotoxicity and applications of antimicrobial peptides for human health. *Biosaf. Health* **2022**, *4*, 118–134. [CrossRef]
96. Zhang, Q.Y.; Yan, Z.B.; Meng, Y.M.; Hong, X.Y.; Shao, G.; Ma, J.J.; Cheng, X.R.; Liu, J.; Kang, J.; Fu, C.Y. Antimicrobial peptides: Mechanism of action, activity and clinical potential. *Mil. Med. Res.* **2021**, *8*, 48. [CrossRef] [PubMed]
97. Lee, J.K.; Luchian, T.; Park, Y. New antimicrobial peptide kills drug-resistant pathogens without detectable resistance. *Oncotarget* **2018**, *9*, 15616–15634. [CrossRef] [PubMed]
98. Ennaas, N.; Hammami, R.; Beaulieu, L.; Fliss, I. Production of antibacterial fraction from Atlantic mackerel (*Scomber scombrus*) and its processing by-products using commercial enzymes. *Food Bioprod. Process.* **2015**, *96*, 145–153. [CrossRef]
99. Wald, M.; Schwarz, K.; Rehbein, H.; Bussmann, B.; Beermann, C. Detection of antibacterial activity of an enzymatic hydrolysate generated by processing rainbow trout by-products with trout pepsin. *Food Chem.* **2016**, *205*, 221–228. [CrossRef]
100. Zhao, W.-H.; Luo, Q.-B.; Pan, X.; Chi, C.-F.; Sun, K.-L.; Wang, B. Preparation, identification, and activity evaluation of ten antioxidant peptides from protein hydrolysate of swim bladders of miiuy croaker (*Miichthys miiuy*). *J. Funct. Foods* **2018**, *47*, 503–511. [CrossRef]
101. Sila, A.; Bougatef, A. Antioxidant peptides from marine by-products: Isolation, identification and application in food systems. A review. *J. Funct. Foods* **2016**, *21*, 10–26. [CrossRef]
102. Chen, Y.; Jin, H.; Yang, F.; Jin, S.; Liu, C.; Zhang, L.; Huang, J.; Wang, S.; Yan, Z.; Cai, X.; et al. Physicochemical, antioxidant properties of giant croaker (*Nibea japonica*) swim bladders collagen and wound healing evaluation. *Int. J. Biol. Macromol.* **2019**, *138*, 483–491. [CrossRef]
103. Atef, M.; Mahdi Ojagh, S. Health benefits and food applications of bioactive compounds from fish byproducts: A review. *J. Funct. Foods* **2017**, *35*, 673–681. [CrossRef]
104. Admassu, H.; Gasmalla, M.A.A.; Yang, R.; Zhao, W. Bioactive peptides derived from seaweed protein and their health benefits: Antihypertensive, antioxidant, and antidiabetic properties. *J. Food Sci.* **2018**, *83*, 6–16. [CrossRef] [PubMed]
105. Suárez-Jiménez, G.M.; Burgos-Hernández, A.; Torres-Arreola, W.; López-Saiz, C.M.; Contreras, C.A.V.; Ezquerra-Brauer, J.M. Bioactive peptides from collagen hydrolysates from squid (*Dosidicus gigas*) by-products fractionated by ultrafiltration. *Int. J. Food Sci. Technol.* **2018**, *54*, 1054–1061. [CrossRef]
106. Sae-leaw, T.; Benjakul, S. Antioxidant activities of hydrolysed collagen from salmon scale ossein prepared with the aid of ultrasound. *Int. J. Food Sci. Technol.* **2018**, *53*, 2786–2795. [CrossRef]
107. Vijayan, D.K.; Perumcherry Raman, S.; Dara, P.K.; Jacob, R.M.; Mathew, S.; Rangasamy, A.; Chandragiri Nagarajarao, R. In vivo anti-lipidemic and antioxidant potential of collagen peptides obtained from great hammerhead shark skin waste. *J. Food Sci. Technol.* **2022**, *59*, 1140–1151. [CrossRef]
108. Qiu, Y.T.; Wang, Y.M.; Yang, X.R.; Zhao, Y.Q.; Chi, C.F.; Wang, B. Gelatin and antioxidant peptides from gelatin hydrolysate of skipjack tuna (*Katsuwonus pelamis*) scales: Preparation, Identification and activity evaluation. *Mar. Drugs* **2019**, *17*, 565. [CrossRef]
109. Nur, S.; Wierson, Y.; Yulia, Y.; Jumaetri Sami, F.; Megawati; Andi, N.A.; Marwati, M.; Gani, S.A. Characterization, antioxidant and α-Glucosidase inhibitory activity of Collagen Hydrolysate from Lamuru (*Caranx ignobilis*) Fishbone. *Sains Malays.* **2021**, *50*, 2329–2341. [CrossRef]
110. Cavinato, M.; Jansen-Durr, P. Molecular mechanisms of UVB-induced senescence of dermal fibroblasts and its relevance for photoaging of the human skin. *Exp. Gerontol.* **2017**, *94*, 78–82. [CrossRef]
111. Li, C.; Fu, Y.; Dai, H.; Wang, Q.; Gao, R.; Zhang, Y. Recent progress in preventive effect of collagen peptides on photoaging skin and action mechanism. *Food Sci. Hum. Wellness* **2022**, *11*, 218–229. [CrossRef]
112. Wang, L.; Jiang, Y.; Wang, X.; Zhou, J.; Cui, H.; Xu, W.; He, Y.; Ma, H.; Gao, R. Effect of oral administration of collagen hydrolysates from *Nile tilapia* on the chronologically aged skin. *J. Funct. Foods* **2018**, *44*, 112–117. [CrossRef]
113. WHO. *WHO European Regional Obesity Report 2022*; WHO: Geneva, Switzerland, 2022.
114. Dethlefsen, L.; McFall-Ngai, M.; Relman, D.A. An ecological and evolutionary perspective on human-microbe mutualism and disease. *Nature* **2007**, *449*, 811–818. [CrossRef] [PubMed]
115. Ejtahed, H.-S.; Angoorani, P.; Soroush, A.-R.; Atlasi, R.; Hasani-Ranjbar, S.; Mortazavian, A.M.; Larijani, B. Probiotics supplementation for the obesity management: A systematic review of animal studies and clinical trials. *J. Funct. Foods* **2019**, *52*, 228–242. [CrossRef]

116. Chotphruethipong, L.; Binlateh, T.; Hutamekalin, P.; Sukketsiri, W.; Aluko, R.E.; Benjakul, S. Hydrolyzed collagen from defatted sea bass skin and its conjugate with epigallocatechin gallate: In vitro antioxidant, anti-inflammatory, wound-healing and anti-obesity activities. *Food Biosci.* **2021**, *43*, 101303. [CrossRef]
117. Song, H.; Zhang, S.; Zhang, L.; Li, B. Ingestion of collagen peptides prevents bone loss and improves bone microarchitecture in chronologically aged mice. *J. Funct. Foods* **2019**, *52*, 1–7. [CrossRef]
118. Venkatesan, J.; Anil, S.; Kim, S.K.; Shim, M.S. Marine fish proteins and peptides for cosmeceuticals: A review. *Mar. Drugs* **2017**, *15*, 143. [CrossRef]
119. Kang, M.C.; Yumnam, S.; Kim, S.Y. Oral intake of collagen peptide attenuates ultraviolet B irradiation-induced skin dehydration in vivo by regulating hyaluronic acid synthesis. *Int. J. Mol. Sci.* **2018**, *19*, 3551. [CrossRef]
120. Varani, J.; Dame, M.K.; Rittie, L.; Fligiel, S.E.; Kang, S.; Fisher, G.J.; Voorhees, J.J. Decreased collagen production in chronologically aged skin: Roles of age-dependent alteration in fibroblast function and defective mechanical stimulation. *Am. J. Pathol.* **2006**, *168*, 1861–1868. [CrossRef]
121. Kumar, A.; Elavarasan, K.; Hanjabam, M.D.; Binsi, P.K.; Mohan, C.O.; Zynudheen, A.A.; Kumar, A.K. Marine collagen peptide as a fortificant for biscuit: Effects on biscuit attributes. *LWT* **2019**, *109*, 450–456. [CrossRef]

foods

MDPI

Article

In Silico Screening of Bioactive Peptides in Stout Beer and Analysis of ACE Inhibitory Activity

Wenhui Tian [1], Cui Zhang [2], Qi Zheng [1], Shumin Hu [2], Weiqiang Yan [1], Ling Yue [1], Zhijun Chen [3], Ci Zhang [1], Qiulian Kong [1,*] and Liping Sun [4,*]

1 Crop Breeding and Cultivation Research Institute, Shanghai Academy of Agricultural Sciences, Shanghai 201403, China; twh9768@163.com (W.T.); zhengqi@saas.sh.cn (Q.Z.); webweifree@163.com (W.Y.); yueling312@sina.com (L.Y.); cizhang6@saas.sh.cn (C.Z.)

2 State Key Laboratory of Biological Fermentation Engineering of Beer, Qingdao 266021, China; zhangcui@tsingtao.com.cn (C.Z.); hsm@tsingtao.com.cn (S.H.)

3 Shanghai Shuneng Irradiation Technology Co., Ltd., Shanghai Academy of Agricultural Sciences, Shanghai 201403, China; 13764023962@126.com

4 Faculty of Food Science and Engineering, Kunming University of Science and Technology, Kunming 650500, China

* Correspondence: qiuliank@aliyun.com (Q.K.); lpsun@kmust.edu.cn (L.S.)

Abstract: Stout beer was selected as the research object to screen angiotensin-converting enzyme (ACE) inhibitory peptides. The peptide sequences of stout beer were identified using ultra-performance liquid chromatography-quadrupole-Orbitrap mass spectrometry with de novo, and 41 peptides were identified with high confidence. Peptide Ranker was used to score the biological activity and six peptides with a score ≥ 0.5 were screened to predict their potential ACE inhibitory (ACEI) activity. The toxicity, hydrophilicity, absorption, and excretion of these peptides were predicted. In addition, molecular docking between the peptides and ACE revealed a significant property of the peptide DLGGFFGFQR. Furthermore, molecular docking conformation and molecular dynamics simulation revealed that DLGGFFGFQR could be tightly bound to ACE through hydrogen bonding and hydrophobic interaction. Lastly, the ACEI activity of DLGGFFGFQR was confirmed using in vitro evaluation and the IC_{50} value was determined to be 24.45 μM.

Keywords: beer polypeptide; angiotensin converting enzyme; virtual screening; molecular docking

Citation: Tian, W.; Zhang, C.; Zheng, Q.; Hu, S.; Yan, W.; Yue, L.; Chen, Z.; Zhang, C.; Kong, Q.; Sun, L. In Silico Screening of Bioactive Peptides in Stout Beer and Analysis of ACE Inhibitory Activity. *Foods* **2024**, *13*, 1973. https://doi.org/10.3390/foods13131973

Academic Editor: Cristina Martínez-Villaluenga

Received: 18 May 2024
Revised: 17 June 2024
Accepted: 19 June 2024
Published: 22 June 2024

1. Introduction

Hypertension is one of the most common chronic diseases with the highest incidence rate and it is also an important risk factor for cardiovascular disease (CVDs) [1,2]. On 19 September 2023, the World Health Organization published its inaugural Global Hypertension Report, which stated that the prevalence of hypertension had steadily increased over the past 2–3 decades, with the number of adults with hypertension doubling from 650 million in 1990 to 1.3 billion in 2019 [3]. The angiotensin-converting enzyme (ACE) plays a pivotal role in the renin-angiotensin system. ACE cleaves the C-terminal histidine and leucine residues of the decapeptide angiotensin I, transforming it into the effective vasoconstrictor angiotensin II and causing an increase in blood pressure [4,5]. Therefore, hypertension can be controlled by inhibiting ACE activity. Food-derived ACE inhibitory (ACEI) peptides are a class of peptides with antihypertensive activity. Unlike traditional antihypertensive drugs, they are non-toxic, have no adverse effects, and do not influence normal blood pressure [6,7]. They have the potential to replace antihypertensive drugs; therefore, the exploration of food-derived ACEI peptides is gradually becoming a research focus [8].

Traditional methods to identify peptide components involve the combination of ultrafiltration separation with purification methods such as volume-exclusion chromatography and ion-exchange chromatography [9]. These techniques can be used to isolate and characterize the individual components that have high activity. However, this method is not only challenging to perform but also provides inefficient results, limiting the number of peptides that can be obtained at a given time. Additionally, there is a risk of losing some highly active components during separation, which hinders high-throughput screening efforts. Proteomics and mass spectrometry techniques have advanced significantly. In silico screening eliminates the need for tedious separation and purification processes and allows batch processing of many polypeptides via rapid screening. A combination of bioinformatics and online databases can be used to screen several peptide sequences for potential bioactivity. This method is rapid, highly efficient and inexpensive, making it especially suitable for screening ACEI peptides [10].

Recently, several ACEI peptides have been identified and evaluated in brewed wine. Wu et al. discovered a polypeptide in Chinese Baijiu for the first time, which indicated an ACE inhibitory effect with an IC_{50} of 446 mmol/L [11]. Guo et al. confirmed the potential antioxidant and active ACEI peptides isolated from Guangdong glutinous rice wine and found significantly increased ACEI activity (30.89%) [12]. Moreover, Yin et al. identified two peptides, LPVGP and FPLQPHQP, from Qingke Baijiu, which had an inhibition with ACE (IC_{50} = 9.05, 5.03 μM) [13]. Beer is one of the most globally consumed fermented beverages that traditionally comprises four ingredients, including malted barley, water, hops, and yeast [14,15]. It is rich in nutrients, has a low alcohol content and can dissolve various peptides, amino acids, vitamins, and some enzymes (either raw or in the process) [16]. Numerous epidemiological studies have reported the protective effects of beer against CVDs [17]. A study reported that light-to-moderate beer drinkers were 57% less likely to develop CVDs than abstainers [18]. Ronan et al. reported that ale, lager, and stout could significantly inhibit platelet-activating factor-induced human platelet aggregation in vitro [19]. It has been observed that beer contains 2–6 mg/mL of protein/peptide [20], primarily derived from the water-soluble protein and insoluble barley protein in the raw material; some peptides are also produced by yeast metabolism during fermentation [21]. However, there has been little research on bioactive peptides in beer.

Stout beer, also known as dark beer, is generally black or brown in color and is prepared using special malts such as burnt and black malt as brewing materials [22,23]. This beer has unique malt and burnt-malt flavors, offering a mellow and slightly sweet taste [24]. Stout beer is rich in nutrients and its amino acid content is 3–4 times higher than that in other types of beers [25]. This study aims to identify peptides in stout beer, screen for bioactive peptides with ACEI activity through in silico screening and verify the activities of the identified peptides in vitro. Furthermore, molecular docking and molecular dynamics simulation (MD) were used to study the interaction between the peptide and ACE. This study could serve as a valuable reference for the development of beer with certain bioactivity.

2. Materials and Methods

2.1. Materials and Reagents

Stout beer was purchased from a market in Kunming, China. Synthesized peptides were obtained from Shanghai Synpeptide Co., Ltd., (Shanghai, China). An ACE activity determination kit was obtained from Sigma-Aldrich (Saint Louis, MO, USA). Trifluoroacetic acid and acetonitrile were purchased from Merck (Darmstadt, Germany).

2.2. Sample Preparation

The stout beer samples were filtered, degassed, and then concentrated from 300 mL to 50 mL. After concentration, the samples were extracted three times with ethyl acetate [26]. The aqueous phase obtained after extraction was subsequently used for peptide analysis.

2.3. Isolation and Identification of Peptides

The peptides in stout beer were analyzed using ultra-performance liquid chromatography-quadrupole-Orbitrap mass spectrometry (UPLC-Q-Orbitrap-MS^2) [27].

The UPLC conditions were as follows: InfinityLab Poroshell 120 EC-C18 (2.1 × 100 mm, 1.9 μm); column temperature of 30 °C; mobile phase A: acetonitrile with 0.1% formic acid; mobile phase B: ultrapure water with 0.1% formic acid. The injection volume was 2 μL and the flow rate was 0.2 mL/min. Gradient elution was 5% A from 0 to 1 min, 10% A from 1 to 2.5 min, 25% A from 2.5 to 12.5 min, 52.5% A from 12.5 to 20 min, 95% A from 20 to 22 min, and 5% A from 22 to 30 min.

The following conditions were used for mass spectrometry: ESI+ ion source; the primary mass spectrometer resolution was 70,000, and the secondary mass spectrum was 35,000; the spray voltage was 3.2 kV; the capillary temperature was 350 °C; the drier temperature was 350 °C; the data acquisition range was m/z 100–1000; the mass spectrometry data scanning mode was full MS-dd MS^2.

The obtained mass data were analyzed using Peaks Studio 8.0 software. Only peptides with an average local confidence (ALC) score ≥ 85% were considered for further analysis.

2.4. Screening of Bioactive Peptides In Silico

In silico digestion was performed to compare the selected peptides with known ACEI peptides in the BIOPEP database (http://www.uwm.edu.pl/biochemia/index.php/en/biopep) (last accessed on 4 April 2024) and the AHTPDB database (http://crdd.osdd.net/raghava/ahtpdb/) (last accessed on 4 April 2024).

Peptide Ranker (http://bioware.ucd.ie/compass/biowareweb/Serverpages/Peptideranker) (last accessed on 8 April 2024) was utilized to predict the possible bioactivities of the peptides. Peptides with a score ≥ 0.5 indicated potential biological activity [28]. In this study, a threshold of 0.5 was established for activity evaluation.

The BIOPEP application (http://www.uwm.edu.pl/Biochemia/biopep/start_biopep.php) (last accessed on 10 April 2024) contains a bioactive peptide sequence database and a program. The program can be used to construct potential bioactivity profiles of peptides, calculating quantitative descriptors to measure their potential values. In this study, the "Activity" and "Profiles of potential biological activity" functions were used to search for potential biological activity and possible active sites.

2.5. Toxicity and Solubility Prediction

ToxinPred (https://webs.iiitd.edu.in/raghava/toxinpred/) (last accessed on 12 April 2024) is a unique and valuable in silico method used to predict the toxicity of peptides and proteins. The Expasy database (https://web.expasy.org/protparam/) (last accessed on 14 April 2024) was used to calculate the hydrophilicity and hydrophobicity indices of the amino acids in peptides, which were assessed and represented using the grand average of hydropathy value (GRAVY). A low GRAVY value indicates strong hydrophilicity, which is conducive to solubility, and vice versa [29].

2.6. Absorption and Excretion Properties Analysis

The prediction of absorption and excretion properties is essential in contemporary drug design and screening. In this study, AdmetSAR (http://lmmd.ecust.edu.cn/admetsar1) (last accessed on 15 April 2024) was used to predict the properties of peptides. Furthermore, the primary evaluation indicators selected in this study indicated the blood–brain barrier (BBB), human intestinal absorption (HIA), and cytochrome P450 inhibitory promiscuity (CYP 450) [30]. The peptide codes were translated into simplified molecular input line entry specifications using PepSMI (https://www.novoprolabs.com/tools) (last accessed on 18 April 2024).

2.7. Molecular Docking

2.7.1. Preparation of Small Molecules

Peptide segment structures were built using the "Build Protein" function in SYBYL-X 2.1.1 software, followed by optimization with parameters set to Max Iterations 5000 and Gradient 0.005 to ensure energy minimization [31].

2.7.2. Preparation of Protein Structure

The crystal structure of ACE (PDBI: 108A) was downloaded from Protein Data Bank (http://www.rcsb.org/pdb) (last accessed on 20 April 2024). The enzyme was processed using the "Surflex-Dock" function in SYBYL-X. The ligand molecule from the original receptor-ligand complex crystal structure was separated from the receptor pocket, water molecules and other small molecules were removed, and the protein was hydrogenated and charged. A docking interface was set within 5 Å of the ligand center.

2.7.3. Molecular Docking

The peptide segments were flexibly docked with ACE using the "Surflex-dock" function in SYBYL-X. Peptides were docked into ACE using molecular visualization. The total score (T-Score), consensus score (C-Score), and number of hydrogen bonds were used to evaluate the docking results. A C-Score $\geq$ 4.0 indicated successful docking, and a T-Score $\geq$ 6.0 indicated superior docking [32]. Based on the evaluation criteria, optimal conformation was selected.

2.7.4. 2D Structure Visualization

The LigPlot program was used to analyze the 2D interaction of the receptor-ligand complex, the interaction between proteins and the ligand complex was determined, and the hydrophobic bond, hydrogen bond and bond length under docking posture were analyzed.

2.8. MD Simulation

The simulation of the peptides and the ACE was performed according to a previous report [33,34]. In the preprocessing of small molecules, Gaussian 16 W was used for the hydrogenation of small molecules, whereas AmberTools 22 was used to impose GAFF on the small molecule force field. Amber99sb-ildn was used as the force field, and a TIP3P water model as the solvent. Na+ ions were used to neutralize the charge. The simulation system adopts the steepest descent method to minimize the energy; the coupling constant was 0.1 ps and the duration was 100 ps. The following conditions were employed: temperature: 300 K, pressure: 1 atm, time: 100 ns, isothermal and isobaric system. The root-mean-square deviation (RMSD), root-mean-square fluctuation (RMSF), the solvent-accessible surface area (SASA), changes in the radius of gyration, and the number of hydrogen bond changes during MD simulation were determined.

2.9. Determination of ACEI Activity In Vitro

ACEI activity determination was determined using an assay kit. Briefly, 10 μL of ACE and 40 μL of the inhibitor were added to the microplate and mixed thoroughly, then 50 μL of the substrate was added to start the reaction. The fluorescence was detected at an excitation wavelength of 320 nm and an emission wavelength of 405 nm at 37 °C for 30 min. The ACE activity of the peptide was calculated as follows:

$$\text{ACEI activity} = [(K_{\text{Std}} - K_{\text{Sample}})/K_{\text{Std}}] \times \text{DF} \times 100\%$$

where K_{Sample} means the slope of the sample curve after subtracting the blank. (RFU/min); K_{Std} means the slope of the standard curve after subtracting the blank (RFU/min); DF means the dilution factor.

2.10. Statistical Analysis

Each experiment was conducted in triplicate, the data were presented as mean ± standard deviation. SYBYL-X 2.1.1 software was utilized for evaluating the molecular docking visualization. LigPlot 1.4.5 software was employed to illustrate the 2D interactions. MD simulation data analysis was performed using GROMACS 2022.3 software.

3. Results and Discussion

3.1. Identification of Peptides

During the brewing process, most proteins are enzymatically metabolized into small molecule polypeptides or single amino acids and are ultimately found in the final beer product. Previous studies were primarily focused on the proteomic composition of beer. Mahya et al. have reviewed the types of proteins in beer [35]. Gianluca and co-authors analyzed the protein and peptide components of two Italian barley malt beers and indicated that Z4-barley protein, ns-LTP1 and two barley albumins are predominantly found in beer [36,37].

In this study, peptides in stout beer were identified using UPLC-Q-Orbitrap-MS^2, and a total of 41 peptides with ALC ≥ 85% were selected (Table 1). The peptides PMAPLPRSGPE and DLGGFFGFQR from Italian beer, LPQQQAQFK, STEWHLD, PPPVHD, LVLPGELAK, LAVMQQQQQQ, PPPVHDTD, PPVPHDTD, LPEDAKVE, HAVSEGTKAVT, LPGELAK, PMAPLPRSGEP, TVSGF, LNFDPNR, LALDTRVG, LDTRVGV, and RVALVY from draft beer, which were previously identified, also appeared in our results [37,38]. The polypeptides present in beer primarily originate from malt protein. Previous studies have shown that the soluble protein content in malt water-soluble protein decreased significantly with an increase in roasting degree [39]. The main raw material of stout beer includes roasted burnt malt, and because roasted stout beer has fewer peptides than other beers [40]. The 41 peptides consisted of 4–14 amino acids, with a molecular weight less than 1600 Da, consistent with the structure of bioactive peptides [41]. Malt typically contains 8–14% protein, and is composed primarily of 18 different amino acids, of which Glu (2.9%), Pro (1.2%) and Leu (1.0%) are the most abundant [38]. Among the 41 peptides identified in this study, 18 contained Glu, 17 contained Pro, and 28 contained Leu, which is consistent with findings from previous studies. Moreover, the obtained peptides were entered into the active peptide database BIOPEP and ACEI peptide database AHTPDB and it was revealed that two of them were the already known peptides LNFDPNR and LPQQQAQFK (IC50 = 38.58, 26.45 μM) [42]. Therefore, the remaining 39 peptides were selected for the current study.

Table 1. Peptides of stout beer.

No	Peptide	RT	Tag Length	ALC (%)	Mass	References
1	ALDTRVGV	15.67	8	94	829.4658	
2	LPQQQAQFK	13.86	9	93	1086.5823	Tian et al. [41]
3	STEWHLD	17.12	7	93	886.3821	Tian et al. [41]
4	TDVRLS	10.05	6	93	689.3708	
5	MFALPVPSQPVD	18.84	12	91	1299.6533	
6	VAVAREVAVG	14.41	10	91	969.5607	
7	FDRLQ	11.86	5	91	677.3497	
8	PPPVHD	4.48	6	91	660.3231	Tian et al. [41]
9	LASQLGLGGLSSL	19.07	13	90	1214.6870	
10	QAAVGGQVVEK	12.17	11	90	1084.5876	
11	KELWSMEK	16.11	8	90	1049.5215	
12	TAAGSL	10.38	6	90	518.2700	
13	LVLPGELAK	18.33	9	90	938.5800	Tian et al. [41]

Table 1. *Cont.*

No	Peptide	RT	Tag Length	ALC (%)	Mass	References
14	EALE	6.20	4	90	460.2169	
15	TALTVVRN	14.20	8	90	872.5079	
16	LAVMQQQQQQ	13.55	10	89	1200.5920	Tian et al. [41]
17	DLGGFFGFQR	18.99	10	89	1142.5508	Picariello et al. [37]
18	SPKMAKNVD	13.08	9	89	988.5012	
19	DNLQGLTKP	15.37	9	89	984.5240	
20	PPPVHDTD	4.46	8	89	876.3977	Tian et al. [41]
21	PPVPHDTD	4.30	8	89	876.3977	Tian et al. [41]
22	ELQTSVR	9.78	7	89	831.4450	
23	LPEDAKVE	13.19	8	88	899.4600	Tian et al. [41]
24	HAVSEGTKAVT	4.73	11	87	1098.5669	Tian et al. [41]
25	SSQGLELSHVFHKS	16.89	14	87	1554.7791	
26	LPGELAK	14.22	7	87	726.4276	Tian et al. [41]
27	AHQQQVPVEVMR	18.09	12	86	1420.7246	
28	PMAPLPRSGEP	15.91	11	86	1150.5803	Picariello et al. [37]
29	TVSGF	15.41	5	86	509.2485	
30	SSLF	18.10	4	86	452.2271	
31	LNFDPNR	15.74	7	86	874.4297	Tian et al. [41]
32	PPPVHDFMNE	4.41	10	86	1181.5176	
33	ELRVR	10.55	5	86	671.4078	
34	EVNNVGQSGLMD	16.34	12	85	1261.5608	
35	ATPCCAEELER	16.22	11	85	1220.5166	
36	VAVARTPTVG	14.57	10	85	969.5607	
37	EAGY	3.79	4	85	438.1750	
38	LALDTRVG	16.39	8	85	843.4814	Tian et al. [41]
39	PMAPLPR	15.47	7	85	780.4316	
40	LDTRVGV	15.14	7	85	758.4286	Tian et al. [41]
41	RVALVY	16.05	6	85	719.4330	Tian et al. [41]

3.2. Process of Selecting Stout Beer Peptides

3.2.1. Activity Score and Prediction

Peptide Ranker is a neural-network-based biopeptide activity prediction server. For each peptide, the score is between 0 and 1 [42]. The higher the peptide ranker score, the greater the possibility of biological activity. The Peptide Ranker scores of 39 peptides are shown in Table 2. The activity evaluation threshold was set at 0.5, based on which six peptides were screened, including DLGGFFGFQR, PMAPLPR, SSLF, PPPVHDFMNE, FDRLQ, and PMAPLPRSGEP with scores of 0.8736, 0.8565, 0.8145, 0.7050, 0.5909, 0.5780, respectively.

The potential biological activities of six peptides were predicted using the "Activity" function of BIOPEP, and the results showed that all peptides were potential ACE inhibitors. The active sites of the selected peptides were predicted using the function "Profiles of potential biological activity" and the results were as follows: GF, FG, GG, LG, FQ in peptide DLGGFFGFQR; PLP, PR, PL, AP, MAP in peptide PMAPLPR; LF in peptide SSLF; PP, DF, PPP in peptide PPPVHDFMNE; RL, LQ in peptide FDRLQ; PLP, PR, PL, GEP, AP, GE, SG, MAP in peptide PMAPLPRSGEP (Table 3). A previous study found that peptides containing Tyr, His, Phe, Arg, and Trp are characteristic of ACEI peptides [43]. The six peptides screened by BIOPEP exhibited these characteristics.

Table 2. Peptide Ranker scores of stout beer.

No.	Peptide	Peptide Ranker Score	No.	Peptide	Peptide Ranker Score
1	DLGGFFGFQR	0.8736	21	SPKMAKNVD	0.1683
2	PMAPLPR	0.8565	22	TAAGSL	0.1628
3	SSLF	0.8145	23	LALDTRVG	0.1545
4	PPPVHDFMNE	0.7050	24	LVLPGELAK	0.1488
5	FDRLQ	0.5909	25	VAVARTPTVG	0.1275
6	PMAPLPRSGEP	0.5780	26	EVNNVGQSGLMD	0.1166
7	LASQLGLGGLSSL	0.4581	27	ALDTRVGV	0.1162
8	SSQGLELSHVFHKS	0.4439	28	RVALVY	0.1076
9	PPPVHD	0.4175	29	ELRVR	0.0990
10	PPPVHDTD	0.2941	30	VAVAREVAVG	0.0900
11	STEWHLD	0.2941	31	QAAVGGQVVEK	0.0833
12	TVSGF	0.2924	32	LPEDAKVE	0.0763
13	MFALPVPSQPVD	0.2698	33	LDTRVGV	0.0728
14	ATPCCAEELER	0.2616	34	TDVRLS	0.0723
15	PPVPHDTD	0.2565	35	HAVSEGTKAVT	0.0701
16	AHQQQVPVEVMR	0.2268	36	LAVMQQQQQQ	0.0597
17	LPGELAK	0.2245	37	TALTVVRN	0.0575
18	DNLQGLTKP	0.2083	38	ELQTSVR	0.0521
19	KELWSMEK	0.1888	39	EALE	0.0483
20	EAGY	0.1687			

Table 3. Bioactivity prediction of active amino acid sequence of peptide.

No.	Peptide	Activity	Active Amino Acid Sequence
1	DLGGFFGFQR	ACE inhibition	GF, FG, GG, LG, FQ
2	PMAPLPR	ACE inhibition	PLP, PR, PL, AP, MAP
3	SSLF	ACE inhibition	LF
4	PPPVHDFMNE	ACE inhibition	PP, DF, PPP
5	FDRLQ	ACE inhibition	RL, LQ
6	PMAPLPRSGEP	ACE inhibition	PLP, PR, PL, GEP, AP, GE, SG, MAP

3.2.2. Evaluation of the Properties of the Selected Peptides

Currently, peptides are being used to treat diseases; however, peptide-based therapy is limited by its toxicity profile. Therefore, it is necessary to predict the toxicity of polypeptides before commencing in-depth studies. ToxinPred is a useful tool to distinguish toxic peptides from the non-toxic ones. Here, the screen peptides were non-toxic, indicating their suitability as food-derived ACEI peptides (Table 4).

Table 4. Evaluated properties of stout beer.

9	Peptide	Toxin	GRAVY	Absorption		Excretion
				BBB	HIA	CYP450 Inhibitory Promiscuity
1	DLGGFFGFQR	Non-Toxin	−0.050	0.9493	0.7473	−0.9658
2	PMAPLPR	Non-Toxin	−0.257	0.9332	−0.5054	−0.9958
3	SSLF	Non-Toxin	1.250	0.9037	0.7437	−0.9888
4	PPPVHDFMNE	Non-Toxin	−0.960	0.7896	0.6805	−0.9657
5	FDRLQ	Non-Toxin	−0.980	0.6429	0.7473	−0.9870
6	PMAPLPRSGEP	Non-Toxin	−0.736	0.9399	−0.4929	−0.9953

Good water solubility is the basis of orally bioactive peptides. The physiological activity of bioactive peptides is closely related to the hydrophobic/hydrophilic properties of the amino acid chains. The GRAVY value reflects the solubility of the peptides and ranges from −2 to 2. Positive values indicate that the peptide is hydrophobic, and negative values indicate hydrophilicity. The GRAVY value of SSLF was 1.250 and that of the other five peptides ranged from −0.050 to −0.980, indicating that SSLF is hydrophobic, whereas DLGGFFGFQR, PMAPLPR, SSLF, PPPVHDFMNE, FDRLQ, PMAPLPRSGEP are hydrophilic.

Absorption and excretion parameters play a crucial role in the identification of food-derived bioactive peptides [44]. Therefore, the properties of the six identified peptides were assessed using the AdmetSAR program (Table 4). The ability of chemicals to cross the BBB is important for the development of central nervous system drugs. Here, it was observed that all six peptides had positive BBB values, suggesting their BBB-penetrating potential and ability to enter the central nervous system. HIA results showed that DLGGFFGFQR, SSLF, PPPVHDFMNE and FDRLQ were HIA+, indicating that they have favorable permeability and absorptive properties. The HIA values of PMAPLPR and PMAPLPRSGEP were negative, indicating that they were not well absorbed after oral administration. CYP 450 is one of the most important enzymes in drug metabolism. Several CYP450 enzymes are involved in the metabolism of various endogenous and exogenous compounds, including active peptides and other dietary products. We analyzed the interactions between the selected peptides and CYP 450 enzymes to determine their metabolic characteristics. The results revealed that CYP 450 enzymes do not affect the efficacy and safety of peptides. Therefore, based on the evaluation of the properties of peptides, we selected three peptides DLGGFFGFQR, PPPVHDFMNE, FDRLQ for further analyses.

3.3. Molecular Docking

Molecular docking is widely used to study molecular interactions and predict binding patterns and forces. To further identify peptides with high ACEI activity, molecular docking was used to simulate and assess the binding capacity of the three peptides with ACE (Table 5). All three peptides had good binding ability with ACE, and the T-Score of DLGGFFGFQR was ≥10. Combining the characteristic findings from the peptide property analysis and the molecular docking score, DLGGFFGFQR was selected from among the three peptides owing to its strong binding ability to ACE.

Table 5. Molecular docking score.

Peptide	T-Score	C-Score	Number of Hydrogen Bonds
DLGGFFGFQR	10.96	4.0	9
PPPVHDFMNE	9.22	4.0	5
FDRLQ	8.98	4.0	11

3.4. Molecular Docking Conformation Analysis

Molecular docking is a computational chemistry method that aims to predict binding patterns and affinity between proteins and small molecules. To understand the interactions between ACE and the peptides, DLGGFFGFQR was docked into the binding site of ACE (Table 6). The ACE active center includes 12 amino acid residues, which were Ala354, Gln281, Glu162, Glu384, Glu411, His353, His383, His387, His513, Lys511, Tyr520, and Tyr523 [45]. The ACE- DLGGFFGFQR binding site complex was predicted by SYBYL-X and the best conformations are shown in Figure 1. Figure 1a shows the surface of the ligand-protein molecule. Figure 1b is a 3D structure model of ACE-DLGGFFGFQR, whereas Figure 1c,d are the 3D and 2D models of the amino acid residue interaction binding mode. It was observed that DLGGFFGFQR formed eight hydrogen bonds with amino acids Cys352, Gln369, Val380, Pro407, His410, Gly408, and Gly414 in ACE, and the residues may enhance the ability of ACE to immobilize peptides and enhance stability, which may lead to increased inhibition. Based on relevant experimental reports and the findings related to the catalytic mechanism of ACE, it could be inferred that the results were similar to those for hemorphins [46]. Furthermore, DLGGFFGFQR formed 27 hydrophobic interactions with ACE and the action sites exhibited were different. Previous research showed that caffeic acid interacts with Asp377, Thr372, Thr166, Cys370, Pro163 at the ACE active site [47], corroborating our observations. This similarity suggested that the peptide was more tightly bound to ACE and may therefore be a more potent inhibitor.

Table 6. Action mode of DLGGFFGFQR and ACE.

Peptide	Interaction Mode	ACE Residue
DLGGFFGFQR	Hydrogen bond	Cys352(3.11Å), Gln369(3.00Å), Val380(2.62Å), Pro407(3.09Å), His410(3.06Å), Gly408(2.33Å), Gly414(3.03Å)
	Hydrophobic interaction	Thr166, Lys343, Val148, Thr150, Asp377, Val351, Pro163, Ala149, Gln160, Trp185, Phe512, Ser147, Tyr146, Thr372, Cys370, Lys368, Ile367, Ala381, Leu161, Ser355, Arg522, Ala356, Gly386, Ile413, Ala412, Met385, Ser526

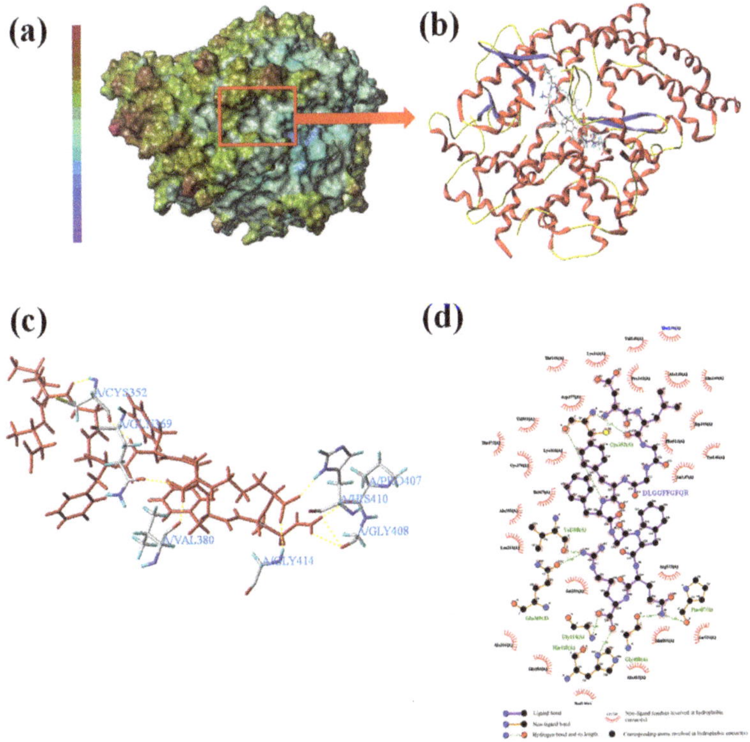

Figure 1. Docking result of DLGGFFGFQR-ACE. (**a**): ligand-protein molecular surface; (**b**): 3D structure of DLGGFFGFQR-ACE; (**c**): 3D binding interactions; (**d**): 2D binding interactions.

3.5. MD Simulation Analysis

MD simulation indicated the interactions between the molecules at the atomic level of microscopic changes for dynamic analysis. The results of MD simulation for the complexes of DLGGFFGFQR and ACE are shown in Figure 2. The change in trend of RMSD of protein and ligand is an important characteristic indicating whether the simulation has reached a state of stability. The effect of DLGGFFGFQR-ACE receptor was more obvious between 0–40 ns and RMSD fluctuated widely, interfering with the overall structure. After 40 ns, RMSD was more stable around 2.5 Å. (Figure 2a).

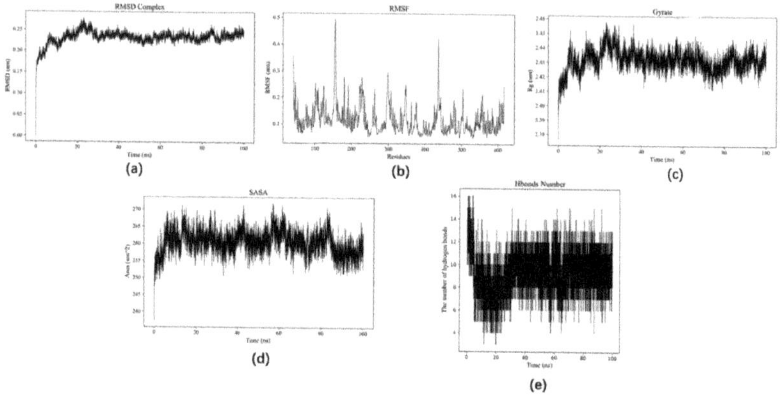

Figure 2. Results of MD simulation for the complexes of DLGGFFGFQR and ACE. (**a**): RMSD changes; (**b**): RMSF changes; (**c**): the radius of gyration changes; (**d**): SASA changes; (**e**): the number of hydrogen bond changes.

RMSF is the mean value of the atomic position changes over time and can be used to characterize the flexibility and motion intensity of protein and amino acids during the simulation process. Amino acid residues with high values of RMSF have high variability and are also the center of the receptor structure region [48]. The amino acid residues in the 130–200 region were significantly affected by the interaction between DLGGFFGFQR and ACE (Figure 2b). At the same time, the RMSF value of this region was significantly higher than that of other regions, indicating that it may be an important area for DLG GFFGFQR to affect ACE and that amino acid fluctuations in this region would affect protein–peptide interactions. The molecular docking result showed that DLGGFFGFQR could form hydrophobic interactions with ACE via amino acid residues including Thr166, Val148, Thr150, Pro163, Ala149, Gln160, Trp185, Ser147, Tyr146, and Leu161. These two findings were in concordance.

The radius of gyration was used to characterize protein structure density. No significant changes in the radius of gyration were noted after the binding of the protein to the peptide (Figure 2c). SASA is used to assess the surface area of a protein molecule exposed to the solution and assess the interactions between them and the stability of the protein structure. The SASA of the ACE receptor decreased after binding (Figure 2d). Hydrogen bonding and hydrophobic interactions play an important role in preserving protein conformation. Figure 2e shows the changes in the number of hydrogen bonds, which tend to be stable after 40 ns.

Based on the above results, it could be concluded that DLGGFFGFQR and ACE had stable interactions. Moreover, the binding of the peptide caused protein compaction and decreased the surface area. Several hydrogen bonds were formed between the small molecules.

3.6. *In Vitro Analysis of ACEI Activity*

In this study, DLGGFFGFQR was screened from stout beer using virtual screening and molecular docking. At the next stage, to further verify the actual inhibitory activity of the screened peptide, DLGGFFGFQR was analyzed to identify its in vitro bioactivity. The IC_{50} of DLGGFFGFQR was 24.45 ± 0.35 μM. Previous studies have reported the isolation of numerous ACEI. The ACEI activity of DLGGFFGFQR was more potent than that of VGLFPSRSF [49] and GRVSNCAA [50] and weaker than that of YGIKVGYAIP [51] and DHSTAVW [52]. Activity verification indicated that the peptide obtained via screening methods exhibited significant ACE inhibition effects. Therefore, this study has demonstrated a rapid screening method for ACEI peptides from stout beer. Further in vitro simulation of oral gastrointestinal digestion is necessary to fully understand the action of peptides.

4. Conclusions

In this study, stout beer was selected to screen for bioactive peptides using a rapid method. Using this method, 41 peptides were identified, and 6 potential biological activity peptides were selected using in silico tools. Through ADMET prediction and molecular docking, DLGGFFGFQR was assessed to have high ACEI activity. Through exploring the hypotensive mechanisms of DLGGFFGFQR by molecular docking and MD simulation analysis, it was found that the peptide could effectively dock with ACE. Finally, the in vitro activity was verified. This study exhibited a useful method to screen for ACEI peptide with excellent biosafety, thereby providing a theoretical basis for further product innovation and research development of functional beer.

Author Contributions: Writing-original draft and data curation, W.T.; resources and funding acquisition, C.Z. (Cui Zhang); validation, Q.Z.; software, S.H.; conceptualization, W.Y.; investigation, L.Y.; data curation, Z.C.; formal analysis, C.Z. (Ci Zhang); project administration and methodology, Q.K.; supervision and writing-review & editing, L.S. All authors have read and agreed to the published version of the manuscript.

Funding: This study was supported by the Open Research Fund of State Key Laboratory of Biological Fermentation Engineering of Beer (Grant No. K202003). Shanghai Agricultural Science and Technology Innovation Program (Grant No. T2023322).

Institutional Review Board Statement: Not applicable.

Informed Consent Statement: Not applicable.

Data Availability Statement: The original contributions presented in the study are included in the article, further inquiries can be directed to the corresponding authors.

Conflicts of Interest: All authors declare that the research was conducted in the absence of any commercial or financial relationships that could be construed as a potential conflict of interest.

References

1. Mills, K.T.; Stefanescu, A.; He, J. The global epidemiology of hypertension. *Nat. Rev. Nephrol.* **2020**, *16*, 223–237. [CrossRef] [PubMed]
2. Yin, R.; Yin, L.; Li, L.; Silva-Nash, J.; Yan, L. Hypertension in China: Burdens, guidelines and policy responses: A state-of-the-art review. *J. Hum. Hypertens.* **2022**, *36*, 126–134. [CrossRef] [PubMed]
3. Kario, K.; Okura, A.; Hoshide, S.; Mogi, M. The WHO Global report 2023 on hypertension warning the emerging hypertension burden in globe and its treatment strategy. *Hypertens. Res.* **2024**, *47*, 1099–1102. [CrossRef] [PubMed]
4. Liu, W.; Zhang, J.; Miyakawa, T.; Li, G.; Gu, R.; Tanokura, M. Antioxidant properties and inhibition of angiotensin-converting enzyme by highly active peptides from wheat gluten. *Sci. Rep.* **2021**, *11*, 5206. [CrossRef] [PubMed]
5. Ngoh, Y.-Y.; Gan, C.-Y. Identification of Pinto bean peptides with inhibitory effects on α-amylase and angiotensin converting enzyme (ACE) activities using an integrated bioinformatics-assisted approach. *Food Chem.* **2018**, *267*, 124–131. [CrossRef] [PubMed]
6. Hartmann, R.; Meisel, H. Food-derived peptides with biological activity: From research to food applications. *Curr. Opin. Biotechnol.* **2007**, *18*, 163–169. [CrossRef] [PubMed]
7. Shukla, P.; Chopada, K.; Sakure, A.; Hati, S. Current Trends and Applications of Food-derived Antihypertensive Peptides for the Management of Cardiovascular Disease. *Protein Pept. Lett.* **2022**, *29*, 408–428. [CrossRef] [PubMed]
8. Ganguly, A.; Sharma, K.; Majumder, K. Food-derived bioactive peptides and their role in ameliorating hypertension and associated cardiovascular diseases. *Adv. Food Nutr. Res.* **2019**, *89*, 165–207. [PubMed]
9. Loas, A.; Pentelute, B.L. Introduction: Peptide Chemistry. *Chem. Rev.* **2020**, *120*, 3049–3050. [CrossRef]
10. Tripathi, N.M.; Bandyopadhyay, A. High throughput virtual screening (HTVS) of peptide library: Technological advancement in ligand discovery. *Eur. J. Med. Chem.* **2022**, *243*, 114766. [CrossRef]

11. Wu, J.; Sun, B.; Zhao, M.; Zheng, F.; Sun, J.; Sun, X.; Li, H.; Huang, M. Discovery of a Bioactive Peptide, an Angiotensin Converting Enzyme Inhibitor in Chinese Baijiu. *J. Chin. Inst. Food Sci. Technol.* **2016**, *16*, 14–20.
12. Guo, J.; Lu, A.; Sun, Y.; Liu, B.; Zhang, J.; Zhang, L.; Huang, P.; Yang, A.; Li, Z.; Cao, Y.; et al. Purification and identification of antioxidant and angiotensin converting enzyme-inhibitory peptides from Guangdong glutinous rice wine. *LWT* **2022**, *169*, 113953. [CrossRef]
13. Yin, Z.; Yan, R.; Jiang, Y.; Feng, S.; Sun, H.; Sun, J.; Zhao, D.; Li, H.; Wang, B.; Zhang, N. Identification of peptides in Qingke baijiu and evaluation of its angiotensin converting enzyme (ACE) inhibitory activity and stability. *Food Chem.* **2022**, *395*, 133551. [CrossRef] [PubMed]
14. Hinojosa-Avila, C.R.; García-Gamboa, R.; Chedraui-Urrea, J.J.T.; García-Cayuela, T. Exploring the potential of probiotic-enriched beer: Microorganisms, fermentation strategies, sensory attributes, and health implications. *Food Res. Int.* **2024**, *175*, 113717.
15. Cao, W.; Baumert, J.L.; Downs, M.L. Compositional and immunogenic evaluation of fractionated wheat beers using mass spectrometry. *Food Chem.* **2020**, *333*, 127379. [CrossRef] [PubMed]
16. Zugravu, C.-A.; Medar, C.; Manolescu, L.S.C.; Constantin, C. Beer and Microbiota: Pathways for a Positive and Healthy Interaction. *Nutrients* **2023**, *15*, 844. [CrossRef]
17. Kang, Q.; Sun, J.; Wang, B.; Sun, B. Wine, beer and Chinese Baijiu in relation to cardiovascular health: The impact of moderate drinking. *Food Sci. Hum. Wellness* **2023**, *12*, 1–13. [CrossRef]
18. Gaetano, G.; Costanzo, S.; Castelnuovo, A.D.; Badimon, L.; Bejko, D.; Alkerwi, A.; Chiva-Blanch, G.; Estruch, R.; Vecchia, C.L.; Panico, S.; et al. Effects of moderate beer consumption on health and disease: A consensus document. *Nutr. Metab. Cardiovasc. Dis.* **2016**, *26*, 443–467. [CrossRef]
19. Lordan, R.; Keeffe, E.; Dowling, D.; Mullally, M.; Heffernan, H.; Tsoupras, A.; Zabetakis, I. The in vitro antithrombotic properties of ale, lager, and stout beers. *Food Biosci.* **2019**, *28*, 83–88. [CrossRef]
20. Watson, H.G.; Decloedt, A.I.; Hemetyck, L.Y.; Landschoot, A.V.; Prenni, J. Peptidomics of an industrial gluten-free barley malt beer and its non-gluten-free counterpart: Characterisation and immunogenicity. *Food Chem.* **2021**, *355*, 129597. [CrossRef]
21. Lodolo, E.J.; Kock, J.L.F.; Axcell, B.C.; Brooks, M. The yeast Saccharomyces cerevisiae-the main character in beer brewing. *FEMS Yeast Res.* **2008**, *8*, 1018–1036. [CrossRef] [PubMed]
22. Merinas-Amo, T.; Martinez-Jurado, M.; Merinas-Amo, R.; Mateo-Fenandez, M.M.; Alonso-Moraga, A. In vivo and in vitro toxicological and biological activities of alpha-acid humulone, isoxanthohumol and stout beer. *Toxicol. Lett.* **2014**, *229*, 179–180. [CrossRef]
23. Liu, X.; Jiang, W.; Zhao, S.; Wang, D.; Tang, K. Analysis of 15 kinds of malt aroma characteristic substances in malt and stout. *Food Ferment. Ind.* **2014**, *8*, 171–177.
24. Liguori, L.; Francesco, G.D.; Orilio, P.; Perretti, G.; Albanese, D. Influence of malt composition on the quality of a top fermented beer. *J. Food Sci. Technol.* **2021**, *58*, 2295–2303. [CrossRef] [PubMed]
25. Negrão, R.; Costa, R.; Duarte, D.; Taveira-Gomes, T.; Mendanha, M.; Soares, R. Angiogenic and Inflammatory activities are modulated in vivo by polyphenol supplemented beer. *FASEB J.* **2010**, *24*, 535.5. [CrossRef]
26. Huo, J.; Sun, B.; Zheng, F.; Sun, J.; Sun, X.; Li, H.; Luo, X.; Huang, M. Identification of a Tripeptide Arg-Asn-His from Chinese Baijiu and Its Antioxidant Activity. *Food Sci.* **2018**, *23*, 126–133.
27. Liu, B.; Zhuang, Y.; Sun, L. Identification and characterization of the peptides with calcium-binding capacity from tilapia (*Oreochromis niloticus*) skin gelatin enzymatic hydrolysates. *Food Chem.* **2020**, *85*, 114–122.
28. Fu, Y.; Liu, Z.; Wang, H.; Zhang, F.; Guo, S.; Shen, Q. Comparison of the generation of α-glucosidase inhibitory peptides derived from prolamins of raw and cooked foxtail millet: In vitro activity, de novo sequencing, and in silico docking. *Food Chem.* **2023**, *411*, 135378. [CrossRef] [PubMed]
29. Uriel, U.-R.; Liceaga, A.M.; Reddivari, L.; Kim, K.-H.; Anderson, J.M. Enzyme kinetics, molecular docking, and in silico characterization of canary seed (*Phalaris canariensis* L.) peptides with ACE and pancreatic lipase inhibitory activity. *J. Funct. Foods* **2022**, *88*, 104892.
30. Ahmad, I.; Mashwani, Z.-R.; Younas, Z.; Yousaf, T.; Ahmad, A.; Vladulescu, C. Antioxidant activity, metabolic profiling, in-silico molecular docking and ADMET analysis of nano selenium treated sesame seed bioactive compounds as potential novel drug targets against cardiovascular disease related receptors. *Heliyon* **2024**, *10*, e27909. [CrossRef]
31. Sun, L.; Chang, W.; Ma, Q.; Zhuang, Y. Purification of Antioxidant Peptides by High Resolution Mass Spectrometry from Simulated Gastrointestinal Digestion Hydrolysates of Alaska Pollock (*Theragra chalcogramma*) Skin Collagen. *Mar. Drugs* **2016**, *14*, 186–200. [CrossRef] [PubMed]
32. Zhuang, Y.; Ma, Q.; Guo, Y.; Sun, L. Protective effects of rambutan (*Nephelium lappaceum*) peel phenolics on H_2O_2-induced oxidative damages in $HepG_2$ cells and D-galactose-induced aging mice. *Food Chem. Toxicol.* **2017**, *108*, 554–562. [CrossRef] [PubMed]
33. Abraham, M.J.; Murtola, T.; Schulz, R.; Páll, S.; Smith, J.C.; Hess, B.; Lindahl, E. GROMACS: High performance molecular simulations through multi-level parallelism from laptops to supercomputers. *SoftwareX* **2015**, *1*, 19–25. [CrossRef]
34. Spoel, D.V.D.; Lindahl, E.; Hess, B.; Groenhof, G.; Mark, A.E.; Berendsen, H.J.C. GROMACS: Fast, flexible, and free. *J. Comput. Chem.* **2005**, *26*, 1701–1718. [CrossRef] [PubMed]
35. Bahmani, M.; Juhász, A.; Bose, U.; Nye-Wood, M.G.; Blundell, M.; Howitt, C.A.; Colgrave, M.L. From grain to malt: Tracking changes of ultra-low-gluten barley storage proteins after malting. *Food Chem.* **2024**, *432*, 137189. [PubMed]

36. Picariello, G.; Bonomi, F.; Iametti, S.; Rasmussen, P.; Pepe, C.; Lilla, S.; Ferranti, P. Proteomic and peptidomic characterisation of beer: Immunological and technological implications. *Food Chem.* **2011**, *124*, 1718–1726. [CrossRef]
37. Picariello, G.; Mamone, G.; Nitride, C.; Addeo, F.; Camarca, A.; Vocca, I.; Gianfrani, C.; Ferranti, P. Shotgun proteome analysis of beer and the immunogenic potential of beer polypeptides. *J. Proteom.* **2012**, *75*, 5872–5882. [CrossRef]
38. Tian, W.; Hu, S.; Zhuang, Y.; Sun, L.; Yin, H. Identification of angiotensin-converting enzyme and dipeptidyl peptidase inhibitory peptides from draft beer by virtual screening and molecular docking. *J. Sci. Food Agric.* **2022**, *102*, 1085–1094.
39. Chen, Y.; Huang, J.; Hu, J.; Yan, R.; Ma, X. Comparative study on the phytochemical profiles and cellular antioxidant activity of phenolics extracted from barley malts processed under different roasting temperatures. *Food Funct.* **2019**, *10*, 2176–2185. [CrossRef]
40. Tian, W.; Sun, L.; Zhang, C.; Hu, S.; Zhuang, Y.; Yin, H. Virtual Screening of Activity Evaluation of Dipeptidyl Peptidase-IV Inhibitory Peptides in White Beer. *Food Sci.* **2022**, *10*, 81–87.
41. Yang, F.; Chen, X.; Huang, M.; Yang, Q.; Cai, X.; Chen, X.; Du, M.; Huang, J.; Wang, S. Molecular characteristics and structure–activity relationships of food-derived bioactive peptides. *J. Integr. Agric.* **2021**, *20*, 2313–2332. [CrossRef]
42. Li, R.; Zhou, X.; Sun, L.; Zhuang, Y. Identification, in silico screening, and molecular docking of novel ACE inhibitory peptides isolated from the edible symbiot *Boletus griseus-Hypomyces chrysospermus*. *LWT* **2022**, *169*, 114008.
43. Ma, M.; Feng, Y.; Miao, Y.; Shen, Q.; Tang, S.; Dong, J.; Zhang, J.Z.H.; Zhang, L. Revealing the Sequence Characteristics and Molecular Mechanisms of ACE Inhibitory Peptides by Comprehensive Characterization of 160,000 Tetrapeptides. *Foods* **2023**, *12*, 1573. [CrossRef] [PubMed]
44. Kumar, S.; Niguram, P.; Bhat, V.; Jinagal, S.; Jairaj, V.; Chauhan, N. Synthesis, molecular docking and ADMET prediction of novel swertiamarin analogues for the restoration of type-2 diabetes: An enzyme inhibition assay. *Nat. Prod. Res.* **2020**, *36*, 2197–2207. [CrossRef] [PubMed]
45. Fan, Y.; Yu, Z.; Zhao, W.; Ding, L.; Zheng, F.; Li, J.; Liu, J. Identification and molecular mechanism of angiotensin-converting enzyme inhibitory peptides from *Larimichthys crocea* titin. *Food Sci. Hum. Wellness* **2020**, *9*, 257–263. [CrossRef]
46. Jobe, A.; Antony, P.; Altabbal, S.; Dhaheri, Y.A.; Vijayan, R. Interaction of hemorphins with ACE homologs. *Sci. Rep.* **2023**, *13*, 3743. [CrossRef] [PubMed]
47. Bare, Y.; Kuki, A.; Tiring, S.S.N.D.; Rophi, A.H.; Krisnamurti, G.C.; Sari, D.R.T. In Silico Study: Prediction the Potential of Caffeic Acid As ACE inhibitor. *El-Hayah J. Biol.* **2019**, *7*, 94–98. [CrossRef]
48. Li, W.; Chen, W.; Wang, J.; Li, Z.; Zhang, Z.; Wu, D.; Yan, M.; Ma, H.; Yang, Y. Structure–Activity Relationship of Novel ACE Inhibitory Undecapeptides from *Stropharia rugosoannulata* by Molecular Interactions and Activity Analyses. *Foods* **2023**, *12*, 3461. [CrossRef] [PubMed]
49. Dong, Y.; Yan, W.; Zhang, Y.; Dai, Z. A novel angiotensin-converting enzyme (ACE) inhibitory peptide from tilapia skin: Preparation, identification and its potential antihypertensive mechanism. *Food Chem.* **2024**, *430*, 137074. [CrossRef]
50. Zhang, Y.; Pan, D.; Yang, Z.; Gao, X.; Dang, Y. Angiotensin I-Converting enzyme (ACE) inhibitory and dipeptidyl Peptidase-4 (DPP-IV) inhibitory activity of umami peptides from *Ruditapes philippinarum*. *LWT* **2021**, *144*, 111265. [CrossRef]
51. Zarei, M.; Ghanbari, R.; Zainal, N.; Ovissipour, R.; Saari, N. Inhibition kinetics, molecular docking, and stability studies of the effect of papain-generated peptides from palm kernel cake proteins on angiotensin-converting enzyme (ACE). *Food Chem. Mol. Sci.* **2022**, *5*, 100147. [CrossRef] [PubMed]
52. Sutopo, C.C.Y.; Aznam, N.; Arianingrum, R.; Hsu, J.-L. Screening potential hypertensive peptides using two consecutive bioassay-guided SPE fractionations and identification of an ACE inhibitory peptide, DHSTAVW (DW_7), derived from pearl garlic protein hydrolysate. *Peptides* **2023**, *167*, 171046. [CrossRef] [PubMed]

nutrients

MDPI

Article

Identification of Novel Peptides in Distillers' Grains as Antioxidants, α-Glucosidase Inhibitors, and Insulin Sensitizers: In Silico and In Vitro Evaluation

Lixin Ding [1,2,3], Xiuqing Zheng [1,2,3], Lei Zhao [4] and Shengbao Cai [1,2,3,*]

1 Faculty of Food Science and Engineering, Kunming University of Science and Technology, Kunming 650500, China; dlx2528@163.com (L.D.); 18388417157@163.com (X.Z.)
2 Yunnan Engineering Research Center for Fruit & Vegetable Products, Kunming 650500, China
3 International Green Food Processing Research and Development Center of Kunming City, Kunming 650500, China
4 Beijing Engineering and Technology Research Center of Food Additives, Beijing Technology and Business University, Beijing 100048, China; zhaolei@th.btbu.edu.cn
* Correspondence: caikmust2013@163.com or caikmust2013@kmust.edu.cn

Abstract: Distillers' grains are rich in protein and constitute a high-quality source of various bioactive peptides. The purpose of this study is to identify novel bioactive peptides with α-glucosidase inhibitory, antioxidant, and insulin resistance-ameliorating effects from distiller's grains protein hydrolysate. Three novel peptides (YPLPR, AFEPLR, and NDPF) showed good potential bioactivities, and the YPLPR peptide had the strongest bioactivities, whose IC_{50} values towards α-glucosidase inhibition, radical scavenging rates of 2,2′-azino-bis (3-ethylbenzothiazoline-6- sulfonic acid) (ABTS) and 2,2-diphenyl-1-picrylhydrazyl (DPPH) were about 5.31 mmol/L, 6.05 mmol/L, and 7.94 mmol/L, respectively. The glucose consumption of HepG2 cells treated with YPLPR increased significantly under insulin resistance condition. Moreover, the YPLPR peptide also had a good scavenging effect on intracellular reactive oxygen species (ROS) induced by H_2O_2 (the relative contents: 102.35% vs. 100%). Molecular docking results showed that these peptides could stably combine with α-glucosidase, ABTS, and DPPH free radicals, as well as related targets of the insulin signaling pathway through hydrogen bonding and van der Waals forces. This research presents a potentially valuable natural resource for reducing oxidative stress damage and regulating blood glucose in diabetes, thereby increasing the usage of distillers' grains peptides and boosting their economic worth.

Keywords: bioactive peptides; molecular docking; UHPLC-ESI-HRMS/MS; insulin resistance; reactive oxygen species

Citation: Ding, L.; Zheng, X.; Zhao, L.; Cai, S. Identification of Novel Peptides in Distillers' Grains as Antioxidants, α-Glucosidase Inhibitors, and Insulin Sensitizers: In Silico and In Vitro Evaluation. *Nutrients* **2024**, *16*, 1279. https://doi.org/10.3390/nu16091279

Academic Editors: Abdelouahed Khalil and Juergen Schrezenmeir

Received: 14 March 2024
Revised: 22 April 2024
Accepted: 23 April 2024
Published: 25 April 2024

1. Introduction

The global incidence of type 2 diabetes mellitus (T2DM), characterized by high blood sugar levels, is steadily rising, posing a significant challenge to the worldwide healthcare system [1]. Prolonged hyperglycemia can lead to severe oxidative stress damage to blood vessels, ultimately resulting in complications such as inflammatory responses and atherosclerosis [2]. In the treatment of T2DM, the key strategies for controlling blood glucose levels involve inhibiting crucial enzymes such as α-glucosidase and improving insulin resistance status; additionally, maintaining normal levels of free radicals in the body can indirectly ameliorate diabetes complications [3–5]. Nowadays, common clinical hypoglycemic drugs include acarbose, voglibose, metformin, and miglitol [6,7]. While these drugs are effective in managing T2DM, they often come with certain toxic side effects and adverse reactions for long-term use of these drugs, with severe cases even leading to patient comas [8]. Hence, the pursuit of safe and effective hypoglycemic compounds and the development of high-efficiency, low-side-effect hypoglycemic functional foods have emerged as a new research focus.

Food-derived bioactive peptides are generally low-molecular-weight compounds that are safe, low in toxicity, and easily absorbed and that exhibit good antibacterial, antihypertensive, hypoglycemic, immunomodulatory, and antioxidant properties [9–11]. Since these peptides have small molecular weights, they are more likely to be absorbed intact by the gastrointestinal tract after oral ingestion, thereby exerting systemic or local effects [12]. Consequently, the study of bioactive peptides has progressively emerged as a prominent research area in recent years. Numerous studies have demonstrated that certain bioactive peptides can regulate blood glucose levels and enhance the management of type 2 diabetes through diverse pathways, such as the inhibition of α-glucosidase activity or the improvement of insulin resistance [13,14]. For instance, Mora et al. (2020) identified the peptides AEEEYPDL and LGVGG from Spanish dry-cured ham, which effectively inhibited α-glucosidase activity [13]. In the study conducted by Hu et al. (2019), water-extracted peptides and proteins were found to alleviate insulin resistance, thereby improving diabetes by enhancing glucose utilization [14].

Distillers' grains, by-products of rice wine production, possess abundant protein and alimentary value. Nevertheless, currently, most distillers' grains are primarily utilized as animal feed, resulting in potential resource wastage. Hence, the purpose of this study is to identify novel bioactive peptides with good α-glucosidase inhibition, antioxidant capacity, and insulin resistance-ameliorating effects from distiller's grains protein hydrolysate and to further reveal the underlying mechanisms. This research may provide new insights into the use of distillers' grains peptides as a valuable natural resource to combat oxidative stress damage and regulate blood glucose levels in diabetes, and the outcomes may hold the potential for enhancing the economic benefits derived from distillers' grains peptides.

2. Materials and Methods

2.1. Chemical and Reagents

Distillers' grains were generously supplied by the Suzhou Bridge Wine & Spirits Co. (Suzhou, China). Peptides YPLPR, AFEPLR, and NDPF (purity $\geq$ 98%) were composed by All Peptide Biotechnology Co., Ltd. (Hangzhou, China). Dulbecco's Modified Eagle Medium (DMEM) medium, embracing 10.0% fetal bovine serum (FBS) and 1.0 mol/L phosphate buffer (PBS) was bought from Gibco (Life Technologies, Inc., Carlsbad, CA, USA). Alkaline protease (2.4 U/g), penicillin-streptomycin (100.0 U/mL penicillin-100.0 mg/mL streptomycin), 3-(4,5-dimethylthiazol-2)-2,5-diphenyltetrazolium bromide (MTT) powder, 2′,7′-dichlorofluorescence yellow diacetate (DCFH DA), and trypsin solution were purchased from Solarbio (Beijing Solarbio Science and Technology Co., Ltd., Beijing, China). The HepG2 cell line was acquired from the Kunming Institute of Animal Science, Chinese Academy of Sciences (Kunming, China). Rosiglitazone, insulin, ABTS, DPPH, α-glucosidase derived from *Saccharomyces cerevisiae* (type I, protein content of at least 30.0 units per milligram), and p-nitrobenzene α-D-glucopyranoside (pNPG, with a minimum purity of 99.0%) were purchased from Sigma-Aldrich (located in Shanghai, China). The glucose kit was purchased from Nanjing Jiancheng Co. (Nanjinng, China). Other reagents applied were all of analytical rating.

2.2. Sample Preparation

Distillers' grains underwent lyophilization using the Christ Alpha 1-2 LD plus lyophilizer and then were finely made into a powder using the Lingdan LD-T300 grinder. The isolation of protein from the distillers' grains followed the way outlined by Wang et al. (2019) (minor modifications) [15,16]. The concrete method of separating proteins is displayed in the Supplementary Materials.

2.3. Composition Analysis of Peptide by UHPLC-ESI-HRMS/MS

The molecular mass and amino acid identity of peptides in rice protein hydrolysate were determined by the Thermo Fisher Ultimate 3000 UHPLC system (Thermo Fisher Scientific, Dreieich, Germany) combined with the Agilent Zorbax SB-C18 column (1.7 μm,

2.1 mm × 100 mm, Dr. Asp 80Maisch, Germany). According to the Tian et al. (2022) method (slightly modified), the mass spectrometry parameters were set [17]. Details of the method are shown in the Supplementary Materials. Peak Studio 8.0 (Bioinformatics Solutions, Waterloo, ON, Canada) was used to analyze molecular mass and amino acid sequence. Peptides with an average local confidence (ALC) score exceeding 85% were deemed to be included.

2.4. Determination of α-Glucosidase Inhibitory Effect

The inhibitory effect on α-glucosidase activity of the protein hydrolysates from distillers' grains or three selected peptides was assessed following the methodology outlined by Zhang et al. (2018) with slight modifications, and the concrete method is shown in the Supplementary Materials [18].

2.5. Assessment of Scavenging Capacity against DPPH and ABTS Radicals

The antioxidant properties of vinasse hydrolysates or three screening peptides were evaluated by DPPH and ABTS free radical scavenging experiments according to the method of Sun et al. (2015) [19]. Detailed information about the test method has been recorded in the Supplementary Materials.

2.6. In Silico Screening of Main Bioactive Peptides in Distillers' Grains Protein Hydrolysates

Peptide bioactivity prediction was conducted using Peptide Ranker (http://distilldeep.ucd.ie/PeptideRanker/, (accessed on 12 October 2023)) and selected peptides with a biological activity score of ≥0.6 for further analysis. It is important to highlight that a higher score indicates a higher likelihood of bioactivity [17]. The detailed method of studying the interaction between distiller's grains protein hydrolysate polypeptide and α-glucosidase by molecular docking technology is noted in the Supplementary Materials. For the pretreatment of peptides, refer to O'Boyle et al. (2011) [20]. Given the unavailability of the 3D structure of α-glucosidase, Saccharomyces cerevisiae isomaltose (PDB Code: 3A4A) from the database RCSB PDB (http://www.rcsb.org/pdb/home/home.do, (accessed on 12 October 2023)) was employed as a substitute. The post-treatment of α-glucosidase and the subsequent molecular docking followed the approach [10]. Based on the results of the molecular docking, peptides with an affinity equal to or superior to WLRL were identified as potential α-glucosidase inhibitory peptides [14]. The structure and interactions of latent α-glucosidase inhibitory peptides were analyzed using PyMOL (version 2.3.1). ToxinPred (http://crdd.osdd.net/raghava/toxinpred/ (accessed on 25 October 2023)) was used to appraise the toxicity of peptides [21]. Additionally, the solubility of peptides was evaluated by an online website (http://www.innovagen.com/proteomicstools, (accessed on 26 October 2023)) [22].

2.7. Peptides Synthesis and Activity Verification

Peptides YPLPR, AFEPLR, and NDPF, derived from distillers' grains protein hydrolysates, were synthesized. The synthesized peptides were stored, and their corresponding activities were verified through in vitro experiments. The α-glucosidase inhibitory effects and free radical scavenging capacities were performed as described above in Sections 2.4 and 2.5.

2.8. Determination of Insulin-Resistance HepG2 Cells on Bioactive Peptides

The MTT method was adopted to assess the impact of samples and insulin on cell viability, helping determine the optimal concentrations of insulin and the samples. The experimental protocol for improving insulin resistance followed previous methods [23,24]. Details are provided in the Supplementary Materials. In this experiment, various groups were established, incorporating the blank control group (K), model group (M), peptide sample groups, and the positive control group (treated with rosiglitazone).

2.9. Evaluation of Scavenging Capacity against Intracellular ROS

According to Liu et al. (2020), the H_2O_2-induced scavenging activity of reactive oxygen species was evaluated [25]. In this experiment, various groups were established, including the blank, sample, positive control, and model groups. Refer to the Supplementary Materials for specific information.

2.10. Molecular Docking to Probe Mechanism of Action

The molecular docking of the peptides YPLPR, AFEPLR, and NDPF with ABTS and DPPH was performed in a manner consistent with Section 2.6. The 3D structures of ABTS (Code: 5360881) and DPPH (Code: 2735032) were obtained from PubChem (https://pubchem.ncbi.nlm.nih.gov/ (accessed on 16 November 2023)). The phosphorylated insulin receptor tyrosine kinase protein (PIRTK protein) and peroxisome proliferator-activated receptor γ (PPARγ) were important targets involved in the insulin signaling pathway [26,27]. The 3D structures of the PIRTK protein (PDB Code: 1IR3) and PPARγ (PDB Code: 2PRG) were obtained from the RCSB PDB database. The grid boxes and center of 1IR3 were $80 \times 80 \times 80$ Å and −24.099, 39.015, 11.431 (X, Y, Z) [26]. The grid boxes of 2PRG had $80 \times 80 \times 80$ Å and grid centers of 50.140, −38.202, and 19.559 (X, Y, Z) [27]. Then, the peptides YPLPR, AFEPLR, and NDPF were docked with proteins. Their results were visualized by PyMOL (version 2.3.1).

2.11. Statistical Analysis

All experiments were conducted at least three times. The data were disposed of and analyzed by Origin 2021 (Origin Lab, Northampton, MA, USA) and SPSS 26 (Chicago, IL, USA) software. The data were represented by a median of 25% and a 75% quartile, and significance was determined by the Kruskal–Wallis test ($p < 0.05$).

3. Results and Discussion

3.1. Identification of the Peptides in Distillers' Grains Protein Hydrolysates

In the process of hydrolysis, the degree of hydrolysis of distiller's grains protein gradually augmented with the increase in time, and after hydrolysis for 2.5 h, the degree of hydrolysis gradually remained steady, indicating that the enzymatic reaction also stabilized. The final hydrolysis degree of alkaline protease was 18.49 ± 0.49%. In previous studies, the final hydrolysis degree of alkaline protease for soybean protein was 19.00% [28], and that for mung bean protein was about 25.00% [29]. In the lump, 20 peptides with ALC scores > 85% were identified among the distillers' grains protein hydrolysate peptides (Table 1). These 20 peptides contained 4–8 amino acids, and their molecular weights ranged from 438–860 Da; all of them were low-molecular-weight peptides. The study of beer and spent grain peptides also showed that the main bioactive peptides were low-molecular-weight peptides with relative molecular masses < 1500 Da [11,17]. The better bioactivity of low molecular mass peptides is mainly due to the fact that they can bind more easily to the active sites of specific enzymes or be more readily absorbed by the gastrointestinal digestive tract and blood vessel system and eventually demonstrate physiological regulatory properties, which also shows that distillers' grains protein hydrolysate peptides may have good biological activity. And in previous studies on rice wine, 99 peptides with ALC scores > 85% were also identified from rice wine [10]. The number of peptides in rice wine was higher than that in distillers' grains protein hydrolysates, mainly because many peptides were further broken down into more peptides during the fermentation of rice wine. The biological activity of distillers' grains protein hydrolysate peptides needs to be further explored and developed.

Table 1. Peptides in distillers' grains protein hydrolysates.

Peptides	ALC (%)	RT (min)	Mass (Da)	Local Confidence (%)
ELELLE	96	2.31	744.3905	99 97 98 94 95 98
RFDR	96	2.2	592.3081	95 96 97 98
PDVGHPM	94	14.64	751.3323	89 97 97 91 96 94 96
YPLPR	92	15.28	644.3646	94 97 96 89 83
VLEPR	91	7.08	612.3595	88 98 97 85 90
AFEPLR	91	15.98	731.3966	90 97 98 85 92 87
FEEL	91	17.76	536.2482	86 93 96 89
WNVN	88	14.76	531.2441	85 79 94 94
EEEF	88	13.81	552.2067	72 94 98 89
FEPLR	88	14.89	660.3595	86 98 81 91 82
WLDY	87	18.69	595.2642	87 89 88 87
ERR	87	1.04	459.2554	96 85 81
FEPL	87	18.25	504.2584	86 94 79 90
LDFEPR	87	17.02	775.3864	88 91 90 96 76 79
FDGVLRGP	86	17	859.4551	89 96 90 95 89 74 79 79
YAGE	86	2.81	438.175	86 85 77 95
NDPF	85	15.77	491.2016	84 87 84 88
RVLEPR	85	4.98	768.4606	82 88 90 95 79 77
WTVN	85	15.35	518.2489	80 82 91 88
WNLN	85	17.09	545.2598	85 77 93 85

RT: retention time.

3.2. α-Glucosidase Inhibitory Effect of Distillers' Grains Protein Hydrolysates

α-glucosidase is a membrane-associated enzyme. Inhibition of α-glucosidase activity can reduce carbohydrate hydrolysis, thus reducing sugar absorption and effectively controlling diabetes. [11]. As shown in Figure 1A, with the increase in concentration, the α-glucosidase inhibitive activity of distiller's grains protein hydrolysate increased, and the α-glucosidase inhibitive activity was 60.88 (58.07, 67.04) % when the concentration of distillers' grains protein hydrolysates was 5 mg/mL, with an IC_{50} value of 3.76 (3.51, 3.83) mg/mL. In the study on rice wine peptides, the α-glucosidase inhibitory activity of 50% ethanol-eluted rice wine peptides had an IC_{50} value of 1.71 (1.65, 1.74) mg/mL [10]. And the IC_{50} value of α-glucosidase depressant activity was 4.94 ± 0.07 mg/mL in the study of soy protein hydrolyzed peptide [15]. The comparison indicated that the distillers' grains protein hydrolysates had α-glucosidase inhibitive activity and deserves more deep investigation and development.

3.3. Scavenging Capacity against DPPH and ABTS Radicals of Distillers' Grains Protein Hydrolysates

The oxidation stress is nearly relevant to diabetes. Reducing the production of free radicals, or directly breaking the free radicals produced by the body, can reverse the tissue damage caused by oxidative stress, thus preventing or delaying diabetes complications. While ABTS and DPPH free radicals are the most common free radicals in the body, the scavenging activities of distillers' grains protein hydrolysates on ABTS and DPPH free radicals are shown in Figure 1B,C. The scavenging activity of distillers' grains protein hydrolysates on ABTS and DPPH radicals increased gradually with increasing concentrations, with IC_{50} values of 0.59 (0.52, 0.64) mg/mL and 1.85 (1.74, 1.86) mg/mL, respectively. This shows that distillers' grains protein hydrolysates have good antioxidant activity.

3.4. Screening of Main Bioactive Peptides in Distillers' Grains Protein Hydrolysates

Peptide Ranker (http://distilldeep.ucd.ie/PeptideRanker/ (accessed on 12 October 2023)) was used to score the probability of biological activity for each peptide sequence, and the outcomes are shown in Table 2. The closer the probability of a peptide being biologically active predicted by Peptide Ranker to 1, the more confident Peptide Ranker is that the peptide is biologically active. Peptides with a score above or equal to 0.6 were more likely to be biologically active [16,17]. Based on the scoring results, ten peptides with scores greater than 0.6 were selected, which were RFDR, PDVGHPM, YPLPR, AFEPLR, FEPLR, WLDY, FEPL, FDGVLRGP, NDPF, and WNLN.

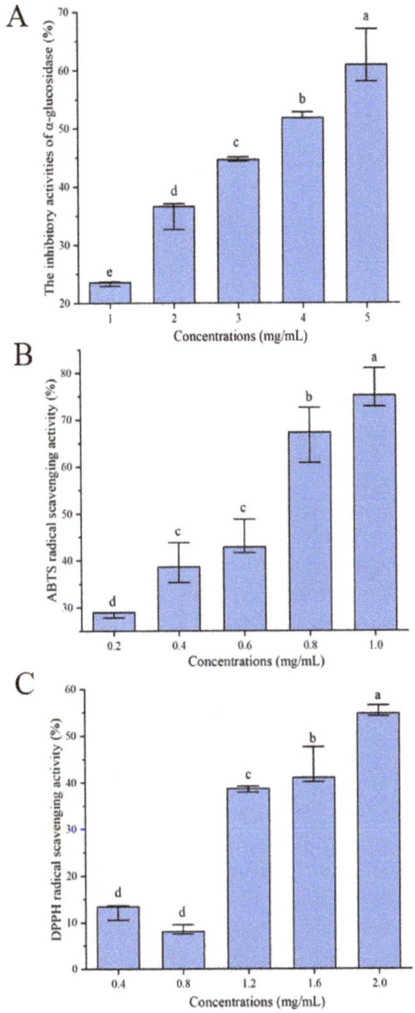

Figure 1. α-Glucosidase inhibitory activities (**A**), ABTS (**B**), and DPPH (**C**) scavenging activities of protein hydrolysate from distillers' grains. The data are displayed as the median (n = 3) with the 25% and 75% quartiles and analyzed using the Kruskal–Wallis test. Significance is represented by different letters ($p < 0.05$).

Due to the fact that α-glucosidase is a more important and directly regulated target in the treatment of diabetes, the potential inhibitory activity of α-glucosidase was used as the main condition of inquiry for further analysis. According to a previous study, the WLRL peptide from soybean protein exhibited a good inhibitory effect toward α-glucosidase [15]. Therefore, this peptide was selected as a positive control to facilitate the analysis of the current study, and its affinity grading after molecular docking with α-glucosidase was −9.1 kcal/mol. Four peptides, YPLPR, AFEPLR, NDPF, and WNLN, had affinities of −10.7, −9.6, −9.1, and −9.2 kcal/mol, respectively, which were better than WLRL, indicating that these peptides may have good α-glucosidase inhibitory potential. Therefore, these peptides can be used as potential functional candidates for α-glucosidase inhibitors for further analysis.

Table 2. Peptides with the main potentially bioactive selected for in silico analyses.

Peptides	Tag Length	Peptide Ranker Score	Affinity (kcal/mol)	Estimated Toxicity	Estimated Solubility
RFDR	4	0.6986	−8.7	Non-toxin	Good
PDVGHPM	7	0.6039	−8.6	Non-toxin	Good
YPLPR	5	0.7972	−10.7	Non-toxin	Good
AFEPLR	6	0.7571	−9.6	Non-toxin	Good
FEPLR	5	0.7106	−9.0	Non-toxin	Good
WLDY	4	0.7981	−8.6	Non-toxin	Poor
FEPL	4	0.7531	−8.8	Non-toxin	Good
FDGVLRGP	8	0.6470	−8.7	Non-toxin	Good
NDPF	4	0.8886	−9.1	Non-toxin	Good
WNLN	4	0.6784	−9.2	Non-toxin	Poor
WLRL (Positive control)	4	0.9210	−9.1	Non-toxin	Poor

The solubility and toxicity of bioactive peptides are closely related to the biological activity of the peptide. Before further application, it plays a large part in analyzing the toxicity of peptides. The Toxin Pred tool is used to predict whether or not peptides are toxic, and it is being utilized more widely to assess numerous bioactive peptides [30]. The prediction results obtained using this tool showed that all the peptides screened in this study were safe and non-toxic. The potential efficacy of bioactive peptides is closely related to the solubility of the peptides, which was assessed by the peptide property calculator Innovagen. The results are shown in Table 2. Among the four peptides obtained by bioactivity scoring, α-glucosidase inhibition binding energy, and toxicity prediction, YPLPR, AFEPLR, and NDPF had good solubility. Therefore, YPLPR, AFEPLR, and NDPF were selected as the main potentially bioactive peptides for further study.

3.5. Activity Verification of Synthetic Novel Peptides

YPLPR had the best inhibitory effect with an IC_{50} value of 5.31 (5.28, 5.35) mmol/L; NDPF had the second inhibitory effect, and AFEPLR had the worst, with IC_{50} values of 6.24 (5.97, 6.35) and 8.69 (8.67, 8.71) mmol/L. These studies have previously identified bioactive peptides with α-glucosidase depressant activity in foods. The peptides LDLQR, AGGFR, and LDNFR from malt were found to exhibit significant α-glucosidase inhibitory activity, with IC_{50} values of 8.59 ± 0.25, 8.66 ± 0.33, and 9.21 ± 0.22 mmol/L, respectively [31]. In contrast, in previous studies on rice wine, α-glucosidase with peptides SSLFR, QFTPR, and FTYPR depressant activity was also discovered, and their IC_{50} values were 7.73 (7.73, 7.74), 9.18 (9.17. 9.18), and 3.57 (3.26, 3.68) mmol/L, respectively [11]. The results of this study, compared with previous studies, all indicated that YPLPR, AFEPLR, and NDPF had good α-glucosidase depressant activity.

The IC_{50} values of ABTS radical scavenging for YPLPR, AFEPLR, and NDPF were 6.05 (5.9, 6.64), >15, and >15 mmol/L, respectively, and the scavenging of DPPH radicals was better than that of ABTS, with IC_{50} values of 7.94 (7.92, 7.96), 13.80 (13.72, 13.94), and 8.71 (8.69, 8.74) mmol/L, respectively. The peptides EAMAPK and AVPYPQ, identified in the research of Pepe et al. (2016), showed antioxidant activity at different concentrations (5–150 mg/mL) [32]. The peptide FTYPR, previously screened from rice wine peptides by molecular simulation, had IC_{50} values of 2.64 (2.63, 2.66) and 6.09 (6.04, 6.82) mmol/L for ABTS radical and DPPH radical scavenging capacity, respectively. It is thus evident that YPLPR, AFEPLR, and NDPF have antioxidant activity; YPLPR has the best antioxidant activity and can be further developed and processed as a potential functional food.

3.6. Improvement of Bioactive Peptides on Insulin Resistance in HepG2 Cells

Insulin is one of the main pathogenic mechanisms of type 2 diabetes mellitus, and the liver is an important site for insulin to function. HepG2 cells are derived from human embryonic hepatocytes and possess many biological characteristics of normal hepatocytes, so HepG2 cells are chosen to establish an insulin resistance model. As the cellular insulin resistance condition arises, the utilization of glucose by the cells decreases, and the consumption of glucose decreases. Therefore, insulin resistance is usually expressed by the consumption of glucose in studies [24]. Using the MTT method, this study explored the effect of different concentrations of insulin on cellular glucose consumption after 24 h of action with HepG2 cells. As shown in Figure 2A, the glucose consumption of the cells gradually decreased as the concentration of insulin increased. When the insulin concentration reached 25.0 µmol/L, the cellular glucose consumption was significantly reduced compared with the group without insulin, which indicated that HepG2 cells developed insulin resistance. And MTT experiments showed that insulin was not cytotoxic at a concentration of 25.0 µmol/L. Therefore, the insulin concentration of 25.0 µmol/L was chosen to establish the insulin resistance model in this experiment. Also, MTT results showed that YPLPR, AFEPLR, and NDPF had no cytotoxicity at 1.0 mmol/mL, so the sample concentrations for subsequent experiments were all 1.0 mmol/L.

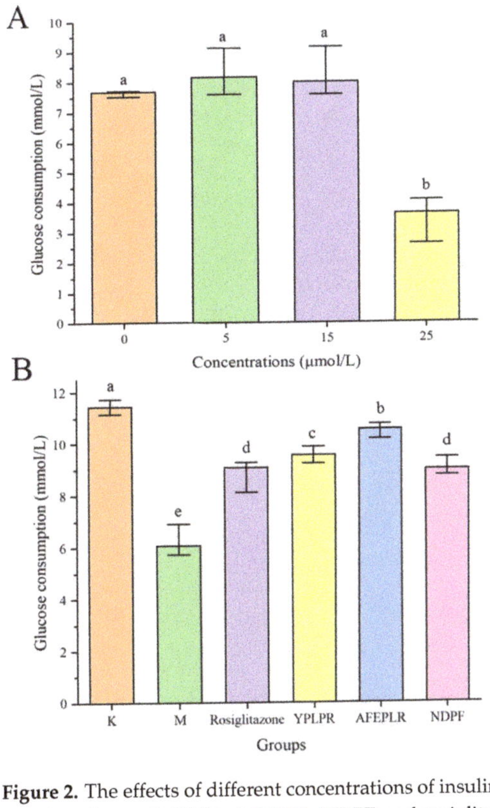

Figure 2. The effects of different concentrations of insulin on glucose consumption of HepG2 cells (**A**); the effects of YPLPR, AFEPLR, NDPF, and rosiglitazone on glucose consumption of HepG2 cells (**B**). Data are presented as the median (n = 6) with the 25% and 75% quartiles; different letters indicated significant differences ($p < 0.05$). (K: control group; M: model group; rosiglitazone: positive control group).

In this study, the effect of YPLPR, AFEPLR, and NDPF on the glucose consumption of HepG2 cells in an insulin resistance state was investigated. As shown in Figure 2B, the glucose consumption of model group M was 6.08 (5.80, 6.81) mmol/L, which was significantly lower than that of blank control group K ($p < 0.05$), indicating that the insulin resistance model of this experiment was successful and the results of the relevant experiments were obtained under the condition that the cells were insulin resistant. The glucose consumption of peptides YPLPR, AFEPLR, and NDPF were 9.56 (9.37, 9.83), 10.56 (10.21, 10.71) and 9.00 (8.80, 9.33) mmol/L, respectively. The glucose consumption of these three peptides was significantly increased compared with model group M; YPLPR and AFEPLR were significantly different from the positive control rosiglitazone group (9.09 (8.29, 9.23) mmol/L) ($p < 0.05$). The efficacy of food-derived peptides to improve insulin resistance has been studied previously, and in a study of walnut-derived peptides, the identified peptides LVRL, LRYL, and VLLALVLLR all increased glucose consumption in HepG2 cells compared to the model group and served to improve insulin resistance [33]. In a study by Gong et al. (2020), specific bioactive peptides SDIPNPIGSE, NPWDQVKR, and QEPVLGPVRGPFP extracted from goat milk casein hydrolysate have demonstrated an improvement in insulin resistance in HepG2 cells exposed to high glucose. This study establishes the potential of goat cheese protein hydrolysate as a food component for enhancing insulin sensitivity [34]. Comparison with these studies indicates that the peptides YPLPR, AFEPLR, and NDPF have strong effects on improving insulin resistance, all of which can promote cellular glucose uptake, and these peptides have some development and utilization value.

3.7. Intracellular ROS Scavenging Activity by Bioactive Peptides

Excessive ROS may attack normal cells or biological macromolecules and undergo oxidative stress, causing severe damage to cellular structures, which causes many diabetic complications and chronic diseases. Accordingly, in this study, the scavenging activities of ABTS and DPPH radicals of bioactive peptides were not only investigated by in vitro experiments, but their antioxidant activity was also further investigated at the cellular level. The cell damage rate of the H_2O_2-induced cell damage model was generally around 50%. As shown in Figure 3A, with the increase in concentration, the cytotoxicity produced by H_2O_2 was enhanced, and the cell viability gradually decreased. When the concentration of H_2O_2 was 1.2 mmol/L, the cell viability rate was 56.38 (50.33, 68.00) %, so the concentration of H_2O_2 of 1.2 mmol/L established the cell damage model in this study. And the intracellular ROS levels in the model group established by 1.2 mmol/L H_2O_2 were significantly higher than those in the blank group ($p < 0.05$), indicating that the oxidative stress model of HepG2 cells induced by 1.2 mmol/L H_2O_2 was successfully established. Consistent with 3.6, YPLPR, AFEPLR, and NDPF were used at 1.0 mmol/L for subsequent experiments.

The intracellular ROS inhibition of the different fractions was represented by flow and histograms. In the flow diagram (Figure 3B), the left-right shift of the peak indicates the magnitude of the fluorescence luminosity, and the more the peak is shifted to the right, the greater the fluorescence intensity. The peak of the model group M was significantly shifted to the right compared with the blank group K, which indicated an increase in fluorescence intensity and an increase in ROS content. As shown in Figure 3C, the relative intracellular ROS content in the M group and the blank group K were 139.46 (135.10, 142.41) % and 101 (95.10, 103.50) %, respectively. The relative intracellular ROS content in the M group was significantly higher than that in the blank group K ($p < 0.05$), which indicated that the oxidative stress model in this batch of experiments was successful. The relative intracellular ROS content in VC, the YPLPR, AFEPLR, and NDPF groups were 92.03 (90.83, 98.11) %, 102.35 (101.38, 103.06) %, 104.11 (103.69, 104.54) % and 95.47 (92.68, 99.37) %, respectively. It can be seen that the intracellular ROS content in the YPLPR, AFEPLR, and NDPF groups was relatively lower than that in the M group; among them, the YPLPR and NDPF intracellular ROS contents were not significantly different from that in the VC group and the blank group ($p > 0.05$). This indicated that the peptides YPLPR, AFEPLR, and NDPF played a good protective role against oxidative stress damage to cells and had a good scavenging effect on reactive oxygen species.

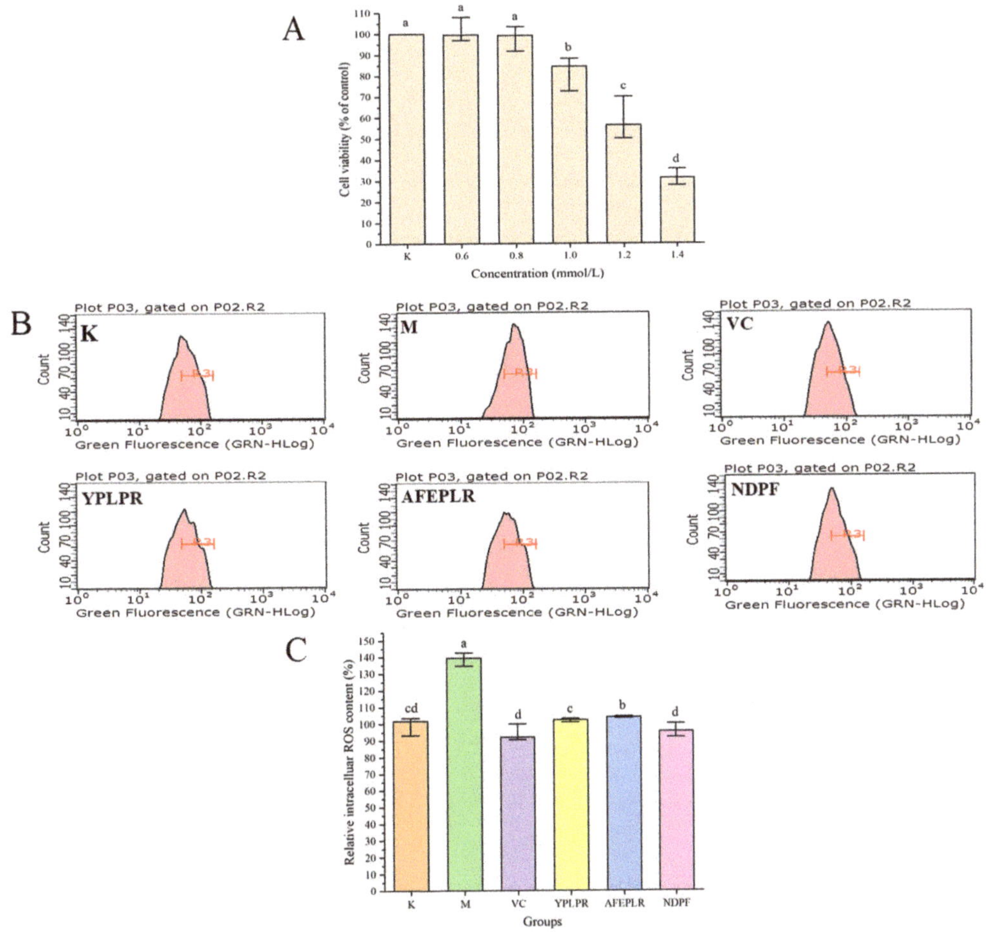

Figure 3. Scavenging of H_2O_2-induced intracellular ROS production by YPLPR, AFEPLR, NDPF, and VC. Effects of different concentrations (**A**) on cell viability; ROS flow chart, R3 represents half the peak width of the blank group K. (**B**); histogram of ROS contents (**C**). Data are displayed as the median (n = 4) with the 25% and 75% quartiles; different letters indicated significant differences ($p < 0.05$). (K: control group; M: model group; VC: positive control group).

3.8. Molecular Docking

3.8.1. The Molecular Docking of Bioactive Peptides with α-Glucosidase

To reveal the possible inhibition mechanisms of YPLPR, AFEPLR, and NDPF towards α-glucosidase, the optimal docking postures between α-glucosidase and three peptides were shown in Figure 4, and the related results are shown in Table 3. Usually, when α-glucosidase binds to the ligand, the active cavity provides a strong hydrophobic environment and multiple hydrogen bond binding sites, which are conducive to the steadiness of the ligand. As shown in Figure 4A,B, YPLPR, AFEPLR, and NDPF can bind to amino acids in the active lumen of α-glucosidase and be completely encapsulated in the α-glucosidase active pocket, forming a relatively stable binding conformation. It has been shown that hydrogen bonding and hydrophobic interactions are important indicators of ligand-receptor binding. Among them, the number and distance of hydrogen bonds are the main drivers of docking ligand–α-glucosidase interaction, where the ligand inhibits α-glucosidase activity by binding to α-glucosidase active amino acids through hydrogen bonds and preventing substrate access to the active pocket [35]. As shown in Figure 4C1,C2 and Table 3, YPLPR formed hydrogen bonds with 9 amino acid residues (Asp215, Ser240, Ser241, Asp242, Glu277, Gln279, Phe314, Arg315, and Asp352) in the α-glucosidase active pocket, and the longest of these bonds was 3.6 Å, the shortest was 1.9 Å. As shown in Figure 4D1,D2 and Table 3, AFEPLR established 11 hydrogen bonds with 7 amino acid residues in the active site of α-glucosidase, with a mean bond distance of 2.4 Å. And it also had hydrophobic interactions with 15 amino acids on α-glucosidase. In contrast, NDPF formed only 3 hydrogen bonds with Lys156, Tyr158, and Asp242 on α-glucosidase, with bond lengths of 2.2 Å, 3.1 Å, and 2.7 Å, respectively. And it had hydrophobic interactions with 8 amino acid residues on α-glucosidase (Figure 4E1,E2). The results indicated that YPLPR, AFEPLR, and NDPF were able to efficiently attach to the active site of α-glucosidase. In previous studies, Asp69, Tyr158, Asp215, Val216, Ser240, Asp242, Arg442, Gln279, Alp411, and Asn415 were identified as the primary active residues responsible for α-glucosidase inhibition [36]. The comprehensive analysis shows that YPLPR interacts with the key residues that inhibit α-glucosidase activity the most, namely Asp215, Ser240, Asp242, and Gln279, which is consistent with the highest YPLPR and α-glucosidase binding scores and the best actual α-glucosidase inhibition experimental activity (3.5 Experimental Results). This result indicates that computer simulation is desirable to be used to assist experimental studies, and it can greatly reduce the cost and time of traditional biological screening, making it an effective method for screening enzyme-inhibiting peptides.

Table 3. The molecular docking results of YPLPR, AFEPLR, and NDPF with α-glucosidase.

	YPLPR	AFEPLR	NDPF
Affinity (kcal/mol)	−10.7	−9.6	−9.1
Number of hydrogen interactions	9	11	3
Amino acid residues involved in hydrogen bonds	Asp 215, Ser 240, Ser 241, Asp 242, Glu 277, Gln 279, Phe 314, Arg 315, Asp 352	Lys 156, Ser 240, Ser 241, His 280, Pro 312, Arg 315, Gln 353	Lys 156, Tyr 158, Asp 242
Number of hydrophobic interactions	10	15	8
Amino acid residues participated in van der waals	Lys 156, Tyr 158, Phe 178, Val 216, Phe 303, Pro 312, Leu 313, Phe 314, Arg 315, Gln 353	Ser 157, Tyr 158, Phe 159, Phe 178, Val 216, Gln 239, Glu 277, His 280, Phe 303, Pro 312, Leu 313, Arg 315, Asp 352, Glu 411, Arg 442	Tyr 158, Phe 178, Glu277, Gln 279, Phe 303, Pro 312, Arg 315, Arg 442

3.8.2. The Molecular Docking of Bioactive Peptides with ABTS and DPPH

The radical scavenging mechanisms of the three peptides YPLPR, AFEPLR, and NDPF on ABTS and DPPH radicals were also investigated by molecular docking. As shown in Figure 5A1,A2, the binding energy of YPLPR to the ABTS radical is −4.2 kcal/mol. ABTS, as a ligand, can form 2 hydrogen bonds with Met7 and Pi bonds with His5 and Pro6 on YPLPR. As shown in Figure 5B1–D2, the binding energies of YPLPR, AFEPLR, and NDPF to the DPPH radical are −4.2, −3.9, and −4.2 kcal/mol, respectively. DPPH as a ligand can form 2 hydrogen bonds with His5 and Pi bonds with Met7 on YPLPR (Figure 5B1,B2). The amino acids of DPPH and AFEPLR are mainly bound by Pi bonds (Leu5) and hydrophobic interactions (Ala1, Phe2, and Glu3) (Figure 5C1,C2). DPPH forms a hydrogen bond with Asp2 on NDPF with a bond distance of 2.5 Å and a Pi bond with Phe4 (Figure 5D1,D2). In summary, the results indicated that amino acid residues belonging to YPLPR, AFEPLR, and NDPF could bind to ABTS and DPPH free radicals through hydrogen bonds, π bonds, or hydrophobic interactions, which made these peptides have good ABTS or DPPH free radical scavenging activity.

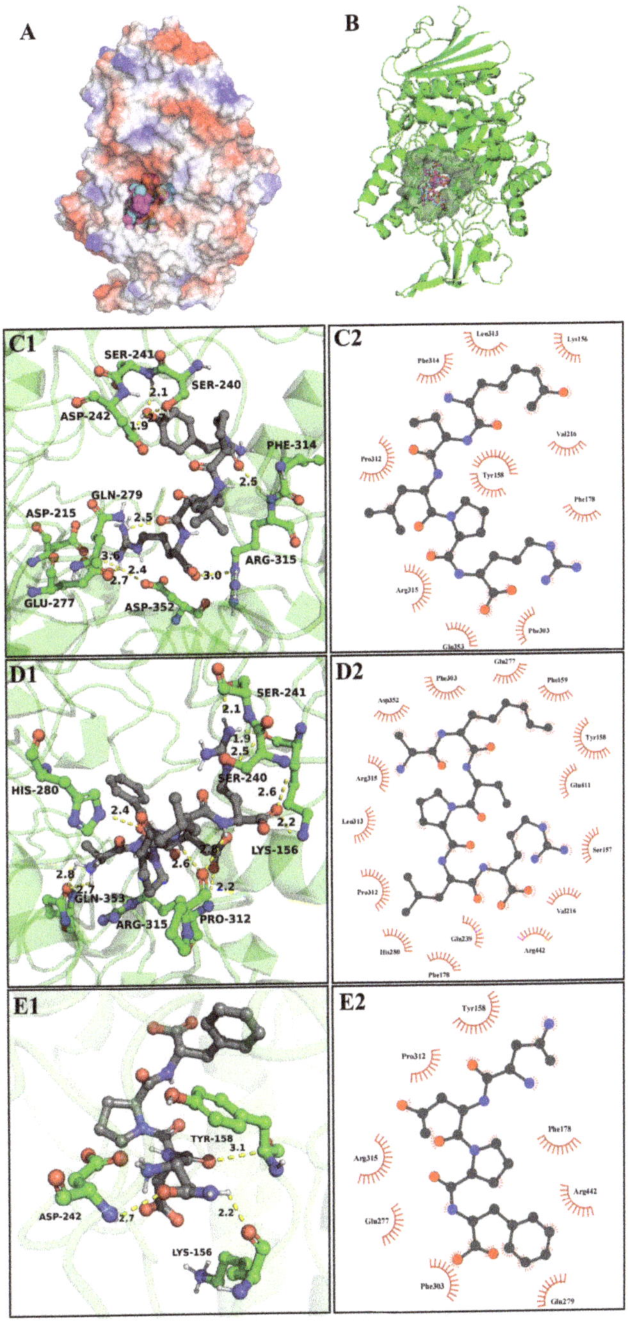

Figure 4. Molecular docking results of YPLPR, AFEPLR, and NDPF with α-glucosidase (PDB Code: 3A4A). The 3D structural surface images of three peptide with α-glucosidase (**A**); active site binding of three peptide with α-glucosidase, (**B**), where YPLPR is in blue, AFEPLR in purple and NDPF in red. The 3D images of the hydrogen bonding of YPLPR (**C1**), AFEPLR (**D1**), and NDPF (**E1**) with the amino acid residues of α-glucosidase (The yellow dotted line represents the hydrogen bond distance, green part is α-glucosidase, and gray part is peptide.), and the 2D images of the hydrophobic interaction of YPLPR (**C2**), AFEPLR (**D2**), and NDPF (**E2**) with the amino acid residues of α-glucosidase.

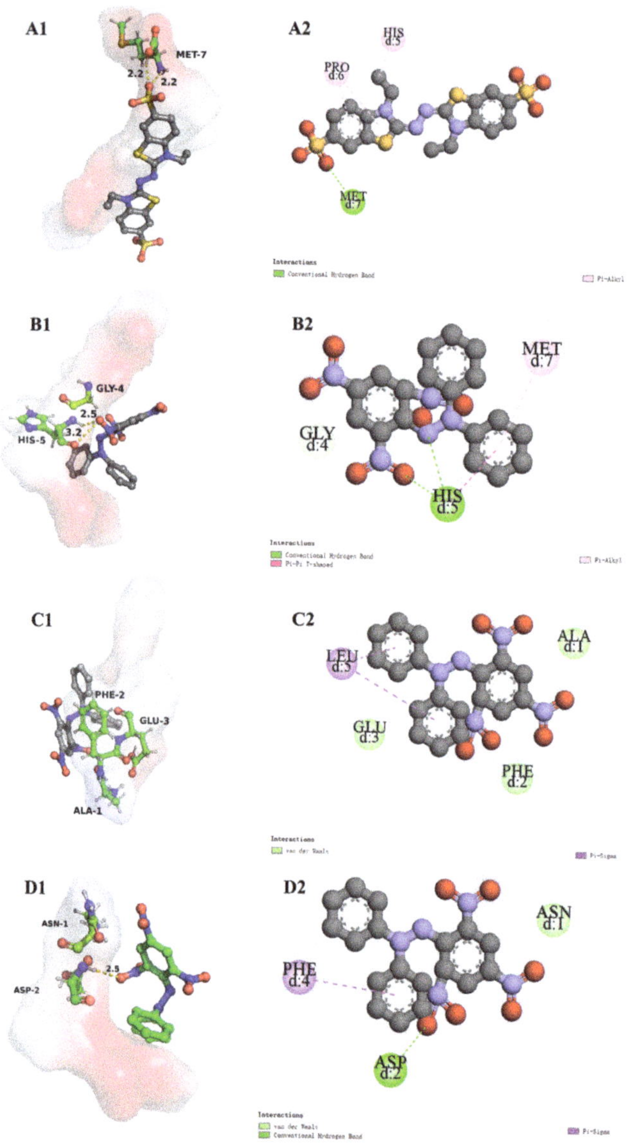

Figure 5. The molecular docking results of YPLPR (**A1**,**A2**) with ABTS and YPLPR (**B1**,**B2**), AFEPLR (**C1**,**C2**), and NDPF (**D1**,**D2**) with DPPH. Peptides are displayed by surface style, ABTS, and DPPH are displayed by rod and ball model. The yellow dotted line represents the hydrogen bond distance.

3.8.3. The Molecular Docking of Bioactive Peptides with Insulin Signaling Pathway

The molecular docking results between YPLPR, AFEPLR, and NDPF and PIRTK protein and PPARγ are shown in Figures 6 and 7. Peptides YPLPR, AFEPLR, and NDPF were able to enter the active cavities of PIRTK protein (Figure 6A,B) and PPARγ (Figure 7A,B) and were completely encapsulated in the active pockets of PIRTK proteins and PPARγ, forming relatively stable binding conformations. For PIRTK protein, the binding energies of YPLPR, AFEPLR, and NDPF to PIRTK protein were −7.2, −7.7, and −7.3 kcal/mol, respectively. YPLPR formed 4 hydrogen bonds with Ser 1006, Glu 1040, Arg 1136, and Asp 1150 on PIRTK protein (Figure 6C1), AFEPLR formed 5 hydrogen bonds with Leu

1002, Ser 1006, Asp 1083, and Arg 1136 on PIRTK protein (Figure 6D1), and NDPF formed 5 hydrogen bonds with Met 1079, Asn 1137, Met 1139, and Asp 1150 on PIRTK protein (Figure 6E1). And their average bond lengths were 2.45 Å, 2.66 Å, and 2.6 Å, respectively. Meanwhile, peptides YPLPR, AFEPLR, and NDPF were also hydrophobic with 7, 12, and 8 amino acid residues on the PIRTK protein, respectively (Figure 6C2,D2,E2). For PPARγ, the binding energies of peptides YPLPR, AFEPLR, and NDPF to PPARγ were −9.2, −8.8, and −8.5 kcal/mol, respectively. YPLPR was able to form hydrogen bonds with amino acid residues Arg288 and Ser 342 on PPARγ (Figure 7C1), AFEPLR was able to form hydrogen bonds with amino acid residues Gln 271, Cys 285, Tyr 327, and Ser 342 on PPARγ (Figure 7D1), and NDPF can form 5 hydrogen bonds with Gln 271, Arg 280, Ser 342, and Glu 343 on PPARγ (Figure 7E1). Peptides YPLPR, AFEPLR, and NDPF also have hydrophobic interactions with 16, 16, and 15 amino acid residues on PPARγ, respectively (Figure 7C2,D2,E2). These strong hydrogen bonding and hydrophobic interactions both stabilize the binding of YPLPR, AFEPLR, and NDPF to PIRTK protein and PPARγ, thus potentially inhibiting the activity of PIRTK protein and PPARγ.

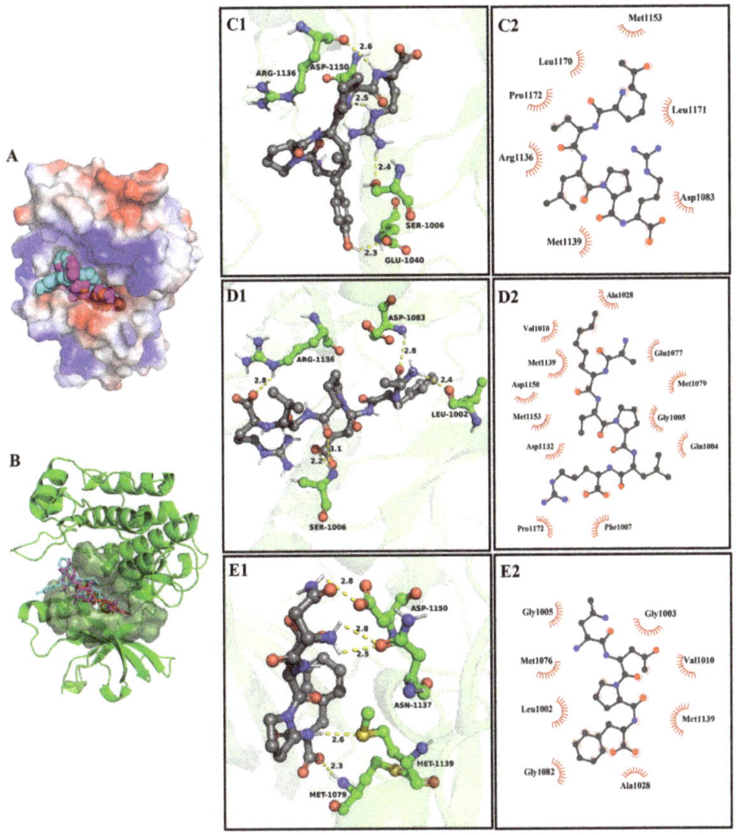

Figure 6. Molecular docking results of YPLPR, AFEPLR, and NDPF with PIRTK protein (PDB Code: 1IR3). The 3D structural surface images of three peptide with receptor protein (**A**); active site binding of three peptide with PIRTK protein (**B**), where YPLPR is in blue, AFEPLR in purple, and NDPF in red. The 3D images of the hydrogen bonding of YPLPR (**C1**), AFEPLR (**D1**), and NDPF (**E1**) with the amino acid residues of PIRTK protein (The yellow dotted line represents the hydrogen bond distance.), and the 2D images of the hydrophobic interaction of YPLPR (**C2**), AFEPLR (**D2**), and NDPF (**E2**) with the amino acid residues of PIRTK protein.

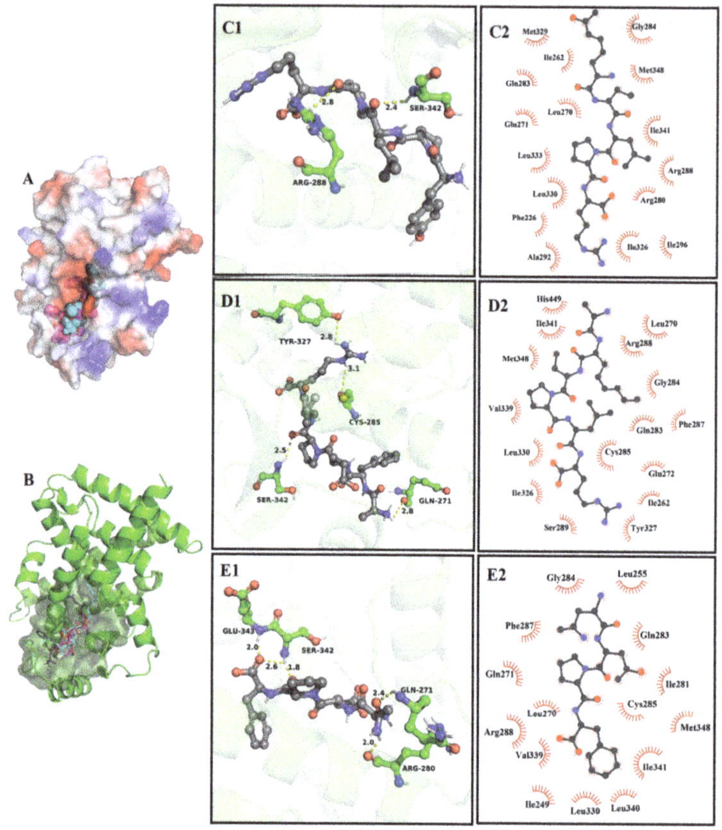

Figure 7. Molecular docking results of YPLPR, AFEPLR, and NDPF with PPARγ (PDB Code: 2PRG). The 3D structural surface images of three peptide with PPARγ (**A**); active site binding of three peptide with receptor protein (**B**), where YPLPR is in blue, AFEPLR in purple, and NDPF in red. The 3D images of the hydrogen bonding of YPLPR (**C1**), AFEPLR (**D1**), and NDPF (**E1**) with the amino acid residues of PPARγ (The yellow dotted line represents the hydrogen bond distance.), and the 2D images of the hydrophobic interaction of YPLPR (**C2**), AFEPLR (**D2**), and NDPF (**E2**) with the amino acid residues of PPARγ.

4. Conclusions

The results showed that a total of 20 peptides with ALC scores > 85% were identified in the distillers' grains peptides obtained by alkaline protease enzymatic digestion, and the molecular weights were concentrated between 438 and 860 Da. The distillers' grains peptides had good α-glucosidase inhibitory activity and antioxidant activity. Bioinformatics analysis showed that peptides YPLPR, AFEPLR, and NDPF were the main potentially bioactive peptides in distillers' grains peptides and the three peptides had good α-glucosidase inhibitory activity, insulin resistance-improving and antioxidant activities. And peptides YPLPR, AFEPLR, and NDPF may exhibit their bioactives by stably binding to α-glucosidase, ABTS, and DPPH radicals, as well as relevant targets of the insulin signaling pathway, through strong hydrogen bonding and van der Waals forces. Generally speaking, these results obtained in the present work may offer a potential natural resource for alleviating oxidative stress damage and controlling blood glucose in diabetes, thereby enhancing the utilization of distillers' grains peptides and augmenting their economic value. However, it should be noted that the methods used in the current study were all in vitro, which may differ from the real environment of the human body; the bioavailability, intestinal absorption, and bioactivities of these peptides need to be further investigated in vivo in the future.

Supplementary Materials: The following supporting information can be downloaded at: https://www.mdpi.com/article/10.3390/nu16091279/s1. Methods: The methods used in this experiment were described in detail.

Author Contributions: Investigation, methodology, data curation, L.D.; investigation, data curation, and writing—original draft, X.Z.; writing—review and editing, methodology, L.Z.; conceptualization, supervision, and funding acquisition, S.C. All authors have read and agreed to the published version of the manuscript.

Funding: The present work was financially supported by the Major Science and Technology Project of the Science and Technology Department of Yunnan Province (Grant Nos. 202102AE090050 and 202202AG050009).

Institutional Review Board Statement: Not applicable.

Informed Consent Statement: Not applicable.

Data Availability Statement: The original contributions presented in the study are included in the article/Supplementary Materials, further inquiries can be directed to the corresponding author due to privacy.

Conflicts of Interest: The authors declare no conflicts of interest.

References

1. Glovaci, D.; Fan, W.; Wong, N.D. Epidemiology of Diabetes Mellitus and Cardiovascular Disease. *Curr. Cardiol. Rep.* **2019**, *21*, 21. [CrossRef]
2. Zhao, C.-C.; Zhang, X.-H.; Chen, J.; Shao, J.-H.; Zhao, Z.-Y.; Tang, Y.-Y. Lignans with α-Glucosidase, Protein Tyrosine Phosphatase 1B, and Aldose Reductase Inhibitory Activities from the Fruits of Viburnum Cylindricum. *Ind. Crops Prod.* **2022**, *178*, 114601. [CrossRef]
3. Kazmi, M.; Zaib, S.; Ibrar, A.; Amjad, S.T.; Shafique, Z.; Mehsud, S.; Saeed, A.; Iqbal, J.; Khan, I. A New Entry into the Portfolio of α-Glucosidase Inhibitors as Potent Therapeutics for Type 2 Diabetes: Design, Bioevaluation and One-Pot Multi-Component Synthesis of Diamine-Bridged Coumarinyl Oxadiazole Conjugates. *Bioorg. Chem.* **2018**, *77*, 190–202. [CrossRef] [PubMed]
4. Oseguera-Toledo, M.E.; Gonzalez de Mejia, E.; Amaya-Llano, S.L. Hard-to-Cook Bean (*Phaseolus vulgaris* L.) Proteins Hydrolyzed by Alcalase and Bromelain Produced Bioactive Peptide Fractions That Inhibit Targets of Type-2 Diabetes and Oxidative Stress. *Food Res. Int.* **2015**, *76*, 839–851. [CrossRef] [PubMed]
5. Paneni, F.; Beckman, J.A.; Creager, M.A.; Cosentino, F. Diabetes and Vascular Disease: Pathophysiology, Clinical Consequences, and Medical Therapy: Part I. *Eur. Heart J.* **2013**, *34*, 2436–2443. [CrossRef]
6. Hedrington, M.S.; Davis, S.N. Considerations When Using Alpha-Glucosidase Inhibitors in the Treatment of Type 2 Diabetes. *Expert. Opin. Pharmacother.* **2019**, *20*, 2229–2235. [CrossRef]
7. Dash, R.P.; Babu, R.J.; Srinivas, N.R. Reappraisal and Perspectives of Clinical Drug–Drug Interaction Potential of α-Glucosidase Inhibitors Such as Acarbose, Voglibose and Miglitol in the Treatment of Type 2 Diabetes Mellitus. *Xenobiotica* **2018**, *48*, 89–108. [CrossRef]
8. Khalid, Z.; Shafqat, S.S.; Ahmad, H.A.; Rehman, H.M.; Munawar, M.A.; Ahmad, M.; Asiri, A.M.; Ashraf, M. Synthesis of 1,2,3-Benzotriazin-4(3H)-One Derivatives as α-Glucosidase Inhibitor and Their in-Silico Study. *Med. Chem. Res.* **2022**, *31*, 819–831. [CrossRef]
9. Wang, X.; Chen, H.; Fu, X.; Li, S.; Wei, J. A Novel Antioxidant and ACE Inhibitory Peptide from Rice Bran Protein: Biochemical Characterization and Molecular Docking Study. *LWT* **2017**, *75*, 93–99. [CrossRef]
10. Zheng, X.; Chi, H.; Ma, S.; Zhao, L.; Cai, S. Identification of Novel α-Glucosidase Inhibitory Peptides in Rice Wine and Their Antioxidant Activities Using in Silico and in Vitro Analyses. *LWT* **2023**, *178*, 114629. [CrossRef]
11. Wei, D.; Fan, W.; Xu, Y. Identification of Water-Soluble Peptides in Distilled Spent Grain and Its Angiotensin Converting Enzyme (ACE) Inhibitory Activity Based on UPLC-Q-TOF-MS and Proteomics Analysis. *Food Chem.* **2021**, *353*, 129521. [CrossRef] [PubMed]
12. Yoshikawa, M.; Fujita, H.; Matoba, N.; Takenaka, Y.; Yamamoto, T.; Yamauchi, R.; Tsuruki, H.; Takahata, K. Bioactive Peptides Derived from Food Proteins Preventing Lifestyle-related Diseases. *BioFactors* **2000**, *12*, 143–146. [CrossRef] [PubMed]
13. Mora, L.; González-Rogel, D.; Heres, A.; Toldrá, F. Iberian Dry-Cured Ham as a Potential Source of α-Glucosidase-Inhibitory Peptides. *J. Funct. Foods* **2020**, *67*, 103840. [CrossRef]
14. Hu, S.; Fan, X.; Qi, P.; Zhang, X. Identification of Anti-Diabetes Peptides from Spirulina Platensis. *J. Funct. Foods* **2019**, *56*, 333–341. [CrossRef]

15. Wang, R.; Zhao, H.; Pan, X.; Orfila, C.; Lu, W.; Ma, Y. Preparation of Bioactive Peptides with Antidiabetic, Antihypertensive, and Antioxidant Activities and Identification of A-glucosidase Inhibitory Peptides from Soy Protein. *Food Sci. Nutr.* **2019**, *7*, 1848–1856. [CrossRef] [PubMed]
16. Xie, Y.; Wang, J.; Li, Z.; Luan, Y.; Li, M.; Peng, X.; Xiao, S.; Zhang, S. Damage Prevention Effect of Milk-Derived Peptides on UVB Irradiated Human Foreskin Fibroblasts and Regulation of Photoaging Related Indicators. *Food Res. Int.* **2022**, *161*, 111798. [CrossRef] [PubMed]
17. Wenhui, T.; Shumin, H.; Yongliang, Z.; Liping, S.; Hua, Y. Identification of in Vitro Angiotensin-converting Enzyme and Dipeptidyl Peptidase IV Inhibitory Peptides from Draft Beer by Virtual Screening and Molecular Docking. *J. Sci. Food Agric.* **2022**, *102*, 1085–1094. [CrossRef]
18. Zhang, X.; Jia, Y.; Ma, Y.; Cheng, G.; Cai, S. Phenolic Composition, Antioxidant Properties, and Inhibition toward Digestive Enzymes with Molecular Docking Analysis of Different Fractions from Prinsepia Utilis Royle Fruits. *Molecules* **2018**, *23*, 3373. [CrossRef]
19. Sun, D.; Huang, S.; Cai, S.; Cao, J.; Han, P. Digestion Property and Synergistic Effect on Biological Activity of Purple Rice (*Oryza sativa* L.) Anthocyanins Subjected to a Simulated Gastrointestinal Digestion in Vitro. *Food Res. Int.* **2015**, *78*, 114–123. [CrossRef]
20. O'Boyle, N.M.; Banck, M.; James, C.A.; Morley, C.; Vandermeersch, T.; Hutchison, G.R. Open Babel: An Open Chemical Toolbox. *J. Cheminform.* **2011**, *3*, 33. [CrossRef]
21. Garcia-Vaquero, M.; Mora, L.; Hayes, M. In Vitro and In Silico Approaches to Generating and Identifying Angiotensin-Converting Enzyme I Inhibitory Peptides from Green Macroalga Ulva Lactuca. *Mar. Drugs* **2019**, *17*, 204. [CrossRef] [PubMed]
22. Ferysiuk, K.; Wójciak, K.M.; Kęska, P. Effect of Willow Herb (*Epilobium angustifolium* L.) Extract Addition to Canned Meat with Reduced Amount. of Nitrite on the Antioxidant and Other Activities of Peptides. *Food Funct.* **2022**, *13*, 3526–3539. [CrossRef] [PubMed]
23. Zhou, M.; Bu, T.; Zheng, J.; Liu, L.; Yu, S.; Li, S.; Wu, J. Peptides in Brewed Wines: Formation, Structure, and Function. *J. Agric. Food Chem.* **2021**, *69*, 2647–2657. [CrossRef] [PubMed]
24. Zhu, D.; Wang, Y.; Du, Q.; Liu, Z.; Liu, X. Cichoric Acid Reverses Insulin Resistance and Suppresses Inflammatory Responses in the Glucosamine-Induced HepG2 Cells. *J. Agric. Food Chem.* **2015**, *63*, 10903–10913. [CrossRef] [PubMed]
25. Liu, X.; Wang, O.; Zhou, J.; Cai, S. Different Phenolic Extracts of Oil Palm Fruits and Caffeic Acid Prevent Palmitic Acid-Induced Lipotoxicity in HepG2 Cells via Improving Mitochondrial Function. *J. Food Qual.* **2020**, *2020*, 8827707. [CrossRef]
26. Chettri, S.; Tamang, S.; Rai, P.; Rai, Y.; Singha, U.K.; Pradhan, K.; Brahman, D.; Sinha, B. Synthesis, Physico-chemical Characterization and Theoretical Exploration of Some 2,4,5-triaryl Imidazole Derivatives. *J. Heterocycl. Chem.* **2023**, *60*, 1394–1415. [CrossRef]
27. da Costa Leite, L.F.C.; Veras Mourão, R.H.; de Lima, M.d.C.A.; Galdino, S.L.; Hernandes, M.Z.; de Assis Rocha Neves, F.; Vidal, S.; Barbe, J.; da Rocha Pitta, I. Synthesis, Biological Evaluation and Molecular Modeling Studies of Arylidene-Thiazolidinediones with Potential Hypoglycemic and Hypolipidemic Activities. *Eur. J. Med. Chem.* **2007**, *42*, 1263–1271. [CrossRef] [PubMed]
28. Mojica, L.; de Mejía, E.G. Optimization of Enzymatic Production of Anti-Diabetic Peptides from Black Bean (*Phaseolus vulgaris* L.) Proteins, Their Characterization and Biological Potential. *Food Funct.* **2016**, *7*, 713–727. [CrossRef]
29. Ying, F.; Lin, S.; Li, J.; Zhang, X.; Chen, G. Identification of Monoamine Oxidases Inhibitory Peptides from Soybean Protein Hydrolysate through Ultrafiltration Purification and in Silico Studies. *Food Biosci.* **2021**, *44*, 101355. [CrossRef]
30. Dai, L.; Kong, L.; Cai, X.; Jiang, P.; Liu, N.; Zhang, D.; Li, Z. Analysis of the Structure and Activity of Dipeptidyl Peptidase IV (DPP-IV) Inhibitory Oligopeptides from Sorghum Kafirin. *J. Agric. Food Chem.* **2022**, *70*, 2010–2017. [CrossRef]
31. Liu, W.; Li, H.; Wen, Y.; Liu, Y.; Wang, J.; Sun, B. Molecular Mechanism for the α-Glucosidase Inhibitory Effect of Wheat Germ Peptides. *J. Agric. Food Chem.* **2021**, *69*, 15231–15239. [CrossRef] [PubMed]
32. Pepe, G.; Sommella, E.; Ventre, G.; Scala, M.C.; Adesso, S.; Ostacolo, C.; Marzocco, S.; Novellino, E.; Campiglia, P. Antioxidant Peptides Released from Gastrointestinal Digestion of "Stracchino" Soft Cheese: Characterization, in Vitro Intestinal Protection and Bioavailability. *J. Funct. Foods* **2016**, *26*, 494–505. [CrossRef]
33. Wang, J.; Wu, T.; Fang, L.; Liu, C.; Liu, X.; Li, H.; Shi, J.; Li, M.; Min, W. Peptides from Walnut (Juglans Mandshurica Maxim.) Protect Hepatic HepG2 Cells from High Glucose-Induced Insulin Resistance and Oxidative Stress. *Food Funct.* **2020**, *11*, 8112–8121. [CrossRef] [PubMed]
34. Gong, H.; Gao, J.; Wang, Y.; Luo, Q.W.; Guo, K.R.; Ren, F.Z.; Mao, X.Y. Identification of Novel Peptides from Goat Milk Casein That Ameliorate High-Glucose-Induced Insulin Resistance in HepG2 Cells. *J. Dairy. Sci.* **2020**, *103*, 4907–4918. [CrossRef] [PubMed]
35. Ding, D.; Du, B.; Zhang, C.; Zaman, F.; Huang, Y. Isolation and Identification of an Antioxidant Collagen Peptide from Skipjack Tuna (*Katsuwonus pelamis*) Bone. *RSC Adv.* **2019**, *9*, 27032–27041. [CrossRef]
36. Mi, S.; Liu, J.; Liu, X.; Fu, Y.; Yi, J.; Cai, S. Inhibitory Effects of Myricetrin and Dihydromyricetin toward α-Glucosidase and Pancreatic Lipase with Molecular Docking Analyses and Their Interaction. *J. Food Qual.* **2021**, *2021*, 9943537. [CrossRef]

foods MDPI

Article

Structure–Activity Relationship of Novel ACE Inhibitory Undecapeptides from *Stropharia rugosoannulata* by Molecular Interactions and Activity Analyses

Wen Li [1,2], Wanchao Chen [1], Jinbin Wang [3], Zhengpeng Li [1], Zhong Zhang [1], Di Wu [1], Mengqiu Yan [1], Haile Ma [2,*] and Yan Yang [1,*]

1 Institute of Edible Fungi, Shanghai Academy of Agricultural Sciences, Shanghai 201403, China; liwen3848@126.com (W.L.); chenwanchao@saas.sh.cn (W.C.); lizp_ln@126.com (Z.L.); zhangz0815@126.com (Z.Z.); vivian218074@163.com (D.W.); yanmengqiu@saas.sh.cn (M.Y.)
2 School of Food & Biological Engineering, Jiangsu University, Zhenjiang 212013, China
3 Institute of Biotechnology Research, Shanghai Academy of Agricultural Sciences, Shanghai 201106, China; wangjinbin2013@126.com
* Correspondence: mhl@ujs.edu.cn (H.M.); yangyan@saas.sh.cn (Y.Y.)

Abstract: Undecapeptide is the central peptide molecule in the peptide base material of *Stropharia rugosoannulata*, and angiotensin-converting enzyme (ACE) plays a crucial role in hypertension. To fully explore the interaction mechanism and ACE-inhibitory activity of long-chain peptides from *Stropharia rugosoannulata*, the binding conformations of twenty-seven undecapeptides with the ACE receptor were revealed by molecule docking. The undecapeptide GQEDYDRLRPL with better receptor binding capacity and higher secondary mass spectral abundance was screened. All amino acid residues except proline in GQEDYDRLRPL interacted with the ACE receptor. GQEDYDRLRPL interfered with the receptor's overall structure, with significant fluctuations in amino acid residues 340–355, including two residues in the receptor's active pockets. The binding constants of GQEDYDRLRPL to the ACE receptors were at the μM level, with a kinetic binding constant of 9.26×10^{-7} M, which is a strong binding, and a thermodynamic binding constant of 3.06×10^{-6} M. Intermolecular interaction were exothermic, enthalpy-driven, and specific binding reactions. GQEDYDRLRPL had an IC_{50} value of 164.41 μmol/L in vitro and superior antihypertensive effects at low-gavage administration in vivo. Obtaining information on the interaction mechanism of ACE-inhibitory undecapeptides from *S. rugosoannulata* with the ACE receptor will help to develop and utilize ACE inhibitors of natural origin.

Keywords: *Stropharia rugosoannulata* undecapeptides; molecular docking; molecular dynamics simulation; molecular thermodynamics reaction; molecular dynamics reaction; ACE inhibition mechanism

Citation: Li, W.; Chen, W.; Wang, J.; Li, Z.; Zhang, Z.; Wu, D.; Yan, M.; Ma, H.; Yang, Y. Structure–Activity Relationship of Novel ACE Inhibitory Undecapeptides from *Stropharia rugosoannulata* by Molecular Interactions and Activity Analyses. *Foods* **2023**, *12*, 3461. https://doi.org/10.3390/foods12183461

Academic Editor: Vermont Punongba Dia

Received: 2 September 2023
Revised: 14 September 2023
Accepted: 14 September 2023
Published: 17 September 2023

1. Introduction

Currently, the number of people with hypertension worldwide is constantly increasing. On 25 August 2021, a research paper titled "Worldwide trends in tension pressure and progress in treatment and control from 1990 to 2019", coauthored by Imperial College London and the World Health Organization, was published in The Lancet [1]. Studies show that the number of adults aged 30 to 79 with hypertension has increased from 650 million to 1.28 billion in the last 30 years. Severe hypertension can cause physical problems like heart, brain, and kidney damage. Patients with malignant hypertension can develop kidney failure and even die quickly. The angiotensin-converting enzyme (ACE) plays a vital role in the induction of elevated blood pressure in vivo and is a drug target for the treatment of hypertension. ACE-inhibitory peptides can inhibit ACE activity and thus have a hypotensive effect. ACE-inhibitory peptides from food sources are considered natural substances and exhibit a high biological activity. They can be obtained by extraction, protein hydrolysis, and fermentation. Unlike antihypertensive drugs, food-derived ACE-inhibitory peptides have few side effects and a significant utilization value for people with hypertension and cardiovascular disease.

Peptides with ACE-inhibitory properties derived from food sources, whether plant or animal-based, are typically found in a mixture of peptide molecules. Isolating and purifying a single peptide fraction can be expensive, and the process is often limited to laboratory settings. Researchers commonly use LC-MS/MS techniques to analyze high-peptide-content mixtures. To predict peptide activity, functional peptide sequences are matched against the published literature. The actual activities of synthesized peptides are then validated through in vitro and in vivo tests. Notably, most peptide products currently available in the industry comprise a mixture of peptide molecules. Therefore, it is essential to determine the composition and conformational relationships of peptide molecules within mixed materials. This information can be used as a reference for the industrial applications of these peptide products.

Stropharia rugosoannulata is a grass rot fungus cultivated on a rice straw substrate. It is known for being rich in protein and peptides. The peptide products that are prepared by hydrolyzing the protein of *S. rugosoannulata* have been found to have excellent ACE-inhibitory activity (Patent No. CN202210297426.2). Despite this potential, the use of *S. rugosoannulata*'s ACE-inhibitory peptide in creating healthy food products has yet to be fully explored. The properties of peptides, such as their sequence, structure, and target receptor recognition, are closely linked to their active properties. Thus, creating a peptide library with different sequences is essential to explore their ACE inhibition mechanism further.

As long-chain peptides, polypeptides have a more complex spatial conformation than oligopeptides, resulting in more active peptide fragments that can be produced after digestion and hydrolysis. A library was constructed to investigate the potential ACE-inhibitory activity of undecapeptides (the main long-chain peptides identified by LC-MS/MS), and the toxicology and absorption activity of the undecapeptides were analyzed. Various techniques, including molecular docking, molecular dynamics simulation, molecular thermodynamics interaction, and molecular dynamics interaction, were employed to study the interaction between the undecapeptides and the ACE receptor. In vitro and in vivo analyses were carried out to verify the hypotensive function of the screened undecapeptide. This study could serve as a reference for developing highly active long-chain peptides.

2. Materials and Methods

2.1. Preparation of the Peptide Base Material of *S. rugosoannulata*

The *S. rugosoannulata* mushroom (mushroom strain NCBI No. SRR14469700) with a concentration of 48 g/L, 2×10^5 U/g alkaline protease added amount of 1% (*w*/*w*) (Beijing Solarbio Science & Technology Co., Ltd., Beijing, China), hydrolysis temperature 42 °C, pH8.5 with synchronous ultrasonic-assisted enzyme hydrolysis conditions (bath ultrasonic power density 120 W/L, ultrasonic frequency 20 kHz, and hydrolysis time 40 min) were carried out to obtain the enzymatic hydrolysis solution of *S. rugosoannulata*. The hydrolysis supernatant was obtained by centrifuging the hydrolysis solution at 9000× *g* for 10 min at 4 °C and ultrafiltrate by the GE 3000 NMWC UF membrane. The permeate was freeze-dried at −70 °C for 48 h to obtain the peptide base material [2]. The method for identifying the peptide sequence of *S. rugosoannulata* was the same as that in the literature [3]. Peptide base materials were desalted, dissolved, centrifuged, and then taken for peptide sequence analysis by LC-MS/MS (Thermo-Obritrap-QE, ThermoFisher Scientific, Shanghai, China). The PEAKS 10.5 software (Bioinformatics Solutions Inc., Waterloo, ON, Canada) was used for peptide database retrieval.

2.2. Construction of *S. rugosoannulata* Undecapeptide Library with ACE Inhibitory Activity

The 82 undecapeptides identified by LC-MS/MS (Supplementary Table S1) were used for screening *S. rugosoannulata* undecapeptides with potential ACE-inhibitory activity. The peptide fragments with ACE-inhibitory activity in undecapeptides were identified by comparing the functional peptide sequence reported in the BIOPEP-UWM (https://biochemia.uwm.edu.pl/biopep-uwm/, accessed on 17 May 2023.). AdmetSAR (http://lmmd.ecust.edu.cn/admetsar1/predict/, accessed on 17 May 2023.) was used to predict the blood–brain barrier penetration, intestinal epithelial Caco-2 cell penetration, human intestinal absorption, and AMES oral toxicity of undecapeptides [4].

2.3. The ACE Inhibition Mechanism Assay of *S. rugosoannulata* Undecapeptides Based on Molecular Docking

Molecular docking is a computer simulation method to explain the binding patterns and interaction mechanisms between biomolecules at the atomic level. The crystal structure of the ACE receptor protein (1O8A) obtained from the PDB database was used as the target protein. MOE2019 (Chemical Computing Group ULC., Montreal, BC, Canada) was used to optimize the energy and structure of the ACE receptor and construct the 3D structure database of undecapeptides. Combining the critical amino acid residue (ALA354, GLU384, TYR523, GLN281, HIS353, HIS513, LYS511, TYR520, GLU162) in functional pockets (S1, S2, S1′) of the ACE receptor reported in the literature [5], the docking scoring value, the number of bonds formed, and the docking energy of undecapeptides to the ACE receptor were used as indicators to select the tightly bound undecapeptide–receptor complexes. Molecular docking was performed with 3 replications to verify the stability of the molecular simulation results. The Site Finder computed and applied the docking sites in the ACE receptor. The London dG score and number of poses were defaulted in MOE. MOE analyzed the interaction modes of the undecapeptides and the ACE receptor.

2.4. The ACE Inhibition Mechanism Assay of *S. rugosoannulata* Undecapeptides Based on Molecular Dynamics Simulation

Molecular dynamics (MD) simulation allows for dynamic analysis of the microscopic changes in intermolecular interactions at the atomic level. After a series of dynamic simulations, the intermolecular interactions system will reach the equilibrium state, and the analysis of the post-equilibrium trajectory can obtain kinetic data to predict the macroscopic properties of the interacting molecules. The MD simulation of undecapeptides and the ACE receptor was performed according to reference [6]. Amber99SB was chosen for the force field, and the TIP3P model was used for the water molecules. The receptor protein–undecapeptide complex system was first placed in the centre of a cubic box, and the distances between them were greater than 1.0 nm. The box was randomly filled with water molecules, and antagonistic sodium ions replaced the water molecules to make the simulated system electrically neutral. The simulated system was optimized using the energy minimization of the fastest decreasing method. Restricted MD simulation was used to equilibrate the systems. MD simulation was performed in an isothermal, isobaric system at a temperature of 300 K and a pressure of 1 atm for 50 ns. Molecular dynamics simulations were also conducted with 3 replications to verify the stability of the molecular simulation results. The conformation, the radius of gyration, and the formation bond changes between undecapeptide and the ACE receptor complex were calculated. MD simulation data analysis was performed by GROMACS 2022.3 software.

2.5. The ACE Inhibition Mechanism Assay of *S. rugosoannulata* Undecapeptide Based on Molecular Interactions

The interaction patterns between undecapeptide and the ACE receptor were analyzed using isothermal titration calorimetry (ITC) and biolayer interferometry (BLI) molecular interaction systems. The selected undecapeptide was synthesized by GL Biochem (Shanghai) Ltd., Shanghai, China.

The ACE receptor protein solution (0.02 mmol) prepared with PBS buffer was injected into the ITC cuvette at 25 °C. The operating program was as follows: stirrer speed, 1000 rpm; 60 s interval between start and first titration; 20 total injections; 2 μL (0.2 mmol) of buffer-prepared peptide solution in a single injection; 150 s interval between two titrations; single injection time, 2 s. Temperature changes in the reaction system were measured, and the ITC Nano analysis software 1.41 was used to calculate the complex thermodynamic binding affinity (K_D), stoichiometry (N), enthalpy (ΔH), and entropy (ΔS) to verify whether the ACE receptor and undecapeptide interacted with each other.

The ACE receptor protein solution (1.0 mg/mL) prepared in PBS buffer was subjected to protein biotinylation (Genemore G-MM-IGT biotinylation kit, BMD Labservice Ltd., Suzhou, China). Protein biotinylation was incubated for 30–60 min and desalted. Four SSA sensors were used to solidify biotinylated proteins, and 4 SSA sensors were used as reference electrodes. Set the baseline equilibration time, loading sample time, and baseline equilibration time of 60 s–800 s–60 s. The peptide solutions were prepared at concentrations of 116.6 μmol/L, 233.2 μmol/L, 466.4 μmol/L, 932.8 μmol/L, 1866 μmol/L, and 3731 μmol/L. A biomolecular interaction cycle program with a baseline equilibration time of 60 s, a binding time of 120 s, and a dissociation time of 120 s was established to perform the interactions between the ACE receptor and undecapeptide at 25 °C. The complex's binding constant (K_{on}), dissociation constant (K_{off}), and kinetic affinity K_D were calculated to determine the interaction modes between the ACE receptor and undecapeptide.

2.6. *The ACE Inhibitory Activity In Vitro and Hypotensive In Vivo Assay of S. rugosoannulata Undecapeptide*

The 0.1 g synthesized peptide was weighed and dissolved in 50 mL of pure water. The solution was diluted step by step to obtain the undecapeptide solutions with concentration gradients of 2.0, 0.4, 0.08, 0.016, and 0.0032 mg/mL. The ACE-inhibitory activity of the undecapeptide solutions was measured by the DOJINDO ACE Kit-WST kit (Shanghai Youlu Biotechnology Co., Ltd., Shanghai, China). The ACE inhibition activity was calculated by detecting the absorbance at 450 nm of 3-Hydroxybutyric acid (3HB), which was produced by ACE-catalyzed 3-Hydroxybutyryl-Gly-Gly-Gly (3HB-GGG). The ACE inhibition activity curve was made with the sample concentration and the inhibition rate, and the sample concentration at 50% inhibition rate (IC_{50}) was calculated.

Eleven-week-old male SHR rats and Wistar rats (SiPeiFu (Suzhou) Biotechnology Co., Ltd., Suzhou, China) weighing 200–250 g were used. After 7 d of acclimatization feeding, SHR rats and Wistar rats were randomly grouped into 6 rats per group for a single gavage administration treatment. The undecapeptide and the commercial antihypertensive drug Benazepril hydrochloride were administered in a gavage dose of 10 mg/kg body weight (the recommended dose of Benazepril hydrochloride). SHR rats with regular saline treatment were used as the blank control group. The systolic blood pressure (SBP) and diastolic blood pressure (DBP) of rats were monitored by the Softron BP system (Softron BP-2010A, Tokyo, Japan) at 0, 2, 4, 6, 8, and 24 h after drug administration. At the end of the treatment, the organ coefficients were calculated (organ weight as a percentage of the body weight of the rat, mg/g). The animal study was reviewed and approved by the Yanxuan Biotechnology Laboratory Animal Welfare & Ethical Committee (Hangzhou) Co., Ltd., Hangzhou, China (YXSW 2201184105).

2.7. *Statistical Analysis of the Data*

The Origin 2019 software was used to plot the variation of the MD simulation parameters. GraphPad Prism 9 software was used for thermodynamics and dynamics data plots obtained from ITC and BLI analysis. The ACE-inhibitory activity in vitro experiment was performed thrice, and the 50% inhibition rate data was calculated. Data from the in vivo tests were presented as means ± standard deviations of 6 measurements.

3. Results

3.1. Undecapeptides with ACE Inhibitory Activity in S. rugosoannulata

A peptide database of 27 undecapeptides with potential ACE-inhibitory activity was constructed. The active ACE-inhibitory peptide fragments were plentiful, with the main ones being PL, RP, LP, and PG, which had high frequencies. The prediction results of admetSAR showed that all 27 undecapeptides were non-toxic (0.667–0.900). The blood–brain barrier penetration (0.824–0.997), Caco-2 cell penetration (0.702–0.875), and human intestinal absorption properties (0.518–0.964) of undecapeptides were good (Supplementary Table S2).

3.2. Molecular Docking Results of the ACE Receptor and Undecapeptides

MOE was used to analyze the active amino acid residues in the ACE receptor (Figure 1A, 133 amino acid residues). The spatial conformation and docking sites of 27 undecapeptides bound to the ACE receptor are shown in Figures 1B–F and S1. It can be seen that the undecapeptide conformations were more stretching in the receptor cavity, and the distribution positions were extensive (the orange dashed box of Figure 1B–F). There were 56 types of amino acid residues in the receptor to which the undecapeptide could bind, and hydrogen bonds and ionic bonds were the central interaction bonds in undecapeptides and the ACE receptor complex (Supplementary Figure S1). The amino acid residues (ALA354, GLU384, TYR523, GLN281, HIS353, HIS513, LYS511, TYR520, GLU162) in the functional pockets (S1, S2, and S1′) of the ACE receptors were used as indicators, and the interaction types and binding energy of undecapeptides are shown in Supplementary Table S3. Except for the residue TYR520 in the active pocket S2, the 27 undecapeptides could form interactions with the other eight critical amino acid residues in varying amounts. The binding energy range was −21.3~−0.5 kcal/mol (the lower the energy of the interaction bonds, the more stable the complex formed). The amino acid residues ALA354, GLU384, HIS353, HIS513, and GLU162 in the receptor formed interactions with more undecapeptides (10–12 undecapeptides), among which GLU384 and GLU162 formed more binding bonds with undecapeptides (19 and 27 bonds, respectively). The bond number formed by LYS511 with undecapeptides was lower than the above two amino acid residues. Still, its total binding energy was lower (−91.6 kcal/mol), second only to that formed by undecapeptide with GLU162 (−99.1 kcal/mol).

The 27 undecapeptides also formed many binding bonds with other amino acid residues in the structural domain of the receptor. LYS118, GLU403, ARG522, and GLU123 were the four primary amino acid residues that formed interactions with undecapeptides. The number of bonds formed by undecapeptides with the four amino acid residues was more than 27, and the total binding energy was −156.7 kcal/mol~−71.7 kcal/mol. It has been reported that HIS383, HIS387, GLU411, and Zn^{2+} were the ACE receptors' catalytic sites and ACE-inhibitory peptides' primary binding sites [5,7]. Although the bond number formed by GLU411 and HIS387 with undecapeptides was low (10 and 11, respectively), the total binding force created by the above two amino acid residues with undecapeptides was strong (binding energy −78.9 kcal/mol and −70.2 kcal/mol, respectively), which contributed a lot to maintaining the stability of undecapeptides and the ACE receptor complex. In addition, undecapeptides also formed many low-energy metal bond/ion forces with zinc ions. The abundance of amino acid residues and zinc ions bound to undecapeptides and the strong interaction force played an essential role in the ACE-inhibitory activity of undecapeptides.

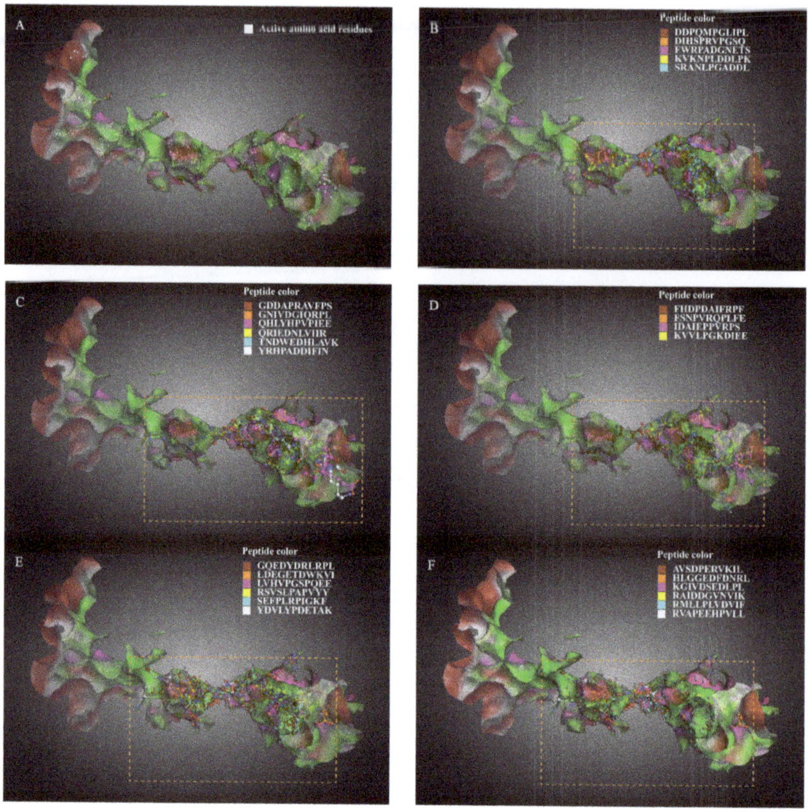

Figure 1. Spatial conformation of the 27 *S. rugosoannulata* undecapeptides bound to the ACE receptor. (**A**) The active cavity of the ACE receptor and the active amino acid residues (white sphere); (**B**) main undecapeptides bound to the amino acid residue ALA354; (**C**) main undecapeptides bound to the amino acid residue GLU384; (**D**) main undecapeptides bound to the amino acid residue HIS353; (**E**) main undecapeptides bound to the amino acid residue LYS511; (**F**) main undecapeptides bound to the amino acid residue GLU162. The orange dashed box showed the interaction region of undecapeptides and the ACE receptor.

Among the 27 undecapeptides, GQEDYDRLRPL had the highest MS abundance (6.80×10^7), and the peptide contained the ACE-inhibitory peptide fragments RL, LRP, PL, RP, GQ, DY, LR, and DR. The peptide bound firmly to critical amino acid residues GLU384, HIS513, LYS511, and GLU162 in S1, S2, and S1′ pockets, forming strong hydrogen bonds and electrostatic interaction forces (−19.8 kcal/mol binding energy with GLU162; −20.5 kcal/mol binding energy with LYS511). The peptide formed low-energy bonds with the critical amino acid residues (LYS118, GLU403, ARG522) and ASP358. The binding energy was −6.4 kcal/mol with ARG522, −9.1 kcal/mol with LYS118, −10.7 kcal/mol with GLU403, and −11.9 kcal/mol with ASP358. The docking amino acid residues from GQEDYDRLRPL mainly included amino nitrogen atoms (G1, Q2, Y5, R7, R9, L11), carbonyl oxygen atoms (D4, Y5, L8), and carboxyl oxygen atoms (E3, D6, L11) (Figure 2). In summary, GQEDYDRLRPL could interact tightly with critical amino acid residues in S1, S2, and S1′ and other key amino acid residues in the receptor structural domain. Except for proline (P), all amino acid residues in GQEDYDRLRPL exerted interactions with the ACE receptor. It was hypothesized that GQEDYDRLRPL might have better ACE-inhibitory activity. Therefore, GQEDYDRLRPL was selected for MD simulation and molecular interactions.

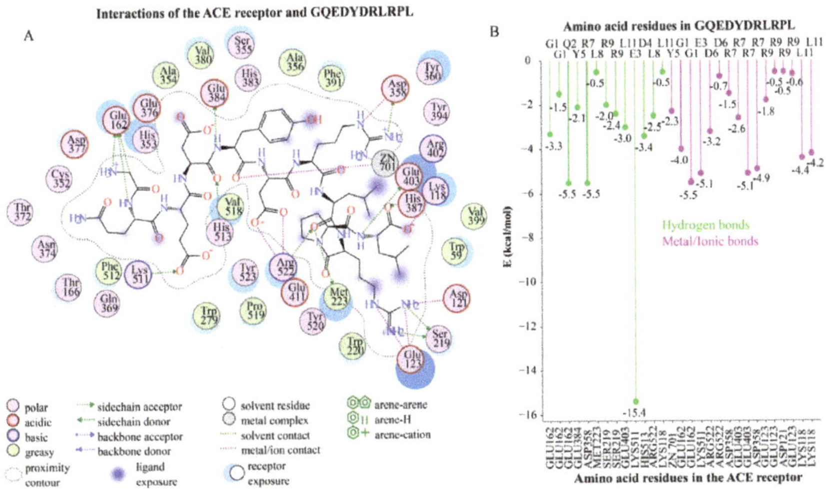

Figure 2. The 2D docking plot of GQEDYDRLRPL and the ACE receptor. (**A**) Docking residues; (**B**) docking bond energy.

3.3. Molecular Dynamics Simulation Results of GQEDYDRLRPL and the ACE Receptor

The results of the MD simulation of GQEDYDRLRPL with the ACE receptor are shown in Figure 3. The root mean square deviation (RMSD) was used to measure the overall change in the structural conformation of the complex during MD simulation and to observe and evaluate the protein stability. The MD simulation system had a more pronounced effect of GQEDYDRLRPL on the ACE receptor between 0 and 15 ns, and the RMSD fluctuated widely, which interfered with the overall structure of the receptor. The RMSD of the system stabilized after 15 ns, and the complex stabilized around 2.5 Å. (Figure 3A).

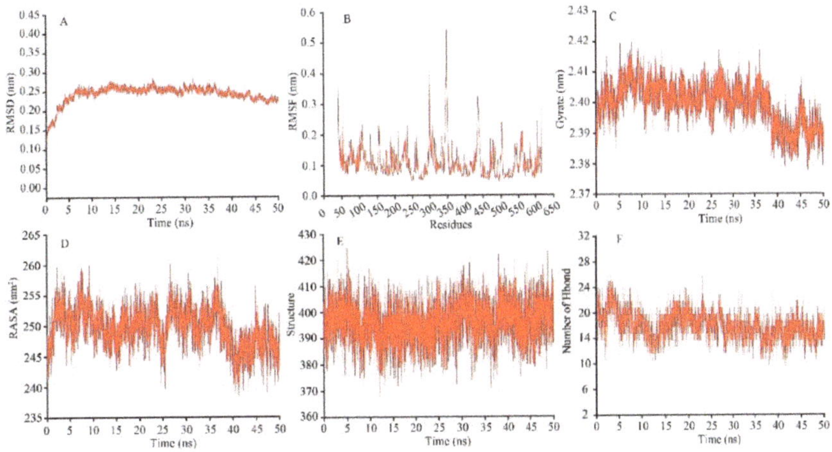

Figure 3. Molecular dynamics simulation results. (**A**) RMSD changes during MD simulation; (**B**) RMSF changes during MD simulation; (**C**) the radius of gyration changes during MD simulation; (**D**) the receptor's solvent-accessible area changes during MD simulation; (**E**) the receptor's secondary structure changes during MD simulation; (**F**) the number of hydrogen bond changes during MD simulation.

The root mean square fluctuation (RMSF) is the deviation of individual amino acid residues from their average position in the receptor over time. It is used to analyze the protein structural fluctuation region. Amino acid residues with high RMSF values have more extraordinary mutability and are the central structural region where peptides affect the receptor. The interaction between GQEDYDRLRPL and the ACE receptor significantly affected amino acid residues 340–355 region. The RMSF in this region was significantly higher than those in other positions, indicating that it was the hot spot region for GQEDYDRLRPL to affect the ACE receptor (Figure 3B), and amino acids fluctuations in this region would affect the state structure of the receptor protein in solution, interfering with protein–peptide interactions. Amino acid residues in this region include MET340, LEU341, GLU342, LYS343, PRO344, THR345, ASP346, GLY347, ARG348, GLU349, VAL350, VAL351, CYS352, HIS353 (in the S2 pocket), ALA354 (in the S1 pocket) and SER355. The molecular docking results showed that GQEDYDRLRPL did not bind to the amino acid residues mentioned above in the final steady-state complex to generate interaction forces. Thus, it is presumed that the amino acid residues bound and dissociated from the GQEDYDRLRPL during the MD simulation, leading to the perturbing of the amino acid residues. This region was also the central region of structural changes in the receptor. Furthermore, the fluctuation of other active amino acid residues with which the peptide interacted was less, indicating that GQEDYDRLRPL and the ACE receptor binding complex had good stability.

The radius of gyration is used to assess the structural compactness of the protein–peptide complex in the MD simulation system, and a smaller radius of gyration indicates a dense structure. The radius of gyration changes during the MD simulation tended to decrease (Figure 3C), favouring the interaction of GQEDYDRLRPL with the ACE receptor. During the 0–50 ns MD simulation process, there were no significant changes in the ACE receptor's solvent-accessible area (Figure 3D) and the secondary structure (Figure 3E). The number of hydrogen bonds within the protein–peptide complex decreased slightly during the 10–15 ns process, which may be related to the influence of GQEDYDRLRPL on the receptor structure. After 15 ns, the number of hydrogen bonds gradually increased and levelled, indicating that the binding of the peptide to the receptor tended to have a stable structure (Figure 3F).

3.4. Molecular Thermodynamics and Dynamics Interactions Results of GQEDYDRLRPL and the ACE Receptors

The ITC interaction results showed that the enthalpy (−440 KJ/mol) and the entropy (−1408 J/mol·K) between GQEDYDRLRPL and the ACE receptor interaction were negative (Gibbs free energy < 0, enthalpy-driven reaction) and the reaction system was spontaneously exothermic. The stable complex was formed by GQEDYDRLRPL interacting with multiple sites on the receptor (N = 2.22). Hydrogen bond interaction and van der Waals forces were the main driving forces for reciprocal binding. The thermodynamic binding affinity K_D value of the receptor and GQEDYDRLRPL was 3.06×10^{-6} M (Figure 4A,B). The ITC analysis results were consistent with the molecular docking results.

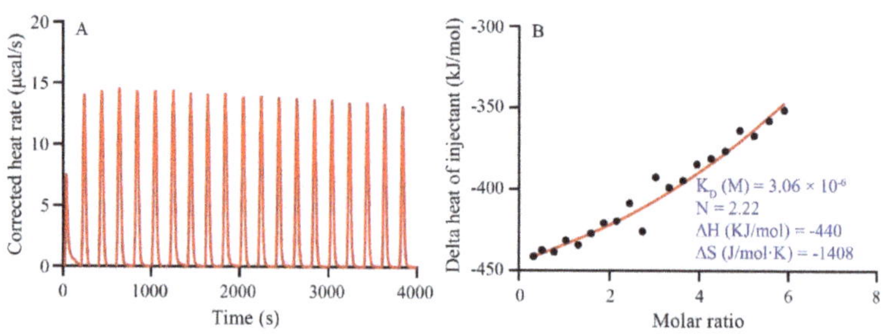

Figure 4. Molecular thermodynamics interactions between GQEDYDRLRPL and the ACE receptor. (**A**) Pulse diagram of heat generated per titration; (**B**) molar heat to molar ratio heat integral fitting plot.

The BLI dynamics interaction results of GQEDYDRLRPL with the ACE receptor at different gradient concentrations showed that the binding constant K_{on} of GQEDYDRLRPL to the ACE receptor was 6.84 × 10^5 (M^{-1} S^{-1}). The primary rate of peptide binding to the receptor was fast, and after 120 s, GQEDYDRLRPL binding to the receptor reached a saturation stage, and the interaction tended to equilibrium. The signal response increased dose-dependently with peptide concentration, and the peptide activity bound to the receptor was better at medium to high concentrations (1866 μmol/L and 3731 μmol/L). The kinetic affinity value K_D of GQEDYDRLRPL to the ACE receptor was 9.26 × 10^{-7} M, a high intensity binding level in protein–peptide molecule interactions. GQEDYDRLRPL was explicitly bound to the ACE receptor (Figure 5). The dynamic results of the BLI analysis were consistent with the ITC thermodynamic results.

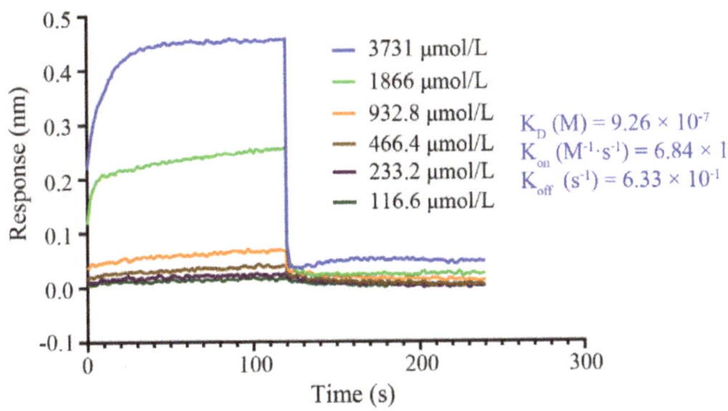

Figure 5. Molecular dynamics interactions between GQEDYDRLRPL and the ACE receptor.

3.5. The ACE Inhibitory Activity In Vitro and Hypotensive Evaluation In Vivo of GQEDYDRLRPL

The in vitro ACE inhibition curve of GQEDYDRLRPL obtained by the kit method was y = 8.4157 ln(x) + 62.597 (R^2 = 0.903), and its IC_{50} value was 164.41 μmol/L. The results of SBP and DBP in spontaneously hypertensive rats (SHR) showed that SHR rats treated with GQEDYDRLRPL and Benazepril hydrochloride by gavage had better blood pressure lowering effects after 2, 4, 6, and 8 h. The most significant blood pressure lowering effect was observed at 8 h. Blood pressure values of the treated groups began to rebound after 8 h. In the GQEDYDRLRPL treatment group, the decrease in SBP was 37.98 ± 1.26 mmHg, and the reduction in DBP was 29.64 ± 1.23 mmHg. In the Benazepril hydrochloride treatment group, the reduction in SBP was 54.94 ± 1.79 mmHg, and the decrease in DBP was 39.17 ± 1.81 mmHg. In the GQEDYDRLRPL gavage-treated group of Wistar rats, the drop in SBP and DBP was 9.11 ± 0.85 mmHg and 2.44 ± 1.20 mmHg, respectively. The effect of GQEDYDRLRPL on the blood pressure of Wistar rats was insignificant. (Figure 6A,B).

The heart organ coefficients in SHR rats showed a significant difference between the Benazepril-hydrochloride-treated group and the blank control group, and there was no significant difference between the GQEDYDRLRPL-treated group and the blank control group. The kidney organ coefficients were significantly lower in the Benazepril hydrochloride treatment group of SHR rats and considerably higher in the GQEDYDRLRPL treatment group of Wistar rats (Figure 6C,D). Benazepril hydrochloride treatment had a more significant effect on the SHR rats' organs. The low body weight in two Wistar rats of the GQEDYDRLRPL gavage-treated groups might be the main reason for the high coefficients of the kidney. In conclusion, GQEDYDRLRPL had a hypotensive effect on SHR rats by gavage treatment and had less impact on the organs of SHR rats. The dosage used for therapy has been lowered to serve as a raw material for the production of hypotensive functional foods.

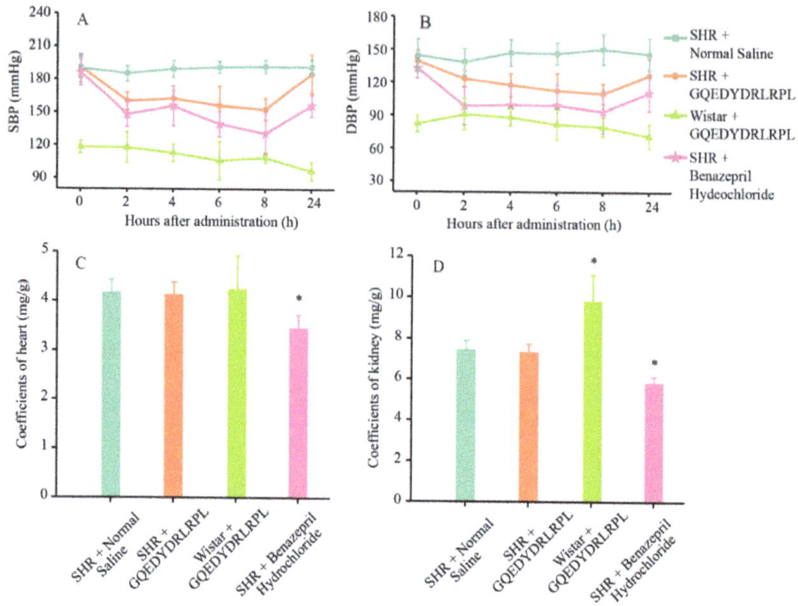

Figure 6. Blood pressure values and organ coefficients of rats at different periods after drug administration. (**A**) Systolic blood pressure; (**B**) diastolic blood pressure; (**C**) heart coefficients; (**D**) kidney coefficients. * indicates a significant difference compared to the blank control group ($p < 0.05$).

4. Discussion

Studies have concluded that elevated blood pressure can put patients at risk for memory loss and Alzheimer's disease [8]. Data from the Blood Pressure Lowering Treatment Trials' Collaboration (BPLTTC) meta-analysis extend findings from an earlier meta-analysis that showed the cardiovascular benefits of blood-pressure-lowering drugs in patients. Over an average of 4 years of follow-up, each 5 mmHg reduction in systolic blood pressure lowered the relative risk of major cardiovascular events by approximately 10% (https://www.medscape.com/viewarticle/936598, accessed on 17 May 2023.). It shows that hypertensive patients need to reduce their blood pressure, and blood pressure regulation in ordinary people also significantly reduces the risk of cardiovascular events. Therefore, active peptide products with blood-pressure-lowering functions will have a vast application space. Using mushrooms to prepare natural bioactive peptides can enhance the value of mushrooms and reduce the production cost of peptides.

Peptide profiling identification [7,9–11], toxicological prediction [12,13], and molecular docking [7,9,11–17] are effective means to obtain peptides with high ACE-inhibitory activity in vitro rapidly. Researchers have reported a variety of food-derived ACE-inhibitory peptides from animal protein sources such as dairy products [9] and seafood [7,12,16,18,19], as well as from soy products [14,20], high-quality crops or low-value plant protein [15,21–24]. The varied sequences of amino acids in peptides have increased the diversity of food-derived peptides. Food-derived ACE-inhibitory peptides were mainly water-soluble and non-toxic. Their inhibited mechanism often includes mixed or competitive mode, binding to ACE's active or inactive sites through hydrogen bonds, electrostatic interactions, and hydrophobic interactions [12–16,25]. The 27 undecapeptides could inhibit ACE activity in a mixed mode by forming hydrogen bonds and electrostatic interaction forces with amino acid residues in the active and inactive sites of the ACE receptor.

MD simulation [18,26], molecular interactions [18,27], simulated digestion in vitro [9–11,20,25,28], and antihypertensive effects evaluation in vivo [14,22,29] effectively verify and reveal the ACE inhibition mechanism. The peptide–receptor complex's interaction mode and conformation changes can help improve our understanding of the inhibition mechanism. Evaluating the activity of the ACE-inhibitory peptide after digestion and absorption can obtain information on the peptide's stability and activation pathway. ACE-inhibitory peptides are usually time-dependent and dose-dependent in regulating blood pressure. The regulatory path involves the renin–angiotensin system (RAS), amelioration of oxidative stress-induced cell damage, activity activation of nitric oxide synthase and gene expression in endothelial cells, and regulation of nitric oxide (NO) and endothelin-1 (ET-1) production [19,22,30–34]. The cross-validation of multiple techniques laid the theoretical foundation for revealing the interaction mechanism of ACE-inhibitory peptide and its hypotensive effects. Molecular docking and molecular dynamics simulations cross-validated that GQEDYDRLRPL can form a tightly structured complex (2.5 Å) with the ACE receptor, with hydrogen bonding interaction driving the peptide molecule to bind to the receptor molecule at multiple sites. Cross-validation of the dynamics and thermodynamic results of the intermolecular interactions confirmed the existence of a moderate affinity force between the biomolecules, and cross-analysis of the in vitro ACE-inhibitory activity and in vivo hypotensive activity confirmed that GQEDYDRLRPL is a novel antihypertensive agent.

The biological activity of peptides is correlated with the type of amino acids in the sequence and its secondary structure. Generally, peptides containing hydrophobic and aromatic amino acid residues at the N- and C-terminus, proline at the second C-terminus position, or those containing arginine and proline had better ACE-inhibitory activity [16,35–37]. Peptides containing negatively charged amino acids were essential for their interaction with ACE and their ACE-inhibitory activity [37]. GQEDYDRLRPL contained hydrophobic amino acid leucine at the C-terminus, arginine, and proline in the peptide sequence and had a negative charge at pH 7.0, which is consistent with the above amino acid composition patterns of the ACE-inhibitory peptides. However, since GQEDYDRLRPL is a long-chain peptide, the spatial position of proline in its secondary structure affects its binding to ACE receptors, as shown by the molecular docking results. Therefore, proline in GQEDYDRLRPL did not directly contribute to the ACE-inhibitory activity of GQEDYDRLRPL.

There is some difference between the ACE-inhibitory activity in vitro and the hypotensive activity in vivo. Peptides with relatively poor ACE-inhibitory activity exhibited better hypotensive activity after digestion in vivo, possibly related to the peptide degradation fragments with a higher ACE-inhibitory activity. GQEDYDRLRPL had a hypotensive effect at lower gavage treatment doses. It is reported that PL was an ACE-inhibitory peptide with an IC_{50} value of 337.3 muM [38]. Virtual digestion of GQEDYDRLRPL by gastrointestinal protease could produce the peptide fragments of GQEDYDR, L, R, and PL, and it is hypothesized that PL peptide fragments also played a role in blood pressure regulation in SHR rats.

5. Conclusions

PL, RP, LP, and PG were the main peptide fragments with ACE-inhibitory activity among the *S. rugosoannulata* undecapeptides. Undecapeptides stretched conformationally in the ACE receptor's active cavity, forming low-energy bonds with the receptor's amino acid residues and zinc ions. The screened undecapeptide, GQEDYDRLRPL, except for proline (P) in the peptide, could bind tightly to the ACE receptor. GQEDYDRLRPL interfered with the overall structure of the ACE receptor, with large fluctuations in the amino acid residues 340–355 region. After 15 ns, the binding complex tended to have a stable structure. The interaction between GQEDYDRLRPL and the ACE receptor was an enthalpy-driven spontaneous exothermic reaction with specific binding. GQEDYDRLRPL had an ACE-inhibitory value (IC_{50}) of 164.41 μmol/L in vitro. At a therapeutic dose of 10 mg/kg body weight by gavage in SHR rats, a significant reduction in blood pressure was observed within 8 h. This study provided a theoretical basis for the *S. rugosoannulata* undecapeptides with ACE-inhibitory activity for producing healthy foods.

Supplementary Materials: The following supporting information can be downloaded at: https://www.mdpi.com/article/10.3390/foods12183461/s1, Figure S1: 2D interaction plots of the 27 S. rugosoannulata undecapeptides to the ACE receptors; Table S1: The 82 undecapeptides of S. rugosoannulata identified by LC-MS/MS; Table S2: Undecapeptides of S. rugosoannulata with ACE-inhibitory activity obtained based on screening and prediction; Table S3: Binding energy of undecapeptides with the active amino acid residues in the ACE receptor's pockets.

Author Contributions: Conceptualization, data curation, writing—original draft., W.L.; software, validation, W.C.; formal analysis, investigation, J.W.; project administration, Z.L.; methodology, validation, Z.Z., D.W. and M.Y.; Resources, Supervision, H.M. and Y.Y. All authors have read and agreed to the published version of the manuscript.

Funding: This research was funded by the Natural Science Foundation of Shanghai, China, grant number 23ZR1426100, and the Agriculture Research System of Shanghai, China, grant number 202309.

Data Availability Statement: The data used to support the findings of this study can be made available by the corresponding author upon request.

Conflicts of Interest: The authors declare no conflict of interest.

References

1. NCD Risk Factor Collaboration. Worldwide trends in hypertension prevalence and progress in treatment and control from 1990 to 2019: A pooled analysis of 1201 population-representative studies with 104 million participants. *Lancet* **2021**, *398*, 957–980. [CrossRef] [PubMed]
2. Li, W.; Chen, W.C.; Ma, H.L.; Wu, D.; Zhang, Z.; Yang, Y. Efficient Preparation of flavor peptides from *Stropharia rugosoannulata* and its ACE inhibitory activity. *J. Chin. Inst. Food Sci. Technol.* **2023**, *23*, 229–240.
3. Li, W.; Chen, W.C.; Ma, H.L.; Wu, D.; Zhang, Z.; Yang, Y. Structural characterization and angiotensin-converting enzyme (ACE) inhibitory mechanism of *Stropharia rugosoannulata* mushroom peptides prepared by ultrasound. *Ultrason. Sonochem.* **2022**, *88*, 106074. [CrossRef] [PubMed]
4. Fu, Y.; Young, J.F.; Løkke, M.M.; Lametsch, R.; Aluko, R.E.; Therkildsen, M. Revalorisation of bovine collagen as a potential precursor of angiotensin I-converting enzyme (ACE) inhibitory peptides based on in silico and in vitro protein digestions. *J. Funct. Foods* **2016**, *24*, 196–206. [CrossRef]
5. Natesh, R.; Schwager, S.L.U.; Sturrock, E.D.; Acharya, K.R. Crystal structure of the human angiotensin-converting enzyme–lisinopril complex. *Nature* **2003**, *6922*, 551–554. [CrossRef]
6. Liu, G.X.; Li, Z.X.; Li, Z.K.; Hao, C.C.; Liu, Y.F. Molecular dynamics simulation and in vitro digestion to examine the impact of theaflavin on the digestibility and structural properties of myosin. *Int. J. Biol. Macromol.* **2023**, *247*, 125836. [CrossRef]
7. Liu, R.; Zhu, Y.H.; Chen, J.; Wu, H.; Shi, L.; Wang, X.Z.; Wang, L.C. Characterization of ACE inhibitory peptides from *Mactra veneriformis hydrolysate* by nano-liquid chromatography electrospray ionization mass spectrometry (Nano-LC-ESI-MS) and molecular docking. *Mar. Drugs* **2014**, *12*, 3917–3928. [CrossRef]
8. Veldsman, M.; Tai, X.Y.; Nichols, T.; Smith, S.; Peixoto, J.; Manohar, S.; Husain, M. Cerebrovascular risk factors impact frontoparietal network integrity and executive function in healthy ageing. *Nat. Commun.* **2020**, *11*, 4340. [CrossRef]
9. Abedin, M.M.; Chourasia, R.; Chiring Phukon, L.; Singh, S.P.; Kumar Rai, A. Characterization of ACE inhibitory and antioxidant peptides in yak and cow milk hard chhurpi cheese of the Sikkim Himalayan region. *Food Chem. X* **2022**, *13*, 100231. [CrossRef]
10. Bhaskar, B.; Ananthanarayan, L.; Jamdar, S. Purification, identification, and characterization of novel angiotensin I-converting enzyme (ACE) inhibitory peptides from alcalase digested horse gram flour. *LWT-Food Sci. Technol.* **2019**, *103*, 155–161. [CrossRef]
11. Yin, Z.T.; Yan, R.Y.; Jiang, Y.S.; Feng, S.B.; Sun, H.L.; Sun, J.Y.; Zhao, D.R.; Li, H.H.; Wang, B.W.; Zhang, N. Identification of peptides in *Qingke baijiu* and evaluation of its angiotensin converting enzyme (ACE) inhibitory activity and stability. *Food Chem.* **2022**, *395*, 133551. [CrossRef] [PubMed]
12. Li, Z.Y.; He, Y.; He, H.Y.; Zhou, W.Z.; Li, M.R.; Lu, A.M.; Che, T.J.; Shen, S.D. Purification identification and function analysis of ACE inhibitory peptide from *Ulva prolifera* protein. *Food Chem.* **2023**, *401*, 134127. [CrossRef] [PubMed]
13. Shao, M.Y.; Wu, H.X.; Wang, B.H.; Zhang, X.; Gao, X.; Jiang, M.Q.; Su, R.H.; Shen, X.R. Identification and characterization of novel ACE inhibitory and antioxidant peptides from *Sardina pilchardus* hydrolysate. *Foods* **2023**, *12*, 2216. [CrossRef] [PubMed]
14. Bhadkaria, A.; Narvekar, D.T.; Nagar, D.P.; Sah, S.P.; Srivastava, N.; Bhagyawant, S.S. Purification, molecular docking and in vivo analyses of novel angiotensin-converting enzyme inhibitory peptides from protein hydrolysate of moth bean (*Vigna aconitifolia* (Jacq.) Marechal) seeds. *Int. J. Biol. Macromol.* **2023**, *230*, 123138. [CrossRef] [PubMed]
15. Fu, Y.; Alashi, A.M.; Young, J.F.; Therkildsen, M.; Aluko, R.E. Enzyme inhibition kinetics and molecular interactions of patatin peptides with angiotensin I-converting enzyme and renin. *Int. J. Biol. Macromol.* **2017**, *101*, 207–213. [CrossRef]
16. Yuan, L.; Sun, L.P.; Zhuang, Y.L. Preparation and identification of novel inhibitory angiotensin-I-converting enzyme peptides from tilapia skin gelatin hydrolysates: Inhibition kinetics and molecular docking. *Food Funct.* **2018**, *9*, 5251–5259.

17. Zheng, Y.J.; Wang, X.Y.; Guo, M.; Yan, X.T.; Zhuang, Y.L.; Sun, Y.; Li, J.R. Two novel antihypertensive peptides identified in millet bran glutelin-2 hydrolysates: Purification, in silico characterization, molecular docking with ace and stability in various food processing conditions. *Foods* **2022**, *11*, 1355. [CrossRef]
18. Feng, X.Z.; Liao, D.K.; Sun, L.X.; Feng, S.Z.; Wu, S.G.; Lan, P.; Wang, Z.F.; Lan, X.D. Exploration of interaction between angiotensin I-converting enzyme (ACE) and the inhibitory peptide from Wakame (*Undaria pinnatifida*). *Int. J. Biol. Macromol.* **2022**, *204*, 193–203. [CrossRef]
19. Wang, C.X.; Song, C.C.; Liu, X.T.; Qiao, B.W.; Song, S.; Fu, Y.H. ACE inhibitory activities of two peptides derived from *Volutharpa ampullacea* perryi hydrolysate and their protective effects on H_2O_2 induced HUVECs injury. *Food Res. Int.* **2022**, *157*, 111402. [CrossRef]
20. Xu, Z.Q.; Wu, C.P.; Sun-Waterhouse, D.X.; Zhao, T.T.; Waterhouse, G.I.N.; Zhao, M.M.; Su, G.W. Identification of post-digestion angiotensin-I converting enzyme (ACE) inhibitory peptides from soybean protein isolate: Their production conditions and in silico molecular docking with ACE. *Food Chem.* **2021**, *345*, 128855. [CrossRef]
21. Chay, S.Y.; Salleh, A.; Sulaiman, N.F.; Abidin, N.Z.; Hanafi, M.A.; Zarei, M.; Saari, N. Blood-pressure lowering efficacy of winged bean seed hydrolysate in spontaneously hypertensive rats, peptide characterization and a toxicity study in Sprague-Dawley rats. *Food Funct.* **2018**, *9*, 1657–1671. [CrossRef] [PubMed]
22. Dong, J.J.; Wang, S.; Yin, X.Y.; Fang, M.; Gong, Z.Y.; Wu, Y.N. Angiotensin I converting enzyme (ACE) inhibitory activity and antihypertensive effects of rice peptides. *Food Sci. Hum. Well.* **2022**, *11*, 1539–1543. [CrossRef]
23. Lai, X.F.; Pan, S.S.; Zhang, W.J.; Sun, L.L.; Li, Q.H.; Chen, R.H.; Sun, S.L. Properties of ACE inhibitory peptide prepared from protein in green tea residue and evaluation of its anti-hypertensive activity. *Process Biochem.* **2020**, *92*, 277–287.
24. Moayedi, A.; Mora, L.; Aristoy, M.C.; Safari, M.; Hashemi, M.; Toldra, F. Peptidomic analysis of antioxidant and ACE-inhibitory peptides obtained from tomato waste proteins fermented using *Bacillus subtilis*. *Food Chem.* **2018**, *250*, 180–187. [CrossRef] [PubMed]
25. Li, Y.; Li, J.R.; Cheng, C.X.; Zheng, Y.J.; Li, H.X.; Zhu, Z.L.; Yan, Y.X.; Hao, W.H.; Qin, N. Study on the in silico screening and characterization, inhibition mechanisms, zinc-chelate activity, and stability of ACE-inhibitory peptides identified in naked oat bran albumin hydrolysates. *Foods* **2023**, *12*, 2268. [CrossRef] [PubMed]
26. Yan, W.L.; Lin, G.M.; Zhang, R.; Liang, Z.; Wu, L.X.; Wu, W.J. Studies on molecular mechanism between ACE and inhibitory peptides in different bioactivities by 3D-QSAR and MD simulations. *J. Mol. Liq.* **2020**, *304*, 112702. [CrossRef]
27. Zhang, B.Y.; Liu, J.B.; Wen, H.D.; Jiang, F.; Wang, E.; Zhang, T. Structural requirements and interaction mechanisms of ACE inhibitory peptides: Molecular simulation and thermodynamics studies on LAPYK and its modified peptides. *Food Sci. Hum. Well.* **2022**, *11*, 1623–1630. [CrossRef]
28. Zhu, Q.A.; Xue, J.W.; Wang, P.; Wang, X.B.; Zhang, J.J.; Fang, X.Z.; He, Z.P.; Wu, F.H. Identification of a novel ACE Inhibitory hexapeptide from camellia seed cake and evaluation of its stability. *Foods* **2023**, *12*, 501. [CrossRef]
29. Zhang, Y.Y.; Zhang, Y.L.; Chen, P.Y.; Shu, F.J.; Li, K.; Qiao, L.S.; Chen, Z.J.; Wang, L.Z. A novel angiotensin-I converting enzyme inhibitory peptide derived from the glutelin of vinegar soaked black soybean and its antihypertensive effect in spontaneously hypertensive rats. *J. Biochem.* **2019**, *166*, 223–230. [CrossRef]
30. He, R.; Wang, Y.J.; Yang, Y.J.; Wang, Z.G.; Ju, X.R.; Yuan, J. Rapeseed protein-derived ACE inhibitory peptides LY, RALP and GHS show antioxidant and anti-inflammatory effects on spontaneously hypertensive rats. *J. Funct. Foods* **2019**, *55*, 211–219. [CrossRef]
31. Li, Y.; Zheng, Y.J.; Zhang, Y.F.; Liu, L.Y.; Zhao, S.L. Purification, characterization, synthesis, in vivo and in vitro antihypertensive activity of bioactive peptides derived from coconut (*Cocos nucifera* L.) cake globulin hydrolysates. *RSC Adv.* **2016**, *6*, 92688–92698. [CrossRef]
32. Lin, K.; Ma, Z.; Ramachandran, M.; De Souza, C.; Han, X.; Zhang, L.W. ACE inhibitory peptide KYIPIQ derived from yak milk casein induces nitric oxide production in HUVECs and diffuses via a transcellular mechanism in Caco-2 monolayers. *Process Biochem.* **2020**, *99*, 103–111. [CrossRef]
33. Zhang, Y.; Pan, D.D.; Yang, Z.C.; Gao, X.C.; Dang, Y.L. Angiotensin I-Converting enzyme (ACE) inhibitory and dipeptidyl Peptidase-4 (DPP-IV) inhibitory activity of umami peptides from *Ruditapes philippinarum*. *LWT-Food Sci. Technol.* **2021**, *144*, 111265. [CrossRef]
34. Zhou, M.J.; Zhao, W.Z.; Xue, W.J.; Liu, J.B.; Yu, Z.P. Potential antihypertensive mechanism of egg white-derived peptide QIGLF revealed by proteomic analysis. *Int. J. Biol. Macromol.* **2022**, *218*, 439–446. [CrossRef] [PubMed]
35. Chen, J.D.; Sun, S.S.; Li, Y.S.; Liu, R. Proteolysis of tilapia skin collagen: Identification and release behavior of ACE-inhibitory peptides. *LWT-Food Sci. Technol.* **2021**, *139*, 110502. [CrossRef]
36. Mirzaei, M.; Mirdamadi, S.; Safavi, M. Structural analysis of ACE-inhibitory peptide (VL-9) derived from *Kluyveromyces marxianus* protein hydrolysate. *J. Mol. Struct.* **2020**, *1213*, 128199. [CrossRef]
37. Zhang, P.; Chang, C.; Liu, H.J.; Li, B.; Yan, Q.J.; Jiang, Z.Q. Identification of novel angiotensin I-converting enzyme (ACE) inhibitory peptides from wheat gluten hydrolysate by the protease of *Pseudomonas aeruginosa*. *J. Funct. Foods* **2020**, *65*, 103751. [CrossRef]
38. Byun, H.G.; Kim, S.K. Structure and activity of angiotensin I converting enzyme inhibitory peptides derived from *Alaskan pollack* skin. *J. Biochem. Mol. Biol.* **2002**, *35*, 239–243. [CrossRef]

molecules

MDPI

Article

Bitter Almond Albumin ACE-Inhibitory Peptides: Purification, Screening, and Characterization In Silico, Action Mechanisms, Antihypertensive Effect In Vivo, and Stability

Nan Qin *, Chao Chen, Najun Zhang, Lulu Song, Yunfei Li, Lili Guo, Rui Liu and Wenfang Zhang

College of Medicine and Food Engineering, Shanxi University of Chinese Medicine, Taiyuan 030619, China; chenhanchen2010@163.com (C.C.); zhangnajun2747@163.com (N.Z.); 18768942280@163.com (L.S.); 15670138865@163.com (Y.L.); cauguolili@163.com (L.G.); ruiliu87@126.com (R.L.); m18935412709@163.com (W.Z.)
* Correspondence: qinnan_qn@sxtcm.edu.cn; Tel.: +86-13753120097

Abstract: Almond expeller is an undeveloped reservoir of bioactive peptides. In the current study, a zinc ion ligand Arg-Pro-Pro-Ser-Glu-Asp-Glu-Asp-Gln-Glu (RPPSEDEDQE) offering a noncompetitive inhibitory effect on ACE (IC_{50}: 205.50 $\mu mol \cdot L^{-1}$) was identified from almond albumin hydrolysates via papain and thermolysin hydrolysis, subsequent chromatographic separation, and UPLC-Q-TOF-MS/MS analysis. Molecular docking simulated the binding modes of RPPSEDEDQE to ACE and showed the formation of hydrogen bonds between RPPSEDEDQE and seven active residues of ACE. Moreover, RPPSEDEDQE could bind to fifteen active sites of ACE by hydrophobic interactions, and link with the His387 and zinc ions of the zinc tetrahedral coordination. Ultraviolet wavelength scanning and Fourier-transformed infrared spectroscopy analysis revealed that RPPSEDEDQE can provide multiple binding sites for zinc ions. However, RPPSEDEDQE cannot bind with any central pocket of ACE, which was evidenced by an inhibition kinetics experiment. Additionally, the zinc-chelating capacity and inhibiting ability against ACE of RPPSEDEDQE were both not significantly reduced by the hydrolysis of gastrointestinal enzymes. A moderate to high dose of RPPSEDEDQE (100–150 $mg \cdot kg\ bw^{-1}$) significantly reduced the systolic and diastolic blood pressure of spontaneous hypertensive rats, but chelation with zinc ions decreased its antihypertensive efficiency. These results indicate that bitter almond albumin peptides may be used for lowering blood pressure.

Keywords: bitter almond albumin peptides; inhibitory mechanisms against ACE; docking mode; zinc coordination patterns; antihypertensive effect

Citation: Qin, N.; Chen, C.; Zhang, N.; Song, L.; Li, Y.; Guo, L.; Liu, R.; Zhang, W. Bitter Almond Albumin ACE-Inhibitory Peptides: Purification, Screening, and Characterization In Silico, Action Mechanisms, Antihypertensive Effect In Vivo, and Stability. *Molecules* **2023**, *28*, 6002. https://doi.org/10.3390/molecules28166002

Academic Editors: Yongliang Zhuang, Hu Hou, Zhipeng Yu and Shengbao Cai

Received: 12 July 2023
Revised: 2 August 2023
Accepted: 8 August 2023
Published: 10 August 2023

1. Introduction

Nowadays, natural antihypertensive peptides have received increasing attention, because approximately 10.8 million people worldwide die from hypertension complications every year, and synthesized antihypertensive drugs usually have side effects such as vomiting, coughing, allergies, and loss of appetite [1]. In comparison, antihypertensive peptides from foods are considered to be more efficient and safe [2]. The inhibition effect on the ACE-Ang II-AT1R axis has been shown to be the main blood-pressure-lowering pathway of antihypertensive drugs [3]. In the ACE-Ang II-AT1R axis, Angiotensin-I-Converting Enzyme (ACE) is found to play a crucial role in the elevation of blood pressure. Previous studies evidenced that the binding center of ACE contained S1, S2, and S′ central pockets that consisted of nine amino acid residues (Lys511, Tyr523, His353, Glu384, His513, Ala354, Glu162, Gln281, and Tyr520), where its catalytic center contained a zinc tetrahedral coordination that consisted of a zinc ion linked with Glu411, His383, and His387 residues [4,5]. Inhibitors of strong binding power with one or more of the nine active sites in the ACE central pocket are found to be competitive inhibitors of ACE. Moreover, peptides that can affect the structure of ACE via binding with other active sites outside the central pockets are shown to offer inhibition activity toward ACE [6]. Additionally, the zinc ions located in the zinc tetrahedral ligand are essential for the catalytic action of ACE [2]. Thus, in theory, peptides that can influence the zinc tetrahedral coordination of ACE should have a potential inhibition capacity against ACE [7]. In the last decades, peptides with inhibition activity against ACE have been extensively isolated from food protein resources [5,8,9]. However, to our best knowledge, evidence that peptides with zinc-chelating ability can inhibit the activity of ACE is currently weak [2].

In addition, substantial evidence suggests that antihypertensive peptides derived from food proteins may be sensitive to digestion and food processing conditions, especially gastrointestinal digestion [10]. Pepsin, trypsin, and other enzymes presenting in the stomach or intestines can change the amino acid sequence of peptides, further cause changes in the electrochemical parameters of peptides, thereby affecting the coordination of peptides with ACE or zinc ions, and lowering their inhibitory ability toward ACE [11–13]. However, few studies offer the effect of gastrointestinal digestion on ACE-inhibitory activity, physicochemical parameters, coordination with zinc ions, and stability of peptides.

Nuts are an important reservoir of antihypertensive peptides. Recently, ACE-inhibitory peptides have been isolated from different nut proteins such as walnut protein, pine nut globulin, almond, and hazelnut protein isolates [14–17]. Almond (*Semen Armeniacae Amarum*) has a high content of protein (around 37 g·100 g^{-1}) [18]. The annual production of almond increased with an increasing worldwide almond yield and demand for apricot oil [19]. Bitter almond protein has a relatively desirable amino acid composition and high emulsifying properties, foam, and gelling properties [20–22]. Bioactive peptides isolated from bitter almond protein hydrolysates were shown to offer health-care functions such as a hypoglycemic effect, antihypertension, and antiradiation activity [23,24]. Zhang & Ye identified a pentapeptide with a high zinc chelating ability from sweet almond expeller amandin hydrolysates [25]. Mirzapour et al. [16] isolated ACE-inhibitory peptides from wild almond proteins. However, there is little data simultaneously referring to the ACE-inhibitory activity and zinc-chelating ability of bitter almond peptides. Our pre-experimental showed that the zinc-chelating ability and inhibition ability toward ACE of bitter almond albumin hydrolysates (BAAHs) were 20.67 $\pm$ 3.58 mg·g^{-1} and 54.95 $\pm$ 4.62%, respectively, indicating that peptides possessing zinc-chelating and ACE-inhibition capacities should be isolated from BAAHs. To improve the antihypertensive function of BAAHs, the aim of this study was to study the isolation, characterization, action mechanism, and zinc ion-chelating capacity of BAAH ACE-inhibitory peptides. Additionally, the physicochemical characteristics, antihypertensive effect in vivo, and stability of BAAH ACE-inhibitory peptides were investigated.

2. Results

2.1. Chromatographic Purification of BAAHs

Figure 1 shows the effects of papain and thermolysin on the hydrolysis degree and ACE-inhibitory activity of BAAHs with an increasing hydrolysis time. From 30 to 150 min, both the hydrolysis degree and ACE-inhibitory activity of BAAHs were increased as the hydrolysis time increased. After 180 min, the hydrolysis degree and ACE-inhibitory activity of BAAHs were both not significantly increased with the increasing hydrolysis time. After hydrolysis for 210 min, the hydrolysis degree and ACE-inhibitory activity of BAAHs were 21.53 $\pm$ 5.78% and 54.95 $\pm$ 4.62% (Figure 1), respectively.

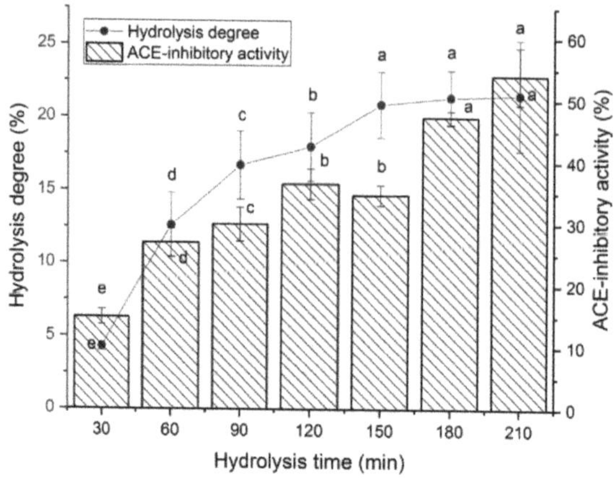

Figure 1. The hydrolysis degree and ACE-inhibitory activity of almond albumin hydrolysates at different hydrolysis times. Different letters (a–e) between the groups marked a statistically significant difference ($p < 0.05$).

Gel chromatography is an effective technique for purification and screening active peptides from parent protein hydrolysates [26,27]. Seven main peaks, separately named as BAAH-1, BAAH-2, BAAH-3, BAAH-4, BAAH-5, BAAH-6, and BAAH-7, were present in the gel chromatographic spectra of BAAHs (Figure 2A). Both the ACE-inhibitory activity (70.48 $\pm$ 2.08% at 1 mg·mL^{-1}) and high zinc-chelating capacity (12.30 $\pm$ 0.90 mg·g^{-1}, Figure 2B) of BAAH-4 were higher than that of the other peaks ($p < 0.05$). After further purification on the Supersil ODS2 column, three peaks, separately named as BAAH-4-A, BAAH-4-B, and BAAH-4-C, were present on the RP-HPLC spectrum of BAAH-4 (Figure 3A). The zinc-chelating ratio of BAAH-4-C (24.44 $\pm$ 1.44 mg·g^{-1}) and its inhibition activity against ACE (78.28 $\pm$ 2.68%) were both higher than those of BAAH-4-A and BAAH-4-B ($p < 0.05$). These results demonstrated that peptide sequences due to both the zinc-chelating ability and inhibition activity against ACE should be identified from BAAH-4-C.

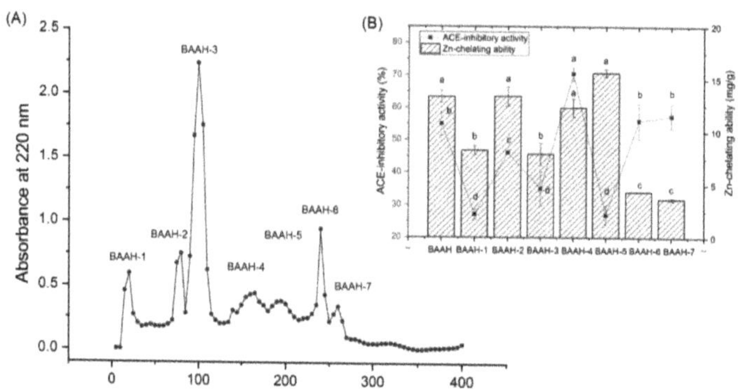

Figure 2. (**A**) The chromatographic separation diagram of almond albumin hydrolysates (BAAHs) on Sephadex G-15 gel column, and (**B**) zinc ion-chelating ability and inhibiting capacity against ACE of the separated main peaks separately named as BAAH-1, BAAH-2, BAAH-3, BAAH-4, BAAH-5, BAAH-6, and BAAH-7. Different letters (a–d) between the groups marked a statistically significant difference ($p < 0.05$).

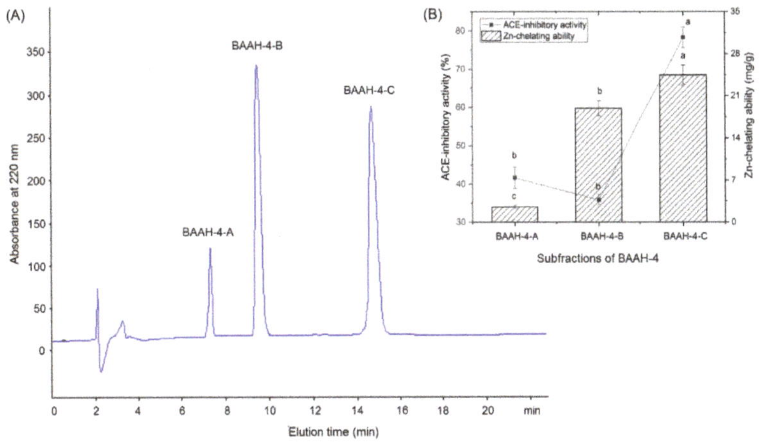

Figure 3. (**A**) The chromatographic separation diagram of BAAH-4 in reverse-phase high-performance liquid chromatography system, and (**B**) the zinc ion-chelating ability and inhibiting capacity against ACE of the obtained main subfractions named as BAAH-4-A, BAAH-4-B, and BAAH-4-C. Different letters (a–c) between the groups marked a statistically significant difference ($p < 0.05$).

2.2. *Identification, Screening, and Inhibitory Effect on ACE of BAAH-4-C Peptides*

The results of ultra-high-performance liquid chromatography and quantum-ultra mass spectrometric determination of BAAH-4-C are shown in Table 1 and Figure 4. Five oligopeptides, Gln-Pro-Pro-Ala-Ala-Ala-Ala-Ala-Ala-Gly-Ala-Gly (QPPAAAAAAGAG), Lys-Thr-Glu-Thr-Gln-Pro (KTETQP), Ser-Pro-Pro-Thr-Ala-Ala-Ala-Ala-Gly-Glu (SPPTAAAAGD), Arg-Pro-Pro-Ser-Glu-Asp-Glu-Asp-Gln-Glu (RPPSEDEDQE), and Thr-Cys-Gly-Ala-Ser (TCGAS) were obtained. However, only RPPSEDEDQE offered potential antihypertension based on the vector machine software score (1.35) calculated by the database AHTpin [28]. Thus, the sequence RPPSEDEDQE was chemically synthesized. The produced RPPSEDEDQE (with purity higher than 99.95%) showed an excellent inhibition activity against ACE (IC_{50}: 205.50 $\mu mol \cdot L^{-1}$, Figure 5A). Additionally, the curve shown in Figure 5A shows that the inhibition activity of RPPSEDEDQE against ACE (y) increases with the increase in the dose (x), and their relationship conformed to the exponential equation: $y = 16.366\ln(x) - 37.156$. RPPSEDEDQE showed a higher inhibitory capacity on ACE compared to peptides EKTFLLYSCPHR identified from camel milk (IC_{50}: 312 $\mu mol \cdot L^{-1}$) and FMRWRDRFL isolated from tombul hazelnut (IC_{50}: 259.7 $\mu mol \cdot L^{-1}$) [17,29], but Captopril offered a much higher inhibition capacity on ACE compared with RPPSEDEDQE ($p < 0.05$) [5]. These results suggested that RPPSEDEDQE was a good inhibitor of ACE.

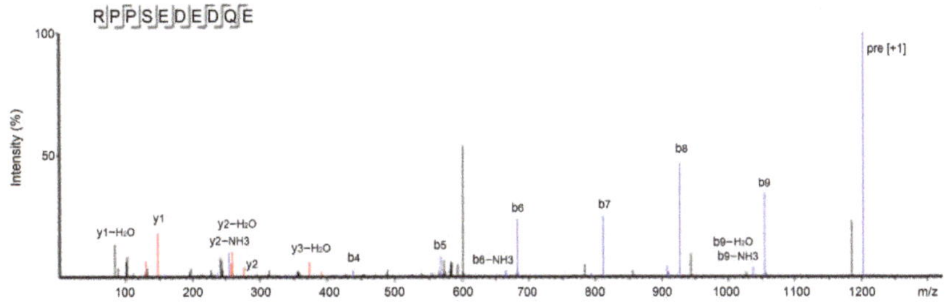

Figure 4. The secondary mass spectrum of peptides RPPSEDEDQE with UPLC-Q-TOF-MS/MS.

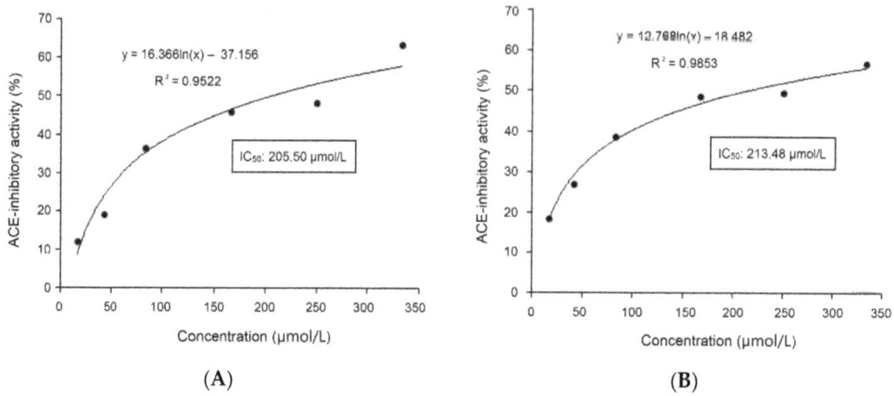

Figure 5. (**A**) The inhibitory effects of RPPSEDEDQE on ACE at different doses, and the fitting equation; and (**B**) the inhibitory effects of RPPSEDEDQE on ACE at different doses, and the fitting equation after the hydrolysis of gastrointestinal enzymes.

Table 1. Identification, in silico screening, chelating ability toward zinc ions, inhibitory effect against ACE, and stability of sequences from bitter almond albumin hydrolysates.

Sequences	RPPSEDEDQE	KTETQP	TCGAS	SPPTAAAAGD	QPPAAAAAAGAG
Molecular mass (Da)	1201.29	535.61	437.52	857.00	766.96
Matched sequence in *Semen Armeniacae Amarum* [a]	P.RPPSEDEDQE.Y	R.KTETQP.-	S.TCGAS.S	F.SPPTAAAAGD.M	G.QPPAAAAAAGAG.R
Vector machine software score [b]	1.35	0.90	−0.58	−0.96	−0.52
Antihypertension predict [c]	AHP	Non-AHP	Non-AHP	Non-AHP	Non-AHP
Inhibitory effect against ACE (IC_{50}: μmol·L^{-1})	205.50	ND	ND	ND	ND
Inhibitory effect against ACE after gastrointestinal digestion (IC_{50}: μmol·L^{-1})	213.48	ND	ND	ND	ND
Chelating ability toward zinc ions (mg·g^{-1})	20.67 ± 3.58 d	19.30 ± 1.46 e	6.44 ± 0.64 f	2.16 ± 0.45 g	1.14 ± 0.01 g
Chelating ability toward zinc ions after gastrointestinal digestion (mg·g^{-1})	18.55 ± 0.95	ND	ND	ND	ND

[a] From National Center for Biotechnology Information (NCBI); [b] Vector machine software score and potential antihypertension were in silico screened with database AHTPIN; [c] AHP: antihypertensive peptide. ND: not measured. Different letters (d–g) between the groups marked a statistically significant difference ($p < 0.05$).

2.3. Chelating Capacity toward Zinc Ions and Physicochemical Characteristics

As shown in Table 2, since RPPSEDEDQE contained a high content of Glu and Asp (50%), which have been shown to be ideal binding residues for metal ions [30], it offered an excellent chelating ability toward zinc ions (20.67 ± 3.58 mg·g^{-1}). The binding capacity of RPPSEDEDQE toward zinc ions was higher than the other four peptide sequences identified in BAAHs and was equal to that of ethylenediamine tetraacetic acid (EDTA, an excellent metal ions chelator) [31]. Additionally, zinc ions can also be bound to the guanidine of Arg at *N*-terminal, carboxyl group of Glu at the *C*-terminal, and the amide bond near the *N*-terminal of RPPSEDEDQE [2].

Physicochemical characteristics such as hydrophilicity, hydrophobicity, and isoelectric points are important for the antihypertensive effect and application of peptides. However, there are a few studies referring to the physicochemical characteristics of peptides [10,32]. As shown in Table 2, RPPSEDEDQE possessed the highest hydrophilicity (1.85) among the five BAAH-4-C peptides. The hydrophilicity of polypeptide was positively correlated with its polarity which was the main factor affecting the chelation between peptides and zinc ions [32]. Therefore, the high hydrophilicity also reflected the high chelation ability of

RPPSEDEDQE toward zinc ions. Since proteins or peptides have a net surface charge of zero and lowest polarity at the isoelectric point (pI), the chelating power of peptides toward metal ions was very low at the pI. Moreover, the polar microenvironment of the peptide may change dramatically at the pI, which affects the binding of the peptide to ACE [33]. Therefore, RPPSEDEDQE and RPPSEDEDQE–zinc chelate should be used away from the pI (3.84, Table 2).

Table 2. In silico analysis of physicochemical characteristics of bitter almond albumin peptides using AHTPIN database.

Peptides	RPPSEDEDQE	KTETQP	TCGAS	SPPTAAAAGD	QPPAAAAAAGAG
Acidic amino acid content (%)	50.00%	16.67%	0.00%	10.00%	0.00%
Hydrophobic amino acid content (%)	20.00%	16.67%	20.00%	60.00%	75.00%
Hydrophobicity	−0.61	−0.09	0.00	−0.01	0.18
Amphiphilicity	0.75	1.59	0.00	0.00	0.00
Hydrophilicity	1.85	−0.66	−0.32	0.09	−0.35
Isoelectric point	3.84	7.25	5.85	3.80	5.88

2.4. *Inhibitory Mechanisms of BAAH Peptides against ACE*

2.4.1. Docking Modes of BAAH Peptides toward ACE Molecule

In this study, the docking modes of BAAH peptides toward ACE were simulated and virtually screened at the molecular level using a molecular modeling tool. The local diagram and global diagram of the best mode (with the highest T-score) are separately shown in Figure 6a,b. Hydrogen bonds were observed between seven active sites of ACE (Lys368, Asp377, Glu376, Pro508, Thr282, Arg522, and Lys454) and RPPSEDEDQE (Figure 6a and Table 3). Hydrogen bonding is a common but very important force between peptides and ACE. Generally, the length of the hydrogen bond was inversely proportional to the force between the peptides and ACE [5,7]. Thus, the short length of the hydrogen bonds between ACE and RPPSEDEDQE (1.94–2.91 Å) suggested that the binding power of RPPSEDEDQE toward ACE was strong. More importantly, the T-score for the best docking mode of BAAH peptides toward ACE (8.06) was above the threshold (6.0), reflecting a strong binding power of RPPSEDEDQE toward ACE. Additionally, hydrophobic interactions were observed between RPPSEDEDQE and fifteen residues in ACE (Table 3). These results were consistent with the high inhibitory effect against ACE (IC_{50}: 205.50 μmol·L^{-1}).

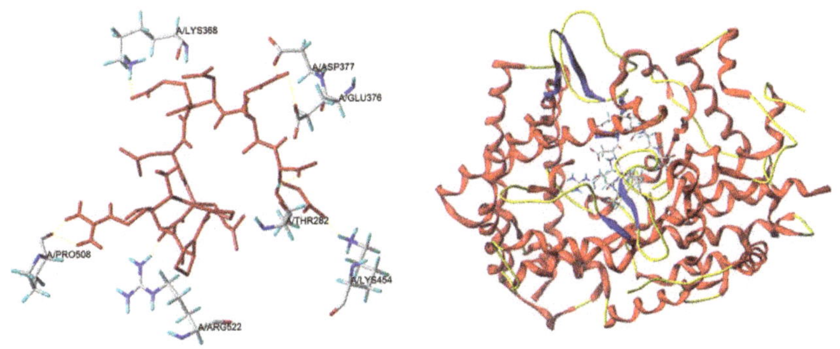

(**a**) (**b**)

Figure 6. The local diagram (**a**) and global diagram (**b**) of the best docking patterns between RPPSEDEDQE and ACE. The structure of ACE was used PDB: 1O8A model.

Table 3. Virtual evaluation on the docking modes of RPPSEDEDQE with ACE, and interaction forces in the best mode.

Ligand	T-Score	C-Score	Interaction Force	Active Sites of ACE and Hydrogen Bond Length
RPPSEDEDQE	8.06	5	Hydrogen bond	Lys368: 1.94 Å; Asp377: 2.91 Å; Glu376: 2.47 Å; Pro508: 2.03 Å; Thr282: 1.97 Å; Arg522: 2.12 Å; Lys454: 2.81 Å
			Hydrophobic interactions	His353, Cys352, Tyr523, Ala354, Ser355, Phe457, Phe527, Asp415, Pro163, Gln369, Cys370, Glu376, Asn374, Ala170, Asn167

T-score is the total score of virtual evaluation, and C-score means the consistency score.

In addition, hydrophobic interactions were also observed between RPPSEDEDQE and His387, a main residue of the zinc tetrahedral coordination in ACE (Table 3). The effects of peptides on the zinc tetrahedral coordination may change ACE's catalytic function [3]. The results in Figure 6 and Table 3 demonstrated that the inhibitory mechanism of BAAH peptides against ACE was mainly the binding of RPPSEDEDQE with ACE, especially the interactions between RPPSEDEDQE and the zinc tetrahedral coordination.

2.4.2. Inhibition Kinetics on ACE

The active residues of ACE that RPPSEDEDQE can dock with via hydrogen bond or hydrophobic interactions did not locate in the binding center of ACE [4], indicating that RPPSEDEDQE offered an uncompetitive inhibitory effect against ACE, which was evidenced by the results of the inhibition kinetics test, as shown in Figure 7. As shown in the Lineweaver–Burk diagram of ACE, the K_m increased with the increasing RPPSEDEDQE dose, but the maximum velocity (V_{max}) decreased when the RPPSEDEDQE dose increased, which was the typical effect of noncompetitive inhibitors [7]. Peptides such as HLNVVHEN and PGSGCAGTDL identified from walnut protein, and LVRYYLLVR from hazelnut protein all showed noncompetitive ACE inhibitory patterns [14,17].

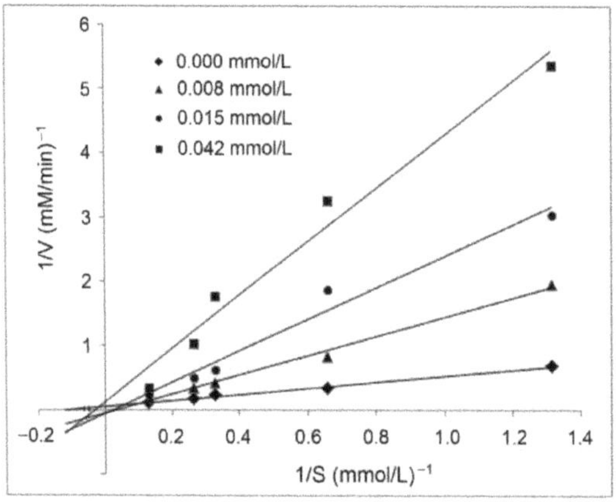

Figure 7. Lineweaver–Burk chart for inhibitory capacity of RPPSEDEDQE toward ACE.

2.4.3. The Binding Mode of RPPSEDEDQE to Zinc Ion

In the current study, ultraviolet wavelength scanning and Fourier-transformed infrared spectroscopy were both employed to analyze the binding mode of RPPSEDEDQE to zinc ion. The ultraviolet adsorption peak of RPPSEDEDQE moved from 199 nm to 206 nm after the zinc chelation (Figure 8A), demonstrating the combination between zinc ions and RPPSEDEDQE [34]. Chelation with zinc ions can cause electronic transitions of peptides' chromophoric groups, thereby resulting in a red- or blue shift of the ultraviolet adsorption peak of peptides [35].

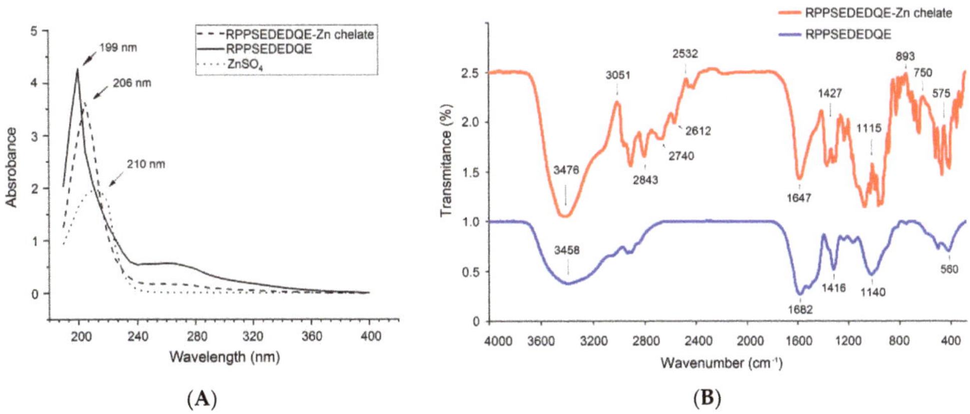

Figure 8. Ultraviolet wavelength scanning profiles (**A**) and Fourier-transformed infrared spectroscopic diagrams (**B**) of RPPSEDEDQE zinc ligands. Pure RPPSEDEDQE was used as comparison.

As shown in Figure 8B, after chelation with zinc ions, the RPPSEDEDQE–zinc chelate showed a different FT-IR spectrum compared with the pure RPPSEDEDQE, suggesting that zinc ions have been linked with RPPSEDEDQE. After the coordination of RPPSEDEDQE with zinc ions, the absorption peaks presented at 3458 cm^{-1} (corresponding to a stretching of –N–H), 1682 cm^{-1} (indicative of the vibration of –C–N), and 886 cm^{-1} (representing the deformation of the amide IV band) [36,37] have separately moved to 3476 cm^{-1}, 1647 cm^{-1}, and 893 cm^{-1}, evidencing the binding between zinc ions and the peptide linkage or amino group of RPPSEDEDQE [33]. Moreover, the redshift from 1416 to 1427 cm^{-1} reflected that an asymmetric bend had happened on –C–O, showing the binding of zinc ions with the carboxyl group of RPPSEDEDQE [32]. Thus, the results in Figure 8A,B show that RPPSEDEDQE can provide multiple binding sites for zinc ions.

In short, the inhibitory mechanisms of RPPSEDEDQE against ACE may include binding with seven active residues of ACE, and coordinating with His387 or the zinc ion of the zinc tetrahedral coordination.

2.5. Stability of RPPSEDEDQE

After gastrointestinal digestion, the inhibitory effect of RPPSEDEDQE on ACE (y) at different doses (x) is shown in Figure 5B. The equation fitting to this curve was $y = 12.768\ln(x) - 18.482$, with an IC_{50} value of 213.48 μmol·L^{-1}, which was equal to that of untreated RPPSEDEDQE (205.50 μmol·L^{-1}, Figure 5A). Moreover, as shown in Table 2, after gastrointestinal digestion, the chelating activity of RPPSEDEDQE toward zinc ions was 18.55 ± 0.95 mg·g^{-1}, which was not different from that of untreated RPPSEDEDQE ($p > 0.5$). It can be obtained that the zinc-chelating ability and ACE-inhibitory capacity of RPPSEDEDQE were both not significantly reduced by the hydrolysis of gastrointestinal proteases ($p > 0.5$).

2.6. Antihypertension In Vivo of RPPSEDEDQE and RPPSEDEDQE–Zn Complexes

As shown in Figure 9A, a moderate to high dose of RPPSEDEDQE (100–150 mg·kg bw^{-1}) significantly decreased the systolic blood pressure of spontaneous hypertensive rats (SHR) after 1 week of oral administration ($p < 0.05$); in comparison, a high dose of RPPSEDEDQE–Zn complexes obviously reduced the systolic blood pressure of SHR after 3 weeks' administration ($p < 0.05$). A low dose of RPPSEDEDQE (50 mg·kg bw^{-1}) and low to middle dose of RPPSEDEDQE–Zn complexes (50–100 mg·kg bw^{-1}) did not show any lowering effect on the systolic blood pressure of SHR. A similar trend was observed in their effects on the diastolic blood pressure of SHR (Figure 9B).

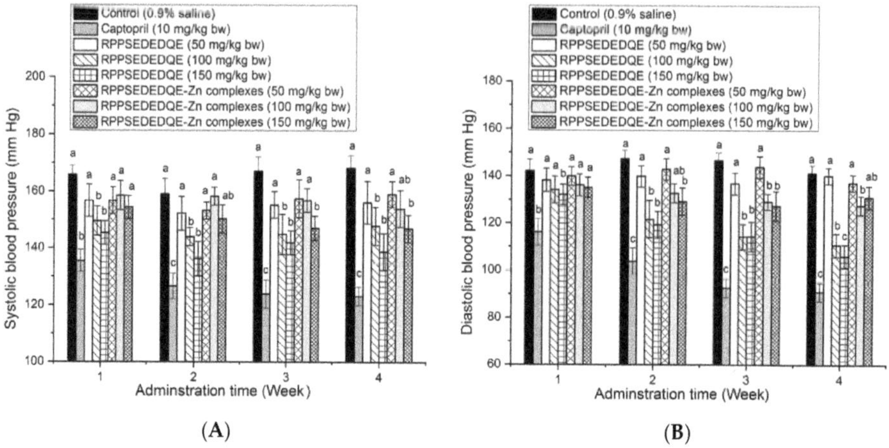

Figure 9. Efficiency of RPPSEDEDQE and RPPSEDEDQE–Zn complexes on the systolic blood pressure (**A**) and diastolic blood pressure (**B**) of spontaneously hypertensive rats. Different letters (a–c) between the groups marked a statistically significant difference ($p < 0.05$).

RPPSEDEDQE remarkably lowered the diastolic blood pressure of SHR at 150 mg/kg body weight after one week of oral administration ($p < 0.05$). RPPSEDEDQE–Zn complexes with a high dose (150 $mg \cdot kg\ bw^{-1}$) decreased the diastolic blood pressure after two weeks of administration. However, a low dose of RPPSEDEDQE and RPPSEDEDQE–Zn complexes did not offer any reducing effect on the diastolic blood pressure of SHR. At the same dose, RPPSEDEDQE's reducing effects on the systolic blood pressure and the diastolic blood pressure were both better than that of RPPSEDEDQE–Zn complexes. Moreover, the reducing efficiency of RPPSEDEDQE and RPPSEDEDQE–Zn complexes on the diastolic and systolic blood pressure were both much lower than that of the positive control (Captopril, a famous antihypertensive drug) ($p < 0.05$) [7].

In short, RPPSEDEDQE was a dose-dependent blood-pressure lowerer, and chelation with zinc ions may hinder its antihypertensive efficiency. The results in Figure 7 showed that the amino group, carboxyl group, and amide bond of RPPSEDEDQE could all chelate with zinc ions. The coordination with zinc ions may affect the interactions between RPPSEDEDQE and ACE, especially the zinc tetrahedral coordination, consequently lowering the antihypertensive effect of RPPSEDEDQE [35]. Therefore, the results of this study suggested that RPPSEDEDQE was not suitable to use as a zinc supplement and blood-pressure-lowering agent at the same time.

3. Discussion

To the best of our knowledge, this is the first study to simultaneously investigate the capacities of bitter almond albumin peptides to inhibit ACE and bind with zinc ions. Previous studies evidenced that thermolysin could quickly cleave the amide bonds linked with hydrophobic amino acid residues, with the released hydrophobic amino acid residues playing an important role in ACE inhibition; whereas papain cleaves multiple sites of polypeptide chains, especially peptide bonds linking with acidic amino acid residues (Glu and Asp), Arg, and aromatic amino acids which was helpful for the release of peptide sequences due to the metal ion-chelating capacity and/or inhibiting ability on ACE [23,38]. Thus, to release the capacities of peptides to inhibit ACE and chelate with zinc ions, bitter almond albumin was hydrolyzed by papain and thermolysin in the current study. Zhang et al. [25] used dual enzymes (flavorzyme and alcalase with a mass ratio of 1:2) to hydrolyze almond amandin, and the hydrolysis degree and zinc-chelating capacity of the hydrolysates were 32.67% ± 4.56% and 14.32 ± 1.32 $mg \cdot g^{-1}$, respectively. These results indicated that papain and thermolysin rather than flavorzyme combined with alcalase could isolate the peptide sequences of the zinc-chelating ability and ACE-inhibitory ability from bitter almond albumin polypeptide chains. A previous study found that papain probably prefers to hydrolyze peptide bonds linking with acidic amino acid residues (Glu and Asp), which was helpful for the release of peptide sequences owing to the metal ion-chelating capacity [17]. Moreover, BAAHs offered a higher inhibiting ability against ACE (53.99 ± 4.62%) than that of wild almond protein hydrolysates (18.26%) at 1 mg·mL^{-1} [16].

A previous study showed that *C*-terminal tripeptides are crucial for the inhibition ability of peptides against ACE. *C*-terminal hydrophobic tripeptides are accepted as the main reason for the excellent inhibiting ability of peptides on ACE [29]. In contrast, the *C*-terminal tripeptides of RPPSEDEDQE (Asp, Gln, and Glu) were all hydrophilic, but it offered a relatively high ACE-inhibitory activity (205.50 µmol·L^{-1}). The main reason was perhaps that the Gln and Glu residues both had a relatively high binding affinity with the active sites of ACE. Peptides with Glu and Gln in the *C*-terminal tripeptide such as ADMIET identified in *Lepidium meyenii*, QTPHQ, and ELHPQ derived from canary seeds also showed high ACE-inhibitory activities [29,30]. Moreover, the *N*-terminal tripeptides of RPPSEDEDQE were Arg, Pro, and Pro which have been shown to have a strong affinity to ACE [7,31]. The peptides RTVFDGEL and RADVFNPR derived from palm kernel glutelin-1 with Arg at the *N*-terminal exhibited an excellent ability to restrain ACE [27].

Recently, most studies have been focusing on the interaction between the peptides and binding sites of ACE, but few data referred to the effects of peptides on the zinc tetrahedral coordination. The results in Figures 6 and 7 evidenced that RPPSEDEDQE can interact with the residue His387 of the zinc tetrahedral coordination and provide multiple chelating sites for zinc ions. Li et al., and Zarei et al., found that symbiotic peptides and palm oil kernel peptides could affect the zinc tetrahedral coordination and offered high ACE inhibition abilities [5,26]. However, the effect of RPPSEDEDQE on the zinc tetrahedral coordination in ACE needs further work.

Although previous studies showed that oligopeptides were sensitive to the protease existing in the gastric or intestinal regions [10,12], RPPSEDEDQE exhibited a relative stability against the hydrolysis of gastrointestinal proteases. One of the reasons may be its high content of Pro residue. It was shown that the rigid ring structure of Pro was helpful for the resistance of peptides to enzymatic digestion [13]. The peptides PPSEPTKL and VEPFPLF that were rich in Pro residue and isolated from bighead carp were found to be stable under the hydrolysis of gastrointestinal enzymes [36]. However, to investigate the changes in the RPPSEDEDQE structure under the hydrolysis of gastrointestinal enzymes, more work is needed.

The results of the current study showed that zinc chelation reduced the in vivo antihypertensive effect of RPPSEDEDQE. It was shown that Mung bean-derived peptides PGSGCAGTDL, camellia peptide GYGYNY, and LYSPLLKYFD from rice bran had dose-dependent antihypertensive effects on SHR, but the addition of zinc ions can reduce their antihypertension [35,39,40]. Zinc chelation may change the coordination of peptides toward ACE, resulting in lower ACE-inhibitory activity and antihypertension [4]. However, Li et al. [5] found that the ACE inhibition ability of symbiotic peptide KYPHVF was not significantly reduced after zinc chelation. Therefore, the specific effect of zinc chelation on the antihypertensive activity of peptides needs further work.

4. Materials and Methods

4.1. Materials and Reagents

Bitter almonds were purchased in Xinglin Garden, Taigu, China. Thermolysin (USP, 1:100,000 U·g^{-1}, where one unit (U) will hydrolyze casein to produce color equivalent to 1.0 µmol of tyrosine per min at pH 7.5 and 37 °C), papain (USP, 1:10,000 U·g^{-1}), pancreatin (USP, 1:3000 U·g^{-1}), and pepsin (USP, 1:100,000 U·g^{-1}) were provided by Yunheng Biotechnology Company, Suzhou, China. ACE and *N*-hippuryl-L-histidyl-L-leucine were purchased from Sigma (St. Louis, MO, USA). Sodium hydroxide, $ZnSO_4·7H_2O$, and other analytical grade chemicals were purchased from Suma Chem. Co. (Shijiazhuang, China).

4.2. Extraction and Proteolysis of Bitter Almond Albumin

Bitter almonds were washed and dried at 50 ± 1 °C for 17 h. The shells of the bitter almonds were removed, and the kernel was ground, and then sieved using a JL-003 sieve (60–mesh, Jujin Screen Factory, Xinxiang, China) [20]. Next, 50 g of bitter almond was mixed with 450 mL of petroleum ether (60–90 °C), and shaken in a THZ-82AB thermostatic oscillator (Nabeichao Shaker Co., Foshan, China) at 40 r/min and 25 °C for 4 h. After filtration, the mud left on the filter paper was dried at 45 ± 1 °C for 3 h to obtain defatted bitter almond. Afterward, the defatted bitter almond (10 g) was thoroughly mixed with 400 mL of distilled water (dH_2O), and incubated in the THZ-82AB thermostatic oscillator (with a rotation speed of 210 r/min) at 45 ± 1 °C for 200 min. Then, the mixture was filtered, and the percolate was centrifuged at 13,000× *g* and 5 °C for 25 min. The supernatant was placed in membranes with a cut-off mass of 7.0 kDa and dialyzed against dH_2O at 4 °C for 36 h. A freeze-dryer (Tingya Centrifuge Factory, Wuhan, China) was employed to dry the dialysis solutions in the membranes, and bitter almond albumin (SAA) was obtained.

As per the method of Zhang et al. [25] with some modifications, in a THZ-82AB thermostatic oscillator with a rotation speed of 170 r/min, SAA (4 g·200 mL^{-1} dH_2O) was hydrolyzed by papain (0.1 g) at 50 °C and pH 7.5 for 120 min. Next, the temperature was decreased to 37 °C, and thermolysin (0.1 g) was added, the proteolysis reaction was continued for 120 min. To maintain the pH value of the reaction solution at 7.5 ± 0.1, the reaction solution was adjusted with dilute 1 mol·L^{-1} of HCl or 1 mol·L^{-1} of NaOH every five minutes. Afterward, the hydrolysis solution was heated in boiling water for 5 min. After centrifugation at 10,480× *g* for 22 min, the supernatant was freeze-dried with a lyophilizer (Tingya Centrifuge Factory, Wuhan, China) to obtain bitter almond albumin hydrolysates (BAAHs). Additionally, the hydrolysis degree of bitter almond albumin was determined using the method of Nielsen et al. [41].

4.3. Chromatographic Separation of BAAHs

The BAAH (0.2 g/100 mL Milli-Q water) was first separated on a gel column (size: 8 × 100 cm, Xihu, Shanghai, China) chromatography filled with Sephadex G-25. The mobile phase was Milli-Q water with an elution speed of 3.2 mL every minute, and the monitored wavelength was 220 nm [28]. The corresponding peak fractions were collected, pooled, freeze-dried, and their inhibition ability against ACE was measured as per the *N*-hippuryl-L-histidyl-L-leucine method [30]. The further separation of subfractions was conducted using an LC-20AT reverse-phase high-performance liquid chromatography instrument (RP-HPLC, Shimadzu, Tokyo, Japan). The subfraction that offered the highest inhibiting ability against ACE was loaded on a Supersil ODS2 chromatographic column (Size: 2.1 × 30 mm, 5 μm; Eilt, Suzhou, China) and eluted by acetonitrile (elution A) and trifluoroacetate solution (1 mL·1000 mL^{-1} ultrapure water, elution B). The RP-HPLC separation was conducted with binary gradient elution at 35 °C for 24 min [7]. The gradient concentration of acetonitrile against elution time was 0–3 min, 5% A; 4–8 min, 5–10% A; 9–15 min, 10–35% A; 13–24 min, 35% A. The corresponding peak fractions were collected and separately freeze-dried, and the fraction offering the highest inhibition ability against ACE was subjected to peptide sequence analysis.

4.4. Inhibition Activity and Kinetics against ACE

The *N*-hippuryl-L-histidyl-L-leucine method was utilized for the determination of the inhibiting capacity of BAAH peptides toward ACE [30]. In a glass test tube, 0.2 mL of ACE solution (dissolved in boric acid buffer, pH 8.3) and 0.2 mL of BAAH peptide aqueous solution (1 mg·mL^{-1}) were added. The mixture in the glass tube was shaken in a water bath at 143 r·min^{-1} and 37 °C. Ten minutes later, 0.65 mL of *N*-hippuryl-L-histidyl-L-leucine (5 mmol·L^{-1}, dissolved in boric acid buffer, pH 8.3) was added. After continuously shaking for 1 h, 0.75 mL of dilute hydrochloric acid was added into the tube. Next, hippuric acid produced in the reaction solution was extracted by ethyl acetate (1.25 mL). The hippuric acid solution (the supernatant) was sucked out with a pipette (DG-1, Dragon TopPette, Beijing,

China) and heated at 70 °C to evaporate the ethyl acetate. The residual hippuric acid was resolved in distilled water, and its absorbance at 228 nm was read. The inhibition capacity of BAAH peptides against ACE was the percentage of the absorbance reduction value at 228 nm to the absorbance of the control group (without BAAH peptide). The concentration of BAAH peptides was defined as IC_{50} when the activity of ACE was inhibited by half. Captopril was used as the positive control.

Furthermore, the Lineweaver–Burk plot of ACE against different concentrations of *N*-hippuryl-L-histidyl-L-leucine (0.13, 0.26, 0.33, 0.66, and 1.32 mmol·L^{-1}) was drawn [30]. Both K_m (Michaelis–Menten constant) and the max reaction velocity (V_m) were calculated to analyze the inhibition kinetic of BAAH peptides (0.008, 0.015, and 0.042 mmol·L^{-1}) against ACE.

4.5. Chelating Ability toward Zinc Ions

The chelating ability of BAAH peptides toward zinc ions was measured using the 4-(β-Pyridinazo)-resorcinol method following the same procedures proved by Ke et al. [36]. In short, 0.75 mmol·L^{-1} of $ZnSO_4$ (dissolved in 0.2 mmol·L^{-1} of HEPES-KOH buffer) and 10 mmol·L^{-1} of 1,4-Dithiothreitol were added into a glass tube. The tube was incubated in the THZ-82AB thermostatic oscillator (195 r·min^{-1} and 25 °C). Five minutes later, the zinc concentration in the glass tube was measured and read as C_0. Next, the BAAH peptides aqueous solution (0.8 mg·mL^{-1}) was added and the mixture was continuously shaken for another 15 min. Then, the zinc concentration in the glass tube was measured and read as C_t. The chelating ability of BAAH peptides toward zinc ions was calculated as follows, and ethylenediamine tetraacetic acid (EDTA) was used as the positive control.

$$\text{Chelating ability towards zinc ions}\left(\text{mg}\cdot\text{g}^{-1}\right) = (C_0 - C_t) \times V/m \quad (1)$$

where V and m represent the volume of the reaction solution (mL) and the mass weight of BAAH peptides (g), respectively.

4.6. Amino Acid Sequence Analysis, Validation, and Screening

First, the lyophilized BAAH subfractions isolated by RP-HPLC were further analyzed on a LaChrom II C_{18} column (3 × 250 mm, 1.9 μm) coupled to an ultra-high-performance liquid chromatography system (ChromasterUltra Rs, HITACHI, Tokyo, Japan) using the same procedures of Ruan et al. [7]. Next, the quadrupole Orbitrap tandem spectrometric determination was conducted on an ACQUITY TQD quantum-ultra mass spectrometer (Waters, Milford, MA, USA) coupled with Peak de novo 8.5 software (BSI, Waterloo, ON, Canada). The specific parameters for the electrospray ionization operation were ion source gas of 60 psi; spray negative voltage of 4300 and positive voltage of 5300 V; mass range from 120 to 1600 m/z; and AGC target value of 1×10^5 [33]. Validation of the BAAH amino acid sequences identified was carried out using the database from the National Biotechnology Information Center (Bethesda, MD, USA). Moreover, the BAAH amino acid sequences identified were in silico screened on the basis of the vector machine software score (SVMS) using the database AHTpin [27]. BAAH sequences that had an SVMS higher than zero were accepted as potential antihypertensive peptides [42].

4.7. Chemical Synthesis, Chelating Capacity toward Zinc Ions, and Physicochemical Characteristics of Potential Antihypertensive Sequences

BAAH amino acid sequences with potential antihypertension were synthesized following the chemical solid procedures in Gangyan Biotech. Co. (Dalian, China). The chemical-synthesized peptides were subjected to inhibition ability against ACE and chelating capacity toward zinc ions determinations. Moreover, the physicochemical characteristics of BAAH amino acid sequences were calculated using in silico tools with database AHTpin [27].

4.8. Docking Modes of BAAH Peptides toward ACE Molecule

Interactions of BAAH peptides toward ACE were determined using a molecular modeling tool (SERFLUX-DOCK SCREEN, SYLYB 2.2.2, Treops Company, Jefferson City, MO, USA) with a three-dimensional structure of ACE (PDB-108A, from Protein Bank) as the docking model. Ligands of ACE and BAAH peptides were virtually screened using multiple scoring functions including Kuntz-D Score, FleX-Chem Score, and Martin-PMF Score, and total score of virtual evaluation (T-score), and consistency score (C-score) were used as the main evaluation index. The T-score of acceptable docking modes should be higher than 6.0. The interaction forces between the interfacing sites of ACE and BAAH peptides including hydrogen bonds and hydrophobic interactions were investigated [7,17].

4.9. Ligand Patterns of Zinc Ions with BAAH Peptides

First, BAAH peptides and zinc ion ligands were prepared with $ZnSO_4 \cdot 7H_2O$ as the zinc ion donor, using the same procedures of Zhang et al. [37]. BAAH peptides were chelated with $ZnSO_4 \cdot 7H_2O$ at a mass ratio of 1:7. The reaction temperature and pH were 52 °C and pH 5.8, respectively [33]. The reaction solution was centrifuged and the supernatant was further precipitated by anhydrous ethanol. Then, the precipitate was freeze-dried and the BAAH peptides' zinc chelation was obtained.

Second, the obtained BAAH peptides' zinc chelation was subjected to ultraviolet wavelength scanning analysis and Fourier-transformed infrared spectroscopy (FT-IR) determination, respectively. In short, ultraviolet wavelength scanning was conducted using an UV759CRT ultraviolet wavelength scanner (Jingke Instrument Lit. Co., Shanghai, China). The range of the scanned wavelength was from 190 to 400 nm [33]. Simultaneously, 3 mg of the BAAH peptides' zinc chelation and dry KBr (0.01 g) were pressed into a translucent sheet. The sheet was used for FT-IR analysis on an FT-730 spectrometer (Jinke Instrument Factory, Shanghai, China) with wavenumbers from 4000 to 400 cm^{-1} [32].

4.10. Stabilities of BAAH Peptides

In a stirring (115 $r \cdot min^{-1}$) water bath at 37 °C, the BAAH peptides' aqueous solution (5 mg/mL, 10 mL) was added into a glass test tube, and the pH value was adjusted to 2.00 ± 0.05 with dilute hydrochloric acid solution (0.5 $mol \cdot L^{-1}$) [13]. Three minutes later, 1.15 mg pepsin (USP, 1:10,000 U/g) was added into the glass tube and mixed thoroughly, and then stirred in the water bath for ninety minutes. Next, the pH value was adjusted to pH 7.00 ± 0.05 with dilute sodium hydroxide solution (0.1 $mol \cdot L^{-1}$), and then 2.25 mg pancreatin (USP, 1:3000 $U \cdot g^{-1}$) was added. The hydrolysis reaction was continued for another 150 min. Then, the glass tube was heated in boiling water to stop the digestion reaction. Afterward, both the chelating ability toward zinc ion, and the inhibition activity of BAAH peptides against ACE were determined using the 4-(β-Pyridinazo)-resorcinol method and *N*-hippuryl-L-histidyl-L-leucine method, respectively [30,36]. Untreated BAAH peptides were used as the blank control.

4.11. Antihypertension In Vivo

The antihypertensive effect was measured as per the modified method of Dong et al. [40]. Male spontaneously hypertensive rats (SHR, 10 weeks, 220 ± 15 g body weight, from Beijing Medicine Yaoli Biotechnology Co., Ltd., Beijing, China) were housed in colony cages maintained at 25 ± 1 °C and 45–55% relative humidity with a 12 h light/dark cycle. Tap water was freely available. The rats were randomly divided into eight groups: negative group, positive group, BAAH peptide low-, middle-, and high-groups, and BAAH–Zn complexes low-, middle-, and high-groups. Each group had four rats. The rats in the negative control group and positive group were gastric intubated with NaCl (0.9%) and Captopril (15 mg/kg/bodyweight once daily), respectively; while the rats in the high-, middle-, and low-dosage groups were orally administrated with 200, 100, and 50 mg of MBAH peptides (or peptide–Zn complexes)/kg/bodyweight once daily, respectively. The oral administration was continued for 28 days. The diastolic and systolic blood pressures of the rats were measured every 7 days using the tail-cuff method with a ZL-620-F noninvasive blood-pressure apparatus (Anhui Yaokun Biotechnology Co., Ltd., Hefei, China). Each rat was measured at least three times. The weight and heart rate of the rats were determined at the same time. This experimental work was approved by the Institutional Animal Care and Use Committee, Shanxi University of Chinese Medicine (No. SNUA230311002). All animals received human care according to the Guidelines Manual for the Care and Use of Laboratory Animals.

4.12. Statistical Analysis

The entire test was repeated three or more times and one-way variance analysis and Duncan's multiple comparison were employed to analyze the difference among results. The difference was considered as significant when the *p* value was less than 0.05.

5. Conclusions

A zinc ion ligand RPPSEDEDQE offering an inhibitory effect on ACE (IC_{50}: 205.50 $\mu mol \cdot L^{-1}$) was isolated from BAAHs using an in vitro test combined with in silico methods. The binding modes of RPPSEDEDQE to the ACE molecule perhaps include three ways: (i) the formation of hydrogen bonds between RPPSEDEDQE and seven active residues of ACE; (ii) RPPSEDEDQE bound to fifteen active residues of ACE via hydrophobic interactions; and (iii) RPPSEDEDQE link with the residue His387 or zinc ion of the zinc tetrahedral coordination in ACE. Inhibitory kinetics analysis evidenced that RPPSEDEDQE was a noncompetitive inhibitor of ACE, because it cannot bind with any central pocket of ACE. RPPSEDEDQE possessed a high hydrophilicity (1.85), but RPPSEDEDQE and RPPSEDEDQE–zinc chelate should be used away from the pI (3.84). Moreover, the zinc-chelating ability and ACE-inhibitory capacity of RPPSEDEDQE were both resistant to the hydrolysis of gastrointestinal enzymes. RPPSEDEDQE (100–150 $mg \cdot kg\ bw^{-1}$) significantly reduced the systolic and diastolic blood pressure of SHRs, but chelation with zinc ions decreased its antihypertensive efficiency. These results indicate that bitter almond albumin peptides may be used for antihypertension. The results of the current study, especially the studies on physicochemical characterization, safety, and effect on the zinc tetrahedral coordination in ACE, can provide a new perspective for studying the action mechanism and application of antihypertensive peptides.

Author Contributions: Investigation, conceptualization, writing—original draft, funding acquisition, N.Q.; investigation, data curation, C.C.; investigation, data curation, N.Z.; investigation, L.S.; validation, Y.L.; writing—review and editing, L.G.; validation, R.L.; writing—review and editing, W.Z. All authors have read and agreed to the published version of the manuscript.

Funding: This research was funded by the Chinese Nutrition Society Fund (CNS-ZD2021-78) and Shanxi University of Chinese Medicine cultivation project (2022TD2008).

Institutional Review Board Statement: The animal study protocol was approved by the Institutional Animal Care and Use Committee, Shanxi University of Chinese Medicine (202303020223-1, 23 February 2023).

Informed Consent Statement: Not applicable.

Data Availability Statement: The data that support the findings of this study are available from the corresponding author upon reasonable request.

Conflicts of Interest: The authors declare no conflict of interest.

Sample Availability: Samples of the bitter almond albumin are available from the authors.

References

1. Duan, X.; Dong, Y.; Zhang, M.; Li, Z.; Bu, G.; Chen, F. Identification and molecular interactions of novel ACE inhibitory peptides from rapeseed protein. *Food Chem.* **2023**, *422*, 136085. [CrossRef]
2. Li, Z.; He, Y.; He, H.; Zhou, W.; Li, M.; Lu, A.; Che, T.; Shen, S. Purification identification and function analysis of ACE inhibitory peptide from *Ulva prolifera* protein. *Food Chem.* **2023**, *345*, 134127. [CrossRef]

3. Wu, J.; Liao, W.; Udenigwe, C.C. Revisiting the mechanisms of ACE inhibitory peptides from food proteins. *Trends Food Sci. Technol.* **2017**, *69*, 214–219. [CrossRef]
4. Pina, A.; Roque, A. Studies on the molecular recognition between bioactive peptides and angiotensin-converting enzyme. *J. Mol. Recognit.* **2009**, *22*, 162–168. [CrossRef]
5. Li, R.; Zhou, X.; Sun, L.; Zhuang, Y. Identification, in silico screening, and molecular docking of novel ACE inhibitory peptides isolated from the edible symbiot Boletus griseus-*Hypomyces chrysospermus*. *LWT-Food Sci. Technol.* **2022**, *169*, 114008. [CrossRef]
6. Yin, Z.; Yan, R.; Jiang, Y.; Feng, S.; Sun, H.; Sun, J.; Zhao, D.; Li, H.; Wang, B.; Zhang, N. Identification of peptides in Qingke baijiu and evaluation of its angiotensin converting enzyme (ACE) inhibitory activity and stability. *Food Chem.* **2022**, *395*, 133551. [CrossRef]
7. Ruan, S.Y.; Sun, L.P.; Sun, X.D.; He, J.L.; Zhuang, Y.L. Novel umami peptides from tilapia lower jaw and molecular docking to the taste receptor T1R1/T1R3. *Food Chem.* **2021**, *362*, 130249. [CrossRef]
8. Lee, S.Y.; Hur, S.J. Antihypertensive peptides from animal products, marine organisms, and plants. *Food Chem.* **2017**, *228*, 506–517. [CrossRef] [PubMed]
9. Nirmal, N.P.; Rajput, M.S.; Rathod, N.B.; Mudgil, P.; Pati, S.; Bono, G.; Nalinanon, S.; Li, L.; Maqsood, S. Structural characteristic and molecular docking simulation of fish protein-derived peptides: Recent updates on antioxidant, anti-hypertensive and anti-diabetic peptides. *Food Chem.* **2023**, *405*, 134737. [CrossRef] [PubMed]
10. Singh, B.P.; Vij, S. In vitro stability of bioactive peptides derived from fermented soy milk against heat treatment, pH and gastrointestinal enzymes. *LWT-Food Sci. Technol.* **2018**, *91*, 303–307. [CrossRef]
11. Wongngam, W.; Hamzeh, A.; Tian, F.; Roytrakul, S.; Yongsawatdigul, J. Purification and molecular docking of angiotensin converting enzyme-inhibitory peptides derived from corn gluten meal hydrolysate and from in silico gastrointestinal digestion. *Process Biochem.* **2023**, *129*, 113–120. [CrossRef]
12. Sahni, P.; Sharma, S.; Surasani, V.K.R. Influence of processing and pH on amino acid profile, morphology, electrophoretic pattern, bioactive potential and functional characteristics of alfalfa protein isolates. *Food Chem.* **2020**, *333*, 127503. [CrossRef] [PubMed]
13. Magouz, O.; Mehanna, N.; Khalifa, M.; Sakr, H.; Gensberger-Reigl, S.; Dalabasmaz, S.; Pischetsrieder, M. Profiles, antioxidative and ACE inhibitory activity of peptides released from fermented buttermilk before and after simulated gastrointestinal digestion. *Innov. Food Sci. Emerg.* **2023**, *84*, 103266. [CrossRef]
14. Tang, H.; Wang, C.; Cao, S.; Wang, F. Novel angiotensin I-converting enzyme (ACE) inhibitory peptides from walnut protein isolate: Separation, identification and molecular docking study. *J. Food Biochem.* **2022**, *46*, e14411. [CrossRef] [PubMed]
15. Liu, X.; Miao, X.; Dan, C.; Li, J.; Min, W. Purification and identification of ACE-inhibiting peptides from wild pine nut peptide fractions (PNPF). *Eur. Food Res. Technol.* **2018**, *244*, 979–988. [CrossRef]
16. Mirzapour, M.; Rezaei, K.; Sentandreu, M.A. Identification of Potent ACE Inhibitory Peptides from Wild Almond Proteins. *J. Food Sci.* **2017**, *82*, 2421–2431. [CrossRef] [PubMed]
17. Song, W.; Fu, J.; Zeng, Q.; Lu, H.; Wang, J.; Fang, L.; Liu, X.; Min, W.; Liu, C. Improving ACE inhibitory activity of hazelnut peptide modified by plastein: Physicochemical properties and action mechanism. *Food Chem.* **2023**, *402*, 134498. [CrossRef] [PubMed]
18. Souza, T.S.P.; Dias, F.F.G.; Koblitz, M.G.B.; Bell, J.M.L.N.M. Effects of enzymatic extraction of oil and protein from almond cake on the physicochemical and functional properties of protein extracts. *Food Bioprod. Process.* **2020**, *122*, 280–290. [CrossRef]
19. Dias, F.F.G.; de Moura Bell, J.M.L.N. Understanding the impact of enzyme-assisted aqueous extraction on the structural, physicochemical, and functional properties of protein extracts from full-fat almond flour. *Food Hydrocolloid.* **2020**, *127*, 107534. [CrossRef]
20. Li, S.G.; Chu, S.; Lu, J.K.; Wang, P.; Ma, M.H. Molecular and structural properties of three major protein components from almond kernel. *J. Food Process. Pres.* **2018**, *42*, e13536. [CrossRef]
21. Ganesh, S.; Ningtyas, D.W.; Prakash, S. Investigating the functionality of enzymatically (transglutaminase and alcalase) treated almond protein isolate. *Food Biosci.* **2020**, *49*, 101914. [CrossRef]
22. Civera, A.; Galan-Malo, P.; Segura-Gil, I.; Mata, L.; Tobajas, A.P.; Sánchez, L.; Pérez, M.D. Development of sandwich ELISA and lateral flow immunoassay to detect almond in processed food. *Food Chem.* **2022**, *371*, 131338. [CrossRef] [PubMed]
23. Zhang, H.; Xue, J.; Zhao, H.; Zhao, X.; Xue, H.; Sun, Y.; Xue, W. Isolation and structural characterization of antioxidant peptides from degreased apricot seed kernels. *J. AOAC Inter.* **2018**, *101*, 1661–1663. [CrossRef] [PubMed]
24. Gu, X.; Gao, T.; Hou, Y.; Li, D.; Fu, L. Identification and characterization of two novel α-glucosidase inhibitory peptides from almond (*Armeniaca sibirica*) oil manufacture residue. *LWT-Food Sci. Technol.* **2020**, *134*, 110215. [CrossRef]
25. Zhang, J.; Ye, Z. Pentapeptide-Zinc Chelate from Sweet Almond Expeller Amandin Hydrolysates: Structural and Physicochemical Characteristics, Stability and Zinc Transport Ability In Vitro. *Molecules* **2022**, *27*, 7936. [CrossRef]
26. Zarei, M.; Ghanbari, R.; Zainal, N.; Ovissipour, R.; Saari, N. Inhibition kinetics, molecular docking, and stability studies of the effect of papain-generated peptides from palm kernel cake proteins on angiotensin-converting enzyme (ACE). *Food Chem. Mol. Sci.* **2022**, *5*, 100147. [CrossRef]
27. Kumar, R.; Chaudhary, K.; Sharma, M.; Nagpal, G.; Chauhan, J.S.; Singh, S.; Gautam, A.; Raghava, G.P. AHTPDB: A comprehensive platform for analysis and presentation of antihypertensive peptides. *Nucleic Acids Res.* **2015**, *43*, D956–D962. [CrossRef] [PubMed]

28. Khakhariya, R.; Basaiawmoit, B.; Sakure, A.A.; Maurya, R.; Bishnoi, M.; Kondepudi, K.K.; Padhi, S.; Rai, A.K.; Liu, Z.; Hati, S. Production and Characterization of ACE Inhibitory and Anti-Diabetic Peptides from Buffalo and Camel Milk Fermented with Lactobacillus and Yeast: A Comparative Analysis with In Vitro, In Silico, and Molecular Interaction Study. *Foods* **2023**, *12*, 2006. [CrossRef]
29. Lin, Z.; Lai, J.; He, P.; Pan, L.; Zhang, Y.; Zhang, M.; Wu, H. Screening, ACE-inhibitory mechanism and structure-activity relationship of a novel ACE-inhibitory peptide from *Lepidium meyenii* (Maca) protein hydrolysate. *Food Biosci.* **2023**, *52*, 102374. [CrossRef]
30. Urbizo-Reyes, U.; Liceaga, A.M.; Reddivari, L.; Kim, K.-H.; Anderson, J.M. Enzyme kinetics, molecular docking, and in silico characterization of canary seed (*Phalaris canariensis* L.) peptides with ACE and pancreatic lipase inhibitory activity. *J. Funct. Foods* **2022**, *88*, 104892. [CrossRef]
31. Abedin, M.M.; Chourasia, R.; Phukon, L.C.; Singh, S.P.; Rai, A.K. Characterization of ACE inhibitory and antioxidant peptides in yak and cow milk hard chhurpi cheese of the Sikkim Himalayan region. *Food Chem. X* **2022**, *13*, 100231. [CrossRef] [PubMed]
32. Sun, R.; Liu, X.; Yu, Y.; Miao, J.; Leng, K.; Gao, H. Preparation process optimization, structural characterization and in vitro digestion stability analysis of Antarctic krill (*Euphausia superba*) peptides-zinc chelate. *Food Chem.* **2021**, *340*, 128056. [CrossRef] [PubMed]
33. Li, Y.; Shi, P.; Zheng, Y.; Guo, M.; Zhuang, Y.; Huo, X. Millet bran protein hydrolysates derived peptides-zinc chelate: Structural characterization, security prediction in silico, zinc transport capacity and stability against different food processing conditions. *J. Food Sci.* **2023**, *88*, 477–490. [CrossRef] [PubMed]
34. Udechukwu, M.C.; Downey, B.; Udenigwe, C.C. Influence of structural and surface properties of whey-derived peptides on zinc-chelating capacity, and in vitro gastric stability and bioaccessibility of the zinc-peptide complexes. *Food Chem.* **2018**, *240*, 1227–1232. [CrossRef]
35. Zheng, Y.J.; Shi, P.Q.; Li, Y.; Zhuang, Y.L.; Zhang, Y.L.; Liu, L.; Wang, W. A novel ACE-inhibitory hexapeptide from camellia glutelin-2 hydrolysates: Identification, characterization and stability profiles under different food processing conditions. *LWT-Food Sci. Technol.* **2021**, *147*, 111682. [CrossRef]
36. Ke, X.; Hu, X.; Li, L.; Yang, X.; Chen, S.; Wu, Y.; Xue, C. A novel zinc-binding peptide identified from tilapia (*Oreochromis niloticus*) skin collagen and transport pathway across Caco-2 monolayers. *Food Biosci.* **2022**, *42*, 101127. [CrossRef]
37. Zhang, Z.; Zhou, F.; Liu, X.; Zhao, M. Particulate nanocomposite from oyster (*Crassostrea rivularis*) hydrolysates via zinc chelation improves zinc solubility and peptide activity. *Food Chem.* **2018**, *258*, 269–277. [CrossRef]
38. Chen, M.; Wang, L.; Zheng, C.; Ma, A.; Hu, K.; Xiang, A.; Sun, Z.; Xie, B.; Xiong, G.; Shi, L.; et al. Novel ACE inhibitory peptides derived from bighead carp (*Aristichthys nobilis*) hydrolysates: Screening, inhibition mechanisms and the bioconjugation effect with graphene oxide. *Food Biosci.* **2023**, *52*, 102399. [CrossRef]
39. Sonklin, G.; Alashi, A.M.; Laohakunjit, N.; Kerdchoechuen, O.; Aluko, R.E. Identification of antihypertensive peptides from mung bean protein hydrolysate and their effects in spontaneously hypertensive rats. *J. Funct. Foods* **2020**, *64*, 103635. [CrossRef]
40. Dong, J.; Wang, S.; Yin, X.; Fang, M.; Gong, Z.; Wu, Y. Angiotensin I converting enzyme (ACE) inhibitory activity and antihypertensive effects of rice peptides. *Food Sci. Hum. Wellness* **2022**, *11*, 1539–1543. [CrossRef]
41. Nielsen, P.M.; Petersen, D.; Dambmann, C. Improved method for determining food protein degree of hydrolysis. *J. Food Sci.* **2001**, *66*, 642–646. [CrossRef]
42. Zaharuddin, N.D.; Barkia, I.; Ibadullah, W.Z.W.; Zarei, M.; Saari, N. Identification, molecular docking, and kinetic studies of six novel angiotensin I converting enzyme (ACE) inhibitory peptides derived from Kenaf (*Hibiscus cannabinus* L.) seed. *Int. J. Biol. Macromol.* **2022**, *220*, 1512–1522. [CrossRef] [PubMed]

foods

MDPI

Article

Preparation and Identification of Peptides with α-Glucosidase Inhibitory Activity from Shiitake Mushroom (*Lentinus edodes*) Protein

Yu Zhang [1,3,4], Yu Chen [1,5], Xinyang Liu [1,6], Wei Wang [1,2,*], Junhong Wang [1], Xue Li [1,3,4] and Suling Sun [1]

1 Institute of Agro-Product Safety and Nutrition, Zhejiang Academy of Agricultural Sciences, Hangzhou 310021, China; zhangy_lk@zaas.ac.cn (Y.Z.)
2 School of Biological and Chemical Engineering, Zhejiang University of Science and Technology, Hangzhou 310012, China
3 Key Laboratory of Agricultural Product Information Traceability, Ministry of Agriculture and Rural Affairs of China, Hangzhou 310021, China
4 Zhejiang Provincial Key Laboratory of Food Safety, Hangzhou 310021, China
5 Collage of Agriculture and Animal Husbandry, Qinghai University, Xining 810016, China
6 College of Wine, North West Agriculture and Forestry University, Xi'an 712199, China
* Correspondence: 122069@zust.edu.cn

Abstract: The shiitake mushroom is the most commonly cultivated edible mushroom in the world, and is rich in protein. This study aims to obtain the peptides with α-glucosidase inhibition activity from shiitake mushroom protein hydrolysate. The conditions of enzymatic hydrolysis of shiitake mushroom protein were optimized by response surface test. The results showed that the optimal conditions were as follows: the E/S was 3390 U/g, the solid–liquid ratio was 1:20, the hydrolysis temperature and time were 46 °C and 3.4 h, respectively, and the pH was 7. The active peptides were separated by gel filtration and identified by LC-MS/MS analysis and virtual screening. The results indicated that fourteen peptides were identified by LC-MS/MS. Among them, four new peptides (EGEPKLP, KDDLRSP, TPELKL, and LDYGKL) with the higher docking score were selected and chemically synthesized to verify their inhibition activity. The IC50 values of EGEPKLP, KDDLRSP, TPELKL, and LDYGKL for α-glucosidase inhibition activity ranged from 452 ± 36 μmol/L to 696 ± 39 μmol/L. The molecular docking results showed that the hydrogen bond and arene–cation bond were the two major interactions between four peptides and 2QMJ. The hydrogen bonds were crucial to the inhibition activity of α-glucosidase. The results indicate the potential of using the peptides from shiitake mushroom protein as functional food with α-glucosidase inhibition activity.

Keywords: shiitake mushroom; α-glucosidase inhibition peptides; identification; virtual screening

Citation: Zhang, Y.; Chen, Y.; Liu, X.; Wang, W.; Wang, J.; Li, X.; Sun, S. Preparation and Identification of Peptides with α-Glucosidase Inhibitory Activity from Shiitake Mushroom (*Lentinus edodes*) Protein. *Foods* **2023**, *12*, 2534. https://doi.org/10.3390/foods12132534

Academic Editor: Ivano De Noni

Received: 1 June 2023
Revised: 24 June 2023
Accepted: 26 June 2023
Published: 29 June 2023

1. Introduction

Diabetes mellitus (DM) is a widespread metabolic disorder that affected 537 million adults globally in 2021 [1]. The number of diabetics is predicted to rise to 643 million by 2030 [2]. Two varieties of DM exist: type I (T1DM) and type II (T2DM). It is estimated that the number of people suffering from T2DM accounts for 90% of the total number of diabetics [3], and T2DM is characterized by hyperglycemia. Chronic damage and dysfunction of various tissues are caused by long-term hyperglycemia, such as cardiovascular disease, retinal damage, chronic kidney disease, and diabetic ketoacidosis [4–6].

The most effective treatment for T2DM is to reduce hyperglycemia, especially postprandial hyperglycemia, by slowing down the carbohydrate metabolism. The membrane-bound α-glucosidase, located in the small intestine's epithelial mucosa, is a major digestive enzyme involved in carbohydrate metabolism. It can cleave the glycoside bonds in carbohydrate to free the glucose [7,8]. Therefore, inhibition of α-glucosidase activity has evolved to be an effective method for T2DM treatment because of its role in delaying the hydrolysis of carbohydrates to inhibit postprandial hyperglycemia [9,10]. Although the synthesized α-glucosidase inhibitors such as acarbose have been widely used for T2DM treatment, their long-term use leads to adverse effects [11,12]. Thus, more and more researchers are trying to look for α-glucosidase inhibitors from natural food as a complementary therapy for T2DM to ameliorate the adverse effects associated with synthetic drugs [13].

Recently, some plant or animal protein hydrolysates with α-glucosidase inhibition activity have been found, such as egg [14,15], bean [16,17], amaranth [18], quinoa [19,20], wheat gluten [21], and black cricket [22]. A variety of α-glucosidase inhibitory peptides were identified from Binglangjiang buffalo casein [10], walnut [23], egg protein [4,24,25], Spirulina platensis [26], Ginkgo biloba seed protein [27], camel whey protein [28], and Changii Radix hydrolysates [29].

Shiitake mushroom (*Lentinus edodes*) is the largest edible mushroom in China and is also the second most commonly cultivated edible mushroom in the world [30–32]. The yield of shiitake mushrooms in China has increased, but there are few processed products, which hinders the development of the shiitake mushroom industry. The shiitake mushroom is widely popular because of its nutritional value due to a variety of nutritional compounds including polysaccharides, protein, fiber, vitamins, and minerals [33–35]. At present, research on shiitake mushrooms mainly focuses on polysaccharides [35], and there are few studies on the development of shiitake mushroom protein. Shiitake mushroom is rich in protein, with a protein content of about 16–22% [34]. In addition, the shiitake mushroom protein contains abundant human essential and non-essential amino acids. Therefore, the shiitake mushroom, as one of the main sources of active peptides, has good prospects and may be crucial to the development of functional foods. The objective of this study is to optimize the preparation conditions of peptides and obtain the hydrolysate with the maximum inhibition activity of α-glucosidase. The peptides from the hydrolysate of shiitake mushroom protein were purified by gel chromatography, and the potential peptides with α-glucosidase inhibition activity were further identified by LC-MS/MS and virtual screening. Then, the relation between the structure of α-glucosidase inhibition peptides and activity was investigated.

2. Materials and Methods

2.1. Materials, Reagents and Instruments

Shiitake mushrooms were acquired from a farmers' market. Neutral protease (50 U/mg), pepsin (1:30,000), alkaline protease (200 U/mg), acid protease (50 U/mg), papain (800 U/mg), and flavor protease (20 U/mg) were purchased from Qiyi Biotechnology (Shanghai, China) Co., Ltd. α-glucosidase (10 U/mg) and 4-nitrophenyl a-D-glucopyranoside (PNPG, 99% purity) were purchased from Sigma-Aldrich (St. Louis, MO, USA). Acetonitrile (HPLC grade, 99.9% purity) was purchased from Fisher Chemical (Hampton, NH, USA). Formic acid (LC-MS grade, 99% purity) was purchased from Thermo Scientific (Waltham, MA, USA). Other reagents were of analytical grade and commercially available.

An Thermo Easy-nanoLC system was connected to a Thermo Scientific QE mass spectrometer, USA Thermo Fisher Scientific. Other devices used were as follows: Analytical balance, Sartorius Scientific Instruments (Beijing) Co., Ltd. (Beijing, China). AMR-100 automatic enzyme label analyzer, Hangzhou Aosheng Instrument Co., Ltd. (Hangzhou, China). MTYK-MI805 pH meter, Beijing Zhonghui Tiancheng Technology Co., Ltd. (Beijing, China). HZQ-B Constant Temperature Culture Shaker, Suzhou Weier Laboratory Supplies Co., Ltd. (Suzhou, China). H-1650 high speed centrifuge, Changsha Xiangyi Centrifuge Instrument Co., Ltd. (Changsha, China). Freeze-drying Instrument, Shanghai Haozhuang Instrument Co., Ltd. (Shanghai, China).

2.2. Peptide Preparation and Condition Optimization

2.2.1. Preparation of Peptides

Extraction of shiitake mushroom protein: A certain quantity of dried shiitake mushroom powders was dissolved in deionized water with the mass–volume ratio of 1:20. The solution pH was adjusted to 9.0 with 1 mol/L NaOH. The mixture was sonicated at 40 KHz and 50 °C for 60 min, then centrifugation was used to separate the supernatant for 15 min at a speed of 4000 rpm. The supernatant pH was adjusted to 3.5 by 1 mol/L HCl. Then, after centrifugation at 5000 rpm/min for 20 min, the precipitates were collected.

Hydrolysate preparation: Two grams of protein powder was dissolved in deionized water according to Section 2.2.2. A certain amount of protease was added according to Section 2.2.2, and mixed once the pH value was adjusted to an ideal level according to Section 2.2.2. The mixture was shaken at 150 rpm/min for a certain period time according to Section 2.2.2. The reaction was terminated in boiling water for 5 min. The pH was adjusted to 7.0 with 1 mol/L NaOH or 1 mol/L HCl according to the hydrolysis pH. Then the supernatant was collected after centrifugation at 5000 rpm/min for 20 min.

2.2.2. Optimization of Single Factor Enzymatic Hydrolysis Conditions

With the E/S of 4000 U/g, the solid–liquid ratio of 20, the pH of 7.0, the hydrolysis temperature of 45 °C, and the hydrolysis time of 4 h as the basic fixed factors, one of the factors was changed and the single factor test was carried out. The gradient design of each factor was as follows: E/S (1000, 2000, 3000, 4000, 5000 U/g), solid–liquid ratio (15, 20, 25, 30, 35), pH (5, 6, 7, 8, 9), hydrolysis time (2, 3, 4, 5, 6), hydrolysis temperature (25, 35, 45, 55, 65 °C). The α-glucosidase inhibition activity of hydrolysate with different conditions was compared.

2.2.3. Response Surface Methodology

The hydrolysis conditions of shiitake mushroom protein were improved using CCD. The inhibitory activity of α-glucosidase acted as the response value. According to the single factor experiment, E/S (A), temperature (B), and time (C) were selected to be optimized for response surface optimization. Table 1 displays the levels of the experimental factors.

Table 1. Parameters and levels of response surface analysis.

Parameters.	**Level**				
	−1.68	**−1**	**0**	**1**	**1.68**
E/S / U/g (A)	1320	2000	3000	4000	4680
Temperature/°C (B)	28.2	35	45	55	61.8
Time/h (C)	1.32	2.00	3.00	4.00	4.68

2.3. *Determination of α-Glucosidase Inhibition Activity*

A quantity of 100 μL α-glucosidase solution (0.2 U/mL) was premixed with 50 μL sample solution in the enzyme label plate, and incubated at 25 °C for 10 min; 50 μL of PNPG (5 mmol/L) solution was added to start reactions, and incubation was continued for another 30 min at 37 °C. The reaction was terminated with 50 μL 0.67 mol/L Na_2CO_3 solution. The absorbance value was measured at 405 nm using an AMR-100 automatic enzyme label analyzer. The results were calculated using the following formula:

$$\text{Inhibitory rate } (\%) = ((A_{control} - A_{sample})/A_{control}) \times 100 \quad (1)$$

2.4. *Purification of α-Glucosidase Inhibition Peptides*

2.4.1. Ultrafiltration

The hydrolysate solution was separated by ultrafiltration membranes of 1 k Da, 3 k Da, and 5 k Da. The fraction was named UF-I (>5 k Da), UF-II (3–5 k Da), UF-III (1–3 k Da), and UF-IV (<1 k Da), and collected separately.

2.4.2. Gel Column Filtration

The fraction F-IV obtained by ultrafiltration was further separated by gel filtration according to Chen et al.'s method [36]. The sample was loaded onto a Sephadex G-10 gel filtration column (2.6 cm × 100 cm) after being dissolved in deionized water. Following that, deionized water was used to elute the peptides at a flow rate of 0.5 mL/min. The eluted fractions were monitored at 280 nm. The desired peak fractions were collected and lyophilized. The α-glucosidase inhibitory rates of elution peaks were determined.

2.5. Identification and Screening of the α-Glucosidase Peptide

The fraction with maximum activity obtained by gel filtration chromatography was analyzed by LC-MS/MS according to the method of Chen et al. [36].

The peptides were diluted in a 20 μL solution of 0.1% formic acid and 5% acetonitrile and separates with an Acclaim PepMap RPLC C18 (75 um i.d. × 150 mm,2 um, 100 Å, nanoViper) linked to a PepMap RSLC C18 in an LC-MS/MS system. The mobile phase was as follows: mobile phase A (0.1% formic acid) and mobile phase B (80% acetonitrile with 0.1% formic acid); the gradient elution was performed with a gradient of 3–99% B (0–34 min) and the flow rate was 500 nL/min.

MS data were acquired over the range from 100 to 1550 m/z in ESI positive mode. Resolution was 120,000, AGC target was 4×10^5, and maximum IT was 50 ms. MS/MS scanning conditions were as follows: resolution was 30,000, AGC target was 1×10^5, maximum IT was 100 ms, TopN was 20, and NCE/stepped NCE was 32.

Peptide sequences were identified using PEAKS Studio X software in combination with a shiitake mushroom protein database search.

2.6. Molecular Docking Analysis

The semi-flexible molecular docking was performed between the identified peptides and the crystal structure of the human α-glucosidase (PDB code: 2QMJ) using MOE software according to Chen et al.'s method [36]. The crystal structure of 2QMJ (PDB DOI: 10.2210/pdb2QMJ/PDB) was downloaded from the PDB database. The structures of 14 peptides identified by LC-MS-MS were drawn using MOE2009 software.

To obtain the receptor molecules required for docking, the water molecules from 2QMJ were eliminated, the molecules were protonated, and energy was minimized. The receptor pocket of the 2QMJ was used as the docking target for molecular docking, and the peptides were used as the ligand. Each coupling was performed 30 times. The peptide with a better docking effect was screened from the obtained results for further analysis.

2.7. Peptide Synthesis

The screened potential α-glucosidase inhibitory peptides were chemically synthesized by China Peptides Co., Ltd. (Shanghai, China). Fmoc solid-phase synthesis was used to obtain the raw peptide. Briefly, chlorine resin was swelled by dichloromethane (DCM). Fmoc-AA (amino acid)-OH, O-benzotriazol-1-yl-tetramethyluronium hexafluorophosphate (HBTU), and N,N-diisopropylethylamine (DIEA) were added to the drained resin for 30 min with a nitrogen bubble reaction to accomplish the condensation reaction. The resin was then washed with DMF to remove Fmoc. After being drained, cutting solution (trifluoroacetic acid (TFA)/H_2O/1,2-ethanedithiol (EDT)/triisopropylsilane (Tis) = 95/1/2/2, $v/v/v/v$) was added to the resin to cut the peptide. The synthetic peptides were purified by HPLC. The molecular weight of the purified peptide was confirmed by mass spectrometry.

2.8. Statistical Analysis

All experiments had three repetitions. Data are expressed as mean ± standard deviation. Statistical significance (Duncan's test, $p < 0.05$) was analyzed by DPS. Response surface data were analyzed using Design-ExpertV8.0.6 software.

3. Results

3.1. Protease Selection

The α-glucosidase inhibition activity of shiitake mushroom protein hydrolysates obtained by hydrolysis with the acid protease, alkaline proteinase, neutral proteinase, flavor proteinase, papain, and pepsin was investigated. The results indicate that the enzyme specificity of the different proteases determines the activity characteristics of the hydrolysate. The α-glucosidase inhibition rates of the hydrolysates obtained with acid protease, alkaline proteinase, neutral proteinase, flavor proteinase, papain, and pepsin were 12.5 ± 0.31%, 14.8 ± 0.35%, 61.5 ± 1.6%, 33.2 ± 0.71%, 10.7 ± 0.28%, and 15.3 ± 0.37%, respectively. Compared with other hydrolysates, the neutral proteinase hydrolysate showed the maximum inhibitory rate against α-glucosidase at the same concentration. Therefore, neutral protease was chosen as the enzyme for hydrolysis.

3.2. Effects of Different Enzyme Concentrations on the Activity of Hydrolysates

The effect of enzyme concentration on the α-glucosidase inhibitory activity of shiitake mushroom protein hydrolysates was investigated. With the increase in the enzyme–substrate ratio (E/S), the α-glucosidase inhibitory rate increased first and then slightly decreased (Figure 1A). The α-glucosidase inhibitory rate of hydrolysate with E/S of 3000 U/g was significantly higher than that of 1000 U/g and 2000 U/g ($p < 0.05$). Therefore, 3000 U/g was selected as the appropriate enzyme concentration.

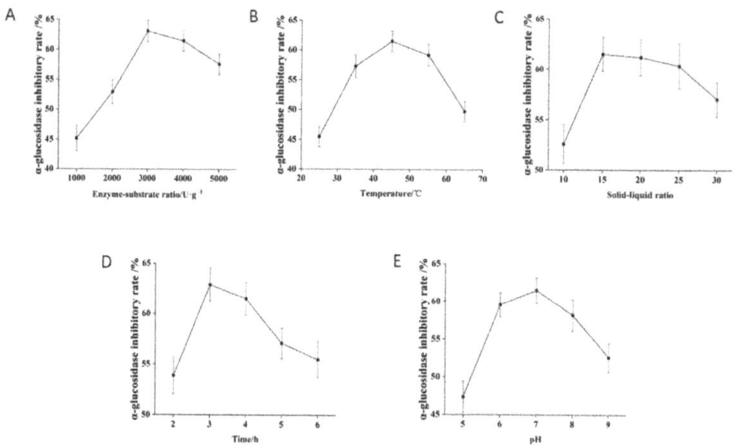

Figure 1. Effects of E/S (**A**), temperature (**B**), solid–liquid ratio (**C**), time (**D**), and pH (**E**) on α-glucosidase inhibition activity.

3.3. Effects of Different Temperature on the Activity of Hydrolysates

Temperature is a key influencing factor for enzymatic reaction. Increasing temperature can speed up the reaction rate, but excessive temperature can cause protease inactivation. The effect of different temperatures on the α-glucosidase inhibition activity of enzymatic hydrolysates was investigated. With the increase in temperature, the inhibitory rate of α-glucosidase first increased and then decreased. When the temperature reached 45 °C, the maximum inhibitory rate of α-glucosidase was attained (Figure 1B). Therefore, 45 °C was chosen as the appropriate condition.

3.4. Effects of Different Substrate Concentrations on the Activity of Hydrolysates

The substrate concentration is a key parameter of enzymatic reaction. When the solid–liquid ratio was 15, the inhibitory rate of enzymatic hydrolysates was lower. When the solid–liquid ratio was 20, the α-glucosidase inhibitory rate of enzymatic hydrolysates increased and then decreased slightly (Figure 1C). It may be that when the solid–liquid ratio was 15, the protein concentration was too high, which affected the fluidity of the system and the contact surface between the enzyme and the substrate. When the solid-liquid ratio reached 20, the hydrolysate had good fluidity. Therefore, the solid–liquid ratio of 20 was chosen as the appropriate condition for enzymatic hydrolysis.

3.5. Effects of Different Enzymolysis Time on the Activity of Hydrolysates

The inhibition activity against α-glucosidase was significantly influenced by the enzymolysis time. With the increase in enzymolysis time, the inhibitory rate of α-glucosidase first increased and reached the highest level at 3 h, and then decreased (Figure 1D), which may be because some active peptides were further hydrolyzed with the increase in time. Therefore, 3 h was chosen as the appropriate condition for enzymatic hydrolysis.

3.6. Effects of Different pH on the Activity of Hydrolysates

With the increase in pH, α-glucosidase inhibitory rates of hydrolysates rose first and then decreased. When pH was 7, the inhibitory rate of α-glucosidase reached the highest level (Figure 1E). There were no significant differences in the α-glucosidase inhibitory rates among pH 6.0, pH 7.0, and pH 8.0. Therefore, pH 7 was selected as the appropriate condition.

3.7. Response Surface Analysis

In the actual process of enzymolysis, various factors may interact with each other. To further investigate the significance of the influence of various factors on the target value, E/S, temperature, and time were taken as independent variables, and α-glucosidase inhibitory rate as the response value, and a CCD experiment was conducted. The results are shown in Table 2. The quadratic polynomial regression equation of α-glucosidase inhibitory rate and E/S (A), enzymolysis temperature (B), and enzymolysis time (C) was obtained as follows:

$$Y = 63.86 + 3.17A + 0.87B + 3.32C - 0.55AB - 0.87AC + 1.10BC - 3.49A^2 - 4.89B^2 - 3.76C^2 \quad (2)$$

Table 2. The experimental results of CCD.

Number	A (E/S u/g)	B (Temperature/°C)	C (Time/h)	Inhibitory Rate/%
1	1.68	0	0	58.2
2	0	0	0	64.3
3	−1	1	−1	43.5
4	0	0	−1.68	48.5
5	−1	−1	−1	43.8
6	1	−1	−1	52.8
7	−1.68	0	0	49.5
8	0	0	0	63.3
9	−1	1	1	55.9
10	0	0	0	64.8
11	1	−1	1	57.3
12	0	0	0	63.1
13	0	0	0	64.5
14	0	0	0	63.2
15	1	1	−1	52.3
16	0	0	1.68	57.7
17	0	−1.68	0	48.5
18	−1	−1	1	49.8
19	0	1.68	0	51.3
20	1	1	1	59.2

According to the variance analysis results (Table 3), the regression model has a high level of significance ($p < 0.0001$). The R^2 was 0.9877, which indicates that the test value and the fitting value had a high correlation. The regression equation model fitted the data well because the p value of the lack of fit was not significant. Consequently, it is possible to forecast the test results using the model. It can be seen from Table 3 and Figure 2 that there were significant interactions between temperature and time, and E/S and time.

Table 3. Variance analysis results of CCD design.

Source	Squares	df	Square	Value	Prob > F
Model	922.97	9	102.55	89.15	<0.0001
A-E/S	136.85	1	136.85	118.97	<0.0001
B-Temperature	10.38	1	10.38	9.03	0.0132
C-Time	150.08	1	150.08	130.46	<0.0001
AB	2.42	1	2.42	2.1	0.1776
AC	6.12	1	6.12	5.32	0.0437
BC	9.68	1	9.68	8.41	0.0158
A^2	175.52	1	175.52	152.58	<0.0001
B^2	344.1	1	344.1	299.13	<0.0001
C^2	203.2	1	203.2	176.65	<0.0001
Residual	11.5	10	1.15		
Lack of Fit	8.69	5	1.74	3.09	0.1207
Pure Error	2.81	5	0.56		
Cor Total	934.48	19			
Model	922.97	9	102.55	89.15	<0.0001
A-E/S	136.85	1	136.85	118.97	<0.0001
B-Temperature	10.38	1	10.38	9.03	0.0132
C-Time	150.08	1	150.08	130.46	<0.0001
AB	2.42	1	2.42	2.1	0.1776
AC	6.12	1	6.12	5.32	0.0437

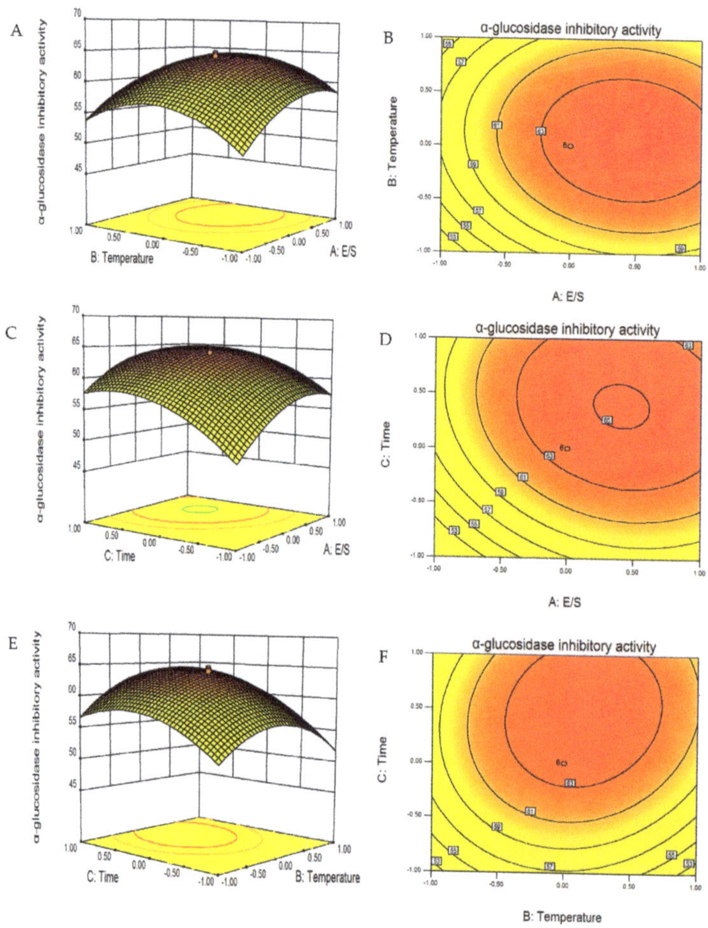

Figure 2. Interaction diagram between factors in the response surface. E/S and temperature (**A**,**B**), E/S and time (**C**,**D**), and temperature and time (**E**,**F**).

The optimal conditions predicted by the model were as follows: the E/S was 3390 U/g, the solid–liquid ratio was 1:20, the hydrolysis temperature and time were 46 °C and 3.4 h, respectively, and the pH was 7. Under optimal conditions, the predicted value of the α-glucosidase inhibitory rate was 65.2%. To verify the reliability of the predictive value of the model, three validation tests were carried out under optimum conditions, and the average α-glucosidase inhibitory rate of hydrolysate was 64.6 ± 1.2%. The result indicates that the optimized parameters of the model were accurate and reliable.

3.8. Separation of Peptides by Ultrafiltration

The molecular weight of the peptide is closely related to the α-glucosidase inhibitory activity. As shown in Figure 3, four fractions were obtained by ultrafiltration from shiitake mushroom protein hydrolysates. The α-glucosidase inhibitory rate of UF-IV was 65.0 ± 3.1%, which was higher than that of the other three fractions. The results indicate that the α-glucosidase inhibitory activity increased with the decrease in molecular weight, which is consistent with the results of previous reports [19,27,37]. Therefore, fraction UF-IV was selected for further purification.

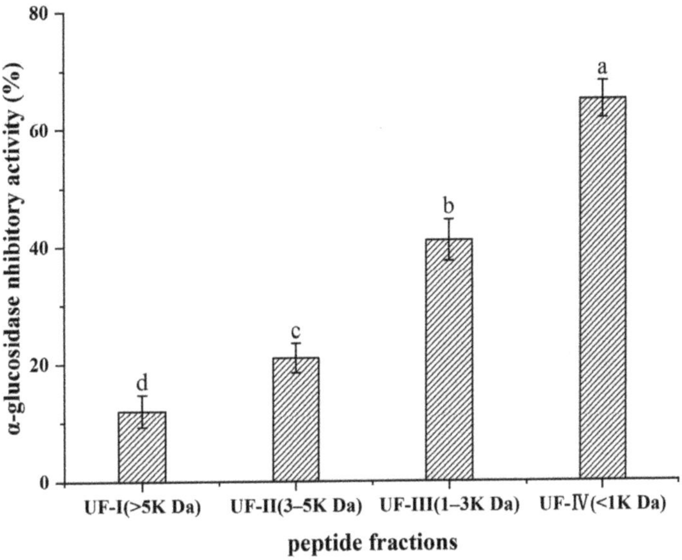

Figure 3. Results of activity of hydrolysate after ultrafiltration. Significant differences are indicated by different letters ($p < 0.05$).

3.9. Separation of Peptides by Gel Filtration

Peptides are frequently purified by gel filtration. The UF-IV fraction was further separated by a gel filtration column (Sephadex G-10). The results are presented in Figure 4. A total of eleven peaks were detected and named as F1–F11. The 11 fractions were collected to study their α-glucosidase inhibitory activity. Fraction F5 exhibited the strongest α-glucosidase inhibition rate. The inhibition rate of F5 was 61.3 ± 0.69%.

3.10. Identification and Screening of α-Glucosidase Inhibitory Peptides

The sequences of fraction F5 were determined by HPLC–MS/MS with de novo sequencing. Fourteen peptides were identified from fraction F5 (score > 80). The molecular weight of the fourteen peptides ranged from 699 Da to 829 Da (Table 4).

The binding energy and binding sites of peptides to receptor proteins are related to their biological activity. To further screen active peptides from these 14 peptides, the peptides docking with the crystal structure of α-glucosidase (2QMJ) were performed using

MOE software. Peptides with binding energy below −14 as well as more than four binding bonds were selected, as shown in Table 5, which were EGEPKLP, KDDLRSP, TPELKL, and LDYGKL (Figure 5). The α-glucosidase inhibition activity of the four peptides was further investigated. EGEPKLP, KDDLRSP, TPELKL, and LDYGKL showed an effective inhibition activity, and their IC50 value was 499 ± 39 μmol/L, 550 ± 37 μmol/L, 452 ± 36 μmol/L, and 696 ± 39 μmol/L, respectively.

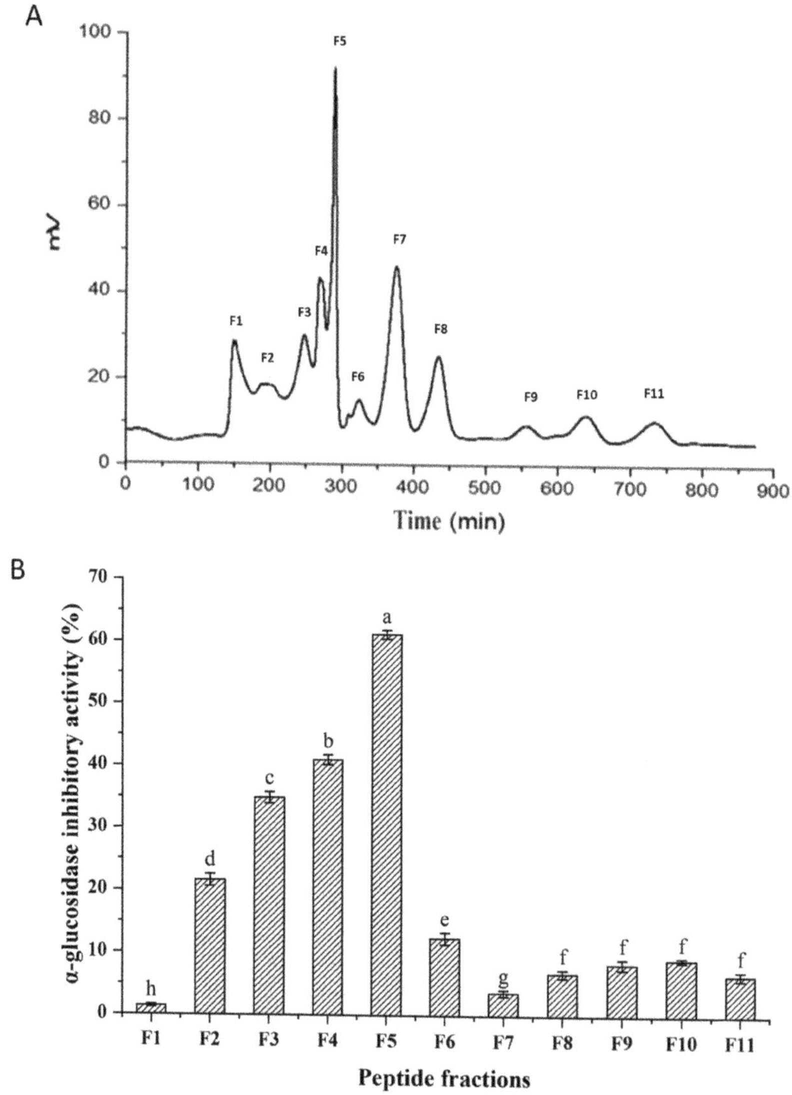

Figure 4. Separation of fraction UF-IV by Sephadex G-10 gel filtration column chromatography. (**A**) Elution profile of the fraction UF-IV; (**B**) α-glucosidase inhibition activity of the fractions separated from UF-IV. Significant differences are indicated by different letters ($p < 0.05$).

Table 4. Peptide sequences identified from fraction F5 (score > 80).

Sequence	Mass	*m/z*	*m/z* Error (ppm)	RT (min)	Length	Score
DVFAHF	734.3387	368.1776	2.5	20.7	6	82
KDDLRSP	829.4293	415.7215	−1.1	11.98	7	83
EDLRLP	741.402	371.7031	−14.1	15.63	6	83
LLAKFE	719.4218	360.7188	1.8	14.72	6	83
EPLEPK	711.3802	356.6971	−0.7	11.21	6	86
LQHLPL	719.433	360.7242	1.3	15.72	6	87
VLSRKL	714.4752	358.2401	−13.2	11.41	6	88
EGEPKLP	768.4017	385.208	−0.4	12.45	7	88
LDYGKL	707.3854	354.699	−2.6	13.24	6	88
TPELKL	699.4167	350.7144	−3.6	11.89	6	88
SPDEPKL	784.3967	393.2057	0.4	11.35	7	90
EEPLPQ	711.3439	356.679	−0.5	12.75	6	92
VVELLK	699.4531	350.7336	−0.6	13.53	6	96
DPEKFP	731.3489	366.6822	1.2	12.79	6	97

Table 5. The peptides with binding energies below −14.

Sequence	E-Score	Site	Molecular Structural Formula
EGEPKLP	−15.42	Lys534, Leu286, Ser521, Ser288	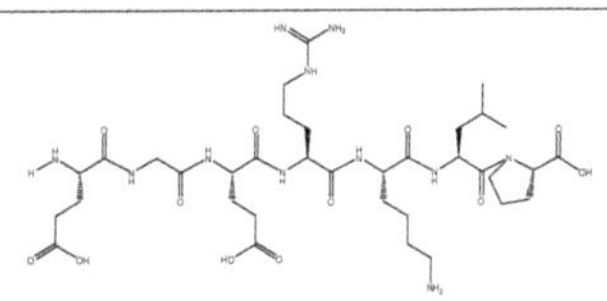
KDDLRSP	−15.83	Lys534, Lys513, His645, Val779, Asp777	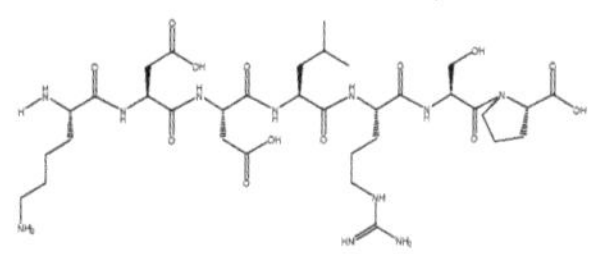
TPELKL	−14.25	Lys534, Lys513, His115, Glu114 (3 binding bonds)	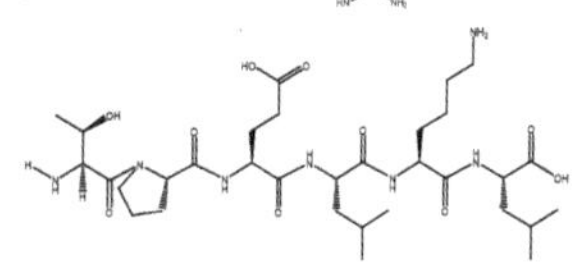
LDYGKL	−16.31	Lys776, Lys513 (2 binding bonds), Asp777 (2 binding bonds)	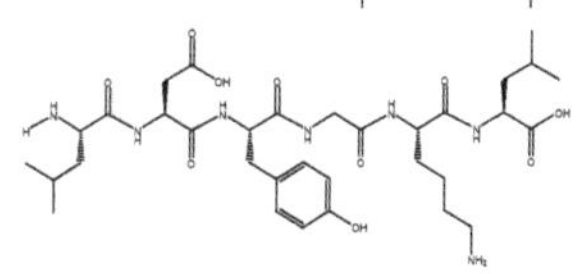

3.11. Molecular Mechanisms of the Active Peptides

The α-glucosidase inhibition activity of these peptides was obtained by molecular binding with residues in the active center. Figure 5 showed that EGEPKLP formed four hydrogen bonds at Lys534, Leu386, ser521, and Ser288 (Figure 6A). KDDLRSP formed five hydrogen bonds at Lys534, Lys513, His645, Val779, and Asp777 of 2QMJ (Figure 6B). TPELKL formed five hydrogen bonds at Lys534, Lys513, and Glu114 and an arene–cation interaction with His115 residue of 2QMJ (Figure 6C). LDYGKL formed five hydrogen bonds at Lys776, Lys513, and Asp777 of 2QMJ (Figure 6D).

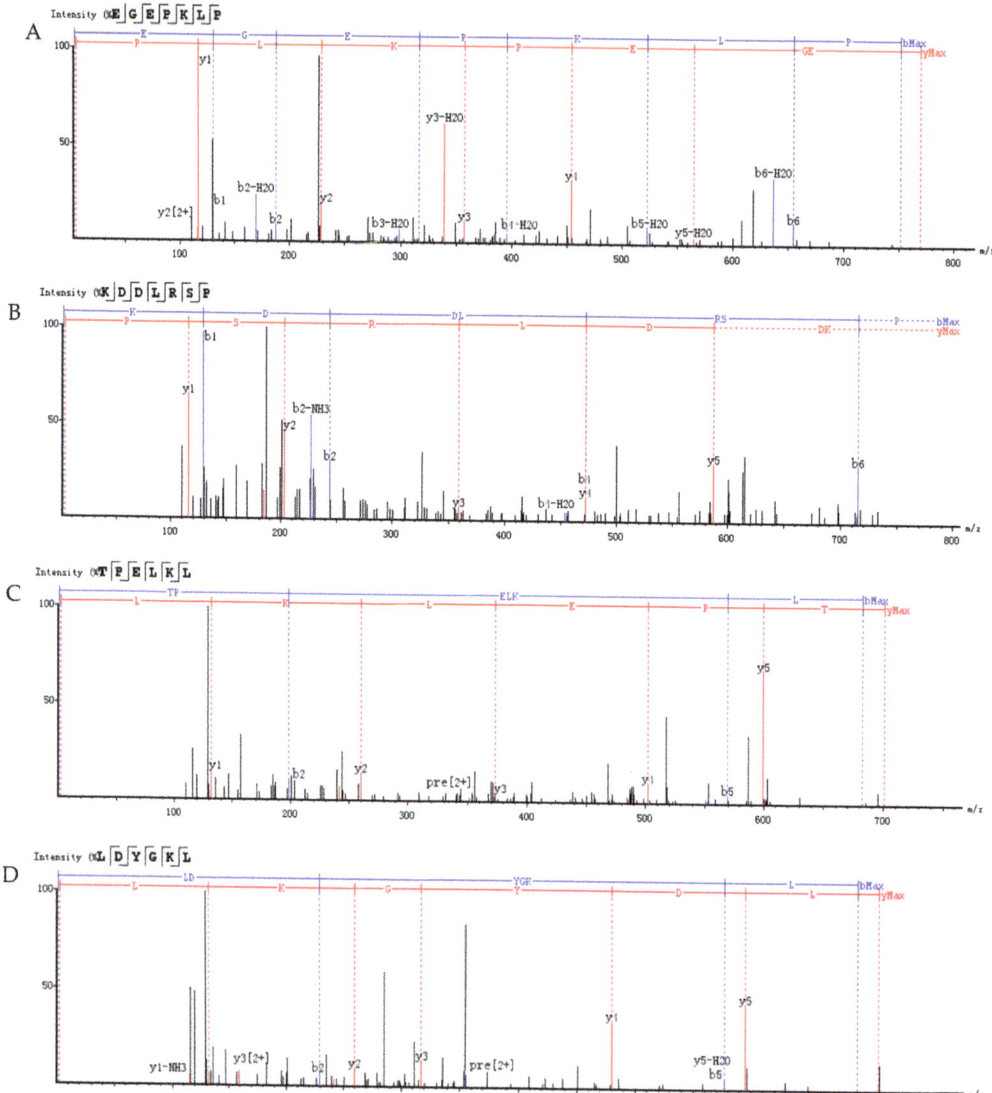

Figure 5. LC-MS-MS chromatogram of F-5. (**A**) EGEPKLP; (**B**) KDDLRSP; (**C**) TPELKL; (**D**) LDYGKL.

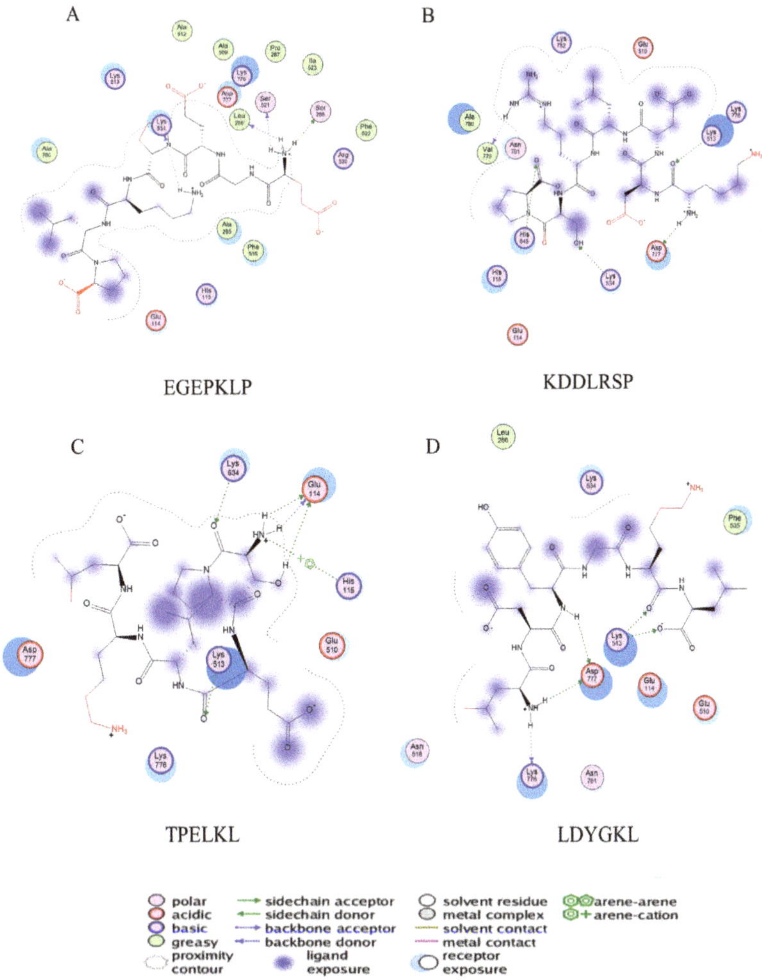

Figure 6. Molecular docking results of peptides. (**A**) EGEPKLP; (**B**) KDDLRSP; (**C**) TPELKL; (**D**) LDYGKL.

4. Discussion

DM is a typical metabolic disease with chronic complications [4–6]. With the increasing incidence of diabetes, the controls of blood glucose and therapeutic complications are faced with severe challenges. Intervention with alpha-glucosidase inhibitors can slow the conversion rate of ingested carbohydrates to glucose, thereby preventing glucose from entering the systemic circulation. Compared with the regulatory mechanism of promoting insulin secretion (biguanides, sulfonylureas, DPP IV inhibitors, GLP-1 agonists, etc.), this intervention pathway does not involve islet cells and can relieve the pressure of insulin metabolism to a certain extent and improve the function of damaged islet beta cells. The α-glucosidase in the small intestine plays an important role in the digestion of carbohydrate [19]. Blocking the enzyme with α-glucosidase inhibitors in the digestive tract will limit the digestion of carbohydrates and reduce postprandial hyperglycemia [17,23].

Recently, many studies have confirmed that α-glucosidase inhibitory peptides prepared from food protein have excellent α-glucosidase inhibitory potential [2]. With their higher bioavailability and lower side effects, the study of α-glucosidase inhibitory peptides is important. Different proteases have unique restriction sites, and thus biological peptides with different amino acid compositions can be produced by different proteases. Ren et al. [38] found that the hydrolysates released by amylopsin, pancrelipase, alcalase, papain, and trypsin from hemp seed protein displayed different α-glucosidase inhibition activity. The hydrolysates prepared by alcalase showed a high α-glucosidase inhibition rate. In this research, the α-glucosidase inhibition rates of the hydrolysates obtained by acid protease, alkaline proteinase, neutral proteinase, flavor proteinase, papain, and pepsin were investigated. The neutral proteinase hydrolysate showed the maximum α-glucosidase inhibition activity. This indicates that the newer short peptides with enhanced α-glucosidase inhibitory activity had been generated in the hydrolysates by neutral protease.

Ultrafiltration and gel chromatography can be used to further isolate and purify active peptides from the hydrolysates. In this study, low molecular weight peptides (MW < 1 k Da) had stronger α-glucosidase inhibition activity than other fractions (>5 k Da, 3–5 k Da, 1–3 k Da), which is consistent with the results of previous reports [27,37]. Wang et al. [27] reported that the hydrolysate from ginkgo biloba seed protein was fractionated into five fractions (<1 k Da, 1–3 k Da, 3–5 k Da, 5–10 k Da, >10 k Da), and the fraction with the MW < 1 k Da showed the highest α-glucosidase inhibition activity. Liu et al. [37] found that the peptide fraction (<1 kDa) of wheat germ had better α-glucosidase inhibition activity than other factions. This may be explained by the fact that the active site on the amino acid residues in the low molecular weight peptides can be exposed to the outside and increase the possibility of interacting with α-glucosidase.

The amino acid composition, molecular weight, and chain length of the peptides influence biological activity of α-glucosidase inhibitory peptides [39–41]. The majority of peptides with α-glucosidase inhibition activity have a short sequence of less than 10 residues [39,40,42]. The amino acid sequences of fraction with the highest activity were determined by HPLC–MS/MS, and 14 peptide sequences (score > 80) were obtained. The molecular weight of these peptides ranged from 699 to 829 Da and the sequence lengths were 6 to 7 amino acids residues. According to previous reports, the short peptide is beneficial to reducing the free energy of peptide-enzyme binding, and improves the inhibitory effect [40]. The identified peptides were further docked to the crystal structure of α-glucosidase. The peptides with lower energy scores and more binding sites may be proposed as active peptides in shiitake mushroom protein hydrolysates. Among these 14 peptides, EGEPKLP, KDDLRSP, TPELKL, and LDYGKL showed lower energy scores and outstanding binding ability to α-glucosidase.

The IC50 values of EGEPKLP, KDDLRSP, TPELKL, and LDYGKL for α-glucosidase inhibitory activity ranged from 452 ± 36 μmol/L to 696 ± 39 μmol/L. In previous studies, the IC50 value of RVPSLM with α-glucosidase inhibitory activity identified from egg white protein was 23.07 μmol/L [4]. The IC50 of α-glucosidase inhibitory peptide KLPGF from albumin was 59.5 ± 5.7 μmol/L [25]. The four peptides (YLGYLEQLLR, TKVIPYVRYL, RNAVPITPTLNR, FALPQYLK) from Binglangjiang buffalo casein had the inhibitory activity of α-glucosidase and their IC50 values were 470 μmol/L, 498 μmol/L, 504 μmol/L, and 543 μmol/L, respectively [10]. LSMSFPPF, MPGPPSD, and VPKIPPP identified from ginkgo biloba seed protein exhibited the α-glucosidase inhibitory activity with the IC50 values of 454.33 ± 32.45 μmol/L, 943.82 ± 73.10 μmol/L, and 1446.81 ± 66.98 μmol/L, respectively [27]. The IC50 values of SPGAGKG and GLAR from germinated chickpea protein were 1.8 mg/mL and 8.7 mg/mL, respectively [43]. Four α-glucosidase inhibitory peptides were identified from dark tea protein with the IC50 values from 0.04 mg/mL to 1.03 mg/mL [44]. Compared with the previous reports, the activity of EGEPKLP, KDDLRSP, TPELKL, and LDYGKL was of a medium level.

The peptides with more hydrophobic amino acids had better inhibitory activity of α-glucosidase [38,45]. In previous reports, hydrophobic amino acids Leu and Pro were commonly found in the sequence of α-glucosidase inhibition peptides, which contributed greatly to the inhibitory activity, such as LR, PFP, PLMLP, KLPGF, RVPSLM, WLRL, SWLRL and LLPLPVLK [4,16,38,45,46]. Basic amino acids (Lys or Arg) in peptides were also important to α-glucosidase inhibition activity, especially Lys or Arg at the N-terminus [40,47]. The amino acid analysis of the four peptides showed that the proportion of Pro and Leu in EGEPKLP, TPELKL, and LDYGKL was 42.9%, 50.0%, and 33.3%, respectively. Although the proportion of proline and leucine is only 28.6% in KDDLRSP, Lys at the N-terminus may contribute to the activity.

Further study of the interactions between the peptide and α-glucosidase showed that EGEPKLP interacted with the 2QMJ through four hydrogen bonds. KDDLRSP and LDYGKL formed five hydrogen bonds with 2QMJ. TPELKL formed five hydrogen bonds and an arene–cation interaction with 2QMJ. The α-glucosidase inhibition activity is mainly dependent on hydrogen bonding, which agrees with the results of previous research [27,48].

5. Conclusions

In this study, the optimal conditions for preparation of α-glucosidase inhibitory peptide from shiitake mushroom were established. Four new α-glucosidase inhibitory peptides (EGEPKLP, KDDLRSP, TPELKL, and LDYGKL) were identified from shiitake mushroom protein hydrolysates using LC-MS/MS and virtual screening. TPELKL exhibited a higher α-glucosidase inhibitory activity than EGEPKLP, KDDLRSP, and LDYGKL. The molecular docking results demonstrated that the hydrogen bond and arene–cation bond were the two major interactions between the four peptides and 2QMJ. The hydrogen bond was crucial to the α-glucosidase inhibition activity. The results suggest that shiitake mushroom may be a reliable source of α-glucosidase inhibitory peptide and provide a feasible strategy for further utilization of shiitake mushroom protein. To improve the possibility of using these peptides as supplements, future studies are required, including examination of the effects of the peptides on other enzymes linked to starch digestion, the antidiabetic mechanism of these peptides, and their safety and stability in vivo.

Author Contributions: Conceptualization, Y.Z.; software, W.W.; methodology and validation, W.W. and Y.Z.; formal analysis and investigation, Y.Z. and Y.C.; resources, Y.Z.; data curation, Y.C.; writing—original draft preparation, X.L. (Xinyang Liu), Y.C. and Y.Z.; writing—review and editing, Y.Z. and W.W.; visualization, X.L. (Xinyang Liu), S.S., X.L. (Xue Li) and J.W.; supervision, Y.Z.; project administration and funding acquisition, Y.Z. and W.W. All authors have read and agreed to the published version of the manuscript.

Funding: This research was funded by Zhejiang Provincial Basic Public Welfare Research Project of China (LGN21C200006), Zhejiang Provincial Natural Sciences Foundation of China (LZ22C200006), and Top young talents of the ten thousand talents program of Zhejiang Province (ZJWR0308016).

Data Availability Statement: The datasets generated for this study are available on request from the corresponding author.

Conflicts of Interest: The authors declare no conflict of interest.

References

1. IDF Diabetes Atlas. International Diabetes Federation 2021. Available online: https://www.diabetesatlas.org (accessed on 5 December 2022).
2. Whiting, D.R.; Guariguata, L.; Weil, C.; Shaw, J. IDF diabetes atlas: Global estimates of the prevalence of diabetes for 2011 and 2030. *Diabetes Res. Clin. Pract.* **2011**, *94*, 311–321. [CrossRef] [PubMed]
3. Evans, J.M.; Newton, R.W.; Ruta, D.A.; MacDonald, T.M.; Morris, A.D. Socio-economic status, obesity and prevalence of Type 1 and Type 2 diabetes mellitus. *Diabet. Med.* **2000**, *17*, 478–480. [CrossRef] [PubMed]
4. Yu, Z.; Yin, Y.; Zhao, W.; Yu, Y.; Liu, B.; Liu, J.; Chen, F. Novel peptides derived from egg white protein inhibiting alpha-glucosidase. *Food Chem.* **2011**, *129*, 1376–1382. [CrossRef]
5. Cole, J.B.; Florez, J.C. Genetics of diabetes mellitus and diabetes complications. *Nat. Rev. Nephrol.* **2020**, *16*, 377–390. [CrossRef] [PubMed]
6. Aronson, D. Hyperglycemia and the pathobiology of diabetic complications. *Adv. Cardiol.* **2008**, *45*, 1–16. [CrossRef]
7. Ponnusamy, S.; Haldar, S.; Mulani, F.; Zinjarde, S.; Thulasiram, H.; RaviKumar, A. Gedunin and Azadiradione: Human Pancreatic Alpha-Amylase Inhibiting Limonoids from Neem (*Azadirachta indica*) as Anti-Diabetic Agents. *PLoS ONE* **2015**, *10*, e0140113. [CrossRef]

8. Konrad, B.; Anna, D.; Marek, S.; Marta, P.; Aleksandra, Z.; Jozefa, C. The Evaluation of Dipeptidyl Peptidase (DPP)-IV, alpha-Glucosidase and Angiotensin Converting Enzyme (ACE) Inhibitory Activities of Whey Proteins Hydrolyzed with Serine Protease Isolated from Asian Pumpkin (*Cucurbita ficifolia*). *Int. J. Pept. Res. Ther.* **2014**, *20*, 483–491. [CrossRef]
9. El-Beshbishy, H.; Bahashwan, S. Hypoglycemic effect of basil (*Ocimum basilicum*) aqueous extract is mediated through inhibition of alpha-glucosidase and alpha-amylase activities: An in vitro study. *Toxicol. Ind. Health* **2012**, *28*, 42–50. [CrossRef]
10. Zhao, Q.; Wei, G.; Li, K.; Duan, S.; Ye, R.; Huang, A. Identification and molecular docking of novel α-glucosidase inhibitory peptides from hydrolysates of Binglangjiang buffalo casein. *LWT* **2022**, *156*, 113062. [CrossRef]
11. Fang, Y.; Wang, S.; Wu, J.; Zhang, L.; Wang, Z.; Gan, L.; He, J.; Shi, H.; Hou, J. The kinetics and mechanism of α-glucosidase inhibition by F5-SP, a novel compound derived from sericin peptides. *Food Funct.* **2017**, *8*, 323–332. [CrossRef]
12. Fujisawa, T.; Ikegami, H.; Inoue, K.; Kawabata, Y.; Ogihara, T. Effect of two alpha-glucosidase inhibitors, voglibose and acarbose, on postprandial hyperglycemia correlates with subjective abdominal symptoms. *Metabolism* **2005**, *54*, 387–390. [CrossRef] [PubMed]
13. Di Stefano, E.; Oliviero, T.; Udenigwe, C.C. Functional significance and structure–activity relationship of food-derived α-glucosidase inhibitors. *Curr. Opin. Food Sci.* **2018**, *20*, 7–12. [CrossRef]
14. Qiu, L.; Deng, Z.; Zhao, C.; Xiao, T.; Weng, C.; Li, J.; Zheng, L. Nutritional composition and proteomic analysis of soft-shelled turtle (*Pelodiscus sinensis*) egg and identification of oligopeptides with alpha-glucosidase inhibitory activity. *Food Res. Int.* **2021**, *145*, 110414. [CrossRef] [PubMed]
15. Acquah, C.; Dzuvor, C.K.O.; Tosh, S.; Agyei, D. Anti-diabetic effects of bioactive peptides: Recent advances and clinical implications. *Crit. Rev. Food Sci. Nutr.* **2022**, *62*, 2158–2171. [CrossRef] [PubMed]
16. Wang, R.; Zhao, H.; Pan, X.; Orfila, C.; Lu, W.; Ma, Y. Preparation of bioactive peptides with antidiabetic, antihypertensive, and antioxidant activities and identification of alpha-glucosidase inhibitory peptides from soy protein. *Food Sci. Nutr.* **2019**, *7*, 1848–1856. [CrossRef]
17. Valencia-Mejia, E.; Batista, K.A.; Fernandez, J.J.A.; Fernandes, K.F. Antihyperglycemic and hypoglycemic activity of naturally occurring peptides and protein hydrolysates from easy-to-cook and hard-to-cook beans (*Phaseolus vulgaris L.*). *Food Res. Int.* **2019**, *121*, 238–246. [CrossRef]
18. Kamal, H.; Mudgil, P.; Bhaskar, B.; Fisayo, A.F.; Gan, C.Y.; Maqsood, S. Amaranth proteins as potential source of bioactive peptides with enhanced inhibition of enzymatic markers linked with hypertension and diabetes. *J. Cereal Sci.* **2021**, *101*, 103308. [CrossRef]
19. Abbasi, S.; Moslehishad, M.; Salami, M. Antioxidant and alpha-glucosidase enzyme inhibitory properties of hydrolyzed protein and bioactive peptides of quinoa. *Int. J. Biol. Macromol* **2022**, *213*, 602–609. [CrossRef]
20. Vilcacundo, R.; Martínez-Villaluenga, C.; Hernández-Ledesma, B. Release of dipeptidyl peptidase IV, α-amylase and α-glucosidase inhibitory peptides from quinoa (*Chenopodium quinoa Willd.*) during in vitro simulated gastrointestinal digestion. *J. Funct. Foods* **2017**, *35*, 531–539. [CrossRef]
21. Mousavi, B.; Azizi, M.H.; Abbasi, S. Antidiabetic bio-peptides of soft and hard wheat glutens. *Food Chem.* **2022**, *4*, 100104. [CrossRef]
22. De Matos, F.M.; De Lacerda, J.T.J.G.; Zanetti, G.; De Castro, R.J.S. Production of black cricket protein hydrolysates with α-amylase, α-glucosidase and angiotensin I-converting enzyme inhibitory activities using a mixture of proteases. *Biocatal. Agric. Biotechnol.* **2022**, *39*, 102276. [CrossRef]
23. Wang, J.; Wu, T.; Fang, L.; Liu, C.; Liu, X.; Li, H.; Shi, J.; Li, M.; Min, W. Anti-diabetic effect by walnut (*Juglans mandshurica Maxim.*)-derived peptide LPLLR through inhibiting α-glucosidase and α-amylase, and alleviating insulin resistance of hepatic HepG2 cells. *J. Funct. Foods* **2020**, *69*, 103944. [CrossRef]
24. Liu, Y.-F.; Oey, I.; Bremer, P.; Carne, A.; Silcock, P. Bioactive peptides derived from egg proteins: A review. *Crit. Rev. Food Sci. Nutr.* **2018**, *58*, 2508–2530. [CrossRef] [PubMed]
25. Yu, Z.; Yin, Y.; Zhao, W.; Liu, J.; Chen, F. Anti-diabetic activity peptides from albumin against alpha-glucosidase and alpha-amylase. *Food Chem.* **2012**, *135*, 2078–2085. [CrossRef]
26. Hu, S.; Fan, X.; Qi, P.; Zhang, X. Identification of anti-diabetes peptides from Spirulina platensis. *J. Funct. Foods* **2019**, *56*, 333–341. [CrossRef]
27. Wang, X.; Deng, Y.; Xie, P.; Liu, L.; Zhang, C.; Cheng, J.; Zhang, Y.; Liu, Y.; Huang, L.; Jiang, J. Novel bioactive peptides from ginkgo biloba seed protein and evaluation of their alpha-glucosidase inhibition activity. *Food Chem.* **2023**, *404*, 134481. [CrossRef]
28. Baba, W.N.; Mudgil, P.; Kamal, H.; Kilari, B.P.; Gan, C.Y.; Maqsood, S. Identification and characterization of novel alpha-amylase and alpha-glucosidase inhibitory peptides from camel whey proteins. *J. Dairy Sci.* **2021**, *104*, 1364–1377. [CrossRef]
29. Liu, L.; Chen, J.; Li, X. Novel peptides with α-glucosidase inhibitory activity from Changii Radix hydrolysates. *Process Biochem.* **2021**, *111*, 200–206. [CrossRef]
30. Jiang, T.; Feng, L.; Wang, Y. Effect of alginate/nano-Ag coating on microbial and physicochemical characteristics of shiitake mushroom (*Lentinus edodes*) during cold storage. *Food Chem.* **2013**, *141*, 954–960. [CrossRef]
31. Huang, Q.; Qian, X.; Jiang, T.; Zheng, X. Effect of chitosan and guar gum based composite edible coating on quality of mushroom (*Lentinus edodes*) during postharvest storage. *Sci. Hortic.* **2019**, *253*, 382–389. [CrossRef]
32. Cui, L.; Liu, Y.; Liu, M.; Ren, M.; Ahmed, A.F.; Kang, W. Identification of phytochemicals from Lentinus edodes and Auricularia auricula with UPLC-Q-Exactive Orbitrap MS. *J. Future Foods* **2022**, *2*, 253–260. [CrossRef]

33. Morales, D.; Rutckeviski, R.; Villalva, M.; Abreu, H.; Soler-Rivas, C.; Santoyo, S.; Iacomini, M.; Smiderle, F.R. Isolation and comparison of α- and β-D-glucans from shiitake mushrooms (*Lentinula edodes*) with different biological activities. *Carbohydr. Polym.* **2020**, *229*, 115521. [CrossRef] [PubMed]
34. Sun, T.; Yao, S.J.; Liu, F.S. A comparative study on nutritional compositions of Lentinula edodes in four main producing regions in China. *Mod. Food Sci. Technol.* **2021**, *37*, 97–103. [CrossRef]
35. Sheng, K.; Wang, C.; Chen, B.; Kang, M.; Wang, M.; Liu, K.; Wang, M. Recent advances in polysaccharides from *Lentinus edodes* (Berk.): Isolation, structures and bioactivities. *Food Chem.* **2021**, *358*, 129883. [CrossRef] [PubMed]
36. Chen, Y.; Zhang, Y.; Qi, Q.; Liang, F.; Wang, N.; Chen, Q.; Li, X.; Sun, S.; Wang, X.; Bai, K.; et al. Preparation and activity evaluation of angiotensin-I converting enzyme inhibitory peptides from protein hydrolysate of mulberry leaf. *Front. Nutr.* **2022**, *9*, 1064526. [CrossRef]
37. Liu, W.; Li, H.; Wen, Y.; Liu, Y.; Wang, J.; Sun, B. Molecular Mechanism for the α-Glucosidase Inhibitory Effect of Wheat Germ Peptides. *J. Agric. Food Chem.* **2021**, *69*, 15231–15239. [CrossRef]
38. Ren, Y.; Liang, K.; Jin, Y.; Zhang, M.; Chen, Y.; Wu, H.; Lai, F. Identification and characterization of two novel α-glucosidase inhibitory oligopeptides from hemp (*Cannabis sativa* L.) seed protein. *J. Funct. Foods* **2016**, *26*, 439–450. [CrossRef]
39. Fadimu, G.J.; Farahnaky, A.; Gill, H.; Olalere, O.A.; Gan, C.Y.; Truong, T. In-Silico Analysis and Antidiabetic Effect of alpha-Amylase and alpha-Glucosidase Inhibitory Peptides from Lupin Protein Hydrolysate: Enzyme-Peptide Interaction Study Using Molecular Docking Approach. *Foods* **2022**, *11*, 3375. [CrossRef]
40. Ibrahim, M.A.; Bester, M.J.; Neitz, A.W.H.; Gaspar, A.R.M. Structural properties of bioactive peptides with alpha-glucosidase inhibitory activity. *Chem. Biol. Drug Des.* **2018**, *91*, 370–379. [CrossRef]
41. Yang, F.-J.; Chen, X.; Huang, M.-C.; Yang, Q.; Cai, X.-X.; Chen, X.; Du, M.; Huang, J.-L.; Wang, S.-Y. Molecular characteristics and structure–activity relationships of food-derived bioactive peptides. *J. Integr. Agric.* **2021**, *20*, 2313–2332. [CrossRef]
42. Obaroakpo, J.U.; Liu, L.; Zhang, S.; Lu, J.; Pang, X.; Lv, J. α-Glucosidase and ACE dual inhibitory protein hydrolysates and peptide fractions of sprouted quinoa yoghurt beverages inoculated with *Lactobacillus casei*. *Food Chem.* **2019**, *299*, 124985. [CrossRef] [PubMed]
43. Chandrasekaran, S.; Gonzalez de Mejia, E. Optimization, identification, and comparison of peptides from germinated chickpea (*Cicer arietinum*) protein hydrolysates using either papain or ficin and their relationship with markers of type 2 diabetes. *Food Chem.* **2022**, *374*, 131717. [CrossRef] [PubMed]
44. Zhao, B.; Su, K.; Mao, X.; Zhang, X. Separation and identification of enzyme inhibition peptides from dark tea protein. *Bioorganic Chem.* **2020**, *99*, 103772. [CrossRef] [PubMed]
45. Sadri, H.; Larki, N.N.; Kolahian, S. Hypoglycemic and hypolipidemic effects of leucine, zinc, and chromium, alone and in combination, in rats with type 2 diabetes. *Biol. Trace Elem. Res.* **2017**, *180*, 246–254. [CrossRef] [PubMed]
46. Kang, M.-G.; Yi, S.-H.; Lee, J.-S. Production and Characterization of a New α-Glucosidase Inhibitory Peptide from *Aspergillus oryzae* N159-1. *Mycobiology* **2013**, *41*, 149–154. [CrossRef]
47. Mudgil, P.; Kilari, B.P.; Kamal, H.; Olalere, O.A.; FitzGerald, R.J.; Gan, C.-Y.; Maqsood, S. Multifunctional bioactive peptides derived from quinoa protein hydrolysates: Inhibition of α-glucosidase, dipeptidyl peptidase-IV and angiotensin I converting enzymes. *J. Cereal Sci.* **2020**, *96*, 103130. [CrossRef]
48. Wei, G.; Zhao, Q.; Wang, D.; Fan, Y.; Shi, Y.; Huang, A. Novel ACE inhibitory, antioxidant and α-glucosidase inhibitory peptides identified from fermented rubing cheese through peptidomic and molecular docking. *LWT* **2022**, *159*, 113196. [CrossRef]

molecules

MDPI

Article

Distribution, Typical Structure and Self-Assembly Properties of Collagen from Fish Skin and Bone

Xuening Zhang [1], Jie Wang [1], Qian Zhang [1], Yan Fan [2,*], Hongwei Zhang [3], Khurshid Ahmad [1] and Hu Hou [1,4,5,6,*]

[1] College of Food Science and Engineering, Ocean University of China, Sansha Road, Qingdao 266404, China; zhangxuening23@163.com (X.Z.); wuaxueya@163.com (J.W.); a1104952517@163.com (Q.Z.); khurshidahmad694@yahoo.com (K.A.)
[2] College of Marine Life Sciences, Ocean University of China, Yushan Road, Qingdao 266003, China
[3] Technology Center of Qingdao Customs, Xinyue Road, Qingdao 266109, China; light04@126.com
[4] Laboratory for Marine Drugs and Bioproducts, Laoshan Laboratory, Qingdao 266237, China
[5] Sanya Oceanographic Institution, Ocean University of China, Sanya 572024, China
[6] Qingdao Institute of Marine Bioresources for Nutrition and Health Innovation, Qingdao 266000, China
* Correspondence: fanyan@ouc.edu.cn (Y.F.); houhu@ouc.edu.cn (H.H.); Tel.: +86-532-60892936 (H.H.)

Abstract: The source and type of collagen are crucial to its application, and both play a decisive role. Collagen was prepared from both tilapia skin and bone and skate skin and cartilage, named as CI-TI-s, CI-TI-b, CI-SK-s, and CII-SK-c, respectively. Types, distributions, structures, and self-assembly of collagen were studied. It showed that yellow collagen fibers from skin arranged longitudinally, while collagen fibers from skate cartilages displayed varying colors. CI-TI-s, CI-TI-b, CI-SK-s, and CII-SK-c showed the typical amide A (3316–3336 cm^{-1}) and amide B (2929–2948 cm^{-1}) in FTIR spectra. CI-TI-b and CII-SK-c showed 218–229 nm of UV absorption, 11.56–12.20 Å of d values in XRD, and 0.12–0.14 of Rpn values in CD. The thermal denaturation temperatures of CI-TI-s and CI-SK-s were 30.7 and 20.6 °C, respectively. The self-assembly of CI-TI-s and CII-SK-c were maximum at pH 7.2 and 7.4–7.6, respectively. The unique collagen peptides of tilapia and skate were GPSGPQGAVGATGPK, PAMPVPGPMGPMGPR, SPAMPVPGPMGPMGPR, GESGPSGPAGPAGPAGVR, SSGPPVPGPIGP-MGPR, GLTGPIGVPGPPGAQGEK, GLAGPQGPR, and GLSGDPGVQGIK, respectively. The unique peptides of type I and type II collagen were GPTGEIGATGLAGAR, GVLGLTGMR, LGLTGMR, GEPGAAGPAGPSGPMGPR, SSGPPVPGPIGPMGPR, and GLSGDPGVQGIK, respectively.

Keywords: collagen; self-assembly; unique peptide; structure

Citation: Zhang, X.; Wang, J.; Zhang, Q.; Fan, Y.; Zhang, H.; Ahmad, K.; Hou, H. Distribution, Typical Structure and Self-Assembly Properties of Collagen from Fish Skin and Bone. *Molecules* **2023**, *28*, 6529. https://doi.org/10.3390/molecules28186529

Academic Editor: Gabriele Rocchetti

Received: 27 July 2023
Revised: 27 August 2023
Accepted: 31 August 2023
Published: 8 September 2023

1. Introduction

Collagen is mainly present in various skins, cartilages, tendons, and bones, and accounts for 30% of the protein content in mammals [1,2]. Type I collagen is the most abundant protein type in bones, skins, scales, and tendons, and it is the principal structural ingredient of connective tissues; this represents about 90% of total human collagen [3]. Type II collagen is usually present in cartilage, which is formed by combining collagen fibers with some proteoglycans [4]. It was reported that in Nile tilapia (*Oreochromis niloticus*), one of the most popularly cultured fish in China, the yields of collagen from skin were above 40% (dry weight) [5]. Skate (*Raja porosa*), a typical marine cartilaginous fish, was widely used in order to extract collagen from its cartilage.

There are many diversities in the function and application for different types of collagen. Type I collagen is often made into sponges and for wound dressing membranes [6,7]. Modified type I collagen is extensively used in bioprinting as a bio-ink due to its high compatibility with human cells [8]. In the pharmaceutical field, collagen I is used for drug delivery as nanospheres and nanoparticles [9]. In the food industry, type I collagen has been widely used as a food additive, as edible film, in drinks, and as a carrier [10]. Type II collagen can promote the differentiation of cartilage cells and improve bone health, and it is especially useful in the treatment of rheumatoid arthritis [11].

The structure and properties of collagen derived from fish have been extensively studied. These studies are important for determining the eventual application of collagen. The collagen α chain shows a repeated Gly-X-Y sequence, in which X and Y are generally Pro and Hyp, respectively [12]. Structural differences of collagen from different sources could not be found by FTIR, XRD, and UV spectroscopy [1]. Zhu et al. [4] found the Rpn of collagen from skate and sturgeon cartilage was 0.19–0.25 in CD, and the collagen exhibited obvious self-assembly with a concentration above 0.3 mg/mL, an adjustment of pH 7.4–7.6, and an NaCl concentration of 120 mmol/L. Zhang et al. [13] found that the denaturation temperatures of collagen from sturgeon scale, skin, and cartilage were 29.6, 26.8, and 36.3 °C, respectively. Romijn et al. [14] estimated the $\chi33/\chi31$-ratio for type I collagen in tendon and type II collagen in cartilage was 1.33 and 1.36, respectively. This was determined by polarization-resolved second harmonic, which cannot be easily used to differentiate collagen types. Therefore, collagen of different types and sources should be determined using other methods. Different collagen peptides could be identified using LC-MS/MS [15]. Zhang et al. [16] found diverse mixtures of collagen type II and I contained 12 specific peptides using LC-MS, which indicated that MS was a possible strategy for collagen type identification. Hu et al. [17] quantified 81 proteins of skipjack, bigeye, and yellowfin tuna by the sequential window acquisition of all theoretical fragment ion mass spectra (SWATH-MS) analysis and screened 14 protein biomarkers with the ability to distinguish the three tuna species.

In this work, collagens from skate and tilapia were extracted by acetic acid and pepsin, and then their structures and properties were assessed by UV, FTIR, CD, XRD, fractional viscosity, and self-assembly. The discovery of unique peptides of sources (tilapia and skate) and collagen types (type I and type II) were conducted by SWATH-MS and orthogonal partial least squares discriminant analysis (OPLS-DA). It could provide a theoretical basis for the application of collagen from different types and sources.

2. Results

2.1. Distribution of Collagen Fibers in TI-s, TI-b, SK-s, and SK-c

VG staining is usually used to analyze the arrangement and distribution of collagen fibers. VG staining of TI-s (a), TI-b (c), SK-s (e), and SK-c (g) is shown in Figure 1. It can be seen that the red collagen fibers are aligned tightly and longitudinally in TI-s, SK-s, and TI-b. The collagen fibers in the SK-c tissue showed a tight reticular structure.

With Sirius red staining under cross-polarized light, collagen I shows strong red or yellow, collagen II displays multiple colors, and collagen III shows weak green [18]. The arrangement and distribution of collagen in TI-s (b), TI-b (d), SK-s (f), and SK-c (h) under polarized light are shown in Figure 1. The collagen of TI-s, TI-b, and SK-s showed bright yellow and red, and might therefore be type I collagen. In addition, there were also some green collagens interlaced with the bright yellow and red, which may be type III collagen. The collagen of SK-c showed weak birefringence and a variety of colors, and was presumed to be type II collagen. Acid-soluble and pepsin-soluble tilapia skin collagen were characterized as type I collagen [19]. Hwang et al. [20] identified that the major collagen in SK-s was type I using SDS–PAGE pattern and peptide maps analysis.

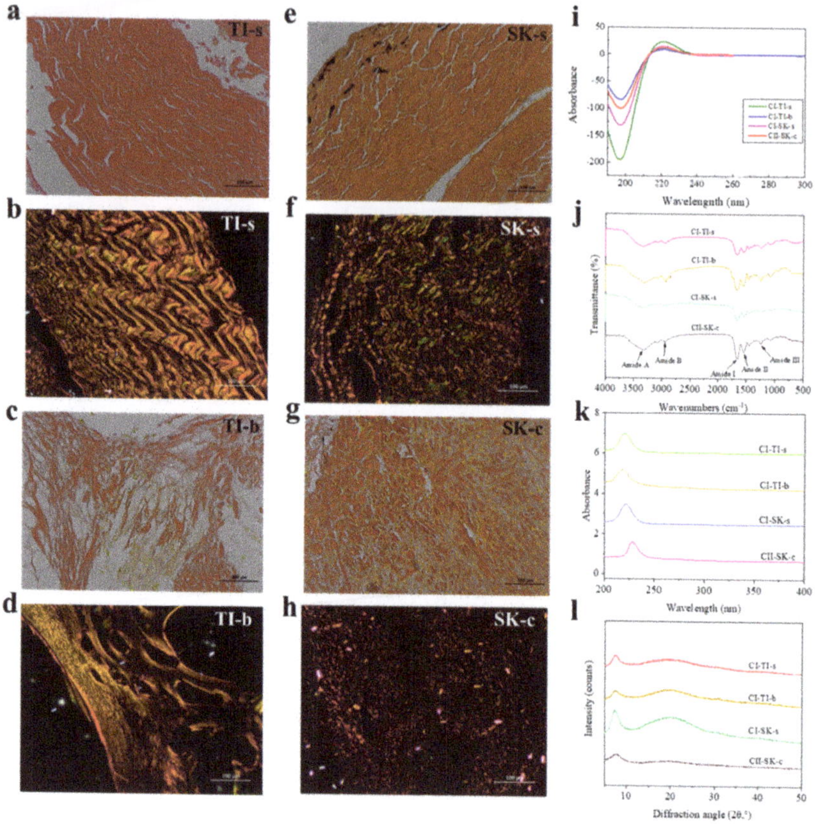

Figure 1. The VG staining and Sirius red staining of collagen in TI−s (**a**,**b**), TI−b (**c**,**d**), SK−s (**e**,**f**), SK−c (**g**,**h**); CD spectra of collagen isolated from tilapia and skate (**i**); FTIR spectra of collagen isolated from tilapia and skate (**j**); UV spectra of collagen isolated from tilapia and skate (**k**); XRD of collagen isolated from tilapia and skate (**l**).

2.2. *Structural Characteristics of CI-TI-s, CI-TI-b, CI-SK-s, and CII-SK-c*

2.2.1. Analysis of the UV Spectra of CI-TI-s, CI-TI-b, CI-SK-s, and CI-SK-c

The UV absorption spectra of collagen isolated from tilapia skins and bones and skate skins and cartilages are shown in Figure 1k. The maximum absorptions of CI-TI-s, CI-TI-b, CI-SK-s, and CII-SK-c were 222, 218, 221, and 229 nm, respectively. Reátegui-Pinedo et al. [1] found that the UV absorption peak of tilapia skin collagen was at 232 nm. The maximum absorption peak of skate cartilage collagen was at 230 nm in UV [4]. Results showed that the collagen of tilapia and skate had a similar UV absorption characteristic. The presence of tryptophan and tyrosine ensured that the maximum UV absorption wavelength of proteins occurred at 275–280 nm [4]. CI-TI-s, CI-TI-b, CI-SK-s, and CII-SK-c did not display clear absorption peaks at 280 nm, showing that the collagen samples contained a small amount of tryptophan and tyrosine. The UV absorption peak of collagen at about 230 nm was related to the $n \rightarrow \Pi^*$ transition of C=O, COOH, and CO-NH_2 groups [4]. The small differences might be related to the amino acid composition. It was found that the tyrosine concentration of skate cartilage collagen was lower than that of tilapia skin collagen [20,21].

2.2.2. FTIR Analysis of CI-TI-s, CI-TI-b, CI-SK-s, and CII-SK-c

The FTIR spectra characteristics of collagen extracted from tilapia and skate are shown in Figure 1j. The amide A band was mainly generated by the N-H stretching vibration [4]; CI-TI-s, CI-TI-b, CI-SK-s, and CII-SK-c were observed at 3336, 3316, 3326, and 3326 cm^{-1}, respectively. The amide B band was associated with the asymmetric stretching vibration of CH_2 [22] and appeared at wavenumbers 2938, 2929, 2935, and 2948 cm^{-1} for CI-TI-s, CI-TI-b, CI-SK-s, and CII-SK-c, respectively. The amide I bands of CI-TI-s (1662 cm^{-1}), CI-TI-b (1660 cm^{-1}), CI-TI-s (1658 cm^{-1}), and CII-TI-c (1652 cm^{-1}) was in accordance with the vibration of C=O groups. The amide II of CI-TI-s (1546 cm^{-1}), CI-TI-b (1551 cm^{-1}), CI-SK-s (1535 cm^{-1}), and CII-SK-c (1550 cm^{-1}) were similar to the results of Zhu et al. [4]. The amide III bands were associated with the N-H and C-N stretching vibration [4], which of CI-TI-s, CI-TI-b, CI-SK-s, and CII-SK-c were observed at 1236, 1238, 1234, and 1238 cm^{-1}, respectively.

2.2.3. CD Spectra Analysis of CI-TI-s, CI-TI-b, CI-SK-s, and CII-SK-c

The CD spectra of CI-TI-s, CI-TI-b, CI-SK-s, and CII-SK-c are shown in Figure 1i. Natural collagen showed a positive peak at 221 nm, a negative absorption at 198 nm, and the intensity ratio of positive and negative absorption peaks (Rpn) of 0.15 [23]. The CD spectra of CI-TI-s, CI-TI-b, CI-SK-s, and CII-SK-c exhibited positive absorption peaks at 221.5, 221.5, 221, and 222 nm, respectively, and negative absorption peaks at 197 nm. The Rpn values were 0.12, 0.12, 0.12, and 0.14, respectively; these values are consistent with the results of Zhu et al. [4]. The CD spectra of collagen from tilapia skins showed a significantly weak positive peak (220 nm) and a strong negative peak (196 nm); the Rpn value of collagen was 0.131 [5].

2.2.4. X-ray Diffraction of CI-TI-s, CI-TI-b, CI-SK-s, and CII-SK-c

The X-ray diffraction (XRD) patterns of CI-TI-s, CI-TI-b, CI-SK-s, and CII-SK-c are shown in Figure 1l. They reveal the two characteristic peaks in the spectra. the first peak, in the range of 5–10°, was associated with the crystal structure inside the collagen [24]. The second peak was about 20° at the diffraction angle (2θ), which was associated with diffuse emission from the internal amorphous region of the collagen [25]. The d values, corresponding to the first peak of CI-TI-s, CI-TI-b, CI-SK-s, and CII-SK-c, were 11.84, 12.20, 12.03, and 11.56 Å, respectively, representing spacing between molecular chains. Sun et al. [23] found d values for the first relatively sharp peak of acid; the pepsin-soluble collagen of tilapia skins were 11.66 Å and 11.90 Å, respectively. The d value, corresponding to the first peak of double-spotted pufferfish skin collagen, was 11.45 Å [26].

2.3. *The O-Glycopeptide Bond of CI-TI-s, CI-TI-b, CI-SK-s, and CII-SK-c*

O-glycosylation, related to the stability of collagen, is formed by the attachment of galactose in its β-configuration to the hydroxyl group of hydroxylysine, which was converted from hydroxylated lysine [4]. The β-elimination reaction (alkali-treatment) can break the O-glycopeptide bond of collagen, which produces more unsaturated amino acids to enhance the UV absorption at 240 nm [27]. The UV spectra before and after β-elimination reaction of CI-TI-s (Figure 2a), CI-TI-b (Figure 2b), CI-SK-s (Figure 2c), and CII-SK-c (Figure 2d) were illustrated in Figure 2. The enhancement of UV absorption after β-elimination reaction of collagen at 240 nm indicated that the O-linked glycopeptide bonds were present, which is consistent with the results of Zhu et al. [4]. O-linked glycopeptide bonds were present in a higher level on CII-SK-c when compared with CI-TI-s, CI-TI-b, and CI-SK-s.

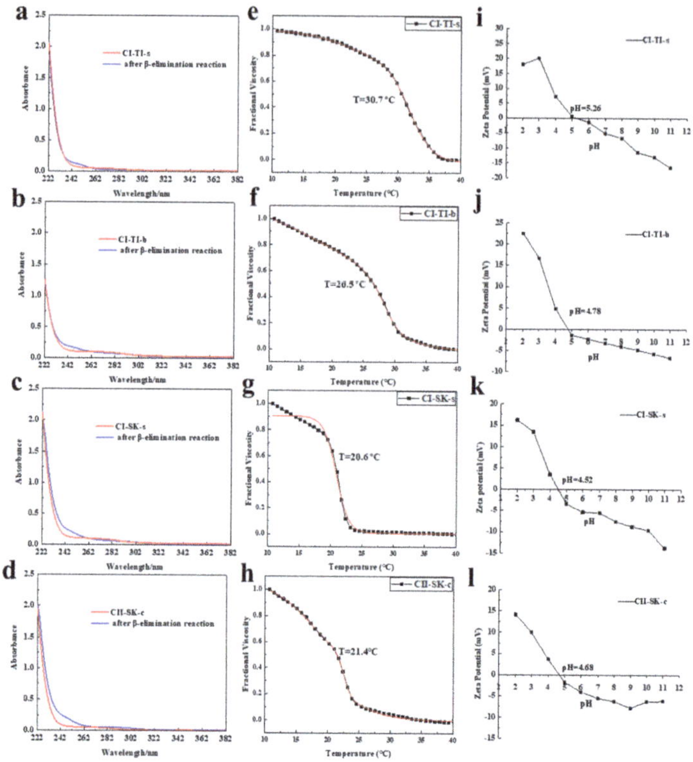

Figure 2. Ultraviolet spectra before and after β elimination reaction, thermal denaturation curves fitting and Zeta potential of CI−TI−s (**a**,**e**,**i**), CI−TI−b (**b**,**f**,**j**), CI−SK−s (**c**,**g**,**k**), and CII−SK−c (**d**,**h**,**l**); The red line were the fractional viscosity fitting curve of collagen.

2.4. Td and Zeta Potential of CI-TI-s, CI-TI-b, CI-SK-s, and CII-SK-c

The fractional viscosity fitting curve of CI-TI-s (Figure 2e), CI-TI-b (Figure 2f), CI-SK-s (Figure 2g), and CII-SK-c (Figure 2h) are shown in Figure 2. The thermal denaturation temperatures (Td) of CI-TI-s, CI-TI-b, CI-SK-s, and CII-SK-c were 30.7, 26.5, 20.6, and 21.4 °C, respectively. The Td of tilapia skins and bone collagen was higher than that of skate skins and cartilage collagen, and CII-SK-c showed a higher Td when compared with that of CI-SK-s. The thermal denaturation temperatures of different collagen can be affected by the temperature in which the species live. Wang et al. [28] found that the Td in the deep sea was lower than that in the shallow sea. It was reported that the Td temperature of collagen was raised as the content of Hyp increased [29]. More proline and hydroxyproline increased the thermal stability of collagen because of a higher density of crosslinks [30], which can also be improved by chitin nanofibers for the usage as scaffolds and wound-dressing materials [31]. Klabukov et al. [32] found that the copolymer had a higher stability and the ability to resist hydrolysis.

Zeta potential can be used to describe the surface charge of colloidal systems, and the isoelectric point (pI) refers to the pH at which the net charge of molecule is zero [33]. The rate of collagen fibril production is largely controlled by electrostatic interactions, and collagen is more conducive to the aggregation of collagen molecules and fibrils formation when the collagen pI is reached [34]. The zeta values of four collagen proteins are shown in Figure 2. The pI of CI-TI-s (Figure 2i), CI-TI-b (Figure 2j), CI-SK-s (Figure 2k), and CII-SK-c (Figure 2l) were 5.26, 4.78, 4.52, and 4.68, respectively, which may be due to the high content of acidic amino acids (glutamate and aspartic acid) in collagen. Ahmed et al. [33] found the pI of collagen might be associated with amino acid composition and amino acid residue sequences, especially on surface domains.

2.5. The Self-Assembly Properties of CI-TI-s, CI-TI-b, CI-SK-s, and CII-SK-c

The effects of concentration, pH, ionic strength, and hyaluronic acid (HA) on the kinetics of collagen self-assembly of CI-TI-s and CII-SK-c are shown in Figure 3. The degree of collagen self-assembly was positively correlated with the concentration of collagen (Figure 3a,b). The higher self-assembly of the CI-TI-s (Figure 3c) was at pH 6.8, and the self-assembly of CII-SK-c (Figure 3d) in the range of pH 7.4–7.6 was higher than that at other pH values, which was similar to the results of Yan et al. [35]. Figure 3e showed that the aggregation rate of CI-TI-s was increased when the sodium chloride concentration reached 80 mmol/L. The aggregation rate of CII-SK-c showed similar characteristics at different ionic strengths (Figure 3f). The self-assembly of CI-TI-s could be found when the HA concentration was in the range of 10% to 20%. However, the self-assembly behavior of CII-SK-c could not be affected by HA (Figure 3g,h). At pH 7.4–7.6 and NaCl concentration of 120 mmol/L, collagen in skate and sturgeon cartilage at concentrations above 0.3 mg/mL exhibited significant self-assembly behavior [4]. In summary, the self-assembly characteristics of CI-TI-s and CII-SK-c were different.

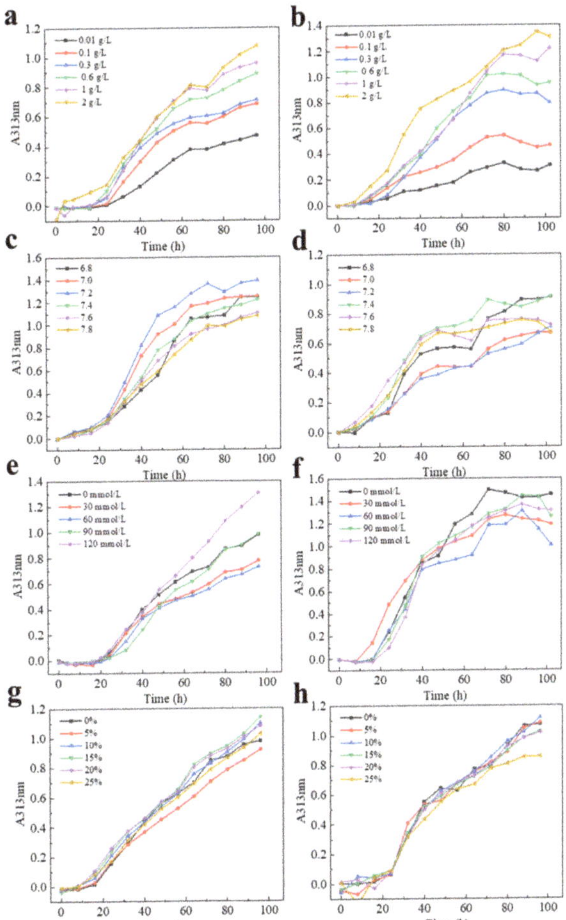

Figure 3. Effects of concentration (**a**,**b**), pH (**c**,**d**), ion strength (**e**,**f**), and HA (**g**,**h**) induce on the kinetic of collagen self-assembly of CI-TI-s and CII-SK-c.

2.6. Discovery of Unique Peptides for C-TI and C-SK

The mass spectrum information of C-TI and C-SK was obtained using DDA collection mode. The DDA data from two samples were searched, and 698 peptides and 17 proteins (C-TI) and 331 peptides and 31 proteins (C-SK) were found at the 95% confidence level. The score plot of OPLS-DA in Figure 4b demonstrated that collagen samples from tilapia were thoroughly separated with those from skate, indicating potential of SWATH-MS based proteomic analysis in discrimination of collagen sources. The OPLS-DA evaluation model parameters consist of R^2X, R^2Y, and Q^2, where R^2X, and R^2Y stand for the explanation rate of the model for the X and Y matrices, respectively, while Q^2 indicates the predictive power of the model. For a perfect categorization, the value of specificity should be near to 1 [36]. R^2X, R^2Y, and Q^2 of the model were 0.712, 0.998, and 0.985, respectively.

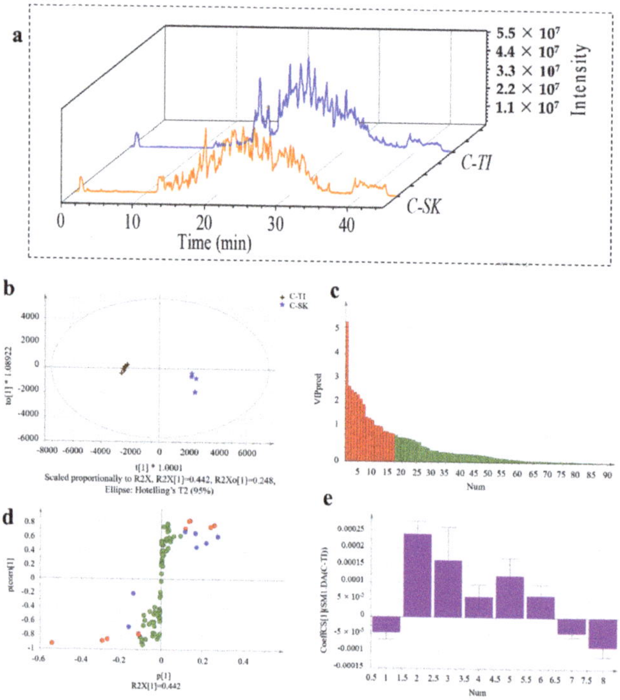

Figure 4. SWATH−MS data analysis of collagen peptides. Representative total ion chromatograms (TICs) with SWATH data of C−TI (blue) and C−SK (orange) (**a**). The OPLS−DA (**b**) (the blue were C−SK and the brown were C−TI; "*" was weights of regression coefficient), VIP (**c**) (the red rectangles were peptides with a VIP value > 1), S−plot (**d**) (the green were VIP value < 1; the blue were VIP value > 1 but |p(corr)| < 0.7; the red were VIP value > 1 and |p(corr)| > 0.7), and Knife cut confidence interval analysis (**e**) for discovering peptide markers of C−TI versus C−SK.

The peptides with a variable importance in the projection (VIP) value higher than 1 were screened. Figure 4c showed that there were 17 peptides with a VIP value > 1 (marked in red) that could be used as potential identification peptides for subsequent analysis. Then the peptides with |p(corr)| > 0.7 in S-plot were selected to the next screen (Figure 4d). The farther from the origin in the S-plot score chart, the greater the contribution of this point to sample classification. Finally, eight specific peptides (marked with purple in Figure 4e) with the jack-knifed confidence intervals in the coefficients plot not passing through zero were identified as potential marker peptides of C-TI and C-SK. Statistical analysis showed that there were significant differences between two samples. To perform the biospecificity check, the bioinformatics tool, Blast, was used to perform a peptide amino acid sequence search of the entire NCBI protein sequence database [37]. A total of eight candidate tagged peptides were submitted to Blast search for biospecificity checks against the entire NCBI non-redundant protein database. The unique peptides of C-TI and C-SK were revealed through the proteomic and chemometrics analysis, including GPSGPQGAVGATGPK, PAMPVPGPMGPMGPR, SPAMPVPGPMGPMGPR, GESGPSGPAGPAGPAGVR, SSGPPVPGPIGPMGPR, GLTGPIGVPGPPGAQGEK, GLAGPQGPR, and GLSGDPGVQGIK (Table 1). The detectability assessment of the potential marker peptides can be performed using the MRM approach. The 81 proteins of skipjack tuna, bigeye tuna, and yellowfin tuna were quantified by SWATH-MS proteomic analysis, and 14 potential protein biomarkers were screened by chemometrics [17]. Results showed that it was feasible to screen unique peptides for C-TI and C-SK using the SWATH-MS proteomic approach.

Table 1. All unique peptides screened by chemometrics.

	Peptide Sequence	*m*/*z*	Protein Source
C-TI/C-SK	data		data
1	GPSGPQGAVGATGPK	640.8333	Type I procollagen alpha 1 chain OS = Okamejei kenojei GN = SkCOL1A1 PE = 2 SV = 1
2	PAMPVPGPMGPMGPR	746.3671	Uncharacterized protein (Fragment) OS = Oreochromis niloticus GN = col1a1 PE = 4 SV = 1
3	SPAMPVPGPMGPMGPR	789.8831	Uncharacterized protein (Fragment) OS = Oreochromis niloticus GN = col1a1 PE = 4 SV = 1
4	GESGPSGPAGPAGPAGVR	760.8762	Uncharacterized protein OS = Oreochromis niloticus GN = LOC100694532 PE = 4 SV = 1
5	SSGPPVPGPIGPMGPR	751.8928	Uncharacterized protein OS = Oreochromis niloticus GN = LOC100694532 PE = 4 SV = 1
6	GLTGPIGVPGPPGAQGEK	816.4412	Uncharacterized protein OS = Oreochromis niloticus GN = LOC100694532 PE = 4 SV = 1
7	GLAGPQGPR	426.7379	Uncharacterized protein (Fragment) OS = Anolis carolinensis GN = COL1A2 PE = 4 SV = 1
8	GLSGDPGVQGIK	564.3064	Uncharacterized protein OS = Gasterosteus aculeatus PE = 4 SV = 1
CI/CII			
1	GPTGEIGATGLAGAR	664.3519	Collagen type I alpha 2 OS = Oreochromis niloticus GN = COL1A2 PE = 2 SV = 1
2	GVLGLTGMR	452.2577	Type I procollagen alpha 1 chain OS = Okamejei kenojei GN = SkCOL1A1 PE = 2 SV = 1
3	LGLTGMR	374.2127	Type I procollagen alpha 1 chain OS = Okamejei kenojei GN = SkCOL1A1 PE = 2 SV = 1
4	GEPGAAGPAGPSGPMGPR	781.8726	Type I procollagen alpha 1 chain OS = Okamejei kenojei GN = SkCOL1A1 PE = 2 SV = 1
5	SSGPPVPGPIGPMGPR	751.8928	Uncharacterized protein OS = Oreochromis niloticus GN = LOC100694532 PE = 4 SV = 1
6	GLSGDPGVQGIK	564.3064	Uncharacterized protein OS = Gasterosteus aculeatus PE = 4 SV = 1

2.7. Discovery of Unique Peptides for Type I and Type II Collagen

The analysis of unique peptides for type I and type II collagen are shown in Figure 5. The mass spectrum information of CI and CII was obtained using the DDA collection mode, and the DDA data from two samples were searched, resulting in 275 peptides and 24 proteins (CI) and 144 peptides and 16 proteins (CII) at the 95% confidence level; these were taken as the quantitative database of SWATH-MS for subsequent analysis. The OPLS-DA analysis was performed to seek marker peptides as differentiators indicating types from collagen I and collagen II. R^2X, R^2Y, and Q^2 of the OPLS-DA evaluation model were 0.895, 0.983, and 0.985, respectively. Figure 5c showed that there were nine peptides with a VIP value > 1 (marked in red). There were seven specific peptides (marked with purple in Figure 5e) with the jack-knifed confidence intervals in the coefficients plot not passing through zero. In addition, unique peptides for CI and CII were identified, including GPTGEIGATGLAGAR, GVLGLTGMR, LGLTGMR, GEPGAAGPAGPSGPMGPR, SSGPPVPGPIGPMGPR, and GLSGDPGVQGIK (Table 1). The results showed the ability to search for unique peptides for type I and type II collagen using the SWATH-MS proteomic approach.

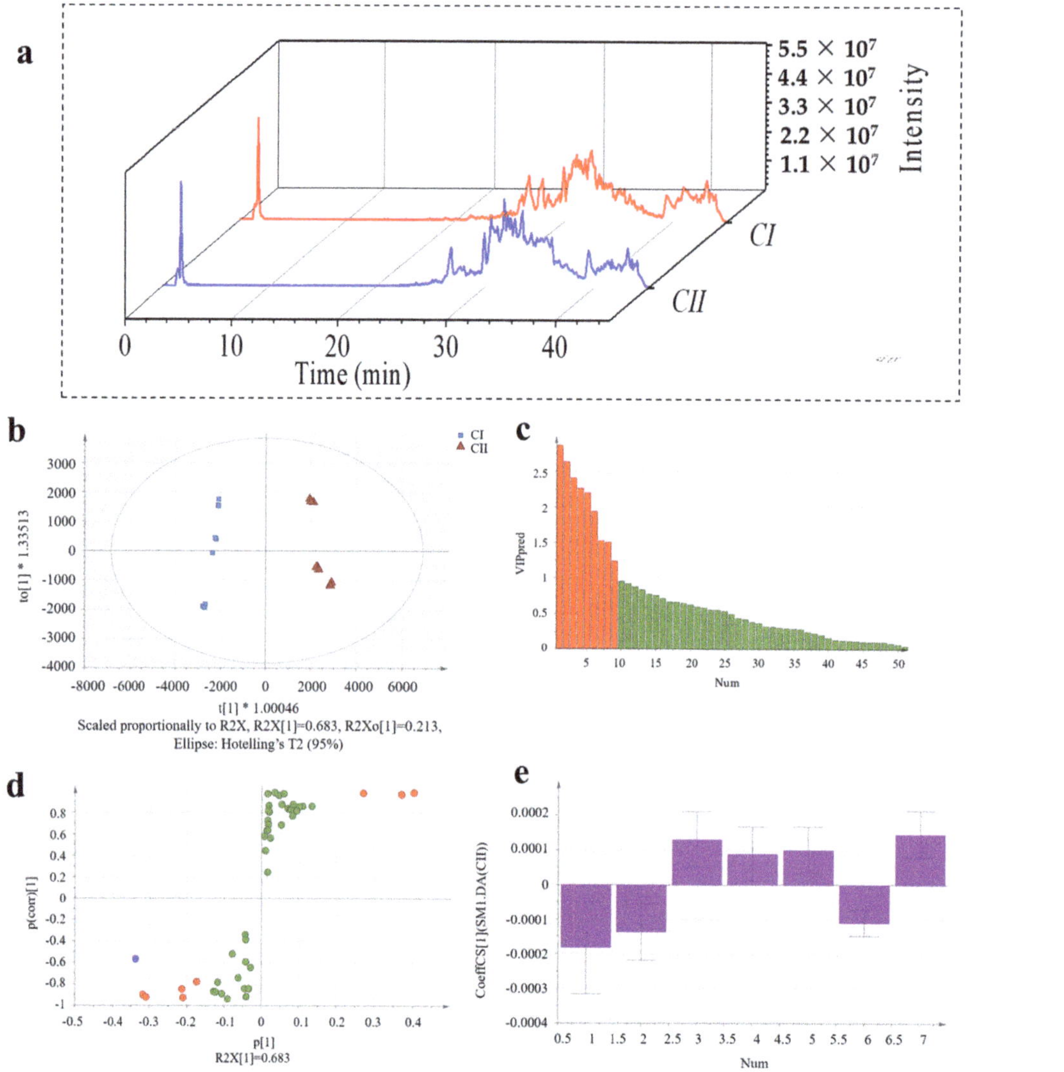

Figure 5. SWATH−MS data analysis of collagen peptides. Representative total ion chromatograms (TICs) with SWATH data of CI (red) and CII (blue) (**a**). The OPLS-DA (**b**) (the blue were CI and the red were CII; "*" was weights of regression coefficient), VIP (**c**) (the red rectangles were peptides with a VIP value > 1), S−plot (**d**) (the green were VIP value < 1; the blue were VIP value > 1 but ∣p(corr)∣ < 0.7; the red were VIP value > 1 and ∣p(corr)∣ > 0.7), and Knife cut confidence interval analysis (**e**) for discovering peptide markers of CI versus CII.

3. Materials and Methods

3.1. Materials

Skins and bones of tilapia (*Oreochromis mossambicus*) and skins and cartilages of skate (*Raja porosa*) were obtained from Nanshan Aquatic Products (Qingdao, China). All raw materials were preserved at −20 °C. Acetonitrile was obtained from Merck (Darmstadt, Germany). Trypsin was obtained from Promega Co., Ltd. (Beijing, China). The pepsin was obtained from Maclean Biochemical Technology Co., Ltd. (Shanghai, China). The other reagents were of analytical grade and bought from Sinopharm Chemical Reagent Co., Ltd. (Shanghai, China).

3.2. *Van Gieson and Picric Acid-Sirius Red Staining*

Tilapia skin (TI-s), tilapia bone (TI-b), skate skin (SK-s), and skate cartilage (SK-c) were treated in 4% formaldehyde for two days and then washed with water. Samples were dehydrated, embedded in paraffin, and then stained with Van Gieson (VG) and Picric acid-Sirius red. Tissues stained with Picric acid-Sirius red were observed using a cross-polarized light microscope and tissues stained with VG were observed using an ordinary light microscope (Ni-E, Nikon Imaging Instrument Sales Co., Ltd., Shanghai, China). For tilapia bones, their pretreatment should be decalcified in a 10% EDTA-Na2 solution for two weeks before dehydration.

3.3. *Preparation of Collagen*

Samples were washed with deionized water (4 °C) after being removed the residual meat with a knife and divided into small pieces (1 cm × 1 cm). They were then soaked for 48 h (1:20, w/v) with the 0.1 mol/L concentration of NaOH. After discarding the non-collagen components, fish skins were retreated with the pre-cold deionized water for a neutral pH. Then the pretreated materials were poured into acetic acid solution (0.5 mol/L) that contained pepsin with the concentration of 0.1% (w/v). After being stirred for 48 h at 4 °C, it was centrifuged (9000× g) for 20 min at 4 °C (2–16KL, Sigma, An Der Unteren Söse 50, Osterode Am Harz, Germany). The collagen was sedimented by adding NaCl to the supernatants until its concentration reached 0.9 mol/L. The collagen was gathered by centrifugation at 6000× g for 30 min and redissolved in 0.5 mol/L acetic acid solution. Then the mixture was dialyzed and lyophilized (VSP62, Marin Christ Co., An Der Unteren Söse 50, Osterode Am Harz, Germany) to obtain tilapia skin collagen (CI-TI-s), skate skin collagen (CI-SK-s), and skate cartilage collagen (CII-SK-c). For tilapia bone collagen (CI-TI-b), the extraction method was similar. Before using acetic acid and pepsin to extract collagen from tilapia bones, the bones needed to be pretreated with EDTA-Na_2 (0.5 mol/L) for 48 h to remove the calcium.

3.4. *Fourier Transform Infrared (FTIR) Spectroscopy Analysis*

FTIR was detected according to the method of Zhu et al. [4]. Freeze-dried collagen was thoroughly mixed, ground with dehydrated potassium bromide, and pressed into sheets. FTIR was measured at 4000–400 cm^{-1} (Nicolet iS 10, Thermo Fisher Scientific, Waltham, MA, USA). It was scanned at the resolution of 2 cm^{-1} 64 times and OMNIC 9.0 software was used for the analysis.

3.5. *Determination of Ultraviolet-Visible Absorption Spectra*

The 0.5 mg/mL samples of collagen were prepared with 0.1 mol/L acetic acid and centrifuged at 9000× g for 5 min at 4 °C. They were scanned by an ultraviolet spectrophotometer (UV-2102 PC, Unico Co., Ltd., Shanghai, China) in the range of 190–400 nm [4].

3.6. *Circular Dichroism (CD) Spectra Analysis*

The 0.5 mg/mL samples of collagen were prepared with 0.05 mol/L acetic acid. After filtering through a 0.22 μm membrane, the solution was measured in the scanning range of 190–260 nm [23].

3.7. Determination of X-ray Diffraction (XRD)

The XRD patterns were determined with Cu Kα radiation applied at 40 kV of voltage. The samples were recorded with a scanning angle (2θ) from 5 ° to 40 ° at a scanning rate of 4 °min^{-1}. The d value can be calculated by using the Bragg's equation as follows [23]:

$$d\ (\text{Å}) = \frac{\lambda}{2\sin\theta}$$

where θ and λ were the Bragg diffraction angle and X-ray wavelength, respectively.

3.8. Thermal Denaturation Temperature (Td) Analysis

The 4‰ collagen was prepared using a 0.1 mol/L acetic acid solution. The viscosity of the solution was analyzed from 4 °C to 40 °C using a rheometer (MCR301, Anton Paar GmbH Co., Ltd., Graz, Austria) at the heating rate of 0.4 °C/min. The Td was computed, at which the fractional viscosity was 0.5 [38]. The specific calculation formula was as follows:

$$F(T) = \frac{\eta_{sp}(T) - \eta_{sp}(40\ ^\circ C)}{\eta_{sp}(4^\circ C) - \eta_{sp}(40\ ^\circ C)}$$

where F was fractional viscosity, η_{sp} was computed by $\frac{\eta-\eta_0}{\eta_0}$, η was the viscosity of the solution, and η_0 was the viscosity of acetic acid (mPa·s).

3.9. Zeta Potential Measurement

Zeta potential was measured according to the method of Yan et al. [35]. The pH of 0.5 mg/mL collagen was adjusted to 2–11 and it was measured using a nano size potentiometer (ZS90, Malvern Instruments, Malvern, UK).

3.10. LC/MS Analysis of Collagen of Different Types and Sources

For the discovery of unique collagen peptides of different sources, collagen from tilapia skins (CI-TI-s), tilapia bones (CI-TI-b), skate skins (CI-SK-s), and skate cartilages (CII-SK-c) were digested with sequencing grade trypsin at 37 °C for 18 h at a protein/trypsin ratio of 25:1 (*w*/*w*). After centrifugation, the digested samples of CI-TI-s and CI-TI-b were mixed in equal volumes, named as C-TI. The digested samples of CI-SK-s and CII-SK-c were mixed in equal volumes, named as C-SK.

For the discovery of unique collagen peptides of different types, the treatment of collagen from cod skins (CI-GA-s), cod bones (CI-GA-b), sturgeon skins (CI-ST-s), and sturgeon cartilages (CII-ST-c) was the same as above. The digested samples of CI-GA-s, CI-TI-s, CI-ST-s, CI-SK-s, CI-GA-b, and CI-TI-b were mixed in equal volumes, named as CI. The digested samples of CII-ST-c and CII-SK-c were mixed in equal volumes, named as CII.

Sample separation was performed by mass spectrometry using an UHPLC-Q/TOF instrument in DDA mode equipped with AdvanceBio Peptide Map column (150 mm × 2.1 mm, 130 Å, 2.7 µm), and it was operated in electrospray positive ion (ESI+) mode with a spray voltage of 5500 V in range of 350–1500 m/z. The data were identified by Protein Pilot software (version 5.0.2, SCIEX) to generate the library for the SWATH analysis. Data from SWATH acquisitions were imported by PeakView software (version 2.1, SCIEX), and then normalized based on the total peak areas sum by MarkerView software (version 1.2.1, SCIEX). Orthogonal partial least squares discriminant analysis (OPLS-DA) was used to screen peptides between CI and CII or C-SK and C-TI in SMICA software (version 14.0, Umetrics). The specificity verification of potential peptide markers was achieved by the Blast searching of the whole NCBI protein database [39].

Concretely, the mass spectrum information was obtained using DDA collection mode, and the Protein Pilot software was used to search the database. In the process of mass spectrometry analysis, the DDA data were searched at the 95% confidence level, which were taken as the quantitative database of SWATH-MS for subsequent analysis. The data obtained by SWATH-MS are quite complex. A further decoding process is necessary for generating meaningful information. The different peptides with a false positive error rate of less than 1% were selected for subsequent OPLS-DA analysis using SIMCA software [40]. To seek marker peptides as differentiators indicating collagen sources, the separated groups were classified as one group and OPLS-DA analysis was then performed.

3.11. Statistical Analysis

Each experiment was repeated at least three times. Statistical analyses were performed using one-way analysis of variance (ANOVA). The probability value of $p < 0.05$ was used as the criterion for significant differences.

4. Conclusions

In this study, the distribution of collagen fibers in tissues of tilapia skin and bone and skate skin fish and cartilage was different. All collagen had a complete triple helix structure and similar UV and X-ray characteristics. Type II collagen had obvious glycopeptide linkage. The Td temperature, Zeta potential, and the optimal self-assembly conditions of collagen from different sources were different. Further studies have shown that unique peptides of different sources and types of collagens could be obtained by SWATH-MS.

Author Contributions: Writing—original draft preparation, writing—review and editing, data curation, investigation, X.Z.; data curation, writing—original draft preparation, investigation, J.W.; writing—review and editing, Y.F.; writing—review and editing, Q.Z.; writing—review and editing, H.Z.; writing—review and editing, K.A.; writing—review and editing, supervision, H.H. All authors have read and agreed to the published version of the manuscript.

Funding: This research was funded by Major Scientific and Technological Innovation Project in Shandong Province, grant number No. 2022CXGC020414; Major Scientific and Technological Innovation Project of Qingdao West Coast New Area, grant number No. ZDKC-2022-01; Taishan Scholar Foundation of Shandong Province, grant number No. tsqn202103033; Marine S&T Fund of Shandong Province for Qingdao Marine Science and Technology Center, grant number No. 2022QNLM030002-4.

Data Availability Statement: The data presented in this study are available on request from the corresponding author.

Conflicts of Interest: The authors declare no conflict of interest.

Sample Availability: Available.

References

1. Reátegui-Pinedo, N.; Salirrosas, D.; Sánchez-Tuesta, L.; Quiñones, C.; Jáuregui-Rosas, S.R.; Barraza, G.; Cabrera, A.; Ayala-Jara, C.; Martinez, R.M.; Baby, A.R.; et al. Characterization of Collagen from Three Genetic Lines (Gray, Red and F1) of *Oreochromis niloticus* (Tilapia) Skin in Young and Old Adults. *Molecules* **2022**, *27*, 1123. [CrossRef] [PubMed]
2. Menezes, M.d.L.L.R.; Ribeiro, H.L.; Abreu, F.d.O.M.d.S.; Feitosa, J.P.d.A.; Filho, M.d.S.M.d.S. Optimization of the Collagen Extraction from Nile Tilapia Skin (*Oreochromis niloticus*) and its Hydrogel with Hyaluronic Acid. *Colloids Surf. B Biointerfaces* **2020**, *189*, 110852. [CrossRef] [PubMed]
3. Fan, X.; Liang, Y.; Cui, Y.; Li, F.; Sun, Y.; Yang, J.; Song, H.; Bao, Z.; Nian, R. Development of Tilapia Collagen and Chitosan Composite Hydrogels for Nanobody Delivery. *Colloids Surf. B Biointerfaces* **2020**, *195*, 111261. [CrossRef] [PubMed]
4. Zhu, L.; Li, J.; Wang, Y.; Sun, X.; Li, B.; Poungchawanwong, S.; Hou, H. Structural Feature and Self-assembly Properties of Type II Collagens from the Cartilages of Skate and Sturgeon. *Food Chem.* **2020**, *331*, 127340. [CrossRef] [PubMed]
5. Ge, B.; Wang, H.; Li, J.; Liu, H.; Yin, Y.; Zhang, N.; Qin, S. Comprehensive Assessment of Nile Tilapia Skin (*Oreochromis niloticus*) Collagen Hydrogels for Wound Dressings. *Marine Drugs* **2020**, *18*, 178. [CrossRef] [PubMed]
6. Naomi, R.; Ridzuan, P.M.; Bahari, H. Current Insights into Collagen Type I. *Polymers* **2021**, *13*, 2642. [CrossRef]
7. Tamer, T.M.; Kenawy, E.R.; Agwa, M.M.; Sabra, S.A.; El-meligy, M.A.; Mohy-Eldin, M.S. Wound Dressing Membranes based on Immobilized Anisaldehyde onto (Chitosan-GA-Gelatin) Copolymer: In-vitro and in-vivo evaluations. *Int. J. Biol. Macromol.* **2022**, *211*, 94–106. [CrossRef]
8. Guo, K.; Wang, H.; Li, S.; Zhang, H.; Li, S.; Zhu, H.; Yang, Z.; Zhang, L.; Chang, P.; Zheng, X. Collagen-Based Thiol-Norbornene Photoclick Bio-Ink with Excellent Bioactivity and Printability. *ACS Appl. Mater. Interfaces* **2021**, *13*, 7037–7050. [CrossRef]
9. Furtado, M.; Chen, L.; Chen, Z.; Chen, A.; Cui, W. Development of Fish Collagen in Tissue Regeneration and Drug Delivery. *Eng. Regen.* **2022**, *3*, 217–231. [CrossRef]

10. Wu, X.; Liu, Y.; Liu, A.; Wang, W. Improved Thermal-stability and Mechanical Properties of Type I Collagen by Crosslinking with Casein, Keratin and Soy Protein Isolate Using Transglutaminase. *Int. J. Biol. Macromol.* **2017**, *98*, 292–301. [CrossRef]
11. Bakilan, F.; Armagan, O.; Ozgen, M.; Tascioglu, F.; Bolluk, O.; Alatas, O. Effects of Native Type II Collagen Treatment on Knee Osteoarthritis: A Randomized Controlled Trial. *Eurasian J. Med.* **2016**, *48*, 95–101. [CrossRef]
12. Al-Shaer, A.; Lyons, A.; Ishikawa, Y.; Hudson, B.G.; Boudko, S.P.; Forde, N.R. Sequence-dependent Mechanics of Collagen Reflect its Structural and Functional Organization. *Biophys. J.* **2021**, *120*, 4013–4028. [CrossRef]
13. Zhang, X.; Ookawa, M.; Tan, Y.; Ura, K.; Adachi, S.; Takagi, Y. Biochemical Characterisation and Assessment of Fibril-forming Ability of Collagens Extracted from Bester *Sturgeon Huso huso × Acipenser ruthenus*. *Food Chem.* **2014**, *160*, 305–312. [CrossRef]
14. Romijn, E.I.; Finnøy, A.; Lilledahl, M.B. Analyzing the Feasibility of Discriminating between Collagen Types I and II using Polarization-resolved Second Harmonic Generation. *J. Biophotonics* **2019**, *12*, e201800090. [CrossRef]
15. Padhi, S.; Chourasia, R.; Kumari, M.; Singh, S.P.; Rai, A.K. Production and Characterization of Bioactive Peptides from Rice Beans using Bacillus Subtilis. *Bioresour. Technol.* **2022**, *351*, 126932. [CrossRef] [PubMed]
16. Zhang, G.; Sun, A.; Li, W.; Liu, T.; Su, Z. Mass Spectrometric Analysis of Enzymatic Digestion of Denatured Collagen for Identification of Collagen Type. *J. Chromatogr. A* **2006**, *1114*, 274–277. [CrossRef] [PubMed]
17. Hu, L.; Zhang, H.; Hu, Z.; Chin, Y.; Zhang, X.; Chen, J.; Liu, D.; Hu, Y. Comparative Proteomics Analysis of Three Commercial Tuna Species through SWATH-MS based Mass Spectrometry and Chemometrics. *Food Control* **2022**, *141*, 109162. [CrossRef]
18. Stephenson, B. A Modified Picro-Sirius Red (PSR) Staining Procedure with Polarization Microscopy for Identifying Collagen in Archaeological Residues. *J. Archaeol. Sci.* **2015**, *61*, 235–243. [CrossRef]
19. Zhang, J.; Jeevithan, E.; Bao, B.; Wang, S.; Gao, K.; Zhang, C.; Wu, W. Structural Characterization, in-vivo Acute Systemic Toxicity Assessment and in-vitro Intestinal Absorption Properties of Tilapia (*Oreochromis niloticus*) Skin Acid and Pepsin Solublilized Type I Collagen. *Process Biochem.* **2016**, *51*, 2017–2025. [CrossRef]
20. Hwang, J.-H.; Mizuta, S.; Yokoyama, Y.; Yoshinaka, R. Purification and Characterization of Molecular Species of Collagen in the Skin of Skate (*Raja kenojei*). *Food Chem.* **2007**, *100*, 921–925. [CrossRef]
21. Hernández-Ruiz, K.L.; López-Cervantes, J.; Sánchez-Machado, D.I.; Campas-Baypoli, O.N.; Quintero-Guerrero, A.A.; de Lourdes Grijalva-Delgado, M.; Chávez-Almanza, A.F. Collagen Peptide Fractions from Tilapia (*Oreochromis aureus Steindachner*, 1864) Scales: Chemical Characterization and Biological Activity. *Food Biosci.* **2023**, *53*, 102658. [CrossRef]
22. Abe, Y.; Krimm, S. Normal Vibrations of Crystalline Polyglycine I. *Biopolymers* **1972**, *11*, 1817–1839. [CrossRef] [PubMed]
23. Sun, L.; Hou, H.; Li, B.; Zhang, Y. Characterization of Acid- and Pepsin-soluble Collagen Extracted from the Skin of Nile Tilapia (*Oreochromis niloticus*). *Int. J. Biol. Macromol.* **2017**, *99*, 8–14. [CrossRef] [PubMed]
24. An, X.; Duan, S.; Jiang, Z.; Chen, S.; Sun, W.; Liu, X.; Sun, Z.; Li, Y.; Yan, M. Role of Chlorogenic Acid and Procyanidin in the Modification of Self-assembled Fibrillar Gel Prepared from Tilapia Collagen. *Polym. Degrad. Stab.* **2022**, *206*, 110177. [CrossRef]
25. Zhu, S.; Gu, Z.; Xiong, S.; An, Y.; Liu, Y.; Yin, T.; You, J.; Hu, Y. Fabrication of a Novel Bio-inspired Collagen–polydopamine Hydrogel and Insights into the Formation Mechanism for Biomedical Applications. *RSC Adv.* **2016**, *6*, 66180–66190. [CrossRef]
26. Gao, X.; He, J.; Chen, J.; Zheng, Y.; Li, Y.; Ye, T. Double-spotted Pufferfish (*Takifugu bimaculatus*) Skin Collagen: Preparation, Structure, Cytocompatibility, Rheological, and Functional Properties. *Arab. J. Chem.* **2023**, *16*, 104402. [CrossRef]
27. Li, T.; Wu, C.-e.; Meng, X.; Fan, G.; Cao, Y.; Ying, R.; Tang, Y. Structural Characterization and Antioxidant Activity of a Glycoprotein Isolated from Camellia Oleifera Abel Seeds against D-galactose-induced Oxidative Stress in Mice. *J. Funct. Foods* **2020**, *64*, 103594. [CrossRef]
28. Wang, L.; An, X.; Yang, F.; Xin, Z.; Zhao, L.; Hu, Q. Isolation and Characterisation of Collagens from the Skin, Scale and Bone of Deep-sea Redfish (*Sebastes mentella*). *Food Chem.* **2008**, *108*, 616–623. [CrossRef]
29. Ikoma, T.; Kobayashi, H.; Tanaka, J.; Walsh, D.; Mann, S. Physical Properties of Type I Collagen Extracted from Fish Scales of *Pagrus Major* and *Oreochromis Niloticas*. *Int. J. Biol. Macromol.* **2003**, *32*, 199–204. [CrossRef]
30. Pati, F.; Adhikari, B.; Dhara, S. Isolation and Characterization of Fish Scale Collagen of Higher Thermal Stability. *Bioresour. Technol.* **2010**, *101*, 3737–3742. [CrossRef]
31. Yano, S.; Yamaguchi, K.; Shibata, M.; Ifuku, S.; Teramoto, N. Photocrosslinked Fish Collagen Peptide/Chitin Nanofiber Composite Hydrogels from Marine Resources: Preparation, Mechanical Properties, and an In Vitro Study. *Polymers* **2023**, *15*, 682. [CrossRef]
32. Klabukov, I.; Tenchurin, T.; Shepelev, A.; Baranovskii, D.; Mamagulashvili, V.; Dyuzheva, T.; Krasilnikova, O.; Balyasin, M.; Lyundup, A.; Krasheninnikov, M.; et al. Biomechanical Behaviors and Degradation Properties of Multilayered Polymer Scaffolds: The Phase Space Method for Bile Duct Design and Bioengineering. *Biomedicines* **2023**, *11*, 745. [CrossRef] [PubMed]
33. Ahmed, R.; Haq, M.; Chun, B.-S. Characterization of Marine Derived Collagen Extracted from the By-products of Bigeye Tuna (*Thunnus obesus*). *Int. J. Biol. Macromol.* **2019**, *135*, 668–676. [CrossRef] [PubMed]
34. Li, Y.; Asadi, A.; Monroe, M.R.; Douglas, E.P. pH Effects on Collagen Fibrillogenesis in Vitro: Electrostatic Interactions and Phosphate Binding. *Mater. Sci. Eng. C* **2009**, *29*, 1643–1649. [CrossRef]
35. Yan, M.; Li, B.; Zhao, X.; Qin, S. Effect of Concentration, pH and Ionic Strength on the Kinetic Self-assembly of Acid-soluble Collagen from Walleye Pollock (*Theragra chalcogramma*) skin. *Food Hydrocoll.* **2012**, *29*, 199–204. [CrossRef]
36. Yi, L.; Dong, N.; Yun, Y.; Deng, B.; Ren, D.; Liu, S.; Liang, Y. Chemometric Methods in Data Processing of Mass Spectrometry-based Metabolomics: A Review. *Anal. Chim. Acta* **2016**, *914*, 17–34. [CrossRef]
37. Camacho, C.; Coulouris, G.; Avagyan, V.; Ma, N.; Papadopoulos, J.; Bealer, K.; Madden, T.L. BLAST+: Architecture and Applications. *BMC Bioinform.* **2009**, *10*, 421. [CrossRef]

38. Muyonga, J.H.; Cole, C.G.B.; Duodu, K.G. Characterisation of Acid Soluble Collagen from Skins of Young and Adult Nile Perch (*Lates niloticus*). *Food Chem.* **2004**, *85*, 81–89. [CrossRef]
39. Buckley, M.; Fraser, S.; Herman, J.; Melton, N.D.; Mulville, J.; Pálsdóttir, A.H. Species Identification of Archaeological Marine Mammals Using Collagen Fingerprinting. *J. Archaeol. Sci.* **2014**, *41*, 631–641. [CrossRef]
40. Gillet, L.C.; Navarro, P.; Tate, S.; Röst, H.; Selevsek, N.; Reiter, L.; Bonner, R.; Aebersold, R. Targeted Data Extraction of the MS/MS Spectra Generated by Data-independent Acquisition: A New Concept for Consistent and Accurate Proteome Analysis. *Mol. Cell. Proteom.* **2012**, *11*, O111.016717. [CrossRef]

antioxidants

MDPI

Article

Partial Purification and Biochemical Evaluation of Protease Fraction (MA-1) from *Mycoleptodonoides aitchisonii* and Its Fibrinolytic Effect

Sung-Ho Lee [1,2,†], Seung-Yub Song [1,2,†], Jun-Hui Choi [3], Seung Kim [3], Hyo-Jeong Lee [3], Jin Woo Park [1,2], Dae-Hun Park [4], Chun-Sik Bae [5] and Seung-Sik Cho [1,2,*]

1 Department of Pharmacy, College of Pharmacy, Mokpo National University, Muan 58554, Republic of Korea; tjdgh0730@naver.com (S.-H.L.); tgb1007@naver.com (S.-Y.S.); jwpark@mnu.ac.kr (J.W.P.)
2 Department of Biomedicine, Health & Life Convergence Sciences, BK21 Four, Biomedical and Healthcare Research Institute, Mokpo National University, Mokpo 58554, Republic of Korea
3 Department of Food Science and Biotechnology, Gwangju University, Gwangju 61743, Republic of Korea; sekai0572@naver.com (J.-H.C.); seungk@gwangju.ac.kr (S.K.); jhshkl@gwangju.ac.kr (H.-J.L.)
4 College of Oriental Medicine, Dongshin University, Naju-si 58245, Republic of Korea; dhj1221@hanmail.net
5 College of Veterinary Medicine, Chonnam National University, 77 Yongbong-ro, Buk-gu, Gwangju 61186, Republic of Korea; csbae210@chonnam.ac.kr
* Correspondence: sjason1@naver.com; Tel.: +82-61-450-2687
† These authors contributed equally to this work.

Abstract: The antioxidative proteolytic fraction, MA-1, was partially purified from *Mycoleptodonoides aitchisonii*. MA-1 was purified to homogeneity using a two-step procedure, which resulted in an 89-fold increase in specific activity and 42.5% recovery. SDS-PAGE revealed two proteins with a molecular weight of 48 kDa. The zymography results revealed proteolytic activity based on the MA-1 band. MA-1 was found to be stable in the presence of Na^+, Ca^{2+}, Fe^{3+}, K^+, and Mg^{2+}. MA-1 was also stable in methanol, ethanol, and acetone, and its enzyme activity increased by 15% in SDS. MA-1 was inhibited by ethylenediaminetetra-acetic acid or ethylene glycol tetraacetic acid and exerted the highest specificity for the substrate, MeO-Suc-Arg-Pro-Tyr-pNA, for chymotrypsin. Accordingly, MA-1 belongs to the family of chymotrypsin-like metalloproteins. The optimum temperature was 40 °C and stability was stable in the range of 20 to 35 °C. The optimum pH and stability were pH 5.5 and pH 4–11. MA-1 exhibited stronger fibrinolytic activity than plasmin. MA-1 hydrolyzed the Aα, Bβ, and γ chains of fibrinogen within 2 h. MA-1 exhibited an antithrombotic effect in animal models. MA-1 was devoid of hemorrhagic activity at a dose of 80,000 U/kg. Overall, our results show that *M. aitchisonii* produces an acid-tolerant and antioxidative chymotrypsin-like fibrinolytic enzyme, and *M. aitchisonii* containing MA-1 could be a beneficial functional material for the prevention of cardiovascular diseases and possible complications.

Keywords: *Mycoleptodonoides aitchisonii*; fibrinolytic enzyme; acidic tolerance antithrombotic; thromboembolism

Citation: Lee, S.-H.; Song, S.-Y.; Choi, J.-H.; Kim, S.; Lee, H.-J.; Park, J.W.; Park, D.-H.; Bae, C.-S.; Cho, S.-S. Partial Purification and Biochemical Evaluation of Protease Fraction (MA-1) from *Mycoleptodonoides aitchisonii* and Its Fibrinolytic Effect. *Antioxidants* **2023**, *12*, 1558. https://doi.org/10.3390/antiox12081558

Academic Editors: Yongliang Zhuang, Hu Hou, Zhipeng Yu and Shengbao Cai

Received: 22 June 2023
Revised: 1 August 2023
Accepted: 2 August 2023
Published: 4 August 2023

1. Introduction

Mycoleptodonoides aitchisonii is an edible mushroom belonging to the family, Climacodontaceae, that is mainly found in India and Japan [1]. This mushroom has a cap that is approximately 3–8 cm in diameter, its color ranges from white to yellow, its surface is smooth, and its stem is short. *M. aitchisonii* grows on the trunks of dead beech trees. Notably, *M. aitchisonii* is an edible mushroom that is commonly used as a natural health product [2].

Recently, Republic of Korea has succeeded in the mass production of *M. aitchisonii* through artificial cultivation [3]. Recently, studies on the biological efficacy of *M. aitchisonii*, the toxicity of single and repeated administration of hot water extracts [4], the anti-asthma [5] and immune-stimulating effects of polysaccharides [6], and the anti-obesity effects [7] of *M. aitchisonii* dried powder, have been conducted. The methanol extract of *M. aitchisonii* has been reported to have an anti-inflammatory effect [8]. Further, the hot water extract has been found to lower blood pressure in animal models [9].

The polysaccharide fractions of *M. aitchisonii* have been reported to increase nerve growth factor (NGF) and catecholamine levels in rat brains [10]. In addition, phenylpentane compounds from *M. aitchisonii* mycelia enhance dopamine release in rat brain striatal slices [11]. In particular, inducing transient bilateral carotid artery occlusion in rats and the oral administration of *M. aitchisonii* or a polysaccharide extract can reduce ischemic damage by affecting monoamine metabolism in the cerebral cortex [12].

The bioactive compounds in mushroom extract include 5-hydroxy-4-(1-hydroxyethyl)-3-methylfuran-2(5H)-one and (3R,4R)-4-((R)-), which are endoplasmic reticulum stress inhibitors. 1-Hydroxyethyl-3-methyldihydrofuran-2(3H)-one, (3S,4R)-5-Phenylpentane-1,3,4-triol [13], and phenylpentane-based volatile substances, such as 1-phenyl-3-pentanol and 1-phenyl-3-pentanone, are known [11].

Although polysaccharides and some compounds of *M. aitchisonii* have been identified and their physiological activities are known, no studies have been conducted on biologically active catalysts. Therefore, in the present study, *M. aitchisonii* was treated with cold water to obtain crude proteins and polysaccharides, and the protease fractions (MA-1) were obtained via gel permeation column chromatography. Partially purified MA-1 was confirmed to affect optimum temperature, pH, stability, metal ions, enzyme inhibitors, and surfactants. In addition, MA-1 was confirmed to exhibit fibrino(geno)lytic activity and act as a chymotrypsin-like metalloprotease. The characteristics of MA-1 were revealed through an evaluation of its blood clot-degrading mechanism and characteristics in vivo.

2. Materials and Methods

2.1. Materials

Bovine fibrinogen, human fibrinogen, human fibrin, bovine thrombin, human thrombin, plasmin, azocasein, tris (hydroxymethyl) aminomethane (Tris), tetramethyl-ethylenediamine (TEMED), sodium dodecyl sulfate-polyacrylamide gel electrophoresis, phenylmethanesulfonyl fluoride (PMSF), ethylenediaminetetra-acetic acid (EDTA), and ethylene glycol tetraacetic acid (EGTA) were purchased from Sigma-Aldrich Co. (Burlington, NJ, USA). The protein markers used for SDS-PAGE were purchased from Fermentas Co. (Burlington, MA, USA). Sephadex resin was purchased from GE Healthcare Co. (Boston, MA, USA). The substrates for the amidolytic assay were purchased from Diapharma (West Chester, OH, USA). Other reagents were of special grade and were purchased commercially.

2.2. Rodents

ICR mice (25 g, 6 weeks of age, male) and Sprague-Dawley (SD) rats (200–300 g, male) were used in the tests. Four mice and four rats were housed in each cage. Mice were reared at 22 $\pm$ 2 °C, with a 12 h light/dark cycle. Feed and water were provided ad libitum. The environment was created to minimize the stress on the animals. All experimental procedures were performed under the guidance of the National Institutes of Health Guide for the Care and Use of Laboratory Animals [13]. It was approved by the Ethics Committee of Chonnam National University (CNU-IACUC-YB-2019-47).

2.3. Preparation of the Crude Extract and Purification of MA-1

M. aitchisonii was supported by the Jeollanamdo Wando Arboretum (Wando, Republic of Korea) and was prepared based on our previous report [5]. The crude protein extract was prepared using 250 g of mycelium powder in 1000 mL filtered sterile water and agitated for 24 h at 10 °C. Insoluble material was removed via filtration and the water extract was freeze-dried. The purification steps were performed under 10 °C. The crude was fractionated via gel permeation chromatography (GPC) using a Sephadex G-100 column (1 $\times$ 100 cm) equilibrated with 10 mM Tris-HCl (pH 7.0) at a flow rate of 2 mL/fraction. Loading volume was 1 mL and the protein amount was 25 mg. The protease activity of the active parts was determined using an azocasein method (at 660 nm) as described below. The protein concentration was determined using the Bradford method (at 595 nm) [14]. Fractions exerting protease activity were concentrated using a Vivaspin (Millipore, Burlington, MA, USA) and used as purified enzymes (MA-1) for further analysis.

2.4. Protease Assay

Protease activity was determined using a previously described method with minor modifications [15]. The mixture containing 50 μL azocasein (2% solution in 10 mM Tris-HCl buffer pH 8.0) and samples (100 μL) were incubated at 40 °C for 60 min and then stopped with 50 μL of trichloroacetic acid (10%). After centrifugation at 13,000× *g* for 10 min, 150 μL of the supernatant was mixed with 300 μL Folin–Ciocalteu's phenol reagent (0.33 M) and 450 μL Na_2CO_3 solution (10%, *w*/*v*), and absorbance was measured at 660 nm. Enzyme activity was converted to a unit that has the ability to produce 1 μg of tyrosine per minute.

2.5. Biochemical Properties of MA-1

Determination of the optimal reaction temperature was performed by comparing enzyme activity at various temperatures under the condition of pH 7.0 (20 to 60 °C). For temperature stability, the remaining activity was measured after leaving the enzyme in the range of 20 to 55 °C degrees for 2 h. The optimal pH for enzyme activity was determined by measuring the residual enzyme activity in various pH conditions (3~11). The pH stability was evaluated by measuring the remaining activity after standing at 35 °C for 2 h under various pH conditions. The effects of metal ions and enzyme inhibitors on protease activity were also assessed using $NaCl_2$, $MgCl_2$, $CaCl_2$, $FeCl_3$, KCl, EDTA, EGTA, and PMSF (1 mM). The effects of organic solvents and surfactants on protease activity were also evaluated with methanol, ethanol, acetone, sodium dodecyl sulfate, triton ×100, and tween 20 (10%, *v*/*v*). The enzyme was mixed with 1 mM of different metal ions and 10% solvents and detergents. Finally, the protease activity was measured using azocasein.

The amidolytic activity based on several chromogenic substrates was determined by mixing 25 μL MA-1 (5 mg/mL) and 100 μL tris buffer (pH 7.5; 10 mM) with 75 μL of substrates (4 mM), such as H-D-Phe-Pip-Arg-pNA (S-2238; for thrombin); H-D-Val-Leu-Lys-pNA (S-2251; for plasmin); H-D-Ile-Pro-Arg-pNA (S-2288; for t-PA); MeO-Suc-Arg-Pro-Tyr-pNA (DPG-586; for chymotrypsin); and Pyro-Glu-Gly-Arg-pNA (DPG-444; for u-PA) in 20 mM Tris-HCl (pH 7.5) in a 1.5 mL tube. The mixtures were incubated at 37 °C for 10 min and the amount of *p*-nitroaniline (or the substrate) released was determined at 405 nm [16].

2.6. Antioxidant Assay

DPPH radical scavenging assay was determined according to Rahman's protocol. Briefly, the sample was added to the DPPH solution (0.4 mM), mixed for 10 min, and measured at 517 nm [17].

The total phenolic content was determined by the Folin reaction [18]. The samples were mixed with Na_2CO_3 solution and Folin–Ciocalteu phenol reagent for 10 min. The mixture was measured at 750 nm. Gallic acid was used as standard. Results were expressed as milligrams of gallic acid equivalents per gram of sample.

The reducing power was evaluated to test the antioxidant properties of the sample. The extract (0.1 mL) was mixed with 0.2 M sodium phosphate buffer (0.5 mL, pH 6.5) and $K_3[Fe(CN)_6]$ (0.5 mL, 1% *w*/*v*), followed by incubation at 50 °C for 20 min, and the reaction was stopped by adding TCA (0.5 mL, 10%). After centrifugation, the supernatant was mixed with distilled water (0.5 mL) and $FeCl_3$ (0.1 mL, 0.1% *w*/*v*), and the absorbance of the mixture was measured at 700 nm. The reducing powers were expressed as vitamin C equivalents [19].

2.7. Fibrinolytic and Fibrinogenolytic Activity Assay

The fibrinolytic activity of MA-1 and the positive control (plasmin from human plasma) were evaluated. MA-1 (1 U) and the control (5 U) were prepared on fibrin plates for the fibrinolytic activity assay. The fibrinolytic activity was analyzed using fibrin plates according to the method described by [20]. The fibrin plates were prepared by pouring the solution comprising 4.5 mg/mL fibrinogen in 10 mM Tris-HCl buffer (pH 7.0), 1.2% agarose, and 0.45 U/mL thrombin into a Petri dish. The solution was incubated for 30 min at room temperature for fibrin clot formation. MA-1 and plasmin (1 and 5 U) were loaded onto a hole (3.5 mm in diameter) of the plate, which was then incubated at 37 °C for 24 h. The fibrinolytic activity was estimated by measuring the diameter of the clear zone around the well.

Quantitative analyses of fibrinogenolytic activity were conducted according to a previously published method, with slight modifications [20]. Briefly, 200 μL of fibrinogen (10 mg/mL) in 10 mM Tris-HCl buffer (pH 7.5) comprising 0.15 M NaCl was incubated at 37 °C with 60 μL MA-1 (0.6 U) for various durations. The incubated sample (10 μL) was withdrawn at the indicated time intervals, then boiled for 5 min to stop the reaction. The resulting degradation products were then analyzed via SDS-PAGE (12% separating gel).

2.8. Ex Vivo Coagulation Assay

Ex vivo coagulation assays were performed according to the method from Broersma et al. [21]. Six experimental groups (ten animals per group) were used: group 1 (saline treatment), groups 2–3 (40,000 and 80,000 U/kg nattokinase), and groups 4–6 (20,000, 40,000, and 80,000 U/kg MA-1). Nattokinase, a well-known thrombolytic enzyme, was selected as the control and purified according to a previously reported method [22]. The test samples (saline, nattokinase, and MA-1) were administered intravenously into the tail vein 30 min before sampling. Blood was collected after treatment on sodium citrate solution (3.8%), and plasma was obtained via centrifugation of the blood samples at 2000 g for 10 min. Activated partial thromboplastin time (APTT) and prothrombin time (PT) was performed according to the instructions provided by the manufacturer (Fisher Diagnostics, Middletown, CT, USA).

2.9. Acute Thromboembolism Animal Model

The collagen and epinephrine-induced thromboembolism model was established in ICR mice according to the method described by Dimmino et al. [23]. Five groups (ten animals/group) were established: group 1 (vehicle), group 2 (collagen + epinephrine), group 3 (nattokinase, 80,000 U/kg + collagen + epinephrine), and groups 4 and 5 (40,000, and 80,000 U/kg MA-1 + a mixture of collagen and epinephrine). One hour before the thrombotic challenge, mice were injected intraperitoneally with nattokinase and MA-1. Thereafter, mice were rapidly injected with 0.1 mL of a mixture of collagen and epinephrine via the tail vein. Mortality or paralysis was recorded for 15 min, and all surviving mice were killed immediately after the experiment.

2.10. $FeCl_3$-Induced Carotid Arterial Thrombus Model

$FeCl_3$-induced arterial thrombosis was established as described previously, with some modifications [24]. Briefly, male SD rats were anesthetized with an intraperitoneal injection of 50 mg/kg pentobarbital. The tested samples (saline, nattokinase (20,000 U/kg) and MA-1 (10,000, 20,000 U/kg)) were administered intravenously 1 min before $FeCl_3$ injury. Then, the right carotid artery segment was exposed by blunt dissection. Two sheets of filter paper (6 mm × 6 mm) were treated with a 4% $FeCl_3$ solution and then placed on opposite sides of the carotid artery in contact with the lateral surface of the vessel for 3 min. After 30 min, the filter paper was removed, and the mice were sacrificed and perfused with 4% paraformaldehyde through cardiac puncture. The secured carotid artery was fixed with 4% paraformaldehyde for 12 h at 4 °C. The obtained carotid arteries were made into frozen sections and H&E staining was performed to evaluate the antithrombotic effect of MA-1. Quantitative evaluation of thrombi was performed using ImageJ 1.46 b image analysis software.

2.11. Hemostasis Assessment

Tail bleeding time was measured according to Kim et al.'s method [24]. Mice were anesthetized and the t-PA (5000 U/kg), nattokinase (80,000 U/kg), and MA-1 (80,000 U/kg) were injected intravenously via the tail veins of mice. After 30 min, the mice were placed on a hot block (37 °C) and the end of the tail was cut, and then immersed in a tube containing saline. The bleeding time was recorded for 30 min.

2.12. Statistical Analysis

Data are expressed as mean ± SD. Statistical significance was calculated using a post hoc test for multiple group comparisons. Differences with p values less than 0.01 and 0.05 were considered statistically significant.

3. Results and Discussion

3.1. Purification of the Protease Fraction, MA-1

We partially purified the protease fraction, MA-1, from *M. aitchisonii* using the steps listed in Table 1. Two proteins were obtained via gel permeation chromatography (GPC, Sephadex G-100) and were found to have similar molecular weights. Approximately 25 mg of crude protein was obtained after cold water extraction. The crude enzyme was subjected to gel permeation chromatography (GPC) on a Sephadex G-100 column to collect fractions that exhibited proteolytic activity (Figure 1B). Finally, the partially purified enzyme (MA-1) was subjected to SDS-PAGE. MA-1 appeared as two bands on the PAGE (Figure 1A). The final yield of the active fraction from 25 mg of crude protein was 0.12 mg (Table 1). The protease activity of the purified MA-1 was verified by gelatin zymography (Figure 1A).

Table 1. Summary of the purification of MA-1.

Step	Total Protein (mg)	Total Activity (U)	Specific Acidity (U/mg)	Fold	Yield (%)
Crude	25	40	1.6	1	100
GPC	0.12	17	141.67	89	0.48

Activity was measured using an azocasein assay. The units of activity were calculated based on a tyrosine standard.

3.2. MA-1 Showed Fibrino(geno)lytic Activity

The fibrinolytic effect of MA-1 on fibrin was analyzed using the fibrin plate method. In Figure 1C, the size of the fibrin cleavage zone after treatment with 1 U MA-1 was similar to that after treatment with 5 U plasmin, suggesting that the fibrinolytic efficiency of MA-1 was greater than that of plasmin. The clear zone owing to MA-1 was 14 mm/U, which was approximately 5-fold higher than that produced by plasmin (17 mm/5 FU) after 24 h of incubation at 37 °C. Additionally, to elucidate the mode of the fibrinogenolytic effect of MA-1, Aα, Bβ, and γ chains were analyzed using SDS-PAGE. As shown in Figure 1D, the Aα-chains were completely cleaved by the enzyme within 5 min and the Bβ-chains were almost completely cleaved within 120 min. Moreover, the enzyme hydrolyzed the γ-chains within 60 min of incubation.

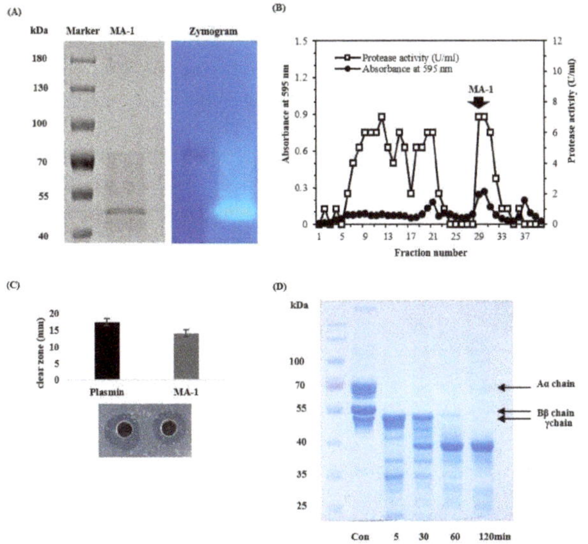

Figure 1. Purification and fibrino(geno)lytic activity of MA-1. (**A**) SDS-PAGE and zymogram of MA-1 in 12% polyacrylamide gel. (**B**) Profiles from Sephadex G-100 column chromatography. (**C**) Fibrinolytic activity on fibrin plate. Fibrin plates composed of 4.5 mg fibrinogen, 0.012 mg agarose, and 0.45 U thrombin in 1 mL of 10 mM Tris-HCl buffer (pH 7.0). Each value represents the mean ± SD of triplicate measurements. (**D**) Fibrinogenolytic pattern analysis of MA-1.

3.3. Biochemical Properties of MA-1

The optimal reaction temperature for MA-1 was determined at various reaction temperatures. The thermal stability of the enzyme was examined via incubation at different temperatures for 2 h. As shown in Figure 2A, MA-1 was active at temperatures between 35 and 45 °C and exhibited maximal activity at 40 °C. The activity of MA-1 was stable at 20–35 °C (Figure 2B). The optimal reaction pH of MA-1 was determined using buffers with various pH values. As shown in Figure 2C, MA-1 was active at pH values of 5.5, 6, and 9. MA-1 was stable in the pH range of 4 to 11 at 4 °C for 2 h. The effect of pH on enzyme activity was also examined. As shown in Figure 2D, MA-1 exhibited a wide stability range from pH 4 to 11.

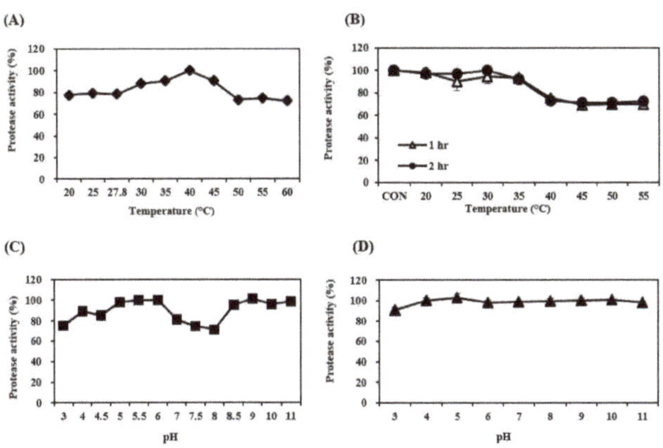

Figure 2. Biochemical properties of the purified enzyme. (**A**) Effect of temperature on protease activity (◆); (**B**) temperature stability (•, △); (**C**) effect of pH on protease activity (■); (**D**) pH stability (▲). Each value represents the mean ± SD of triplicate measurements.

The effects of several metal ions, protease inhibitors, organic solvents, and surfactants are summarized in Table 2. The enzyme was inhibited by 1 mM ethylenediaminetetraacetic acid (EDTA) and ethyleneglycol tetraacetic acid (EGTA), which are inhibitors of metalloproteases. However, MA-1 was not inhibited by PMSF, a protease inhibitor that reacts with proteases like chymotrypsin, trypsin, and papain. To further investigate the characteristics of MA-1, its amidolytic activity was evaluated using various chromogenic substrates (Figure 3). The enzyme exhibited the highest specificity for the MeO-Suc-Arg-Pro-Tyr-pNA substrate for chymotrypsin, whereas MA-1 exhibited weak activity for the substrates t-PA, thrombin, plasmin, and u-PA. These findings suggest that MA-1 has hydrolytic characteristics similar to those of serine proteases or serine metalloproteases.

Table 2. Biochemical properties of MA-1.

	Relative Activity (%)
Control	100 ± 1.9
Metal/inhibitors (1 mM)	
Na^{+}	102.6 ± 4.1
Ca^{2+}	99.4 ± 1.3
Fe^{3+}	95.9 ± 2.3
K^{+}	95.1 ± 2.0
Mg^{2+}	99.4 ± 3.4
EDTA	75.7 ± 3.2
EGTA	78.0 ± 1.6
PMSF	100.6 ± 2.4
Solvents/surfactants (10%)	
Methanol	98.9 ± 2.2
Ethanol	99.7 ± 1.7
Acetone	100.1 ± 2.0
SDS	116.2 ± 1.6
Glycerol	95.3 ± 2.5
Triton X-100	66.4 ± 2.2
Tween 20	82.4 ± 2.0

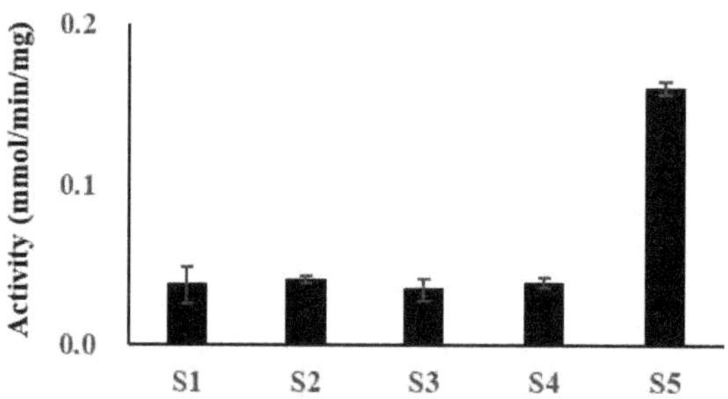

Figure 3. Amidolytic activity of MA-1. The amidolytic activity with several chromogenic substrates was determined by mixing MA-1 and various substrates at 37 °C for 10 min. The amount of *p*-nitroaniline (or substrate) released was determined at 405 nm. Each value represents the mean ± SD of triplicate measurements. S1 (H-D-Ile-Pro-Arg-pNA for t-PA); S2 (H-D-Val-Leu-Lys-pNA for plasmin); S3 (Pyro-Glu-Gly-Arg-pNA for u-PA); S4 (MeO-Suc-Arg-Pro-Tyr-pNA for chymotrypsin); and S5 (H-D-Phe-Pip-Arg-pNA for thrombin).

The effects of metal ions, solvents, and surfactants on enzyme activity were examined. MA-1 was incubated with 1.0 mM of metal ions, 10% of solvents, and 10% of surfactants and residual enzyme activity was measured. MA-1 was not inhibited by the metal ions. In the various solvent tests, 10% (v/v) ethanol, methanol, and acetone did not decrease the enzyme activity; however, Triton X-100 (approximately 34.6% reduction), glycerol (approximately 4.7% reduction), and Tween 20 (approximately 17.6% reduction) decreased the activity. SDS increased enzyme activity (approximately 16.2%) (Table 2). The DPPH radical scavenging ability of MA-1 was 82.83% at a concentration of 5 mg/mL, and ascorbic acid, a control, showed 85.92% at a concentration of 25 μg/mL. The reductive activity of MA-1 expressed as vitamin C equivalents was 20.16 ± 1.28 μg/100 μg protein.

3.4. MA-1 Showed Excellent Anti-Coagulant Effect in an Ex Vivo Coagulation Model

By measuring the APTT and PT, we determined the coagulant factors (extrinsic and intrinsic) of MA-1 that exhibited anticoagulant effects. The ex vivo test was conducted by dividing mice into the control and experimental groups (groups treated with MA-1). The effects of MA-1 (20,000–80,000 U/kg) on the APTT and PT were evaluated ex vivo. MA-1 delayed the APTT in a dose-dependent manner. At a dose of 80,000 U/kg, MA-1 prolonged the APTT by 3.38 fold (120.8 ± 4.8 s) compared with the control (35.5 ± 1.8 s) (Table 3). However, MA-1 did not affect the PT compared with the control (12.3 ± 0.4 s). Our findings indicate that MA-1 may possess anticoagulant properties by prolonging the common and/or intrinsic coagulation pathways.

Table 3. Ex vivo anticoagulant activity of MA-1.

		APTT (s)	PT (s)
Control		35.5 ± 1.8	12.3 ± 0.4
MA-1 (U/kg)	20,000	46.1 ± 3.0	12.2 ± 0.4
	40,000	67.0 ± 3.3 *	12.0 ± 0.5
	80,000	120.8 ± 4.8 *	11.5 ± 0.5

Each value represents the mean ± SD of at least 3 independent experiments. One-way ANOVA followed by post hoc Tukey's test. * $p < 0.01$, compared with the saline-treated control group.

3.5. MA-1 Showed Effects of Inhibiting Clot Formation and Delaying Bleeding in Animal Models of Thrombosis and Hemostasis

In the present study, we experimented with the acute thromboembolism model and a $FeCl_3$-induced arterial thrombosis model to evaluate the antithrombotic activity of MA-1. Collagen with epinephrine treatment caused 100% death or paralysis in the vehicle-treated group. Nattokinase (80,000 U/kg/day) was used as a positive control and exhibited a 50% protection rate against death and paralysis and the MA-1-treated group displayed 30% and 40% protection rates when administered 40,000 U/kg/day and 80,000 U/kg/day.

The $FeCl_3$-induced arterial thrombosis model was employed as another animal model in this study. The percentage occlusion of the right carotid artery was determined to investigate the antithrombotic activity of MA-1 and nattokinase as positive controls. Figure 4 shows the qualitative and quantitative percentages of occlusion of the right carotid. As shown in Figure 4B, the $FeCl_3$-induced vehicle-treated group displayed perfect occlusion compared with the control group (Figure 4A). As shown in Figure 4C, the positive control had an occlusion inhibition percentage of 24.1 ± 2.95% ($p < 0.01$) due to nattokinase, while the MA-1-treated group had percentages of 10.29 ± 1.42% and 16.62 ± 1.15% ($p < 0.01$) compared to the $FeCl_3$ + saline-treated group (Figure 3D,E). Based on our findings, the antithrombotic activity of MA-1 was confirmed in the $FeCl_3$-induced arterial thrombosis model and compared with that of nattokinase. Although the MA-1 exhibited antithrombotic activity compared with the vehicle, its occlusion inhibition was lower than that of nattokinase.

To assess the effect of MA-1 on hemostasis, saline, nattokinase (80,000 U/kg), MA-1 (80,000 U/kg), and t-PA (5000 U/kg) were administered and the tail bleeding time was measured after cutting off the tip. Hemostasis was achieved at approximately 410, 1620, 570, and 660 s with saline, t-PA, nattokinase, and MA-1, respectively (Table 4). MA-1 displayed a similar or slightly longer hemostasis time than nattokinase, increasing the hemostasis time by 1.6 times compared to that of the control group treated with saline. However, compared to t-PA, which increased the hemostasis time by approximately 4 times compared to the control, the hemostasis time of the experimental group treated with MA-1 was confirmed to be relatively short.

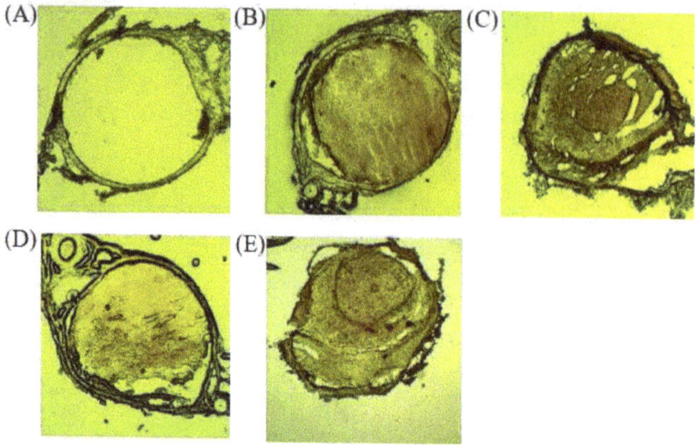

Figure 4. Antithrombotic effect of MA-1. The effects of MA-1 on $FeCl_3$-induced carotid arterial thrombus formation in rats were determined. The carotid arteries of rats were occluded with saline (**A**); $FeCl_3$ + saline (**B**); $FeCl_3$ + nattokinase ((**C**), 20,000 U/kg/day); $FeCl_3$ + MA-1 ((**D**,**E**); 10,000 and 20,000 U/kg/day).

Table 4. Effect of MA-1 on tail bleeding time.

	Bleeding Time (s)
Vehicle	410 ± 20
t-PA	1620 ± 50 *
Nattokinase	570 ± 20 *
MA-1	660 ± 40 *

Bleeding time was estimated using the tail transection model after intravenous injection of vehicle (saline), t-PA (5000 U/kg), nattokinase (80,000 U/kg), or MA-1 (80,000 FU/kg). Each value represents the mean ± SD (n = 4). One-way ANOVA followed by post hoc Tukey's test. * $p < 0.01$, compared with the vehicle group.

4. Discussion

Since the report of the anti-inflammatory efficacy of the *M. aitchisonii* extract in the 1990s, research on its biological activity and active ingredients has been steadily progressing. In the last decade, toxicity and bioactivity evaluations of the functional materials of *M. aitchisonii* extracts have been conducted. Hot water extracts have been reported to be closely associated with immune regulation, improvement of brain function, antihypertensive effects, and improvement of ischemic damage [1,5–7,12,13]. Recently, the hot water extracts of artificially grown *M. aitchisonii* have been reported to contain a high polysaccharide content of 30% or more. Therefore, the need to develop *M. aitchisonii* as a functional health material for ingestion as a dry powder or aqueous extract has increased. To date, no reports have described the purification and functionality of functional catalysts in the water-soluble fractions of *M. aitchisonii*. Therefore, we partially purified the proteolytic enzyme obtained from *M. aitchisonii* at a low temperature to investigate its functionality.

Intravascular thrombosis, the accumulation of intravascular fibrin, is caused by the accumulation of fibrin in blood vessels. Thrombosis causes cardiovascular disease and myocardial infarction, which is statistically known to cause more than 17 million deaths annually [25]. In general, injections are commonly used for the treatment of thrombosis, and research on fibrinolytic agents that can be administered orally is ongoing [26,27]. Thrombosis is closely associated with fibrin clot formation. Fibrin is degraded by plasmin (EC 3.4.21.7), and inactive plasminogen is converted into active plasmin by the tissue-type plasminogen activator (tPA) (EC 3.4.21.73). Urokinase (EC 3.4.21.73), streptokinase (EC 3.4.99.22), and tissue-type plasminogen activator (tPA) are typical fibrinolytic agents [28].

The fibrinolytic agents used for medical purposes have a strong clot-dissolving action, but have a limited route of administration and side effects, such as bleeding. Therefore, studies are being actively conducted to reduce the side effects and develop materials that can be administered orally. Attempts have been made to develop fibrinolytic enzymes and materials containing fibrinolytic enzymes via daily oral intake as food sources to prevent intravascular and/or cardiovascular diseases.

Thrombolytic enzymes have been found in microorganisms [29], fungi [30], marine organisms [31], and leeches [32]. In particular, Bacillus Nattokinase produced by bacteria in the genus can be administered orally and is known to have excellent fibrinolytic activity [33]. Therefore, in this study, fibrinolytic enzymes were compared with nattokinase and MA-1 in vivo to assess thrombolysis.

We prepared a water extract of dried *M. aitchisonii* at a low temperature (10 °C) and freeze-dried this extract. A highly active proteolytic enzyme fraction (MA-1) was then obtained through one-step column purification. Using sodium dodecyl sulfate-polyacrylamide gel electrophoresis, two protein mixtures of similar sizes (approximately 48 kDa) were identified. Fibrinolytic enzymes with a size similar to that of MA-1 can be found in some microorganisms. *Enterobacterium serratia* E–15 is reported to induce fibrinolysis and can distinguish and dissolve dead and damaged tissues without harming cells. Serralysin, a proteolytic enzyme of less than 50 kDa [34,35], *Shewanella* sp. IND20, and *Psuedoalteromonas* sp. IND11 produce thrombolytic enzymes with molecular weights of 44 kDa and 64 kDa, and exhibit direct coagulolytic activity and proenzyme plasminogen (PLG) activating ability [36,37].

MA-1 exhibited optimal proteolytic activity at pH 5.5 and 40 °C, and high activity even at around pH 9, which might be due to the different physicochemical characteristics of the two proteins in MA-1. MA-1 was stable when left for 2 h at pH 5, and 20–35 °C. When MA-1 was left for 2 h, it exhibited more than 90% residual activity at pH 3 to 11. These results suggest that MA-1 is composed of two types of proteins and the MA-1 fraction must be further subdivided through detailed separation.

MA-1 was not significantly affected by the presence of Na^+, Ca^{2+}, Fe^{3+}, K^+, or Mg^{2+}. In addition, MA-1 was not affected by organic solvents, such as methanol, ethanol, or acetone. However, SDS increased the enzyme activity of MA-1. EDTA and EGTA are metal chelators, and if an enzyme is inhibited by EDTA or EGTA, it is classified as a metalloenzyme [38]. The enzymatic activity of MA-1 was found to be inhibited by EDTA and EGTA. In general, proteases inhibited by PMSF are classified as serine proteases; however, MA-1 did not exhibit any change in enzymatic activity when treated with PMSF. Nonetheless, when MeO-Suc-Arg-Pro-Tyr-pNA, which has substrate specificity for chymotrypsin among the serine proteases, was treated with MA-1, MA-1 effectively degraded MeO-Suc-Arg-Pro-Tyr-pNA. These results suggest that MA-1 is a serine metalloprotease with chymotrypsin-like activity.

MA-1 exhibited fibrino(geno)lytic activity that was approximately five times stronger than that of plasmin. In fact, based on the fibrinogenolytic assay, the Aα, Bβ, and γ-chains were confirmed to be completely degraded within 120 min. Fibrinogen subchain specificity tended to differ for each enzyme. For example, WRL101 derived from *Bacillus* strain has substrate specificity for Bβ-chain [39]. The AprE127 enzyme derived from the *Bacillus* strain had substrate specificity similar to that of MA-1. AprE127 displayed strong Aα-chain decomposition ability and weak Bβ and γ-chain decomposition ability at 80 min [40].

Based on the results of the in-tube MA-1 assay, we investigated the efficacy of MA-1 in vivo. First, mice were intravenously (IV) treated with 20,000–80,000 U/kg of MA-1, and blood was collected to confirm anticoagulant activity. MA-1 was confirmed to delay the APTT. In the 80,000 U/kg administration group, APTT delayed coagulation by more than 120 s, and MA-1 was found to be the coagulation factor of the endogenous pathway (I, II, V, X, XII, XI, IX, VIII, prekallikrein, high molecular weight kininogen, and HMWK), and significantly delayed blood coagulation [41].

To confirm the ability of MA-1 to eliminate intravascular thrombosis, an experimental rat model was applied [42]. $FeCl_3$ was added to the carotid arteries to induce thrombosis.

Nattokinase from *Bacillus subtilis natto* was used as the control because it is a representative fibrinolytic enzyme after various administrations [33]. Nattokinase displayed an antithrombotic effect of 24.0 ± 2.9% when a dose of 80,000 U/kg was administered, while MA-1 displayed an antithrombotic effect of 16.6 ± 1.2% when administered at a dose of 80,000 U/kg. Several mushroom-derived proteases have been reported. Kim et al. purified a thrombolytic protease derived from *Cordyceps militaris* and reported its characteristics; however, its activity was not identified in vivo. Thus, the antithrombotic efficacy of MA-1 could not be considered [43].

Tania et al. compared the antithrombotic effects of four types of mushrooms (*Lentinula edodes, Pleurotus ostreatus, Pleurotus eryngii,* and *Agrocybe aegerita*) but presented only fibrin(ogen)olytic results [44]. Xiao et al. only revealed the physicochemical characteristics of thrombolytic enzymes, indicating limitations in the potential development of mushroom-derived antithrombotic materials [45].

The effect of MA-1 was analyzed using a thromboembolism-induced mouse model [46]. Thromboembolisms are caused by slow blood flow, excessive blood clotting, and vascular damage [47]. In the present study, blood circulation disorders were induced in mice via the administration of epinephrine and collagen. Nattokinase (80,000 U/kg/day) was also employed as a control. Based on the results, nattokinase exhibited a protective effect of 50%, while MA-1 displayed a protective effect of 40%. There are no reports on the thromboembolism inhibitory effects of mushroom-derived thrombolytic enzymes. Therefore, further studies on the amino acid sequence of MA-1, mushroom-derived proteins, and the development of an oral/injectable thrombosis prevention material based on this analysis are needed in the future.

To predict the side effects of MA-1, a tail bleeding test was performed using tPA and nattokinase as controls. For antithrombotic agents with rapid effects, delayed coagulation is a disadvantage. As shown in Table 4, tPA caused a coagulation delay of 1600 s. Notably, the clotting times of nattokinase (570 s) and MA-1 (660 s) were more than 1000 s faster than those of tPA. Previously, we reported that C142 (540 s), isolated from *Bacillus* strain C142, had the same coagulation delay time as nattokinase (550 s) [22]. Although MA-1 displayed a longer coagulation delay than C142, the bleeding side effects were expected to be similar to those of tPA.

We isolated a fibrinolytic enzyme fraction (MA-1) from *M. aitchisonii* and investigated its physicochemical properties. MA-1 exhibited excellent fibrinolytic activity compared to the control (plasmin and nattokinase) based on the animal experiments conducted. Although MA-1 is not composed of a single protein, our findings provide a clue to the development of mushroom-derived fibrinolytic enzymes. In future studies, amino acid sequence identification of MA-1 proteins, treatment of thrombotic diseases via different routes of administration (oral, injection), and evaluation of the curative and preventive efficacy are needed.

5. Conclusions

Antioxidative protease fraction, MA-1, was purified from *M. aitchisonii.* Both partially purified proteins were found to have similar molecular weights. This enzyme more efficiently degrades fibrin clots via direct fibrino(geno)lysis than it degrades plasmin. MA-1 hydrolyzed the Aα-, Bβ-, and γ-chains of fibrinogen within 2 h. The enzyme may be a chymotrypsin-like metalloprotease based on substrate specificity assay, and the mode of fibrino(geno)lysis. We revealed the ex vivo and in vivo antithrombotic effects of MA-1 via two different mechanisms: anticoagulation activity due to the inhibition of the activation of coagulation factors, and/or anti-clotting activity due to the inhibition of platelet aggregation and blood clot formation. These findings suggest that *M. aitchisonii* containing MA-1 can be a beneficial functional material for preventing cardiovascular diseases and their possible complications.

Author Contributions: Conceptualization, S.-S.C., D.-H.P. and C.-S.B.; investigation, S.-H.L., S.-Y.S. and J.-H.C.; data curation, J.W.P., S.K. and H.-J.L.; writing—original draft preparation, S.-S.C. All authors have read and agreed to the published version of the manuscript.

Funding: This work was supported by the National Research Foundation of Korea (NRF) grant funded by the Korean government (MSIT) (No. 2022R1A5A8033794).

Institutional Review Board Statement: Animal experiments are approved by the Ethics Committee of Chonnam National University (CNU-IACUC-YB-2019-47).

Informed Consent Statement: Not applicable.

Data Availability Statement: Should any raw data files be needed they are available from the corresponding author upon reasonable request.

Conflicts of Interest: The authors declare no conflict of interest.

References

1. Okuyama, S.; Terashima, T.; Kawamura, Y.; Yokogoshi, H. Enhancing effect of *Mycoleptodonoides aitchisonii* on synthesis of nerve growth factor and releasing dopamine in the rat brain. *Nutr. Neurosci.* **2004**, *7*, 41–47. [CrossRef] [PubMed]
2. Chandrasekaran, G.; Oh, D.S.; Shin, H.J. Versatile applications of the culinary-medicinal mushroom *Mycoleptodonoides aitchisonii* (Berk.) Maas G. (Higher Basidiomycetes): A review. *Int. J. Med. Mushrooms* **2012**, *14*, 395–401. [CrossRef] [PubMed]
3. Kim, Y.; Jung, B.M.; Wi, A.J.; Park, W.S.; Bang, M.; Park, D.H.; Seo, J.-W.; Oh, D.-S. Growth characters and harvest time for the artificial cultivation of *Mycoleptodonoides aitchisonii*. *J. Mushrooms* **2015**, *13*, 114–118. [CrossRef]
4. Lee, S.Y.; Park, D.H.; Kim, J.Y.; Lee, Y.J.; Oh, D.S.; Bang, M.A.; Seo, J.W. Genotoxicity and Single-Treatment Toxicity Evaluation of *Mycoleptodonoides aitchisonii* (Agaricomycetes) Water Extract. *Int. J. Med. Mushrooms* **2017**, *19*, 717–726. [CrossRef]
5. Lee, S.Y.; Bae, C.S.; Seo, J.H.; Cho, S.S.; Bae, M.S.; Oh, D.S.; Park, D.H. *Mycoleptodonoides aitchisonii* suppresses asthma via Th2 and Th1 cell regulation in an ovalbumin-induced asthma mouse model. *Mol. Med. Rep.* **2018**, *17*, 11–20. [CrossRef]
6. Xu, H.; Cho, S.S.; Lee, S.Y.; Oh, D.S.; Lim, S.K.; Park, D.H. Immune-Stimulating Effects of *Mycoleptodonoides aitchisonii* (Agaricomycetes) Water Extract via TNF-α and IFN-γ. *Int. J. Med. Mushrooms* **2017**, *19*, 809–815. [CrossRef]
7. Lee, M.R.; Begum, S.; Oh, D.S.; Wee, A.J.; Yun, B.S.; Sung, C.K. Ameliorating Effect of *Mycoleptodonoides aitchisonii* on High-fat Diet-induced Obese Mice. *Prev. Nutr. Food Sci.* **2014**, *19*, 69–74. [CrossRef]
8. Yasukawa, K.; Kanno, H.; Kaminaga, T.; Takido, M.; Kasahara, Y.; Kumaki, K. Inhibitory Effect of Methanol Extracts from Edible Mushroom on TPA-induced Ear Oedema and Tumour Promotion in Mouse Skin. *Phytother. Res.* **1996**, *10*, 367–369. [CrossRef]
9. Sato, T.; Takeuchi, A.; Ishida, T.; Obara, K.; Hirose, T.; Takai, K.; Fujino, K.; Sakamoto, Y. Antihypertensive effect of an aqueous extract from fruit body of *Mycoleptodonoides aitchisonii* in spontaneously hypertensive rats. *Pharmacometrics* **2001**, *61*, 177–183.
10. Okuyama, S.; Lam, N.V.; Hatakeyama, T.; Terashima, T.; Yamagata, K.; Yokogoshi, H. *Mycoleptodonoides aitchisonii* affects brain nerve growth factor concentration in newborn rats. *Nutr. Neurosci.* **2004**, *7*, 341–349. [CrossRef]
11. Okuyama, S.; Sawasaki, E.; Yokogoshi, H. Conductor compounds of phenylpentane in *Mycoleptodonoides aitchisonii* mycelium enhance the release of dopamine from rat brain striatum slices. *Nutr. Neurosci.* **2004**, *7*, 107–111. [CrossRef] [PubMed]
12. Okuyama, S.; Tashiro, K.; Lam, N.V.; Hatakeyama, T.; Terashima, T.; Yokogoshi, H. Effect of the edible mushroom *Mycoleptodonoides aitchisonii* on transient global ischemia-induced monoamine metabolism changes in rat cerebral cortex. *J. Med. Food* **2012**, *15*, 96–99. [CrossRef] [PubMed]
13. Choi, J.H.; Suzuki, T.; Okumura, H.; Noguchi, K.; Kondo, M.; Nagai, K.; Hirai, H.; Kawagishi, H. Endoplasmic reticulum stress suppressive compounds from the edible mushroom *Mycoleptodonoides aitchisonii*. *J. Nat. Prod.* **2014**, *77*, 1729–1733. [CrossRef] [PubMed]
14. Kruger, N.J. The Bradford Method For Protein Quantitation. In *The Protein Protocols Handbook*; Walker, J.M., Ed.; Humana Press: Totowa, NJ, USA, 2009; pp. 17–24.
15. Paggi, R.A.; Giménez, M.I.; De Castro, R.E. Proteolytic Activity Assays in Haloarchaea. In *Archaea: Methods and Protocols*; Methods in Molecular Biology; Clifton, N.J., Ed.; Springer: New York, NY, USA, 2022; Volume 2522, pp. 319–330. [CrossRef]
16. Choi, J.H.; Kim, S. Fibrinolytic and Thrombolytic Effects of an Enzyme Purified from the Fruiting Bodies of Boletus pseudocalopus (Agaricomycetes) from Korea. *Int. J. Med. Mushrooms* **2021**, *23*, 47–57. [CrossRef] [PubMed]
17. Rahman, M.M.; Islam, M.B.; Biswas, M.; Khurshid Alam, A.H.M. In vitro antioxidant and free radical scavenging activity of different parts of Tabebuia pallida growing in Bangladesh. *BMC Res. Notes* **2015**, *8*, 621. [CrossRef]
18. Kupina, S.; Fields, C.; Roman, M.C.; Brunelle, S.L. Determination of Total Phenolic Content Using the Folin-C Assay: Single-Laboratory Validation, First Action 2017.13. *J. AOAC Int.* **2018**, *101*, 1466–1472. [CrossRef]
19. Seo, J.H.; Kim, J.E.; Shim, J.H.; Yoon, G.; Bang, M.A.; Bae, C.S.; Lee, K.J.; Park, D.H.; Cho, S.S. HPLC analysis, optimization of extraction conditions and biological evaluation of *Corylopsis coreana* Uyeki Flos. *Molecules* **2016**, *21*, 94. [CrossRef]
20. Simkhada, J.R.; Cho, S.S.; Mander, P.; Choi, Y.H.; Yoo, J.C. Purification, biochemical properties and antithrombotic effect of a novel Streptomyces enzyme on carrageenan-induced mice tail thrombosis model. *Thromb. Res.* **2012**, *129*, 176–182. [CrossRef]

21. Broersma, R.J.; Kutcher, L.W.; Heminger, E.F.; Krstenansky, J.L.; Marshall, F.N. Antithrombotic activity of a novel C-terminal hirudin analog in experimental animals. *Thromb. Haemost.* **1991**, *65*, 377–381. [CrossRef]
22. Choi, J.H.; Kim, J.E.; Kim, S.; Yoon, J.; Park, D.H.; Shin, H.J.; Lee, H.J.; Cho, S.S. Purification and partial characterization of a low molecular fibrinolytic serine metalloprotease C142 from the culture supernatant of Bacillus subtilis C142. *Int. J. Biol. Macromol.* **2017**, *104*, 724–731. [CrossRef]
23. DiMinno, G.; Silver, M.J. Mouse antithrombotic assay: A simple method for the evaluation of antithrombotic agents in vivo. Potentiation of antithrombotic activity by ethyl alcohol. *J. Pharmacol. Exp. Ther.* **1983**, *225*, 57–60.
24. Kim, D.W.; Choi, J.H.; Park, S.E.; Kim, S.; Sapkota, K.; Kim, S.J. Purification and characterization of a fibrinolytic enzyme from Petasites japonicus. *Int. J. Biol. Macromol.* **2015**, *72*, 1159–1167. [CrossRef] [PubMed]
25. Wang, C.; Du, M.; Zheng, D.; Kong, F.; Zu, G.; Feng, Y. Purification and Characterization of Nattokinase from Bacillus subtilis Natto B-12. *J. Agric. Food Chem.* **2009**, *57*, 9722–9729. [CrossRef] [PubMed]
26. Suzuki, Y.; Kondo, K.; Matsumoto, Y.; Zhao, B.-Q.; Otsuguro, K.; Maeda, T.; Tsukamoto, Y.; Urano, T.; Umemura, K. Dietary supplementation of fermented soybean, natto, suppresses intimal thickening and modulates the lysis of mural thrombi after endothelial injury in rat femoral artery. *Life Sci.* **2003**, *73*, 1289–1298. [CrossRef] [PubMed]
27. Hsu, R.-L.; Lee, K.-T.; Wang, J.-H.; Lee, L.Y.L.; Chen, R.P.Y. Amyloid-Degrading Ability of Nattokinase from Bacillus subtilis Natto. *J. Agric. Food Chem.* **2009**, *57*, 503–508. [CrossRef] [PubMed]
28. Wang, S.-L.; Wu, Y.-Y.; Liang, T.-W. Purification and biochemical characterization of a nattokinase by conversion of shrimp shell with Bacillus subtilis TKU007. *New Biotechnol.* **2011**, *28*, 196–202. [CrossRef]
29. Sharma, C.; Osmolovskiy, A.; Singh, R. Microbial Fibrinolytic Enzymes as Anti-Thrombotics: Production, Characterisation and Prodigious Biopharmaceutical Applications. *Pharmaceutics* **2021**, *13*, 1880. [CrossRef]
30. Bezerra, V.H.S.; Cardoso, S.L.; Fonseca-Bazzo, Y.; Silveira, D.; Magalhães, P.O.; Souza, P.M. Protease Produced by Endophytic Fungi: A Systematic Review. *Molecules* **2021**, *26*, 7062. [CrossRef]
31. Zhao, L.; Lin, X.; Fu, J.; Zhang, J.; Tang, W.; He, Z. A Novel Bi-Functional Fibrinolytic Enzyme with Anticoagulant and Thrombolytic Activities from a Marine-Derived Fungus Aspergillus versicolor ZLH-1. *Mar. Drugs* **2022**, *20*, 356. [CrossRef]
32. Sawyer, R.T. Thrombolytics and Anticoagulants from Leeches. *Bio/Technology* **1991**, *9*, 513–518. [CrossRef]
33. Weng, Y.; Yao, J.; Sparks, S.; Wang, K.Y. Nattokinase: An Oral Antithrombotic Agent for the Prevention of Cardiovascular Disease. *Int. J. Mol. Sci.* **2017**, *18*, 523. [CrossRef] [PubMed]
34. Miyata, K.; Maejima, K.; Tomoda, K.; Isono, M. Serratia Protease: Part I. Purification and General Properties of the Enzyme. *Agric. Biol. Chem.* **1970**, *34*, 310–318. [CrossRef]
35. Bhagat, S.; Agarwal, M.; Roy, V. Serratiopeptidase: A systematic review of the existing evidence. *Int. J. Surg.* **2013**, *11*, 209–217. [CrossRef] [PubMed]
36. Rosen, G.; Naor, R.; Kutner, S.; Sela, M.N. Characterization of fibrinolytic activities of Treponema denticola. *Infect. Immun.* **1994**, *62*, 1749–1754. [CrossRef]
37. Vijayaraghavan, P.; Prakash Vincent, S.G. A low cost fermentation medium for potential fibrinolytic enzyme production by a newly isolated marine bacterium, *Shewanella* sp. IND20. *Biotechnol. Rep.* **2015**, *7*, 135–142. [CrossRef]
38. Sharma, M.; Gat, Y.; Arya, S.; Kumar, V.; Panghal, A.; Kumar, A. A Review on Microbial Alkaline Protease: An Essential Tool for Various Industrial Approaches. *Ind. Biotechnol.* **2019**, *15*, 69–78. [CrossRef]
39. Su, P.; Dong, H.; Yiel, L.; Ook, K.; Jae, J.; Choi, N.-S. Identification of fibrinogen-induced nattokinase WRL101 from Bacillus subtilis WRL101 isolated from Doenjang. *Afr. J. Microbiol. Res.* **2013**, *7*, 1983–1992. [CrossRef]
40. Frias, J.; Toubarro, D.; Fraga, A.; Botelho, C.; Teixeira, J.; Pedrosa, J.; Simões, N. Purification and Characterization of a Thrombolytic Enzyme Produced by a New Strain of Bacillus subtil. *J. Microbiol. Biotechnol.* **2021**, *31*, 327–337. [CrossRef]
41. Cui, N.; Cai, H.; Li, Z.; Lu, Y.; Wang, G.; Lu, A. Tigecycline-induced coagulopathy: A literature review. *Int. J. Clin. Pharm.* **2019**, *41*, 1408–1413. [CrossRef]
42. Li, W.; Nieman, M.; Sen Gupta, A. Ferric Chloride-induced Murine Thrombosis Models. *J. Vis. Exp. JoVE* **2016**, *115*, e54479. [CrossRef]
43. Kim, J.S.; Sapkota, K.; Park, S.E.; Choi, B.S.; Kim, S.; Nguyen, T.H.; Kim, C.S.; Choi, H.S.; Kim, M.K.; Chun, H.S.; et al. A fibrinolytic enzyme from the medicinal mushroom Cordyceps militaris. *J. Microbiol.* **2006**, *44*, 622–631. [PubMed]
44. Petraglia, T.; Latronico, T.; Liuzzi, G.M.; Fanigliulo, A.; Crescenzi, A.; Rossano, R. Edible Mushrooms as Source of Fibrin(ogen)olytic Enzymes: Comparison between Four Cultivated Species. *Molecules* **2022**, *27*, 8145. [CrossRef] [PubMed]
45. Liu, X.-L.; Zheng, X.-Q.; Qian, P.-Z.; Kopparapu, N.-K.; Deng, Y.-P.; Nonaka, M.; Harada, N. Purification and Characterization of a Novel Fibrinolytic Enzyme from Culture Supernatant of Pleurotus ostreatus. *J. Microbiol. Biotechnol.* **2014**, *24*, 245–253. [CrossRef] [PubMed]
46. Diaz, J.A.; Obi, A.T.; Myers, D.D., Jr.; Wrobleski, S.K.; Henke, P.K.; Mackman, N.; Wakefield, T.W. Critical review of mouse models of venous thrombosis. *Arterioscler. Thromb. Vasc. Biol.* **2012**, *32*, 556–562. [CrossRef] [PubMed]
47. Kebede, B.; Ketsela, T. Magnitudes of Risk Factors of Venous Thromboembolism and Quality of Anticoagulant Therapy in Ethiopia: A Systematic Review. *Vasc. Health Risk Manag.* **2022**, *18*, 245–252. [CrossRef] [PubMed]

foods

MDPI

Review

Extraction, Identification, and Quantification of Polyphenols from the *Theobroma cacao* L. Fruit: Yield vs. Environmental Friendliness

Juan Manuel Silva [1,2], Fernanda Peyronel [2,*], Yinan Huang [2], Carlos Eugenio Boschetti [1] and Maria G. Corradini [2,3,*]

1 Institute of Biotechnological and Chemical Processes (IPROByQ-CONICET), National University of Rosario (UNR), Rosario 2000, SF, Argentina; js.juanmanuelsilva@gmail.com (J.M.S.); ceboschetti@gmail.com (C.E.B.)
2 Food Science Department, University of Guelph, Guelph, ON N1G 2W1, Canada; yh699@cornell.edu
3 Arrell Food Institute, University of Guelph, Guelph, ON N1G 2W1, Canada
* Correspondence: fsvaikau@uoguelph.ca (F.P.); mcorradi@uoguelph.ca (M.G.C.); Tel.: +1-519-824-4120 (ext. 56422) (F.P.); +1-519-824-4120 (ext. 53344) (M.G.C.)

Abstract: The cacao fruit is a rich source of polyphenols, including flavonoids and phenolic acids, which possess significant health benefits. The accurate identification and quantification of these bioactive compounds extracted from different parts of the cacao fruit, such as pods, beans, nibs, and cacao shells, require specific treatment conditions and analytical techniques. This review presents a comprehensive comparison of extraction processes and analytical techniques used to identify and quantify polyphenols from various parts of the cacao fruit. Additionally, it highlights the environmental impact of these methods, exploring the challenges and opportunities in selecting and utilizing extraction, analytical, and impact assessment techniques, while considering polyphenols' yield. The review aims to provide a thorough overview of the current knowledge that can guide future decisions for those seeking to obtain polyphenols from different parts of the cacao fruit.

Keywords: *Theobroma cacao* L.; cacao; cocoa; polyphenols; extraction; analytical methods; green chemistry; sustainability; yield; penalty points

Citation: Silva, J.M.; Peyronel, F.; Huang, Y.; Boschetti, C.E.; Corradini, M.G. Extraction, Identification, and Quantification of Polyphenols from the *Theobroma cacao* L. Fruit: Yield vs. Environmental Friendliness. *Foods* **2024**, *13*, 2397. https://doi.org/10.3390/foods13152397

Academic Editors: Yongliang Zhuang, Hu Hou, Zhipeng Yu and Shengbao Cai

Received: 13 June 2024
Revised: 22 July 2024
Accepted: 23 July 2024
Published: 29 July 2024

1. Introduction

Cocoa products attract considerable scientific interest due to the health claims derived from their high content of bioactive compounds, notably phenolic compounds, praised for their antioxidant capacity. The cacao pod and its parts, i.e., cacao pod husks (CPHs), nibs, and cacao bean shells, have surfaced as promising sources of bioactive compounds.

The content and composition of polyphenols of the cacao pod depend not only on factors directly affecting the feedstock, such as tree genotype, growing conditions, and the ripeness degree of cacao pods at the time of harvesting, but also on the processing methods employed at both the farm and the industry level. The quantification of polyphenols is also contingent upon the methodology used for their efficient extraction from the feedstock and the subsequent analytical techniques utilized for measurement.

The *Theobroma cacao* L. tree, indigenous to the Amazonian rainforest, is cultivated globally, particularly in regions within 20 degrees north and south of the Equator, with significant production occurring in West Africa, Southeast Asia, and Central and South America. The fruit consists of an outer cacao pod husk (CPH) that surrounds the seeds (cacao beans), intermingled with a white mucilage. Cacao beans comprise a cotyledon and a coat or cacao shell (Figure 1). Conventionally, "cacao" refers to the unprocessed parts of the *Theobroma cacao* L. fruit, while "cocoa" typically denotes processed parts, a distinction adhered to herein.

Figure 1. (**A**) Longitudinal cut of a *Theobroma cacao* fruit open in half showing the cacao pod husk, seeds, and mucilage. (**B**) Fermented, sun-dried, and roasted cacao beans. (**C**) Cacao bean shell and nib.

Processing of cocoa beans at the farm typically involves discarding the CPHs and fermenting the mucilage together with the beans, a step crucial for flavor, color, and polyphenol content modulation. Fermentation causes cell destruction in the cotyledon, facilitating the release of polyphenols from storage cells [1]. Once the polyphenols are not protected by the cell, they are exposed to pH and temperature changes, as well as many enzymes present in the cotyledon. The total polyphenol content in fermented and dried whole cocoa beans has been reported in the range of 12–18% (d.w.) [2,3]. Several studies have shown that fermentation notably decreases polyphenol levels when compared with non-fermented beans [4–6]. Certain polyphenols may increase post-fermentation, influenced by genetic factors such as anatomical features of the beans [7]. Fermentation is immediately followed by sun-drying the beans before they are stored for transportation to the processing industries.

Drying and storage of the beans have a minor influence on the polyphenol content. However, the roasting process of the cacao beans in the industry, which is the initial step in the manufacture of chocolate, significantly impacts the polyphenol levels. The thermolabile molecular structures of polyphenols are affected by high and prolonged roasting temperatures. Roasting temperatures ranging from 70 to 150 °C for 15 to 45 min have been shown to reduce total polyphenol content in cacao feedstock [4,8,9]. Milder roasting treatments (~100 °C) decrease flavanol content by 18%, while more intense roasting (~150 °C) results in a 50% reduction compared to non-roasted beans [4]. Moreover, bean provenance has also been correlated to polyphenol thermal stability. For example, flavanols in cocoa beans from the Ivory Coast were less affected by roasting than those from Java [4]. Furthermore, it was observed that roasting at particular temperatures causes certain polyphenols to increase at the expense of others [10]. Migration of polyphenols from the bean interior to its shell occurs during roasting, rendering the shell a potential source of ingredients for food applications.

This review used the Eco-Scale penalty point system [11] to compare protocols for the extraction, identification, and quantification of polyphenolic compounds across different cacao fruit feedstocks. This penalty point system was employed to determine the potential impact the method and the analytical techniques have on human health as well as the environment. This review shows that polyphenol yield is highly dependent on the methods used for extraction, identification, and quantification, which also play a role in the sustainability of a methodology.

Readers can utilize this information to identify environmentally friendly extraction procedures and analytical quantification methods for characterizing polyphenols in cacao feedstocks. Additionally, if the reader is interested in implementing an appropriate scale-up procedure, this review should offer valuable insights into the efficacy and sustainability of industrial extractions and monitoring processes.

2. Molecular Structure of the Polyphenols in Cacao Feedstock

The health benefits of polyphenols are primarily attributed to their molecular structure and degree of polymerization [12–14]. The majority of cocoa polyphenols (>90%) occur as polymers, with monomers only accounting for 5–10% [15].

Polyphenols are classified into different groups according to their molecular structure. The number of aromatic rings and the structural elements that link these rings together determine which category a polyphenol belongs to. Figure 2 shows the names of the main polyphenols present in cacao, while Supplementary Figure S1 presents details of their chemical structures.

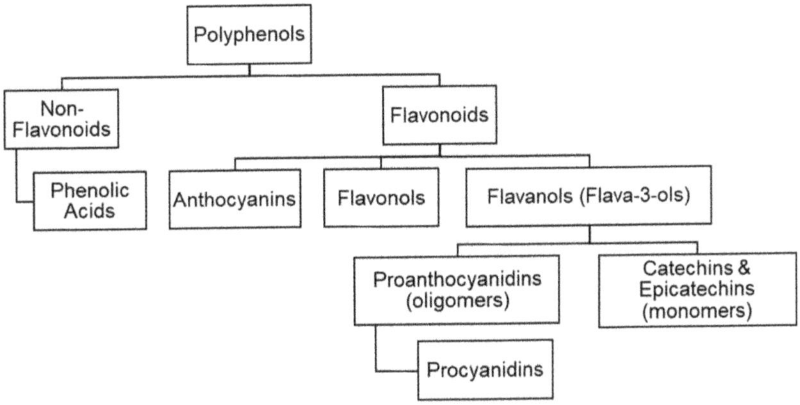

Figure 2. Classification of polyphenols in the *Theobroma cacao* fruit based on their chemical structure.

Flavonoids (Figure S1A), which include flavanols (also called flava-3-ols), anthocyanins (Figure S1B), and flavonols (Figure S1C), are the largest group of phenolic compounds in cacao and cocoa byproducts [16]. Flavanols are the principal polyphenols in cocoa beans, followed by anthocyanins and flavonols like tannins [17,18]. Among the non-flavonoid compounds in cacao, phenolic acids are the most commonly found [19,20].

Catechin (trans-isomer) and epicatechin (cis-isomer) are the most abundant flavanols. Both of them have two stereo-isomers, i.e., (+)-catechin, (−)-catechin, (+)-epicatechin, and (−)-epicatechin (Figure S1D). The most important are (−)-epicatechin, which constitutes 30–40% of total polyphenols in cacao, and (+)-catechin, in the form of monomers or oligomers such as proanthocyanidins (Figure S1D) [2,21,22]. Proanthocyanidins are classified according to their degree of polymerization (DP); DP equal to 1 refers to monomers, 2–10 are considered oligomers, and >10 are polymers [23]. Procyanidins are classified into types A and B based on their stereo-configuration and the link between monomers. The B-type is the predominant form in cocoa products [16,24], while gallocatechin and epigallocatechin are present in smaller amounts [25–27]. For instance, the epicatechin content of cocoa shells varies from 0.31–2.24 mg g^{-1} across 12 different bean types [28].

Anthocyanins are pigments known for their antioxidant and antimicrobial activities [16,20] (Figure S1E). This group of polyphenols constitutes ~4% of the total polyphenols in cocoa products. However, the fermentation process contributes to the degradation of some of them, e.g., cyanidin-3-galactoside and cyanidin-3-arabinoside [2,29]. Among the flavonols, which are glycosylated compounds similar to the anthocyanins, quercetin-O-glycosides, such as quercetin-3-O-galactoside, quercetin-3-O-arabinoside, and quercetin-3-O-glucoside, have been identified in cocoa beans [30,31].

Non-flavonoid polyphenols are present in small quantities in cacao beans [32]. Phenolic acids (Figure S1F) such as N-caffeoyl-3-O-hydroxytyrosine (clovamide), N-p-coumaroyltyrosine (deoxyclovamide), cinnamoyl-L-aspartic acid, and caffeoyl-glutamic acid (see Figure S6) have been reported in the different cacao fruit parts. These compounds, as well as procyanidins, contribute to the flavor of both unfermented cacao beans and roasted cocoa nibs [18].

3. Green Chemistry and the Eco-Scale

The purpose of applying green chemistry principles to extraction and analytical techniques is to eliminate, as much as possible, solvents, processes, and byproducts that can negatively impact human health and the environment [33].

The Eco-Scale [11,34] evaluates a process's overall impact based on the potential pollution of reagents/solvents and equipment energy consumption using penalty points (PP). The higher the total PP, the more harmful and less green the process. In this review, the PP associated with reported polyphenol extraction and identification/quantification methods, considering the amount of reagent or solvent used and equipment energy consumption, are reported and compared. PP are assigned based on the Globally Harmonized System of Classification and Labeling of Chemicals (GHS) pictograms and signal words. The Eco-Scale also considers waste generated, and its post-treatment, which contributes to the total PP for the method. The Eco-Scale's merit is that the whole process's environmental impact becomes quantifiable, allowing easy comparison between different methods.

This review reports and summarizes the penalty points (PP) for all the surveyed extraction and identification steps for polyphenols from the cacao feedstocks. PPs calculated for (1) pre-treatment and extraction steps, and (2) identification or quantification steps are presented in independent tables. Equation (1) was utilized to estimate the total PP (PP_T) for each step [11,34]:

$$PP_T = PP_C + PP_E \tag{1}$$

where PP_C are the penalty points accrued from the amount and type of chemicals used, and PP_E are those contributed from the consumption of energy.

Table 1 shows the criteria used to determine the PPs from the chemicals used and the consumed energy during each step [11,34].

Table 1. Eco-Scale penalty points (PPs) assigned to reagents/solvents and equipment used in this review as described by [11] for sample pre-treatment, extraction, and analytical methods.

Criteria				Penalty Points (PP) Assigned
Reagent/Solvent Used and Attributes			Equipment Used	
Amount Used (mL or g)	Risk: Number of Pictograms in Label	Risk: Signal Word in Label	Energy Consumption (kWh)	
0	0	No signal word	<0.1	**0**
<10	1	"Warning"	0.1–1.5	**1**
10–100	2	"Hazard"	>1.5	**2**
>100	3			**3**

The PP associated with a particular chemical (i.e., reagent or solvent) employed during a selected process (pre-treatment, extraction, identification, and quantification) were calculated by multiplying the amount used (PP = 0, 1, 2, or 3) by the PP due to the reagent/solvent risk. The reagent/solvent risk was calculated by multiplying the PP associated with the number of pictograms on the chemical label (PP = 0, 1, 2, or 3) by the PP related to the type of signal words displayed on the label (PP = 0, 1, or 2), as shown in Table 1. The pictograms and signal words should comply with the Globally Harmonized System of Classification and Labeling of Chemicals (GHS). The assignment of PP was carried out on a per-pictogram basis, meaning that if two pictograms were present, two PP were assigned. Supplementary Table S1 provides a list of common reagents and solvents often reported in polyphenol extractions and identification, along with their corresponding reagent/solvent risk PP. This information is helpful in selecting greener alternatives to process and characterize cacao and cocoa products.

The primary criterion used to determine the penalty points associated with the use of equipment, as indicated in Table 1, is energy consumption. PP and energy consumption of equipment commonly used for the extraction and identification of polyphenols are provided in Supplementary Table S2.

4. Polyphenols Extraction Methods

Separating polyphenols from vegetal matrices typically involves a sequential process comprising pre-treatment, extraction or soaking, and sample concentration. Pre-treatment involves preparatory techniques aimed at enhancing the polyphenol yield before extraction/soaking of the feedstock. In this review, the term feedstock encompasses all parts of the cacao fruit, including CPHs, beans, cacao shells, and nibs. During extraction or soaking, a solvent is utilized to facilitate the migration of polyphenols from the solid matrix to the liquid phase. Solid residues are then separated from the solvent-containing polyphenol extract through filtration or centrifugation. Sample concentration is frequently attained through solvent evaporation, while commercialization often necessitates extract purification or isolation. However, this review focused primarily on comparing extraction and identification/quantification procedures; therefore, sample concentration and purification were excluded from the analysis.

4.1. Pre-Treatment of the Samples

The selection of suitable pre-treatment steps is critical when dealing with labile compounds such as polyphenols, which degrade, oxidize, polymerize, or form complexes over time. Several pre-treatments used to increase polyphenols' extraction yield from cacao feedstocks have been reported in the literature. These pre-treatments can be categorized based on their primary mechanism into: (i) increasing the surface area exposed to the solvent, (ii) water removal, (iii) lipid removal, and (iv) loosening intracellular compounds in the feedstock.

Size reduction of the feedstock, for example, by milling, increases the surface area exposed to the solvent, thus enhancing polyphenol yields. Milling can be achieved using a centrifugal or a cutting mill [35–37] or by crushing cacao parts with a knife mill [4,37]. Sieving after milling is typically performed to obtain a monodisperse particle-size distribution ranging from 0.5 mm to 1 mm [38–40]. However, friction during milling might increase the temperature, which can promote the complexation of procyanidins with matrix components and hinder their subsequent extraction. Drying the samples before milling can prevent caking, facilitate the breaking of the cell structure [41–43], and preclude undesirable enzymatic reactions [44].

Drying is necessary for raw CPHs or cacao beans due to their high water content. Fermented cocoa beans can be sun-dried, eliminating the need for additional drying. Freeze- and air-drying are the preferred methods to reduce water content, especially for raw beans or pod husks [45–47]. Freeze-drying is less disruptive than air-drying for heat-labile compounds like polyphenols, and air-drying requires longer times and higher temperatures (>100 °C), leading to the degradation of some phenolic compounds. Microwave-, freeze-, and hot-air-drying were compared for CPH [48], with hot-air-drying at 60 °C for 24 h resulting in a reduction of ~60% of total polyphenols, flavanols, and flavonoids compared to both freeze- and microwave-drying.

The defatting process involves soaking the milled–dried solid material in solvent multiple times, leading to higher yields of polyphenols, the targeted analytes [39,42,47]. Commonly used solvents are hexane and petroleum ether, although less common ones include heptane and dichloromethane [41,49,50]. Soxhlet defatting results in greater fat extraction efficiency but requires longer running times and higher energy consumption [26,46]. In contrast, mechanical defatting, such as pressing, does not use a solvent and is more environmentally friendly. Kobori et al. [51] reported that mechanical defatting of cocoa powder improved total polyphenol and procyanidin content by 14% and 19%, respectively, compared to extractions of non-defatted cocoa powder.

Pulsed electric field (PEF) pre-treatments release intracellular compounds, like polyphenols, by permeabilizing the cell membranes. Since PEF is a non-thermal treatment, it does not affect thermolabile compounds like polyphenols. Barbosa-Pereira et al. [28] applied PEF to pre-treat cocoa bean shells immersed in water. These authors showed that by adequately selecting the parameters of the PEF (e.g., duration of pulse and strength), the extractability of polyphenols increased by about 30%.

Table 2 summarizes the various pre-treatment methods utilized for cacao feedstocks, including pod husks, beans, nibs, and shells, along with their respective states (e.g., raw, sun-dried).

Table 2. Summary of the pre-treatment and extraction methods for different feedstock from the *Theobroma cacao* L. fruit and the computed penalty points (PPs) obtained by using Supplementary Tables S1 and S2.

Material	State	Extraction Conditions				PP	Ref.
		Pre-Treatment Step		Extraction Step			
		Reagents	Equipment	Reagents/Solvents	Equipment		
Pod Husk	Raw	-	Electric oven, mill	Methanol	Mixer, shaker, rotary evaporator, fridge	19	[43]
		-	Blender, dryer, microwave, freezer	Acetone, acetic acid, water	Freezer, shaker, centrifuge	17	[48]
Beans	Raw	Water, hexane	Freeze-dryer, grinder, dryer	Water, methanol, acetic acid	US bath, shaker, centrifuge	23	[49]
	Fermented; Sundried	Hexane	Incubator, dryer	Methanol, water	Centrifuge, freezer, shaker	21	[52]
		Petroleum ether	-	Water, acetic acid, acetonitrile	Blender, US bath, centrifuge, heater, fridge	25	[36]
		Hexane	Grinder, centrifuge	Acetone, acetic acid, water	US bath, centrifuge, vortex, freezer	23	[53]
	Fermented; Sundried; Roasted	Dry ice, n-hexane	Mill, fridge	Water; DMSO, acetone, diatomaceous earth	PLE	20	[4]
		Hexane, nitrogen	Mill	Acetone; acetic acid; water	US bath, vortex, centrifuge	19	[39]
Nibs	Fermented; Sundried	Hexane	Shaker	Methanol, water, acetone	US bath, centrifuge, rotary evaporator, thermostatic bath	36	[54]
		n-hexane	Mill, centrifuge, freeze-dryer	Acetone, water	Centrifuge	26	[23]
	Fermented; Sundried; Roasted	Petroleum ether	Grinder, centrifuge, fridge, thermostat	Water	Homogenizer, shaking water bath	10	[55]
		Heptane	Centrifuge, roaster	Acetone, acetic acid, water, nitrogen	Centrifuge PHWE, rotary evaporator, US bath, freezer	34	[49]
		Hexane	Centrifuge grinder, shaker	Acetic acid, acetone, water	Rotary evaporator, centrifuge, freeze-dryer	40	[56]

Table 2. *Cont.*

Material	State	Extraction Conditions				PP	Ref.
		Pre-Treatment Step		Extraction Step			
		Reagents	Equipment	Reagents/Solvents	Equipment		
Shells	Raw	-	Freeze-dryer, blender	Ethanol	US bath, rotary evaporator, thermostatic bath	16	[46]
	Fermented; Sundried; Roasted	Hexane	Grinder, orbital shaker	Water, acetic acid, acetone	Rotary evaporator, centrifuge, vortex; freeze-dryer	38	[57]
		Hexane	Mill, Soxhlet, freeze-dryer	Water	Centrifuge	18	[47]
		-	Grinder	CO_2	SFE, freezer	12	[58]
		-	Grinder	Ethanol	PLE; fridge, rotary evaporator	15	[58]
		-	PEF, mill, grinder, centrifuge, vacuum	Ethanol	Centrifuge, orbital shaker, freezer	18	[29]

The listed penalty points (PP) in Table 2 are the sum of those associated with each pre-treatment and extraction method, calculated using Equation (1) and employing the techniques documented in the scientific literature. The environmental impact assessed using the Eco-Scale based on the information in Table 2 revealed that milling is an efficient and speedy process, with energy consumption penalties from 0 to 2 PP, depending on the equipment employed (Supplementary Table S2). Although freeze- and oven-drying are straightforward steps, they can result in 2 to 4 PPs due to their energy consumption. Defatting, on the other hand, requires solvents such as hexane or petroleum ether, resulting in a substantial increase in PPs (8 to 12 PP) when using between 10 and 100 mL. Pulsed electric field (PEF) is an eco-friendly pre-treatment when using water as the soaking medium, with energy consumption being the only contributor to the PPs, resulting in a single PP [28]. However, substituting methanol for water can increase the PPs to ~12 when using 10 to 100 mL of solvent (Supplementary Table S2). Lower penalty points are always preferred as they indicate a more efficient and cost-effective method. The selection of appropriate pre-treatment methods is crucial as it can significantly impact the final penalty point counts. Therefore, careful consideration and evaluation should be given when deciding which pre-treatment method(s) to utilize in order to retain the efficiency of the extraction with minimal environmental impact.

4.2. Extraction Procedures

The extraction process aims to achieve the highest possible yield of polyphenols by optimizing the processing conditions. Solid–liquid extraction methods have traditionally been used to obtain polyphenols from cacao feedstock. These methods involve immersing the material in a chosen solvent for a specific time. The yield can be increased by facilitating mass transfer through forced convection (i.e., stirring), increasing the temperature, or applying ultrasound during the extraction process. On the other hand, supercritical fluid extraction is gaining popularity due to its high yields and its classification as a green technique, owing to the utilization and recycling of a supercritical fluid. Table 2 compares the conditions reported in the literature for extracting polyphenols from various cacao feedstocks, along with their corresponding penalty points (PP).

Solid–liquid extraction techniques that use solvents other than water are categorized as non-green techniques. A "greener method" incurs fewer PPs compared to other methods. Ideally, a green technique would not incur any PP, but such a technique has not been developed yet. Solvent selection during extraction plays a crucial role in determining the

PPs associated with the technique. Both methanol and ethanol have higher PP values than acetone (6 vs. 4 PP). In contrast, water incurs no penalty points. The pH of the solvents can also contribute to PP, depending on the type and quantity of reagent used. For example, formic acid, acetic acid, and hydrochloric acid are each assigned 4 PP. The volume of solvent used in the extraction process can also significantly impact the reagent PP value. Increasing the solvent volume from 1 to 10 mL can double the latter PP value, as presented in Table 1.

To select an appropriate extraction method, both the extraction yield and the associated penalty points should be considered. Focusing solely on the latter can lead to misleading choices. In some cases, a method with higher PPs may still be more efficient in terms of yield than a method with lower PP. For instance, Nsor-Atindana et al. [59] extracted total polyphenols from cocoa bean shells using water (0 PP), acetone (4 PP), ethanol (6 PP), and methanol (6 PP), obtaining yields of 17.2, 42, 23, and 25 mg eq. g^{-1} (d.w.), respectively. In this case, selecting a method based solely on penalty points may lead to a lower yield since acetone, which has a moderate PP contribution, allowed for doubling the extraction efficacy. Hernández-Hernández et al. [26] compared the effectiveness of five solvent combinations, including water, acidified water, methanol–water 80:20 (v/v), methanol–acidified water, and ethanol–acidified-water, in combination with acetone–water 70:30 (v/v), to extract total polyphenols from raw cacao nibs. The reagent risk penalty points for all the listed solvents were 0, 4, 6, 11, and 14, respectively, and the corresponding yields were 5.7, 9.4, 14.6, 20.4, and 49.5 mg eq. g^{-1} (d.w.). Therefore, a balance between penalty points and yield should be considered when selecting an extraction method, as solely focusing on penalty points can result in an ineffective or non-profitable extraction with low yields or a high-yield method that generates waste requiring additional remedial steps.

Due to its environmentally friendly status, water has been extensively explored as a viable solvent for extracting polyphenols. Camu et al. [35] and Manzano et al. [45] extracted defatted cocoa nibs and cocoa shells, respectively, with water, obtaining modest yields of ~0.025 and 6 mg eq. g^{-1} d.w., respectively. Using water as a solvent is highly desirable when pursuing a greener methodology; however, this often results in lower extraction yields. Consequently, when transitioning to greener alternatives (e.g., switching from methanol to water as an extraction solvent), the steps to increase yield should be considered [26]. The use of physical or mechanical treatments can increase the efficacy of water during the extraction process, as long as the selected treatments result in lower penalty points compared to more efficient solvents.

Among these treatments, ultrasound and pressurized assisted extractions can help release polyphenols from the feedstock into the solvent. Ultrasound-assisted extraction (UAE) applies ultrasound while the feedstock is being soaked in the solvent. It is recognized as a green technology due to its low energy consumption, i.e., $\leq$1.5 kWh (see Table S2), and lack of requirements for high temperature and additional solvents [37,41,46,47,49,54,60,61]. Using ultrasound increases polyphenol extraction efficiencies by ~30% compared to soaking alone [54]. Although UAE can be performed at room temperature, increasing the temperature during the water-soaking step can increase the total polyphenol content by 15% [46].

Pressurized liquid extraction (PLE) is a technique that uses elevated temperatures and pressure to rapidly extract compounds from a sample. The feedstock and solvent are placed in a pressurized vessel, facilitating extraction. When water is used as the extraction solvent, the process is referred to as pressurized hot water extraction (PHWE), sub-critical water extraction, or superheated water extraction [62]. PLE is a low-cost method that achieves high yields without the need for harmful solvents or pre-treatment steps, such as defatting. Plaza et al. [49] compared the efficiency of PHWE at 125 °C with UAE at 30 °C for extracting polyphenols from cocoa nibs, beans, and chocolate. The PHWE method resulted in a ~6-fold increase in total polyphenol yield compared to UAE, particularly for procyanidins in nibs and beans. However, there was no significant difference in yield between the two methods for chocolate samples.

Supercritical fluid extraction (SFE) is a promising alternative to conventional liquid–solid techniques that is gaining increasing attention due to its potential as a green methodology. SFE utilizes a solvent at supercritical conditions, where its pressure and temperature are modulated to achieve the supercritical state, resulting in the separation of the extractant from the feedstock. The use of a supercritical fluid, characterized by gas-like viscosity and diffusivity, facilitates the mass transfer of analytes from the matrix to the fluid and its circulation within confined spaces, combined with a liquid-like density that imparts high solvation power to the fluid [63]. The most commonly used supercritical fluid is CO_2, which presents a low reagent risk (2 PP—See Table S1). SFE also offers the advantage of CO_2 recyclability, making it an eco-friendly method that eliminates the need for waste treatment and minimizes environmental contamination. Other reagents, such as water (0 PP) and ethanol (5 PP), can be added as co-solvents. SFE has demonstrated consistent and superior extraction yields of polyphenols in comparison to other techniques. For instance, the SFE with CO_2 of cocoa beans has been reported to yield 43.3–64.2 mg eq. g^{-1} (d.w.) [59]. In a study by Mazzutti et al. [58], SFE with CO_2 and hexane extraction were compared in terms of total extracted polyphenols from fermented and roasted cocoa shells. Although the yields were similar (4.0 vs. 4.5 mg eq. g^{-1} d.w.), the reagent risk associated with hexane (8 PP) outweighed the low risk of CO_2 (2 PP), which underscores the growing popularity of SFE as an extraction alternative.

Table 2 summarizes not only the type and conditions of the pre-treatments but also those of the extraction methods reported. The utilization of solvents such as methanol, acetone, or acetic acid in extraction procedures can result in elevated PP values when compared to solvents like water or ethanol. Even when using 10–100 mL of such solvents, an increase of 8 to 12 PP can be observed. In addition, the use of sophisticated equipment like rotary evaporators, pressurized vessels, or freeze-dryers incurs a penalty of 2 PP each due to their reagent requirements, which, coupled with their energy consumption, can increase the PP to 3 or more. However, it is crucial to note that selecting an extraction method should not solely be based on the penalty points incurred, as higher PP values may not necessarily reflect higher efficiency in terms of polyphenol yield, as has been already mentioned for the pre-treatments and will be further discussed in this review.

4.3. Optimization of the Extraction Procedure

The quest for environmentally friendly chemical approaches has led to the exploration of two distinct pathways, which are not mutually exclusive. The first approach involves replacing, recycling, or eliminating harmful solvents, reagents, or energy-intensive processes. The second approach optimizes outcomes such as extraction yields by systematically exploring processing conditions using an adequate design-of-experiments (DOE) approach. This strategy can lead to a method that is more efficient and eco-friendlier. The DOE approach consists of two main steps: factor screening and optimization. The response surface methodology (RSM) is commonly used to optimize the factors and their values to achieve the best possible outcome. This optimization technique has been successfully applied to the extraction of polyphenols from cocoa products by identifying suitable pre-treatment conditions [28], selecting appropriate temperature, time, and reagent concentrations during traditional extractions [43,46], and identifying the most effective temperature, pressure, and reagent concentrations in an SFE-CO_2 process [49]. Although significant improvements have been achieved through appropriate DOE and optimization techniques, most previous studies selected extraction conditions based solely on prior reported work, without considering sustainability performance parameters (PP). The integration of PP in the optimization process remains largely unexplored but holds great potential for further improving the eco-friendliness of chemical processes.

5. Polyphenol Identification and Quantification Procedures

Accurately quantifying phenolic compounds is essential for comparing the yields obtained by various extraction methods and determining their selectivity towards different phenolic components. The quantification techniques can be broadly categorized into two groups: spectrophotometric- and chromatographic-based methods. These techniques enable researchers to precisely measure the concentration of phenolic compounds in a sample, thereby facilitating the identification and characterization of these bioactive compounds.

5.1. Spectrophotometric Methods

UV–Vis absorbance spectroscopy is widely utilized for the identification and quantification of polyphenols. This technique is based on the capability of certain compounds, referred to as chromophores, to absorb light. In the absence of natural chromophores in the sample, it is necessary to prepare the sample to elicit a detectable color response by promoting a reaction between the compound of interest, such as a polyphenol, and a suitable chromophore. The observed absorbance is then compared to a calibration curve obtained using a standard reagent, which is used to determine the quantity of polyphenols present in the sample. The polyphenol concentration is usually expressed as milligrams of calibration standard equivalents per gram of sample, usually based on dry weight. Only a small amount of the standard is required for this analysis. However, the use of specific reagents, such as gallic acid and epicatechin, incurs penalty points related to their reagent risk PP, as outlined in the Supplementary Table S1.

Several assays can produce a color cue from polyphenols, including Folin–Ciocalteu (F–C), vanillin, aluminum chloride, 4-(dimethylamino) cinnamaldehyde (DMAC), and acid butanol. However, the F–C assay is the most used due to its simplicity and comprehensive results. Nevertheless, the assay's limited specificity toward polyphenols is a significant drawback since it reacts with several oxidation substrates in the sample, including but not limited to phenols [64,65]. Upon reacting with the F–C reagent in a Na_2CO_3 solution, phenolic compounds form blue complexes, which absorb at 740–765 nm [66]. The kinetics and efficacy of the reaction depend on factors such as reagent volume and concentration, temperature, and environmental light exposure. The total polyphenol content is determined from a standard curve of gallic acid, catechin, or (−)-epicatechin. Ramirez-Sanchez et al. [67] reported that (−)-epicatechin demonstrated more sensitivity than gallic acid and is preferred as a standard since it is present in all cacao and cocoa products. The penalty points associated with this method on the Eco-Scale are relatively low and typically amount to <5 PP, with 1 PP attributed to the amount of reagents (<10 mL), up to 3 PP for reagent risk, and 1 PP for the use of a spectrophotometer.

The quantification of total flavonoids is commonly carried out through the aluminum chloride ($AlCl_3$) colorimetric assay [68]. The extract is typically treated with a solution of sodium nitrite and sodium hydroxide to enhance the method's selectivity, followed by the addition of an $AlCl_3$ solution [69]. The presence of $NaNO_2$ in an alkaline medium allows the determination of rutin, luteolin, and catechins. Conversely, the absence of $NaNO_2$ only permits the quantification of flavonols and luteolin. The absorbance is measured at 500–510 nm, and the reaction time is approximately 15–30 min [70]. Standards such as catechin [28], rutin [40,70], (−)-epicatechin [38], and quercetin [71] are commonly employed. The addition of $NaNO_2$ increases the selectivity of the assay, but also contributes to 6 PPs on the Eco-Scale, resulting in a total of 16 PPs. Thus, it is recommended to avoid using $NaNO_2$ unless its selectivity is required for the analysis.

Total flavanols can be determined using the 4-(dimethylamino) cinnamaldehyde (DMAC) method. This method employs a DMAC reagent that specifically reacts with (−)-epicatechin, (+)-catechin, epigallocatechin, and gallocatechin, producing a green coloration that can be measured at 640 nm [72]. A DMAC in HCl/ethanol solution is utilized, with 10% *w*/*v* of HCl being preferred to achieve shorter reaction times. Replacing ethanol with water as the solvent slows the reaction with DMAC, and thus, it is not recommended [48]. Epicatechin is commonly used as the calibration standard [48,57]. However, it is worth noting that the DMAC method has a relatively high score of 11 PP.

The vanillin assay is a method used to quantify flavonols instead of total polyphenols. Flavonols react with vanillin (2.4% w/v) and methanol under acidic conditions (HCl), and the intensity of the resulting red coloration is measured at a wavelength of 500 nm [73,74]. The absorbance of the derivatized sample is compared to a calibration curve typically prepared with catechin. To increase the stability of the analyte/vanillin complex, the reaction should be protected from light. The reaction should be carried out at a constant temperature of 25 °C for 15 min to minimize variability. Still, this step is not associated with high energy consumption and carries no penalty points. The sensitivity of the method is improved when the hydrochloric acid (HCl) is diluted with methanol instead of water at a ratio of 30% v/v [75]. However, due to the need to optimize the method for better sensitivity, particularly the additional risk associated with the reagents, this method carries 11 PP. Replacing hydrochloric acid with sulphuric acid [42] reduces the PP from 11 to 9 and was found to render a better catalyst [76]. Any reduction in penalty points is considered an improvement, even if it is only a reduction of 2 PP.

When it is necessary to determine the total proanthocyanidin content of an extract, the acid butanol assay is employed. This assay involves dissolving a Fe(III) salt, such as $NH_4Fe(SO_4)_2$ x12 H_2O, in a concentrated HCl–butanol solution (usually in a ratio of 1:4 or 1:6) [23,55]. Under acidic conditions, leucoanthocyanins are depolymerized, and the corresponding anthocyanidins' absorbance is measured at 550 nm [77,78]. Cyanidin is the most commonly used standard in this method. The method has a penalty score of 6 due to the use of low-risk reagents and solvents with a volume of less than 10 mL.

As presented above, the quantification of polyphenols using spectrophotometric techniques requires the absorption of derived compounds using UV–Vis light. Additionally, near-infrared and fluorescence spectroscopy are popular techniques for polyphenol quantification. Near-infrared spectroscopy (NIRS) is a rapid, reliable, and non-destructive technique that does not require sample derivatization or solvent use. Consequently, this technique is considered environmentally friendly, or "green", compared to other spectrophotometric techniques. NIRS can detect specific functional groups within polyphenols at different wavelengths in the near-infrared range (12,500 to 4000 cm^{-1} or 800 to 2500 nm), including methyl, methylene, and ethylene moieties with the first overtone of O–H and –CH stretching vibration, which are assigned to catechins and epicatechins [79,80]. To analyze samples using NIRS, they must be dehydrated and ground beforehand. The penalty points for NIRS are associated with the sampling preparation and actual testing, such as the energy consumption of the drier, grinder, and NIR equipment. However, since these processes are rapid and have low energy requirements, the penalty points for this technique are low. NIRS is suitable for determining the total procyanidin content in cacao beans, which were tested unaltered and showed a content range of 0.6–19 mg eq. g^{-1} d.w [81]. One of the main limitations of this technique is that a reference set should be generated using alternative techniques such as absorbance spectroscopy or HPLC, and the values of the reference set should be entered in the corresponding software along with the collected NIR spectra for each sample. To generate a reliable reference set, 60 to 100 samples are required, as recommended by the manufacturer. However, the environmental impact of this procedure is small, as it only needs to be performed once for similar samples, after which the established database can be used for years without the need to collect information again.

Similar to NIRS, fluorescence spectroscopy (FS) techniques allow determining polyphenol composition without adding reagents or compound derivatization since they rely on the auto-fluorescence of the polyphenols present in the sample. If the sample requires a dilution step of the extract to avoid inner filter effects, this could contribute to additional PP. Ramirez-Sanchez and co-workers [68] used steady-state fluorescence spectroscopy to detect (−)-epicatechin in microsamples from cacao seeds and cocoa products. They compared the results with the Folin–Ciocalteu (F–C) method and found that the polyphenol quantification was lower than that obtained using the reference method (F–C). However, measuring the autofluorescence of the samples is a fast and reliable green methodology with values of 1 PP due to energy use (Supplementary Table S2). Their findings demonstrate the potential of fluorescence spectroscopy for determining polyphenols, specifically (−)-epicatechin, in cacao seeds and cocoa products.

5.2. Chromatographic Methods

Spectrophotometric techniques are useful for easy, quick, and economical screening of extracts from cacao and cocoa products. Still, due to the complexity of the cocoa matrix and lack of specificity, the results may not provide a thorough identification of the present compounds. As an alternative, chromatographic techniques coupled to different detectors permit the separation and identification of individual molecules based on their molecular weight, stereo-chemistry, and polarity, constituting a more accurate technique for evaluating polyphenols from cacao and cocoa products. However, some phenols might react unpredictably with spectrophotometric assay reagents, leading to erroneous results, further emphasizing this method's limitations.

Liquid Chromatography

Liquid chromatography (LC) techniques, including high-performance liquid chromatography (HPLC), ultra-HPLC (UHPLC), and ultra-performance liquid chromatography (UPLC), are often chosen for the identification and quantification of polyphenols. The main differences among them are the required time and solvent consumption. UPLC and UHPLC can attain higher pressures than HPLC, resulting in higher flow rates and shorter runs. Liquid chromatography (LC) incurs penalty points due to the energy equipment consumption and the amount and risk of solvents used.

Analyzing polyphenols in cocoa products requires proper identification and quantification, which can be achieved through a wide range of detectors. The multi-wavelength UV–Vis diode array detector (DAD) coupled with an HPLC system is the most popular used [23,35,43,48,54,55,82,83]. Photodiode array detectors (PDA) have also been utilized to detect specific polyphenols in cocoa bean shells, like 5-caffeoylquinic acid and epicatechin [27]. Fluorescence detectors (FD) have also been used to study cacao products [47,49,51,84], but their main drawback is co-elution of peaks due to a lack of detector resolution. Mass spectroscopy (MS) detection systems are used to overcome co-elution and selectivity problems. Out of the several ionization sources [85], electrospray ionization (ESI) is the most popular for analyzing phenolic compounds in cocoa products. The most widely used analyzers for polyphenol detection are quadrupole (Q) and time-of-flight (TOF). Coupling LC equipment with one mass spectrometer, or up to three, allows for detailed analysis of polyphenols [85–87], as seen in studies analyzing (−)-epicatechin, (+)-catechin, (−)-catechin, epicatechin, catechin, procyanidin B1, procyanidin B2, procyanidin B5, and procyanidin dimers in unroasted and roasted cocoa beans [4]. The use of MS detectors has its advantages, such as sensitivity and mass range, but it also incurs extra costs and energy consumption, resulting in higher PP.

On the other hand, gas chromatography (GC) can also be used to separate and quantify polyphenols. This technique requires an extra step, i.e., compound derivatization (methylation and acetylation), since it can only measure volatiles. Lack of derivatization led to the unsuccessful detection of polyphenols in cocoa bean shells even after several steps of extraction [63]. This step is time-consuming and involves the use of additional reagents that can impact the sustainability of the method.

UPLC coupled with MS detectors is preferred for analyzing polyphenols in cocoa products due to its shorter analysis time, higher peak efficiency, higher resolution, and higher sensitivity than HPLC. A study comparing the efficiency of HPLC with NP column and UPLC with RP column coupled to a triple quadrupole found that UPLC [24] reduced analysis time by a factor of seven (from 80 to 12.5 min) and allowed the detection and quantification of oligomers from trimer to monomers, which was not possible with HPLC. On the other hand, the environmental impact of chromatographic methods is directly related to the duration of the analysis process, the solvent selection, and the quantity used. The amount of waste generated is proportional to the flow, and longer processes are often used due to the type of equipment or column properties. Therefore, selecting a method with a lower flow and shorter duration can significantly reduce the environmental impact.

Table 3 shows the PPs associated with the identification and quantification of polyphenols based on selected procedures reported in the literature.

Table 3. Summary of the identification and quantification methods for different feedstock from *Theobroma cacao* L. and the computed penalty points obtained by using Supplementary Tables S1 and S2.

Material	State	Identification and Quantification of Polyphenols				PP	Ref.
		Technique *1	Reagents/Solvents *2	Standard	Equipment		
Pod Husk	Raw	S	F–C	Gallic acid	UV–Vis spectrometer	4	[43,48]
		C	MeOH, water, acetonitrile, H_3PO_4	Gallic, vanillic, caffeic, ferulic, ellagic acid	HPLC–DAD	27	[43]
		C	Formic acid, methanol, water	Gallic acid, catechin, quercetin, epicatechin, p-coumaric A., protocatechuic A	HPLC–DAD	18	[48]
Beans	Raw	S	F–C	Gallic acid	UV–Vis spectrometer	4	[79]
			-	-	NIRS	4	[79]
			F–C	Ferulic acid	UV–Vis spectrometer	5	[88]
		C	-	-	Electronic tongue	4	[79]
	Fermented, Sundried	S	F–C	Gallic acid	UV–Vis spectrometer	4	[52,53]
			F–C	Epicatechin	UV–Vis spectrometer	4	[36]
		C	Water, acetonitrile, acetic acid	Epicatechin, catechin	HPLC–FLD	26	[36]
			Dichloromethane, water, methanol, acetic acid	Epicatechin	HPLC–UV	24	[53]
	Fermented, Sundried, Roasted	S	F–C	Gallic acid	UV–Vis spectrometer	4	[39]
		C	Borate buffer, water, hydroxypropyl-y-cyclodextrin, sodium hydroxide, acetonitrile, formic acid	Epicatechin, catechin	UHPLC–UV–QqQ	24	[4]
			Acetic acid, methanol	Epicatechin, catechin, catechin gallate, gallocatechin, epigallocatechin	Capillary electrophoresis	20	[39]
			Acetic acid, methanol		HPLC–DAD	26	[39]
			Dichloromethane, methanol, water, acetic acid		HPLC–UV–QqQ	25	[39]

Table 3. *Cont.*

Material	State	Identification and Quantification of Polyphenols				PP	Ref.
		Technique *1	Reagents/Solvents *2	Standard	Equipment		
Nibs	Fermented, Sundried	S	F–C	Gallic acid	UV–Vis spectrometer	4	[54]
			F–C	Epicatechin	UV–Vis spectrometer	4	[38]
		C	Water, formic acid, ethanol, acetonitrile	Epicatechin, catechin	HPLC–DAD	20	[54]
			Ethyl acetate, butanol, water, 2-propanol, Sephadex LH-20, propanol, acetone	Catechin, epicatechin, PA B2, B3, B4, PA C1, o-arabinoside, cinnamtannin A2, quercetin, quercetin-3-o-glycoside	Semi-preparative SCPC, HPLC–ESI–Q	52	[38]
	Fermented, Sundried, Roasted	S	F–C	Gallic acid	UV–Vis spectrometer	4	[49,55]
		C	Water, formic acid, methanol, acetonitrile	Catechin, epicatechin, procyanidin B1, Procyanidin B2	HPLC–DAD	20	[55]
			Ammonium formate, acetonitrile, methanol, nitrogen, formic acid, acetic acid, sodium formate	Catechin, epicatechin, procyanidin B2	HPLC–DAD–ECD–CAD, HPLC–DAD–MS, HPLC–FLD	42	[49]
Bean Shells	Raw	S	Aluminum chloride	Rutin	UV–Vis spectrometer	11	[46]
		C	Ethanol, water, formic acid, acetonitrile	Procyanidin B2, epicatechin	UPHLC–Q–TOF	22	[46]
	Fermented, Sundried, Roasted	S	F–C	Gallic acid	UV–Vis spectrometer	4	[28,45,58]
			Ethanol, HCl, DMAC	Epicatechin	UV–Vis Spectrometer	12	[57]
		C	Water, methanol, formic acid	Procyanidin B2, epicatechin, catechin, theobromine, caffeine	UPLC–ESI–QqQ	12	[57]
			Helium		GC–MS	4	[58]
			Water, formic acid, methanol	5-caffeoylquinic acid, epicatechin, caffeine, theobromine	HPLC–PDA	18	[28]

*1 S = spectrophotometric; C = chromatographic. *2 F–C= Folin–Ciocalteau reagents.

In general, UV–Vis spectrophotometric techniques using Folin–Ciocalteau reagents and gallic acid as a standard resulted in 4 PP, which is relatively low. Resourcing to more specific and comprehensive methods, such as the chromatographic technique using HPLC–DAD with methanol, water, acetonitrile, and H_3PO_4 as mobile phases, significantly increases the environmental impact (27 PP); robustness and comprehensiveness come at a cost. The selection of the mobile phases for chromatographic techniques, such as HPLC–FLD with water, acetonitrile, and acetic acid, only has a mild effect on the sustainability of the method, obtaining 26 PP. Table 3 shows that all the chromatographic techniques reported tend to have higher penalty points than the spectrophotometric techniques, regardless of their type or operating wavelengths (e.g., NIR, UV–Vis, or fluorescence). This can be attributed to the length of the measurement (minutes for spectroscopy vs. hours for chromatography) and the requirement of reagents other than the standards. Hence, mindfully aiming for the resolution and specificity required during the identification and quantification of polyphenols should be considered when choosing a method, as they can impact the overall cost and sustainability of the analysis.

6. Polyphenol Yield Comparison

Table 4 allows for a comparison of the environmental impact and the yield of the technique.

Table 4. Summary of penalty points for different cacao feedstock due to extraction, analysis, and total polyphenol yield measured using non-destructive techniques.

Material	State	PP		Yield	Ref.
		Extraction	Analysis	Total Phenol Content (meq g^{-1})	
Pod Husk	Raw	19	4	151	[43]
		17	4	5–20	[48]
Beans	Raw	7	5	0.08–0.12	[88]
	Fermented, Sundried	21	4	0.6–6	[52]
		25	4	0.2–0.3	[35]
		23	4	40–120	[61]
	Fermented, Sundried, Roasted	16	4	30–70	[39]
Nibs	Fermented, Sundried	36	4	140	[54]
		26	4	80–120	[38]
	Fermented, Sundried, Roasted	34	4	10–35	[49]
Bean Shells	Raw	16	11	7.4	[46]
	Fermented, Sundried, Roasted	48	12	1–4	[57]
		25	4	43	[58]
		18	4	6	[47]
		18	4	21–55	[28]

The availability of extraction techniques based on significantly different principles, which also require different levels of pre-treatment, determined that the PP associated with the extraction steps covered a broader range, from 7 to 48, than the analytical steps. The identification and quantification of the polyphenols are often performed by a more limited pool of techniques, which also is reflected in their PP, which ranged from 4–12. Although in some studies, a high penalty point value for extraction corresponded to a high polyphenol yield, this was not always the rule. For instance, the highest yields (i.e., 140 meq g^{-1}) were obtained from fermented and sundried nibs, with 36 PP for the extraction. However, a similar yield (about 120 meq g^{-1}) could be achieved using extractions with less environmental impact, e.g., 23 PP for fermented and sundried beans. This suggests that the correlation between penalty points and polyphenol yield is not always straightforward and that a mindful selection of extraction conditions and feedstocks can result in high yields. It should be noted that there can be other factors at play, such as the nature of the polyphenols in each material and their accessibility for extraction. Therefore, further studies are required to comprehensively understand the relationship between extraction efficiency and polyphenol yield. Table 4 is intended to guide the reader in understanding the role that extraction and identification might have on the environment to inform future decisions.

7. Conclusions

Cacao feedstocks are abundant sources of polyphenols known to possess numerous health benefits. To achieve efficient extraction and accurate identification with minimal environmental impact, careful consideration of pre-treatment, extraction, and analytical determination conditions is required. Pre-treatment methods, such as milling and drying, as well as the use of pulsed electric fields, have been shown to effectively release intracellular compounds, depending on the specific feedstock and targeted analytes. Pre-treatments that are rapid and involve low energy consumption are preferred from a sustainability perspective, while other techniques, such as defatting, which requires solvents and their posterior elimination, have a considerable environmental impact and should not be prioritized. The transition from traditional solid–liquid extraction methods to supercritical fluid extraction is gaining popularity and should be supported due to its high yields and green classification. Using water and only one solvent (ethanol) together with some mechanical pre-treatments can lead to only 10 PP for fermented, sundried, and roasted nibs [77]. Using supercritical fluid extraction with CO_2 can lead to 12 PP for fermented and sundried cocoa shells [64]. When more than two organic solvents are used, the PP normally range from 17 to 40 (Table 2).

Polyphenol identification and quantification are commonly performed using spectrophotometric methods, with UV–Vis absorbance spectroscopy being widely used. Several assays, including the most popular Folin–Ciocalteu (F–C), can be successfully applied to quantify total polyphenols. Although the penalty points derived from the reagent risks can affect the environmental standing of the method, in general, the penalty points associated with the spectrometric methods are relatively low, ranging from 4 to 12 PP (Table 4).

For more accurate identification of polyphenols in cacao and cocoa products, chromatographic techniques such as HPLC, UHPLC, and UPLC, coupled with different detectors or mass spectrometry, are the preferred methods. The selection of solvents and their separation conditions will play a crucial role in determining the environmental impact of the technique. It is noteworthy that the analysis duration, solvent selection, and the quantity employed carry substantial environmental implications. Consequently, it is imperative to consider the associated penalty points while selecting the appropriate method. The PPs associated with the chromatographic techniques ranged from 12 to 52 (Table 4).

The PP cannot be analyzed on their own, as the polyphenol yield is the goal of the extraction and identification/quantification. In this regard, the higher yields were obtained when methanol was used to extract from cacao husk [43], or a combination of hexane, acetone, and acetic acid to extract from the beans [62] and a combination of hexane, methanol, and acetone to extract from the nibs [54]. Those three methods carried PP of 19, 23, and 36, respectively, with a yield of 151, 40–120, and 140 meq g^{-1}, respectively. The highest yield from cacao shells was obtained when a combination of ethanol and water was used in the extraction, giving 19 PPs and a yield of 21–55 meq g^{-1}.

Mazzutti et al. [58] showed the lowest PPs in the extraction, as it used a combination of water and ethanol, but it gave a polyphenol yield of 43 meq g^{-1} in the cacao husks. The tables presented herein (Tables 2–4 and Supplementary Tables S1 and S2) serve the purpose of consolidating dispersed information found in the literature. This compilation aids readers in the judicious selection of pre-treatment, extraction, and analytical methodologies, contingent upon the unique attributes of the chosen feedstock and the specific analytes of interest. Additionally, these tables provide insight into penalty point considerations and total polyphenol yields. The users are encouraged to deliberately choose extraction and analysis procedures and conditions to attain maximal yield while concurrently mitigating environmental repercussions. The food industry is always looking for new materials that can be used to enhance already existing food products or to develop new ones. Cacao byproducts, such as husks and shells, are unavoidable in the chocolate industry. Polyphenols obtained from parts of the cacao fruit should be viewed as possible food enrichers. We believe that this review can help the industry make educated choices when the time comes to enrich their products.

Supplementary Materials: The following supporting information can be downloaded at: https://www.mdpi.com/article/10.3390/foods13152397/s1, Figure S1: (A) Schematic view of the flavonoid basic structure; (B) Main anthocyanins found in cacao; (C) Commonly found flavonols in cacao; (D) Common flavanols found in cacao; (E) Type B Procyanidins commonly found in cacao; (F) Structural formulas of phenolic acids commonly found in cacao; Table S1. Reagent/solvents reported and their penalty points calculated as proposed by the Eco-Scale in which the PP from the pictogram are multiplied by the PP attributed to the hazard word; Table S2. Equipment reported, showing kWh usage for the listed working conditions from which the energy consumption penalty points (PP) were calculated using the criteria in Table 2.

Author Contributions: Conceptualization: J.M.S., F.P. and M.G.C.; formal analysis, J.M.S., F.P. and Y.H.; resources, M.G.C.; data curation, J.M.S., F.P. and Y.H.; writing—original draft preparation, J.M.S. and F.P.; writing—review and editing, J.M.S., F.P., C.E.B. and M.G.C., supervision, F.P., C.E.B. and M.G.C.; funding acquisition, C.E.B. and M.G.C. All authors have read and agreed to the published version of the manuscript.

Funding: This research was funded by NSERC for the Discovery Program RGPIN-2019-04995.

Institutional Review Board Statement: Not applicable.

Informed Consent Statement: Not applicable.

Data Availability Statement: No new data were created or analyzed in this study. Data sharing is not applicable to this article.

Acknowledgments: J.M.S. thanks Consejo Nacional de Investigaciones Científicas y Técnicas, Argentina (CONICET) for his fellowship, and the Emerging Leaders in the Americas Program (ELAP) for the financial support of his stay in Canada. C.E.B. thanks the CONICET. F.P. thanks the Food Science Department at the University of Guelph. M.G.C. thanks NSERC for the Discovery Program RGPIN-2019-04995 and the Arrell Food Institute at the University of Guelph.

Conflicts of Interest: The authors declare no conflicts of interest.

References

1. Afoakwa, E.O.; Paterson, A.; Fowler, M.; Ryan, A. Flavor Formation and Character in Cocoa and Chocolate: A Critical Review. *Crit. Rev. Food Sci. Nutr.* **2008**, *48*, 840–857. [CrossRef] [PubMed]
2. Aprotosoaie, A.C.; Luca, S.V.; Miron, A. Flavor Chemistry of Cocoa and Cocoa Products-An Overview. *Compr. Rev. Food Sci. Food Saf.* **2016**, *15*, 73–91. [CrossRef] [PubMed]
3. Jalil, A.M.M.; Ismail, A. Polyphenols in Cocoa and Cocoa Products: Is There a Link between Antioxidant Properties and Health? *Molecules* **2008**, *13*, 2190–2219. [CrossRef] [PubMed]
4. Kothe, L.; Zimmermann, B.F.; Galensa, R. Temperature Influences Epimerization and Composition of Flavanol Monomers, Dimers and Trimers during Cocoa Bean Roasting. *Food Chem.* **2013**, *141*, 3656–3663. [CrossRef] [PubMed]
5. Misnawi; Jinap, S.; Jamilah, B.; Nazamid, S. Effects of Incubation and Polyphenol Oxidase Enrichment on Colour, Fermentation Index, Procyanidins and Astringency of Unfermented and Partly Fermented Cocoa Beans. *Int. J. Food Sci. Technol.* **2003**, *38*, 285–295. [CrossRef]
6. Albertini, B.; Schoubben, A.; Guarnaccia, D.; Pinelli, F.; Della Vecchia, M.; Ricci, M.; Di Renzo, G.C.; Blasi, P. Effect of Fermentation and Drying on Cocoa Polyphenols. *J. Agric. Food Chem.* **2015**, *63*, 9948–9953. [CrossRef]
7. Trognitz, B.; Cros, E.; Assemat, S.; Davrieux, F.; Forestier-Chiron, N.; Ayestas, E.; Kuant, A.; Scheldeman, X.; Hermann, M. Diversity of Cacao Trees in Waslala, Nicaragua: Associations between Genotype Spectra, Product Quality and Yield Potential. *PLoS ONE* **2013**, *8*, e54079. [CrossRef] [PubMed]
8. Hurst, W.J.; Krake, S.H.; Bergmeier, S.C.; Payne, M.J.; Miller, K.B.; Stuart, D.A. Impact of Fermentation, Drying, Roasting and Dutch Processing on Flavan-3-Ol Stereochemistry in Cacao Beans and Cocoa Ingredients. *Chem. Cent. J.* **2011**, *5*, 53. [CrossRef]
9. Cienfuegos-Jovellanos, E.; Del Mar Quiñones, M.; Muguerza, B.; Moulay, L.; Miguel, M.; Aleixandre, A. Antihypertensive Effect of a Polyphenol-Rich Cocoa Powder Industrially Processed to Preserve the Original Flavonoids of the Cocoa Beans. *J. Agric. Food Chem.* **2009**, *57*, 6156–6162. [CrossRef] [PubMed]

10. Payne, M.J.; Hurst, W.J.; Miller, K.B.; Rank, C.; Stuart, D.A. Impact of Fermentation, Drying, Roasting, and Dutch Processing on Epicatechin and Catechin Content of Cacao Beans and Cocoa Ingredients. *J. Agric. Food Chem.* **2010**, *58*, 10518–10527. [CrossRef] [PubMed]
11. Gałuszka, A.; Migaszewski, Z.M.; Konieczka, P.; Namieśnik, J. Analytical Eco-Scale for Assessing the Greenness of Analytical Procedures. *Trend Anal. Chem.* **2012**, *37*, 61–72. [CrossRef]
12. Santos-Buelga, C.; Scalbert, A. Proanthocyanidins and Tannin-like Compounds - Nature, Occurrence, Dietary Intake and Effects on Nutrition and Health. *J. Sci. Food Agric.* **2000**, *80*, 1094–1117. [CrossRef]
13. King, R.E.; Bomser, J.A.; Min, D.B. Bioactivity of Resveratrol. *Compr. Rev. Food Sci. Food Saf.* **2006**, *5*, 65–70. [CrossRef]
14. Hu, S.J.; Kim, B.Y.; Baik, M.Y. Physicochemical Properties and Antioxidant Capacity of Raw, Roasted and Puffed Cacao Beans. *Food Chem.* **2016**, *194*, 1089–1094. [CrossRef] [PubMed]
15. Khan, N.; Nicod, N.M. Biomarkers of Cocoa Consumption. In *Chocolate and Health*; Paoletti, R., Poli, A., Conti, A., Visioli, F., Eds.; Springer Science & Business Media: Berlin, Germany, 2012; pp. 33–40.
16. Tsao, R. Chemistry and Biochemistry of Dietary Polyphenols. *Nutrients* **2010**, *2*, 1231–1246. [CrossRef] [PubMed]
17. Sanbongi, C.; Osakabe, N.; Natsume, M.; Takizawa, T.; Gomi, S.; Osawa, T. Antioxidative Polyphenols Isolated from *Theobroma cacao*. *J. Agric. Food Chem.* **1998**, *46*, 454–457. [CrossRef]
18. Tomás-Barbéran, F.; Borges, G.; Crozier, A. Phytochemicals in Cocoa and Flavan-3-Ol Bioavailability. In *Teas, Cocoa and Coffee: Plant Secondary Metabolites and Health*; Wiley-Blackwell: Hoboken, NJ, USA, 2011; pp. 193–217, ISBN 9781444334418.
19. Fayeulle, N.; Vallverdu-Queralt, A.; Meudec, E.; Hue, C.; Boulanger, R.; Cheynier, V.; Sommerer, N. Characterization of New Flavan-3-Ol Derivatives in Fermented Cocoa Beans. *Food Chem.* **2018**, *259*, 207–212. [CrossRef] [PubMed]
20. Oracz, J.; Zyzelewicz, D.; Nebesny, E. The Content of Polyphenolic Compounds in Cocoa Beans (*Theobroma cacao* L.), Depending on Variety, Growing Region, and Processing Operations: A Review. *Crit. Rev. Food Sci. Nutr.* **2015**, *55*, 1176–1192. [CrossRef] [PubMed]
21. Niemenak, N.; Rohsius, C.; Elwers, S.; Omokolo Ndoumou, D.; Lieberei, R. Comparative Study of Different Cocoa (*Theobroma cacao* L.) Clones in Terms of Their Phenolics and Anthocyanins Contents. *J. Food Compos. Anal.* **2006**, *19*, 612–619. [CrossRef]
22. Calderón, A.I.; Wright, B.J.; Hurst, W.J.; Van Breemen, R.B. Screening Antioxidants Using LC-MS: Case Study with Cocoa. *J. Agric. Food Chem.* **2009**, *57*, 5693–5699. [CrossRef] [PubMed]
23. Pedan, V.; Fischer, N.; Rohn, S. Extraction of Cocoa Proanthocyanidins and Their Fractionation by Sequential Centrifugal Partition Chromatography and Gel Permeation Chromatography. *Anal. Bioanal. Chem.* **2016**, *408*, 5905–5914. [CrossRef] [PubMed]
24. Cheynier, V. Polyphenols in Foods Are More Complex than Often Thought. *Am. J. Clin. Nutr.* **2005**, *81*, 223S–229S. [CrossRef] [PubMed]
25. Ortega, N.; Romero, M.P.; Macià, A.; Reguant, J.; Anglès, N.; Morelló, J.R.; Motilva, M.J. Comparative Study of UPLC-MS/MS and HPLC-MS/MS to Determine Procyanidins and Alkaloids in Cocoa Samples. *J. Food Compos. Anal.* **2010**, *23*, 298–305. [CrossRef]
26. Hernández-Hernández, C.; Morales-Sillero, A.; Fernández-Bolaños, J.; Bermúdez-Oria, A.; Morales, A.A.; Rodríguez-Gutiérrez, G. Cocoa Bean Husk: Industrial Source of Antioxidant Phenolic Extract. *J. Sci. Food Agric.* **2019**, *99*, 325–333. [CrossRef]
27. Papillo, V.A.; Locatelli, M.; Travaglia, F.; Bordiga, M.; Garino, C.; Coïsson, J.D.; Arlorio, M. Cocoa Hulls Polyphenols Stabilized by Microencapsulation as Functional Ingredient for Bakery Applications. *Food Res. Int.* **2019**, *115*, 511–518. [CrossRef] [PubMed]
28. Barbosa-Pereira, L.; Guglielmetti, A.; Zeppa, G. Pulsed Electric Field Assisted Extraction of Bioactive Compounds from Cocoa Bean Shell and Coffee Silverskin. *Food Bioproc. Technol.* **2018**, *11*, 818–835. [CrossRef]
29. De Taeye, C.; Eyamo Evina, V.J.; Caullet, G.; Niemenak, N.; Collin, S. Fate of Anthocyanins through Cocoa Fermentation. Emergence of New Polyphenolic Dimers. *J. Agric. Food Chem.* **2016**, *64*, 8876–8885. [CrossRef] [PubMed]
30. Tomas-Barberán, F.A.; Cienfuegos-Jovellanos, E.; Marín, A.; Muguerza, B.; Gil-Izquierdo, A.; Cerdá, B.; Zafrilla, P.; Morillas, J.; Mulero, J.; Ibarra, A.; et al. A New Process to Develop a Cocoa Powder with Higher Flavonoid Monomer Content and Enhanced Bioavailability in Healthy Humans. *J. Agric. Food Chem.* **2007**, *55*, 3926–3935. [CrossRef] [PubMed]
31. Jonfia-Essien, W.A.; Navarro, N. Effect of Storage Management on Free Fatty Acid Content in Dry Cocoa Beans. In Proceedings of the 10th International Working Conference on Stored Product Protection, Estoril, Portugal, 27 June 2008.
32. Mazor Jolić, S.; Radojčic Redovnikovic, I.; Marković, K.; Ivanec Šipušić, D.; Delonga, K. Changes of Phenolic Compounds and Antioxidant Capacity in Cocoa Beans Processing. *Int. J. Food Sci. Technol.* **2011**, *46*, 1793–1800. [CrossRef]
33. Keith, L.H.; Gron, L.U.; Young, J.L. Green Analytical Methodologies. *Chem. Rev.* **2007**, *107*, 2695–2708. [CrossRef] [PubMed]
34. Gałuszka, A.; Migaszewski, Z.; Namieśnik, J. The 12 Principles of Green Analytical Chemistry and the Significance: Mnemonic of Green Analytical Practices. *Trends Anal. Chem.* **2013**, *50*, 78–84. [CrossRef]
35. Bordiga, M.; Locatelli, M.; Travaglia, F.; Coïsson, J.D.; Mazza, G.; Arlorio, M. Evaluation of the Effect of Processing on Cocoa Polyphenols: Antiradical Activity, Anthocyanins and Procyanidins Profiling from Raw Beans to Chocolate. *Int. J. Food Sci. Technol.* **2015**, *50*, 840–848. [CrossRef]
36. Camu, N.; De Winter, T.; Addo, S.K.; Takrama, J.S.; Bernaert, H.; De Vuyst, L. Fermentation of Cocoa Beans: Influence of Microbial Activities and Polyphenol Concentrations on the Flavour of Chocolate. *J. Sci. Food Agric.* **2008**, *88*, 2288–2297. [CrossRef]
37. Grillo, G.; Boffa, L.; Binello, A.; Mantegna, S.; Cravotto, G.; Chemat, F.; Dizhbite, T.; Lauberte, L.; Telysheva, G. Analytical Dataset of Ecuadorian Cocoa Shells and Beans. *Data Brief.* **2019**, *22*, 56–64. [CrossRef] [PubMed]

38. Vriesmann, L.C.; de Mello Castanho Amboni, R.D.; De Oliveira Petkowicz, C.L. Cacao Pod Husks (*Theobroma cacao* L.): Composition and Hot-Water-Soluble Pectins. *Ind. Crops Prod.* **2011**, *34*, 1173–1181. [CrossRef]
39. Ioannone, F.; Di Mattia, C.D.; De Gregorio, M.; Sergi, M.; Serafini, M.; Sacchetti, G. Flavanols, Proanthocyanidins and Antioxidant Activity Changes during Cocoa (*Theobroma cacao* L.) Roasting as Affected by Temperature and Time of Processing. *Food Chem.* **2015**, *174*, 256–262. [CrossRef] [PubMed]
40. Karim, A.A.; Azlan, A.; Amin, I.; Hashim, P.; Abdullah, N.A. Antioxidant Properties of Cocoa Pods and Shells. *Malays. Cocoa J.* **2014**, *8*, 49–56.
41. D'Souza, R.N.; Grimbs, S.; Behrends, B.; Bernaert, H.; Ullrich, M.S.; Kuhnert, N. Origin-Based Polyphenolic Fingerprinting of *Theobroma cacao* in Unfermented and Fermented Beans. *Food Res. Int.* **2017**, *99*, 550–559. [CrossRef] [PubMed]
42. Carrillo, L.C.; Londoño, J.; Gil, A. Comparison of Polyphenol, Methylxanthines and Antioxidant Activity in *Theobroma cacao* Beans from Different Cocoa-Growing Areas in Colombia. *Food Res. Int.* **2014**, *60*, 273–280. [CrossRef]
43. Teboukeu, G.B.; Djikeng, F.T.; Klang, M.J.; Ndomou, S.H.; Karuna, M.S.L.; Womeni, H.M. Polyphenol Antioxidants from Cocoa Pods: Extraction Optimization, Effect of the Optimized Extract, and Storage Time on the Stability of Palm Olein during Thermoxidation. *J. Food Process Preserv.* **2018**, *42*, 1–11. [CrossRef]
44. Dávila, J.A.; Hernandez, V.; Rosenberg, M.; Cardona, C.A. Design of Processes for Supercritical Extraction and Microencapsulation of Antioxidants from Fruits. In *Handbook of Supercritical Fluids, Fundamentals, Properties and Applications*; Osborne, J., Ed.; Nova Science Publisher: New Dehli, India, 2014; pp. 229–256.
45. Manzano, P.; Hernández, J.; Quijano-Avilés, M.; Barragán, A.; Chóez-Guaranda, I.; Viteri, R.; Valle, O. Polyphenols Extracted from *Theobroma cacao* Waste and Its Utility as Antioxidant. *Emir. J. Food Agric.* **2017**, *29*, 45–50. [CrossRef]
46. Yusof, A.H.M.; Gani, S.S.A.; Zaidan, U.H.; Halmi, M.I.E.; Zainudin, B.H. Optimization of an Ultrasound-Assisted Extraction Condition for Flavonoid Compounds from Cocoa Shells (*Theobroma cacao*) Using Response Surface Methodology. *Molecules* **2019**, *24*, 711. [CrossRef] [PubMed]
47. Payne, M.J.; Hurst, W.J.; Stuart, D.A.; Ou, B.; Fan, E.; Ji, H.; Kou, Y. Determination of Total Procyanidins in Selected Chocolate and Confectionery Products Using DMAC. *J. AOAC Int.* **2010**, *93*, 89–96. [CrossRef] [PubMed]
48. Valadez-Carmona, L.; Plazola-Jacinto, C.P.; Hernández-Ortega, M.; Hernández-Navarro, M.D.; Villarreal, F.; Necoechea-Mondragón, H.; Ortiz-Moreno, A.; Ceballos-Reyes, G. Effects of Microwaves, Hot Air and Freeze-Drying on the Phenolic Compounds, Antioxidant Capacity, Enzyme Activity and Microstructure of Cacao Pod Husks (*Theobroma cacao* L.). *Innov. Food Sci. Emer. Technol.* **2017**, *41*, 378–386. [CrossRef]
49. Plaza, M.; Oliveira, D.; Nilsson, A.; Turner, C. Green and Efficient Extraction Method to Determine Polyphenols in Cocoa and Cocoa Products. *Food Anal. Methods* **2017**, *10*, 2677–2691. [CrossRef]
50. Zhong, J.L.; Muhammad, N.; Gu, Y.C.; Yan, W.D. A Simple and Efficient Method for Enrichment of Cocoa Polyphenols from Cocoa Bean Husks with Macroporous Resins Following a Scale-up Separation. *J. Food Eng.* **2019**, *243*, 82–88. [CrossRef]
51. Kobori, K.; Maruta, Y.; Mineo, S.; Shigematsu, T.; Hirayama, M. Polyphenol-Retaining Decaffeinated Cocoa Powder Obtained by Supercritical Carbon Dioxide Extraction and Its Antioxidant Activity. *Foods* **2013**, *2*, 462–477. [CrossRef] [PubMed]
52. Vázquez-Ovando, A.; Molina-Freaner, F.; Nuñez-Farfán, J.; Betancur Ancona, D.; Salvador-Figueroa, M. Classification of Cacao Beans (*Theobroma cacao* L.) of Southern Mexico Based on Chemometric Analysis with Multivariate Approach. *Eur. Food Res. Technol.* **2015**, *240*, 1117–1128. [CrossRef]
53. Di Mattia, C.; Martuscelli, M.; Sacchetti, G.; Scheirlinck, I.; Beheydt, B.; Mastrocola, D.; Pittia, P. Effect of Fermentation and Drying on Procyanidins, Antiradical Activity and Reducing Properties of Cocoa Beans. *Food Bioproc. Technol.* **2013**, *6*, 3420–3432. [CrossRef]
54. Quiroz-Reyes, C.N.; Aguilar-Mendez, M.A.; Ramırez-Ortız, M.E.; Ronquillo-De Jesus, E. Comparative Study of Ultrasound and Maceration Techniques for the Extraction of the Polyphenols from Cocoa Beans. *Rev. Mex. Ing. Quim.* **2013**, *12*, 11–18.
55. Quiroz-Reyes, C.N.; Fogliano, V. Design Cocoa Processing towards Healthy Cocoa Products: The Role of Phenolics and Melanoidins. *J. Funct. Foods* **2018**, *45*, 480–490. [CrossRef]
56. Ortega, N.; Romero, M.P.; MacIà, A.; Reguant, J.; Anglès, N.; Morelló, J.R.; Motilva, M.J. Obtention and Characterization of Phenolic Extracts from Different Cocoa Sources. *J. Agric. Food Chem.* **2008**, *56*, 9621–9627. [CrossRef] [PubMed]
57. Okiyama, D.C.G.; Soares, I.D.; Cuevas, M.S.; Crevelin, E.J.; Moraes, L.A.B.; Melo, M.P.; Oliveira, A.L.; Rodrigues, C.E.C. Pressurized Liquid Extraction of Flavanols and Alkaloids from Cocoa Bean Shell Using Ethanol as Solvent. *Food Res. Int.* **2018**, *114*, 20–29. [CrossRef] [PubMed]
58. Mazzutti, S.; Rodrigues, L.G.G.; Mezzomo, N.; Venturi, V.; Ferreira, S.R.S. Integrated Green-Based Processes Using Supercritical CO_2 and Pressurized Ethanol Applied to Recover Antioxidant Compouds from Cocoa (*Theobroma cacao*) Bean Hulls. *J. Supercrict Fluids* **2018**, *135*, 52–59. [CrossRef]
59. Nsor-Atindana, J.; Zhong, F.; Mothibe, K.J.; Bangoura, M.L.; Lagnika, C. Quantification of Total Polyphenolic Content and Antimicrobial Activity of Cocoa (*Theobroma cacao* L.) Bean Shells. *Pak. J. Nutr.* **2012**, *11*, 574–579. [CrossRef]
60. Grillo, G.; Boffa, L.; Binello, A.; Mantegna, S.; Cravotto, G.; Chemat, F.; Dizhbite, T.; Lauberte, L.; Telysheva, G. Cocoa Bean Shell Waste Valorisation; Extraction from Lab to Pilot-Scale Cavitational Reactors. *Food Res. Int.* **2019**, *115*, 200–208. [CrossRef] [PubMed]

61. Di Mattia, C.; Martuscelli, M.; Sacchetti, G.; Beheydt, B.; Mastrocola, D.; Pittia, P. Effect of Different Conching Processes on Procyanidin Content and Antioxidant Properties of Chocolate. *Food Res. Int.* **2014**, *63*, 367–372. [CrossRef]
62. Mustafa, A.; Turner, C. Pressurized Liquid Extraction as a Green Approach in Food and Herbal Plants Extraction: A Review. *Anal. Chim. Acta* **2011**, *703*, 8–18. [CrossRef] [PubMed]
63. Shi, J.; Xue, S.J.; Ma, Y.; Jiang, Y.; Ye, X.; Yu, D. Green Separation Technologies in Food Processing: Supercritical-CO_2 Fluid and Subcritical Water Extraction. In *Green Technologies in Food Production and Processing—Food Engineering Series*; Boye, J.I., Arcand, Y., Eds.; Springer: Boston, MA, USA, 2012; pp. 273–294.
64. Singleton, V.L.; Orthofer, R.; Lamuela-Raventos, R.M. Analysis of Total Phenols and Other Oxidation Substrates and Antioxidants by Means of Folin-Coicalteu Reagent. *Methods Enzymol.* **1999**, *299*, 152–178.
65. Sánchez-Rangel, J.C.; Benavides, J.; Heredia, J.B.; Cisneros-Zevallos, L.; Jacobo-Velázquez, D.A. The Folin-Ciocalteu Assay Revisited: Improvement of Its Specificity for Total Phenolic Content Determination. *Anal. Methods* **2013**, *5*, 5990–5999. [CrossRef]
66. Singleton, V.L.; Rossi, J.A., Jr. Colorimetry of Total Phenolics With Phosphomolybdic-Phosphotungstic Acid Reagents. *Am. J. Enol. Vitic.* **1965**, *16*, 144–158. [CrossRef]
67. Ramirez-Sanchez, I.; Maya, L.; Ceballos, G.; Villarreal, F. Fluorescent Detection of (-)-Epicatechin in Microsamples from Cacao Seeds and Cocoa Products: Comparison with Folin-Ciocalteu Method. *J. Food Compos. Anal.* **2010**, *23*, 790–793. [CrossRef] [PubMed]
68. Shraim, A.M.; Ahmed, T.A.; Rahman, M.M.; Hijji, Y.M. Determination of Total Flavonoid Content by Aluminum Chloride Assay: A Critical Evaluation. *LWT* **2021**, *150*, 111932. [CrossRef]
69. Pękal, A.; Pyrzynska, K. Evaluation of Aluminium Complexation Reaction for Flavonoid Content Assay. *Food Anal. Methods* **2014**, *7*, 1776–1782. [CrossRef]
70. Zhishen, J.; Mengcheng, T.; Jianming, W. The Determination of Flavonoid Contents in Mulberry and Their Scavenging Effects on Superoxide Radicals. *Food Chem.* **1999**, *64*, 555–559. [CrossRef]
71. Chang, C.-C.; Yang, M.-H.; Wen, H.-M.; Chern, J.-C. Estimation of Total Flavonoid Content in Propolis by Two Complementary Colometric Methods. *J. Food Drug Anal.* **2002**, *10*, 178–182. [CrossRef]
72. Mcmurrough, I.; Mcdowell, J. Chromatographic Separation and Automated Analysis of Flavanois. *Anal. Biochem.* **1978**, *91*, 92–100. [CrossRef] [PubMed]
73. Swain, T.; Hillis, W.E. The Phenolic Constituents of Prunus Domestica. I.—The Quantitative Analysis of Phenolic Constituents. *J. Sci. Food Agric.* **1959**, *10*, 63–68. [CrossRef]
74. Goldstein, J.L.; Swain, T. Changes in Tannins Un Ripening Fruits. *Phytochemistry* **1963**, *2*, 371–383. [CrossRef]
75. Broadhurst, R.B.; Jones, W.T. Analysis of Condensed Tannins Using Acidified Vanillin. *J. Sci. Food Agric.* **1978**, *29*, 788–794. [CrossRef]
76. Schofield, P.; Mbugua, D.M.; Pell, A.N. Analysis of Condensed Tannins: A Review. *Anim. Feed. Sci. Technol.* **2001**, *91*, 21–40. [CrossRef]
77. Dalzell, S.A.; Kerven, G.L. A Rapid Method for the Measurement of Leucaena Spp Proanthocyanidins by the Proanthocyanidin (Butanol/HCl) Assay. *J. Sci. Food Agric.* **1998**, *78*, 405–416. [CrossRef]
78. Nel, A.P. Tannins and Anthocyanins: From Their Origin to Wine Analysis - A Review. *S Afr. J. Sci.* **2018**, *39*, 1–20. [CrossRef]
79. Huang, X.; Teye, E.; Sam-Amoah, L.K.; Han, F.; Yao, L.; Tchabo, W. Rapid Measurement of Total Polyphenols Content in Cocoa Beans by Data Fusion of NIR Spectroscopy and Electronic Tongue. *Anal. Methods* **2014**, *6*, 5008–5015. [CrossRef]
80. Álvarez, C.; Pérez, E.; Cros, E.; Lares, M.; Assemat, S.; Boulanger, R.; Davrieux, F. The Use of near Infrared Spectroscopy to Determine the Fat, Caffeine, Theobromine and (-)-Epicatechin Contents in Unfermented and Sun-Dried Beans of Criollo Cocoa. *J. Near Infrared Spectrosc.* **2012**, *20*, 307–315. [CrossRef]
81. Whitacre, E.; Oliver, J.; Van Den Broek, R.; Van Engelen, P.; Kremers, B.; Van Der Horst, B.; Stewart, M.; Jansen-Beuvink, A. Predictive Analysis of Cocoa Procyanidins Using Near-Infrared Spectroscopy Techniques. *J. Food Sci.* **2003**, *68*, 2618–2622. [CrossRef]
82. Nsor-Atindana, J.; Zhong, F.; Mothibe, K.J. In Vitro Hypoglycemic and Cholesterol Lowering Effects of Dietary Fiber Prepared from Cocoa (*Theobroma cacao* L.) Shells. *Food Funct.* **2012**, *3*, 1044–1050. [CrossRef] [PubMed]
83. Hernández-Hernández, C.; Viera-Alcaide, I.; Morales-Sillero, A.M.; Fernández-Bolaños, J.; Rodríguez-Gutiérrez, G. Bioactive Compounds in Mexican Genotypes of Cocoa Cotyledon and Husk. *Food Chem.* **2018**, *240*, 831–839. [CrossRef] [PubMed]
84. Robbins, R.J.; Leonczak, J.; Li, J.; Johnson, J.C.; Collins, T.; Kwik-Uribe, C.; Schmitz, H.H. Determination of Flavanol and Procyanidin (by Degree of Polymerization 1-10) Content of Chocolate, Cocoa Liquors, Powder(s), and Cocoa Flavanol Extracts by Normal Phase High-Performance Liquid Chromatography: Collaborative Study. *J. AOAC Int.* **2012**, *95*, 1153–1160. [CrossRef] [PubMed]
85. Fulcrand, H.; Mané, C.; Preys, S.; Mazerolles, G.; Bouchut, C.; Mazauric, J.P.; Souquet, J.M.; Meudec, E.; Li, Y.; Cole, R.B.; et al. Direct Mass Spectrometry Approaches to Characterize Polyphenol Composition of Complex Samples. *Phytochemistry* **2008**, *69*, 3131–3138. [CrossRef] [PubMed]
86. Ajila, C.M.; Brar, S.K.; Verma, M.; Tyagi, R.D.; Godbout, S.; Valéro, J.R. Extraction and Analysis of Polyphenols: Recent Trends. *Crit. Rev. Biotechnol.* **2011**, *31*, 227–249. [CrossRef] [PubMed]

87. La Barbera, G.; Capriotti, A.L.; Cavaliere, C.; Montone, C.M.; Piovesana, S.; Samperi, R.; Zenezini Chiozzi, R.; Laganà, A. Liquid Chromatography-High Resolution Mass Spectrometry for the Analysis of Phytochemicals in Vegetal-Derived Food and Beverages. *Food Res. Int.* **2017**, *100*, 28–52. [CrossRef] [PubMed]
88. Othman, A.; Ismail, A.; Ghani, N.A.; Adenan, I. Antioxidant Capacity and Phenolic Content of Cocoa Beans. *Food Chem.* **2007**, *100*, 1523–1530. [CrossRef]

antioxidants

Article

RSM- and ANN-Based Multifrequency Ultrasonic Extraction of Polyphenol-Rich *Sargassum horneri* Extracts Exerting Antioxidative Activity via the Regulation of MAPK/Nrf2/HO-1 Machinery

Ahsan Javed [1], Md Badrul Alam [1,2], Marufa Naznin [3,4], Raees Ahmad [3,4], Chang Hyung Lee [5], Sunghwan Kim [3,4,*] and Sang-Han Lee [1,2,*]

[1] Department of Food Science and Biotechnology, Graduate School, Kyungpook National University, Daegu 41566, Republic of Korea; ahsanjaved@knu.ac.kr (A.J.); mbalam@knu.ac.kr (M.B.A.)
[2] Food and Bio-Industry Research Institute, Inner Beauty/Antiaging Center, Kyungpook National University, Daegu 41566, Republic of Korea
[3] Department of Chemistry, Kyungpook National University, Daegu 41566, Republic of Korea; naznin@knu.ac.kr (M.N.); ahmed@knu.ac.kr (R.A.)
[4] Mass Spectroscopy Converging Research Center, Green Nano Materials Research Center, Kyungpook National University, Daegu 41566, Republic of Korea
[5] Bio-MAX Institute, Seoul National University, Seoul 08826, Republic of Korea; changhyung@snu.ac.kr
* Correspondence: sunghwank@knu.ac.kr (S.K.); sang@knu.ac.kr (S.-H.L.)

Abstract: *Sargassum horneri* (SH) is widely consumed as a healthy seaweed food in the Asia–Pacific region. However, the bioactive components contributing to its biological activity remain unknown. Herein, we optimized multifrequency ultrasonic-assisted extraction conditions to achieve higher antioxidant activity using a response surface methodology and an artificial neural network. High-resolution mass spectrometry (HRMS; negative mode) was used to tentatively identify the secondary metabolites in the optimized SH extract, which were further tested against oxidative stress in RAW264.7 cells. Additionally, the identified compounds were analyzed in silico to determine their binding energies with the Keap1 protein (4L7B). We identified 89 compounds using HRMS, among which 19 metabolites (8 polyphenolics, 2 flavonoids, 2 lignans, 2 terpenes, 2 tannins, 2 sulfolipids, and 1 phospholipid) were putatively reported for the first time in SH. The in vitro results revealed that optimized SH extract inhibited oxidative stress via the Nrf2/MAPKs/HO-1 pathway in a dose-dependent manner. This result was validated by performing in silico simulation, indicating that sargaquinoic acid and glycitein-7-*O*-glucuronide had the highest binding energies (−9.20 and −9.52 Kcal/mol, respectively) toward Keap1 (4L7B). This study offers a unique approach for the scientific community to identify potential bioactive compounds by optimizing the multivariant extraction processing conditions, which could be used to develop functional and nutraceutical foods.

Keywords: *Sargassum horneri*; multifrequency ultrasonic-assisted extraction; secondary metabolites; antioxidantive activity

Citation: Javed, A.; Alam, M.B.; Naznin, M.; Ahmad, R.; Lee, C.H.; Kim, S.; Lee, S.-H. RSM- and ANN-Based Multifrequency Ultrasonic Extraction of Polyphenol-Rich *Sargassum horneri* Extracts Exerting Antioxidative Activity via the Regulation of MAPK/Nrf2/HO-1 Machinery. *Antioxidants* **2024**, *13*, 690. https://doi.org/10.3390/antiox13060690

Academic Editors: Dimitrios Stagos, Yongliang Zhuang, Hu Hou, Zhipeng Yu and Shengbao Cai

Received: 9 April 2024
Revised: 9 May 2024
Accepted: 30 May 2024
Published: 4 June 2024

1. Introduction

Reactive oxygen and nitrogen species (ROS/RNS) play a significant role in maintaining cellular homeostasis, and an imbalanced redox status or uncontrolled ROS/RNS levels induce the production of numerous oxidative stress-linked ailments, including cancer, inflammation, and cardiovascular diseases [1]. Hydrogen peroxide (H_2O_2) is a widely used inducer that triggers oxidative damage and stress as it undergoes Fenton's reaction with Fe^{2+} ions, leading to the formation of a highly reactive -OH radical. Although macrophages defend cell homeostasis from numerous infectious agents, they produce ROS and RNS upon stimulation, causing epigenetic alterations that contribute to the development of chronic illness [2]. Therefore, activated macrophage models are widely used to propose functional diets through a multi-targeted strategy. Phytochemicals with intrinsic antioxidant activity can directly or indirectly activate a wide range of protective signaling cascades and may be employed to treat oxidative stress-related diseases [3]. As a result, understanding and confirming natural chemical actions, as well as identifying the underlying molecular pathways, are critical for determining their potential medicinal relevance.

Many antioxidant enzymes and detoxifying agents, especially heme oxygenase-1 (HO-1) are relied upon in the induction of nuclear factor erythroid 2-related factor 2 (Nrf2) [4]. Under normal conditions, cytosolic Kelch-like ECH-associated protein 1 (Keap1) causes the degradation of Nrf2 through the ubiquitin–proteasome system. Oxidative stress or xenobiotic challenge can prevent Nrf2 degradation by modifying the reactive cysteine residue of Keap1, leading to the translocation of Nrf2 to the nucleus and its binding to antioxidant-related elements (AREs) in the promoter regions of antioxidant and cytoprotective genes [5]. Furthermore, Nrf2 nuclear translocation is dependent on the activation of mitogen-activated protein kinases (MAPKs), phosphatidylinositol 3-kinase/Akt (PI3K/AKT), and protein kinase C (PKC) [6]. *Sargassum horneri* (SH) is a widely used foodstuff in the Asia and Pacific region due to its well-known attributes. It possesses numerous bioactive ingredients including polyphenols, polysaccharides, lignans, and terpenes, reflecting its antimicrobial, antioxidative, antidiabetic, and anticancer characteristics as a food resource. So far, various marine bioactive ingredients have been isolated from SH and studied against different diseases such as diabetes, cancer, and neurological disorders [7]. Additionally, Jayawardena et al. [8] demonstrated that ethanolic SH extracts helped attenuate the fine dust-caused inflammation. Likewise, Han et al. [9], isolated the (−)-lolidolide and determined its anti-inflammatory activity, which involves activating the Nrf2/HO-1 signaling in α-MSH-stimulated HaCaT cells.

In any experiment, the first and most crucial phase is the recovery and purification of bioactive compounds. Artificial neural networks (ANNs) and response surface methodology (RSM) are widely used statistical tools for the optimization of the multivariant parameter to replace non-conventional methods; they also provide the predicted results, thus helping the optimization [10–12]. Likewise, in silico approaches, especially molecular docking simulation, are gaining the attention of the scientific community to screen out receptor–ligand interactions in the suppression or activation of specific proteins [13]. To the best of our knowledge, the multifrequency ultrasonic extraction optimization of antioxidant activity using two sophisticated statistical tools (ANNs and RSM), the profiling of secondary metabolites using high-resolution mass spectrometry (HRMS), and the quantification of the binding energies of identified compounds using molecular docking simulation have never been performed. Our study focused on optimizing the extraction conditions to achieve maximum antioxidant activity and profiling the bioactive compounds responsible for those biological activities using HRMS. Additionally, in vitro and in silico processes were conducted to attenuate the oxidative stress via Nrf2/MAPKs/HO-1 cascade using the RAW264.7 cells and determine the binding energies of the potentially identified compounds using molecular docking.

2. Materials and Methods

2.1. Sample Collection and Preparation

In the middle of April 2021, *Sargassum horneri* (SH) samples were collected from the coastal areas of the Republic of Korea and verified by the scientific officer at the Department of Oceanography, Kyungpook National University, Republic of Korea. Samples were washed and dried at 37 °C, and multifrequency ultrasonic-assisted extraction (MUAE) was performed as briefed by Vázquez-Rodríguez et al. [13] with slight modifications. Briefly, powdered SH (10 g) was soaked for 1 h before extraction with 150 mL of solvent according to the RSM experimental design Table 1. The sample-to-solvent ratio was 1:15 for MUAE. Based on the RSM–Box–Behnken design (BBD) model, MUAE was performed using an ultrasonic bath (Elma Schmidbauer GmbH, Singen, Germany) at 26, 35, and

40 kHz [14]. The MUAE samples were kept at 25 °C for 30 min to increase the extractability of bioactive components, filtered (Whatman No. 1; Schleicher & Schuell, Keene, NH, USA), and lyophilized using a freeze dryer (Il-Shin Biobase, Goyang, Republic of Korea). Samples were then stored at −20 °C in the polyethylene bags for further analysis.

2.2. Experimental Designs Using Statistical Models

The RSM–BBD model followed the methodology of Javed et al. [15]. In short, four independent variables—time (X_1), temperature (X_2), ethanol concentration (X_3), and ultrasonic frequency (X_4)—were checked against four experimental responses—total phenolic content (TPC, Y1), total flavonoid content (TFC, Y_2), 1-diphenyl-2-picrylhydrazyl (DPPH, Y_3), and 1-diphenyl-2-picrylhydrazyl (ABTS, Y_4). ANOVA and quadratic polynomial equations were used to calculate the significance level of the fitted model. After developing the RSM statistical model, the ANN model was constructed using experimental values to check the accuracy of the two models. The ANN model conditions were developed by following the methodology illustrated by Javed et al. [16]. In short, ANNs comprised hidden layers (tan and purplin) with cascade-forward and feed-forward networks. Two different functions, that is, Levenberg–Marquardt back-propagation (trainLM) and Broyden–Fletcher–Goldfarb–Shanno (BFGS) were applied to construct the model using deep learning toolbox in MATLAB R2020a (MathWorks, Natick, MA, USA). For accurate model construction, 65% of experimental data was used for input, 20% of experimental data was used for testing, and 15% of experimental data was used for validation. The hit and trial method was applied to minimize the mean square error, MSE. After constructing the RSM and ANN models, both models were compared using 4 different parameters elaborated by Javed et al. [15], that is, the absolute average deviation (AAD), the standard error of prediction (SEP), the coefficient of determination (R^2), and the root mean square error (RMSE). To calculate the optimized condition for the ANN model, a further genetic algorithm (GA) was applied. In the GA model, ANN data were used as a fitness function, and the highest results for the target variables were determined. The mutation function, initial population size, crossover percentage, and evolutionary algebra were selected based on the current situation while the other parameters of GA were kept at their default settings (Table S1).

2.3. Antioxidant Activities

SH extracts were analyzed thrice for their TPC and TFC using the Folin–Ciocalteu test and the aluminum chloride colorimetric method, respectively, by following the method described by Javed et al. [15]. Regression equations (TPC: $y = 0.0526x + 0.0020$; $R^2 = 0.9856$, TFC: $y = 0.0\ 336x + 0.0021$; $R^2 = 0.9894$) were derived from calibration curves. TPC and TFC were expressed in gallic acid and catechin equivalents ((mg)/dry weight sample (g)), respectively. ABTS and DPPH radical scavenging assays were analyzed using ascorbic acid as a standard as per Alam et al. [3]. The calibration curves were used to calculate the regression equations for DPPH ($y = 0.0555x + 0.0123$; $R^2 = 0.9879$) and ABTS ($y = 0.0369 + 0.0133$; $R^2 = 0.9823$).

Table 1. Comparison of Box–Behnken design (BBD) for independent variables against corresponding target responses (experimental), RSM and ANN predicted values.

Run No	Parameter				TPC (mg GAE/g)			TFC (mg CAE/g)			DPPH (% Inhibition)			ABTS (% Inhibition)		
	X_1 (%)	X_2 (min)	X_3 (°C)	X_4 kHz	Experimental Value	RSM Predicted	ANN Predicted	Experimental Value	RSM Predicted	ANN Predicted	Experimental Value	RSM Predicted	ANN Predicted	Experimental Value	RSM Predicted	ANN Predicted
1	50	50	20	35	47.78 ± 3.25	49.27	49.55	35.77 ± 1.52	38.84	35.39	19.24 ± 1.62	20.10	19.26	33.07 ± 1.19	32.65	33.09
2	50	35	60	26	55.85 ± 3.54	54.93	58.95	20.08 ± 3.14	24.99	20.06	23.58 ± 1.78	22.85	22.63	35.01 ± 1.74	30.84	35.54
3	50	20	40	40	57.62 ± 2.87	60.47	60.48	23.49 ± 2.87	24.88	23.61	26.59 ± 2.17	25.46	26.74	32.11 ± 1.76	34.76	32.92
4	50	50	60	35	71.80 ± 1.98	71.23	71.96	36.54 ± 2.36	34.91	38.89	30.36 ± 1.15	29.09	29.69	32.01 ± 1.66	34.34	32.16
5	30	50	40	35	65.33 ± 4.25	67.46	67.04	22.04 ± 3.14	23.66	27.68	25.65 ± 2.17	25.54	25.69	49.29 ± 1.39	45.89	49.31
6	50	50	40	40	66.40 ± 2.64	61.25	64.13	32.71 ± 2.77	33.02	32.96	26.04 ± 1.94	23.66	24.69	32.16 ± 1.58	33.16	32.43
7	50	50	40	26	65.33 ± 3.14	64.58	64.30	28.57 ± 3.35	23.77	28.84	25.65 ± 1.81	26.75	25.69	38.04 ± 1.95	38.24	37.65
8	50	35	60	40	65.95 ± 1.64	66.07	64.15	35.81 ± 2.31	33.71	35.46	26.33 ± 0.37	26.31	25.37	33.79 ± 2.75	31.37	33.11
9	70	50	40	35	53.03 ± 2.88	55.90	55.18	57.45 ± 2.87	58.90	59.83	17.46 ± 1.75	19.28	15.56	20.03 ± 3.88	20.35	18.57
10	30	35	40	40	63.56 ± 4.35	63.72	63.08	22.04 ± 2.70	19.80	24.23	23.58 ± 1.44	24.24	23.82	43.83 ± 3.65	42.28	42.29
11	50	35	40	35	66.25 ± 1.61	67.37	67.87	25.55 ± 1.93	26.78	30.05	28.19 ± 1.52	29.34	28.38	44.22 ± 2.44	44.87	43.17
12	30	20	40	35	67.65 ± 3.47	65.86	65.63	22.67 ± 3.40	23.95	22.34	26.01 ± 1.85	25.18	25.24	38.21 ± 3.36	35.48	39.26
13	30	35	60	35	64.40 ± 2.95	66.65	65.95	24.76 ± 1.88	27.94	24.88	21.68 ± 1.17	24.44	21.63	34.40 ± 2.38	37.80	34.26
14	30	35	40	26	69.76 ± 1.98	68.74	70.00	22.87 ± 2.67	20.52	25.6	31.06 ± 1.99	29.73	31.05	32.27 ± 2.35	34.60	31.39
15	50	20	60	35	58.12 ± 3.16	55.24	58.67	35.92 ± 3.79	33.78	34.02	20.91 ± 2.33	19.38	20.16	29.95 ± 2.01	30.32	29.34
16	50	20	40	26	50.57 ± 2.75	54.26	52.15	28.22 ± 3.02	24.01	28.7	15.95 ± 1.34	17.74	15.03	27.22 ± 1.58	28.33	25.49
17	70	20	40	35	50.17 ± 2.97	49.13	49.92	47.20 ± 2.75	48.33	48.85	14.40 ± 2.20	15.52	14.6	24.73 ± 2.98	25.73	25.71
18	70	35	40	26	47.69 ± 3.11	46.29	47.69	41.30 ± 1.79	42.90	41.44	13.06 ± 1.12	11.74	12.79	24.24 ± 3.45	25.97	22.54
19	30	35	20	35	61.70 ± 3.77	60.00	62.25	27.95 ± 2.78	26.50	26.29	28.75 ± 2.30	27.62	28.86	36.34 ± 2.15	38.31	35.5
20	50	35	40	35	68.01 ± 2.58	67.37	67.87	26.78 ± 2.35	26.78	30.05	31.41 ± 1.60	29.34	28.38	45.85 ± 2.64	44.87	43.17
21	50	20	20	35	57.71 ± 3.79	56.89	58.75	27.12 ± 1.97	29.68	27.4	25.08 ± 1.89	25.69	23.98	34.02 ± 2.78	31.64	31.77
22	50	35	20	26	56.82 ± 4.20	57.25	57.09	24.74 ± 2.78	29.61	26.32	22.46 ± 2.64	22.99	23.22	31.65 ± 2.25	30.47	33.00
23	70	35	40	40	54.69 ± 3.13	54.19	53.26	48.86 ± 1.95	53.73	48.9	21.18 ± 1.52	21.85	20.19	22.24 ± 2.46	19.62	22.75
24	70	35	60	35	53.98 ± 2.65	56.01	54.85	58.59 ± 2.60	56.39	55.7	20.18 ± 1.39	21.00	20.03	20.33 ± 2.62	20.85	18.74
25	70	35	20	35	44.28 ± 2.95	42.35	46.88	64.49 ± 3.10	57.66	62.51	18.22 ± 2.29	15.14	17.55	20.88 ± 2.35	19.97	19.10
26	50	35	20	40	46.44 ± 3.57	48.98	48.97	33.21 ± 3.16	31.01	30.82	21.93 ± 1.67	24.15	21.26	28.34 ± 3.69	31.29	31.17
27	50	35	40	35	67.85 ± 4.10	67.37	67.87	28.02 ± 1.30	26.78	30.05	28.41 ± 1.71	29.34	28.38	44.55 ± 2.45	44.87	43.17

2.4. High-Resolution Mass Spectroscopy (HRMS)

Electrospray ionization tandem mass spectrometry (ESI-MS/MS) analysis was conducted as per the procedures of Choi et al. [17]. In brief, optimized SH extracts were injected using (15 μL/min) into the ESI of the Q-Exactive Orbitrap mass spectrometer (Thermo Fisher Scientific Inc., San Jose, CA, USA) using a 500 μL graduated syringe. The setting of the ESI-MS was as follows: the mode was selected as negative; full width at half maximum was kept at a mass resolution of 140,000; the flow rates of sweep gas and sheath gas were kept at 0 and 5, respectively, whereas the temperature was kept at 320 °C and the spray voltage was set at 4.20 kV. For a parent peak, fragmentations were conducted at three different collision energies (CEs = 10, 30, and 40). The other condition of the MS/MS experiment was kept the same except for the auxiliary and sheath gas rates, which were 10, respectively, and an S-lens Rf level of 50. Xcalibur 3.1 and Foundation 3.1 (Thermo Fisher Scientific Inc., Rockford, IL, USA) were used to analyze and process peak identification and spectral data. Fragments obtained from the negative mode were verified by comparing with the online databases, i.e., METLIN, HMDB, and FoodB.

2.5. Cell Viability Assay

Cell viability was assessed via MTT assay, following Alam et al. [5]. Briefly, H_2O_2 pretreated Raw264.7 cells (American Type Culture Collection, Manassas, VA, USA) were treated with various concentrations of optimized SH extracts (5–20 mg/mL) and gallic acid (50 μg/mL). Cells were incubated for 18 h at 37 °C and media was removed using suction. Cells were incubated again for 30 min after treatment with 100 μL of MTT (10%) solution. After the suction of the MTT solution, 100% dimethyl sulfoxide (DMSO) was added, and the optical density was read at 590 nm using a microplate reader (Victor3, PerkinElmer, Waltham, MA, USA).

2.6. Reactive Oxygen Species (ROS) Analysis

ROS produced as an outcome of oxidative damage were analyzed using the methodology of Alam et al. [5], using the DCFH-DA procedure. In short, seeded RAW 264.7 cells were pretreated with H_2O_2, incubated for 30 min, and treated with optimized SH extracts (5–10 mg/mL) and GA (50 μg/mL). Raw264.7 cells were washed with PBS, treated with DCFH-DA (25 μM), and incubated for 30 min at 37 °C. The fluorescence intensity was analyzed at 485 nm (excitation) and 528 nm (emission) using a Victor fluorescence microplate reader (PerkinElmer) to calculate the ROS generation rate.

2.7. Western Blotting and Cell Lysate Preparation

After treatment, cells were disintegrated with RIPA buffer to collect the proteins. The collected proteins were then treated with 5X SDS-PAGE (3M Science, Seoul, Republic of Korea) sample buffer and denatured at 95 °C for 12 min. Proteins were analyzed by 10% SDS gel electrophoresis for 95 min, and protein bands were transferred onto nitrocellulose membranes for 2 h. Skim milk (5%) and bovine serum albumin (5%) were added to the membrane, which was then incubated overnight at 4 °C with a first antibody. After incubation, membranes were washed and treated with the secondary antibody (anti-goat and anti-rabbit IgG). Bands were detected using a chemiluminescence system (PerkinElmer) and identified based on their molecular weights [5].

2.8. Ligand Retrieving and Protein Preparation

The structures of putatively identified compounds were retrieved online using the PubChem database. The energy minimization of all the structures and Keap1 (4L7B) protein was performed with the Chimera software (version 1.5), using the procedure of Diniyah et al. [18].

2.9. Molecular Docking Study

Among the 89 identified compounds, 55 compounds (phenolic, flavonoids, terpenes, lignans, and tannins) were used to analyze their binding affinity with Keap1 protein (4L7B) using Auto-Dock Vina 4.2.621 through DockingApp's interface. All the water molecules and ligands were removed from the initial structure and polar hydrogen atoms were added before the docking. After carefully analyzing the literature, we made the size of the grid box 25 Å × 25 Å × 25 Å with its center at position x = 2.4, y = 2.8, and z = −29.21. ΔG values were calculated to measure each compound's root-mean-square deviation (RMSD) [18]. Compounds with an RMSD of ≤9.0 were further compared with the MolDock scores using MolDock software (version 7.0.0). The same grid box information was used with the MolDock optimizer algorithm to obtain the MolDock score. Pymol and Discovery Studio were used to visualize the 2D and 3D interaction between the protein amino acid and ligands.

2.10. Statistical Analysis

The experimental results were statistically analyzed using Design Expert 11 (version 8.0.6, STAT-EASE, Inc., Minneapolis, MN, USA), MATLAB R2020a software (MathWorks), and OriginPro software (OriginPro® 2021b SR1 v9.8.5.204). All results are reported as the mean standard deviation of three independent experiments (n = 3), with at least three replicates for each sample in each experiment. The value of $p < 0.05$ was considered statistically significant, and different letters represent statistically different means.

3. Results

3.1. Design Matrix and Fitting RSM Modeling

Optimizing the UAE technique using different frequencies in this experiment aimed to increase the extractability of bioactive ingredients with minimum resource utilization. The experimental values for the 27 runs are listed in Table 1. The target responses of experimental UAE exhibited various values for TPC (44.28–71.80 mg GAE/g), TFC (20.08–64.49 mg CAE/g), DPPH activity (13.063–31.41% inhibition), and ABTS activity (20.03–49.29% inhibition) against independent variables (X_1, X_2, X_3, and X_4). The model fitness can be determined using the model significance and the lack of fitness, which should be non-significant, and, in both cases, our model fulfills the requirement of the well-fitted model as shown in Table S2A–D. The RSM- and ANN-predicted models were significant to the target response model in Table 1; although, when compared to each run, various targeted responses are non-significant to the predicted RSM and ANN model run numbers (Table 1). Similarly, in the RSM model, for TPC, the concentration, time, and temperature showed significant effects, whereas, in the case of TFC and ABTS, only the concentration showed a significant effect. For DPPH, concentration and time exhibited a significant trend (Table S2). Likewise, for TPC, the interactions of time × temperature and temperature × frequency and the squares of temperature, time, concentration, and frequency also exhibited a significant effect. In the case of TFC, the squares of concentration and temperature showed a significant effect in the RSM model. For DPPH, the interactions of concentration × frequency and time × temperature and the squares of concentration and time exhibited a significant effect. In ABTS, a significant effect was observed in the squares of concentration, time, temperature, and frequency.

In this study, the R^2 values of each MUAE value of all target responses (R^2 = 0.91–0.93) were higher than the acceptable range ($R^2 \geq 0.80$), indicating that the model is well fitted as observed in Table S3A–D. Likewise, adequate precision represents a signal-to-noise ratio, and higher than four is desirable for a well-fitted model. In the current study, the ratio was between 11.04 and 13.04, suggesting that the model was well fitted and had appropriate signals to traverse the design space.

RSM contour plots were created to better understand the relationship between the independent and target response variables (Figure 1A–D). The highest values for TPC (71.80 mg GAE/g) and TFC (64.49 mg CAT/g) were observed in run no. 4 (X_1 = 50%, X_2 = 50 min, X_3 = 60 °C, and X_4 = 35 kHz) and run no. 25 (X_1 = 70%, X_2 = 35 min, X_3 = 20 °C, and X_4 = 40 kHz), respectively. A DPPH value was achieved in run no. 20 (X_1 = 50%, X_2 = 35 min, X_3 = 40 °C, and X_4 = 35 kHz), and the maximum ABTS value was achieved in run no. 5 (X_1 = 30%, X_2 = 50 min, X_3 = 40 °C, and X_4 = 35 kHz), respectively.

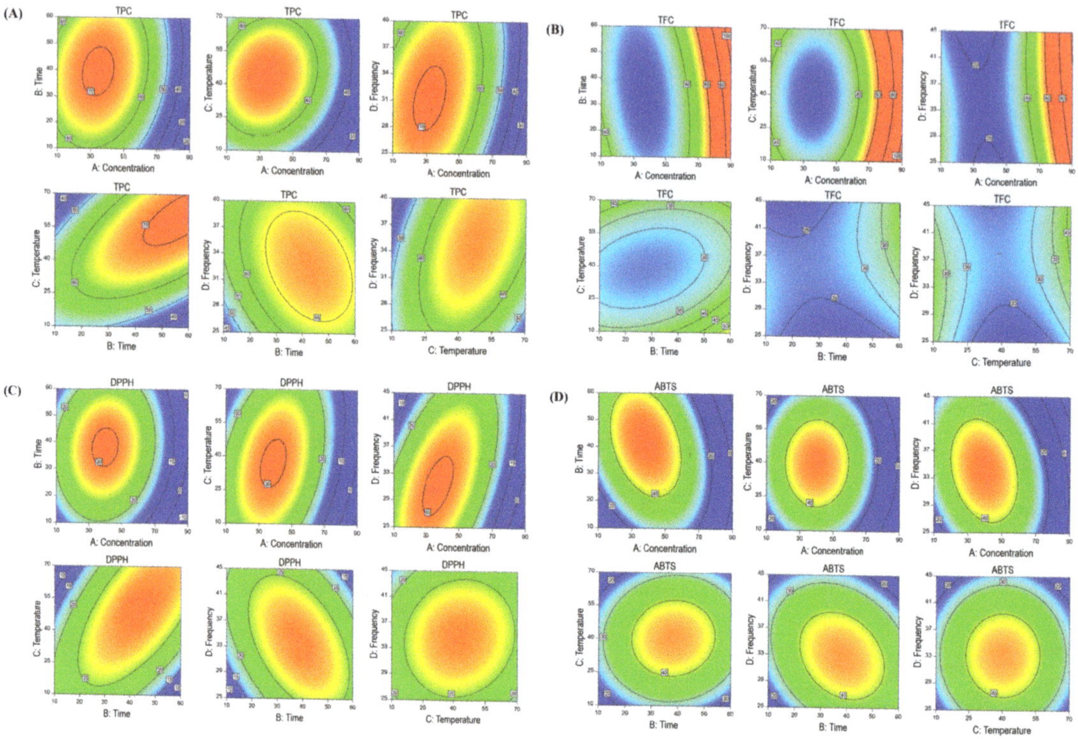

Figure 1. RSM interaction plots of SH extraction on ethanol concentration, time, temperature, and ultrasound frequency using BBD design for TPC (**A**), TFC (**B**), DPPH (**C**), and ABTS (**D**).

Figure 2 illustrates the optimal ANN architecture topology for UAE conditions. The hit and trial method was adopted to minimize the mean error, and an ANN topology of 4–10–4 was sufficient to optimize the ANN model as the neuron optimization was conducted on up to 20 neurons to find the best fitness values (lower MSE). The results indicated that, for all the dependent variables, the lowest MSE value was calculated for ten neurons, which was then further checked via the validation performance and histogram error as depicted in Figure 2.

Table S4 shows the comparison in terms of the prediction abilities and the ANN model reflected better predicting ability than RSM. Table S5 exhibits the optimized model conditions for both the RSM and ANN models. After optimizing the condition, we repeated the experiment with slight parameter changes (X_1: 50%, X_2: 30 min, X_3: 35 °C, and X_4: 35 kHz) to validate the optimized results. The optimized SH extract exhibited TPC (66.34 mg GAE/g), TFC (43.63 mg CAT/g), DPPH (29.11% of inhibition), and ABTS (46.77% of inhibition), respectively. The validation condition also showed that the ANN-optimized model conditions are better in terms of prediction.

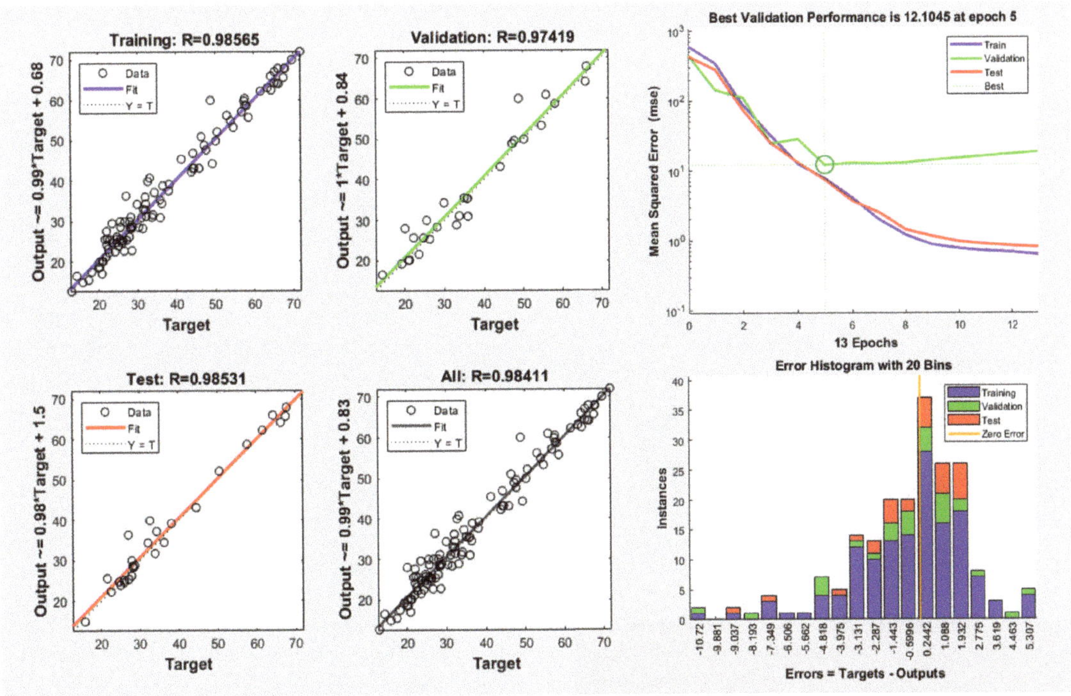

Figure 2. Optimal topology and trained model performance curves with epoch numbers for the dependent variables of the developed ANN model.

3.2. SH Extract Metabolite Profiling Using High-Resolution LC-MS/MS

Secondary metabolites in the UAE-optimized SH extracts were further tentatively identified using the negative mode of ESI-MS/MS equipment. Table 2 indicates the presence of 89 compounds, including 19 metabolites (8 phenolics, 2 flavonoids, 2 lignans, 2 terpenes, 2 tannins, 2 sulfolipids, and 1 phospholipid) that were tentatively reported for the first time in SH-negative mode using MS data from the parent ion mass, recognized fragmentation patterns for the given classes of compounds, and neutral mass loss, paired with comparisons of the existing literature and searches in online databases.

3.2.1. Phenolic Acids and Flavonoids

The literature showed that phenolic acids lose certain fragments such as methyl (15 Da), hydroxyl (18 Da), and carboxyl (44 Da) during the collision, which helped in determining the compounds [3]. Compounds **1–14** have been tentatively identified in *Sargassum fusiforme* based on previously reported fragmentation behavior data (Table 1) [15–19]. Interestingly, 5-(3′,5′-Dihydroxyphenyl)-γ-valerolactone 3-*O*-glucuronide and coumaroylquinic acid were tentatively acknowledged for the first time in optimized MUAE SH extract. Compounds **15–17** were previously reported in *Centroceras* sp., *Ecklonia* sp., and *U. fasciata* [20]. These compounds were tentatively identified for the first time in the SH. Compounds **18–22** were previously reported in *Dasya* sp., *Centroceras* sp., and *Sargassum* sp., respectively [21–23]. Compounds **21** and **22** putatively reported for the first time in the SH.

Each subgroup of polyphenols has a distinct fragmentation behavior in MS/MS, including flavonoids. Typically, C-ring bonds cleave (RDA process), resulting in A or B rings, and sometimes portions of the C ring [5]. Similarly, a few small neutral ions, such as CO (−28 Da), C_2H_2O (−42 Da), CO_2 (−44 Da), and 2CO (−56 Da), might also be yielded during the disintegration. In the previous literature, depending on the fragmentation behavior, compounds **23–28** were identified as flavonoid glycosides epigallocatechin, epicatechin, dalbergin, catechin, dihydrobiochanin A, and glycitein 7-*O*-glucuronide. Dihydrobiochanin A (**27**) was previously discovered in *Grateloupia* sp., *Ascophyllum nodosum*, green seaweeds, and *Ecklonia* sp., respectively. Likewise, glycitein 7-*O*-glucuronide (**28**) was reported in *Centroceras* sp., but both compounds were putatively identified in SH for the first time in this study (Table 2).

3.2.2. Lignans and Terpenes

Few main fragmented ions of [M–H-15], [M–H-18], [M–H-44], and [M–H-46] were produced in dibenzylbutyrolactones as a result of a loss of CH_3, H_2O, or CO_2, and also in the combination of CO and H_2O. However, butanediol lignans generate [M–H-48] ions by breaking at the β-position as a combined loss of CHO and H_2O from the diol structure [24,25]. Based on the monoisotopic mass [M–H]–, fragmentation behavior in mass spectroscopy, and previous studies, compounds **29–33** were recognized as lignan molecules and previously reported in *Sargassum* sp., *Ecklonia* sp., and sea vegetables [15–17,19]. Among lignans, compounds **32** and **33** were tentatively reported for the first time in the SH.

Compounds **34–37** were previously reported in *Sargassum fusiforme, Codium* sp., and *Desmarestia antarctica*, respectively [15,19]. Sargaquinoic acid (compound **38**) and sargahydroquinoic acid (compound **39**) were previously identified in *Sargassum serratifolium, Sargassum fusiforme,* and *Sargassum yezoense* [15,26,27]. Compounds **40–42** were identified as dihydroactinidiolide, humulene epoxide II, and isoamijiol based on the mass fragmentation behavior previously reported for *Fucus virsoides* [28]. Compound **43** was identified as 3,5-dihydroxy-6,7-megastigmadien-9-one and was previously reported in the genus *Ulva*. Compounds **40** and **43** were putatively reported for the first time in the SH. Compounds **44–47** were identified as mojabanchromanol, fallahydroquinone, fallaquinone, and 1-H-indol-6-Carbaldehde and were previously identified in *Sargassum horneri* [29,30].

3.2.3. Tannins, Sulfolipids, and Phospholipids

In SH, compounds **48–51** were identified as phlorotannins in various species of *Sargassum* based on the fragment behavior [31]. Compounds **52** and **53** were previously identified in the *Sargassum horneri* as 1-*O*-(11-hexadecenoly)-3-*O*-(6′-sulpho-α-D-quinovopyranosyl) glycerol and 1-*O*-hexadecanoyl-3-*O*-(6′-sulfo-α-D-quinovopyranosyl) glycerol, as they display activity in vitro and in vivo during the regulation of neuroinflammation [32]. Meanwhile, compounds **54** and **55** were identified as dioxinodehydroeckol and dibenzodioxin-1,3,6,8-tetraol, respectively, and these compounds were tentatively reported for the first time in SH.

Compounds **56–58** were identified as sulfolipids and phospholipids depending on their specific fragmentation patterns. These compounds have previously been reported in *Fucus vesiculosus* and *Sargassum fusiforme* (brown algae) [15,33]. Sulfolipids and phospholipids in SH are reported here for the first time.

3.2.4. Carboxylic Acid, Fatty Acids, Sugars, and Other Compounds

Compounds **59–68** were previously reported in a variety of seaweeds and were identified as carboxylic acids based on their fragmentation patterns and previously reported in *Sargassum fusiforme.* Compounds **69–80** were fatty acids, compounds **81–83** were sugars, and compounds **84–89** were other compounds that were also previously reported by other seaweeds, especially *sargassum* sp. [3,15,34–38], as shown in Table 2.

Table 2. Putatively identified bioactive compounds in optimized *Sargassum horneri* extract by ESI-MS/MS.

Group	No	Compound Name	EF	OM *m/z′*	CM *m/z′*	MS/MS (Negative Mode)	Reference
Phenolic compounds	1.	Ferulic acid	$C_{10}H_{10}O_4$	193.0499	193.0500	177, 149, 133	[15–17,19]
	2.	Sinapic acid	$C_9H_8O_3$	163.0393	163.0395	119	
	3.	Protocatechuic acid	$C_7H_6O_4$	153.0186	153.0186	135, 109	
	4.	Caffeic acid	$C_9H_8O_4$	179.0344	179.0344	161, 135, 117	
	5.	Syringic acid	$C_9H_{10}O_5$	197.0449	197.0450	181, 153, 125	
	6.	Salicylic acid	$C_7H_6O_3$	137.0236	137.0244	93	
	7.	Cinnamoyl glucose	$C_{15}H_{18}O_7$	309.0974	309.0962	147, 131, 103	
	8.	*p*-Hydroxybenzaldehyde	$C_7H_6O_2$	121.0287	121.0289	92, 77	
	9.	Hydroxytyrosol 4-*O*-glucoside #	$C_{14}H_{20}O_8$	315.1083	315.1079	153, 123	
	10.	5-(3′,4′-Dihydroxyphenyl)-valeric acid	$C_{11}H_{14}O_4$	209.0816	209.0813	165, 149	
	11.	5-(3′,5′-Dihydroxyphenyl)-γ-valerolactone 3-*O*-glucuronide #	$C_{17}H_{20}O_{10}$	383.0972	383.0978	205, 191	
	12.	Quinic acid	$C_7H_{12}O_6$	191.0560	191.0555	147, 129	
	13.	Syringin	$C_{17}H_{24}O_9$	371.1356	371.1345	353, 209, 191	
	14.	Coumaroylquinic acid #	$C_{16}H_{18}O_8$	337.0910	337.0965	293, 191, 163	
	15.	3-Sinapoylquinic acid #	$C_{18}H_{22}O_{10}$	397.1140	397.1171	381, 191, 129	[20]
	16.	Hydroxyferulic acid #	$C_{10}H_{10}O_5$	209.0456	209.0477	191, 177	
	17.	Caffeoyl tartaric acid #	$C_{13}H_{12}O_9$	311.0408	311.0446	293, 267, 179	
	18.	Vanillic acid 4-sulfate	$C_8H_8O_7S$	246.9918	246.9904	203, 167.03	[21,22]
	19.	Vanillic acid	$C_8H_8O_4$	167.0343	167.0344	151, 123.04	
	20.	3,4-Dihydroxyphenylglycol	$C_8H_{10}O_4$	169.0500	169.0493	151, 123	
	21.	2-Hydroxy-2-phenylacetic acid #	$C_8H_8O_3$	151.0401	151.0498	107	
	22.	Isopropyl 3-(3,4-dihydroxyphenyl)-2-hydroxypropanoate #	$C_{12}H_{16}O_5$	239.0929	239.0925	173, 123	
Flavonoids and derivatives	23.	Epigallocatechin	$C_{15}H_{14}O_7$	305.0652	305.0636	287, 137, 125	[15,16,23–25]
	24.	Epicatechin	$C_{15}H_{14}O_6$	289.0706	289.0715	245, 109, 125	
	25.	Dalbergin	$C_{16}H_{12}O_4$	267.0710	267.0666	251, 223, 197	
	26.	Catechin	$C_{15}H_{14}O_6$	289.0718	289.0682	245, 205, 139	
	27.	Dihydrobiochanin A #	$C_{16}H_{14}O_5$	285.0817	285.0763	270	
	28.	Glycitein 7-*O*-glucuronide #	$C_{22}H_{20}O_{11}$	459.0927	459.0923	283, 267	
Lignans	29.	Arctigenin	$C_{21}H_{24}O_6$	371.1468	371.1452	355, 327, 311	[15,16,19]
	30.	Isohydroxymatairesinol	$C_{20}H_{21}O_7$	373.1264	373.1266	355, 343, 327, 311	
	31.	Conidendrin	$C_{20}H_{20}O_6$	355.1174	355.1200	340, 311	
	32.	Deoxyschisandrin #	$C_{24}H_{32}O_6$	415.2127	415.2120	402, 347, 316, 301	
	33.	Secoisolariciresinol #	$C_{20}H_{26}O_6$	361.1629	361.1772	343, 331, 315	
Terpenes	34.	Carnosol	$C_{20}H_{26}O_4$	329.1762	329.1749	285, 267	
	35.	Carnosic acid	$C_{20}H_{28}O_4$	331.1921	331.1902	287, 269	
	36.	Loliolide	$C_{11}H_{16}O_3$	195.1021	195.1021	161, 179, 133, 105	
	37.	Isololiolide	$C_{11}H_{16}O_3$	195.1021	195.1013	161, 179, 133, 105	

Table 2. *Cont.*

Group	No	Compound Name	EF	OM *m/z'*	CM *m/z'*	MS/MS (Negative Mode)	Reference
Terpenes	38.	Sargahydroquinoic acid	$C_{27}H_{38}O_4$	425.2680	425.2691	381	[15,19,20,27,28]
	39.	Sargaquinoic acid	$C_{27}H_{36}O_4$	423.2526	423.2579	379	
	40.	Dihydroactinidiolide #	$C_{11}H_{16}O_2$	179.1071	179.1072	135, 121	
	41.	Humulene epoxide II	$C_{15}H_{24}O$	219.1751	219.1748	201	
	42.	(−)-Isoamijiol	$C_{20}H_{32}O_2$	303.2331	303.2319	261, 243, 225	
	43.	3,5-Dihydroxy-6,7-megastigmadien-9-one #	$C_{13}H_{20}O_3$	223.1327	223.1334	205, 187, 163	[30,31]
	44.	Mojabanchromanol	$C_{27}H_{36}O_4$	423.2558	423.2576	379	
	45.	Fallahydroquinone	$C_{27}H_{40}O_4$	427.2854	427.2878	383	
	46.	Fallaquinone	$C_{27}H_{38}O_4$	425.2691	425.2740	381	
	47.	1-H-indol-6-carbaldehde	C_9H_7NO	144.0455	144.0441	-	
Tannins	48.	Trifuhalol-A	$C_{18}H_{14}O_{10}$	389.0513	389.0496	123, 125, 139, 245, 263, 265, 353	[15,31]
	49.	Phloroglucinol	$C_6H_6O_3$	125.0237	125.0238	97	
	50.	Fucophlorethol	$C_{36}H_{26}O_{14}$	680.1154	680.1179	610,601, 495, 469, 229	
	51.	Eckol	$C_{18}H_{12}O_9$	371.0403	371.0403	335, 317, 246, 229, 140	
	52.	1-*O*-(11-hexadecenoly)-3-*O*-(6′-sulpho-α-D-quinovopyranosyl) glycerol	$C_{25}H_{45}O_{11}S$	553.2643	535.2678	317, 299, 253, 225, 207	[32]
	53.	1-*O*-Hexadecanoyl-3-*O*-(60-sulfo-α-D-quinovopyranosyl) glycerol	$C_{25}H_{47}O_{11}S$	555.2824	555.2834	317, 299, 255, 224, 207	
	54.	Dioxinodehydroeckol #	$C_{18}H_{10}O_9$	369.0249	369.0284	238,195, 167, 112	
	55.	Dibenzodioxin-1,3,6,8-tetraol #	$C_{12}H_7O_6$	246.9919	246.9909	203, 121	[15,33]
Sulfolipids and phospholipids	56.	Sulfolipid #	$C_{25}H_{47}O_{11}S$	555.2848	555.2822	299, 255, 225, 206, 164	
	57.	Sulfolipid (SQMG (14:0) #	$C_{23}H_{43}O_{11}S$	527.2535	527.2500	225	
	58.	Phospholipid #	$C_{25}H_{44}NO_7P$	500.2733	500. 2701	303, 259, 196	
Carboxylic acids	59.	Fumaric acid	$C_4H_4O_4$	115.005	115.0021	71	[3,15,34–38]
	60.	Threonic acid	$C_4H_8O_5$	135.0290	135.0299	117, 91, 72	
	61.	Gentisic acid	$C_7H_6O_4$	153.0187	153.0178	152, 108, 81	
	62.	Kainic acid	$C_{10}H_{15}NO_4$	212.0922	212.0916	168, 194, 150	
	63.	Mannuronic acid	$C_6H_{10}O_7$	193.0353	193.0340	175, 103, 72	
	64.	Diethyl phthalate	$C_{12}H_{14}O_4$	221.0818	221.0843	193, 177, 149, 121	
	65.	Phthalic acid	$C_8H_6O_4$	165.0188	165.0178	121 26, 119, 58	
	66.	3-Oxooctanoic acid	$C_8H_{14}O_3$	157.0863	157.0855	139, 113, 97	
	67.	D-Glucaric acid derivative	$C_{12}H_{14}O_{10}$	317.0544	317.0536	209	
	68.	Azelaic acid	$C_9H_{16}O_4$	187.0969	188.0961	187, 124, 169, 111	

Table 2. *Cont.*

Group	No	Compound Name	EF	OM m/z'	CM m/z'	MS/MS (Negative Mode)	Reference
Fatty acids	69.	Caprylic acid	$C_8H_{15}O_2$	143.1070	143.1062	125, 99, 59	[3,15,34–38]
	70.	Vaccenic acid	$C_{18}H_{34}O_2$	281.2486	281.2473	263, 223, 163, 71	
	71.	Palmitic acid	$C_{16}H_{32}O_2$	255.2330	255.2316	237, 211, 197	
	72.	Myristic acid	$C_{14}H_{28}O_2$	227.2015	227.2002	209, 183, 179	
	73.	Octadecendioic acid	$C_{18}H_{32}O_4$	311.223	311.2217	299, 269, 251, 223	
	74.	α-Linoleic acid	$C_{18}H_{32}O_2$	279.2331	279.2315	261, 235, 233	
	75.	13-keto-9Z,11E-Octadecadienoic acid	$C_{18}H_{30}O_3$	293.2125	293.2109	275, 195, 113	
	76.	10-Oxooctadecanoic acid	$C_{18}H_{34}O_3$	297.2436	297.2420	279, 209, 141, 127	
	77.	10,16-Dihydroxy-palmitic acid	$C_{16}H_{32}O_4$	287.2227	287.2213	269, 257, 239, 185	
	78.	2,4-Decadienal	$C_{10}H_{16}O$	151.1119	151.1113	133, 123, 119, 93	
	79.	Lauric acid	$C_{12}H_{24}O_2$	199.1697	199.1689	181, 155	
	80.	Vernolic acid	$C_{18}H_{32}O_3$	295.2278	295.2263	277, 251, 195, 127	
Sugars	81.	D-Galactose	$C_6H_{12}O_6$	179.0568	179.0548	161, 143, 113, 101	
	82.	Mannitol	$C_6H_{14}O_6$	181.0712	181.0703	165, 147, 129, 111	
	83.	Gluconic acid	$C_6H_{12}O_7$	195.0517	195.0498	177, 159, 129, 98	
Other compounds	84.	3,4-Dihydroxybenzaldehyde	$C_7H_6O_3$	137.0237	137.0244	108	
	85.	4-Hydroxyphenyl acetate	$C_8H_8O_3$	151.0394	151.0401	106	
	86.	$PGF_{2\alpha}$	$C_{20}H_{34}O_5$	353.2335	353.2320	317, 273, 235, 127	
	87.	Dinor $PGF_{2\alpha}$	$C_{18}H_{29}O_5$	325.1954	325.2007	307, 199, 183, 171, 129	
	88.	Threonyl-histidylglutamic acid	$C_{15}H_{23}N_5O_7$	384.1518	384.1514	-	
	89.	Dihydroxyphenylalanine	$C_9H_{10}O_7NS$	276.0185	276.0172	259, 231, 215, 196, 179, 150, 135	

EF: elemental formula; OM: observed mass; CM: calculated mass; (-): Negative mode; # First-time identification.

3.3. Attenuation of H_2O_2-Induced Cellular Oxidative Stress by SH

H_2O_2 is widely accepted as a model to induce oxidative stress and assess how that stress affects biological systems in cells and tissues. Figure 3A depicted that after treating RAW264.7 cells with H_2O_2, it induced cell death, which was revered in pretreated gallic acid and with SH in a dose-dependent fashion. Likewise, the SH sample also attenuated cellular ROS in a dose-dependent manner like that of gallic acid (50 µg/mL), as shown in Figure 3B. Catalase, superoxide dismutase-1 (SOD1), and other enzymes are considered as first-line in attenuating oxidative stress and in sustaining the cellular redox environment [39]. As shown in Figure 3C, treatment with H_2O_2 caused a significant surge in oxidative stress, which was reversed by triggering the CAT and SOD with the GA and SH pretreatment in a dose-dependent manner.

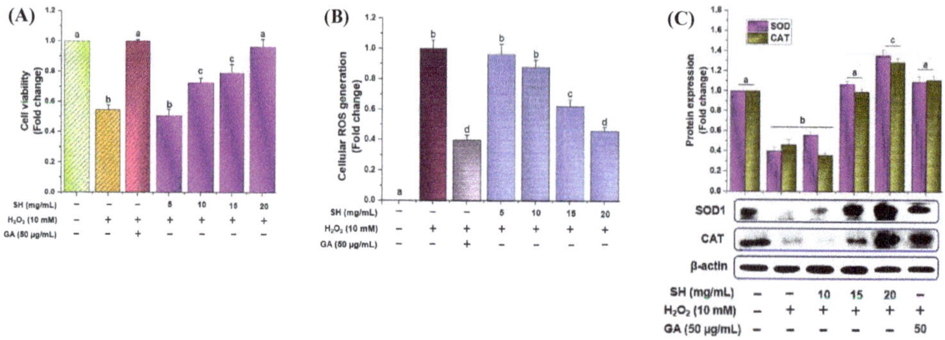

Figure 3. Effect of optimized SH extract on cell viability, ROS generation, and primary enzymes treated with and without H_2O_2 in RAW264.7 cells. Cells were pretreated with SH and gallic acid for 12 h, followed by treatment with or without 10 mM of H_2O_2 for 6 h. Cell viability with H_2O_2 (**A**). The generation of cellular ROS was evaluated using the DCFH-DA method (**B**); the protein expression of SOD1 and catalase was measured with Western blot analysis (**C**), and the relative protein expression was quantified using Image J software (Version 1.54i 03). Differences in the alphabetic letters represent statistical significance ($p < 0.05$) to one another.

3.4. *Nrf2 Regulation via SH Induction of Phase II Enzymes*

To regulate the phase II enzyme (HO-1), Nrf2 translocation into the nucleus is considered an important part after disintegration from the Keap1 protein. Thus, to verify whether SH can increase the phase II antioxidant enzymes through Nrf2 degradation from Keap1, we treated RAW264.7 cells with SH extract. As depicted in Figure 4A, SH dose-dependently increased the dislocation of Nrf2 from the cytoplasm to the nucleus similar to gallic acid. Because of Nrf2 translocation, the HO-1 level increases after treatment with the optimized SH extracts as illustrated in Figure 4B. This trial illustrates that SH may disrupt/degrade the Nrf2 by Keap1, resulting in the upregulation of HO-1 concentration.

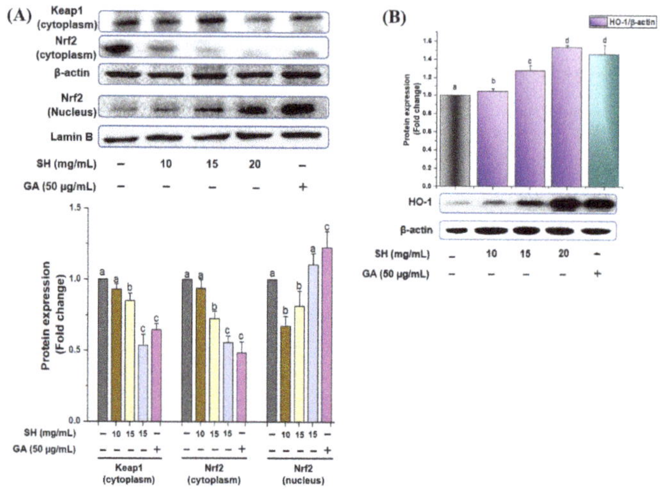

Figure 4. Activation of phase II antioxidant enzymes via the regulation of Nrf2. Optimized SH extract dose-dependently decreases the Keap1 and facilitates the translocation of Nrf2 protein expression into the nucleus (**A**), resulting in the upregulation of phase II enzyme expression (**B**) using immunoblotting assay. The relative protein expression was quantified using Image J software. Differences in the alphabetic letters represent statistical significance ($p < 0.05$) to one another. Gray: non treated; purple: sample treated; green: gallic acid treated.

3.5. MAPK Activation via SH and Regulation of HO-1

The previous literature has shown that the phosphorylation of MAPKs, including JNK, ERK, and p38, can regulate the HO-1 expression in various cell types. Thus, Western blot analysis was conducted to determine the signaling pathways participating in the regulation of HO-1 expression by treating RAW264.7 cells with SH extract. As shown in Figure 5A, SH extract enhanced the phosphorylation of JNK and ERK at 45 min; however, ERK did not show any result regarding the phosphorylation. To further verify whether ERK and JNK modulate the HO-1 expression, we treated the cells with the ERK (U0123) and JNK (SP600125) inhibitor before stimulation with SH. As depicted in Figure 5B, both JNK and ERK inhibitors downregulate the HO-1 expression, which is reversed in the SH treatment. This indicates that ERK and JNK phosphorylation might be involved in the regulation of HO-1 in RAW 264.7 cells. To further verify whether either MAPK/Nrf2/HO-1 cascade has a significant role in regulating oxidative damage, pretreated with SH extracts, Raw264.7 cells were further treated with the MAPK-specific (U0123; SP600125) inhibitors to analyze the production of ROS. As shown in Figure 5C, H_2O_2 induced the production of ROS, which was downregulated via the SH extract. However, when treated with specific inhibitors, the trend was again reversed, which reflects the involvement of MAPK pathways in oxidative stress.

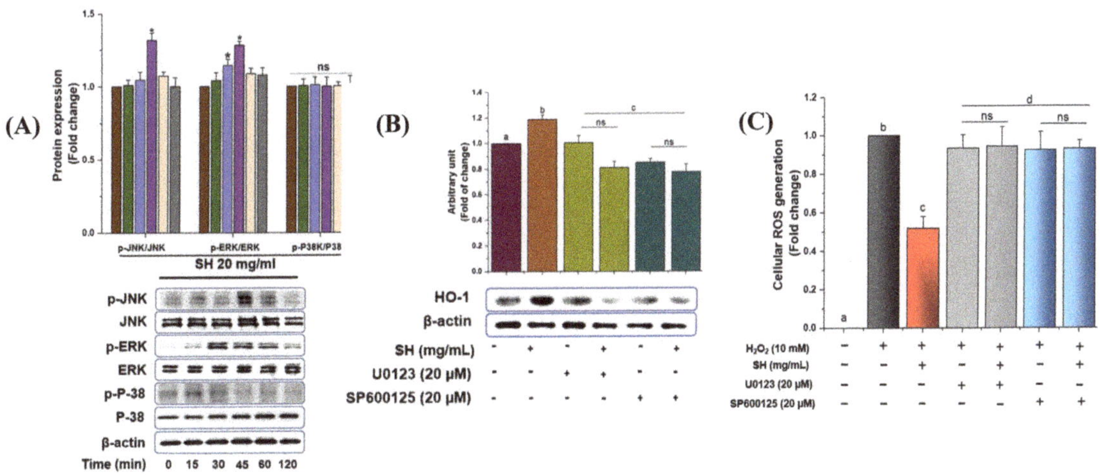

Figure 5. Activation of ERK, p38, and JNK via SH results in Nrf2 translocation. RAW 264.7 cells were treated with SH (20 mg/mL) at various times, and kinase activity was determined with immunoblot assay (**A**). Cells were treated with optimized SH extract and specific inhibitors, SP600125 (JNK inhibitor) and U0123 (ERK), for 1 h, and HO-1 protein levels were analyzed via Western blot analysis. Red: 0 min; green: 15 min; blue: 30 min; purple: 45 min, paste: 60 min, and gray: 120 min (**B**); the production of ROS; red: non-treated; orange: sample; yellow: sample with ERK inhibitor, green: sample with JNK inhibitor. (**C**) and the relative protein expression were quantified using Image J software gray: model control, red: SH; light gray: SH+ERK inhibitor, and sky:SH+JNK inhibitor. Differences in the alphabetic letters represent statistical significance ($p < 0.05$) to one another.

3.6. Docking Results

Numerous studies have revealed that molecular docking is an extremely useful tool for studying ligand–protein interactions during receptor activation or inhibition to better understand the mechanism of action [18]. Based on previous metabolite profiling and Western blot results, we hypothesized that tentatively identified compounds have the ability to attenuate oxidative stress. In order to prove this hypothesis, we performed a molecular docking simulation to identify compounds and calculate the binding affinity of bioactive compounds with Keap1 protein (4L7B) using AutoDoc Vina [40] and MolDock [41]. Fifty-five compounds were checked for the binding energy calculation (Table S6), and those bioactive compounds that showed more binding energy (cutoff value ≤ -9) in AutoDoc Vina were further validated and compared in MolDock. Among them, five compounds exhibited the highest energy, and sargaquinoic acid glycitein 7-*O*-glucuronide (Table 3) exhibited the highest binding affinity in AutoDoc Vina (−9.20 Kcal/mol; 9.52 Kcal/mol) and MolDock (−155.842; −148.975), respectively.

Table 3. AutoDoc Vina and MolDock comparative binding energy scores for selected compounds.

Identified Compounds	MolDock Score	Chimera (Kcal/mol)
Glycitein 7-*O*-glucuronide	−148.975	−9.5
Sargaquinoic acid	−155.842	−9.2
Carnosol	−108.911	−9.0
Conidendrin	−149.703	−9.0
Eckol	−140.451	−9.1
Isohydroxymatairesinol	−150.118	−9.1

4. Discussion

An unbalanced and uncontrolled production of RNS and ROS results in oxidative stress/damage progression in a cellular environment by reacting with proteins, lipids, and DNA [3]. In order to attenuate the uncontrolled ROS/RNS generation, phytochemicals with intrinsic antioxidants play a significant role by regulating the cellular protective signaling cascades [5]. Numerous trials have been conducted in order to find the appropriate bioactive agent; however, marine seaweeds, especially marine brown algae, are still underutilized plant resources. *Sargassum horneri* (SH) is widely used in the Asia and Pacific region due to its biological and pharmacological properties, as this species contains various bioactive compounds [14]. The literature showed that few studies have been conducted to isolate some certain bioactive compounds ((-)-lolidolide, catechin, epicatechin, etc.), which exhibited anti-inflammatory, anti-oxidative activities [14–16]. Although, until today, no study has been conducted to optimize the multifrequency ultrasonic extraction conditions to achieve higher antioxidant activity followed by the identification of bioactive compounds using HRMS in the optimized extract or further check the underlying mechanism of action using in vitro and in silico assays to attenuate the oxidative stress. Keeping these limitations in view, the current trial was conducted to achieve the highest antioxidant activity using sophisticated statistical techniques (RSM and ANNs) to optimize the multifrequency ultrasonic extraction and identification of bioactive compounds in the optimized extracts. Furthermore, the attenuation of oxidative stress was assessed via Nrf2/MAPKs/HO-1 cascade followed by checking the binding energies of identified compounds with Keap1 protein (4L7B) using molecular docking software (AutoDoc and MolDock).

To optimize the extraction conditions, the RSM is frequently used as an appropriate statistical technique for modeling complicated multivariate processes wherein responses are affected by several factors, such as time, temperature, equipment intensity, and power [16]. The statistical significance of the fitted RSM model can be calculated using the ANOVA results, which also contained other important model parameters, such as the regression coefficient (β), coefficient of variation (CV), adjusted correlation factor (R^2), F-value, *p*-value, adequate precision, model significance, and lack of fit. In the current trial, in the RSM model, for TPC extraction, the concentration (X_1); time (X_2); temperature (X_3); the interactions of time (X_2) × temperature (X_3) and temperature (X_3) × frequency (X_4); and the squares of temperature ($X_3{}^2$), time ($X_2{}^2$), concentration ($X_1{}^2$), and frequency ($X_4{}^2$) exhibited a significant effect. In the case of the TFC response, the concentration (X_1) and the squares of concentration ($X_1{}^2$) and temperature ($X_2{}^2$) were significantly affected by the TFC extraction. For DPPH, the concentration (X_1); time (X_2); the interactions of concentration (X_1) × frequency (X_4) and time (X_2) × temperature (X_3); and the squares of concentration ($X_1{}^2$) and time ($X_2{}^2$) exhibited a significant effect, whereas, in the ABTS response, the concentration (X_1) and squares of concentration ($X_1{}^2$), time (X_2), temperature (X_3), and frequency ($X_4{}^2$)

also exhibited a significant effect on the ABTS content. Similarly, the CV value, coefficient of variation (CV), and adjusted correlation factor (R^2) are also in the range (Tables S2–S4), as expected, which was further verified via the previous results of Alshammari et al. [40]; they optimized the heat reflex extraction conditions of Ajwa dates and calculated the R^2, F-values, p-values, and, most importantly, the model significance criteria and lack of fit as 'non-significant'. Recently, Javed et al. [16] optimized the microwave-assisted extraction conditions of *Sargassum fusiforme* (a subspecies of *Sargassum*), which supports our results; however, the MUAE approach reflects the better extraction conditions by keeping the extraction temperature lower. Similarly, Choi et al. [19] determined the optimal heat reflux extraction conditions for the recovery of bioactive ingredients from *Nypa fruticans* Wurmb. using RSM and ANN models, which also supports the current results. However, the current trial is better in terms of the optimizing factors and extraction method because Choi et al. [17] used three factors (time, temperature, and ethanolic concentration) and used a conventional method to extract the bioactive compounds.

ANNs have also emerged as practical and powerful nonlinear computational tools for complicated nonlinear processes owing to their superior learning and predictive modeling capabilities [15]. Principally, ANNs are based on the human central nervous system (CNS), where a complex network of interconnected neurons is capable of computing in response to input data [14]. The ANN model was trained at a topology of 4–10–4 and showed better validation performance and less histogram error as depicted in Figure 2. After optimizing the RSM and ANN models, the prediction abilities of both models were determined and compared using numerous statistical metrics, such as the absolute average deviation (AAD), root mean square error (RMSE), R^2 (coefficient of determination), and standard error of prediction (SEP) [40,41]. The ANN model prediction abilities are better than that of RSM as exhibited in Table S4. The present outcomes are consistent with the results of Choi et al. [19] as they also optimized the extraction condition for *Nypa fruticans* Wurmb and reported that the ANN model is better in terms of prediction. Alshammari et al. [40] also reported on the accuracy of the ANN model in terms of prediction in comparison with RSM prediction, which is also consistent with our present results. Although both studies predicted the values for only three independent factors (time, temperature, and solvent concentration), in the current study, four independent factors were used.

Hydrogen peroxide (H_2O_2) is widely accepted as a model to induce cellular oxidative to investigate to assess how oxidative stress affects biological systems in both cells and tissues [42]. The current trial depicted that optimized extract dose-dependently attenuates ROS generation by regulating the CAT and SOD; see Figure 3A–C. The present outcomes showed that SH helps in promoting the regulation of enzyme proteins, which in turn attenuate oxidative damage by maintaining the cellular redox balance. The literature has shown that plants/food enriched with bioactive compounds, especially polyphenolic compounds, enhance the SOD1 and CAT levels, which results in mitigating oxidative stress. In the previous study, we also optimized the *Sargassum fusiforme* ethanolic extract and determined its activity against oxidative stress, which downregulates oxidative stress by enhancing the SOD and CAT enzymes. Likewise, various investigations also illustrate that the administration of brown algae, especially *Sargassum* sp., helps in reducing oxidative damage and stress. Numerous studies have exhibited that polyphenolics, such as protocatechuic acid, vanillic acid, gallic acid, catechin, and epicatechin (tentatively detected in the SH sample), can upregulate the endogenous antioxidant system, leading to cellular protection from oxidative stress [3,5,15,16]. Consequently, it is assumed that the surge in the first-line antioxidant enzymes via SH might be due to the presence of phenolics and flavonoids, which could be present in higher concentrations during the optimization in the current study. The literature has shown that the presence of bioactive compounds, especially gallic acid and loliolide, scavenged the free radical chain reaction by transferring hydrogen atoms [43]. Likewise, ethanolic, methanolic, and acetone extracts, etc., and extracted polysaccharides were used to check the antioxidant activity or underlying mechanism of action without quantifying the bioactive compounds, but the current trial tries to show the possible compounds that could be behind attenuating the activity against diseases.

The literature has exhibited that phase II antioxidant enzymes, especially heme oxygenase-1 (HO-1), impart an important role in downregulating the ROS and are regulated by the Nrf2, a central controller of ARE-driven antioxidant gene expression. In a normal state, Nrf2 activity is strictly managed by the Kelch-like ECH-associated protein 1 (Keap1) as an adaptor protein for Cullin-3 (Cul3)-dependent E3 ubiquitin ligase enzyme, which manages the Nrf2 degradation and ubiquitination [3,5]. Optimized SH extract helps in the translocation of Nrf2 into the nucleus, which results in the activation of HO-1 (phase II enzyme protein) in a dose-dependent manner (Figure 4A,B). Numerous previous studies have indicated that *Sargassum* sp., including *Sargassum fusiforme* and *Sargassum serratifolium*, actively upregulate HO-1 expression through the regulation of Nrf2 [44]. Likewise, in our previous study, we conducted the metabolite profiling of *Sargassum fusiforme,* which exhibited the presence of various polyphenolic compounds including protocatechuic acid, vanillic acid, gallic acid, and naringenin, which were also present in SH extract because of same *Sargassum* genus (Table 2). Thus, processing conditions might alter the concentrations of the bioactive compounds that need to be carefully selected during the compound's extraction or separation.

Previously, studies have exhibited that MAPK phosphorylation could be involved in the regulation of phase II enzyme HO-1. SH extract upregulates the HO-1 levels dose-dependently, which was further verified by treating the RAW264.7 cells with the specific inhibitors ERK (U0123) and JNK (SP600125) (Figure 5A–C). Previous studies have illustrated that the presence of certain bioactive compounds, including fucosterol, loliolide, sargachromenol, gallic acid, 3-Hydroxy-5,6-epoxy-β-ionone, etc., detected or isolated from SH resulted in the attenuation of oxidative stress and inflammation [14,15,45]. Furthermore, Alam et al. [5] checked the activity of *Nymphaea nouchali* (Burm. f) stem extracts along with metabolite profiling using a high-resolution mass spectrometer and reported different bioactive compounds, including gallic acid, naringin, epicatechin, catechin, 5-*O*-caffeoylquinic acid, 3-Feruloylquinic, etc., resulting in the attenuation of oxidative stress through the regulation of MAPK/NRF2/HO-1/ROS signaling. In the current trial, we use a novel approach to extract the bioactive compounds, including lignans, terpenes, tannins, etc., which have not been previously reported together in a trial [3,5], which could be the reason for the stronger antioxidative stress activity exhibited by SH.

Today, in silico approaches are attracting more attention as these modern computational and experimental methodologies may be combined more effectively than a battery of laboratory experimental analysis [46,47]. For this purpose, MolDoc and AutoDoc molecular docking simulations were carried out on the 55 selected and identified compounds against the Keap1 protein (4L7B). Among them, five compounds exhibited the highest energy, and sargaquinoic acid glycitein 7-*O*-glucuronide (Table 3) exhibited the highest binding affinity in AutoDoc Vina (−9.20 Kcal/mol; 9.52 Kcal/mol) and MolDoc (−155.842; −148.975), respectively. According to the literature, one Nrf2 molecule is attached to a Keap1 dimer through two distinct motifs (DLG and ETGE) in the neh2 region via a "hinge and latch" mechanism. The ETGE and DLG motifs form a β-turn structure through the electrostatic interaction among glutamate and acidic aspartate with Arg380, Arg415, and Arg483 in the Kelch region of Keap1 [48]. Therefore, these three amino acids gained more importance during the ligand–protein interaction. Li et al. [30] used the egg-derived tri-peptide proteins to inhibit the direct interaction of Keap1 with Nrf2 using molecular docking, and the results showed that DKK and DDW proteins bind with Arg380, Asn382, Arg415, Arg482, and Ser508, respectively. Likewise, Adelusi et al. [48] determined the quantum mechanics, dynamics, and docking experiments to identify the Keap1 inhibitors, and among 50 antioxidant compounds, maslinic acid exhibited the highest binding energy (−10.6 kJ/mol). Among different amino acid residue interactions, maslinic acid forms van der Waals interactions with Arg415 of the Keap1 protein. Recently, Diniyah et al. [17] also determined the binding ability of four bioactive compounds, i.e., gallic acid, coumaric acid, epicatechin, and catechin, for Keap1 protein, and the results showed that these four compounds also form bonds with different amino acids, especially Arg415. In the current experiment, sargaquinoic acid forms one hydrogen interaction with Arg483, whereas glycitein 7-*O*-glucuronide forms a pi–pi interaction with Arg415 (Figure 6A–D), which is parallel with the previous studies [17].

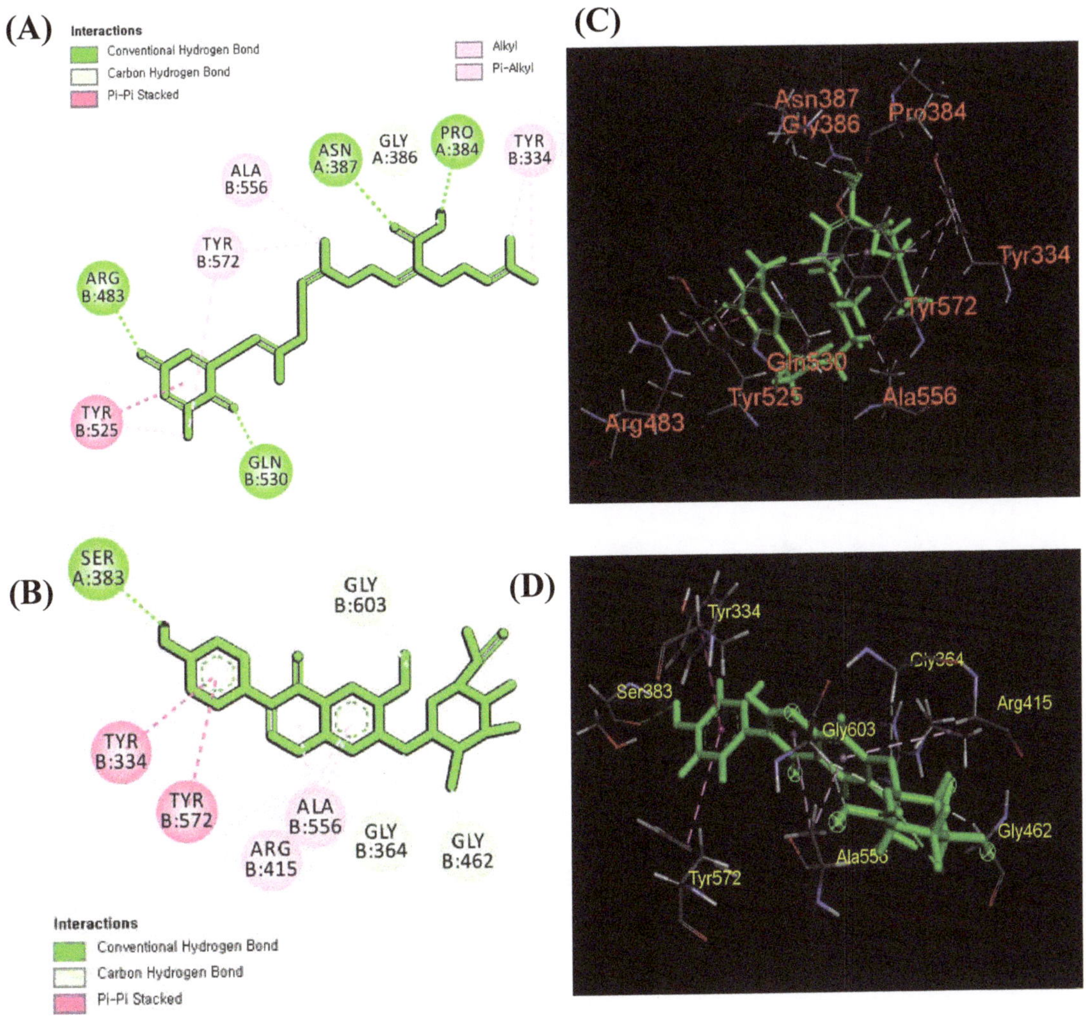

Figure 6. Molecular interaction results of sargaquinoic acid with Keap1 protein (4L7B). The 2D molecular interaction of sargaquinoic acid (**A**) and of glycitein 7-*O*-glucuronide (**B**) and the 3D molecular interaction of sargaquinoic acid (**C**) and of glycitein 7-*O*-glucuronide (**D**) with surrounding Keap1 protein amino acids.

5. Conclusions

After optimizing the extraction condition, 89 bioactive compounds were reported in optimized SH extract, and among them, 19 compounds, including phenolic acid, flavonoids, terpenes, lignan, etc., were tentatively reported for the first time in SH. The in vitro results indicated that optimized SH extract attenuates the oxidative stress in a dose-dependent fashion via the Nrf2/MAPK/HO-1 signaling pathway due to the presence of active ingredients, including gallic acid, epicatechin, catechin, 5-*O*-caffeoylquinic acid, and 3-feruloylquinic, etc. For further validation, in silico simulation was conducted on the Keap1 protein using 55 bioactive compounds and the result indicated that sargaquinoic acid and glycitein 7-*O*-glucuronide could potentially bind with Keap1, as exhibited by higher binding energy (9.20 Kcal/mol; 9.52 Kcal/mol), by forming a hydrogen bond with Arg483 and Arg415. The current investigation provides an alternative statistical technique and supports a preferred extraction technology for identifying important bioactive compounds that could be otherwise present in lower concentrations under non-optimized conditions. Because temperature, time, and solvent concentration greatly affect extraction yield, multivariant optimization is a promising approach instead of single-factor optimization. This trial provides the basis for the scientific community to isolate/extract compounds that could be used in broad commercial applications as promising ingredients for the development of functional foods and nutraceuticals. However, further in vitro and in vivo research should be conducted to check the toxicity and mechanism of action of identified compounds, especially sargaquinoic acid and glycitein 7-*O*-glucuronide, against oxidative stress and related diseases. Furthermore, the dynamics and kinetics of these compounds should also be checked using YASARA to visualize protein interaction to validate our in silico results.

Supplementary Materials: The following supporting information can be downloaded at https://www.mdpi.com/article/10.3390/antiox13060690/s1: Table S1: Setting parameters of genetic algorithm used in the model optimization of SH; Table S2: ANVOA analysis for developed TPC, TFC, DPPH, and ABTS models; Table S3: Fit statistics values of independent variables and corresponding target responses; Table S4: Comparison of the prediction abilities of the RSM and ANN models; Table S5: RSM- and ANN-optimized model condition; and Table S6: Binding affinity of identified bioactive compounds for Keap1 protein (4L7B).

Author Contributions: A.J. and M.N. performed the experiments. A.J., M.B.A., R.A., C.H.L., S.K. and S.-H.L. designed the research and analyzed the data. A.J., M.B.A., S.K. and S.-H.L. wrote the paper. A.J., R.A. and S.-H.L. revised the manuscript. All authors have read and agreed to the published version of the manuscript.

Funding: This study was supported by the National Research Foundation of Korea (NRF) and funded by the Ministry of Science and ICT [grant numbers 2020R1A2C2011495, 2021R1IA1A01058062, and RS-2023-00278670].

Institutional Review Board Statement: Not applicable.

Informed Consent Statement: Not applicable.

Data Availability Statement: The data will be available upon request.

Acknowledgments: We thank all the laboratory members for their help during the experiments. I would also like to thank the Korean government for providing a Ph.D. scholarship (GKS).

Conflicts of Interest: The authors declare that they have no known competing financial interests or personal relationships that could have appeared to influence the work reported in this paper.

References

1. Banerjee, J.; Das, A.; Sinha, M.; Saha, S. Biological efficacy of medicinal plant extracts in preventing oxidative damage. *Oxid. Med. Cell. Longev.* **2018**, *1–2*, 7904349. [CrossRef] [PubMed]
2. Castaneda, O.A.; Lee, S.C.; Ho, C.T.; Huang, T.C. Macrophages in oxidative stress and models to evaluate the antioxidant function of dietary natural compounds. *J. Food Drug Anal.* **2017**, *25*, 111–118. [CrossRef] [PubMed]
3. Alam, M.B.; Ahmed, A.; Islam, S.; Choi, H.-J.; Motin, M.A.; Kim, S.; Lee, S.-H. Phytochemical Characterization of Dillenia indica L. bark by paper spray ionization-mass spectrometry and evaluation of its antioxidant potential against t-BHP-induced oxidative stress in RAW 264.7 cells. *Antioxidants* **2020**, *9*, 1099. [CrossRef]
4. Taguchi, K.; Motohashi, H.; Yamamoto, M. Molecular mechanisms of the Keap1–Nrf2 pathway in stress response and cancer evolution. *Genes Cells* **2011**, *16*, 123–140. [CrossRef]
5. Alam, M.B.; Naznin, M.; Islam, S.; Alshammari, F.H.; Choi, H.-J.; Song, B.-R.; Lee, S.-H. High Resolution Mass Spectroscopy-Based Secondary Metabolite Profiling of *Nymphaea nouchali* (Burm. f) Stem Attenuates Oxidative Stress via Regulation of MAPK/Nrf2/HO-1/ROS Pathway. *Antioxidants* **2021**, *10*, 719. [CrossRef]

6. Baird, L.; Dinkova-Kostova, A.T. The cytoprotective role of the Keap1–Nrf2 pathway. *Arch. Toxicol.* **2011**, *85*, 241–272. [CrossRef] [PubMed]
7. Javed, A.; Hussain, M.B.; Tahir, A.; Waheed, M.; Anwar, A.; Shariati, M.A.; Plygun, S.; Laishevtcev, A.; Pasalar, M. Pharmacological applications of phlorotannins: A comprehensive review. *Curr. Drug Discov. Technol.* **2021**, *18*, 282–292. [CrossRef] [PubMed]
8. Jayawardena, T.U.; Sanjeewa, K.K.A.; Fernando, I.P.S.; Ryu, B.M.; Kang, M.C.; Jee, Y.; Lee, W.W.; Jeon, Y.J. *Sargassum horneri* (Turner) C. Agardh ethanol extract inhibits the fine dust inflammation response via activating Nrf2/HO-1 signaling in RAW 264.7 cells. *BMC Complement Altern. Med.* **2021**, *18*, 249. [CrossRef]
9. Han, E.-J.; Fernando, I.P.S.; Kim, H.-S.; Lee, D.-S.; Kim, A.; Je, J.-G.; Seo, M.-J.; Jee, Y.-H.; Jeon, Y.-J.; Kim, S.-Y.; et al. (−)-Loliolide Isolated from *Sargassum horneri* Suppressed Oxidative Stress and Inflammation by Activating Nrf2/HO-1 Signaling in IFN-γ/TNF-α-Stimulated HaCaT Keratinocytes. *Antioxidants* **2021**, *10*, 856. [CrossRef] [PubMed]
10. Aung, T.; Kim, S.J.; Eun, J.B. A hybrid RSM-ANN-GA approach on optimisation of extraction conditions for bioactive component-rich laver (*Porphyra dentata*) extract. *Food Chem.* **2022**, *366*, 130689. [CrossRef]
11. Yahya, H.S.M.; Abbas, T.; Amin, N.A.S. Optimization of hydrogen production via toluene steam reforming over Ni–Co supported modified-activated carbon using ANN coupled GA and RSM. *Int. J. Hydrogen Energy* **2021**, *46*, 24632–24651. [CrossRef]
12. Devi, S.A.; Joseph, J.; Rajkumar, J. In silico approach of antibacterial compounds from mangrove—Avicennia marina through docking analysis. *Biomed Res.* **2015**, *26*, 50.
13. Vázquez-Rodríguez, B.; Gutiérrez-Uribe, J.A.; Antunes-Ricardo, M.; Santos-Zea, L.; Cruz- Suárez, L.E. Ultrasound-assisted extraction of phlorotannins and polysaccharides from Silvetia compressa (*Phaeophyceae*). *J. Appl. Phycol.* **2020**, *32*, 1441–1453. [CrossRef]
14. Ez Zoubi, Y.; Fadil, M.; Bousta, D.; El Ouali Lalami, A.; Lachkar, M.; Farah, A. Ultrasound-Assisted Extraction of Phenolic Compounds from Moroccan *Lavandula stoechas* L.: Optimization Using Response Surface Methodology. *J. Chem.* **2021**, *1*, 11. [CrossRef]
15. Javed, A.; Alam, M.B.; Naznin, M.; Shafique, I.; Kim, S.; Lee, S.H. Tyrosinase inhibitory activity of Sargassum fusiforme and characterisation of bioactive compounds. *Phytochem. Anal.* **2023**, *7*, 1–15. [CrossRef] [PubMed]
16. Javed, A.; Naznin, M.; Alam, B.; Fanar, A.; Song, B.; Kim, S.; Lee, S.H. Metabolite Profiling of Microwave-Assisted Sargassum fusiforme Extracts with Improved Antioxidant Activity Using Hybrid Response Surface Methodology and Artificial Neural. *Antioxidants* **2022**, *11*, 2246. [CrossRef]
17. Choi, H.J.; Naznin, M.; Alam, M.B.; Javed, A.; Alshammari, F.H.; Kim, S.; Lee, S.H. Optimization of the extraction conditions of Nypa fruticans Wurmb. using response surface methodology and artificial neural network. *Food Chem.* **2021**, *381*, 132086. [CrossRef]
18. Diniyah, N.; Alam, M.B.; Javed, A.; Alshammari, F.; Choi, H.J.; Lee, S.H. In silico and docking studies on the binding activities of Keap1 of antioxidant compounds in non-oilseed legumes. *Arab. J. Chem.* **2023**, *16*, 104414. [CrossRef]
19. Islam, S.; Alam, M.B.; Ahmed, A.; Lee, S.; Lee, S.-H.; Kim, S. Identification of secondary metabolites in Averrhoa carambola L. bark by high-resolution mass spectrometry and evaluation for α-glucosidase, tyrosinase, elastase, and antioxidant potential. *Food Chem.* **2020**, *332*, 127377. [CrossRef]
20. Zhong, B.; Robinson, N.A.; Warner, R.D.; Barrow, C.J.; Dunshea, F.R.; Suleria, H.A.R. LC-ESI-QTOF-MS/MS characterization of seaweed phenolics and their antioxidant potential. *Mar. Drugs* **2020**, *18*, 331. [CrossRef]
21. Ghallab, D.S.; Shawky, E.; Ibrahim, R.S.; Mohyeldin, M.M. Comprehensive metabolomics unveil the discriminatory metabolites of some Mediterranean Sea marine algae in relation to their cytotoxic activities. *Sci. Rep.* **2022**, *2*, 8094. [CrossRef]
22. Agregán, R.; Munekata, P.E.S.; Franco, D.; Dominguez, R.; Carballo, J.; Lorenzo, J.M. Phenolic compounds from three brown seaweed species using LC-DAD–ESI-MS/MS. *Food Res. Int.* **2017**, *99*, 979–985. [CrossRef] [PubMed]
23. Islam, S.; Alam, M.B.; Ann, H.-J.; Park, J.-H.; Lee, S.-H.; Kim, S. Metabolite Profiling of Manilkara zapota L. Leaves by High-Resolution Mass Spectrometry Coupled with ESI and APCI and In Vitro Antioxidant Activity, α-Glucosidase, and Elastase Inhibition Assays. *Int. J. Mol. Sci.* **2021**, *22*, 132. [CrossRef] [PubMed]
24. Lomakool, S.; Ruangrit, K.; Jeerapan, I.; Tragoolpua, Y.; Pumas, C.; Srinuanpan, S.; Duangjan, K. Biological activities and phytochemicals profiling of different cyanobacterial and microalgal biomass. *Biomass Convers. Biorefin.* **2021**, *13*, 4195–4211. [CrossRef]
25. Xu, W.H.; Zhao, P.; Wang, M.; Liang, Q. Naturally occurring furofuran lignans: Structural diversity and biological activities. *Nat. Prod. Res.* **2019**, *33*, 1357–1373. [CrossRef] [PubMed]
26. Kim, S.N.; Choi, H.Y.; Lee, W.; Park, G.M.; Shin, W.S.; Kim, Y.K. Sargaquinoic acid and sargahydroquinoic acid from Sargassum yezoense stimulate adipocyte differentiation through PPARα/γ activation in 3T3-L1 cells. *FEBS Lett.* **2008**, *582*, 3465–3472. [CrossRef] [PubMed]
27. Jerković, I.; Cikoš, A.M.; Babić, S.; Čižmek, L.; Bojanić, K.; Aladić, K.; Jokić, S. Bioprospecting of less-polar constituents from endemic brown macroalga fucus virsoides J. Agardh from the adriatic sea and targeted antioxidant effects in vitro and in vivo (Zebrafish model). *Mar. Drugs* **2021**, *19*, 235. [CrossRef] [PubMed]
28. Herath, K.H.I.N.M.; Kim, H.J.; Jang, J.-H.; Kim, H.-S.; Kim, H.J.; Jeon, Y.-J.; Jee, Y. Mojabanchromanol Isolated from *Sargassum horneri* Attenuates Particulate Matter Induced Inflammatory Responses via Suppressing TLR2/4/7-MAPK Signaling in MLE-12 Cells. *Mar. Drugs* **2020**, *18*, 355. [CrossRef] [PubMed]

29. Rushdi, M.I.; Abdul-Rehman, I.A.M.; Sabar, H.; Attia, E.Z.; Abdelraheem, W.M.; Madkour, H.A.; Hassan, H.M.; Elmaidomy, A.H.; Adbelmohsen, U.R. Pharmacological and natural products diversity of the brown algae genus Sargassum. *RSC Adv.* **2020**, *10*, 24951–24972. [CrossRef] [PubMed]
30. Li, Y.; Fu, X.; Duan, D.; Liu, X.; Xu, J.; Gao, X. Extraction and Identification of Phlorotannins from the Brown Alga, Sargassum fusiforme (Harvey) Setchell. *Mar. Drugs* **2017**, *15*, 49. [CrossRef]
31. Cho, J.H.; Kim, D.E.; Lee, J.S.; Seo, M.S.; Kim, M.E.; Lee, J.S. *Sargassum horneri* (Turner) C. Agardh Extract Regulates Neuroinflammation In Vitro and In Vivo. *Curr. Issues Mol. Biol.* **2022**, *44*, 5416–5426. [CrossRef]
32. Buedenbender, L.; Astone, F.A.; Tasdemir, D. Bioactive Molecular Networking for Mapping the Antimicrobial Constituents of the Baltic Brown Alga Fucus vesiculosus. *Mar. Drugs* **2021**, *18*, 311. [CrossRef] [PubMed]
33. El-Sheekh, M.M.; El-Shenody, R.A.E.K.; Bases, E.A.; El Shafay, S.M. Comparative assessment of antioxidant activity and biochemical composition of four seaweeds, rocky bay of abu qir in alexandria, Egypt. *Food Sci. Technol.* **2021**, *41*, 29–40. [CrossRef]
34. Grina, F.; Ullah, Z.; Kaplaner, E.; Moujahid, A.; Eddoha, R.; Nasser, B.; Essamadi, A. In vitro enzyme inhibitory properties, antioxidant activities, and phytochemical fingerprints of five Moroccan seaweeds. *S. Afr. J. Bot.* **2020**, *128*, 152–160. [CrossRef]
35. Jørgensen, K.; Olesen, P.T. Kainic acid in the seaweed *Palmaria palmata* (dulse). *Food Addit. Contam. Part B Surveill* **2018**, *11*, 198–200. [CrossRef] [PubMed]
36. Maheswari, M.U.; Reena, A. Phytochemical Profiling of the Red Seaweed, Halymenia dilatata by GC-MS Analysis. *Int. J. Pharm. Sci. Res.* **2017**, *8*, 167–172.
37. Pereira, A.G.; Fraga-Corral, M.; Garcia-Oliveira, P.; Lourenço-Lopes, C.; Carpena, M.; Prieto, M.A.; Simal-Gandara, J. The use of invasive algae species as a source of secondary metabolites and biological activities: Spain as case-study. *Mar. Drugs* **2021**, *19*, 178. [CrossRef] [PubMed]
38. Park, C.; Lee, H.; Hong, S.H.; Kim, J.H.; Park, S.K.; Jeong, J.W.; Kim, G.Y.; Hyun, J.W.; Yun, S.J.; Kim, B.W.; et al. Protective effect of diphlorethohydroxycarmalol against oxidative stress-induced DNA damage and apoptosis in retinal pigment epithelial cells. *Cutan. Ocul. Toxicol.* **2019**, *38*, 298–308. [CrossRef]
39. Troot, O.; Olson, A.J. Software News and Update AutoDock Vina: Improving the Speed and Accuracy of Docking with a New Scoring Function, Efficient Optimization, and Multithreading. *J. Comput. Chem.* **2010**, *31*, 455–461. [CrossRef]
40. Thomsen, R.; Christensen, M.H. MolDock: A new technique for high-accuracy molecular docking. *J. Med. Chem.* **2006**, *49*, 3315–3321. [CrossRef]
41. Alshammari, F.; Alam, M.B.; Naznin, M.; Javed, A.; Kim, S.; Lee, S.-H. Profiling of Secondary Metabolites of Optimized Ripe Ajwa Date Pulp (*Phoenix dactylifera* L.) Using Response Surface Methodology and Artificial Neural Network. *Pharmaceuticals* **2023**, *16*, 319. [CrossRef]
42. Rezaei, M.; Valipour, M.; Valipour, M. Modelling Evapotranspiration to Increase the Accuracy of the Estimations Based on the Climatic Parameters. *Water Conserv. Sci. Eng.* **2016**, *1*, 197–207. [CrossRef]
43. Tan, J.; Cui, P.; Ge, S.; Cai, X.; Li, Q.; Xue, H. Ultrasound assisted aqueous two-phase extraction of polysaccharides from Cornus officinalis fruit: Modeling, optimization, purification, and characterization. *Ultrason. Sonochem.* **2016**, *84*, 105966. [CrossRef] [PubMed]
44. Ransy, C.; Vaz, C.; Lombès, A.; Bouillaud, F. Use of H_2O_2 to Cause Oxidative Stress, the Catalase Issue. *Int. J. Mol. Sci.* **2021**, *21*, 9149. [CrossRef] [PubMed]
45. Park, C.; Jeong, J.-W.; Lee, D.-S.; Yim, M.-J.; Lee, J.M.; Han, M.H.; Kim, S.; Kim, H.-S.; Kim, G.-Y.; Park, E.K.; et al. Sargassum serratifolium Extract Attenuates Interleukin-1β-Induced Oxidative Stress and Inflammatory Response in Chondrocytes by Suppressing the Activation of NF-κB, p38 MAPK, and PI3K/Akt. *Int. J. Mol. Sci.* **2018**, *19*, 2308. [CrossRef] [PubMed]
46. Choi, K.S.; Shin, T.S.; Chun, J.; Ahn, G.; Han, E.J.; Kim, M.J.; Kim, J.B.; Kim, S.H.; Kho, K.H.; Kim, D.H.; et al. Sargahydroquinoic acid isolated from Sargassum serratifolium as inhibitor of cellular basophils activation and passive cutaneous anaphylaxis in mice. *Int. Immunopharmacol.* **2022**, *105*, 108567. [CrossRef] [PubMed]
47. Dinkova-Kostova, A.T.; Fahey, J.W.; Kostov, R.V.; Kensler, T.W. KEAP1 and done? Targeting the NRF2 pathway with sulforaphane. *Trends Food Sci Technol.* **2017**, *69*, 257–269. [CrossRef] [PubMed]
48. Adelusi, T.I.; Hammed, M.A.; Idrus, M.O.; Oyedele, Q.K.; Adedotun, I.O. Molecular dynamics, quantum mechanics and docking studies of some Keap1 inhibitors—An insight into the atomistic mechanisms of their antioxidant potential. *Heliyon* **2021**, *7*, e07317. [CrossRef]

molecules

Article

EGCG Suppresses Adipogenesis and Promotes Browning of 3T3-L1 Cells by Inhibiting Notch1 Expression

Yinghao Wang [1,2,†], Chunfeng Li [1,3,†], Wenyuan Peng [1,3], Jun Sheng [1], Chengting Zi [1,3,*] and Xiaoyun Wu [1,2,*]

1 Key Laboratory of Puer Tea Science, Ministry of Education, Yunnan Agricultural University, Kunming 650201, China; wyhh0518@163.com (Y.W.); 15154962737@163.com (C.L.); u1n3fj9u3v@163.com (W.P.); shengj@ynau.edu.cn (J.S.)
2 Department of Science, Yunnan Agricultural University, Kunming 650201, China
3 College of Food Science and Technology, Yunnan Agricultural University, Kunming 650201, China
* Correspondence: zichengting@126.com (C.Z.); wuxiaoyun7979@gmail.com (X.W.)
† These authors contributed equally to this work.

Abstract: Background: With the changes in lifestyle and diet structure, the incidence of obesity has increased year by year, and obesity is one of the inducements of many chronic metabolic diseases. Epigallocatechin gallate (EGCG), which is the most abundant component of tea polyphenols, has been used for many years to improve obesity and its complications. Though it has been reported that EGCG can improve obesity through many molecular mechanisms, EGCG may have many mechanisms yet to be explored. In this study, we explored other possible mechanisms through molecular docking and in vitro experiments. Methods: AutoDock Vina was selected for conducting the molecular docking analysis to elucidate the interaction between EGCG and Notch1, while molecular dynamics simulations were employed to validate this interaction. Then, the new regulation mechanism of EGCG on obesity was verified with in vitro experiments, including a Western blot experiment, immunofluorescence experiment, oil red O staining, and other experiments in 3T3-L1 adipocytes. Results: The molecular docking results showed that EGCG could bind to Notch1 protein through hydrogen bonding. In vitro cell experiments demonstrated that EGCG can significantly reduce the sizes of lipid droplets of 3T3-L1 adipocytes and promote UCP-1 expression by inhibiting the expression of Notch1 in 3T3-L1 adipocytes, thus promoting mitochondrial biogenesis. Conclusions: In this study, molecular docking and in vitro cell experiments were used to explore the possible mechanism of EGCG to improve obesity by inhibiting Notch1.

Keywords: EGCG; Notch1; UCP-1; obesity; mitochondria

Citation: Wang, Y.; Li, C.; Peng, W.; Sheng, J.; Zi, C.; Wu, X. EGCG Suppresses Adipogenesis and Promotes Browning of 3T3-L1 Cells by Inhibiting Notch1 Expression. *Molecules* **2024**, *29*, 2555. https://doi.org/10.3390/molecules29112555

Academic Editors: Yongliang Zhuang, Hu Hou, Zhipeng Yu and Shengbao Cai

Received: 29 April 2024
Revised: 25 May 2024
Accepted: 27 May 2024
Published: 29 May 2024

1. Introduction

Obesity is now one of the most common metabolic diseases, affecting more than 600 million people worldwide [1], and long-term obesity leads to an increased risk of diabetes [2], hyperuricemia [3], and cardiovascular disease [4]. At present, the main reason leading to obesity is excessive dietary intake and lack of physical activity, resulting in the excessive accumulation of lipids. The heat production of brown adipocytes and the browning of white adipocytes are new treatment strategies for obesity, and the browning of white fat has attracted attention in recent years due to its significant effect on fat reduction [5]. The increased expression of uncoupling protein 1 (UCP-1) mRNA and mitochondrial development in white fat are conducive to browning and can increase mitochondrial density and cell respiration [6].

Molecular docking is a computer simulation technique that simulates the interaction between a molecule and a target protein at the atomic level and calculates the affinity to evaluate the binding [7]. Molecular dynamics simulation is widely used in pharmacology and biomedical fields to study the interaction between ligands and receptors as well as conformational changes [8].

EGCG, a widely studied polyphenol, is the most abundant polyphenol in green tea [9]. It has anti-tumor [10], anti-obesity [11], anti-diabetes [12] and anti-inflammatory [13] effects. In recent years, many studies have demonstrated that EGCG can treat obesity through multiple signaling pathways. Li et al. found that EGCG can inhibit lipid accumulation in obese mice by activating AMPK [11]. Impaired bile acid (BA) synthesis leads to impaired lipid metabolism; Sheng et al. found that EGCG could alleviate obesity by regulating the bile acid signaling pathway and intestinal flora using metabolomics analysis [14]. Obesity also leads to neuroinflammation and energy metabolism disorders. Zhou et al. found that the supplementation of EGCG significantly inhibited obesity induced by a high-fat diet by enhancing brown adipose tissue (BAT) thermogenesis and reduced hypothalamic inflammation by down-regulating the hypothalamic STAT3 signaling pathway [15]. Additionally, EGCG also has significant anti-lipogenesis activity in 3T3-L1 cells [16]. Choi et al. also found that the anti-obesity effect of EGCG required the up-regulation of Beclin1-dependent autophagy and lipid catabolism in white adipose tissue (WAT) [17]. These results suggest that EGCG, as a natural polyphenol compound, has multi-target and multi-pathway characteristics in vivo due to its polyhydroxyl structure [18]. Therefore, EGCG may have many mechanisms to improve obesity yet to be explored.

Notch1, as an inflammatory marker protein, exerts regulatory effects on obesity development. Previous studies have demonstrated that the activation of Notch1 signaling promotes adipocyte proliferation [19], while the modulation of the Notch1 signal in adipose tissue macrophages is crucial for regulating inflammation and metabolism [20]. Moreover, inhibiting Notch1 expression has been shown to significantly enhance brown fat thermogenesis, promote white fat browning, and ameliorate obesity [21–23]. Although EGCG can improve inflammation by inhibiting Notch1 and has significant effects on obesity [11,24], it is still uncertain whether EGCG in adipose tissue can have the same metabolic-promoting and obesity-improving effects by targeting Notch1.

This study aims to investigate the potential of EGCG in inhibiting Notch1 expression, promoting adipocyte browning, and improving obesity through the establishment of a molecular docking model, conducting molecular dynamics experiments, and utilizing in vitro cell models.

2. Results

2.1. Molecular Docking

Molecular docking is a theoretical method employed to investigate the interaction between receptors and ligands, enabling the prediction of the binding mode and molecular mechanism [25]. In this study, AutoDock was utilized to dock EGCG with Notch1. Degree analysis was conducted using AutoDock Vina-1.5.6 software, revealing a binding energy of −6.9 kcal/mol between Notch1 and EGCG. The binding mode of EGCG at the pocket of Notch1 is depicted in Figure 1A,B. The visual examination of the molecular docking results between EGCG and Notch1 was performed using PyMOL-1.8.6 software, demonstrating that EGCG forms hydrogen bonds with sites TYP482, GLY481, ASP492, and GLU493 on the Notch1 protein (Figure 1C). Table 1 presents the binding free energy values for the EGCG-Notch1 complex: total binding free energy (-144.317 ± 1.15 kcal/mol), van der Waals energy (-184.34 ± 1.16 kcal/mol), electrostatic energy (-32.03 ± 0.79 kcal/mol), polar solvation energy (89.13 ± 106 kcal/mol), and SASA (solvent-accessible surface area) energy (-17.06 ± 0.9 kcal/mol). These findings highlight the crucial role played by van der Waals interactions in forming the EGCG-Notch1 complex.

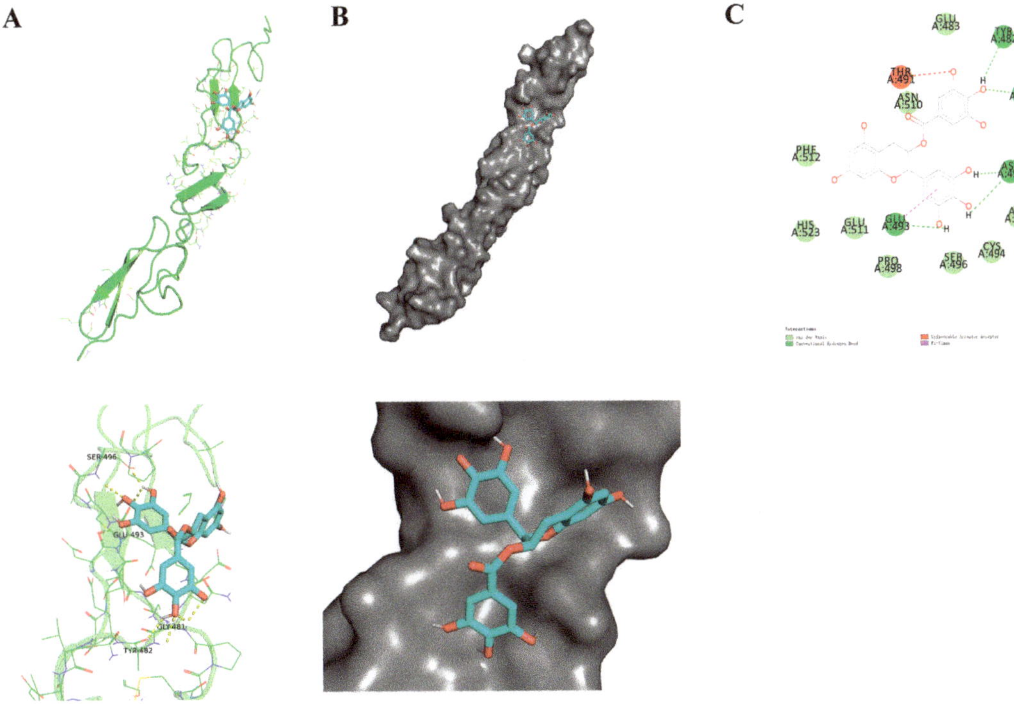

Figure 1. Molecular docking model of EGCG and Notch1. (**A**) The 3D model of the EGCG interaction with Notch1; (**B**) EGCG-docked Notch1 structure depicted and EGCG zoom-in binding pocket. (**C**) The 2D model of the EGCG interaction with Notch1.

Table 1. Calculated MM-PBSA binding free energy in kcal/mol for the Notch1-EGCG complex.

Energy Contribution	Value (kcal/mol)
ΔG_{vdw} [a]	−184.34 ± 1.16
ΔG_{Electr} [b]	−32.03 ± 0.79
ΔG_{Polar} [c]	89.13 ± 1.06
ΔG_{SASA} [d]	−17.06 ± 0.90
ΔG_{bind} [e]	−144.30 ± 1.15

[a] ΔG_{vdw} is van der Waals energy; [b] ΔG_{Electr} is electrostatic energy contribution; [c] ΔG_{Polar} is polar contributions between the solute and solvent to the solvation energy; [d] ΔG_{SASA} is non-polar solvation energy using the solvent accessible surface area; [e] and ΔG_{bind} is the total free binding energy.

2.2. Molecular Dynamics (MD) Simulation

The stability of the ligand-protein system was analyzed at different time scales using MD simulations to further investigate the interaction between Notch1 and EGCG. Molecular dynamics simulations were performed for 30 ns using the GROMAC 5.1.2 software package [26]. Figure 2A demonstrates that the α-C RMSD values of free Notch1 are comparable to those of the Notch1-EGCG complex. However, after 25 ns, there is a larger fluctuation in α-C RMSD values for Notch1, while the α-C RMSD values for the Notch1-EGCG complex exhibit smaller fluctuations, indicating enhanced stability upon the binding of EGCG to Notch1. The Rg value reflects the compactness of a complex (Figure 2B), where compared to Notch1 alone, a decrease in Rg is observed after 20 ns for Noch1-EGCG, suggesting tighter structure formation in the EGCG-Notch1 complex. The trajectory-based RMSF plot of the EGCG-Notch1 complex was utilized to analyze results from 20 ns to 30 ns (Figure 2C), with exceptional stability among interacting residues indicated by an

RMSF value of 0.15 Å. The SASA value of the complex indicates the solvent polarizability energy of molecular interactions, as illustrated in Figure 2D. It is significantly higher for Notch1 binding with EGCG compared to Notch1 alone, suggesting that upon binding with EGCG, the structure of Notch1 undergoes contraction and forms a more stable complex. As depicted in Figure 2E, hydrogen bonding between EGCG and Notch1 occurs from the initial stages and gradually intensifies within the range of 0~7. Furthermore, as shown in Figure 2F, the calculated distance between EGCG and Notch1 reaches a minimum value of 0.16 nm. These results further validate the findings from docking studies, emphasizing the crucial role played by hydrogen bonding in mediating interactions between the EGCG-Notch1 complex.

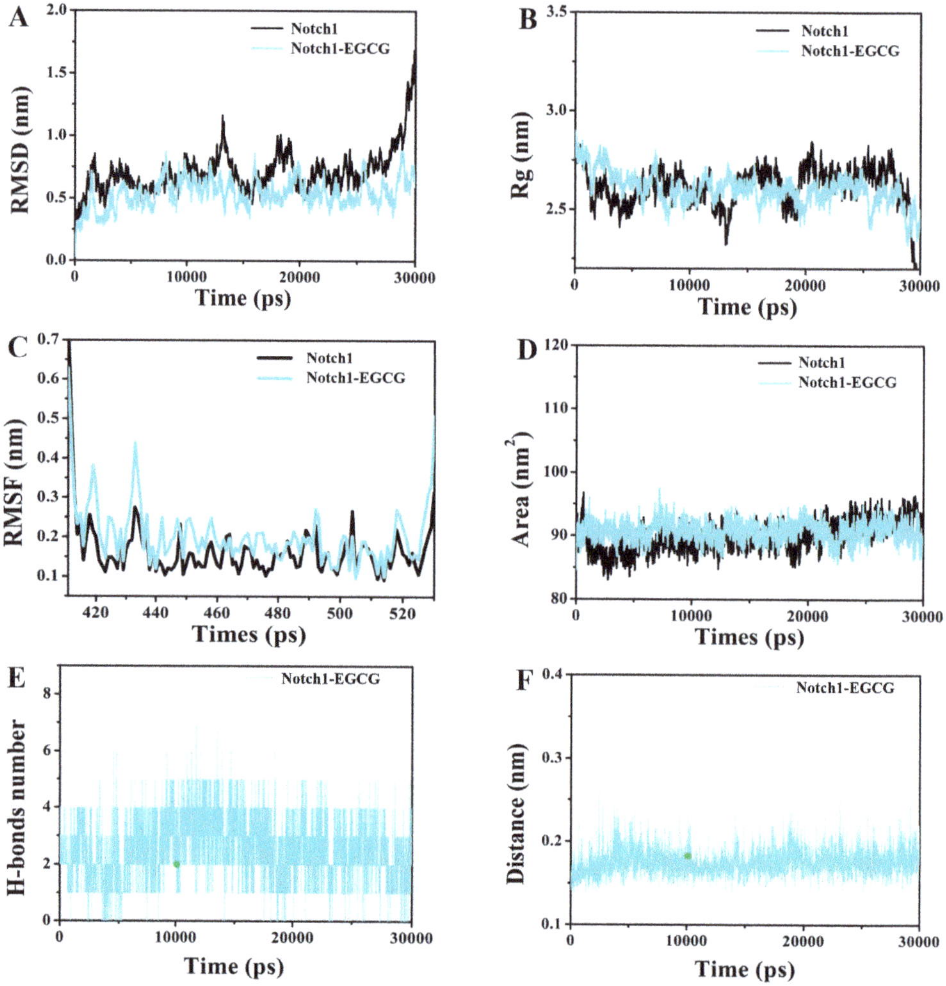

Figure 2. Molecular dynamics simulation of EGCG with Notch1 using the GROMACS package and the 54a7 force field at 300 K during 30 ns simulation times. (**A**) The RMSD of carbon alpha plot for Notch1 and the EGCG-Notch1 complex. (**B**) The Rg plot for Notch1 and the EGCG-Notch1 complex. (**C**) The RMSF plot for Notch1 and the EGCG-Notch1 complex. (**D**) The SASA plot for Notch1 and the EGCG-Notch1 complex. (**E**) The number of hydrogen bonds formed between EGCG and Notch1 during the simulation. (**F**) The distance of hydrogen bond formed between EGCG and Notch1.

2.3. EGCG Structural Formula, EGCG Cytotoxicity to 3T3-L1 Cells

Before starting the experiment, we investigated the cytotoxicity of EGCG to 3T3-L1 cells through cytotoxicity tests, and it was found that EGCG concentrations (2.5–80 μM) did not produce cytotoxicity in 3T3-L1 cells (Figure 3B).

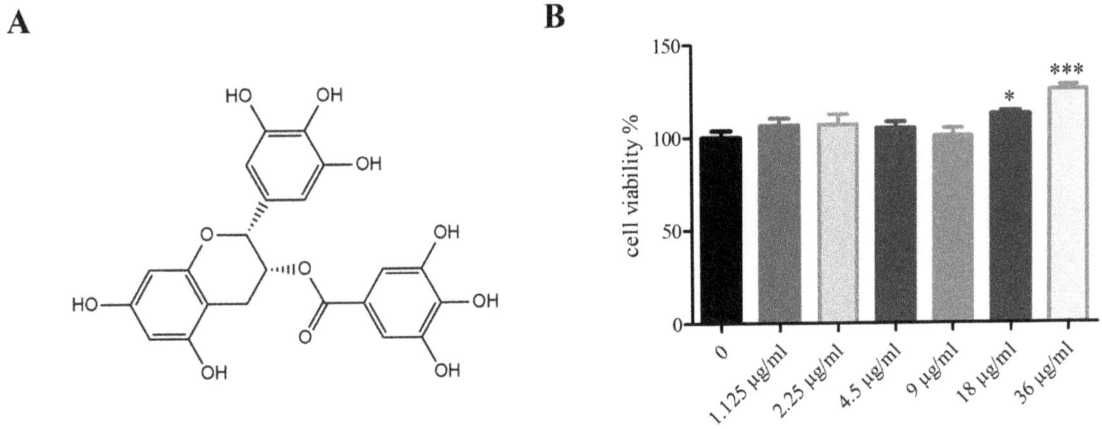

Figure 3. The structural formula of EGCG and the cytotoxicity of EGCG to 3T3-L1 cells. (**A**) Structural formula of EGCG. (**B**) MTT showed the cytotoxicity of EGCG at different concentrations in 3T3-L1 cells. The values are expressed as means ± SEM (n = 6). * $p < 0.05$ and *** $p < 0.001$ compared with the 0 μM EGCG.

2.4. Effect of EGCG on Lipid Synthesis during 3T3-L1 Cell Differentiation

PPARγ and PPARα are key proteins in fatty acid synthesis [27], and the expressions of PPARγ and PPARα gradually increased with the differentiation stages of 3T3-L1 cells. After EGCG treatment, the expressions of PPARγ and PPARα at each stage were significantly inhibited in a dose-dependent manner (Figure 4A). After the treatment of 3T3-L1 cells with different doses of EGCG, the lipid droplet sizes of 3T3-L1 cells were significantly reduced in a dose- and time-dependent manner (Figure 4B). These results suggested that EGCG treatment significantly inhibited lipid synthesis during 3T3-L1 cell differentiation and reduced lipid droplet sizes.

2.5. Effects of EGCG Treatment on Notch1 and UCP1 Expression in 3T3-L1 Cells

UCP-1 is one of the marker proteins of mitochondrial genesis [28]. Previous studies have shown that the increased expression of UCP-1 can significantly inhibit fat synthesis, and the expression of Notch1 is also related to UCP-1 [21]. After the EGCG treatment of mature 3T3-L1, the expression level of Notch1 decreased significantly with the increasing treatment time; in contrast, UCP-1 increased significantly with the increasing EGCG treatment time (Figure 5A). UCP-1 immunofluorescence and mitochondrial staining also showed the same results; EGCG significantly enhanced the expression of UCP-1 in 3T3-L1 cells and promoted mitochondrial genesis (Figure 5B). These results suggested that EGCG could promote UCP-1 expression and mitochondrial genesis by inhibiting Notch1 expression.

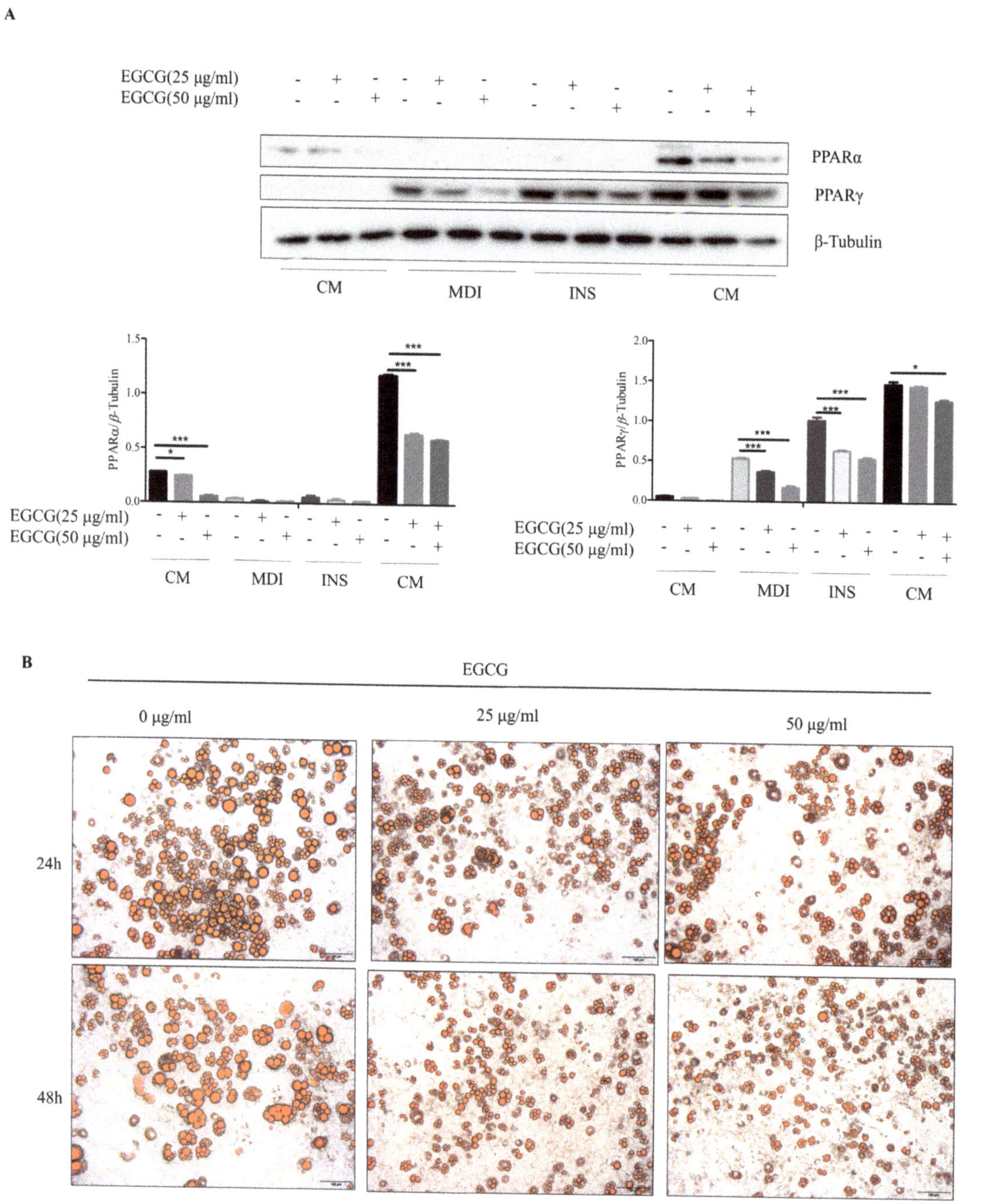

Figure 4. Effect of EGCG on lipid synthesis during 3T3-L1 cell differentiation. (**A**) Western blot analysis showed the expression levels of PPARγ and PPARα protein in 3T3-L1 cells treated with different concentrations of EGCG at different differentiation stages. (**B**) Oil red O staining in 3T3-L1 cells treated with different concentrations of EGCG at 24 h and 48 h. The values are expressed as means ± SEM (n = 3). * $p < 0.05$ and *** $p < 0.001$ compared with the other group.

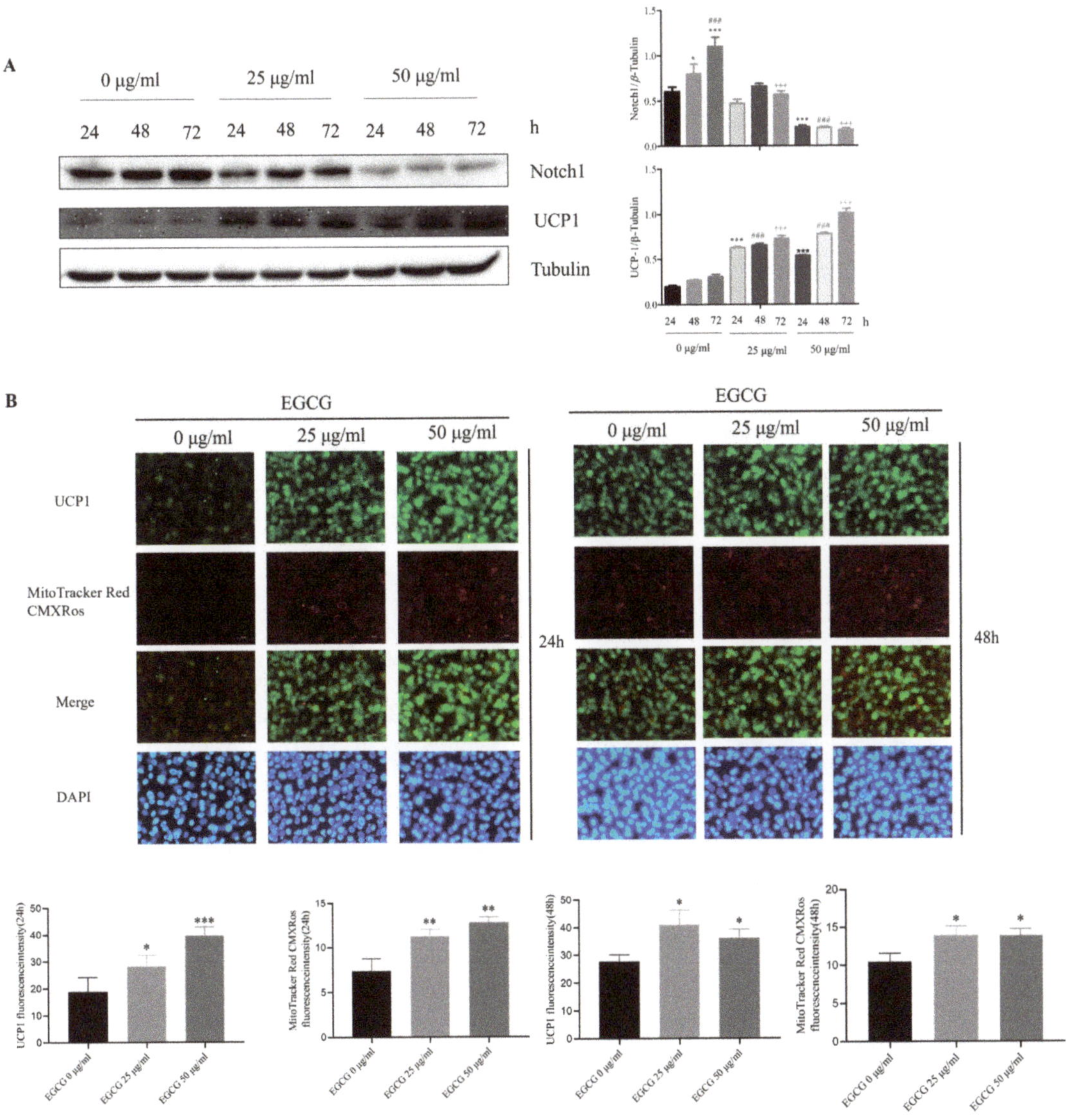

Figure 5. Effects of EGCG treatment on Notch1 and UCP-1 expression in 3T3-L1 cells. (**A**) Western blot analysis showed Notch1 and UCP-1 protein levels in 3T3-L1 cells treated with different concentrations of EGCG for 24 h. (**B**) UCP-1 immunofluorescence staining (green). Mitochondrial fluorescence staining (red) micrographs showed UCP-1 expression and mitochondrial biogenesis in 3T3-L1 cells treated with different concentrations of EGCG for 24 h. The values are expressed as means ± SEM (n = 3). * $p < 0.05$, ** $p < 0.01$ and *** $p < 0.001$ compared with the 0 μg/mL-24 h group, ### $p < 0.001$ compared with the 0 μg/mL-48 h group, and +++ $p < 0.001$ compared with the 0 μg/mL-72 h group.

3. Discussion

With the improvement in living standards and the changes in lifestyles, the number of obese people has increased year by year, and obesity is accompanied by various complications, which has attracted more and more attention around the world. Obesity is also a major factor in many chronic metabolic diseases, including atherosclerosis [29], non-alcoholic fatty liver disease [30], and chronic kidney disease [31]; therefore, improving obesity can effectively prevent a variety of chronic metabolic diseases. Meanwhile, studies have shown that inhibiting GATA3 in adipocytes can significantly inhibit adipose differentiation and improve insulin resistance and T2DM induced by obesity [32]. At present, the treatment for obesity is mainly dependent on changes in lifestyles and energy intake reduction, with drug intervention usually used in severe cases [33]. In recent years, with the popularity of the concept of food and medicine homology, people have paid more and more attention to seeking effective ingredients from food to prevent chronic diseases [34]. Many studies have shown that many foods with the same origin as food and medicine can improve obesity. Wang et al. found that puerarin in pueraria can prevent obesity caused by a high-fat diet by regulating intestinal flora [35]. *Moringa oleifera*, which is a homologous food and medicine, has also been shown to alleviate obesity caused by a high-fat diet by regulating intestinal flora [36]. Tea, as one of the most commonly consumed drinks in the world, contains many potent ingredients that can improve obesity [37,38]. As the tea variety containing the most tea polyphenols, green tea has been proven to have a significant improvement effect on obesity by many studies [39,40] and is related to a variety of effective components contained in it, including EGCG, ECG, EC, etc. [41], and there may be synergistic effects. In addition, while green tea extract produces effects, various polyphenols contained in it may be mutually complementary or have the same effect. Chendi Zhan et al. found that EGCG and EGC both have the effect of improving Alzheimer's disease [42], while EGCG and ECG can significantly inhibit melanoma proliferation [43]. However, as the most abundant and effective of tea polyphenols, EGCG [44] plays a key role in the improvement of obesity and other metabolic diseases by green tea extract [45,46]. As the most abundant component of tea polyphenols, EGCG has been shown to improve many diseases, including cancer, arthritis, diabetes, and colitis [47–51]. Many studies have proven that EGCG has an improving effect on obesity [11,52]. The purpose of this study is to explore some possible mechanisms of EGCG on obesity through the combination of molecular docking, molecular dynamics, and in vitro experiments. The results showed that EGCG could inhibit adipocyte fat synthesis, promote mitochondrial biogenesis, and promote white fat browning by inhibiting Notch1 expression.

Studies have shown that fat cell hypertrophy promotes cell rupture through physical reasons, thereby causing inflammation [53] and that the ectopic accumulation of lipids, mainly in the liver, is the main cause of insulin resistance and inflammation [54,55]. A previous study also showed that the occurrence of inflammation not only reduced the action of insulin but also decreased the heat production of brown adipocytes and beige adipocytes [56]. The enhanced thermogenic activity of brown fat and the browning of white adipocytes can prevent obesity and related metabolic disorders. However, excessive obesity would aggravate local inflammation of white adipocytes and brown adipocytes, impair energy consumption and glucose absorption, and, thus, directly change the thermogenic activity of brown adipocytes and the browning process of white adipocytes [57]. Notch1 is a key target of inflammation [58], and is associated with many inflammatory pathways. Studies have shown that inhibition of the Notch1-PI3K/AKT signaling pathway can inhibit inflammation and immune response induced by nitric oxide [59]. Notch1 has also been extensively studied as a key receptor for EGCG in many diseases [60–62]. These results and studies further suggest that the improvement of EGCG on obesity is strongly related to Notch1. To further confirm our hypothesis, we used molecular docking to demonstrate that EGCG is hydrogen bonded to the Notch1 protein. The results of molecular dynamics indicate that the complex structure formed by EGCG and Notch1 is more compact compared to Notch1. EGCG has been proven to have an improving effect on obesity. We first used the oil red O experiment for verification, and the results showed that EGCG treatment for 24 h and 48 h could significantly reduce the sizes of lipid droplets in 3T3-L1 adipocytes in a dose-dependent manner. EGCG also significantly improved lipid accumulation during 3T3-L1 cell differentiation. Our results showed that EGCG treatment could significantly inhibit the expression of PPARγ and PPARα proteins during the differentiation of 3T3-L1 cells. These results indicate that EGCG has a significant ameliorative effect on obesity.

UCP-1, as a marker protein for mitochondria, is also a marker protein for the browning of white fat [28]. We have previously discussed the effects of inflammation on brown fat thermogenesis and the browning of white adipocytes and discussed the role of Notch1 in this process. Due to the close relationship between EGCG and Notch1, we demonstrated in vitro experiments that EGCG treatment can significantly inhibit the expression of Notch1 in 3T3-L1 adipocytes while enhancing the expression of UCP-1. Mitochondrial staining can clearly express mitochondrial biogenesis and is also one of the signs of the browning of white adipocytes [63]. Our results show that EGCG treatment can significantly enhance mitochondrial genesis and UCP-1 expression in 3T3-L1 adipocytes. These results show that EGCG can enhance UCP-1 expression and mitochondrial biogenesis by inhibiting Notch1 expression, thereby promoting the browning of white adipocytes to achieve the purpose of ameliorating obesity (Figure 6). This study used molecular simulation docking and in vitro experiments to verify that EGCG inhibits the expression of Notch1 by binding to Notch1, thus promoting the browning of adipocytes. However, as a polyphenol, EGCG has been proven to have a variety of biological activities and biological targets. It has been found that EGCG can improve obesity in various ways, including reducing inflammation levels [15] and activating the AMPK signaling pathway [11], and it has been proven that EGCG can prevent obesity by up-regulating adipocyte autophagy [64]. Therefore, EGCG has many effects on improving obesity and promoting the browning of white fat.

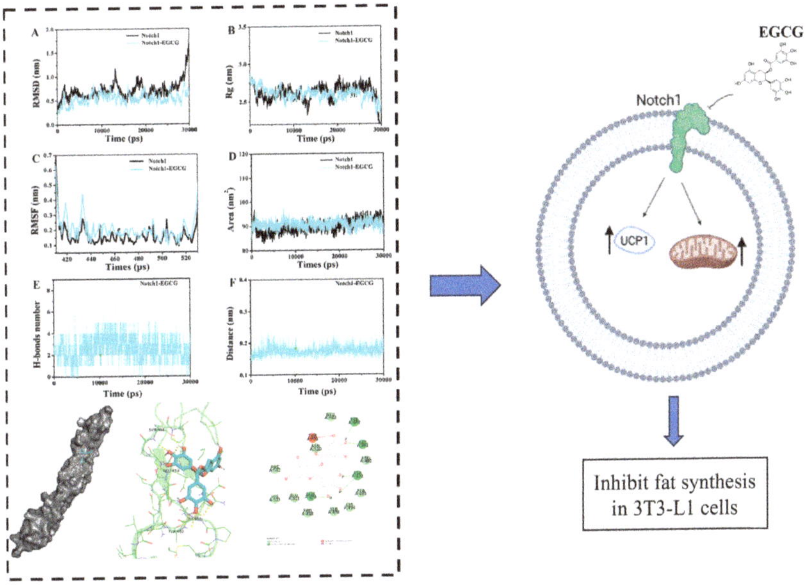

Figure 6. In summary, combined with molecular dynamics simulation and molecular simulation docking, EGCG promotes mitochondrial biogenesis and UCP-1 expression in 3T3-L1 adipocytes by inhibiting Notch1 expression, thereby promoting white adipocytes browning and improving obesity. Image created with Biobender.com.

4. Materials and Methods

4.1. Molecular Docking

The predicted target proteins were molecular-docked with EGCG. The proteins were downloaded from the PDB database (http://www.rcsb.org). The water molecules and ligands were removed using PyMOL, and then the target proteins were imported into AutoDockTools 1.5.6 for hydroprocessing. EGCG was imported as a ligand and then simulated docking with the target proteins and stored in PDBQT format [65]. The results were visualized with PyMOL and Discovery Studio 2019 Client.

4.2. Molecular Dynamics (MD) Simulation

The MD simulations of Notch1 and EGCG-Notch1 complex were conducted using the GROMACS 5.1.2 software package, employing the GROMACS 54a7 force field [66,67]. The solvation of Notch1 and EGCG-Notch1 complex systems was performed with a simple point charge (SPC) water model, followed by their placement within a cubic box filled with water molecules [68]. To maintain a distance of 1 nm from the edge of the box, counterions (Na^+) were added to neutralize the surface charges of both complexes. Subsequently, energy minimization was carried out for 50,000 steps to optimize the prepared systems. Equilibration was achieved through normal volume and temperature (NVT) ensemble at 300 K for 100 ps, followed by density equilibration under normal pressure and temperature (NPT) conditions at 300 K over another 100 ps duration. Finally, an MD simulation was performed for a total time span of 30 ns with a time step size set as 2.0 fs. Various structural properties including root mean square deviation (RMSD), root mean square fluctuation (RMSF), radius of gyration (Rg), solvent accessible surface area (SASA), hydrogen bond formation between ligand and protein along with its distance measurement, and principal component analysis (PCA) were calculated to analyze system dynamics. Additionally, the secondary structure analysis of the EGCG-Notch1 complex employed a defined secondary structure assignment algorithm provided by the do_dssp tool.

4.3. Cell Culture and Differentiation

The 3T3-L1 murine preadipocytes were purchased from Kunming Cell Bank, Kunming Institute of Zoology, Chinese Academy of Sciences. The 3T3-L1 preadipocytes were cultured using Dulbecco's Modified Eagle Medium (DMEM) and supplemented with 10% fetal bovine serum (FBS, Gibco) containing penicillin G sodium salt 100 U/mL. Streptomycin sulfate, in a 0.1 mg/mL volume, (Meilunbio, Dalian, China) was cultured in a 5% CO_2 incubator at 37 °C. After 5 days of culture, 3T3-L1 cells were treated with 10% FBS and 0.5 mmol/L 3-isobutyl-1-methylxanthine (IBMX, Yuanye, Shanghai, China), 2.5 mmol/L Dexamethasone (Dex, Sigma-Aldrich, St. Louis, MI, USA), and 8.7 mmol/L insulin were cultured in DMEM (MDI) medium for 2 days. Then, the DMEM (INS) medium containing 10% FBS and 10 mmol/L insulin was replaced and cultured for 2 days. Subsequently, DMEM (CM) medium containing 10% FBS was replaced every 2 days until 3T3-L1 cells differentiated and matured.

4.4. Western Blot Analysis

Cell proteins were extracted using RIPA buffer (Solarbio, Beijing, China). The protein concentration was detected using the BCA method. Proteins were isolated using 10% SDS-PAGE electrophoresis. It was then transferred to a polyvinylidene fluoride (PVDF) membrane (Merk Millipore, Burlington, MA, USA), sealed with 5% skim milk powder at room temperature for 1 h, followed by primary antibody at 4 °C overnight. It was washed with TBST solution 3 times, 5 min each time, and incubated with 2 antioxidants containing horseradish peroxidase at room temperature for 1 h. Finally, the protein bands were detected using supersensitive enhanced chemiluminescence substrates and quantified using ImageJ 1.46v.

4.5. Oil Red O Staining

After the drug treatment, the cells were washed twice with PBS, fixed at room temperature with 4% paraformaldehyde for 20 min, washed twice with PBS, and stained with Oil Red O solution for 15 min. The photomicrograph was obtained through a microscope (OLYMPUS CKX41, Shanghai, China).

4.6. Cellular Immunofluorescence

The 3T3-L1 adipocytes were treated with EGCG and then subjected to immunofluorescence assay. After fixation with 4% paraformaldehyde and blocking with 5% skim milk powder, the primary antibody was incubated at 4 °C overnight, followed by a FITC-coupled secondary antibody incubated at room temperature for 1 h Finally, after being fixed at room temperature with a DAPI sealer. Images were observed and obtained using a fluorescence microscope (Leica DM2000, Solms, Germany).

4.7. Statistical Analysis

Data are presented as the means ± standard error of the mean (SEM). One-way analysis of variance (ANOVA) and *t*-tests were performed using the SPSS 13.0 software package. A value of $p < 0.05$ was considered to be statistically significant.

5. Conclusions

According to the key role of inflammation in the development of obesity, EGCG has a resistant activity. And, we found that EGCG could bind to Notch1 through molecular docking and molecular dynamics experiments. Moreover, in vitro experiments, it was proved that EGCG could promote the mitochondrial biogenesis and UCP-1 expression of adipocytes by inhibiting Notch1 expression, so as to promote the browning of white adipocytes and improve obesity.

Author Contributions: Conceptualization, X.W. and J.S.; methodology, Y.W., C.L. and W.P.; software, Y.W. and C.L.; validation, X.W.; formal analysis, C.Z.; data curation, Y.W. and C.L.; writing-original draft preparation, Y.W., C.L. and X.W.; writing—review and editing, X.W. and C.Z.; funding acquisition, X.W. and C.Z. All authors have read and agreed to the published version of the manuscript.

Funding: This work was supported by the Natural Science Foundation of China (No. 81760667). The Yunnan Province Agricultural Bisic Research Joint Foundation (202101BD070001-028; 202301BD070001-097), and the Yunnan Ten Thousand Talents Plan Young & Elite Talents Project (YNWR-QNB J-2020-178). The Xingdian Talent Plan of Yunnan Province (Yunling Scholars Project).

Institutional Review Board Statement: Not applicable.

Informed Consent Statement: Not applicable.

Data Availability Statement: All data used in this study are included in this article.

Conflicts of Interest: The authors declare no conflicts of interest.

References

1. Mohammed, M.S.; Sendra, S.; Lloret, J.; Bosch, I. Systems and WBANs for Controlling Obesity. *J. Healthc. Eng.* **2018**, *2018*, 1564748. [CrossRef] [PubMed]
2. Al-Sulaiti, H.; Diboun, I.; Agha, M.V.; Mohamed, F.F.S.; Atkin, S.; Dömling, A.S.; Elrayess, M.A.; Mazloum, N.A. Metabolic signature of obesity-associated insulin resistance and type 2 diabetes. *J. Transl. Med.* **2019**, *17*, 348. [CrossRef] [PubMed]
3. Gong, M.; Wen, S.; Nguyen, T.; Wang, C.; Jin, J.; Zhou, L. Converging Relationships of Obesity and Hyperuricemia with Special Reference to Metabolic Disorders and Plausible Therapeutic Implications. *Diabetes Metab. Syndr. Obes.* **2020**, *13*, 943–962. [CrossRef] [PubMed]
4. Tamura, Y. The Role of Zinc Homeostasis in the Prevention of Diabetes Mellitus and Cardiovascular Diseases. *J. Atheroscler. Thromb.* **2021**, *28*, 1109–1122. [CrossRef] [PubMed]
5. Cheng, L.; Wang, J.; Dai, H.; Duan, Y.; An, Y.; Shi, L.; Lv, Y.; Li, H.; Wang, C.; Ma, Q.; et al. Brown and beige adipose tissue: A novel therapeutic strategy for obesity and type 2 diabetes mellitus. *Adipocyte* **2021**, *10*, 48–65. [CrossRef] [PubMed]
6. Cheng, L.; Zhang, S.; Shang, F.; Ning, Y.; Huang, Z.; He, R.; Sun, J.; Dong, S. Emodin Improves Glucose and Lipid Metabolism Disorders in Obese Mice via Activating Brown Adipose Tissue and Inducing Browning of White Adipose Tissue. *Front. Endocrinol.* **2021**, *12*, 618037. [CrossRef] [PubMed]
7. Meng, X.Y.; Zhang, H.X.; Mezei, M.; Cui, M. Molecular docking: A powerful approach for structure-based drug discovery. *Curr. Comput. Aided Drug Des.* **2011**, *7*, 146–157. [CrossRef]
8. Wu, X.; Xu, L.Y.; Li, E.M.; Dong, G. Application of molecular dynamics simulation in biomedicine. *Chem. Biol. Drug Des.* **2022**, *99*, 789–800. [CrossRef]
9. Kochman, J.; Jakubczyk, K.; Antoniewicz, J.; Mruk, H.; Janda, K. Health Benefits and Chemical Composition of Matcha Green Tea: A Review. *Molecules* **2020**, *26*, 85. [CrossRef]
10. Lin, Y.; Jiang, M.; Chen, W.; Zhao, T.; Wei, Y. Cancer and ER stress: Mutual crosstalk between autophagy, oxidative stress and inflammatory response. *Biomed. Pharmacother.* **2019**, *118*, 109249. [CrossRef]
11. Li, F.; Gao, C.; Yan, P.; Zhang, M.; Wang, Y.; Hu, Y.; Wu, X.; Wang, X.; Sheng, J. EGCG Reduces Obesity and White Adipose Tissue Gain Partly through AMPK Activation in Mice. *Front. Pharmacol.* **2018**, *9*, 1366. [CrossRef] [PubMed]

12. Luc, K.; Schramm-Luc, A.; Guzik, T.J.; Mikolajczyk, T.P. Oxidative stress and inflammatory markers in prediabetes and diabetes. *J. Physiol. Pharmacol.* **2019**, *70*, 809–824. [CrossRef] [PubMed]
13. Mokra, D.; Joskova, M.; Mokry, J. Therapeutic Effects of Green Tea Polyphenol (-)-Epigallocatechin-3-Gallate (EGCG) in Relation to Molecular Pathways Controlling Inflammation, Oxidative Stress, and Apoptosis. *Int. J. Mol. Sci.* **2022**, *24*, 340. [CrossRef] [PubMed]
14. Sheng, L.; Jena, P.K.; Liu, H.X.; Hu, Y.; Nagar, N.; Bronner, D.N.; Settles, M.L.; Bäumler, A.J.; Wan, Y.Y. Obesity treatment by epigallocatechin-3-gallate-regulated bile acid signaling and its enriched Akkermansia muciniphila. *FASEB J.* **2018**, *32*, fj201800370R. [CrossRef] [PubMed]
15. Zhou, J.; Mao, L.; Xu, P.; Wang, Y. Effects of (-)-Epigallocatechin Gallate (EGCG) on Energy Expenditure and Microglia-Mediated Hypothalamic Inflammation in Mice Fed a High-Fat Diet. *Nutrients* **2018**, *10*, 1681. [CrossRef] [PubMed]
16. Oruganti, L.; Reddy Sankaran, K.; Dinnupati, H.G.; Kotakadi, V.S.; Meriga, B. Anti-adipogenic and lipid-lowering activity of piperine and epigallocatechin gallate in 3T3-L1 adipocytes. *Arch. Physiol. Biochem.* **2023**, *129*, 1152–1159. [CrossRef] [PubMed]
17. Choi, C.; Song, H.D.; Son, Y.; Cho, Y.K.; Ahn, S.Y.; Jung, Y.S.; Yoon, Y.C.; Kwon, S.W.; Lee, Y.H. Epigallocatechin-3-Gallate Reduces Visceral Adiposity Partly through the Regulation of Beclin1-Dependent Autophagy in White Adipose Tissues. *Nutrients* **2020**, *12*, 3072. [CrossRef] [PubMed]
18. Chen, D.; Wan, S.B.; Yang, H.; Yuan, J.; Chan, T.H.; Dou, Q.P. EGCG, green tea polyphenols and their synthetic analogs and prodrugs for human cancer prevention and treatment. *Adv. Clin. Chem.* **2011**, *53*, 155–177. [CrossRef] [PubMed]
19. Gan, L.; Liu, Z.; Wu, T.; Feng, F.; Sun, C. αMSH promotes preadipocyte proliferation by alleviating ER stress-induced leptin resistance and by activating Notch1 signal in mice. *Biochim. Biophys. Acta Mol. Basis Dis.* **2017**, *1863*, 231–238. [CrossRef]
20. Siouti, E.; Salagianni, M.; Manioudaki, M.; Pavlos, E.; Klinakis, A.; Galani, I.E.; Andreakos, E. Notch signaling in adipose tissue macrophages prevents diet-induced inflammation and metabolic dysregulation. *Eur. J. Immunol.* **2024**, *54*, e2350669. [CrossRef]
21. Bi, P.; Shan, T.; Liu, W.; Yue, F.; Yang, X.; Liang, X.R.; Wang, J.; Li, J.; Carlesso, N.; Liu, X.; et al. Inhibition of Notch signaling promotes browning of white adipose tissue and ameliorates obesity. *Nat. Med.* **2014**, *20*, 911–918. [CrossRef] [PubMed]
22. Wang, H.; Yu, L.; Wang, J.; Zhang, Y.; Xu, M.; Lv, C.; Cui, B.; Yuan, M.; Zhang, Y.; Yan, Y.; et al. SLC35D3 promotes white adipose tissue browning to ameliorate obesity by NOTCH signaling. *Nat. Commun.* **2023**, *14*, 7643. [CrossRef] [PubMed]
23. Lee, S.G.; Chae, J.; Woo, S.M.; Seo, S.U.; Kim, H.J.; Kim, S.Y.; Schlaepfer, D.D.; Kim, I.S.; Park, H.S.; Kwon, T.K.; et al. TGFBI remodels adipose metabolism by regulating the Notch-1 signaling pathway. *Exp. Mol. Med.* **2023**, *55*, 520–531. [CrossRef] [PubMed]
24. Xie, H.; Sun, J.; Chen, Y.; Zong, M.; Li, S.; Wang, Y. EGCG Attenuates Uric Acid-Induced Inflammatory and Oxidative Stress Responses by Medicating the NOTCH Pathway. *Oxidative Med. Cell. Longev.* **2015**, *2015*, 214836. [CrossRef] [PubMed]
25. Chen, G.; Seukep, A.J.; Guo, M. Recent Advances in Molecular Docking for the Research and Discovery of Potential Marine Drugs. *Mar. Drugs* **2020**, *18*, 545. [CrossRef] [PubMed]
26. Çapan, İ.; Shehu, A.; Sert, Y.; Çelik, İ.; Erol, M.; Koca, İ.; Servi, S. Synthesis, molecular docking, molecular dynamics and evaluation of Drug-Likeness properties of the fused N-Formyl pyrazoline substituted new dehydroepiandrosterone derivatives. *J. Biomol. Struct. Dyn.* **2023**, *41*, 2492–2503. [CrossRef] [PubMed]
27. Xie, Z.; Gao, G.; Wang, H.; Li, E.; Yuan, Y.; Xu, J.; Zhang, Z.; Wang, P.; Fu, Y.; Zeng, H.; et al. Dehydroabietic acid alleviates high fat diet-induced insulin resistance and hepatic steatosis through dual activation of PPAR-γ and PPAR-α. *Biomed. Pharmacother.* **2020**, *127*, 110155. [CrossRef] [PubMed]
28. Teruel, T.; Hernandez, R.; Benito, M.; Lorenzo, M. Rosiglitazone and retinoic acid induce uncoupling protein-1 (UCP-1) in a p38 mitogen-activated protein kinase-dependent manner in fetal primary brown adipocytes. *J. Biol. Chem.* **2003**, *278*, 263–269. [CrossRef] [PubMed]
29. Csige, I.; Ujvárosy, D.; Szabó, Z.; Lőrincz, I.; Paragh, G.; Harangi, M.; Somodi, S. The Impact of Obesity on the Cardiovascular System. *J. Diabetes Res.* **2018**, *2018*, 3407306. [CrossRef]
30. Polyzos, S.A.; Kountouras, J.; Mantzoros, C.S. Obesity and nonalcoholic fatty liver disease: From pathophysiology to therapeutics. *Metabolism* **2019**, *92*, 82–97. [CrossRef]
31. Jiang, Z.; Wang, Y.; Zhao, X.; Cui, H.; Han, M.; Ren, X.; Gang, X.; Wang, G. Obesity and chronic kidney disease. *Am. J. Physiol. Endocrinol. Metab.* **2023**, *324*, E24–E41. [CrossRef] [PubMed]
32. El-Arabey, A.A.; Abdalla, M. GATA3 as an immunomodulator in obesity-related metabolic dysfunction associated with fatty liver disease, insulin resistance, and type 2 diabetes. *Chem. Biol. Interact.* **2022**, *366*, 110141. [CrossRef] [PubMed]
33. Wadden, T.A.; Tronieri, J.S.; Butryn, M.L. Lifestyle modification approaches for the treatment of obesity in adults. *Am. Psychol.* **2020**, *75*, 235–251. [CrossRef]
34. Chen, J. Essential role of medicine and food homology in health and wellness. *Chin. Herb. Med.* **2023**, *15*, 347–348. [CrossRef] [PubMed]
35. Wang, L.; Wu, Y.; Zhuang, L.; Chen, X.; Min, H.; Song, S.; Liang, Q.; Li, A.D.; Gao, Q. Puerarin prevents high-fat diet-induced obesity by enriching Akkermansia muciniphila in the gut microbiota of mice. *PLoS ONE* **2019**, *14*, e0218490. [CrossRef] [PubMed]
36. Li, L.; Ma, L.; Wen, Y.; Xie, J.; Yan, L.; Ji, A.; Zeng, Y.; Tian, Y.; Sheng, J. Crude Polysaccharide Extracted From Moringa oleifera Leaves Prevents Obesity in Association With Modulating Gut Microbiota in High-Fat Diet-Fed Mice. *Front. Nutr.* **2022**, *9*, 861588. [CrossRef] [PubMed]

37. Abiri, B.; Amini, S.; Hejazi, M.; Hosseinpanah, F.; Zarghi, A.; Abbaspour, F.; Valizadeh, M. Tea's anti-obesity properties, cardiometabolic health-promoting potentials, bioactive compounds, and adverse effects: A review focusing on white and green teas. *Food Sci. Nutr.* **2023**, *11*, 5818–5836. [CrossRef]
38. Liu, X.; Hu, G.; Wang, A.; Long, G.; Yang, Y.; Wang, D.; Zhong, N.; Jia, J. Black Tea Reduces Diet-Induced Obesity in Mice via Modulation of Gut Microbiota and Gene Expression in Host Tissues. *Nutrients* **2022**, *14*, 1635. [CrossRef] [PubMed]
39. Ohishi, T.; Fukutomi, R.; Shoji, Y.; Goto, S.; Isemura, M. The Beneficial Effects of Principal Polyphenols from Green Tea, Coffee, Wine, and Curry on Obesity. *Molecules* **2021**, *26*, 453. [CrossRef]
40. Jurgens, T.M.; Whelan, A.M.; Killian, L.; Doucette, S.; Kirk, S.; Foy, E. Green tea for weight loss and weight maintenance in overweight or obese adults. *Cochrane Database Syst. Rev.* **2012**, *12*, Cd008650. [CrossRef]
41. Chakrawarti, L.; Agrawal, R.; Dang, S.; Gupta, S.; Gabrani, R. Therapeutic effects of EGCG: A patent review. *Expert Opin. Ther. Pat.* **2016**, *26*, 907–916. [CrossRef] [PubMed]
42. Zhan, C.; Chen, Y.; Tang, Y.; Wei, G. Green Tea Extracts EGCG and EGC Display Distinct Mechanisms in Disrupting Aβ(42) Protofibril. *ACS Chem. Neurosci.* **2020**, *11*, 1841–1851. [CrossRef] [PubMed]
43. Du, B.X.; Lin, P.; Lin, J. EGCG and ECG induce apoptosis and decrease autophagy via the AMPK/mTOR and PI3K/AKT/mTOR pathway in human melanoma cells. *Chin. J. Nat. Med.* **2022**, *20*, 290–300. [CrossRef] [PubMed]
44. Kim, H.S.; Quon, M.J.; Kim, J.A. New insights into the mechanisms of polyphenols beyond antioxidant properties; lessons from the green tea polyphenol, epigallocatechin 3-gallate. *Redox Biol.* **2014**, *2*, 187–195. [CrossRef] [PubMed]
45. Suzuki, Y.; Miyoshi, N.; Isemura, M. Health-promoting effects of green tea. *Proc. Jpn. Acad. Ser. B Phys. Biol. Sci.* **2012**, *88*, 88–101. [CrossRef] [PubMed]
46. Mehmood, S.; Maqsood, M.; Mahtab, N.; Khan, M.I.; Sahar, A.; Zaib, S.; Gul, S. Epigallocatechin gallate: Phytochemistry, bioavailability, utilization challenges, and strategies. *J. Food Biochem.* **2022**, *46*, e14189. [CrossRef] [PubMed]
47. Almatroodi, S.A.; Almatroudi, A.; Khan, A.A.; Alhumaydhi, F.A.; Alsahli, M.A.; Rahmani, A.H. Potential Therapeutic Targets of Epigallocatechin Gallate (EGCG), the Most Abundant Catechin in Green Tea, and Its Role in the Therapy of Various Types of Cancer. *Molecules* **2020**, *25*, 3146. [CrossRef] [PubMed]
48. Zhu, T.; Li, M.; Zhu, M.; Liu, X.; Huang, K.; Li, W.; Wang, S.X.; Yin, Y.; Li, P. Epigallocatechin-3-gallate alleviates type 2 diabetes mellitus via β-cell function improvement and insulin resistance reduction. *Iran. J. Basic Med. Sci.* **2022**, *25*, 483–488. [CrossRef] [PubMed]
49. Yang, E.J.; Lee, J.; Lee, S.Y.; Kim, E.K.; Moon, Y.M.; Jung, Y.O.; Park, S.H.; Cho, M.L. EGCG attenuates autoimmune arthritis by inhibition of STAT3 and HIF-1α with Th17/Treg control. *PLoS ONE* **2014**, *9*, e86062. [CrossRef]
50. Wu, Z.; Huang, S.; Li, T.; Li, N.; Han, D.; Zhang, B.; Xu, Z.Z.; Zhang, S.; Pang, J.; Wang, S.; et al. Gut microbiota from green tea polyphenol-dosed mice improves intestinal epithelial homeostasis and ameliorates experimental colitis. *Microbiome* **2021**, *9*, 184. [CrossRef]
51. Luz, J.R.D.D.; López, J.A.; Ferreira, M.P.; de Sousa, R.M.; Silva, S.V.E.; Almeida, M.D.G.; Araujo-Silva, G. In Vitro Antithrombotic, Antitumor and Antiangiogenic Activities of Green Tea Polyphenols and Its Main Constituent Epigallocatechin-3-gallate. *Processes* **2023**, *11*, 76. [CrossRef]
52. Sun, X.; Dey, P.; Bruno, R.S.; Zhu, J. EGCG and catechin relative to green tea extract differentially modulate the gut microbial metabolome and liver metabolome to prevent obesity in mice fed a high-fat diet. *J. Nutr. Biochem.* **2022**, *109*, 109094. [CrossRef] [PubMed]
53. Monteiro, R.; Azevedo, I. Chronic inflammation in obesity and the metabolic syndrome. *Mediat. Inflamm.* **2010**, *2010*, 289645. [CrossRef]
54. Xu, X.; Poulsen, K.L.; Wu, L.; Liu, S.; Miyata, T.; Song, Q.; Wei, Q.; Zhao, C.; Lin, C.; Yang, J. Targeted therapeutics and novel signaling pathways in non-alcohol-associated fatty liver/steatohepatitis (NAFL/NASH). *Signal Transduct. Target. Ther.* **2022**, *7*, 287. [CrossRef] [PubMed]
55. Xu, L.; Nagata, N.; Ota, T. Glucoraphanin: A broccoli sprout extract that ameliorates obesity-induced inflammation and insulin resistance. *Adipocyte* **2018**, *7*, 218–225. [CrossRef] [PubMed]
56. Omran, F.; Christian, M. Inflammatory Signaling and Brown Fat Activity. *Front. Endocrinol.* **2020**, *11*, 156. [CrossRef]
57. Villarroya, F.; Cereijo, R.; Gavaldà-Navarro, A.; Villarroya, J.; Giralt, M. Inflammation of brown/beige adipose tissues in obesity and metabolic disease. *J. Intern. Med.* **2018**, *284*, 492–504. [CrossRef]
58. Fazio, C.; Piazzi, G.; Vitaglione, P.; Fogliano, V.; Munarini, A.; Prossomariti, A.; Milazzo, M.; D'Angelo, L.; Napolitano, M.; Chieco, P.; et al. Inflammation increases NOTCH1 activity via MMP9 and is counteracted by Eicosapentaenoic Acid-free fatty acid in colon cancer cells. *Sci. Rep.* **2016**, *6*, 20670. [CrossRef]
59. Villegas, S.N.; Gombos, R.; García-López, L.; Gutiérrez-Pérez, I.; García-Castillo, J.; Vallejo, D.M.; Da Ros, V.G.; Ballesta-Illán, E.; Mihály, J.; Dominguez, M. PI3K/Akt Cooperates with Oncogenic Notch by Inducing Nitric Oxide-Dependent Inflammation. *Cell Rep.* **2018**, *22*, 2541–2549. [CrossRef]
60. Jin, H.; Gong, W.; Zhang, C.; Wang, S. Epigallocatechin gallate inhibits the proliferation of colorectal cancer cells by regulating Notch signaling. *Onco Targets Ther.* **2013**, *6*, 145–153. [CrossRef]
61. Kwak, T.W.; Park, S.B.; Kim, H.J.; Jeong, Y.I.; Kang, D.H. Anticancer activities of epigallocatechin-3-gallate against cholangiocarcinoma cells. *OncoTargets Ther.* **2017**, *10*, 137–144. [CrossRef] [PubMed]

62. Wang, Y.N.; Wang, J.; Yang, H.N.; Zhang, B.L.; Zhang, P.; Sun, P.Y.; Zhang, N.; Wang, Y.; Sheng, J.; Wang, X.J.; et al. The oxidation of (-)-epigallocatechin-3-gallate inhibits T-cell acute lymphoblastic leukemia cell line HPB-ALL via the regulation of Notch1 expression. *RSC Adv.* **2020**, *10*, 1679–1684. [CrossRef] [PubMed]
63. Lin, C.; Chen, J.; Hu, M.; Zheng, W.; Song, Z.; Qin, H. Sesamol promotes browning of white adipocytes to ameliorate obesity by inducing mitochondrial biogenesis and inhibition mitophagy via β3-AR/PKA signaling pathway. *Food Nutr. Res.* **2021**, *65*, 7577. [CrossRef] [PubMed]
64. Kim, S.N.; Kwon, H.J.; Akindehin, S.; Jeong, H.W.; Lee, Y.H. Effects of Epigallocatechin-3-Gallate on Autophagic Lipolysis in Adipocytes. *Nutrients* **2017**, *9*, 680. [CrossRef] [PubMed]
65. Trott, O.; Olson, A.J. AutoDock Vina: Improving the speed and accuracy of docking with a new scoring function, efficient optimization, and multithreading. *J. Comput. Chem.* **2010**, *31*, 455–461. [CrossRef]
66. Poger, D.; Van Gunsteren, W.F.; Mark, A.E. A new force field for simulating phosphatidylcholine bilayers. *J. Comput. Chem.* **2010**, *31*, 1117–1125. [CrossRef] [PubMed]
67. Schmid, N.; Eichenberger, A.P.; Choutko, A.; Riniker, S.; Winger, M.; Mark, A.E.; van Gunsteren, W.F. Definition and testing of the GROMOS force-field versions 54A7 and 54B7. *Eur. Biophys. J.* **2011**, *40*, 843–856. [CrossRef]
68. Chandar, N.B.; Efremenko, I.; Silman, I.; Martin, J.M.L.; Sussman, J.L. Molecular dynamics simulations of the interaction of Mouse and Torpedo acetylcholinesterase with covalent inhibitors explain their differential reactivity: Implications for drug design. *Chem. Biol. Interact.* **2019**, *310*, 108715. [CrossRef]

foods

Article

Preparation and Characterization of a Novel *Longzhua mushroom* Polysaccharide Hydrogel and Slow-Release Behavior of Encapsulated Rambutan Peel Polyphenols

Lingxin Zhao, Jiapeng Li, Yangyue Ding and Liping Sun *

Faculty of Food Science and Engineering, Kunming University of Science and Technology, Kunming 650500, China; zhao13001662637@163.com (L.Z.); ljp134518@163.com (J.L.); dingyangyue77@163.com (Y.D.)
* Correspondence: lpsun@kmust.edu.cn

Abstract: Natural polyphenols have drawbacks such as instability and low bioavailability, which can be overcome by encapsulated slow-release systems. Natural polymer hydrogels are ideal materials for slow-release systems because of their high biocompatibility. In this study, Longzhua mushroom polysaccharide hydrogel (LMPH) was used to encapsulate rambutan peel polyphenols (RPP) and delay their release time to improve their stability and bioavailability. The mechanical properties, rheology, stability, swelling properties, water-holding capacity, RPP loading, and slow-release behavior of LMPH were investigated. The results showed that LMPH has adequate mechanical and rheological properties, high thermal stability, excellent swelling and water-holding capacity, and good self-healing behavior. Increasing the polysaccharide content not only improved the hardness (0.17–1.13 N) and water-holding capacity of LMPH (90.84–99.32%) but also enhanced the encapsulation efficiency of RPP (93.13–99.94%). The dense network structure slowed down the release of RPP. In particular, LMPH5 released only 61.58% at 48 h. Thus, a stable encapsulated slow-release system was fabricated using a simple method based on the properties of LMPH. The developed material has great potential for the sustained release and delivery of biologically active substances.

Keywords: *Longzhua mushroom*; mushroom polysaccharide; hydrogel; slow-release; rambutan peel polyphenols

Citation: Zhao, L.; Li, J.; Ding, Y.; Sun, L. Preparation and Characterization of a Novel *Longzhua mushroom* Polysaccharide Hydrogel and Slow-Release Behavior of Encapsulated Rambutan Peel Polyphenols. *Foods* **2024**, *13*, 1711. https://doi.org/10.3390/foods13111711

Academic Editor: Barry J. Parsons

Received: 27 April 2024
Revised: 22 May 2024
Accepted: 27 May 2024
Published: 29 May 2024

1. Introduction

Rambutan (*Nephelium lappaceum* L.) is an important tropical commercial fruit that can be eaten fresh or processed. The peel, rich in phenolic compounds, is discarded as a by-product during processing [1–3]. Rambutan peel polyphenols have high antioxidant and anti-inflammatory activities [4] but are susceptible to degradation, instability, and low bioavailability [5–8]. Encapsulation delivery systems have been used to improve the chemical stability of natural bioactives (e.g., polyphenols), particularly through the application of hydrogel systems [9].

Hydrogel is a substance with a three-dimensional network structure [10], capable of absorbing and holding large amounts of water, and has adjustable mechanical and chemical properties [11]. In particular, because of their porous structure, hydrogels are employed in the food industry as ideal materials for the encapsulation and delivery of bioactive substances [12,13]. At present, many researchers prefer to use natural polymers to prepare hydrogels [14]. The biocompatibility, accessibility, and renewability of natural polymers (e.g., polysaccharides) have led to their increased appeal as biomaterials compared to synthetic polymers [15].

Polysaccharides are macromolecular natural polymers. The hydroxyl and carboxyl groups on the polysaccharide molecular chain give the hydrogel excellent water retention properties [16,17]. In addition, natural polysaccharides can bind polyphenols through hydrogen bonding and hydrophobic interactions and immobilize them in the porous network of the hydrogel for slow release [18,19].

Natural polysaccharides are derived from various sources, including animals, plants, microorganisms, and algae [20]. These polysaccharides, including chitosan, starch, cellulose, and lignin, have been extensively used to prepare natural hydrogels [21]. In recent years, fungal polysaccharide resources have gained increased attention because of their rich availability and safety [22]. Mushroom is a higher edible fungus [23]. Mushroom polysaccharide is a non-animal source, so the risk of disease transmission is limited [24]. Moreover, mushroom polysaccharides have high bioactivity and gelling properties and, thus, have promising applications in tissue engineering and encapsulation [25,26]. Most studies have described mushroom polysaccharides mixed with other substances to construct gel systems. For example, polysaccharides isolated from *Agaricus blazei* improved the gelling ability of Pluronic® F127 systems [27]. In the study by Zhang et al. [24], *Pleurotus tuber-regium* was compounded with xanthan gum to construct a hydrogel for encapsulation. The bio-hydrogels prepared by mixing the polysaccharides of *Hericium erinaceus* and hydrolyzed extracts of *Paulownia elongata x fortunei* showed high stability [28]. In addition, some studies on single-component polysaccharide hydrogels have mostly been conducted by changing the pH and adding ions to make them gelatinous; for example, *Poria cocos* polysaccharides were acidified to make them gelatinous by Li et al. [20]. These methods have the potential to affect the structural and functional activity of the encapsulated substance. Therefore, there is an urgent need for a natural polysaccharide that can autonomously form weak gels to construct hydrogel systems.

At present, reports on the study and application of this single-component polysaccharide and its gelling properties are rare. Longzhua mushroom is a new variety obtained by controlling the growth and developmental conditions of *Auricularia polytricha* to make it look like a claw. Longzhua mushroom has the same rich polysaccharide content as *Auricularia polytricha*, and we hypothesize that its polysaccharides have gelling properties similar to those of *Auricularia polytricha*.

The aim of this study was to construct a simple, green, safe and stable single-component hydrogel as an encapsulation delivery system, using Longzhua mushroom polysaccharide (LMP) to slow down the release rate of rambutan peel polyphenols (RPP). The mechanical properties, microstructure, self-healing ability, stability, water-holding capacity, and swelling properties of the hydrogel were investigated. We hope that this study will provide a novel, safe, and easily accessible material for single-component hydrogels and expand the research related to LMP.

2. Materials and Methods

2.1. Materials

Fresh Longzhua mushroom was purchased from the wild mushroom market in Kunming, Yunnan Province, China. Fresh rambutan was purchased from the fruit market in Kunming. Anhydrous ethanol was obtained from Zhiyuan Co., Ltd. (Tianjin, China). Folin phenol reagent was purchased from Macklin Biochemical Science and Technology Co., Ltd. (Shanghai, China). Methylene blue was obtained from Aladdin Biochemical Technology Co. (Shanghai, China). FT-IR grade KBr was purchased from Aladdin Reagent (Shanghai, China). Na_2CO_3 was purchased from Macklin Biochemical Science and Technology Co., Ltd. (Shanghai, China). All reagents were of analytical grade.

2.2. Preparation of Longzhua mushroom Polysaccharide (LMP) Hydrogel

The LMP extraction process was based on the method of Sun et al. [29]. Longzhua mushroom powder was soaked in 95% ethanol (1:15, w/v), then ultrasonicated twice for 30 min each time. The mixtures were centrifuged at 5000 rpm for 10 min, and the sediment was retained. Ultra-pure water (900 mL) was added to the sediment and extracted at a constant speed of 1000 rpm at 95 °C for 3 h. The mixture was then centrifuged at 5000 rpm for 15 min, and the supernatant was collected and concentrated under reduced pressure.

The solution was mixed with four volumes of anhydrous ethanol and precipitated for 12 h. The precipitates were collected and lyophilized to obtain LMP.

LMP (100, 200, 300, 400, and 500 mg) and 10 mL of ultra-pure water were weighed and heated at 80 °C for 10 min while stirring and then stored at 4 °C. The hydrogels were named LMPH1–LMPH5, with the polysaccharide content ranging from 1% to 5% (w/v).

2.3. Texture Properties of LMPH

The hardness, springiness, cohesiveness, chewiness, and resilience of the hydrogel samples were measured using a texture analyzer (TA-XT PlusC, Stable Micro Systems, Godalming, UK) in TPA mode according to a previous method, with some modifications [30]. A P/36R probe was used for compression. The pre-test, test, and post-test speed was 1 mm/s. The determination time interval was 1.00 s, the compression distance was 5 mm, and the thixotropic force was 1.0 g. All measurements were conducted a minimum of five times.

2.4. Rheological Properties of LMPH

Rheological analyses were performed with a modular advanced rheometer (MCR 102, Anton Par, Graz, Austria) [31]. A 25-mm-diameter pressure plate (PP25) was used as the suspension measurement system at 25 °C, and the measurement gap was set to 1 mm.

2.4.1. Viscosity Sweep

A shear rate of 0.1 to 100 s^{-1} at 25.0 °C was used to determine the apparent viscosity of all samples, and a total of 41 data points were collected.

2.4.2. Strain Sweep

The linear viscoelastic range (LVR) was determined by adjusting the strain in the range of 0.01–1000% at a 1 Hz frequency. A total of 30 data points were collected. To avoid breaking the gels, subsequent rheological tests were performed within 1% of the strain.

2.4.3. Frequency Sweep

Frequency sweep tests ranged from 0.1 to 100 rad/s. A total of 21 data points were collected. The results of the elastic (G′) and viscous (G″) moduli were recorded.

2.5. Scanning Electron Microscope (SEM)

The internal structure of LMPH was observed using a scanning electron microscope (Helios 5 CX, Thermo Scientific, Waltham, MA, USA). Before observation, the samples were lyophilized using a freeze dryer, then cut into small pieces with a sterilized scalpel, and spluttered with gold [32]. Images were observed and recorded at different magnifications.

2.6. Self-Restoring Characteristic

2.6.1. Macro Self-Recovery Performance

Macro self-recoveries refer to the Wei et al. [33] method and make some modifications. Two hydrogels of different colors were prepared, one stained with methylene blue and the other left untreated. The hydrogels were left at room temperature and observed for self-healing without any external stimuli, and after 6 h, the hydrogels were lifted with forceps and recorded with a camera.

2.6.2. Micro Self-Recovery Performance

The continuous step-strain test was performed at 1 Hz in the LVR (1% strain) and out of the LVR (1000% strain). The hydrogel was held for 150 s in each scanning segment, and the test was repeated five times at 25.0 °C [31].

The FTIR spectra of the samples were obtained by the KBr press method on a NICOLET iS50 Fourier transform infrared spectrometer (Thermo Scientific, USA). All spectra were recorded from 4000 to 400 cm^{-1} with 64 scans and a 1 cm^{-1} resolution.

2.7. Stability of LMPH

2.7.1. Differential Scanning Calorimetry (DSC)

The thermal properties of LMPH were measured using a differential scanning calorimeter (DSC214, NETZSCH, Selb, Germany). LMPH (10 mg) was weighed into an aluminum pan, and an empty aluminum pan was used as a reference [34]. The temperature was increased from 20 °C to 200 °C at a rate of 10 °C/min, with a nitrogen flow rate of 40 mL/min [35].

2.7.2. Freeze–Thaw Stability

The hydrogel was heated at 80 °C for 15 min to prepare a solution and expanded at 25 °C overnight. An empty tube was weighed. The solution was transferred into the empty tube. The tubes and solution were weighed and stored at −20 °C for 20 h, then thawed at 25 °C for 4 h as a freeze–thaw cycle. The entire experiment comprised five freeze–thaw cycles. At the end of each thawing cycle, LMPH was centrifuged at 5000 rpm for 10 min, and the separated water was weighed. Equation (1) was used to calculate the water separation rate.

$$\text{Water separation rate} = \frac{M_2}{(M_1 - M)} \times 100\% \quad (1)$$

where M is the weight of the empty tube, M_1 is the total weight of the sample and centrifuge tube, and M_2 is the weight of water separated after centrifugation.

2.8. Water Holding Capacity and Swelling Properties of LMPH

2.8.1. Swelling Property

The swelling behavior of LMPH was investigated by gravimetric analysis. Hydrogels with different mass fractions were prepared in a 5 mL centrifugal tube. After vacuum freeze-drying, the dried LMPH was weighed 100 mg and placed on a 200-mesh filter. The LMPH was then immersed in ultrapure water and removed at 0.5, 1, 2, 3, 6, 9, 12, 24, 36, and 48 h, and any water present on the surface was quickly dried and then weighed. The expansion rate was calculated using Equation (2):

$$\text{Swelling ratio} = \frac{(W_1 - W_0)}{W_0} \times 100\% \quad (2)$$

where W_1 and W_0 are the weight of the swollen hydrogel at the predetermined time interval and the initial weight of the lyophilized hydrogel, respectively.

2.8.2. Water Holding Capacity

LMPH (20 mL) was prepared in a 50 mL centrifuge tube and then placed at room temperature for 2 h. The hydrogel was then centrifuged at 5000 rpm for 15 min. The water layer was removed from the gel surface with a dry filter paper until a constant weight was obtained. The WHC was calculated using Equation (3):

$$\text{WHC} = \frac{(M_2 - M_0)}{(M_1 - M_0)} \times 100\% \quad (3)$$

where M_2, M_1, and M_0 are respectively the weight of the centrifuge tube containing hydrogel after water removal, the initial weight of the centrifuge tube containing hydrogel, and the weight of the blank centrifuge tube.

2.9. In Vitro Sustained Release Behavior

2.9.1. Encapsulation Efficiency (EE)

Rambutan peel polyphenols (RPP) powder from our previous work [36]. RPP suspension was prepared by taking 10 mg of RPP dispersed in 10 mL of ultrapure water. LMP powder (100, 200, 300, 400, 500 mg) was added to 10 mL of RPP suspension and stirred at 80 °C for 15 min to completely dissolve the LMP powder to form five hydrogel samples of embedded RPP. The hydrogel was washed with 20 mL of ultrapure water, and 0.5 mL of the eluate was added with 2.5 mL of 10% Folin phenol, vortexed, and left to stand for 5 min, followed by the addition of 2 mL of Na_2CO_3, and the reaction was carried out at the dark place for 1 h. The polyphenol content of the washings was determined at 765 nm [37]. The encapsulation efficiency of rambutan polyphenol was determined using Equation (4):

$$EE = \left(1 - \frac{A_2}{A_1}\right) \times 100\% \quad (4)$$

where A_1 and A_2 are absorbance values of RPP solution and cleaning solution, respectively.

2.9.2. In Vitro Release Behavior

The freeze-dried hydrogel containing RPP was immersed in 50 mL of deionized water at 37 °C, and the oscillation was maintained at 100 rpm during the experiment. The solution (2 mL) was removed at 0.5, 1, 2, 3, 6, 9, 12, 24, 36, and 48 h, and replaced with the same volume of deionized water to maintain a constant volume [38]. The collected samples were analyzed based on the standard curve to determine the released amount of RPP. The cumulative release of RPP was calculated using Equation (5):

$$\text{Cumulative Release} = \frac{C_t V_0 + \sum_1^{t-1} V_s C_t}{m} \times 100\% \quad (5)$$

where C_t is the concentration of RPP in the solution at the time of sampling (mg/L); V_s is the volume of each sample (L); V_0 is the total volume of the solution (L); m is the drug-carrying capacity of the hydrogel (mg).

2.10. Statistical Analysis

All analyses were performed in triplicate. Means, standard deviations, and graphs were obtained using Origin 2023. Data sets were evaluated using one-way ANOVA (Duncan test) with a significance of $p < 0.05$. Statistical analyses were performed using the SPSS 21 package for Windows.

3. Results and Discussions

3.1. Texture Property

Texture is an important index to evaluate the structural stability of hydrogels. The texture measurements of LMPH are shown in Table 1. The hardness of LMPH increased substantially from 0.17 N (LMPH1) to 1.13 N (LMPH5). As a result of the increase in the content of Longzhua mushroom polysaccharide, crosslinking increased, which caused the hydrogel to form numerous hydrogen bonds and a dense network structure, resulting in an increase in hardness [32,39]. Generally, the structural integrity of a hydrogel is directly proportional to its hardness; the higher the hardness of a hydrogel, the higher its structural integrity [40]. The structural integrity of LMPH5 was significantly higher than that of LMPH1.

Chewiness can be a comprehensive response to the work performed by the chewing process of the sample, chewiness is positively correlated with hardness, the greater the hardness, the greater the work performed by chewing, so chewiness increases with increasing hardness (from 8.59 N to 82.55 N), which also reflects the structural stability of the LMPH is proportional to the polysaccharide content.

Cohesiveness is used to measure the internal bonding strength of the final structure of a hydrogel [41]. The cohesiveness of LMPH gradually increased, indicating an improvement in the internal hydrogel bond strength, which allowed the sample to withstand secondary deformation. LMPH5 exhibited the strongest cohesion (1.13). This indicates that an increase in the polysaccharide content promotes greater entanglement and the formation of stronger networks, leading to greater resistance to deformation [42]. The hydrogels prepared by Li et al. [43] had a maximum hardness of up to 0.9 N, a maximum chewiness of up to 3 N, and a maximum cohesiveness of up to 0.8, compared to which the hardness, chewiness, and cohesiveness of LMPH were significantly higher, which demonstrates that LMPH possesses better textural properties.

Table 1. Textural property of LMPH.

Sample	Hardness (N)	Chewiness (N)	Cohesiveness	Springiness (mm)	Resilience
LMPH1	0.17 ± 0.02^{d}	8.59 ± 0.62^{d}	0.92 ± 0.06^{b}	0.55 ± 0.04^{a}	0.26 ± 0.04^{b}
LMPH2	0.26 ± 0.01^{d}	15.53 ± 1.89^{d}	1.05 ± 0.06^{a}	0.56 ± 0.06^{a}	0.27 ± 0.02^{b}
LMPH3	0.61 ± 0.05^{c}	39.03 ± 6.28^{c}	1.07 ± 0.04^{a}	0.60 ± 0.11^{a}	0.31 ± 0.03^{ab}
LMPH4	0.90 ± 0.06^{b}	62.29 ± 4.39^{b}	1.09 ± 0.03^{a}	0.63 ± 0.03^{a}	0.35 ± 0.04^{a}
LMPH5	1.13 ± 0.18^{a}	82.55 ± 8.39^{a}	1.13 ± 0.09^{a}	0.66 ± 0.06^{a}	0.27 ± 0.05^{b}

Each value represents the mean ± SD ($n = 5$). Values with different letters in the same column differ significantly ($p < 0.05$).

Springiness is used to describe the ability of hydrogels to bounce back. In general, an increase in hardness and cohesiveness leads to an increase in plasticity and rigidity, consequently resulting in reduced springiness of the hydrogels. However, the springiness of LMPH did not decrease significantly with increasing hardness. This result suggests that polysaccharide molecules form more hydrogen bonds within the hydrogel, resulting in a stronger hydrogel structure [44]. Resilience indicates the ability of a sample to bounce back during the first compression process. The recovery of LMPH5 decreased substantially compared with that of LMPH4 (0.35 to 0.27). Because of excessive crosslinking inside the hydrogel, the strong intermolecular interactions limit the movement of the molecular chains, resulting in longer sample recovery times. There have been cases where excessive crosslinking has reduced mechanical properties in similar experiments [45].

3.2. *Rheology Analysis*

Rheological properties are crucial for the stability and viscoelastic behavior of hydrogels. The viscosity sweep of the LMPH is shown in Figure 1A. All gels showed substantial shear thinning. This shear-dependent behavior can be explained by the disruption of the low-energy intermolecular bonds between the polysaccharide chains, which is typical of weak gels [46]. This feature is common in plant polysaccharides, such as *Tremella* polysaccharide gel and *Dendrobium officinale* polysaccharide gel [47,48], which are pseudoplastic fluids. All pseudoplastic hydrogels exhibit shear-thinning properties. Shear-thinning systems have been widely utilized in tissue engineering and the delivery of bioactive substances [49]. In addition, the viscosity of samples increased as the proportion of the polysaccharide increased (i.e., LMPH5 was the most viscous), which suggests that LMPH forms a dense network structure. The increase in hydrogen bonding with increasing polysaccharide concentration leads to the tight connection between the polysaccharide molecules, restricting their movement and thus increasing the viscosity of LMPH [50–52].

As shown in Figure 1B, G′ was constant in the range of 0.1–10%. To obtain more information about the gel structure and mechanical properties, frequency sweep tests were performed within the LVR (1%). Figure 1C shows that LMPH2–LMPH5 exhibited G′ > G″ in the frequency range of 0.1–100 rad/s, suggesting the presence of an interconnected network structure and indicating excellent stability against mechanical stress [53]. However, when the angular frequency increased, LMPH1 behaved as G″ > G′, and the hydrogel exhibited a viscous behavior. This result is similar to that of tanδ (Figure 1D). As the angular frequency increases, LMPH1 conforms to $\tan\delta > 1$, and the LMPH1 system exhibits the characteristics of liquid viscosity. This phenomenon is related to the slippage effect caused by the small-layer synergistic effect of partial sample decomposition [54,55]. In addition, the G′ of LMPH5 was the highest, which means that when the hydrogel resisted deformation, most of the deformation energy was stored in the LMPH5. Moreover, the network structure of this hydrogel is denser, resulting in increased deformation resistance, which is consistent with the texture measurement. The results in Figure 1D are similar to those in Figure 1C, LMPH2–LMPH5 as tanδ < 1, the system mainly exhibits solid elastic properties [56].

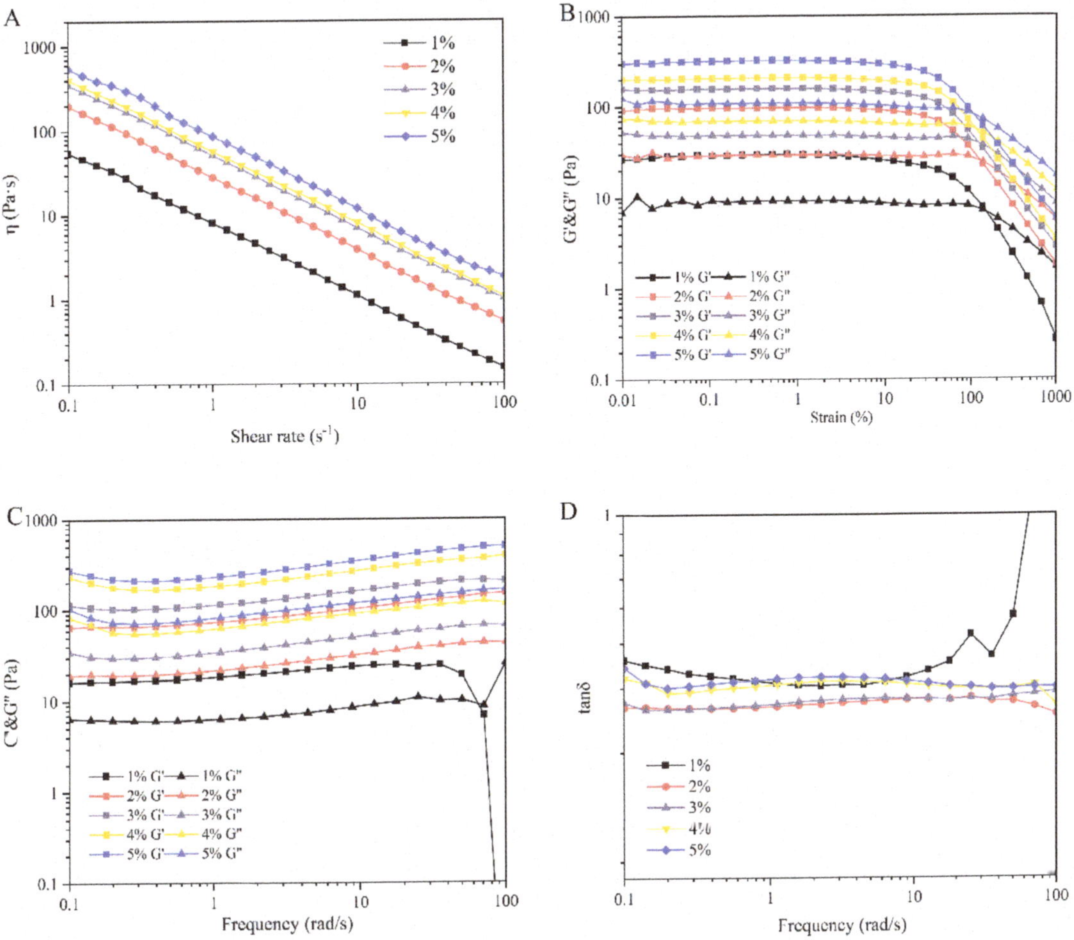

Figure 1. Rheological properties of LMPH. (**A**) Viscosity sweep; (**B**) LVR; (**C**) frequency sweep; (**D**) loss factor.

3.3. The Self-Healing Ability of LMPH

3.3.1. Macroscopic Morphology of LMPH Self-Healing Process

The macroscopic self-healing properties of LMPH are shown in Figure 2A. The methylene blue-treated hydrogel and the untreated hydrogel were cut into two pieces and spliced together. After 2 h at room temperature, the boundary between the two hydrogels became blurred. After 6 h, the hydrogel did not break when picked up with tweezers. This suggests that the hydrogen bonding between the polysaccharide molecules contributes to the self-healing properties of the hydrogel rather than simple adhesion occurring at the interface of hydrogel fragments [57].

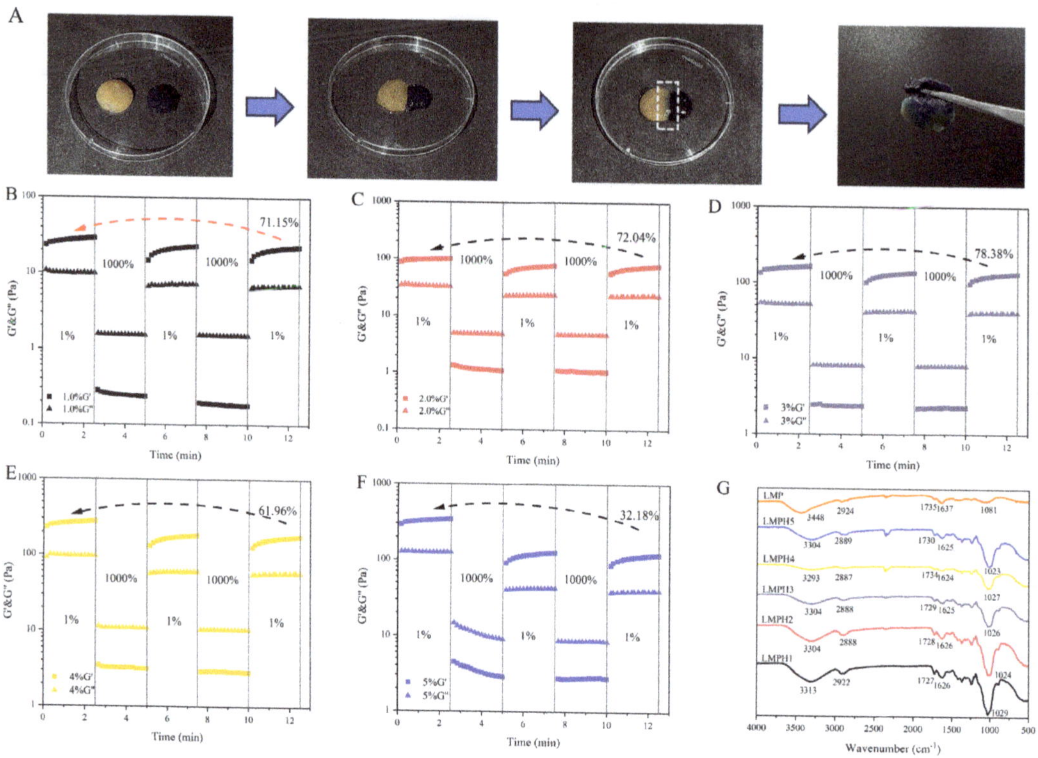

Figure 2. Self-healing properties of LMPH. (**A**) Macroscopic self-healing behavior of cut hydrogels at room temperature; (**B–F**) continuous step-strain tests of the hydrogel; (**G**) FT-IR spectroscopy of LMP and LMPH.

3.3.2. Time Sweep

The self-healing ability of LMPH was tested by changing the shear strain. As shown in Figure 2B–F, the rheological behavior of LMPH within the LVR (1% strain) was elastic (G′ > G″), whereas, at larger strain scans (1000% strain), LMPH shifted to a viscous behavior (G′ < G″). When the shear stress returned to 1%, G′ recovered and the hydrogel showed elastic properties again, suggesting recovery of the gel network and demonstrating the good self-healing ability of LMPH [58]. However, the hydrogel could not reach the original G′ value. The self-healing abilities of LMPH1–LMPH5 were different. LMPH3 exhibited the best self-healing performance, and a 78.38% recovery was reached after two consecutive structural failures. LMPH5 showed the worst recovery (only 32.18%) compared to the Poria polysaccharide hydrogel prepared by Li et al. (minimum 62.25%) [20]. This is because its higher polysaccharide concentration promotes higher entanglement and stronger network formation; thus, it limits chain motion and requires a longer time to bring the system back to equilibrium [42].

3.3.3. FT-IR

The FT-IR spectra of LMP and LMPH are shown in Figure 2G. The peaks at 3448, 2924, 1735, 1637, and 1081 cm^{-1} were attributed to the O-H stretching vibration, C-H stretching vibration, COO- stretching vibration, and alcohol hydroxyl variable angle vibration of LMP [59]. Compared with LMP, the characteristic peaks in the spectrum of LMPH all appear at positions with a low vibration frequency. The peaks of LMP at 3448, 1735, and 1081 cm^{-1} are shifted to positions around 3304, 1730, and 1027 cm^{-1} in LMPH.

The positions of these peaks are redshifted, indicating the presence of hydrogen-bonding interaction forces within the hydrogel [60–62]. The formation of hydrogen bonds reduces the density of the bonding electron cloud, and stretching vibration absorption shifts to a low wavenumber [63]. LMPH is mainly formed by hydrogen-bonding crosslinking. As a result of hydrogen-bonding interactions and complexation between the polysaccharide molecules in the network, the polymer chains on the surface of the hydrogel wounds are constantly moving and re-crosslinking, thus restoring the hydrogel network at the molecular level [64,65]. Thus, we hypothesize that the self-healing ability of hydrogels is caused by the presence of intermolecular hydrogen bonding and complexation [66].

3.4. SEM

Considering the slow-release application of hydrogels in this study, knowledge of their microstructure can provide information on the synergistic relationship between their swelling behavior and release behavior [67]. Figure 3 shows the network structure of LMPH. The hydrogel exhibits a porous structure, and the pore size is in the range of 50–200 μm. The porous structure of LMPH can provide a larger surface area and pore capacity, which is conducive to the adsorption, storage, and release of substances, making LMPH ideal for loading and embedding bioactive substances for slow release [68,69]. From LMPH1 to LMPH5, the network structure gradually becomes more organized and denser. Nevertheless, if the network structure of the hydrogel is too ordered and dense, it will limit the penetration and diffusion of water molecules, reduce swelling, and slow down the release rate of bioactive substances [70].

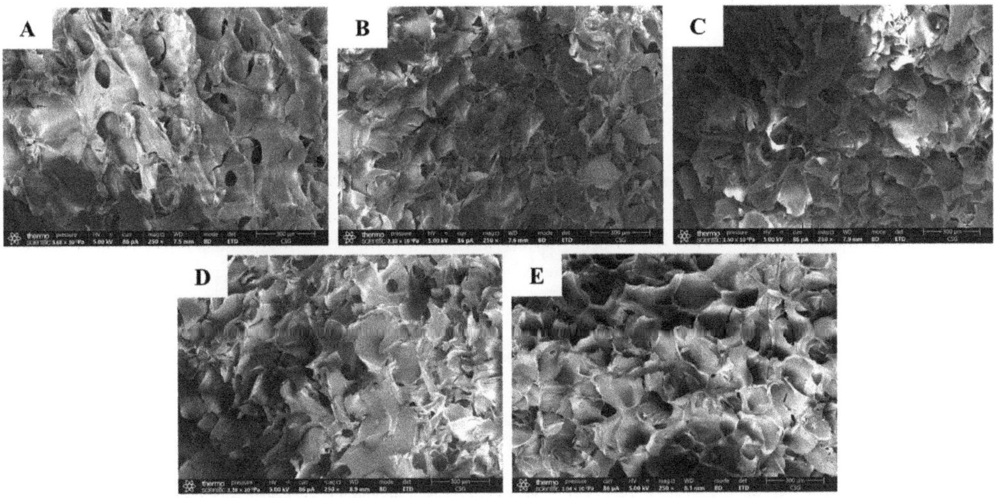

Figure 3. SEM images of LMPH. (**A**) LMPH1; (**B**) LMPH2; (**C**) LMPH3; (**D**) LMPH4; (**E**) LMPH5.

3.5. Water Hold Capacity and Swelling Rate

3.5.1. WHC

The WHC of a hydrogel represents its ability to retain water under external forces [71,72]. As shown in Figure 4A, the lowest WHC of LMPH was 90.84%, and the highest WHC of the hydrogel prepared by He et al. [71] was only 75.57%, indicating that the water-holding property of LMPH was superior. With the increase of polysaccharide content, the WHC also increased, and the WHC of LMPH5 increased significantly to 99.32%.

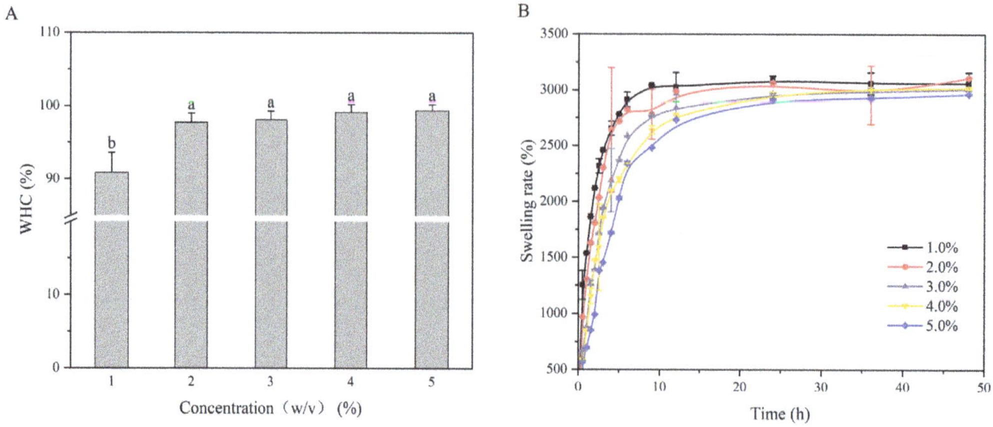

Figure 4. Water retention and swelling properties of LMPH. (**A**) WHC; (**B**) swelling rate. Different lower-case letters represent significant differences between samples.

The number and density of gel networks are two key factors affecting WHC. The WHC of LMPH1 is low because of the loose gel networks, whereas LMPH5 has the highest WHC because it has the highest polysaccharide content, contains more hydroxyl groups, and has high hydrophilicity [73]. A high polysaccharide content results in the formation of a dense network structure of hydrogen bonds through interactions, and a large amount of water is immobilized by physical interception [74].

3.5.2. Swelling Rate

Swelling capacity is another key property of hydrogels as it reflects their ability to adsorb large amounts of water and is reflective of their internal structure [75]. As shown in Figure 4B, LMPH reached swelling equilibrium within 15 h, and the swelling ratio exceeded 2500%. As the polysaccharide content increased, the swelling rate decreased from 3129.94% (LMPH1) to 2965.88% (LMPH5). The hydrogen bonding interactions between –OH groups in the polysaccharide molecules prevent water from entering the interior of the hydrogel, resulting in a low swelling ratio [76]. This outcome was consistent with Y. Wang et al. [70], who reported that a dense crosslink density hindered the diffusion of water, as demonstrated by the water-holding capacity test.

3.6. Stability

3.6.1. Differential Scanning Calorimetry (DSC)

The thermal stability of the LMPH with different polysaccharide contents was analyzed by DSC. The LMPH exhibited an endothermic peak (Figure 5), indicating that melting and decomposition took place. The endothermic peak at approximately 110 °C is related to the thermal decomposition of polysaccharides, including the hydrogen bond between galacturonic acid units and the conformational change, decarboxylation of side-chain groups, and carbon in the ring [77]. The temperature at which LMPH1 and LMPH3 began to decompose increased from 91.3 °C to 93.4 °C, and the peak temperature of decomposition increased from 108.8 °C to 116.1 °C, respectively. The higher temperature required for decomposition demonstrated that the increase in the content of the polysaccharide enhanced the thermal stability of LMPH [78]. The higher values imply higher stability caused by stronger bonds within the chains due to bond formation, loss of internal water, or other reasons [79,80]. In summary, LMPH shows high thermal stability, ensuring the stable and sustained release of bioactives at high temperatures.

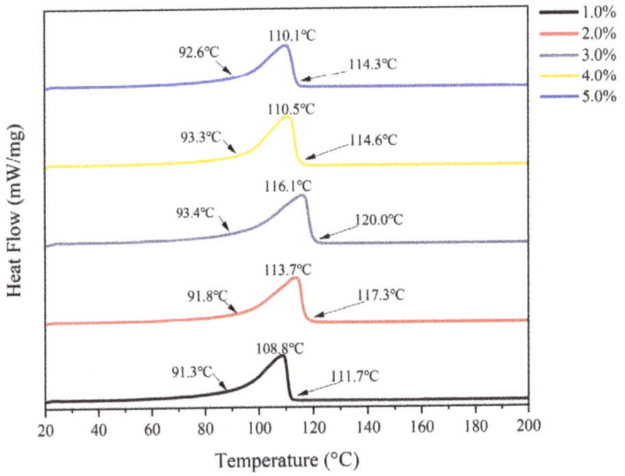

Figure 5. DSC curves of LMPH.

3.6.2. Freeze–Thaw Stability

The experimental results are summarized in Table 2. As the freezing and thawing cycles increased, the water separation rate of LMPH first increased and then decreased. The low water separation rate in Cycle 1 is due to the initial formation of the gel network of LMPH, which is less affected by the freeze–thaw cycle. However, as the number of freeze–thaw cycles increases, the water separation rate of LMPH increases, indicating that the freeze–thaw cycle causes great damage to the network structure of LMPH [81]. The formation and melting of ice crystals during freezing and thawing disrupt the network structure of the hydrogel, resulting in the disruption of the hydrogen bonds between the polysaccharide molecules [82]. For Cycle 3, the water separation rate of LMPH1 reached 34.65%. Then, the water separation rate started to decrease, and the water separation rate of LMPH1 was 12.25% in Cycle 5. The freeze–thaw stability of LMPH was significantly better compared with the prepared ginkgo polysaccharide hydrogels (with a minimum water separation rate of not less than 70%) of Zhang et al. [34]. This shows that the network structure of LMPH is gradually improved, the freeze–thaw stability of the hydrogel is gradually improved, and the infiltration and separation of water are reduced.

Table 2. The water separation rate (%) of LMPH in each cycle.

Concentration	Water Separation Rate (%)				
	Cycle 1	Cycle 2	Cycle 3	Cycle 4	Cycle 5
LMPH1	1.26 ± 0.09	12.0 ± 1.89	34.65 ± 2.33 [a]	21.73 ± 1.42 [a]	12.25 ± 1.26
LMPH2	ND	1.61 ± 0.14	21.64 ± 2.18 [b]	4.57 ± 0.72 [b]	ND
LMPH3	ND	ND	10.97 ± 0.11 [c]	1.21 ± 0.21 [c]	ND
LMPH4	ND	ND	4.28 ± 0.12 [d]	0.28 ± 0.03 [c]	ND
LMPH5	ND	ND	0.57 ± 0.19 [d]	ND	ND

Note: Different lowercase letters in the same column indicate significant difference ($p < 0.05$). ND: Not detected.

The higher the polysaccharide content in LMPH, the better its freeze–thaw stability. The freeze–thaw stability is related to viscosity [83]. LMPH5 exhibited the highest freeze–thaw stability because it had the highest viscosity. The hydrogen bonds in the polysaccharide molecules increase the viscosity of the gel and the strength and stability of the gel network. The high viscosity limits the separation of water [84,85]. In general, the denser the gel network, the stronger the water-binding capacity. This result is consistent with the rheology results (static viscosity sweep tests).

3.7. In Vitro Release Behavior

3.7.1. Encapsulation Efficiency (EE)

As shown in Figure 6A, the EEs were 93.13% (LMPH1), 98.89% (LMPH2), 99.24% (LMPH3), 99.56% (LMPH4), and 99.94% (LMPH5). EE is proportional to the polysaccharide content in LMPH. A high EE is related to the porous structure of the hydrogel, which adsorbs more bioactive molecules. As the polysaccharide content increases, the stronger the interactions of hydrogen bonds formed between polysaccharide molecules, the more ordered the network structure is formed, and the more ordered network structure can provide more contact and immobilization points for RPP. RPP was captured by the network structure of the hydrogel. When the hydrogel is rinsed with ultrapure water, it is difficult for water molecules to enter the hydrogel with high polysaccharide concentration and for RPP molecules to separate from the dense network structure. As a result, the polysaccharide concentration of the hydrogel is higher, and the encapsulation efficiency is higher. By tuning the porous structure and network organization of the gels, higher encapsulation efficiency can be achieved, expanding the potential of hydrogels for nutrient delivery applications.

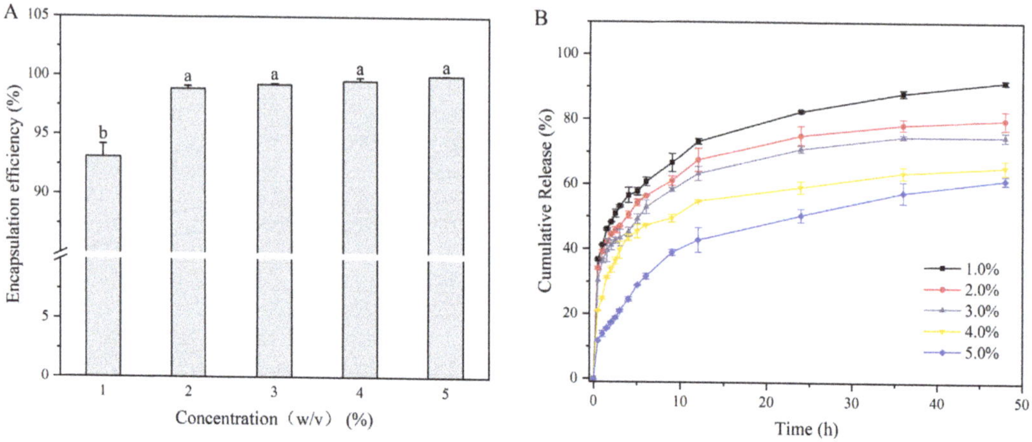

Figure 6. Encapsulation efficiency and release behavior of RPP in vitro. (**A**) Encapsulation efficiency of RPP; (**B**) in-vitro release behavior. Different lower-case letters represent significant differences between samples.

3.7.2. In Vitro Release Behavior of Polyphenols from Rambutan Peel Polyphenols

Based on the above-mentioned porous structure and swelling properties of LMPH, this study used LMPH as a slow-release material and rambutan peel polyphenol as a model bioactive substance. Figure 6B shows the sustained release of rambutan peel polyphenols by LMPH within 48 h. The rapid release trend of polyphenols during the first 5 h was attributed to the inability of the cross-linked network of the hydrogel to control surface drug release when the hydrogel was in a release medium that allowed free initial dissolution of polyphenols. With time, the release of polyphenols gradually increased. LMPH1 showed the highest cumulative polyphenol release rate (91.70%). LMPH5 had the lowest cumulative release rate (61.58%). Just as high-concentration polysaccharide hydrogels prevented water molecules from entering the gel interior in swelling experiments and hydrogels trapped water molecules in water retention experiments, the dense network structure limited interchain repulsion, slowed the penetration of water molecules into the interior of the hydrogel, and allowed for better sequestration of polyphenols [86]; a result consistent with those of M. Li et al. [43], where the release rate of substances inside the hydrogel was reduced due to a denser network structure inside the hydrogel, and the dense network structure negatively affected the release behavior.

4. Conclusions

In this study, a novel one-component hydrogel was successfully prepared by heating using polysaccharides from Longzhua mushroom. Compared with the common methods of preparing hydrogels, which usually require the addition of an inducer to make them gel or the compounding of two or more substances to form a gel, the preparation method in this study is simple and easy to use. Then, Longzhua mushroom polysaccharide hydrogel (LMPH) was used to study the encapsulation and slow release of rambutan peel polyphenols (RPP). The rheological and textural properties of the LMPH were characterized. All the hydrogels showed adequate viscoelasticity and textural properties. LMPH exhibited good self-healing properties driven by hydrogen bonding. LMPH has a three-dimensional porous network structure, and its water-holding capacity (WHC) and swelling rate are related to its porous structure. The WHC and swelling properties of the hydrogel were correlated with the polysaccharide content of LMPH, with WHC being directly proportional to the polysaccharide content and swelling rate being inversely proportional to the polysaccharide content. LMPH was evaluated for the slow release of RPP, which lasted for 48 h. These results suggest that the novel LMPH has adequate mechanical and self-healing properties as well as stability and thus has a wide range of potential applications in the slow release of nutrients and delivery of bioactive substances.

Author Contributions: Conceptualization, L.Z., Y.D. and L.S.; methodology, J.L.; investigation, L.Z. and J.L.; data curation, L.Z. and J.L.; writing—original draft preparation, L.Z.; writing—review and editing, Y.D. and L.S.; visualization, L.Z.; supervision, Y.D. and L.S.; funding acquisition, Y.D. and L.S. All authors have read and agreed to the published version of the manuscript.

Funding: This work was supported by the Yunnan Major Scientific and Technological Project (Grant No. 202202AG050009) and the Yunnan Fundamental Research Project (Grant No. 202201BE070001-055).

Data Availability Statement: The original contributions presented in the study are included in the article, further inquiries can be directed to the corresponding author.

Conflicts of Interest: The authors declare no conflict of interest.

References

1. Solís-Fuentes, J.A.; Camey-Ortíz, G.; Hernández-Medel, M.d.R.; Pérez-Mendoza, F.; Durán-de-Bazúa, C. Composition, phase behavior and thermal stability of natural edible fat from rambutan (*Nephelium lappaceum* L.) seed. *Bioresour. Technol.* **2010**, *101*, 799–803. [CrossRef] [PubMed]
2. Hernández-Hernández, C.; Aguilar, C.N.; Rodríguez-Herrera, R.; Flores-Gallegos, A.C.; Morlett-Chávez, J.; Govea-Salas, M.; Ascacio-Valdés, J.A. Rambutan (*Nephelium lappaceum* L.): Nutritional and functional properties. *Trends Food Sci. Technol.* **2019**, *85*, 201–210. [CrossRef]
3. Phuong, N.N.M.; Le, T.T.; Dang, M.Q.; Van Camp, J.; Raes, K. Selection of extraction conditions of phenolic compounds from rambutan (*Nephelium lappaceum* L.) peel. *Food Bioprod. Process.* **2020**, *122*, 222–229. [CrossRef]
4. Xiao, J.; Liu, B.; Zhuang, Y. Effects of rambutan (*Nephelium lappaceum*) peel phenolics and Leu-Ser-Gly-Tyr-Gly-Pro on hairless mice skin photoaging induced by ultraviolet irradiation. *Food Chem. Toxicol.* **2019**, *129*, 30–37. [CrossRef]
5. Heleno, S.A.; Martins, A.; Queiroz, M.J.R.P.; Ferreira, I.C.F.R. Bioactivity of phenolic acids: Metabolites versus parent compounds: A review. *Food Chem.* **2015**, *173*, 501–513. [CrossRef]
6. Hu, B.; Liu, X.; Zhang, C.; Zeng, X. Food macromolecule based nanodelivery systems for enhancing the bioavailability of polyphenols. *J. Food Drug Anal.* **2017**, *25*, 3–15. [CrossRef]
7. Siddiqui, I.A.; Adhami, V.M.; Bharali, D.J.; Hafeez, B.B.; Asim, M.; Khwaja, S.I.; Ahmad, N.; Cui, H.; Mousa, S.A.; Mukhtar, H. Introducing Nanochemoprevention as a Novel Approach for Cancer Control: Proof of Principle with Green Tea Polyphenol Epigallocatechin-3-Gallate. *Cancer Res.* **2009**, *69*, 1712–1716. [CrossRef]
8. Li, Z.; Jiang, H.; Xu, C.; Gu, L. A review: Using nanoparticles to enhance absorption and bioavailability of phenolic phytochemicals. *Food Hydrocoll.* **2015**, *43*, 153–164. [CrossRef]
9. Yang, M.; Zhao, S.; Zhao, C.; Cui, J.; Wang, Y.; Fang, X.; Zheng, J. Caseinate-reinforced pectin hydrogels: Efficient encapsulation, desirable release, and chemical stabilization of (−)-epigallocatechin. *Int. J. Biol. Macromol.* **2023**, *230*, 123298. [CrossRef]
10. Yegappan, R.; Selvaprithiviraj, V.; Amirthalingam, S.; Jayakumar, R. Carrageenan based hydrogels for drug delivery, tissue engineering and wound healing. *Carbohydr. Polym.* **2018**, *198*, 385–400. [CrossRef] [PubMed]
11. Münster, L.; Capáková, Z.; Fišera, M.; Kuřitka, I.; Vícha, J. Biocompatible dialdehyde cellulose/poly(vinyl alcohol) hydrogels with tunable properties. *Carbohydr. Polym.* **2019**, *218*, 333–342. [CrossRef]

12. Yan, S.; Wu, S.; Zhang, J.; Zhang, S.; Huang, Y.; Zhu, H.; Li, Y.; Qi, B. Controlled release of curcumin from gelatin hydrogels by the molecular-weight modulation of an oxidized dextran cross-linker. *Food Chem.* **2023**, *418*, 135966. [CrossRef]
13. Do, N.H.N.; Huynh, T.N.A.; Le, T.X.; Ha, A.C.; Le, P.K. Encapsulation of *Triphasia trifolia* extracts by pH and thermal dual-sensitive chitosan hydrogels for controlled release. *Carbohydr. Polym.* **2023**, *320*, 121264. [CrossRef]
14. Samadian, H.; Maleki, H.; Allahyari, Z.; Jaymand, M. Natural polymers-based light-induced hydrogels: Promising biomaterials for biomedical applications. *Coord. Chem. Rev.* **2020**, *420*, 213432. [CrossRef]
15. Hu, X.; Wang, Y.; Zhang, L.; Xu, M. Construction of self-assembled polyelectrolyte complex hydrogel based on oppositely charged polysaccharides for sustained delivery of green tea polyphenols. *Food Chem.* **2020**, *306*, 125632. [CrossRef]
16. Yu, Q.; Chen, W.; Zhong, J.; Huang, D.; Shi, W.; Chen, H.; Yan, C. Purification, structural characterization, and bioactivities of a polysaccharide from *Coreopsis tinctoria*. *Food Front.* **2022**, *3*, 736–748. [CrossRef]
17. Manzoor, A.; Dar, A.H.; Pandey, V.K.; Shams, R.; Khan, S.; Panesar, P.S.; Kennedy, J.F.; Fayaz, U.; Khan, S.A. Recent insights into polysaccharide-based hydrogels and their potential applications in food sector: A review. *Int. J. Biol. Macromol.* **2022**, *213*, 987–1006. [CrossRef]
18. Peers, S.; Montembault, A.; Ladavière, C. Chitosan hydrogels incorporating colloids for sustained drug delivery. *Carbohydr. Polym.* **2022**, *275*, 118689. [CrossRef]
19. Obaroakpo, J.U.; Liu, L.; Zhang, S.; Lu, J.; Liu, L.; Pang, X.; Lv, J. In vitro modulation of glucagon-like peptide release by DPP-IV inhibitory polyphenol-polysaccharide conjugates of sprouted quinoa yoghurt. *Food Chem.* **2020**, *324*, 126857. [CrossRef]
20. Li, W.; Fang, K.; Yuan, H.; Li, D.; Li, H.; Chen, Y.; Luo, X.; Zhang, L.; Ye, X. Acid-induced *Poria cocos* alkali-soluble polysaccharide hydrogel: Gelation behaviour, characteristics, and potential application in drug delivery. *Int. J. Biol. Macromol.* **2023**, *242*, 124383. [CrossRef]
21. Srivastava, N.; Richa; Roy Choudhury, A. Recent advances in composite hydrogels prepared solely from polysaccharides. *Colloids Surf. B Biointerfaces* **2021**, *205*, 111891. [CrossRef]
22. Zhang, M.; Li, Y.; Wang, W.; Yang, Y.; Shi, X.; Sun, M.; Hao, Y.; Li, Y. Comparison of physicochemical and rheology properties of Shiitake stipes-derived chitin nanocrystals and nanofibers. *Carbohydr. Polym.* **2020**, *244*, 116468. [CrossRef]
23. Wang, W.; Tan, J.; Nima, L.; Sang, Y.; Cai, X.; Xue, H. Polysaccharides from fungi: A review on their extraction, purification, structural features, and biological activities. *Food Chem. X* **2022**, *15*, 100414. [CrossRef]
24. Zhang, R.; Tao, Y.; Xu, W.; Xiao, S.; Du, S.; Zhou, Y.; Hasan, A. Rheological and controlled release properties of hydrogels based on mushroom hyperbranched polysaccharide and xanthan gum. *Int. J. Biol. Macromol.* **2018**, *120*, 2399–2409. [CrossRef]
25. Ma, X.; Yang, M.; He, Y.; Zhai, C.; Li, C. A review on the production, structure, bioactivities and applications of *Tremella polysaccharides*. *Int. J. Immunopathol. Pharmacol.* **2021**, *35*, 20587384211000541. [CrossRef]
26. Li, D.; Wei, Z.; Sun, J.; Xue, C. Tremella polysaccharides-coated zein nanoparticles for enhancing stability and bioaccessibility of curcumin. *Curr. Res. Food Sci.* **2022**, *5*, 611–618. [CrossRef]
27. Menezes, T.M.F.; Campelo, M.d.S.; Lima, A.B.N.; Câmara Neto, J.F.; Saraiva, M.M.; de Sousa, J.A.C.; Gonzaga, M.L.d.C.; Leal, L.K.A.M.; Ribeiro, M.E.N.P.; Ricardo, N.M.P.S.; et al. Effects of polysaccharides isolated from mushrooms (*Lentinus edodes* Berk or *Agaricus blazei* Murill) on the gelation of Pluronic® F127. *Colloids Surf. A Physicochem. Eng. Asp.* **2022**, *642*, 128684. [CrossRef]
28. Rodríguez-Seoane, P.; Torres, M.D.; González-Muñoz, M.J.; Sinde-Stompel, E.; Domínguez, H. Formulation of bio-hydrogels from *Hericium erinaceus* in *Paulownia elongata x fortunei* autohydrolysis aqueous extracts. *Food Bioprod. Process.* **2021**, *128*, 12–20. [CrossRef]
29. Sun, L.; Su, X.; Zhuang, Y. Preparation, characterization and antiglycation activities of the novel polysaccharides from *Boletus snicus*. *Int. J. Biol. Macromol.* **2016**, *92*, 607–614. [CrossRef]
30. Wang, M.; Yin, Z.; Sun, W.; Zhong, Q.; Zhang, Y.; Zeng, M. Microalgae play a structuring role in food: Effect of spirulina platensis on the rheological, gelling characteristics, and mechanical properties of soy protein isolate hydrogel. *Food Hydrocoll.* **2023**, *136*, 108244. [CrossRef]
31. Hou, Y.; Zhao, J.; Yin, J.; Geng, F.; Nie, S. The synergistic gelation of *Dendrobium officinale* polysaccharide (Dendronans) with xanthan gum and its rheological and texture properties. *Food Hydrocoll.* **2023**, *141*, 108674. [CrossRef]
32. Lin, R.; Xu, R.; Chen, H.; Liu, B.; Yuan, C.; Guo, L.; Cui, B.; Fang, Y. Dual cross-linked starch hydrogel for eugenol encapsulation and the formation of hydrogen bonds on textural hydrogel. *Carbohydr. Polym.* **2023**, *316*, 121044. [CrossRef]
33. Wei, Z.; Yang, J.H.; Liu, Z.Q.; Xu, F.; Zhou, J.X.; Zrínyi, M.; Osada, Y.; Chen, Y.M. Novel Biocompatible Polysaccharide-Based Self-Healing Hydrogel. *Adv. Funct. Mater.* **2015**, *25*, 1352–1359. [CrossRef]
34. Zhang, Z.; Chen, Z.; Zhang, C.; Kang, W. Physicochemical properties and biological activities of Tremella hydrocolloids. *Food Chem.* **2023**, *407*, 135164. [CrossRef]
35. Khalifa, I.; Li, M.; Mamet, T.; Li, C. Maltodextrin or gum Arabic with whey proteins as wall-material blends increased the stability and physiochemical characteristics of mulberry microparticles. *Food Biosci.* **2019**, *31*, 100445. [CrossRef]
36. Tan, J.; Ma, Q.; Li, J.; Liu, Q.; Zhuang, Y. Bioavailability and Antioxidant Activity of Rambutan (*Nephelium lappaceum*) Peel Polyphenols during in Vitro Simulated Gastrointestinal Digestion, Caco-2 Monolayer Cell Model Application, and Colonic Fermentation. *J. Agric. Food Chem.* **2023**, *71*, 15829–15841. [CrossRef]
37. Bhat, M.A.; Rather, R.A.; Shalla, A.H. Texture and rheological features of strain and pH sensitive chitosan-imine graphene-oxide composite hydrogel with fast self-healing nature. *Int. J. Biol. Macromol.* **2022**, *222*, 3129–3141. [CrossRef]

38. Chang, A.; Ye, Z.; Ye, Z.; Deng, J.; Lin, J.; Wu, C.; Zhu, H. Citric acid crosslinked sphingan WL gum hydrogel films supported ciprofloxacin for potential wound dressing application. *Carbohydr. Polym.* **2022**, *291*, 119520. [CrossRef]
39. Stenner, R.; Matubayasi, N.; Shimizu, S. Gelation of carrageenan: Effects of sugars and polyols. *Food Hydrocoll.* **2016**, *54*, 284–292. [CrossRef]
40. Mahajan, P.; Bera, M.B.; Panesar, P.S. Modification of Kutki millet (*Panicum sumatrense*) starch properties by the addition of amino acids for the preparation of hydrogels and its characterization. *Int. J. Biol. Macromol.* **2021**, *191*, 9–18. [CrossRef]
41. Teng, L.Y.; Chin, N.L.; Yusof, Y.A. Rheological and textural studies of fresh and freeze-thawed native sago starch–sugar gels. II. Comparisons with other starch sources and reheating effects. *Food Hydrocoll.* **2013**, *31*, 156–165. [CrossRef]
42. Kongjaroen, A.; Methacanon, P.; Seetapan, N.; Fuongfuchat, A.; Gamonpilas, C.; Nishinari, K. Effects of dispersing media on the shear and extensional rheology of xanthan gum and guar gum-based thickeners used for dysphagia management. *Food Hydrocoll.* **2022**, *132*, 107857. [CrossRef]
43. Li, M.; Cui, H.; Lou, W. *Millettia speciosa* Champ cellulose-based hydrogel as a novel delivery system for *Lactobacillus paracasei*: Its relationship to structure, encapsulation and controlled release. *Carbohydr. Polym.* **2023**, *316*, 121034. [CrossRef]
44. Cui, T.; Wu, Y.; Ni, C.; Sun, Y.; Cheng, J. Rheology and texture analysis of gelatin/dialdehyde starch hydrogel carriers for curcumin controlled release. *Carbohydr. Polym.* **2022**, *283*, 119154. [CrossRef]
45. Lin, X.; Zhao, X.; Xu, C.; Wang, L.; Xia, Y. Progress in the mechanical enhancement of hydrogels: Fabrication strategies and underlying mechanisms. *J. Polym. Sci.* **2022**, *60*, 2525–2542. [CrossRef]
46. Xu, X.; Xu, J.; Zhang, Y.; Zhang, L. Rheology of triple helical Lentinan in solution: Steady shear viscosity and dynamic oscillatory behavior. *Food Hydrocoll.* **2008**, *22*, 735–741. [CrossRef]
47. Feng, J.; Tian, H.; Chen, X.; Cai, X.; Shi, X.; Wang, S. Interaction between fish gelatin and tremella polysaccharides from aqueous solutions to complex coacervates: Structure and rheological properties. *Food Hydrocoll.* **2023**, *138*, 108439. [CrossRef]
48. Zhu, H.; Xu, L.; Wang, J.; Zhang, Z.; Xu, X.; Yang, K.; Sun, P.; Liao, X.; Cai, M. Rheological behaviors of ethanol-fractional polysaccharides from *Dendrobium officinale* in aqueous solution: Effects of concentration, temperature, pH, and metal ions. *Food Hydrocoll.* **2023**, *137*, 108311. [CrossRef]
49. Nair, R.; Roy Choudhury, A. Synthesis and rheological characterization of a novel shear thinning levan gellan hydrogel. *Int. J. Biol. Macromol.* **2020**, *159*, 922–930. [CrossRef]
50. Hu, Y.; Li, C.; Tan, Y.; McClements, D.J.; Wang, L. Insight of rheology, water distribution and in vitro digestive behavior of starch based-emulsion gel: Impact of potato starch concentration. *Food Hydrocoll.* **2022**, *132*, 107859. [CrossRef]
51. Lin, Y.; An, F.; He, H.; Geng, F.; Song, H.; Huang, Q. Structural and rheological characterization of pectin from passion fruit (*Passiflora edulis f. flavicarpa*) peel extracted by high-speed shearing. *Food Hydrocoll.* **2021**, *114*, 106555. [CrossRef]
52. Wen, Y.; Che, Q.T.; Kim, H.W.; Park, H.J. Potato starch altered the rheological, printing, and melting properties of 3D-printable fat analogs based on inulin emulsion-filled gels. *Carbohydr. Polym.* **2021**, *269*, 118285. [CrossRef] [PubMed]
53. Fitzsimons, S.M.; Tobin, J.T.; Morris, E.R. Synergistic binding of konjac glucomannan to xanthan on mixing at room temperature. *Food Hydrocoll.* **2008**, *22*, 36–46. [CrossRef]
54. Ferraro, G.; Fratini, E.; Sacco, P.; Asaro, F.; Cuomo, F.; Donati, I.; Lopez, F. Structural characterization and physical ageing of mucilage from chia for food processing applications. *Food Hydrocoll.* **2022**, *129*, 107614. [CrossRef]
55. Li, L.; Yan, J.-N.; Lai, B.; Wang, C.; Sun, J.-J.; Wu, H.-T. Rheological properties of chia seed gum extracted by high-speed shearing and its comparison with commercial polysaccharides. *Food Hydrocoll.* **2023**, *144*, 108936. [CrossRef]
56. Shao, P.; Qin, M.; Han, L.; Sun, P. Rheology and characteristics of sulfated polysaccharides from chlorophytan seaweeds Ulva fasciata. *Carbohydr. Polym.* **2014**, *113*, 365–372. [CrossRef] [PubMed]
57. Han, W.; Chen, C.; Yang, K.; Wang, H.; Xia, H.; Zhao, Y.; Teng, Y.; Feng, G.; Chen, Y.M. Hyaluronic acid and chitosan-based injectable and self-healing hydrogel with inherent antibacterial and antioxidant bioactivities. *Int. J. Biol. Macromol.* **2023**, *227*, 373–383. [CrossRef] [PubMed]
58. Zhou, P.; Wen, L.; Ai, T.; Liang, H.; Li, J.; Li, B. A novel emulsion gel solely stabilized by the hot water extracted polysaccharide from psyllium husk: Self-healing plays a key role. *Food Hydrocoll.* **2022**, *130*, 107718. [CrossRef]
59. Tong, C.; Jiang, S.; Ye, D.; Li, K.; Liu, J.; Zeng, X.; Wu, C.; Pang, J. Enhanced mechanical property and freeze-thaw stability of alkali-induced heat-set konjac glucomannan hydrogel through anchoring interface effects of carboxylated cellulose nanocrystals. *Food Hydrocoll.* **2023**, *142*, 108812. [CrossRef]
60. Li, S.; Wang, L.; Zheng, W.; Yang, G.; Jiang, X. Rapid Fabrication of Self-Healing, Conductive, and Injectable Gel as Dressings for Healing Wounds in Stretchable Parts of the Body. *Adv. Funct. Mater.* **2020**, *30*, 2002370. [CrossRef]
61. Ren, W.; Qiang, T.; Chen, L. Recyclable and biodegradable pectin-based film with high mechanical strength. *Food Hydrocoll.* **2022**, *129*, 107643. [CrossRef]
62. Wang, C.; Jiang, X.; Kim, H.-J.; Zhang, S.; Zhou, X.; Chen, Y.; Ling, H.; Xue, Y.; Chen, Z.; Qu, M.; et al. Flexible patch with printable and antibacterial conductive hydrogel electrodes for accelerated wound healing. *Biomaterials* **2022**, *285*, 121479. [CrossRef] [PubMed]
63. Needham, P. Hydrogen bonding: Homing in on a tricky chemical concept. *Stud. Hist. Philos. Sci. Part A* **2013**, *44*, 51–65. [CrossRef]
64. Cao, J.; He, G.; Ning, X.; Wang, C.; Fan, L.; Yin, Y.; Cai, W. Hydroxypropyl chitosan-based dual self-healing hydrogel for adsorption of chromium ions. *Int. J. Biol. Macromol.* **2021**, *174*, 89–100. [CrossRef] [PubMed]

65. Yang, X.; Liu, G.; Peng, L.; Guo, J.; Tao, L.; Yuan, J.; Chang, C.; Wei, Y.; Zhang, L. Highly Efficient Self-Healable and Dual Responsive Cellulose-Based Hydrogels for Controlled Release and 3D Cell Culture. *Adv. Funct. Mater.* **2017**, *27*, 1703174. [CrossRef]
66. Qi, X.; Tong, X.; You, S.; Mao, R.; Cai, E.; Pan, W.; Zhang, C.; Hu, R.; Shen, J. Mild Hyperthermia-Assisted ROS Scavenging Hydrogels Achieve Diabetic Wound Healing. *ACS Macro Lett.* **2022**, *11*, 861–867. [CrossRef] [PubMed]
67. Deka, R.; Sarmah, J.K.; Baruah, S.; Dutta, R.R. An okra polysaccharide (*Abelmoschus esculentus*) reinforced green hydrogel based on guar gum and poly-vinyl alcohol double network for controlled release of nanocurcumin. *Int. J. Biol. Macromol.* **2023**, *234*, 123618. [CrossRef] [PubMed]
68. Ridolfo, R.; Tavakoli, S.; Junnuthula, V.; Williams, D.S.; Urtti, A.; van Hest, J.C.M. Exploring the Impact of Morphology on the Properties of Biodegradable Nanoparticles and Their Diffusion in Complex Biological Medium. *Biomacromolecules* **2020**, *22*, 126–133. [CrossRef] [PubMed]
69. Wang, W.; Shi, D.; Zhang, Y.; Li, W.; Li, F.; Feng, H.; Ma, L.; Yang, C.; Peng, Z.; Song, G.; et al. An injectable hydrogel based on hyaluronic acid prepared by Schiff base for long-term controlled drug release. *Int. J. Biol. Macromol.* **2023**, *245*, 125341. [CrossRef]
70. Wang, Y.; Yang, M.; Zhao, Z. Facile fabrication of self-healing, injectable and antimicrobial cationic guar gum hydrogel dressings driven by hydrogen bonds. *Carbohydr. Polym.* **2023**, *310*, 120723. [CrossRef]
71. He, X.; Zhao, H.; Xu, Y.; Yi, S.; Li, J.; Li, X. Synergistic effects of oat β-glucan combined with ultrasound treatment on gel properties of silver carp surimi. *Ultrason. Sonochemistry* **2023**, *95*, 106406. [CrossRef] [PubMed]
72. Qin, L.; Fu, Y.; Yang, F.; Chang, Z.; Zou, C.; Gao, H.; Jiang, D.; Jia, C. Effects of polysaccharides autoclave extracted from Flammulina velutipes mycelium on freeze-thaw stability of surimi gels. *LWT* **2022**, *169*, 113941. [CrossRef]
73. Liu, M.; Chen, G.; Zhang, H.; Yu, Q.; Mei, X.; Kan, J. Heat-induced inulin-gluten gel: Insights into the influences of inulin molecular weight on the rheological and structural properties of gluten gel to molecular and physicochemical characteristics. *Food Hydrocoll.* **2021**, *111*, 106397. [CrossRef]
74. Wan, L.; Yang, Z.; Cai, R.; Pan, S.; Liu, F.; Pan, S. Calcium-induced-gel properties for low methoxyl pectin in the presence of different sugar alcohols. *Food Hydrocoll.* **2021**, *112*, 106252. [CrossRef]
75. Ma, Z.; Yang, X.; Ma, J.; Lv, J.; He, J.; Jia, D.; Qu, Y.; Chen, G.; Yan, H.; Zeng, R. Development of the mussel-inspired pH-responsive hydrogel based on *Bletilla striata* polysaccharide with enhanced adhesiveness and antioxidant properties. *Colloids Surf. B Biointerfaces* **2021**, *208*, 112066. [CrossRef] [PubMed]
76. Ge, Y.; Wu, J.; Pang, M.; Hu, D.; Li, Z.; Wang, X.; Sun, L.; Chen, X.; Yao, J. Novel carboxymethyl chitosan/N-acetylneuraminic acid hydrogel for the protection of *Pediococcus pentosaceus*. *Food Res. Int.* **2022**, *156*, 111355. [CrossRef] [PubMed]
77. Lv, D.; Chen, F.; Yang, X.; Yin, L.; Rashid, M.T.; Li, Y. Spontaneous gelation behaviors and mechanism of *Ficus awkeotsang* Makino pectin. *Int. J. Biol. Macromol.* **2023**, *247*, 125712. [CrossRef] [PubMed]
78. Zhang, D.; Chen, D.; Patel, B.; Campanella, O.H. Pectin as a natural agent for reinforcement of pea protein gel. *Carbohydr. Polym.* **2022**, *298*, 120038. [CrossRef]
79. Milan, E.P.; Martins, V.C.A.; Horn, M.M.; Plepis, A.M.G. Influence of blend ratio and mangosteen extract in chitosan/collagen gels and scaffolds: Rheological and release studies. *Carbohydr. Polym.* **2022**, *292*, 119647. [CrossRef]
80. Trębacz, H.; Szczęsna, A.; Arczewska, M. Thermal stability of collagen in naturally ageing and in vitro glycated rabbit tissues. *J. Therm. Anal. Calorim.* **2018**, *134*, 1903–1911. [CrossRef]
81. Zou, J.; Wang, L.; Sun, G. Sustainable and Reusable Gelatin-Based Hydrogel "Jelly Ice Cubes" as Food Coolant. II: Ideal Freeze–Thaw Conditions. *ACS Sustain. Chem. Eng.* **2021**, *9*, 15365–15374. [CrossRef]
82. Zhang, B.; Bai, B.; Pan, Y.; Li, X.-M.; Cheng, J.-S.; Chen, H.-Q. Effects of pectin with different molecular weight on gelatinization behavior, textural properties, retrogradation and in vitro digestibility of corn starch. *Food Chem.* **2018**, *264*, 58–63. [CrossRef] [PubMed]
83. Yuan, C.; Xu, D.; Cui, B.; Wang, Y. Gelation of κ-carrageenan/Konjac glucommanan compound gel: Effect of cyclodextrins. *Food Hydrocoll.* **2019**, *87*, 158–164. [CrossRef]
84. Gadkari, P.V.; Tu, S.; Chiyarda, K.; Reaney, M.J.T.; Ghosh, S. Rheological characterization of fenugreek gum and comparison with other galactomannans. *Int. J. Biol. Macromol.* **2018**, *119*, 486–495. [CrossRef] [PubMed]
85. Liu, Y.; Lei, F.; He, L.; Xu, W.; Jiang, J. Comparative study on the monosaccharides of three typical galactomannans hydrolyzed by different methods. *Ind. Crops Prod.* **2020**, *157*, 112895. [CrossRef]
86. Zavareh, H.S.; Pourmadadi, M.; Moradi, A.; Yazdian, F.; Omidi, M. Chitosan/carbon quantum dot/aptamer complex as a potential anticancer drug delivery system towards the release of 5-fluorouracil. *Int. J. Biol. Macromol.* **2020**, *165*, 1422–1430. [CrossRef]

foods

Article

Effects of Different Extraction Methods on the Structural and Functional Properties of Soluble Dietary Fibre from Sweet Potatoes

Liuqing Yang, Rongan Zhu, Ning Zhang, Wenya Zhao and Chuyan Wang *

School of Biology, Food and Environment, Hefei University, Hefei 230601, China; yangliuqing0928@163.com (L.Y.); zhurongan0304@163.com (R.Z.); 17333140632@163.com (N.Z.); zhaowenya0317@163.com (W.Z.)
* Correspondence: wangchuyan19@163.com

Abstract: In this study, hot water treatment (WT), ultrasonic treatment (UT), ultrasonic-sodium hydroxide treatment (UST), ultrasonic-enzyme treatment (UET), and ultrasonic-microwave treatment (UMT) were used to treat sweet potatoes. The structural, physicochemical, and functional properties of the extracted soluble dietary fibres (SDFs) were named WT-SDF, UT-SDF, UST-SDF, UET-SDF, and UMT-SDF, respectively. Scanning electron microscopy (SEM), Fourier transform infrared spectroscopy (FT-IR), X-ray diffraction (XRD), thermal properties, and Brunauer–Emmett–Teller (BET) analysis were employed. The structural results indicated that the UST-SDF exhibited the best thermal stability, highest crystallinity, and maximum specific surface area. Moreover, compared to hot water extraction, ultrasonic extraction, or ultrasonic extraction in combination with other methods, enhanced the physicochemical and functional properties of the SDF, including extraction yield, water-holding capacity (WHC), oil-holding capacity (OHC), glucose adsorption capacity (GAC), glucose dialysis retardation index (GDRI), sodium cholate adsorption capacity (SCAC), cholesterol adsorption capacity (CAC), nitrite ion adsorption capacity (NIAC), and antioxidant properties. Specifically, the UST-SDF and UMT-SDF showed better extraction yield, WHC, OHC, GAC, CAC, SCAC, and NIAC values than the other samples. In summary, these results indicate that UST and UMT could be applied as ideal extraction methods for sweet potato SDF and that UST-SDF and UMT-SDF show enormous potential for use in the functional food industry.

Keywords: sweet potatoes; SDF; structural properties; physicochemical properties; functional properties

Citation: Yang, L.; Zhu, R.; Zhang, N.; Zhao, W.; Wang, C. Effects of Different Extraction Methods on the Structural and Functional Properties of Soluble Dietary Fibre from Sweet Potatoes. *Foods* **2024**, *13*, 2395. https://doi.org/10.3390/foods13152395

Academic Editors: Marina Russo and Rafael Guillén Bejarano

Received: 15 May 2024
Revised: 24 July 2024
Accepted: 24 July 2024
Published: 29 July 2024

1. Introduction

Sweet potato (*Ipomoea batatas* (L.) Lam) belongs to the *Convolvulaceae* family and is considered vital worldwide. It is rich in nutrients such as starch, protein, phosphorus, calcium, iron, carotene, vitamins, thiamine, riboflavin, and niacin [1]. Sweet potatoes generally contain 8% dietary fibre (DF), and are considered a high-quality food resource with balanced nutrition. Sweet potato resources are abundant; however, deep processing levels are low, mainly focusing on sweet potato starch production. China is the largest producer of sweet potatoes in the world, with fresh sweet potato yields alone exceeding 52 million tonnes in 2019. However, the short storage period and insufficient utilisation have caused a large amount of sweet potato resources to be wasted [2]. Therefore, further effective utilisation of sweet potato resources for deep processing to prepare high-quality DF can maximise resource value.

DF is a widely indigestible nutrient obtained from plant cell walls and is the seventh most important nutrient in living organisms [3]. Dietary fibre is classified into two types based on its capacity to dissolve in water: soluble and insoluble dietary fibre (SDF/IDF) [4]. Compared to IDF, SDF displays better characteristics [5], such as superior solubility, water-/oil-holding capacity, the prevention and treatment of colon cancer, and lower blood cholesterol content [6,7]. SDF is an excellent fat substitute that promotes the flavour of different foods, such as meatballs, dairy products, and baked goods [8,9]. Sweet potatoes are rich in SDF and have high economic value. Therefore, extracting SDF from sweet potatoes could reduce the waste of biological resources.

Recently, SDF extraction methods have included physical, chemical, biological, and composite procedures. Different extraction methods can result in SDFs with different structural, physicochemical, and functional properties [10]. Among these, ultrasound-assisted extraction is widely used to obtain bioactive compounds from various natural sources [11]. Ultrasonic extraction is based on the principle that ultrasound propagates through a liquid medium to induce acoustic cavitation. These cavitation bubbles rupture on the surface of plant materials, generating shock waves and shear forces, thereby damaging plant cell walls and facilitating the further extraction of target analytes [12]. Sodium hydroxide treatment can result in the swelling of plant cell walls and β-elimination of major components, as well as the alteration of cellulose micro-morphology [13]. Microwave treatment can increase intracellular pressure and damage cell walls in fresh tissues, thereby facilitating the release of active ingredients [14]. However, to the best of our knowledge, no study has compared the effects of ultrasound treatment in combination with sodium hydroxide, enzymatic, and microwave methods on the characteristics of SDF from sweet potatoes.

In this study, hot water treatment (WT), ultrasonic treatment (UT), ultrasonic-sodium hydroxide treatment (UST), ultrasonic-enzyme treatment (UET), and ultrasonic-microwave treatment (UMT) were used to extract SDF from sweet potatoes (WT-SDF, UT-SDF, UST-SDF, UET-SDF, and UMT-SDF). The structural, physicochemical, and functional properties of the extracted WT-SDF, UT-SDF, UST-SDF, UET-SDF, and UMT-SDF were compared to screen for a better method for extracting SDF from sweet potatoes.

2. Materials and Methods

2.1. Materials

Sweet potatoes (watermelon red) were acquired from Sixian (Suzhou, China). These were peeled, cleaned, diced, and dried at 60 °C until the moisture content was below 10%, followed by crushing and screening via a 60-mesh sifter to obtain sweet potato powder. Papain (800,000 U/g) was provided by Beijing Solarbio Science & Technology Co., Ltd., Beijing, China. Cellulase (50 U/g) was provided by Shanghai Yuanye Bio-Technology Co., Ltd., Shanghai, China. Analytical-grade NaOH, 95% ethanol, glucose, cholesterol, sodium gallate, and other chemicals used in the experiments were purchased from Hefei Hongyuan Trading Co., Ltd., Hefei, China.

2.2. SDF Extraction

2.2.1. Preparation of WT-SDF

Sweet potato powder (10 g) was added to distilled water (300 mL) and incubated at 90 °C for an hour in an Electro-Thermostatic Water Cabinet (DK-8D, Shanghai Jing Hong Laboratory Instrument Co., Ltd., Shanghai, China). The mixture was continuously stirred at room temperature for 2 h. After centrifugation for 18 min at 4000 rpm, the supernatant was precipitated with four volumes of 95% ethanol for 12 h, and the mixture was then centrifuged at 4000 rpm for 5 min. Finally, the sediment was collected and freeze-dried to obtain WT-SDF.

2.2.2. Preparation of UT-SDF, UST-SDF, UET-SDF, and UMT-SDF

Sweet potato powder (10 g) was added to distilled water (300 mL), and the pH was adjusted to 4.5 using phosphate buffer. Papain (0.6%, m/v) was added to this mixture and stirred continuously at 60 °C for 40 min, followed by inactivation in a 100 °C water bath for 10 min. Papain-treated samples were collected for subsequent extraction.

This mixture was immersed below the water level in an ultrasonic instrument (L-UCS-15L, LABGIC, Beijing, China) for ultrasonic extraction at an ultrasonic power of 270 W at 40 °C for 30 min. After the treatment described in Section 2.2.1, UT-SDF was obtained.

To obtain UST-SDF, the sample was mixed and stirred with a NaOH (5%, v/v) solution at 40 °C for 30 min under an ultrasonic power of 270 W [15]. The remaining steps were similar to those described in Section 2.2.1. To prepare UET-SDF, the sample was mixed with cellulase (0.6%, m/v) at 40 °C for 30 min under an ultrasound power of 270 W. The remaining procedures were similar to those described in Section 2.2.1. To obtain UMT-SDF, the sample was stirred at 40 °C for 30 min under an ultrasound power of 270 W and then microwaved for 4 min at 630 W in a microwave oven (P70F20CL-DG(B0), Galanz, Foshan, China). The remaining steps are the same as those presented in Section 2.2.1.

2.3. *Structural Characterization*

2.3.1. Scanning Electron Microscopy (SEM)

The morphology of the SDFs was determined by the method described by Du et al. [16], using a scanning electron microscope (SEM; Regulus 8230, HITACHI, Tokyo, Japan) at 3.0 kV. The SDF samples were placed on double-sided tape and evenly sprayed with a gold layer. Micrographs were captured at 30,000× magnification.

2.3.2. Fourier Transform Infrared Spectroscopy (FT-IR)

The FT-IR spectra of the SDFs were recorded using a Fourier Transform Infrared Spectrometer (Nicolet iS5, Thermo Fisher Scientific, Waltham, MA, USA) within the range of 400–4000 cm^{-1} with a resolution of 4 cm^{-1} and 32 scans. Briefly, the samples (10 mg) were blended with KBr powder (1 g), followed by being ground, and tableted using a tablet press (GS01150, Specac, Kent, UK) at a pressure of 2 tons.

2.3.3. Thermogravimetric Analysis (TGA)

The thermal properties of the SDFs were measured using a thermogravimetric analyser (STA6000, PerkinElmer, Waltham, MA, USA). The sample (20 mg) was heated from 0 °C to 800 °C at a heating rate of 10 °C/min with a flow rate of 20 mL/min under a high-purity N_2 atmosphere.

2.3.4. X-ray Diffraction (XRD)

The XRD patterns of the SDFs were obtained using an X-ray diffractometer (Ultima IV, Rigaku Corporation, Akishima-shi, Japan). The diffraction angle (2θ) was scanned from 5° to 70° at a scanning rate of 2°/min.

2.3.5. Brunauer–Emmett–Teller (BET) Surface Area

The specific surface areas and pore sizes of the SDFs were measured according to the procedures described by Ma et al. [17] using an analyser (ASAP 2460, Micromeritics, Norcross, GA, USA). Nitrogen adsorption/desorption isotherms were obtained over a relative pressure (P/P_0) ranging from approximately 10^{-5} to 0.995.

2.4. *Physicochemical Properties*

2.4.1. Colour Analysis

The colour indices of the SDFs, including L* (lightness), b* (redness), and a* (yellowness) values were detected using a colour reader (CR-10, Konica Minolta, Tokyo, Japan).

2.4.2. Water-Holding Capacity (WHC)

The SDF samples (W_1, 1 g) were added to distilled water (25 mL) at room temperature for 2 h. After 15 min of centrifugation at 4000 rpm by a centrifuge (JW-2032, Anhui Jiawen Instrument Equipment Co., Ltd., Hefei, China), the residue was immediately collected and weighed (W_2). The WHC was calculated as follows:

$$\text{WHC (g/g)} = (W_2 - W_1)/W_1 \quad (1)$$

2.4.3. Oil-Holding Capacity (OHC)

The SDF samples (M_1, 1 g) were added to soybean oil (25 mL) at room temperature for 2 h. Following centrifugation at 4000 rpm for 15 min, the supernatant (free oil) was removed. The residue was weighed (M_2). The OHC was calculated as follows:

$$\text{OHC (g/g)} = (M_2 - M_1)/M_1 \quad (2)$$

2.5. Functional Properties

2.5.1. Glucose Adsorption Capacity (GAC)

GAC was measured based on Niu et al.'s method with some modifications [18]. The SDF samples (M, 0.1 g) were mixed with 5 mL of a glucose solution (1 mg/mL). The mixture was positioned at room temperature for 2 h and centrifuged at 4000 rpm for 18 min. The concentration of glucose in the supernatant was measured using the DNS method, and GAC was calculated as follows:

$$\text{GAC (mg/g)} = (M_1 - M_2)/M \quad (3)$$

where M_1 and M_2 are the weights of glucose in solution before and after adsorption, respectively, and M is the weight of the SDF sample.

2.5.2. Glucose Dialysis Retardation Index (GDRI)

The GDRI was determined in accordance with Qi et al.'s method with some modifications [19]. The SDF samples (0.2 g) were dispersed in 10 mL of glucose solution (1 mM) through a dialysis membrane with a cut-off molecular weight of 1000 D, and incubated in distilled water (200 mL) at 37 °C for 300 min. At 15, 30, 60, 90, 120, 150, 180, 240, and 300 min, 1 mL of this mixture was removed. The DNS method was adopted for measuring the glucose concentration, with the blank group as the control. The GDRI was calculated as follows:

$$\text{GDRI (\%)} = (1 - C_1/C_0) \times 100\% \quad (4)$$

where C_1 and C_0 are glucose concentrations in the sample and blank groups, respectively.

2.5.3. Sodium Cholate Adsorption Capacity (SCAC)

SCAC was measured by the method reported by Benitez et al. [20]. The SDF samples (M, 0.2 g) were placed in a 50 mL centrifuge tube with 0.2 g of sodium cholate. The mixture was mixed with 100 mL of NaCl solution (0.15 M), and then incubated at a 37 °C water bath for 3 h. Following 18 min of centrifugation at 4000 rpm, the supernatant (1 mL) was transferred to a glass tube with a plug. Subsequently, the mixture was added to 45% H_2SO_4 (6 mL) and 0.3% furfural (1 mL), followed by a 65 °C water bath for 30 min. Finally, the absorbance was measured at 620 nm using a UV-Vis spectrophotometer (UV-1780, Shimadzu Corporation, Suzhou, China) and quantified with the standard curve. SCAC was calculated by the following equation:

$$\text{SCAC (mg/g)} = (M_1 - M_2)/M \quad (5)$$

where M_1 and M_2 are the weights of sodium cholate in solution before and after adsorption, respectively, and M is the weight of the SDF sample.

2.5.4. Cholesterol Adsorption Capacity (CAC)

CAC was detected using Gan et al.'s procedure with some modifications [21]. Yolk (10 mL) was mixed with distilled water (90 mL) and stirred to form an emulsion. The SDF samples (M, 0.1 g) were mixed with the emulsion (5 mL), and the pH was adjusted to 7.0. The mixture was shaken at room temperature for 2 h and centrifuged at 4000 rpm for 15 min. The supernatant (1 mL) was collected and diluted in glacial acetic acid (9 mL). Then, the diluent (0.4 mL) was taken to measure the cholesterol concentration at 550 nm and was quantified with the standard curve. An emulsion of diluted eggs without SDF was used as the blank. CAC was calculated by the equation below:

$$\text{CAC (mg/g)} = (M_1 - M_2)/M \quad (6)$$

where M_1 and M_2 are the weights of the cholesterol in solution before and after adsorption, respectively, and M is the weight of the SDF sample.

2.5.5. Nitrite Ion Adsorption Capacity (NIAC)

NIAC was evaluated using Zhu et al.'s process [22]. The SDF samples (M, 0.1 g) were added to 5 mL of $NaNO_2$ solution (20 μg/mL). The pH was adjusted to 2.0 to imitate the gastric environment. The mixture was incubated at room temperature for 2.5 h and centrifuged at 4500 rpm for 12 min. Subsequently, the supernatant (0.3 mL) was transferred into a tube and deionised water (1.7 mL) was added. Then, 2 mL of p-aminobenzene sulfonic acid (4 μg/mL) together with 1 mL of hydrochloride naphthodiamide (2 μg/mL) was added. After 30 min of reaction in the dark, the concentration of $NaNO_2$ was measured at 538 nm and quantified using a standard curve. NIAC was measured through the following equation:

$$\text{NIAC } (\mu g/g) = (M_1 - M_2)/M \quad (7)$$

where M_1 and M_2 are the weights of the $NaNO_2$ in solution before and after adsorption, respectively, and M is the weight of the SDF sample.

2.5.6. Antioxidant Properties

Freeze-dried SDF samples (0.2 g) were transferred to test tubes and 50% ethanol (10 mL) was added. This mixture was shaken in a 60 °C water bath for 4 h, and centrifuged at 4500 rpm for 15 min. The supernatants (2 mL) were collected and mixed with $ABTS^{\cdot +}$ working solution (0.5 mL). The mixture was incubated in the dark for an hour. The $ABTS^{\cdot +}$ radical scavenging activity was measured by the following equation:

$$\%\text{ scavenging activity } (\%) = \frac{A_0 - A_1}{A_0} \times 100\% \quad (8)$$

where A_1 and A_0 are the absorbance (734 nm) of the sample and control groups (2.5 mL of $ABTS^{\cdot +}$ working solution), respectively.

2.6. *Statistical Analysis*

Each test was repeated thrice, and the data were presented as mean ± standard deviation after being statistically processed. SPSS software (version 26.0) was used for *t*-test analysis of intergroup deviation. Differences were considered statistically significant at $p < 0.05$.

3. Results and Discussion

3.1. *Structural Properties*

3.1.1. SEM Observation

SEM was used to observe the morphological characteristics of the SDF samples (Figure 1). There were significant differences in the morphology of the WT-SDF, UT-SDF, UST-SDF, UET-SDF, and UMT-SDF. The WT-SDF particles aggregated into sheets with smooth surfaces. After UT, the UT-SDF particles connected to form a sheet-like structure with small particles adhering to the surface, and the texture became loose and swollen. Compared to the WT-SDF and UT-SDF, the structures of the UST-SDF, UET-SDF, and UMT-SDF shrank and connected to a network with more unevenly sized pores. The UST-SDF particles became tighter and more disordered, which was probably associated with the strong corrosiveness of NaOH. The UMT-SDF had larger pores and looser textures than the UET-SDF.

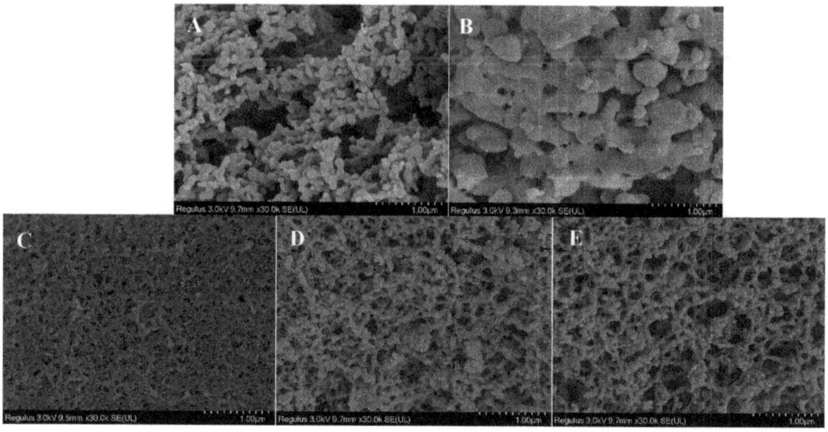

Figure 1. SEM images of WT-SDF (**A**), UT-SDF (**B**), UST-SDF (**C**), UET-SDF (**D**), and UMT-SDF (**E**) with a magnification of 30,000× at 1 μm.

3.1.2. FT-IR Analysis

The FT-IR spectra of the SDF samples extracted by various methods were recorded within the range from 400 cm^{-1} to 4000 cm^{-1}. As shown in Figure 2, several similar characteristic peaks are observed in the spectra of the five SDF samples. The strong and wide peak at 3384 cm^{-1} correlates with the -OH stretching of hydrogen bonds with polysaccharides [18]. Compared with other samples, the UST-SDF exhibited a peak intensity reduction at 2930 cm^{-1}, which is a typical absorption peak of polysaccharide compounds, owing to the C-H stretching vibration of the methylene group of the polysaccharides [23,24]. This could probably be caused by the breakage of bonds in the molecule by sodium hydroxide treatment. The absorption peak at 1649 cm^{-1} might be an asymmetric vibration of C=O in the carboxyl group [25]. The peaks at 1200–1400 cm^{-1} were caused by the angular vibration of C-H, which is the characteristic absorption peak of sugars. The spectra of the five SDF samples showed slight differences in this region, especially that of the UST-SDF. The strong absorption peaks at 1154 cm^{-1} and 1025 cm^{-1} are related to the stretching vibration of C-O in C-O-C and the angular vibration of O-H in C-O-H, respectively, which are also characteristic absorption peaks of polysaccharides. Additionally, the peak at 930 cm^{-1} corresponding with the stretching vibration of β-glycosidic linkages was observed in all samples except the UST-SDF sample, demonstrating the degradation of polysaccharide chains caused by sodium hydroxide [26]. The positions of the characteristic absorption peaks in the FT-IR spectra showed no significant changes; however, the intensities of some peaks changed slightly, which may be due to the redistribution of cellulose caused by the different treatment methods.

3.1.3. TGA

Figure 3 shows the TGA results for the SDF samples. The curves of the five SDF samples exhibited similar trends, and were divided into three temperature ranges. In the first set of temperature ranges (20 °C–200 °C), the curve changed relatively smoothly, and the weight of the sample was decreased slightly, mainly because of the loss of dehydration [27]. In the second set of temperature ranges (200 °C–400 °C), the maximum weight loss was observed, which might be related to the pyrolysis of polysaccharides and hemicellulose [28]. In the final set of temperature ranges (400 °C–800 °C), the weight loss of the SDF samples slowed, mainly due to carbon decomposition. In addition, the residual masses of the WT-SDF, UT-SDF, UST-SDF, UET-SDF, and UMT-SDF were 17.514%, 20.938%, 30.209%, 9.669%, and 18.717%, respectively, indicating better thermal stability of the UST-SDF. Overall, the SDF samples were relatively stable below 200 °C.

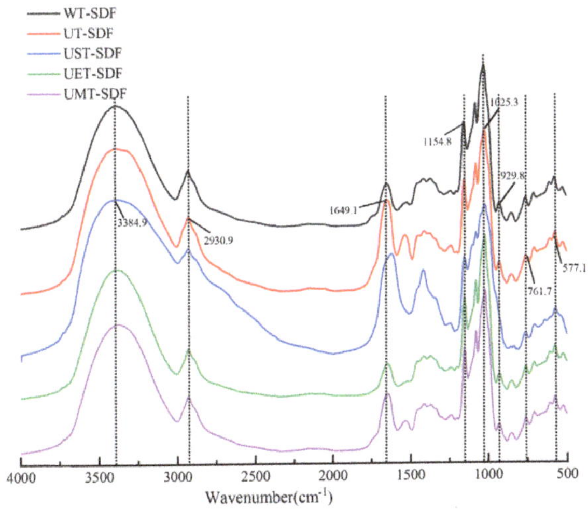

Figure 2. FT-IR spectra of WT-SDF, UT-SDF, UST-SDF, UET-SDF, and UMT-SDF.

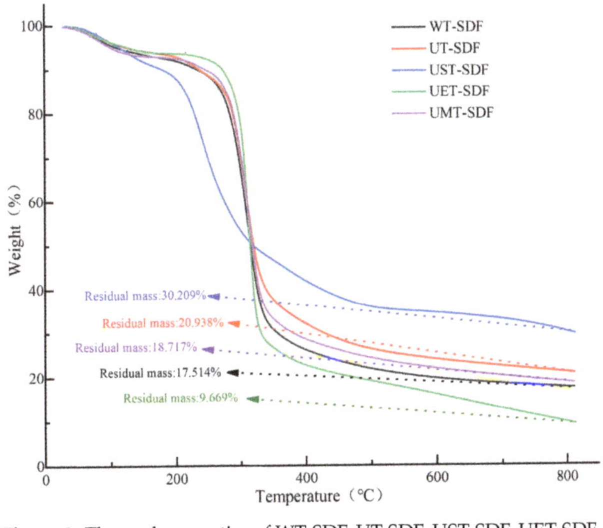

Figure 3. Thermal properties of WT-SDF, UT-SDF, UST-SDF, UET-SDF, and UMT-SDF.

3.1.4. XRD Analysis

XRD was employed to characterise the crystalline morphologies of the polymers. As shown in Figure 4, all the SDF samples had typical crystalline peaks at 13.1° and 19.9°, whereas the prominent diffraction peak was around 19.9°, which is considered a typical structure of cellulose type I. All the SDF samples showed similar XRD patterns, indicating that these extraction methods did not significantly alter the crystal structure of the SDF. Among all the samples, the UST-SDF had the highest crystallinity, followed by the WT-SDF. The crystal structure of SDF is mainly maintained by hydrogen bonds formed between adjacent hydroxyl groups [29]. UT, cellulase hydrolysis, and microwave treatment can result in the degradation of the glycosidic bonds in SDF, resulting in a lower crystallinity of UT-SDF, UET-SDF, and UMT-SDF. The UST-SDF had small diffuse diffraction peaks near 26.9° and 38.3°, as evidenced by the presence of partial crystallisation [8], which is consistent with the UST-SDF having the highest crystallinity of all the SDF samples. This might be due to the combined treatment with ultrasound and alkali, which makes the molecular rearrangement more regular. The UMT-SDF had the lowest crystallinity, suggesting that its crystal structure was severely damaged by the ultrasonic microwaves.

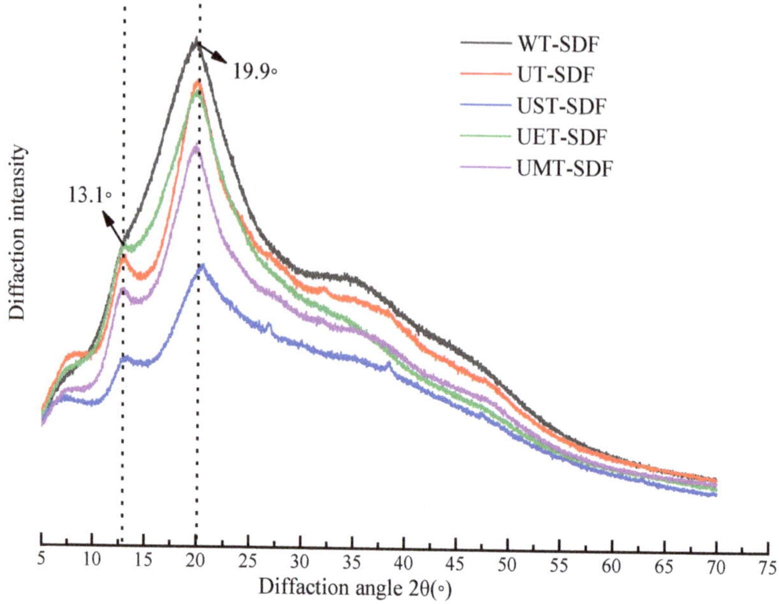

Figure 4. XRD of WT-SDF, UT-SDF, UST-SDF, UET-SDF, and UMT-SD.

3.1.5. BET Surface Area and Pore Size

The specific surface areas and pore sizes of the SDFs are listed in Table 1. The specific surface areas of the WT-SDF, UT-SDF, UST-SDF, UET-SDF, and UMT-SDF were 7.95, 26.66, 37.86, 10.57, and 31.38 m^2/g, respectively, with pore sizes of 17.25, 28.33, 26.16, 15.33, and 27.94 nm, respectively. Compared to the WT-SDF, the specific surface area of the other SDF samples increased, and the effect of ultrasonic-assisted NaOH or microwave treatment on the specific surface area increase of the SDF was more obvious than that of UT alone. Moreover, the nitrogen adsorption–desorption isotherms of all the SDF samples were depicted as a V-type isotherm with an H3-type hysteresis loop (Figure 5). In the higher p/p_0 region, the rate of isotherm ascent increased, and H3-type hysteresis loops were formed, indicating the presence of mesopores and stacking in the SDFs.

Table 1. Specific surface area, pore size, colours, and extraction yield of WT-SDF, UT-SDF, UST-SDF, UET-SDF, and UMT-SDF.

Sample	Specific Surface Area (m^2/g)	Pore Size (nm)	L*	a*	b*	Extraction Yield (g/100 g)
WT-SDF	7.95	17.25	84.53 ± 0.46 [b]	12.27 ± 0.19 [a]	13.20 ± 0.14 [b]	3.80 ± 0.20 [e]
UT-SDF	26.66	28.33	83.23 ± 2.44 [c]	11.06 ± 0.08 [b]	11.83 ± 0.34 [c]	6.27 ± 0.25 [d]
UST-SDF	37.86	26.16	72.20 ± 1.69 [e]	4.77 ± 0.05 [e]	17.83 ± 0.29 [a]	10.03 ± 0.61 [a]
UET-SDF	10.57	15.33	88.73 ± 0.95 [a]	8.93 ± 0.19 [c]	10.57 ± 0.29 [e]	7.83 ± 0.12 [c]
UMT-SDF	31.38	27.94	83.20 ± 0.50 [d]	5.07 ± 0.12 [d]	10.67 ± 0.38 [d]	8.30 ± 0.30 [b]

Values marked by the different letters (a–e) above are significantly different ($p < 0.05$).

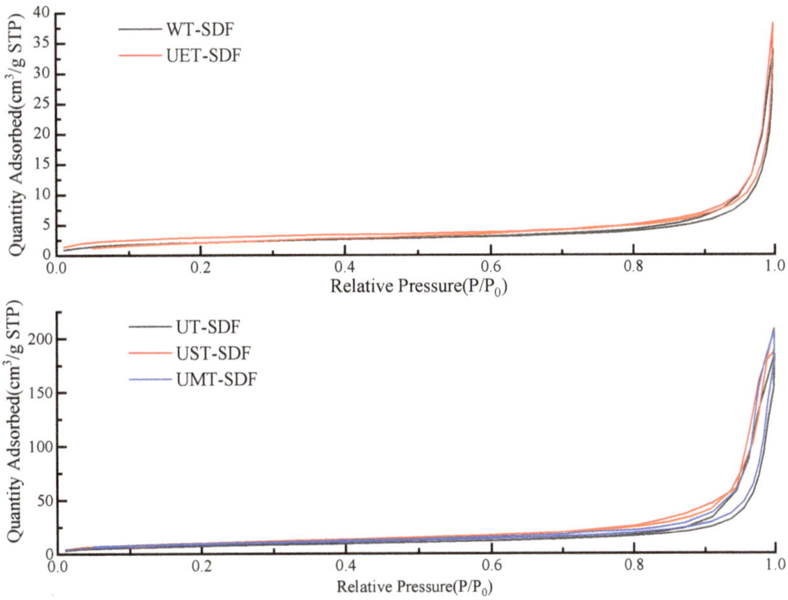

Figure 5. N_2 Adsorption–desorption of WT-SDF, UT-SDF, UST-SDF, UET-SDF, and UMT-SDF.

3.2. Physicochemical Properties

3.2.1. Extraction Yield and Colours of SDFs

The extraction yield and colour parameters (L*, a*, and b*) of the sweet potato SDFs are presented in Table 1. Compared to the WT-SDF, the yields of the UT-SDF (6.27 ± 0.25%), UST-SDF (10.03 ± 0.61%), UET-SDF (7.83 ± 0.12%), and UMT-SDF (8.30 ± 0.30%) were increased, which might be due to the possibility of ultrasound energy increasing the intracellular pressure, rupturing the cell wall and facilitating the further extraction of SDF.

Colour is an essential property that consumers perceive and that often affects their preferences. The L* (white) parameter of the UET-SDF elevated obviously ($p < 0.05$), and the b * (yellow) parameter of the UET-SDF decreased significantly ($p < 0.05$), indicating that the structure of naturally occurring pigments might be degraded under enzymatic conditions. Compared to the other groups, the UST-SDF samples had the lowest L* and a* (red), and the highest b* values, which is likely to be because the sodium hydroxide treatment caused a browning reaction in the SDF [30]. However, the dark colour of the SDF may affect its utilisation in food.

3.2.2. WHC

The WHC of the sweet potato SDFs subjected to different treatments is shown in Figure 6A. SDF with a higher WHC prevents food from shrinking and alters its viscosity [31]. As shown in Figure 6A, the UT-SDF, UST-SDF, UET-SDF, and UMT-SDF had higher WHC values than the WT-SDF, indicating that ultrasound alone or in combination with other methods efficiently enhanced the hydration properties of the SDF. The main reason for this is that the porous microstructure and larger specific surface area are both beneficial for the interaction between water and fibres [28]. Compared to the UT-SDF, the WHC of the UET-SDF decreased, which was related to a decrease in the specific surface area and pore size caused by cellulase hydrolysis.

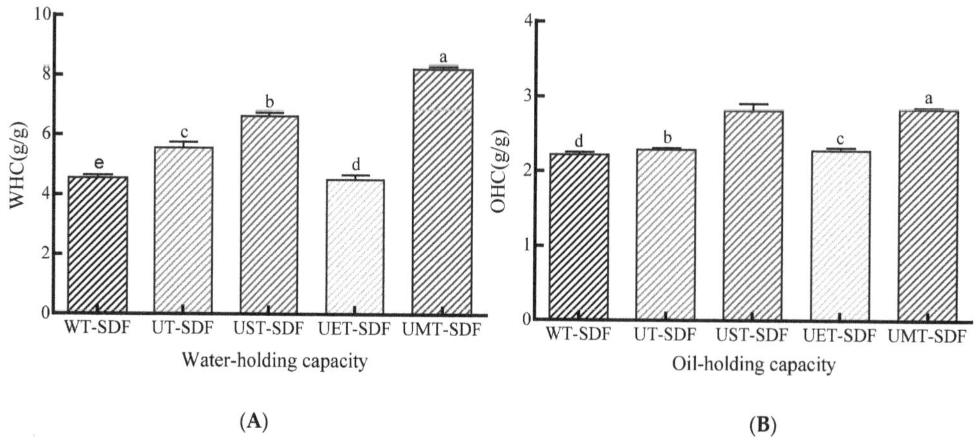

Figure 6. The WHC (**A**) and OHC (**B**) of WT-SDF, UT-SDF, UST-SDF, UET-SDF, and UMT-SDF. Values marked by the different letters (a–e) above are significantly different ($p < 0.05$).

3.2.3. OHC

The OHC of SDF is of great significance for food applications, including avoiding fat loss during cooking [32]. As presented in Figure 6B, the OHC of the WT-SDF, UT-SDF, and UET-SDF displayed little difference and was markedly lower than those of the UMT-SDF ($p < 0.05$) and UST-SDF. The UMT-SDF exhibited the highest OHC, followed by the UST-SDF, although this difference was not of significance ($p > 0.05$), which might be attributed to the exposure of more active groups and an elevation in the specific surface area after the alkaline and microwave treatments.

3.3. Functional Properties

3.3.1. GAC

SDF can combine with glucose in intestinal fluid, thereby reducing postprandial blood glucose levels. GAC is a significant functional property of SDF and is often used to estimate its capacity to postpone glucose absorption. As depicted in Figure 7A, all the SDF samples effectively adsorbed glucose. Compared with hot water extraction, ultrasound treatment effectively enhanced the GAC of the SDF, whereas ultrasound combined with other methods had a better enhancement effect. The UMT-SDF exhibited the strongest adsorption capacity (48.00 ± 0.01 mg/g), which was consistent with the research showing that SDFs with a higher WHC and larger surface area had more of a chance to contact and adsorb glucose [33]. These results suggest that UMT-SDFs exert a strong hypoglycaemic effect in vitro.

3.3.2. GDRI

The GDRI has been used as an effective indicator to evaluate the effects of DFs on delayed glucose absorption in the gastrointestinal tract. As shown in Figure 7B, the inhibitory capacity of the SDF on glucose diffusion decreased continuously with dialysis time, and the final glucose concentration reached equilibrium. The GDRI of the UST-SDF and UMT-SDF was relatively high, possibly due to their increased surface area, which led to the physical hindrance of glucose molecules and stronger interception of glucose in the network structure formed by the fibres [19].

3.3.3. SCAC

Sodium cholate is an endogenous emulsifier that plays an important role in lipid digestion and absorption. Therefore, the inhibition of sodium gallate production can effectively reduce fat absorption in the small intestine [34]. Figure 8A shows that the

UMT-SDF and UST-SDF had a higher SCAC, which is consistent with their higher OHC, whereas the UST-SDF had a lower SCAC than the UT-SDF. A higher OHC indicates a stronger affinity between the SDF and oil molecules. Therefore, UMT-SDF has a stronger ability to adsorb sodium cholate, effectively reducing the accumulation of bile salts in the circulation in the liver and kidneys.

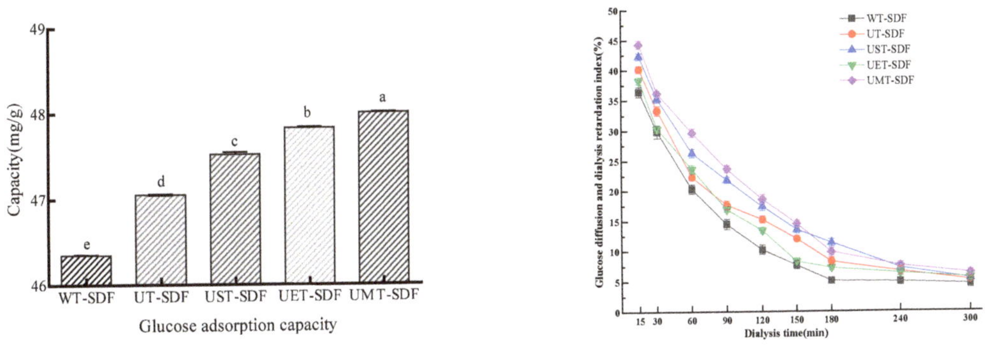

(**A**) (**B**)

Figure 7. The GAC (**A**) and GDRI (**B**) of WT-SDF, UT-SDF, UST-SDF, UET-SDF, and UMT-SDF. Values marked by the different letters (a–e) above are significantly different ($p < 0.05$).

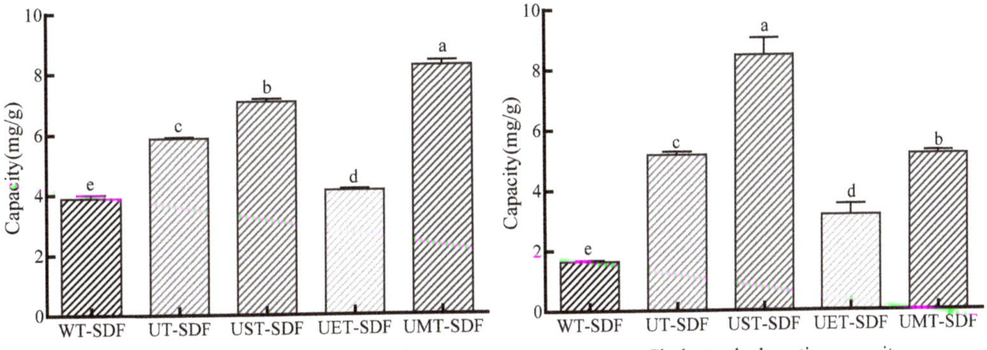

(**A**) (**B**)

Figure 8. The SCAC (**A**) and CAC (**B**) of WT-SDF, UT-SDF, UST-SDF, UET-SDF, and UMT-SDF. Values marked by the different letters (a–e) above are significantly different ($p < 0.05$).

3.3.4. CAC

CAC is another essential functional property of SDF that reduces serum cholesterol levels and protects cardiovascular function [35]. The CAC of the SDFs is shown in Figure 8B; the WT-SDF had the lowest CAC. Meanwhile, the UMT-SDF and UST-SDF displayed higher CAC values, which might have been caused by their loose structures and larger specific surface areas. Yan et al. [36] found that SDF with more pores could provide more sites for the binding of water and cholesterol, thus possessing a strong CAC. The high SCAC and CAC values of the UMT-SDF and UST-SDF show their potential as ingredients in hypolipidaemic foods.

3.3.5. NIAC

Nitrite can react with secondary and tertiary amines/amides to produce nitrosamines with strong carcinogenicity in gastric acid environments. Therefore, the NIAC of SDF is a meaningful functional property. As shown in Figure 9, the adsorption capacity of the SDFs for nitrite ions in a simulated gastric environment was investigated (pH = 2). The results showed that the NIAC of the UMT-SDF was significantly higher than that of the other samples ($p < 0.05$), which might be due to the UMT-SDF having a better structure and a massive explosion of functional groups to interact with more nitrite ions.

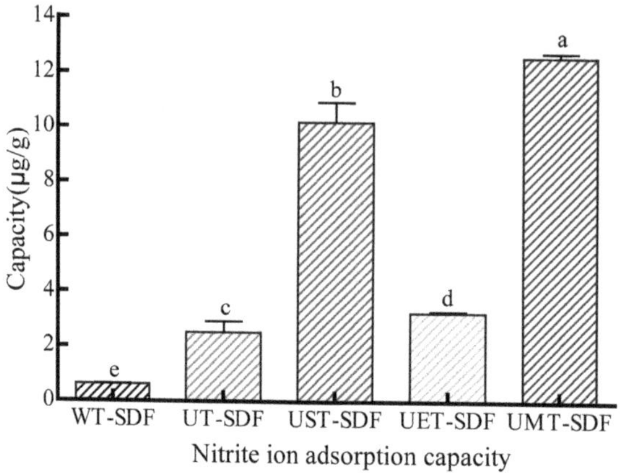

Figure 9. The NIAC of WT-SDF, UT-SDF, UST-SDF, UET-SDF, and UMT-SDF. Values marked by the different letters (a–e) above are significantly different ($p < 0.05$).

3.3.6. Antioxidant Properties

The ABTS cationic radical model has been widely used to detect T-AOC in SDF [37]. The antioxidant activity was investigated using $ABTS^{\cdot+}$ radical scavenging activity. Free radical scavenging activities in $ABTS^{\cdot+}$ are shown in Figure 10. All the samples displayed superior antioxidant activity. The UT-SDF showed the strongest $ABTS^{\cdot+}$ radical scavenging activity, followed by the UMT-SDF ($p > 0.05$), whereas the UET-SDF showed the lowest activity. Hence, the present study identified sweet potatoes as an important source of antioxidant DF.

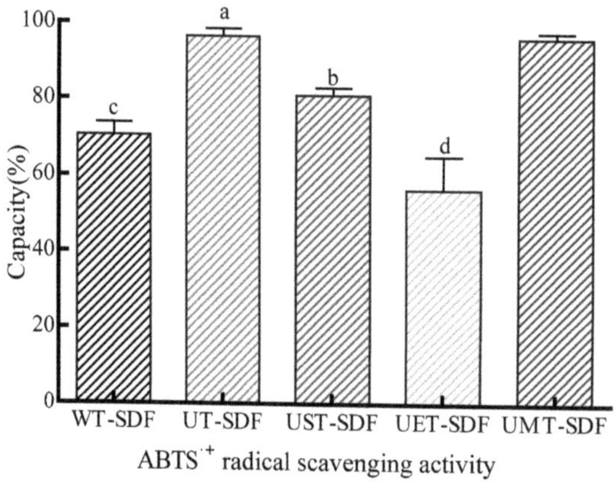

Figure 10. Effects of WT-SDF, UT-SDF, UST-SDF, UET-SDF, and UMT-SDF on radical $ABTS^{\cdot+}$ scavenging activity. Values marked by the different letters (a–d) above are significantly different ($p < 0.05$).

4. Conclusions

This study employed five methods to extract SDFs from sweet potatoes and investigate their structural, physicochemical, and functional properties. The UST-SDF exhibited the best thermal stability, highest crystallinity, maximum specific surface area, and highest extraction yield. Compared with the WT-SDF and UT-SDF, the UST-SDF and UMT-SDF showed better oil-holding and water-holding, glucose adsorption, sodium cholate, cholesterol adsorption, and nitrite ion absorption capacities, whereas the functions of the UET-SDF were not significantly improved. The UT-SDF had the best antioxidant properties, followed by the UST-SDF and UMT-SDF. Overall, UST-SDF and UMT-SDF have enormous potential for application in the functional food industry.

Author Contributions: Conceptualization, methodology, data curation, L.Y. and R.Z.; software, investigation, data curation, L.Y., N.Z. and W.Z.; writing—original draft preparation, writing—review and editing, supervision, funding acquisition, L.Y. and C.W. All authors have read and agreed to the published version of the manuscript.

Funding: This work was financially supported by the University Natural Science Research Project of Anhui Educational Committee (2022AH051786), and the Anhui Provincial Excellent Scientific Research and Innovation Team (2023AH010050).

Institutional Review Board Statement: Not applicable.

Informed Consent Statement: Not applicable.

Data Availability Statement: The original contributions presented in the study are included in the article, further inquiries can be directed to the corresponding author.

Conflicts of Interest: The authors declare no conflicts of interest.

References

1. Hussein, A.E.; Emmanuel, O.; Chinyere, A.; Majekodunmi, R.O. Utilization of sweet potato starches and flours as composites with wheat flours in the preparation of confectioneries. *Afr. J. Biotechnol.* **2015**, *14*, 17–22. [CrossRef]
2. Vannini, M.; Marchese, P.; Sisti, L.; Saccani, A.; Mu, T.; Sun, H.N.; Celli, A. Integrated Efforts for the Valorization of Sweet Potato By-Products within a Circular Economy Concept: Biocomposites for Packaging Applications Close the Loop. *Polymers* **2021**, *13*, 1048. [CrossRef] [PubMed]
3. Liu, S.; Jia, M.Y.; Chen, J.J.; Wang, H.S.; Dong, R.H.; Nie, S.P.; Xie, M.Y.; Yu, Q. Removal of bound polyphenols and its effect on antioxidant and prebiotics properties of carrot dietary fiber. *Food Hydrocoll.* **2019**, *93*, 284–292. [CrossRef]
4. Jia, M.Y.; Chen, J.J.; Liu, X.Z.; Xie, M.Y.; Nie, S.P.; Chen, Y.; Xie, J.H.; Yu, Q. Structural characteristics and functional properties of soluble dietary fiber from defatted rice bran obtained through Trichoderma viride fermentation. *Food Hydrocoll.* **2019**, *94*, 468–474. [CrossRef]
5. Huang, S.Y.; He, Y.W.; Zou, Y.P.; Liu, Z. Modification of insoluble dietary fibres in soya bean okara and their physicochemical properties. *Int. J. Food Sci. Technol.* **2016**, *50*, 2606–2613. [CrossRef]
6. Madenci, A.B.; Bilgiçli, N.; Türker, S. Effects of dietary fibre and antioxidant-rich ingredients on some quality characteristics of fresh and dry pasta. *Qual. Assur. Saf. Crops* **2018**, *10*, 315–324. [CrossRef]
7. Moczkowska, M.; Karp, S.; Niu, Y.; Niu, Y.; Kurek, M.A. Enzymatic, enzymatic-ultrasonic and alkaline extraction of soluble dietary fibre from flaxseed-A physicochemical approach. *Food Hydrocoll.* **2019**, *90*, 105–112. [CrossRef]
8. Dong, W.J.; Wang, D.D.; Hu, R.S.; Long, Y.Z.; Lv, L.S. Chemical composition, structural and functional properties of soluble dietary fiber obtained from coffee peel using different extraction methods. *Food Res. Int.* **2020**, *136*, 109497. [CrossRef] [PubMed]
9. Niu, X.; Shi, Q.K.; Zhao, C.B.; Yin, H.H.; Xu, X.Y.; Wu, Y.Z.; Cao, Y.; Zhang, H.; Liu, J.S. Effects of ultrasonic modification on the physicochemical properties and structure of oat dietary fiber. *Food Sci.* **2020**, *41*, 130–136.
10. Peerajit, P.; Chiewchan, N.; Devahastin, S. Effects of pretreatment methods on health-related functional properties of high dietary fiber powder from lime residues. *Food Chem.* **2012**, *132*, 1891–1898. [CrossRef]
11. Seguenka, B.; Nascimento, L.H.D.; Feiden, T.; Fernandes, I.A.; Magro, J.D.; Junges, A.; Valduga, E.; Steffens, J. Ultrasound-assisted extraction and concentration of phenolic compounds from jabuticaba sabará (*Plinia peruviana* (Poir.) Govaerts) peel by nanofiltration membrane. *Food Chem.* **2024**, *453*, 139690. [CrossRef]
12. İşçimen, E.M.; Hayta, M. Optimisation of ultrasound assisted extraction of rice bran proteins: Effects on antioxidant and antiproliferative properties. *Qual. Assur. Saf. Crops* **2018**, *10*, 165–174. [CrossRef]
13. Backes, E.; Kato, C.G.; Corrêa, R.C.G.; Moreira, R.F.P.M.; Peralta, R.A.; Barros, L.; Ferreira, I.G.F.R.; Zanin, G.M.; Bracht, A.; Peralta, R.M. Laccases in food processing: Current status, bottlenecks and perspectives. *Trends Food Sci. Technol.* **2021**, *115*, 445–460. [CrossRef]

14. Wei, E.W.; Yang, R.; Zhao, H.P.; Wang, P.H.; Zhao, S.Q.; Zhai, W.C.; Zhang, Y.; Zhou, H.L. Microwave-assisted extraction releases the antioxidant polysaccharides from seabuckthorn (*Hippophae rhamnoides* L.) berries. *Int. J. Biol. Macromol.* **2019**, *123*, 280–290. [CrossRef]
15. Zhang, W.M.; Zeng, G.G.; Pan, Y.G.; Chen, W.X.; Huang, W.X.; Chen, H.M.; Li, Y.S. Properties of soluble dietary fiber polysaccharide from papaya peel obtained through alkaline or ultrasound-assisted alkaline extraction. *Carbohyd. Polym.* **2017**, *172*, 102–112. [CrossRef] [PubMed]
16. Du, X.J.; Wang, L.; Huang, X.; Jing, H.J.; Ye, X.; Gao, W.; Bai, X.P.; Wang, H.X. Effects of different extraction methods on structure and properties of soluble dietary fiber from defatted coconut flour. *LWT-Food Sci. Technol.* **2021**, *143*, 111031. [CrossRef]
17. Ma, S.; Ren, B.; Diao, Z.J.; Chen, Y.W.; Qiao, Q.L.; Liu, X.B. Physicochemical properties and intestinal protective effect of ultra-micro ground insoluble dietary fibre from carrot pomace. *Food Funct.* **2016**, *7*, 3902–3909. [CrossRef] [PubMed]
18. Niu, Y.G.; Li, N.; Xia, Q.; Hou, Y.W.; Xu, G.N. Comparisons of three modifications on structural, rheological and functional properties of soluble dietary fibers from tomato peels. *LWT-Food Sci. Technol.* **2018**, *88*, 56–63. [CrossRef]
19. Qi, J.; Li, Y.; Masamba, K.G.; Shoemaker, C.F.; Zhong, F.; Majeed, H.; Ma, J.G. The effect of chemical treatment on the in vitro hypoglycemic properties of rice bran insoluble dietary fiber. *Food Hydrocoll.* **2016**, *52*, 699–706. [CrossRef]
20. Benitez, V.; Rebollo-Hernanz, M.; Hernanz, S.; Chantres, S.; Aguilera, Y.; Martin-Cabrejas, M.A. Coffee parchment as a new dietary fiber ingredient: Functional and physiological characterization. *Food Res. Int.* **2019**, *122*, 105–113. [CrossRef]
21. Gan, J.P.; Huang, Z.Y.; Yu, Q.; Peng, G.Y.; Chen, Y.; Xie, J.H.; Nie, S.P.; Xie, M.Y. Microwave assisted extraction with three modifications on structural and functional properties of soluble dietary fibers from grapefruit peel. *Food Hydrocoll.* **2020**, *101*, 105549. [CrossRef]
22. Zhu, Y.; Chu, J.X.; Lu, Z.X.; Lv, F.X.; Bie, X.M.; Zhang, C.; Zhao, H.Z. Physicochemical and functional properties of dietary fiber from foxtail millet (Setaria italic) bran. *J. Cereal Sci.* **2018**, *79*, 456–461. [CrossRef]
23. Ma, M.M.; Mu, T.H. Modification of deoiled cumin dietary fiber with laccase and cellulase under high hydrostatic pressure. *Carbohyd. Polym.* **2016**, *136*, 87–94. [CrossRef] [PubMed]
24. Xu, H.G.; Jiao, Q.; Yuan, F.; Gao, Y.X. In vitro binding capacities and physicochemical properties of soluble fiber prepared by microfluidization pretreatment and cellulase hydrolysis of peach pomace. *LWT-Food Sci. Technol.* **2015**, *63*, 677–684. [CrossRef]
25. Wang, H.O.; Liu, S.H.; Zhou, X.J.; Yang, X.Y.; Gao, Q.; Tanokura, M.; Xue, Y.L. Treatment with hydrogen peroxide improves the physicochemical properties of dietary fibres from Chinese yam peel. Int. *J. Food Sci. Technol.* **2020**, *55*, 1289–1297. [CrossRef]
26. Hua, M.; Lu, J.X.; Qu, D.; Liu, C.; Zhang, L.; Li, S.S.; Chen, J.B.; Sun, Y.S. Structure, physicochemical properties and adsorption function of insoluble dietary fiber from ginseng residue: A potential functional ingredient. *Food Chem.* **2019**, *286*, 522–529. [CrossRef] [PubMed]
27. Liu, J.; Wen, X.Y.; Zhang, X.Q.; Pu, H.M.; Kan, J.; Jin, C.H. Extraction, characterization and in vitro antioxidant activity of polysaccharides from black soybean. *Int. J. Biol. Macromol.* **2015**, *72*, 1182–1190. [CrossRef] [PubMed]
28. Liu, Y.L.; Zhang, H.B.; Yi, C.P.; Quan, K.; Lin, B.P. Chemical composition, structure, physicochemical and functional properties of rice bran dietary fiber modified by cellulase treatment. *Food Chem.* **2021**, *342*, 128352. [CrossRef] [PubMed]
29. Gil-López, D.I.L.; Lois-Correa, J.A.; Sánchez-Pardo, M.E.; Domínguez-Crespo, M.A.; Torres-Huerta, A.M.; Rodríguez-Salazar, A.E.; Orta-Guzmán, V.N. Production of dietary fibers from sugarcane bagasse and sugarcane tops using microwave-assisted alkaline treatments. *Ind. Crops Prod.* **2019**, *135*, 159–169. [CrossRef]
30. Nasseri, R.; Ngunjiri, R.; Moresoli, C.; Yu, A.P.; Yuan, Z.S.; Xu, C.B. Poly (lactic acid)/acetylated starch blends: Effect of starch acetylation on the material properties. *Carbohyd. Polym.* **2020**, *229*, 115453. [CrossRef]
31. Elleuch, M.; Bedigian, D.; Roiseux, O.; Besbes, S.; Blecker, C.; Attia, H. Dietary fibre and fibre-rich by-products of food processing: Characterisation, technological functionality and commercial applications: A review. *Food Chem.* **2011**, *124*, 411–421. [CrossRef]
32. Schneeman, B.O. Fiber, inulin and oligofructose: Similarities and differences. *J. Nutr.* **1999**, *129* (Suppl. S7), 1424S. [CrossRef] [PubMed]
33. Miehle, E.; Hass, M.; Bader-Mittermaier, S.; Eisner, P. The role of hydration properties of soluble dietary fibers on glucose diffusion. *Food Hydrocoll.* **2022**, *131*, 107822. [CrossRef]
34. Siddiqui, H.; Sultan, Z.; Yousuf, O.; Malik, M.; Younis, K. A review of the health benefits, functional properties, and ultrasound-assisted dietary fiber extraction. *Bioact. Carbohydr. Diet. Fibre* **2023**, *30*, 100356. [CrossRef]
35. Nsor-Atindana, J.; Zhong, F.; Mothibe, K.J. In vitro hypoglycemic and cholesterol lowering effects of dietary fiber prepared from cocoa (*Theobroma cacao* L.) shells. *Food Funct.* **2012**, *3*, 1044–1050. [CrossRef]
36. Yan, X.G.; Ye, G.; Chen, Y. Blasting extrusion processing: The increase of soluble dietary fiber content and extraction of soluble-fiber polysaccharides from wheat bran. *Food Chem.* **2015**, *180*, 106–115. [CrossRef]
37. Li, H.; Wang, X.; Xiong, Q.; Yu, Y.; Peng, L.C. Sulfated modification, characterization, and potential bioactivities of polysaccharide from the fruiting bodies of Russula virescens. *Int. J. Biol. Macromol.* **2020**, *154*, 1438–1447. [CrossRef]

molecules

MDPI

Article

Dual Enzymolysis Assisted by Acrylate or Phosphate Grafting: Influences on the Structural and Functional Properties of Jujube Residue Dietary Fiber

Yitao Zhao, Jianguo Xu *, Yajun Zheng *, Qi Li, Yihao Huang, Meichen Zong and Wangjin Guo

Food Science College, Shanxi Normal University, Taiyuan 030619, China; zhaoyitao30@163.com (Y.Z.); 18803470377@163.com (Q.L.); yourshyh@163.com (Y.H.); 15648635631@163.com (M.Z.); 19581580969@139.com (W.G.)

* Correspondence: xjg71@163.com (J.X.); zyj_coconut@163.com (Y.Z.); Tel.: +86-13015425996 (J.X.); Tel.: +86-15735742642 (Y.Z.)

Abstract: Jujube residue is an abundant and low-cost dietary fiber resource, but its relatively lower hydration and functional properties limit its utilization as an ingredient of functional food. Thus, cellulase and hemicellulase hydrolysis, enzymatic hydrolysis assisted by phosphate grafting (EPG), and enzymatic hydrolysis assisted by acrylate grafting (EAG) were used to improve the functional properties of jujube residue dietary fiber (JRDF) in this study. The results evidenced that these modifications all increased the porosity of the microstructure of JRDF and increased the soluble fiber content, surface area, and hydration properties, but reduced its brightness ($p < 0.05$). Moreover, JRDF modified by enzymolysis combined with acrylate grafting offered the highest extractable polyphenol content, oil, sodium cholate, and nitrite ion sorption abilities. Meanwhile, JRDF modified via enzymolysis assisted by phosphate grafting showed the highest soluble fiber content (23.53 $g \cdot 100\ g^{-1}$), water-retention ability (12.84 $g \cdot g^{-1}$), viscosity (9.37 cP), water-swelling volume (10.80 $mL \cdot g^{-1}$), and sorption ability of copper (II) and lead (II) ions. Alternatively, JRDF modified with cellulase hydrolysis alone exhibited the highest glucose adsorption capacity (21.9 $g \cdot 100\ g^{-1}$) at pH 7.0. These results indicate that EPG is an effective way to improve the hypolipidemic effects of JRDF, while EAG is a good choice to enhance its hydration and hypoglycemic properties.

Keywords: jujube residue dietary fiber; dual enzymatic hydrolysis; phosphate grafting; acrylate grafting; functional capacities

Citation: Zhao, Y.; Xu, J.; Zheng, Y.; Li, Q.; Huang, Y.; Zong, M.; Guo, W. Dual Enzymolysis Assisted by Acrylate or Phosphate Grafting: Influences on the Structural and Functional Properties of Jujube Residue Dietary Fiber. *Molecules* **2024**, *29*, 478. https://doi.org/10.3390/molecules29020478

Academic Editor: Patricia Morales Gómez

Received: 30 November 2023
Revised: 6 January 2024
Accepted: 16 January 2024
Published: 18 January 2024

1. Introduction

Today, 537 million adults have diabetes worldwide, and there will be 1.31 billion diabetics by 2050 [1]. A main pathogeny for diabetes is a lack of dietary fiber (DF), and insufficient DF can also cause other serious harm to the body, including disorders of sugar and lipid metabolism, obesity, cardiovascular and cerebrovascular diseases, and colon cancer [2]. DF is a kind of polysaccharide that cannot be digested and absorbed by the human gastrointestinal tract. But some DFs can be fermented and utilized by microorganisms in the large intestine [3]. Recently, the role of DFs in prevention and adjuvant treatment of diabetes, hyperlipidemia, and hypertension has been evidenced in a large number of clinical trials [4]. Moreover, DFs are widely used in foods as thickening and water-retaining agents, texture improvers, emulsifiers, and clarifying agents [5]. In addition, DF is also available for use as a low-cost and renewable adsorbent in the purification of drinking water and the removal of contaminants (such as heavy metals, nitrite, and sulfides) from food [6]. Previous studies demonstrated that the hypolipidemic, hypoglycemic, and harmful-substance-removing activities of DFs were closely correlated with their microstructure, surface area, hydration properties, and adsorption capacities [6,7]. Some functional groups of DFs such as free hydroxyls, and the carboxyl, phenolic acid, and uronic acid groups improve their bounding affinity with water, oil, sugar, and metal ions, but phenolic and uronic acids are mainly in the form of conjugated mono-, di-, or oligosaccharides of fibers [8].

DF is classified as water insoluble (IDF) and soluble (SDF). Both IDF and SDF play an important role in the functional properties of DFs, and in an ideal DF, the SDF content could be no less than 30% [2]. Cereal and fruit byproducts such as rice bran, millet bran, and orange residue are potential supplements for human DF [5]. However, the application of cereal and fruit fibers as ingredients of functional foods is limited due to their low SDF (around 4 g·100 g^{-1}) and relatively poor hydration and adsorption properties [9]. Thus, certain physical, biological, and chemical methods such as alkaline treatment, high-pressure extrusion, ultrasonic treatment, enzymolysis, fermentation, and hydroxypropylation have been used to improve the hydration and adsorption properties of DFs [3,10,11]. Of these, phosphate and acrylate grafting are both shown to effectively improve the functional properties of starch and other fibers in food industry [5]. However, data regarding the composite effects of these methods, especially biological methods assisted by chemical ones on DFs, are scare.

Red jujube (*Ziziphus jujuba* Mill.) is a traditional species endemic to China. It has been cultivated in China for more than 4000 years [12]. Jujube residue is the main byproduct of jujube juice and offers a potential DF resource that is in abundant supply. Recently, jujube residue has received increasing attention because it contains a relatively high content of fiber (15 g·100 g^{-1}). However, the soluble fiber content is only around 5 g·100 g^{-1} [13]. Cellulose and hemicellulose are the main components of jujube residue dietary fiber (JRDF), accounting for approximately 50.3 and 34.6 g·100 g^{-1}, respectively [14]. JRDF has a considerable water-retention capacity (around 4 g·g^{-1}), but its hydration properties and glucose, oil, and nitrite sorption capacities are poor [15–17]. Thus, the application of jujube residue as an ingredient in foods is rare. However, in recent decades, certain physical, chemical, and biological methods, along with acrylate grafting, have proven to effectively improve the physicochemical and functional properties of plant fibers [18–20]. To our best knowledge, there have been few studies on the synergistic influences of physical, chemical, and biological modifications on JRDF. Therefore, the purpose of this study was to investigate the influences of dual enzymatic hydrolysis assisted by acrylate grafting or phosphate grafting on the functional properties and the relevant structural, hydration, and adsorption properties of JRDF, with the aim to expand the potential applications of JRDF as as ingredient of functional foods.

2. Results and Discussion

2.1. Chemical Component of JRDF

The phosphate-grafting and acrylate-grafting degrees of JRDF-EPG and JRDF-EAG were 2.36 ± 0.33% and 3.57 ± 0.17%, respectively, demonstrating that phosphate and propenyl ester groups were introduced into the JRDF after enzymatic hydrolysis and phosphate grafting and acrylate grafting. As shown in Table 1, the three modifications (cellulase and hemicellulase hydrolysis alone, enzymolysis assisted by acrylate grafting, and enzymolysis assisted by phosphate grafting) showed no remarkable influences on the fat, protein, and ash contents of the JRDF ($p > 0.05$) but offered significantly improved SDF contents ($p < 0.05$), suggesting that these modifications all enhanced the hydrophilicity of the JRDF. In comparison, the IDF content of JRDF was obviously reduced after these modifications ($p < 0.05$), corresponding to its decreased cellulose, hemicellulose, and lignin contents. Cellulase and hemicellulase hydrolysis can crack the glucosidic bonds in cellulose and hemicellulose [19], exposing more hydrophilic groups and thereby reducing the IDF content of the JRDF.

Table 1. Compositions of jujube residue dietary fiber (JRDF), JRDF treated with cellulase and hemicellulase hydrolysis (JRDF-E), and JRDF modified via enzymolysis assisted by phosphate grafting (JRDF-EPG) or acrylate grafting (JRDF-EAG).

Constituent	Jujube Residue	JRDF	JRDF-E	JRDF-EAG	JRDF-EPG
Moisture ($g \cdot 100\ g^{-1}$)	6.34 ± 0.36 a	7.47 ± 0.20 a	5.38 ± 0.19 a	5.99 ± 0.24 a	6.46 ± 0.17 a
Fat ($g \cdot 100\ g^{-1}$)	2.07 ± 0.09 a	1.28 ± 0.09 b	1.54 ± 0.02 b	1.12 ± 0.09 b	2.08 ± 0.09 a
Ash ($g \cdot 100\ g^{-1}$)	1.72 ± 0.07 a	2.19 ± 0.21 a	1.86 ± 0.08 a	2.42 ± 0.09 a	2.05 ± 0.22 a
Protein ($g \cdot 100\ g^{-1}$)	1.94 ± 0.09 a	1.38 ± 0.08 a	1.39 ± 0.07 a	1.74 ± 0.08 a	1.89 ± 0.09 a
Total dietary fiber ($g \cdot 100\ g^{-1}$)	18.23 ± 1.37 b	75.38 ± 2.28 a	81.37 ± 2.84 a	80.93 ± 4.82 a	80.09 ± 3.58 a
Insoluble dietary fiber ($g \cdot 100\ g^{-1}$)	14.92 ± 0.32 d	69.72 ± 4.52 a	66.43 ± 3.34 a	62.95 ± 1.79 b	56.56 ± 4.95 c
Soluble dietary fiber ($g \cdot 100\ g^{-1}$)	3.31 ± 0.08 e	5.66 ± 0.11 d	14.94 ± 1.18 c	17.98 ± 2.56 b	23.53 ± 2.32 a
Extractable phenols ($g \cdot 100\ g^{-1}$)	0.76 ± 0.05 d	0.83 ± 0.01 c	1.14 ± 0.08 b	1.56 ± 0.11 a	1.23 ± 0.02 b
Cellulose ($g \cdot 100\ g^{-1}$)	51.13 ± 3.64 a	53.56 ± 3.47 a	42.19 ± 3.25 b	39.38 ± 2.77 b	40.99 ± 3.67 b
Lignin ($g \cdot 100\ g^{-1}$)	12.52 ± 1.33 a	12.53 ± 0.55 a	9.85 ± 0.08 a	10.79 ± 0.34 a	10.73 ± 1.05 a
Hemicellulose ($g \cdot 100\ g^{-1}$)	33.54 ± 0.45 a	37.45 ± 4.39 a	26.48 ± 2.73 b	28.79 ± 2.66 b	20.97 ± 0.39 c

Different letters (a–e) on the same line represent a significant difference ($p < 0.05$).

Both JRDF-EPG and JRDF-EAG offered higher SDF contents than that of JRDF-E ($p < 0.05$), evidencing that enzymolysis assisted by phosphate grafting and acrylate grafting was more effective in improving the hydrophilicity of JRDF than enzymolysis alone. The phosphate and propenyl ester groups introduced after these modifications were also responsible for the higher SDF contents of JRDF-EPG and JRDF-EAG, respectively. The phosphate group was a polar group and had high hydrophilicity, while the propenyl ester group had both hydrophilicity and hydrophobicity [21]. JRDF-EPG showed a higher SDF content than JRDF-EAG ($p < 0.05$) because the introduced phosphate group had higher polarity than the propenyl ester group [18]. Furthermore, the SDF content of JRDF-EPG ($23.53\ g \cdot 100\ g^{-1}$) was similar to that of oat bran fiber ($22.05\ g \cdot 100\ g^{-1}$) [20], indicating its potential for applications as a functional food ingredient.

Additionally, JRDF-E, JRDF-EPG, and JRDF-EAG all showed higher extractable polyphenol contents than JRDF ($p < 0.05$), mainly due to the hydrolysis of cellulose by cellulase. Cellulase and hemicellulase caused the degradation of fibers and the release of more phenolic compounds [21,22], significantly surpassing the extractable phenol content of JRDF. JRDF-EAG showed the highest polyphenol content. Previous studies found that the introduced propenyl ester groups could increase the bounding polyphenol content of DFs [23], which may have been the case here.

2.2. Particle Sizes and Colors of JRDFs

As shown in Table 2, JRDF-E, JRDF-EAG, and JRDF-EPG all exhibited smaller particle sizes ($D_{3,2}$: 59.38–95.23 μm) but larger surface areas ($86.44–167.35\ m^2 \cdot kg^{-1}$) than those of JRDF ($p < 0.05$). The main reason was that enzymatic hydrolysis can degrade glycosidic bonds and crack the cell well of JRDF [24], leading to a decrease in its particle size. Moreover, during phosphate grafting and acrylate grafting, heating and alkaline treatment can also cause the degradation of polysaccharide chains and thus an increased surface area of JRDF. JRDF-EAG showed the largest surface area ($167.35 \pm 4.42\ m^2 \cdot kg^{-1}$) and the smallest particle size (59.38 ± 3.35 μm), followed by JRDF-EPG and JRDF-E. A bigger surface area indicates that the chance of DFs touching oil and water molecules is increased, which is conducive to the adsorption capacity of DFs [7].

Table 2. Particle size distribution, colors, and adsorption properties of JRDF, JRDF-E, JRDF-EPG, and JRDF-EAG.

Properties		JRDF	JRDF-E	JRDF-EAG	JRDF-EPG
$D_{3,2}$ (μm)		116.47 ± 4.71 b	76.71 ± 2.07 d	59.38 ± 3.35 e	95.23 ± 4.05 c
Surface area ($m^2 \cdot kg^{-1}$)		64.53 ± 3.74 d	105.15 ± 4.74 b	167.35 ± 4.42 a	86.44 ± 2.95 c
L^*		53.8 ± 1.02 a	40.75 ± 3.13 b	36.95 ± 2.44 b	38.56 ± 3.34 b
a^*		7.95 ± 0.23 b	10.57 ± 0.32 a	11.75 ± 0.37 a	9.36 ± 0.27 a
b^*		11.32 ± 0.26 c	14.26± 0.26 b	19.54 ± 0.37 a	15.08 ± 1.42 b
ΔE		Control	13.63 b	19.13 a	15.76 a b
Water-retention ability ($g \cdot g^{-1}$)		6.68 ± 0.24 c	7.17 ± 0.36 b c	10.96 ± 0.49 b	12.84 ± 0.37 a
Water-swelling volume ($mL \cdot g^{-1}$)		6.05 ± 0.04 c	8.42 ± 0.16 b	8.80 ± 0.34 b	10.80 ± 0.26 a
Viscosity (cP)		2.01 ± 0.27 c	5.56 ± 0.08 b	7.00 ± 0.44 b	9.37 ± 0.24 a
Oil sorption ability ($g \cdot g^{-1}$)		0.78 ± 0.09 c	1.24 ± 0.11 b c	3.73 ± 0.23 a	1.89 ± 0.03 b
Sodium cholate sorption ability ($g \cdot g^{-1}$)		10.78 ± 0.46 c	10.27 ± 2.21 c	30.86 ± 1.24 a	20.46 ± 1.85 b
Glucose sorption ability ($\mu mol \cdot g^{-1}$)	pH 2.0	6.49 ± 0.26 c	14.43 ± 0.36 a	9.26 ± 1.02 b	13.75 ± 0.45 a
	pH 7.0	10.17 ± 0.38 c	24.94 ± 2.02 a	14.52 ± 0.52 b	21.69 ± 1.17 a
NO^- sorption ability ($\mu g \cdot g^{-1}$)	pH 2.0	7.36 ± 0.38 d	11.43 ± 0.23 c	21.64 ± 1.95 a	16.54 ± 1.25 b
	pH 7.0	2.48 ± 0.08 c	6.32 ± 0.05 b	16.85 ± 0.41 a	9.69 ± 0.32 b
Equilibrium sorption amount of Cu^{2+} ($mg \cdot g^{-1}$)		9.58 ± 0.25 c	14.54 ± 0.43 b	19.62 ± 1.37 a	21.41 ± 0.27 a
Equilibrium sorption amount of Pb^{2+} ($mg \cdot g^{-1}$)		13.42 ± 0.35 c	20.45 ± 0.75 b	18.32 ± 0.85 b	25.64 ± 1.22 a

Different letters (a–e) on the same line represent a significant difference ($p < 0.05$).

Moreover, the L^* value (representing brightness) of the JRDF was reduced while the a^* and b^* values (indicative of redness and yellowness, respectively) were both increased after these mixed modifications ($p < 0.05$), indicating that a browning reaction probably occurred in the JRDF. JRDF-EAG showed the highest ΔE (19.13, Table 2) among the modified JRDFs, likely because heating (70 °C) and sodium hydroxide (6.25 $mol \cdot L^{-1}$) treatment during acrylate grafting can both cause a browning reaction for DFs [25]. However, a darker color may affect the JRDFs' utilization in foods.

2.3. *Structural Characteristics*

2.3.1. Surface Microstructure

JRDF offered an irregular microstructure (Figure 1a), while JRDF-E, JRDF-EAG, and JRDF-EPG all showed microstructures with more fragmentation and pores (Figure 1b–d). After enzymolysis and chemical modifications, the polysaccharide chains and cell walls of the JRDF were destroyed, and chemical components such as fat, cellulose, and lignin were removed, resulting in more fragmented and porous microstructures [26]. A more porous and fragmented microstructure means DFs have a larger surface area and a higher sorption capacity in water and oil [27]. Our findings supported those of previous studies, which showed that enzymolysis, acrylate grafting, and hydroxypropylation can all make the microstructure of fibers more fragmented or porous [18,20].

2.3.2. Fourier-Transformed Infrared Spectroscopy

Although the FT-IR spectrum of JRDF was similar to that of JRDF-E, JRDF-EAG, and JRDF-EPG, there were visible differences among them (Figure 2). The broad peak located at around 3340 cm^{-1} (representative of the stretching vibration of O–H) in the spectrum of JRDF moved to 3349, 3342, and 3336 cm^{-1} in the spectra of JRDF-E, JRDF-EAG, and JRDF-EPG, respectively, suggesting that the O–H bond of furfural acid was changed during these mixed modifications [28]. Compared to JRDF, a new peak at around 880 cm^{-1} (corresponding to the stretching vibration of the β-glycosidic bond) appeared in the spectra of JRDF-E, JRDF-EAG, and JRDF-EPG, evidencing the degradation of polysaccharide chains caused by ultrasound and enzymatic hydrolysis [29]. In the case of JRDF-EAG, the peaks at 1742 cm^{-1} (corresponding to asymmetric bending of the acetyl group) and 1100 cm^{-1} (representing the stretching vibration of C–O–C) transferred to 1752 and 1085 cm^{-1}, respectively; and the peak at 1525 cm^{-1} (indicative of the stretching vibration of –C=O) transferred to 1519 cm^{-1}, evidencing the introduction of the propenyl ester group [19]. Moreover, a new peak at around 2395 cm^{-1} (corresponding to the stretching vibration of the phosphate group) appeared in the spectrum of JRDF-EPG [10,30], demonstrating that the phosphate group had been grafted with JRDF. Similar results were obtained by Djordjević et al. [31]. In general, these results evidenced that the three modifications changed the chemical bonds of JRDF. However, the binding patterns and sites of the phosphate and acetyl groups with JRDF molecules require further investigation.

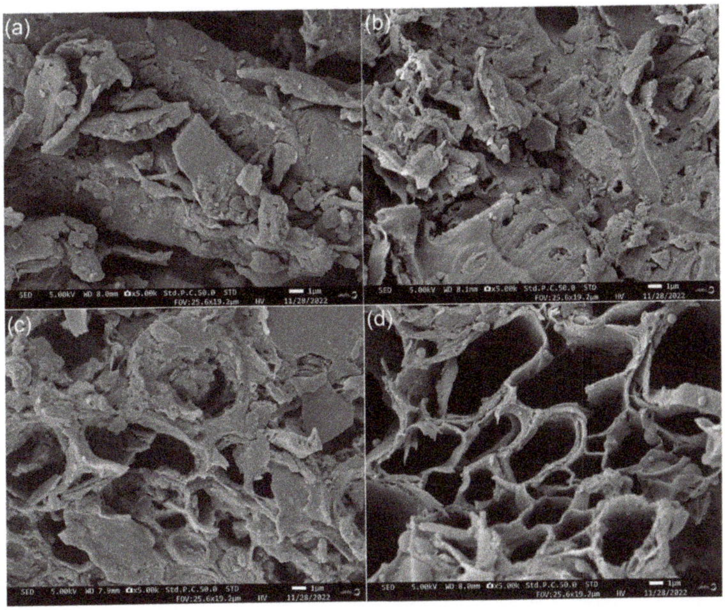

Figure 1. Scanning electron micrographs of JRDF (**a**), JRDF-E (**b**), JRDF-EAG (**c**), and JRDF-EPG (**d**) with a magnification of 5000× at 1 μm. JRDF, jujube residual dietary fiber; JRDF-E, JRDF modified with cellulase and hemicellulase hydrolysis; JRDF-EAG, JRDF modified with cellulase and hemicellulase hydrolysis assisted by acrylate grafting; and JRDF-EPG, JRDF modified with cellulase and hemicellulase hydrolysis assisted by phosphate grafting.

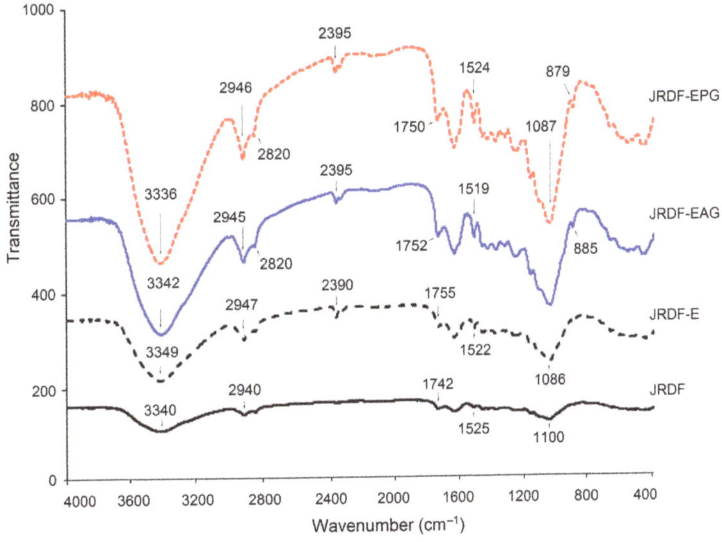

Figure 2. Fourier-transformed infrared spectra of JRDF, JRDF-E, JRDF-EAG, and JRDF-EPG.

2.3.3. X-ray Diffraction Profiles

As shown in Figure 3A, the JRDFs all offered a typical X-ray diffraction profile of fibers with two intensity peaks at 18.4° and 22.2° (corresponding to the cellulose crystal structures I and II, respectively) [31]. The crystallinities of JRDF-E, JRDF-EPG, and JRDF-EAG were lower than that of JRDF ($p < 0.05$), indicating that enzymatic hydrolysis combined with acrylate grafting or phosphate grafting destroyed the cellulose crystal structure of the JRDF. The crystal structure of the fibers was mainly maintained by hydrogen bonds formed between adjacent hydroxyl groups [10]. Cellulase and hemicellulase hydrolysis can both cause degradation of the glucosidic bonds of fibers, and the heating and alkaline treatments during acrylate and phosphate grafting can break down hydrogen bonds, leading to a lower crystallinity of JRDF-E, JRDF-EAG, and JRDF-EPG [19,32]. A similar trend was observed by other previous studies [31,33].

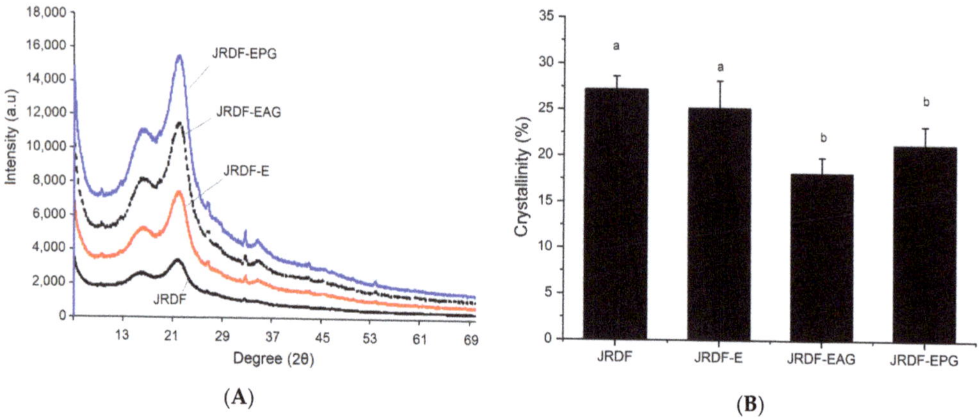

Figure 3. X-ray diffractions (**A**) and crystallinities (**B**) of JRDF, JRDF-E, JRDF-EAG, and JRDF-EPG. Different letters on the columns mean significant difference ($p < 0.05$).

2.4. *Hydration Properties*

High hydration properties including the WRA, WSV, and viscosity allow fibers to rapidly interact with water molecules; expand their volume in water; and thus inhibit the diffusion of oil, heavy metal ions, or sugar in water or food, promoting the aggregation of these materials and facilitating the adsorption of DFs [34,35]. As shown in Table 2, JRDF-E, JRDF-EAG, and JRDF-EPG all exhibited a higher WRA, WSV, and viscosity than JRDF ($p < 0.05$), evidencing that dual enzymolysis alone or combined with acrylate grafting or phosphate grafting all effectively improved the hydration properties of the JRDF. The main reason was the higher soluble contents of JRDF-E, JRDF-EAG, and JRDF-EPG (Table 1). After enzymatic hydrolysis combined with acrylate grafting or phosphate grafting, the polarity of the JRDF was significantly increased, thereby improving the interactions between the JRDF and water molecules. Moreover, the more fragmented and porous surface microstructures and bigger surface areas of JRDF-EAG and JRDF-EPG (Figure 1c,d, Table 1) also contributed to their high WRA, viscosity, and WSV because a porous microstructure and bigger surface area are both conducive to the interactions between water and fibers [24]. Furthermore, the introduced propenyl ester and phosphate groups enhanced the steric hindrance between fiber chains [18], thereby improving the WSV of the JRDF.

JRDF-EPG showed the highest WSV (10.80 $mL \cdot g^{-1}$), WRA (12.84 $g \cdot g^{-1}$), and viscosity (9.37 cP) among all the JRDFs. After phosphate grafting, the introduced phosphate groups had high polarity and a strong affinity with water molecules [36], significantly increasing the viscosity, WSV, and WRA of JRDF. Since the polarity of the acrylate group was lower than that of the hydroxypropyl and phosphate groups, JRDF-EAG had a lower WSV and WRA than those of JRDF-EPG ($p < 0.05$). With that said, the viscosity of JRDF-EAG was higher than that of JRDF ($p < 0.05$); previous studies showed that introducing the acrylate group could enhance the resistance of fiber fluid and increase the viscosity of fibers [23]. In addition, the WSV, WRA, and viscosity of JRDF-EPG were all higher than those of JRDF-E ($p < 0.05$), suggesting that enzymolysis combined with phosphate grafting is a better choice to increase the hydration properties of JRDF than enzymolysis alone. Previous studies also showed that acetylation and phosphate grafting improved the hydration properties of millet fiber and corn starch and thus expanded their applications in the food industry [21,25].

2.5. Functional Properties

2.5.1. Adsorption Capacity of Oil on JRDFs

As shown in Table 2, JRDF-EAG exhibited the highest OSA (3.73 ± 0.23 g·g^{-1}), followed by JRDF-EPG; meanwhile, JRDF-E showed a lower OSA than that of JRDF, although this difference was not significant ($p > 0.05$). After acrylate grafting, the introduced propenyl ester group significantly increased the affinity of JRDF-EAG for oil molecules [25]. Previous studies also found that esterification remarkably increased the OSA of DFs [18,37]. In addition, the high WSV (8.80 mL·g^{-1}), large surface area (167.35 m^2·kg^{-1}, Table 2), and porous microstructure (Figure 1c) were all conducive to the OSA of JRDF-EAG. The higher WSV indicates that JRDF-EAG had a larger expansion volume in water, which increased the contact area between JRDF-EAG and oil [26]. Moreover, JRDF-EPG showed a higher OSA than JRDF ($p < 0.05$). One reason for this finding was the phosphate groups introduced after phosphate grafting, which could promote the formation of a network-like structure between polysaccharide chains, assisting JRDF's adsorption of oil [18]. Additionally, other supporting factors were the high WSV and the porous microstructure of JRDF-EPG (Table 2 and Figure 1d). These results demonstrated that JRDFs adsorbed oil via chemical and physical adsorption.

2.5.2. Sorption Ability of Sodium Cholate

Sodium cholate is an endogenous emulsifier and plays an important role in digestion and the absorption of lipid. Therefore, inhibiting the production of sodium cholate can effectively reduce the absorption of fat in the small intestine [3]. The results in Table 2 show that both JRDF-EAG and JRDF-EPG offered higher SASCs than that of JRDF ($p < 0.05$), which was in accordance with their higher OSA. The higher OSA indicated that the DFs had a stronger affinity with oil molecules. After acrylate grafting, the propenyl ester group that was introduced significantly increased the hydrophobicity of JRDF and thus enhanced its adsorbing ability for oil and sodium cholate. Moreover, the large surface area and porous microstructure contributed to the SASC of JRDF-EAG. In the case of JRDF-EPG, the phosphate group that was introduced increased the SDF content and hydrophilicity of JRDF, which was conducive to the affinity of DFs for sodium cholate, an endogenous emulsifier with hydrophobicity and hydrophilicity. Furthermore, the SASC of JRDF-E was lower than that of JRDF-EAG and JRDF-EPG ($p < 0.05$), demonstrating that acrylate grafting and phosphate grafting, rather than cellulase and hemicellulase hydrolysis, enhanced the SASC of JRDF. A similar trench was observed by Zheng et al. [18]. The high OSA and SASC of JRDF-EPG highlight its potential for application as an ingredient in hypolipidemic foods [35].

2.5.3. Sorption Ability of Glucose

As was shown in Table 2, the GSAs of all JRDFs at pH 7.0 were higher than those at pH 2.0, indicating that JRDFs had a higher ability to adsorb glucose in the intestine than in the stomach because the acidic environment in the latter was not conducive to the dissociation of carboxylic and phenolic acid, which have a strong affinity for glucose [8]. JRDF-E and JRDF-EPG both offered higher GSAs (21.69–24.94 µmol·g^{-1}) than JRDF ($p < 0.05$), which can probably be attributed to their higher SDF contents and WSV, more porous microstructures, and larger surface areas (Table 1, Figure 1b–d, and Table 2). The higher WSV and larger surface areas indicate that the JRDFs had more of a chance to contact and adsorb glucose, while a porous microstructure meant that the JRDF had a higher ability to rupture glucose [34]. More importantly, the phosphate group that was introduced enhanced the hydrophilicity of JRDFs and thus improved their chemical adsorption of glucose. Accordingly, the GSAs of JRDF-E and JRDF-EPG were 2.45 and 2.13 times that of JRDF. Likewise, previous studies also found that enzymatic hydrolysis and phosphate grafting improved the GSAs of coconut cake fiber and Chia fiber [32,33]. In contrast, JRDF-EAG had a lower GSA than that of JRDF, although this difference was not significant ($p > 0.05$), highlighting the reducing minimal influence of acrylate grafting on the GSA of JRDF. The relatively high GSAs of JRDF-E and JRDF-EPG suggested their potential applications as ingredients of hypoglycemic foods [29].

2.5.4. NISA

The JRDFs all exhibited higher NISAs at pH 2.0 than at pH 7.0, suggesting that nitrite ions were more easily adsorbed by JRDFs in the stomach. In an acidic environment, nitrite ions are easily transferred into nitrogen oxide, which can be easily adsorbed by organic acids in DFs [27]. JRDF-E, JRDF-EPG, and JRDF-EAG all exhibited higher NISAs (11.43–21.64 $\mu mol \cdot g^{-1}$, Table 2) than that of JRDF ($p < 0.05$), mainly due to their higher phenolic acid contents (Table 1). JRDF-EAG offered the highest NISA (21.64 ± 1.95 $\mu mol \cdot g^{-1}$) at pH 2.0, corresponding to it having the highest phenol content (Table 1). Moreover, the more porous microstructures (Figure 1b–d), larger WSVs, and smaller particle sizes (Table 2) all indicate that these modified JRDFs had a higher capacity to adsorb nitrite ions [34]. Additionally, the phosphate and propenyl ester groups that were introduced both had a relatively strong affinity for nitrite ions [18].

2.5.5. Copper Ion (II) Sorption Ability

Excess copper ions can cause damage to the liver and kidneys [38]. Figure 4 depicts the isotherm adsorption kinetics of copper ions (II) on JRDFs. JRDF, JRDF-E, JRDF-EAG, and JRDF-EPG all showed a time-dependent Cu^{2+} sorption capacity, reaching an equilibrium sorption amount at around 80 min. Furthermore, JRDF-E, JRDF-EAG, and JRDF-EPG all showed higher ESCIs than that of JRDF ($p < 0.05$) (Table 2), evidencing that enzymatic hydrolysis alone or with phosphate grafting or acrylate grafting increases the adsorption ability of copper ions on JRDF. One reason for this is that enzymolysis causes the degradation of cellulose and lignin and thus the release of more functional groups such as uronic, carboxyl, and phenolic groups, which have a relatively strong affinity with copper ions [26]. In addition, the phosphate and ester groups that were introduced remarkably increased the negative charges of the JRDFs [18], thus increasing their ESCIs. JRDF-EPG exhibited the highest ESCI (24.14 ± 0.27 $mg \cdot g^{-1}$) because the introduced phosphate group has a strong chelating ability with copper ions [36]. Meanwhile, the ESCIs of JRDF-EAG and JRDF-EPG were higher than that of JRDF-E ($p < 0.05$), evidencing that enzymolysis assisted by phosphate grafting or acrylate grafting was more effective at increasing the ESCI of JRDF than enzymolysis alone.

2.5.6. Lead Ion (II) Sorption Capacity

Lead pollution can damage the human nervous system and induce mania, coma, and even death [27]. As shown in Figure 5, JRDF, JRDF-E, JRDF-EAG, and JRDF-EPG all showed a time-dependent sorption of lead ions (II) within 5–80 min and reached an equilibrium sorption value at around 65 min. Moreover, the ESLIs of JRDF-E, JRDF-EAG, and JRDF-EPG were higher than that of JRDF ($p < 0.05$, Table 2), mainly due to their larger WSV and surface area (Table 2), more porous microstructure (Figure 1b–d), and higher polyphenol contents (Table 1), because a larger WSV, larger surface area, and porous microstructure indicate that the fibers had a greater chance and capacity to capture lead ions [35]. In addition, the phosphate and propenyl ester groups that were introduced all increased the negative charge of JRDF, thereby improving its ESLI. JRDF-EPG showed the highest ESLI (25.37 ± 0.32 $mg \cdot g^{-1}$), which can mainly be attributed to the high metal chelation ability of the introduced phosphate group [36]. The ESLIs of JRDF-EAG and JRDF-EPG (20.45–25.37 mg·g^{-1}) were higher than those of acrylate-grafted banana fiber (10.45 mg·g^{-1}) [23], corn starch (15.32 mg·g^{-1}) [27], and palm kernel fiber (19.8 mg·g^{-1}) [38] but lower than that of activated carbon (51.15 mg·g^{-1}) [31], suggesting their potential applications to remove heavy metals from foods.

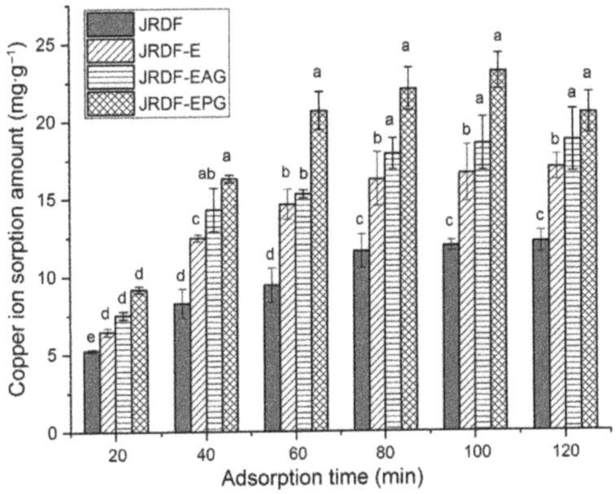

Figure 4. Isothermal adsorption kinetics of copper ions (II) on JRDF, JRDF-E, JRDF-EAG, and JRDF-EPG. Different lowercase letters (a–e) on the bars evidence the difference is significant ($p < 0.05$).

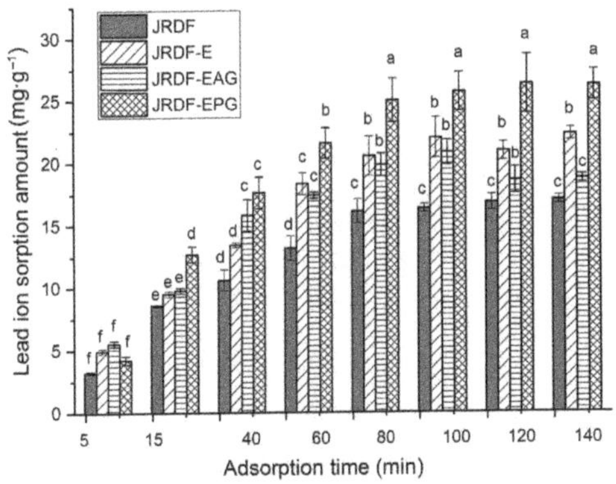

Figure 5. Isothermal adsorption kinetics of lead ions (II) on JRDF, JRDF-E, JRDF-EAG, and JRDF-EPG. Different lowercase letters (a–f) on the bars evidence that the difference is significant ($p < 0.05$).

3. Materials and Methods

3.1. Materials and Reagents

Craw jujube residue was donated by Sanduo Juice Co., Ltd., Yushe, China. Hemicellulase (2.0 × 10^4 U·g^{-1}), cellulase (Trichoderma Vride G, 3.0 × 10^5 U·g^{-1}), amyloglucosidase (Aspergillus niger, 1.0 × 10^5 U·g^{-1}), and α-amylase (Bacillus licheniformis) were all purchased from Yeyan Biotechnology Corporation (Shuzhou, China). Sodium sulfate, acrylic acidisopropenyl, sodium hydroxide, and other analytical reagents were purchased from Shanghuiyuan Biotech (Shijiazhuang, China).

3.2. Preparation of JRDF

Briefly, craw jujube residue was dried, ground, and sieved using a ZJ-IIC sieve (with an aperture of 150 μm, Zhuji Sieve Factory, Zhuji, China) [6]. The jujube residue powder (18 g) was dispersed in deionized water (dH$_2$O, 360 mL) and then hydrolyzed by 0.2 g of α-amylase (at pH 7.0 and 65 °C for 90 min), 0.2 g of papain (at pH 7.0 and 50 °C for

120 min), and 0.15 g of glucoamylase (at pH 4.0 and 60 °C for 120 min) in sequence. After heating at 100 °C for 12 min, the mixture was filtered, and the residue on the filter paper was collected and dried (50 °C, 3 h) to obtain jujube residue dietary fiber (JRDF).

3.3. Dual Enzymatic Hydrolysis of JRDF

JRDF (20 g) and 260 mL of phosphate buffer (pH 4.5, 0.1 mol·L^{-1}) were added into a glass flask and mixed thoroughly [33]. Thirty minutes later, cellulase (75 U·g^{-1}) was added, and the flask was shaken using an EZC-004H water bath oscillator (Shanyu Oscillator Factory, Nantong, China) with a temperature of 55 °C and shaking rate of 195 r·min^{-1}. Two hours later, the mixture was adjusted to pH 6.5 with the addition of NaOH (0.1 mol·L^{-1}) and hemicellulase (25 U·g^{-1} JRDF). The reaction was continued at a temperature of 35.0 °C and shaking rate of 195 r·min^{-1} for 2 h. Then, the mixture was adjusted to pH 7.0 and heated at 100 °C for 5 min. After filtration, the residue was dehydrated using a GFGZ-II dryer (Huaxing Inst. Fac., Hengyang, China) with a heating temperature of 50 °C. Six hours later, jujube residue dietary fiber treated with cellulase and hemicellulase hydrolysis (JRDF-E) was obtained.

3.4. Acrylate Grafting of JRDF-E

Sodium hydroxide (6.25 mol·L^{-1}, 24 mL), dH_2O (180 mL), and acrylic acid (15 mL) were mixed in a triangular flask and shaken in the EZC-004H oscillator at 65 r·min^{-1} and 25 °C [23]. Forty minutes later, 2 g of JRDF-UE was added, and suspension was continuously shaken at 65 r·min^{-1} for ten minutes. Next, thiosalicylic acid (58.4 mmol·L^{-1}, 1.5 mL) and potassium thiosulfate (74 mmol·L^{-1}, 2.25 mL) were added in sequence. The reaction mixture was shaken at 195 r·min^{-1} and 70 °C. Three hours later, the suspension was filtered through a nylon sieve with an aperture of 100 μm, and the residue was washed using anhydrous ethanol (30 mL) in triplicate. After heating at 50 °C for 8 h, JRDF modified via enzymolysis assisted by acrylate grafting (JRDF-EAG) was obtained. The grafting degree was determined by following the same procedures as Rani et al. [23].

3.5. Phosphate Grafting of JRDF-E

JRDF-E (25 g) was dispersed in 250 mL of dH_2O and stirred in the EZC-004H oscillator at 145 r·min^{-1} and 25 °C [18]. Fifteen minutes later, the suspension was adjusted to pH 11, and then 3 g sodium trimetaphosphate and sodium tripolyphosphate (0.3 g) were added. Next, the reaction suspension was shaken at 145 r·min^{-1} and 47 °C in the oscillator. Three hours later, the suspension was adjusted to pH 7.0 and then filtered with Waterman filter paper. The residue was rinsed with dH_2O in triplicates (20 mL every time) and dried in the DHG-9053A blast drying oven with a heating temperature of 50 °C. Six hours later, JRDF modified via enzymolysis assisted by phosphate grafting (JRDF-EPG) was obtained. The grafting degree was defined as the increment in phosphorus content in JRDF-EPG after phosphate grafting [39].

3.6. Chemical Composition Measurement

The moisture, fat, protein, and ash contents of JRDF, JRDF-E, JRDF-EPG, and JRDF-EAG were separately determined using the methods AOAC.920.39, AOAC.92.05, AOAC.955.04, and AOAC.924.05 (AOAC, 2000). The acid-insoluble lignin, neural detergent fiber, and acid detergent fiber were measured as per the procedure from Souza et al. [40] to calculate the lignin, cellulose, and hemicellulose contents, respectively. Moreover, the total dietary fiber (TDF) and IDF contents were determined using the AOAC.991.43 method (AOAC, 2000), and the SDF content was calculated from them [11].

3.7. Particle Size Analysis

A nanometer laser particle size analyzer (WINNER 2309A, Jinan Winner particle Technology Co., Ltd., Jinan, China) with a refractive index of 1.33 was employed to determine the $D_{3,2}$ (Sauter mean diameter, μm) and specific surface area of JRDFs [18].

3.8. Color Analysis

Color indexes including L (the lightness of DF), b (the redness), and a (the yellowness) values were measured with an NH810 spectrocolorimeter (Three-NH Colorimeter Co., Shenzhen, China) with D65 light source. The color difference (ΔE) was calculated with untreated JRDF as the control:

$$\Delta E = \sqrt{(L - L_0)^2 + (a - a_0)^2 + (b - b_0)^2} \quad (1)$$

where L_0, a_0, and b_0 are the color indexes of the untreated JRDF.

3.9. Structural Characteristics

3.9.1. Surface Microstructure Analysis

Approximately 2 mg of JRDFs was placed on a preparation table and coated with a 10 nm gold layer [22]. A JEOL-JSM-7500F scanning electron microscope (Japan Electronics Co., Ltd., Showima city, Japan) was used to scan the microstructures of the samples with an acceleration voltage of 10 kV. Micrographs were captured at a magnification of 2000× and scale bar of 5 μm.

3.9.2. Fourier-Transformed Infrared Spectroscopy (FT-IR)

An FT-752 FT-IR spectrometer (Jinke Instrument Factory, Shanghai, China) was employed for FT-IR analysis with a scanning range of 4000–400 cm^{-1}, according to the same procedures used by Zhang et al. [22].

3.9.3. X-ray Diffraction Investigation

An X-ray diffraction investigation was performed on an Ultima IV-185 X-ray powder diffractometer (Rigaku Corporation of Japan, Tokyo, Japan) according to the procedures used by Liu et al. [24]. The difference between the lowest and the maximum intensities per the maximum intensity was defined as the crystallinity (%) of a sample.

3.10. Hydration Properties Analysis

Hydration properties including the water-retaining ability (WRA) and water-swelling volume (WSV) of JRDFs were determined according to the methods used by Zheng et al. [18].

3.11. Viscosity

Viscosity determination was administrated on a RAV-I5 viscometer (Yebei Viscosity Instrument Factory, Zhuhai, China). The concentration of JRDF was 2.5 g 100 mL, and the shear rate was 600 s^{-1}.

3.12. Functional Properties

3.12.1. Oil Sorption Ability

The soy oil sorption ability (OSA) of JRDFs was conducted according to the same procedures from Zheng and Li [33]. Briefly, JRDFs (1 g) were mixed with soybean oil in a centrifugal tube and left for 1 h at 25 °C. The mixture was then centrifuged at 1500× g for 10 min, the supernatant decanted, and the pellet was recovered by filtering through a linen mesh. OSA is expressed as follows:

$$\text{ORA (g/g)} = (\text{pellet weight} - \text{dry weight})/\text{dry weight} \quad (2)$$

3.12.2. Sodium Cholate Sorption Ability

The sodium cholate sorption ability (SCSA) of JRDFs was conducted according to the same procedures from Peerajit et al. [41]. Briefly, JRDFs (2 g) were mixed with 100 mL of sodium cholate (1 mg·mL^{-1}, pH 7.0) and incubated in a shaking water bath at 37 °C for 2 h. Afterward, the suspension was centrifuged at 3000× g for 10 min, and then 1 mL of mixture was collected and mixed with 6 mL of 45% sulfuric acid and 1 mL of 0.3% furaldehyde. After incubation at 65 °C for 30 min, the absorbance at 620 nm was measured. SCSA was calculated as follows:

$$\text{SCSA}\left(\text{g}\cdot\text{g}^{-1}\right) = (C_0 - C_t) \times V/W \quad (3)$$

where C_t and C_0 were the sodium cholate concentrations of the reaction solution (1 $\text{mg}\cdot\text{mL}^{-1}$) before and after the adsorption, respectively; V is the volume of reaction solution; and W is the weight of JRDFs.

3.12.3. Glucose Sorption Ability

In conical flasks, JRDFs (1 g) were dispersed in 100 mL of glucose solutions with different concentrations (0.1, 0.2, 0.3, 0.4, and 0.5 $\text{g}\cdot\text{L}^{-1}$) and then adjusted to pH 2.0 or pH 7.0, respectively [18]. The suspensions in the flasks were shaken at 25 °C in the EZC-004H oscillator with a shaking rate of 145 $\text{r}\cdot\text{min}^{-1}$. Three hours later, the suspensions were filtered through a nylon cloth via aperture of 100 mesh. The filtrate liquid was used for glucose content determination with the phenol–sulfuric acid method [42]. The glucose sorption ability (GSA) is defined as follows:

$$\text{GSA}\left(\text{mg}\cdot\text{g}^{-1}\right) = (C_0 - C_S) \times V/\ W \quad (4)$$

where V, W, C_0, and C_S are the glucose solution volume, dry weight of JRDFs, and glucose concentrations before and after adsorption, respectively.

3.12.4. Nitrite Ions' Sorption Ability

One gram of JRDFs was dispersed in 200 mL of $NaNO_2$ (2 $\mu\text{g}\cdot\text{mL}^{-1}$) and adjusted to pH 2.0 or 7.0 with acetic acid (0.25 $\text{mol}\cdot\text{L}^{-1}$); then, they were shaken in the EZC-004H oscillator at 135 $\text{r}\cdot\text{min}^{-1}$ and 35 °C for 120 min [18]. Next, the suspensions were filtered, and the filtrate liquid was collected. The N-1-naphthylethylenediamine dihydrochloride method was used for the nitrite concentration determination [29]. After adsorption, the reduction in nitrate concentration according to the weight of JRDFs was defined as the nitrite ion sorption ability (NISA).

3.12.5. Copper Ion (II) Sorption Capacity

JRDFs (0.5 g) were dispersed in 125 mL of $CuSO_4$ (325 $\mu\text{g}\cdot\text{mL}^{-1}$) and then shaken at 165 $\text{r}\cdot\text{min}^{-1}$ and 25 °C for 10, 20, 40, 60, 80, and 100 min in the EZC-004H oscillator [18]. Next, the suspension was filtered through a nylon cloth (120 mesh). The filtrate liquid was pooled and used for copper ion (II) concentration determination with the KI–soluble starch titration method [43]. The equilibrium sorption of copper ions (II) (ESCI) on JRDFs was calculated as follows:

$$\text{ESCI}\left(\text{mg}\cdot\text{g}^{-1}\right) = 125 \times (C_0 - C_S)/\ W \quad (5)$$

where W, C_0, and C_S are the dry weight of JRDFs, initial Cu^{2+} concentration, and Cu^{2+} concentration after the adsorption, respectively.

3.12.6. Lead Ion (II) Sorption Ability

Isothermal adsorption kinetics of lead ions (Pb^{2+}) on JRDFs was investigated according to the modified procedures from Rani et al. [23]. Briefly, JRDFs (1 g) were dispersed in 220 mL of lead nitrate (70 $\text{mg}\cdot\text{L}^{-1}$) and then shaken in the EZC-004H oscillator at 175 $\text{r}\cdot\text{min}^{-1}$ and 35 °C for 5, 20, 35, 50, 65, 80, 95, and 110 min. Next, the suspensions were filtered using a nylon cloth (120 mesh), and then the filtrate liquid was used for Pb^{2+}

concentration determination with the flame atomic absorption spectrometry method [27]. The equilibrium sorption of lead ions (II) (ESLI) on JRDFs was calculated as follows:

$$\text{ECLI}\left(\text{mg}\cdot\text{g}^{-1}\right) = (C_0 - C_S) \times V/M \quad (6)$$

where V, W, C_0, and C_S are the $Pb(NO_3)_2$ solution volume, dry weight of JRDFs, and the initial and final Pb^{2+} concentrations, respectively.

3.13. Statistical Analysis

Each test was performed at least three times. Statistical analysis was performed using the SPSS program (Ver.24.0, SPSS Inc., Chicago, IL, USA), and Duncan's multiple comparison was used for analysis of variance between data, with a significance level of 95% ($p < 0.05$).

4. Conclusions

Dual enzymatic (cellulase and hemicellulase) hydrolysis alone and enzymatic hydrolysis separately combined with phosphate grafting and acrylate grafting all made the microstructure of JRDF more porous and increased the SDF content, surface area, and hydration properties of JRDF ($p < 0.05$). Enzymolysis combined with acrylate grafting significantly improved the polyphenol content and hydrophobicity of JRDF and thus enhanced its oil, sodium cholate, and nitrite ion sorption abilities. Meanwhile, enzymolysis assisted by phosphate grafting remarkably improving the hydrophilicity of JRDF, resulting in higher SDF content; viscosity; and glucose, copper, and lead ion sorption amounts of JRDF ($p < 0.05$). Therefore, dual enzymolysis assisted by phosphate grafting is an effective way to expand the application of JRDF as an ingredient of hypoglycemic foods or a heavy-metal adsorbent, while dual enzymolysis assisted by acrylate grafting can expand the applications of JRDF as an ingredient of hypolipidemic foods. However, further work is required to improve our understanding of the in vitro functional properties of these modified JRDFs.

Author Contributions: Investigation, writing—original draft, data curation, Y.Z. (Yitao Zhao); Investigation, data curation, funding acquisition, J.X.; conceptualization, writing—original draft, funding acquisition, Y.Z. (Yajun Zheng); investigation, Q.L.; writing—review and editing, Y.H.; validation, M.Z.; validation, W.G. All authors have read and agreed to the published version of the manuscript.

Funding: This research was funded by the Natural Science Foundation of Shanxi Province, China (202203021221139), and the Shanxi graduate education innovation plan, China (2023XJG012).

Institutional Review Board Statement: Not applicable.

Informed Consent Statement: Not applicable.

Data Availability Statement: The data presented in this study are available on request from the corresponding author. The data are not publicly available due to the data are contained within the article.

Conflicts of Interest: The authors declare no conflicts of interest.

References

1. GBD 2021 Diabetes Collaborators. Global, regional, and national burden of diabetes from 1990 to 2021, with projections of prevalence to 2050: A systematic analysis for the Global Burden of Disease Study 2021. *Lancet* **2023**, *402*, 203–234. [CrossRef] [PubMed]
2. Qin, W.; Sun, L.; Miao, M.; Zhang, G. Plant-sourced intrinsic dietary fiber: Physical structure and health function. *Trends Food Sci. Technol.* **2021**, *118*, 341–355. [CrossRef]
3. Siddiqui, H.; Sultan, Z.; Yousuf, O.; Malik, M.; Younis, K. A review of the health benefits, functional properties, and ultrasound-assisted dietary fiber extraction. *Bioact. Carbohydr. Diet. Fibre* **2023**, *30*, 100356. [CrossRef]
4. Torbica, A.; Radosavljević, M.; Belović, M.; Djukić, N.; Marković, S. Overview of nature, frequency and technological role of dietary fibre from cereals and pseudocereals from grain to bread. *Carbohyd. Polym.* **2022**, *290*, 119470. [CrossRef] [PubMed]

5. Elleuch, M.; Bedigian, D.; Roiseux, O.; Besbes, S.; Blecker, C.; Attia, H. Dietary fibre and fibre-rich by-products of food processing: Characterisation, technological functionality and commercial applications: A review. *Food Chem.* **2011**, *124*, 411–421. [CrossRef]
6. Harahap, M.; Perangin-Angin, Y.A.; Purwandari, V.; Goei, R.; Tok, A.L.Y.; Gea, S. Acetylated lignin from oil palm empty fruit bunches and its electrospun nanofibres with PVA: Potential carbon fibre precursor. *Heliyon* **2023**, *9*, 14556. [CrossRef] [PubMed]
7. Huang, Y.L.; Ma, Y.S.; Tsai, Y.H.; Chang, S.K.C. In vitro hypoglycemic, cholesterol-lowering and fermentation capacities of fiber-rich orange pomace as affected by extrusion. *Int. J. Biol. Macromol.* **2019**, *124*, 796–801. [CrossRef] [PubMed]
8. Peng, G.Y.; Gan, J.P.; Dong, R.H.; Chen, Y.; Xie, J.H.; Huang, Z.Y.; Gu, Y.; Huang, D.; Yu, Q. Combined microwave and enzymatic treatment improve the release of insoluble bound phenolic compounds from the grapefruit peel insoluble dietary fiber. *LWT—Food Sci. Technol.* **2021**, *149*, 111905. [CrossRef]
9. Dong, R.; Liao, W.; Xie, J.; Chen, Y.; Peng, G.; Xie, J.; Sun, N.; Liu, S.; Yu, C.; Yu, Q. Enrichment of yogurt with carrot soluble dietary fiber prepared by three physical modified treatments: Microstructure, rheology and storage stability. *Innov. Food Sci. Emerg. Technol.* **2022**, *75*, 102901. [CrossRef]
10. Gil-López, D.I.L.; Lois-Correa, J.A.; Sánchez-Pardo, M.E.; Domínguez-Crespo, M.A.; Torres-Huerta, A.M.; Rodríguez-Salazar, A.E.; Orta-Guzmán, V.N. Production of dietary fibers from sugarcane bagasse and sugarcane tops using microwave-assisted alkaline treatments. *Ind. Crops Prod.* **2019**, *135*, 159–169. [CrossRef]
11. Zhu, L.; Yu, B.; Chen, H.; Yu, J.; Yan, H.; Luo, Y.; He, J.; Huang, Z.; Zheng, P.; Mao, X.; et al. Comparisons of the micronization, steam explosion, and gamma irradiation treatment on chemical composition, structure, physicochemical properties, and in vitro digestibility of dietary fiber from soybean hulls. *Food Chem.* **2022**, *366*, 130618. [CrossRef]
12. Rashwan, A.K.; Karim, N.; Shishir, M.R.I.; Bao, T.; Lu, Y.; Chen, W. Jujube fruit: A potential nutritious fruit for the development of functional food products. *J. Func. Food.* **2020**, *75*, 104205. [CrossRef]
13. Cai, W.; Tang, F.; Guo, Z.; Zhang, Q.; Zhao, X.; Ning, M.; Shan, C. Effects of pretreatment methods and leaching methods on jujube wine quality detected by electronic senses and HS-SPME–GC–MS. *Food Chem.* **2020**, *330*, 127330. [CrossRef] [PubMed]
14. Lin, X.M.; Ji, X.L.; Wang, M.; Yin, S.; Peng, Q. An alkali-extracted polysaccharide from Zizyphus Jujuba cv. Muzao: Structural characterizations and antioxidant activities. *Int. J. Biol. Macromol.* **2019**, *136*, 607–615. [CrossRef] [PubMed]
15. Ji, X.; Hou, C.; Yan, Y.; Shi, M.; Liu, Y. Comparison of structural characterization and antioxidant activity of polysaccharides from jujube (*Ziziphus jujuba* Mill.) fruit. *Int. J. Biol. Macromol.* **2020**, *149*, 1008–1018. [CrossRef]
16. Wang, Y.; Liu, X.; Zhang, J.; Liu, G.; Liu, Y.; Wang, K.; Yang, M.; Cheng, H.; Zhao, Z. Structural characterization and in vitro antitumor activity of polysaccharides from *Zizyphus jujuba* cv. Muzao. *RSC Adv.* **2015**, *5*, 7860–7867. [CrossRef]
17. Bai, L.; Zhang, H.; Liu, Q.; Zhao, Y.; Cui, X.; Guo, S.; Zhang, L.; Ho, C.T.; Bai, N. Chemical characterization of the main bioactive constituents from fruits of Ziziphus jujube. *Food Funct.* **2016**, *7*, 2870–2877. [CrossRef]
18. Zheng, Y.; Xu, B.; Shi, P.; Tian, H.; Li, Y.; Wang, X.; Wu, S.; Liang, P. The influences of acetylation, hydroxypropylation, enzymatic hydrolysis and crosslinking on improved adsorption capacities and in vitro hypoglycemic properties of millet bran dietary fibre. *Food Chem.* **2022**, *368*, 130883. [CrossRef]
19. Backes, E.; Kato, C.G.; Corrêa, R.C.G.; Moreira, R.F.P.M.; Peralta, R.A.; Barros, L.; Ferreira, I.G.F.R.; Zanin, G.M.; Bracht, A.; Peralta, R.M. Laccases in food processing: Current status, bottlenecks and perspectives. *Trends Food Sci. Technol.* **2021**, *115*, 445–460. [CrossRef]
20. Kanwar, P.; Yadav, R.B.; Yadav, B.S. Influence of chemical modification approaches on physicochemical andstructural properties of dietary fiber from oat. *J. Cereal Sci.* **2023**, *111*, 103688. [CrossRef]
21. Shaikh, M.; Haider, S.; Ali, T.M.; Hasnain, A. Physical, thermal, mechanical and barrier properties of pearl millet starch films as affected by levels of acetylation and hydroxypropylation. *International J. Biol. Macromol.* **2019**, *124*, 209–219. [CrossRef]
22. Zhang, S.; Dong, Z.; Shi, J.; Yang, C.; Fang, Y.; Chen, G.; Chen, H.; Tian, C. Enzymatic hydrolysis of corn stover lignin by laccase, lignin peroxidase, and manganese peroxidase. *Bioresour. Technol.* **2022**, *361*, 127699. [CrossRef] [PubMed]
23. Rani, K.; Gomathi, T.; Vijayalakshmi, K.; Saranya, M.; Sudha, P.N. Banana fiber Cellulose Nano Crystals grafted with butyl acrylate for heavy metal lead (II) removal. *Int. J. Biol. Macromol.* **2019**, *131*, 461–472. [CrossRef]
24. Liu, Y.; Zhang, H.; Yi, C.; Quan, K.; Lin, B. Chemical composition, structure, physicochemical and functional properties of rice bran dietary fiber modified by cellulase treatment. *Food Chem.* **2021**, *342*, 128352. [CrossRef]
25. Nasseri, R.; Ngunjiri, R.; Moresoli, C.; Yu, A.; Yuan, Z.; Xu, C. Poly(lactic acid)/acetylated starch blends: Effect of starch acetylation on the material properties. *Carbohydr. Polym.* **2020**, *229*, 115453. [CrossRef]
26. Zadeike, D.; Vaitkeviciene, R.; Degutyte, R.; Bendoraitiene, J.; Rukuiziene, Z.; Cernauskas, D.; Svazas, M.; Juodeikiene, G. A comparative study on the structural and functional properties of water-soluble and alkali-soluble dietary fibers from rice bran after hot-water, ultrasound, hydrolysis by cellulase, and combined pre-treatments. *Int. J. Food Sci. Technol.* **2021**, *57*, 1137–1149. [CrossRef]
27. Kim, S.R.; Park, J.Y.; Park, E.Y. Effect of ethanol, phytic acid and citric acid treatment on the physicochemical and heavy metal adsorption properties of corn starch. *Food Chem.* **2024**, *431*, 137167. [CrossRef]
28. Hazarika, B.J.; Sit, N. Effect of dual modification with hydroxypropylation and cross-linking on physicochemical properties of taro starch. *Carbohydr. Polym.* **2016**, *140*, 269–278. [CrossRef]
29. Hua, M.; Lu, J.; Qu, D.I.; Liu, C.; Zhang, L.; Li, S.; Chen, J.; Sun, Y. Structure, physicochemical properties and adsorption function of insoluble dietary fiber from ginseng residue: A potential functional ingredient. *Food Chem.* **2019**, *286*, 522–529. [CrossRef]

30. Akinterinwa, A.; Oladele, E.; Adebayo, A.; Gurgur, E.; Iyanu, O.O.; Ajayi, O. Cross-linked-substituted (esterified/etherified) starch derivatives as aqueous heavy metal ion adsorbent: A review. *Water Sci. Technol.* **2020**, *82*, 1–26. [CrossRef]
31. Djordjević, M.; Šereš, Z.; Maravić, N.; Šćiban, M.; Šoronja-Simović, D.; Djordjević, M. Modified sugar beet pulp and cellulose-based adsorbents as molasses quality enhancers: Assessing the treatment conditions. *LWT—Food Sci. Technol.* **2021**, *150*, 111988. [CrossRef]
32. Zohaib, H.; Muhammad, I.; Haseeb, A.M.; Kamran, K.M. Ultrasound-Assisted Modification of Insoluble Dietary Fiber from Chia (*Salvia hispanica* L.) Seeds. *J. Food Qual.* **2021**, *2021*, 5035299. [CrossRef]
33. Zheng, Y.J.; Li, Y. Physicochemical and functional properties of coconut (*Cocos nucifera* L.) cake dietary fibres: Effects of cellulase hydrolysis, acid treatment and particle size distribution. *Food Chem.* **2018**, *257*, 135–142. [CrossRef] [PubMed]
34. Miehle, E.; Hass, M.; Bader-Mittermaier, S.; Eisner, P. The role of hydration properties of soluble dietary fibers on glucose diffusion. *Food Hydrocoll.* **2022**, *131*, 107822. [CrossRef]
35. Raji, Z.; Karim, A.; Karam, A.; Khalloufi, S. A review on the heavy metal adsorption capacity of dietary fibers derived from agro-based wastes: Opportunities and challenges for practical applications in the food industry. *Trends Food Sci. Technol.* **2023**, *137*, 74–91. [CrossRef]
36. Haq, F.; Mehmood, S.; Haroon, M.; Kiran, M.; Waseem, K.; Aziz, T.; Farid, A. Role of starch based materials as a bio-sorbents for the removal of dyes and heavy metals from wastewater. *J. Polym. Environ.* **2022**, *30*, 1730–1748. [CrossRef]
37. Asadpour, R.; Yavari, S.; Kamyab, H.; Ashokkumar, V.; Chelliapan, S.; Yuzir, A. Study of oil sorption behaviour of esterified oil palm empty fruit bunch (OPEFB) fibre and its kinetics and isotherm studies. *Environ. Technol. Innov.* **2021**, *22*, 101397. [CrossRef]
38. Prabha, R.T.; Udayashankara, T.H. Adsorption of copper metal ions from aqueous solution using rice husk and groundnut shell. *Int. J. Sci. Res.* **2014**, *8*, 705–709. [CrossRef]
39. Zheng, Y.; Li, J.; Wang, X.; Guo, M.; Cheng, C.; Zhang, Y. Effects of three biological combined with chemical methods on the microstructure, physicochemical properties and antioxidant activity of millet bran dietary fibre. *Food Chem.* **2023**, *411*, 135503. [CrossRef]
40. Souza, N.F.; Pinheiro, J.A.; Brígida, A.I.S.; Morais, J.P.S.; Filho, M.M.S.; Rosa, M.F.R. Fibrous residues of palm oil as a source of green chemical buildingblocks. *Ind. Crops Prod.* **2016**, *94*, 480–489. [CrossRef]
41. Peerajit, P.; Chiewchan, N.; Devahastin, S. Effects of pretreatment methods on health-related functional properties of high dietary fiber powder from lime residues. *Food Chem.* **2022**, *132*, 1891–1898. [CrossRef]
42. Daou, C.; Zhang, H. Functional and physiological properties of total, soluble, and insoluble dietary fibers derived from defatted rice bran. *J. Food Sci. Technol.* **2014**, *51*, 3878–3885. [CrossRef] [PubMed]
43. Ho, Y.S.; Ofomaja, A.E. Kinetic studies of copper ion adsorption on palm kernel fibre. *J. Hazard. Mater.* **2006**, *137*, 1796–1802. [CrossRef] [PubMed]

foods

MDPI

Article

Transcriptome Analysis Reveals the Mechanism of Quinoa Polysaccharides Inhibiting 3T3-L1 Preadipocyte Proliferation

Cong Teng [1], Shengyuan Guo [2,3], Ying Li [1,*] and Guixing Ren [2,3,*]

1 Institute of Agro-Product Processing, Jiangsu Academy of Agricultural Sciences, Nanjing 210014, China; tengcong95@163.com
2 College of Food and Bioengineering, Chengdu University, Chengdu 610106, China; guoshengyuan217@163.com
3 College of Life Science, Shanxi University, Taiyuan 030006, China
* Correspondence: hijoly@163.com (Y.L.); renguixing@sxu.edu.cn (G.R.)

Abstract: Quinoa is a highly nutritious and biologically active crop. Prior studies have demonstrated that quinoa polysaccharides exhibit anti-obesity activity. This investigation confirmed that quinoa polysaccharides have the ability to inhibit the growth of 3T3-L1 preadipocytes. The objective of transcriptome research was to investigate the mechanism of quinoa water-extracted polysaccharides and quinoa alkaline-extracted polysaccharides that hinder the growth of 3T3-L1 preadipocytes. There were 2194 genes that showed differential expression between untreated cells and those treated with high concentrations of quinoa water-extracted polysaccharides (QWPHs). There were 1774 genes that showed differential expression between untreated cells and those treated with high concentrations of quinoa alkaline-extracted polysaccharides (QAPHs). Through gene ontology and KEGG pathway analysis, 20 characteristic pathways are found significantly enriched between the untreated group and the QAPH and QWPH groups. These pathways include the NOD-like receptor, Hepatitis C, and the PI3K-Akt signaling pathway. Atp13A4 and Gbgt1 have been identified as genes that are upregulated and downregulated in both the untreated group and the QWPH group, as well as in the untreated group and the QAPH group. These findings establish a theoretical foundation for exploring quinoa polysaccharides as an anti-obesity agent.

Keywords: transcriptome; quinoa; polysaccharides; 3T3-L1; cell proliferation

Citation: Teng, C.; Guo, S.; Li, Y.; Ren, G. Transcriptome Analysis Reveals the Mechanism of Quinoa Polysaccharides Inhibiting 3T3-L1 Preadipocyte Proliferation. *Foods* **2024**, *13*, 2311. https://doi.org/10.3390/foods13152311

Academic Editor: Barry Parsons

Received: 17 May 2024
Revised: 17 July 2024
Accepted: 18 July 2024
Published: 23 July 2024

1. Introduction

Quinoa (*Chenopodium quinoa Willd.*) is a kind of pseudocereal native to the Andes of South America, and contains protein, polysaccharides, unsaturated fatty acids, minerals, and saponins [1]. Recent studies have shown that quinoa has antioxidant [2], immunomodulatory [3], and anti-cancer activities in vitro and in vivo [4]. Polysaccharide is an important class of biological macromolecules, which participate in and mediate the regulation of various cell life phenomena. Many non-starch polysaccharides exhibit anti-obesity effects [5]. Quinoa polysaccharides, as important active substances in quinoa, have exhibited multiple biological activities [6].

Obesity is an independent metabolic disorder that brings about a more complex set of health problems, including type II diabetes, cancer, hypertension, and other cardiovascular diseases. Hyperplastic adipogenesis and adipocyte hypertrophy are important causes of obesity [7]. Obesity caused by the excessive accumulation of adipose tissue is one notable health issue [8]. The increase in obesity cells is caused by cell division and proliferation. Some polysaccharides are used to control the proliferation and differentiation of fat cells, thereby achieving anti-obesity effects [9]. 3T3-L1 preadipocytes are widely used in the study of anti-obesity drug application because they can differentiate mature fat cells in vitro [10]. However, their strong proliferation capacity seems to be ignored. The proliferation of preadipocytes is associated with the formation of adipocytes, and obesity can be treated by inhibiting the proliferation of preadipocytes [11]. Previous research reported that quinoa polysaccharide significantly inhibited 3T3-L1 preadipocyte differentiation to achieve an anti-obesity effect [12]. Elucidating the antiproliferative mechanism of quinoa polysaccharides on 3T3-L1 preadipocytes can clarify the signaling pathway, which provides a basis for quinoa anti-obesity functional food. In addition, anti-obesity drugs have some side effects. Quinoa polysaccharides are safer compared to anti-obesity drugs. The mechanism that quinoa polysaccharides use for inhibiting the proliferation of 3T3-L1 preadipocytes deserves an in-depth study.

There were several prior studies conducted to analyze the transcriptome sequencing of adipocyte and tumor cells in order to identify functional genes. In Ying's study, transcriptome analysis was used to reveal that Neurotrophin, VEGF, and the P53 signaling pathway were identified and related to phycocyanin's inhibitory effects on SKOV-3 cells [13]. Smilax glabra Roxb (SGF) can induce the expression of genes involved in triglyceride storage, fatty acid β-oxidation, and mitochondrial biogenesis [14]. Other related transcriptional changes are related to the AMPK/PGC-1α signaling and inflammation and PI3K/AKT signaling pathways.

This study was conducted to investigate the impact of quinoa polysaccharides on the proliferation of 3T3-L1 preadipocytes. High-throughput sequencing technology was used to analyze transcripts of different concentrations of quinoa water-extracted polysaccharides (QWPs), as well as quinoa alkaline-extracted polysaccharide (QAP)-treated and untreated groups at the transcriptome level. The mechanism used by quinoa polysaccharides for regulating 3T3-L1 preadipocyte proliferation was elaborated. It is helpful to provides a basis for the development of a quinoa anti-obesity agent.

2. Materials and Methods

2.1. Sample and Reagents

Quinoa was obtained from the Institute of Crop Sciences, the Chinese Academy of Agricultural Sciences. Dulbecco's modified Eagle's medium (DMEM), Penicillin–Streptomycin 10,000 U/mL, fetal bovine serum (FBS), Trypsin–EDTA Solution, and α-amylase were purchased from Sigma-Aldrich (St. Louis, MO, USA). Ethanol, N-hexane, sodium hydroxide, HCl, chloroform, and N-butanol were of analytical grade, and they were purchased from Beijing Chemistry Industry Group Co., Ltd. (Beijing, China).

2.2. Quiona Polysaccharide Preparation

Methods of polysaccharide extraction and purification were used according to previous research [2,12]. Quinoa seeds were milled and sieved over 60 mesh, then soaked in N-hexane for 6 h to remove lipids and pigments. The QWP extraction method was employed as follows. We skimmed quinoa powder and extracted it twice at 1:10 with distilled water (90 °C, 4 h). The QAP extraction method was employed as follows. Sodium hydroxide (1.25 mol/L) was added to the pre-treated skimmed quinoa powder at 1:10 (*w*/*v*) and extracted twice for 3 h. The extract was mediumed to pH 7.0 and centrifuged (4000× *g*, 20 min, 4 °C) using a centrifuge (H1850R, Cence, Changsha, China). The supernatant was combined with ethanol for 12 h at a ratio of 1:5 at 4 °C, then centrifuged (5000× *g*, 15 min, 4 °C) to collect sediment. The precipitate was then deproteinized by using the Sevage method. To remove starch and perform dialysis, the fraction was fully treated with α-amylase, and dialysis was performed. The QAP and QWP fractions were obtained by freeze-drying using a freeze-drier (LGJ-12, Songyuan, China) from Beijing Songyuan Huaxing Technology Development Co., Ltd. The QAP and QWP fractions were separated and purified by an ion-exchange column and a sephadex column. In brief, the eluting components were dialyzed and freeze-dried to obtain a column chromatography separation of DEAE Sepharose Fast Flow. Further purification was carried out using the Sephacryl S-300 system to obtain a new component.

2.3. Measurement of Anti-Proliferation Activity

3T3-L1 preadipocytes were purchased from the Chinese Academy of Sciences Shanghai Cell Bank (Shanghai, China), cultured in DMEM supplemented with 4.5 g L^{-1} D-glucose, 10% FBS, and 1% penicillin at 37 °C containing 5% CO_2. The cytotoxic effects of QAP and QWP on 3T3-L1 cells were measured following the previous method [15]. The 3T3-L1 adipocyte cells were seeded in a 96-well plate (5×10^4 cells/well). After 18 h, the medium was replaced with DMEM containing 100 μL samples at varying concentrations. After 24 h, the culture medium was replaced with 100 μL of MTT solution (1 mg/mL). The plates were then incubated for an additional 4 h at 37 °C and 5% CO_2. The MTT solution was removed, and 100 μL of dimethylsulfoxide (DMSO) was added for 30 min. The absorbance was measured at 570 nm. For the anti-proliferation assay, the 3T3-L1 adipocyte cells were cultivated in a 96-well plate (2.5×10^4 cells/well). After 18 h, the medium was replaced with fresh DMEM containing 100 μL of samples at varying concentrations. Following a 72 h incubation at 37 °C in an environment with 5% CO_2, the cells were subjected to staining with MTT for a duration of 30 min. Cell proliferation was quantified by measuring the absorbance at 570 nm using a multi-plate reader (Synergy H1, BioTek, Winooski, VT, USA).

2.4. Effect of Quinoa Polysaccharides on Transcriptional Level of 3T3-L1 Preadipocytes

The high-throughput sequencing platform (NovaSeq 6000, Illumina, Shanghai, China) was utilized for transcriptome sequencing [16]. FastQC (0.11.9) was used to analyze the data for quality control, and the high-quality sequences were compared with the reference genome in the Ensembl database to obtain the expression level of each gene [17]. Differentially expressed genes (DEGs) between untreated and quinoa polysaccharide-treated 3T3-L1 cell lines were identified. The condition was the differential expression multiple |log2foldchange| > 1, significance p-value < 0.05. Volcano and MA plots of DEGs were plotted using the ggplots2 software (3.4.1) package in R-Project. The distance was calculated using the Euclidean method, and the clustering was performed using the hierarchical clustering longest distance method. We mapped all genes that showed differential expression using the GO database. The term enriched by DEGs was calculated using hypergeometric distribution, with the whole genome as a reference. The count of DEGs at various levels of the Kyoto Encyclopedia of Genes and Genomes (KEGG) pathway was counted at various levels, and the primary metabolic and signaling pathways associated with these DEGs were identified. The pathways for significant enrichment of differential genes were calculated in the context of the whole genome. The hypergeometric distribution method was used to calculate the p-value ($p < 0.05$).

2.5. Quantitative RT-PCR (qPCR)

Q-PCR analysis was performed using total RNA extracted from 3T3-L1 preadipocytes that were cultured for 48 h. The reverse transcription process utilized random hexamer primers and was conducted using a reverse transcription system (TaKaRa, RR036A, Beijing, China). Gene-specific primers were generated using Primer Premier software (6.25), and GAPDH was employed as the internal reference gene for quantifying gene expression in qRT-PCR (Table 1).

Table 1. Primers and sequences used in qRT-PCR analysis.

Gene	Primers	Sequences (5′→3′)
Atp13A4	Atp13A4-F	CACGTATGGGCACATTGTGTC
	Atp13A4-R	TGAGACCAAATGCGCTGTTTA
Gbgt1	Gbgt1-F	TGGGTGTATCTTGAGAACTGGC
	Gbgt1-R	GTACTGTGACCATACCACGGG
GAPDH	GAPDH-F	GGGAGCCAAAAGGGTCATCA
	GAPDH-R	TGATGGCATGGACTGTGGTC

2.6. Data Statistics

All data were measured in three replicates (means ± SD), and the analysis of variance (ANOVA) was followed by Duncan's multiple-range test. The statistical analyses were conducted using SPSS version 22.0. Statistical significance was defined as $p < 0.05$.

3. Results and Discussion

3.1. Anti-Proliferation Effect of Quinoa Polysaccharides on 3T3-L1 Preadipocytes

The cytotoxic and anti-proliferation effects of QWP and QAP on the 3T3-L1 preadipocytes were measured. QWP and QAP had no cytotoxic effects on 3T3-L1 preadipocytes within 0.25 to 4 mg/mL (Figure 1A). QAP showed stronger inhibitory activity than QWP at high concentrations, except at 0.25 mg/mL. At 4 mg/mL, QAP inhibited the proliferation of 3T3-L1 preadipocytes by 38.94%, which was higher than that of QWP (29.05%) (Figure 1B). In Figure 1C, after being treated with QWP and QAP, the number of 3T3-L1 preadipocytes was significantly reduced (Figure 1C).

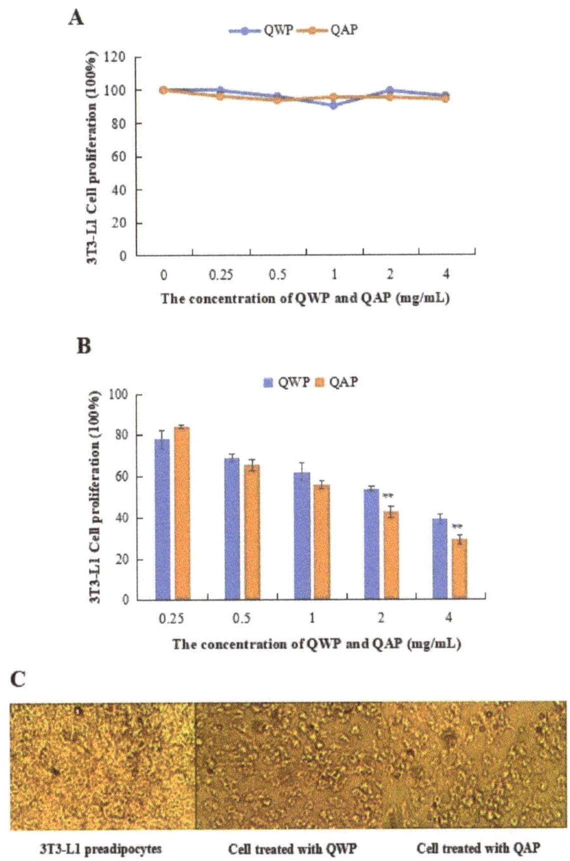

Figure 1. Anti-proliferative activity analysis of quinoa polysaccharides. (**A**) Cytotoxic effect of QWP and QAP on 3T3-L1 adipocytes. (**B**) QWP and QAP dramatically suppressed cell growth in the anti-proliferation experiment. (**C**) Topography of 3T3-L1 adipocytes and QWP- and QAP-treated samples. ** $p < 0.01$ shows significant differences. QAP represents quinoa alkaline-extracted polysaccharides. QWP represents quinoa water-extracted polysaccharides.

In the 3T3-L1 preadipocyte differentiation experiment, treatment with quinoa polysaccharides could inhibit PPARγ, C/EBPα, C/EBPβ, and C/EBPδ [12]. In addition, previous studies found that quinoa polysaccharide showed an inhibition effect on the proliferation of SMMC 7721 cells [18]. However, the mechanism of inhibiting cell proliferation is not clear and is worthy of further study. Apoptosis is considered a kind of programmed cell death that can lead to characteristic changes, including blistering, cell contraction, nuclear fragmentation, chromatin condensation, and chromosome DNA fragmentation [19]. Cordyceps polysaccharide inhibits the expression of Cyclin E and CDK2, promotes cell apoptosis, and inhibits the proliferation of Hela cells [20]. Mushroom polysaccharides are rich in β-glucan, which can decrease the levels of lipid hydrogen peroxide in 3T3-L1 cells and effectively stimulate lipogenesis and lipolysis [21].

In order to further explore the global intracellular regulation of 3T3-L1 cells and reveal their anti-obesity function and possible molecular mechanisms, 3T3-L1 cells in the control group, 2 mg/mL, and 4 mg/mL QWP- and QAP-treated groups were collected and isolated for subsequent transcriptome analysis. Adipose tissue is an important endocrine organ in the body. It can secrete numerous adipokines to regulate energy metabolism, glucose and lipid metabolism, vascular homeostasis, and immune and reproductive processes in the body. As an important active ingredient, it regulates a variety of biological processes, including the regulation of blood pressure, immune responses, neurotransmitter transmission, and cell proliferation. We investigated the effects of two quinoa polysaccharides on the proliferation of 3T3-L1 preadipocytes, and the results showed that quinoa polysaccharides could inhibit the proliferation of 3T3-L1 preadipocytes, and QAP had a better inhibitory effect than QWP. The 3T3-L1 cell differentiation process is widely used in the study of adipocyte differentiation. However, there are few studies on the molecular mechanism of bioactive components affecting 3T3-L1 cells by transcriptomic technology. 3T3-L1 preadipocytes are widely used as cell models for studying obesity. Ginsenoside Rg1 could improve the phosphorylation level of AMP-activated protein kinase (AMPK) in 3T3-L1 cells to inhibit adipogenesis and reduce intracellular lipid content [22].

3.2. *Analysis of Transcriptome Sequencing Results*

In total, 48,878,089, 49,409,051, 55,533,208, 51,156,189, and 53,245,847 clean reads were recognized from QAPHs, QAPLs, QWPHs, QWPLs, and the control group (CT), respectively. The extracted total RNA is shown in Figure S1. The electrophoresis results of all RNA samples showed two bands of 28S and 18S. In addition, the values of RIN and 28S/18S in all RNA samples were more than 9.8 and 1.6 (Table S1). All these results showed that the RNA extracted in this experiment is of high quality and can be used in subsequent experiments. The mapping ratios were both over 95.8%, respectively, showing significant levels of gene expression (Table S2). The base content distribution is generally used to determine the presence of separation between A and T, G and C. And it is basically stable throughout the entire sequencing process, showing a horizontal line. The use of 6 bp random primers during reverse transcription synthesis of cDNA can lead to normal variations that result in a specific preference for the nucleotide composition of the initial locations (Figure S2). Between CT and QWPHs, 2194 genes were significantly expressed, including 653 upregulated and 1541 downregulated. Between CT and QAPHs, there were 1774 DEGs, including 728 upregulated and 1046 downregulated (Figures 2–4). The number of DEGs was determined by summing the number of genes in each colored circle. The Venn plot illustrates the common DEGs between the two control groups (Figure 3A).

There were three levels in the expression analysis of the transcriptome, including gene, transcript, and exon expression levels [23]. Diverse splicing variants of a single gene can give rise to distinct biological consequences. The FPKM density distribution offers a comprehensive examination of the expression patterns of all genes in the sample. Most genes exhibit moderate expression levels, with a smaller fraction being expressed at either a low or high level. Our results correspond to these descriptions (Figure S3).

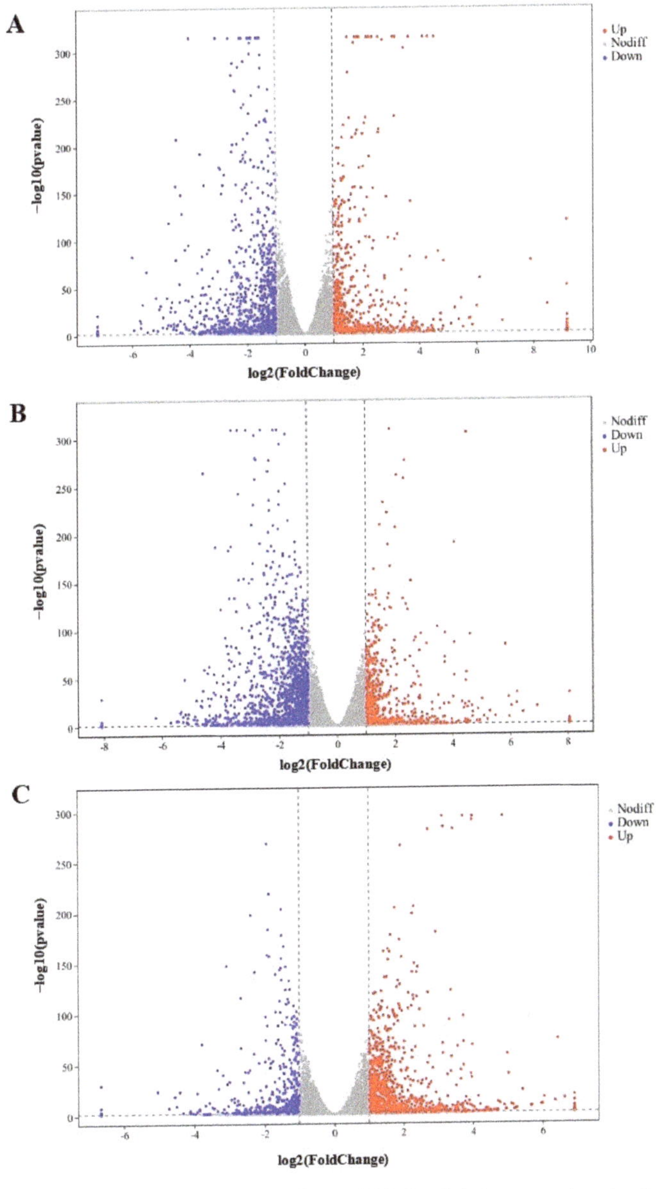

Figure 2. Genes that are expressed at variable levels between various treatment groups. The horizontal axis represents the logarithmic changes in gene expression among different groups. (**A**) CT vs. QAPHs; (**B**) CT vs. QWPHs; (**C**) QAPHs vs. QWPHs. The discrepancy was more pronounced when the adjusted p-value was less and the −log10 (adjusted p-value) was larger. The figure shows that the vertical bar represents a difference threshold that is 1.5 times higher, while the horizontal bar represents a threshold of statistical significance at $p < 0.05$. Distinct color splashes represented various genes, with gray dots indicating genes without notable variation, red dots indicating significantly upregulated genes, and blue dots relating to downregulated genes. CT represents the control group. QAPH represents high concentrations of quinoa alkaline-extracted polysaccharides. QWPH represents high concentrations of quinoa water-extracted polysaccharides.

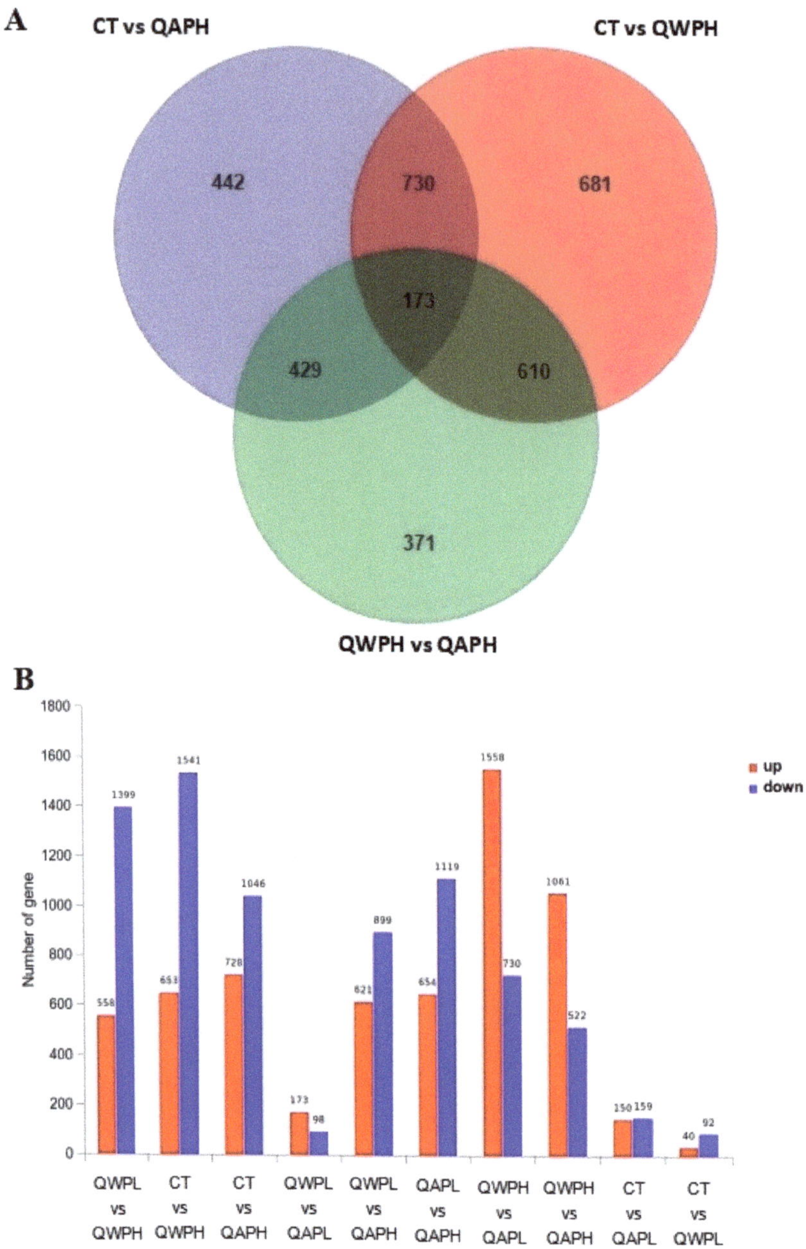

Figure 3. (**A**) Venn diagram illustrating the number of genes differing in overlap among comparison groups. (**B**) The upregulation and downregulation of genes.

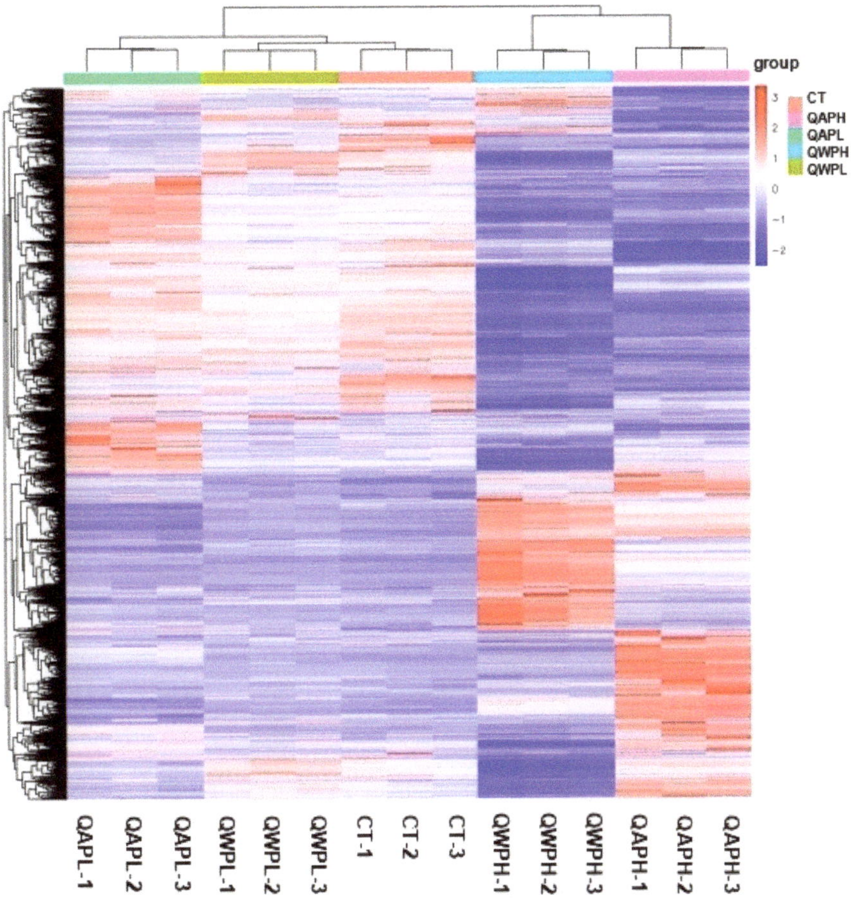

Figure 4. Heat map of mRNA transcripts showing hierarchical clustering of altered mRNA transcripts in 3T3-L1 preadipocytes with or without polysaccharide (QAP and QWP) treatment. Up- and downregulated genes are in red and blue, respectively. QAP represents quinoa alkaline-extracted polysaccharides. QWP represents quinoa water-extracted polysaccharides.

In this study, the correlation coefficients between the biologically replicated samples are higher than 0.9, indicating that the correlation between the samples is high (Figure S4). In addition, PCA analysis shows high similarity between samples. Recently, transcriptome sequencing technology has developed rapidly and is widely used to identify cell engineering. Designing a transcriptome-guided cell engineering system requires a large number of samples to accurately predict key genes.

3.3. Differential Expression Analysis

Obesity is a complex biological process. A better understanding of various signaling molecules will help elucidate the mechanisms that inhibit the proliferation and development of 3T3-L1 cells. Previous transcriptome sequencing showed that Isocitrate dehydrogenase 2 (Idh2), fatty acid synthase (Fasn), Lpin1, and lipoproteinlipase (Lpl) were affected by all-trans retinoic acid and made to undergo significant changes, indicating that these genes may be inhibiting the differentiation of 3T3-L1 cells and play a key role [24]. In this study, ten DEGs were identified after QAP and QWP treatments compared with the blank group. Fold changes of ATPase 13A4 (ATP13A4) between CT and QAPLs, CT and

QAPHs were 7.231 and 574.614. It is worth noting that ATP13A4 expression decreased after treatment with QWP. Fold changes of ATPase 13A4 (ATP13A4) between CT and QAPLs, CT and QAPHs were 3.579 and 65.734. ATP13A4 is a cation-transporting P5-type ATPase that is linked to neurodevelopmental disorders. Studies have shown that the loss of ATP13A4 is related to neuropsychiatric phenotypes [25]. In this study, after quinoa polysaccharide treatment, the expression of ATP13A4 increased, indicating that QWP and QAP can inhibit the symptoms of obesity caused by neurodevelopmental disorders. The Globoside alpha-1,3-N-acetylgalactosaminyltransferase 1 (Gbgt1) gene is involved in encoding forssman glycosphingolipid synthase, and its presence leads to the expression of FORS1 glycosphingolipids in human red blood cells [26]. The Gbgt1 expression of 3T3-L1 cells treated with QAP and QWP was lower than that in the untreated group. Fold changes of GBGT1 between CT and QAPLs, CT and QAPHs were 0.92 and 0.06. Fold changes of GBGT1 between CT and QWPLs, CT and QWPHs were 0.604 and 0.024. Moreover, the expression of GBGT1 was more inhibited after high-concentration quinoa polysaccharide treatment than after low-concentration quinoa polysaccharide treatment (Tables 2 and 3).

Table 2. Fold changes of DEG treated with QAP.

Gene id	Name	CT vs. QAPLs	CT vs. QAPHs	QAPLs vs. QAPHs
ENSMUSG00000038094	ATP13A4	7.231	574.614	79.74
ENSMUSG00000051228	Nyx	2.656	69.157	26.1
ENSMUSG00000057933	Gsta2	1.478	37.748	25.636
ENSMUSG00000021062	Rab15	1.697	24.484	24.484
ENSMUSG00000111709	Gm3776	1.176	28.313	24.15
ENSMUSG00000018656	Tcaf3	1.02	22.454	22.121
ENSMUSG00000026012	Cd28	1.026	20.365	19.944
ENSMUSG00000063851	Rnf183	12.741	242.082	19.058
ENSMUSG00000038963	Slco4a1	1.429	25.326	17.766
ENSMUSG00000056457	Prl2c3	1.02	17.553	17.26
ENSMUSG00000041559	SLRR2E	0.739	0.019	0.026
ENSMUSG00000020676	Ccl11	0.86	0.033	0.038
ENSMUSG00000027220	Syt13	0.798	0.036	0.045
ENSMUSG00000040170	Fmo2	0.611	0.029	0.048
ENSMUSG00000037206	Islr	0.862	0.046	0.053
ENSMUSG00000078922	Tgtp1	0.347	0.019	0.667
ENSMUSG00000042436	Mfap4	0.813	0.047	0.058
ENSMUSG00000026829	Gbgt1	0.92	0.06	0.06
ENSMUSG00000020053	Igf1	0.734	0.045	0.061
ENSMUSG00000034009	Rxfp1	0.264	0.016	0.062

Table 3. Fold changes of DEG treated with QWP.

id	Name	CT vs. QWPLs	CT vs. QWPHs	QWPLs vs. QWPHs
ENSMUSG00000059383	Gfral	1.733	75.096	43.574
ENSMUSG00000020646	Mboat2	2.039	74.532	36.663
ENSMUSG00000038094	Atp13A4	3.579	65.734	18.422
ENSMUSG00000039691	Tspan10	3.079	62.112	20.235
ENSMUSG00000030827	Fgf21	6.198	61.422	9.954
ENSMUSG00000029797	Sspo	2.387	56.769	23.88
ENSMUSG00000040026	Saa3	7.253	44.579	6.188
ENSMUSG00000026822	Lcn2	2.594	39.139	15.152
ENSMUSG00000009356	Lpo	3.028	37.007	12.247
ENSMUSG00000034402	Kcnh5	2.638	33.928	12.916

Table 3. Cont.

id	Name	CT vs. QWPLs	CT vs. QWPHs	QWPLs vs. QWPHs
ENSMUSG00000031364	Grpr	0.679	0.004	0.005
ENSMUSG00000022468	Endou	0.855	0.019	0.023
ENSMUSG00000024810	Il33	0.85	0.024	0.029
ENSMUSG00000027238	Frmd5	0.908	0.028	0.031
ENSMUSG00000048368	Omd	0.806	0.029	0.036
ENSMUSG00000031554	Adam5	0.956	0.036	0.037
ENSMUSG00000085584	Rtl9	0.884	0.033	0.038
ENSMUSG00000024059	Clip4	0.889	0.034	0.038
ENSMUSG00000026829	Gbgt1	0.604	0.024	0.04
ENSMUSG00000048782	Insc	0.538	0.022	0.041

3.4. GO Analysis of DEGs

Between the CT and QAPL treatment groups, nineteen categories and one category were included in the biological process and cellular component, respectively, in which the response to the interferon-beta pathway showed the most significant difference (Figure 5A). The comparison between the CT and QAPH treatment groups revealed that there were four categories included in the cellular component, fourteen categories included in the biological process, and two categories included in the molecular function. Among these categories, the extracellular region exhibited the most significant difference (Figure 5B). Between the CT and QWPL treatment groups, sixteen categories, three categories, and one category were included in biological processes, cellular components, and molecular function. Defense response was the most significant pathway (Figure 6A). Between the CT and QWPH treatment groups, three, three, and fourteen categories were included in the cellular component, molecular function, and biological process. It was found that the extracellular matrix was the most significant change pathway (Figure 6B). The GO enrichment analysis of DEGs revealed that the cellular process in 3T3-L1 cells was the function most significantly affected by quinoa polysaccharides.

3.5. KEGG Pathway Analysis of DEGs

Pathway significance enrichment analysis identified the involvement of many genes in biochemical metabolic and signal transduction pathways. To further understand the biological functions involved in the differential genes, KEGG analysis was performed on these 2194 differential genes.

Through the significant enrichment analysis of the pathway, the metabolic and signal transduction pathways involved in differential gene involvement were identified. There were 212 metabolic pathways with significant expression differences identified in the CT and QAPLs groups, 35 of which showed the most significant differences ($p < 0.05$, Figure 7A). A total of 302 metabolic pathways with significant expression differences were identified between the CT and QAPHs, 35 of which showed the most significant differences ($p < 0.05$, Figure 7B). A total of 84 metabolic pathways with significant expression differences were identified between the CT and QWPLs, 12 of which showed the most significant differences ($p < 0.05$, Figure 8A). A total of 310 metabolic pathways with significant expression differences were identified between the CT and QWPHs, 57 of which showed the most significant differences ($p < 0.05$, Figure 8B). Significantly enriched pathways included Herpes simplex infection, NOD-like receptor signaling pathway, Hepatitis C, Influenza A, and PI3K-Akt signaling pathway, which is associated with human diseases, immunology pathways metabolism, genetic information processing, immunity, apoptosis, proliferation, and adhesion. The compound has the potential to trigger programmed cell death in 3T3-L1 preadipocytes by blocking the activity of nuclear factor kappa-B (NF-κB) and modifying the mitogen-activated protein kinases (MAPKs) pathway [11]. The effect of tuckahoe flavonoids was determined on 3T3-L1 cells by transcriptome technology, and the sample had no cytotoxic effect at a concentration of 250 μg/mL [14]. A total of 1529 genes

(657 upregulated and 872 downregulated) were significantly differentially expressed after cocos flavonoids treatment. The GO enrichment analysis revealed a significant enrichment in extracellular matrix tissue, aging, and inflammatory responses between blank and sample tissues. It was found that the key pathways of all genes mostly include the lysosome, focal adhesion, TGF-β signaling pathway, calcium signaling route, ECM-receptor interaction, and PI3K-Akt signaling pathway, as identified using KEGG analysis. This result is consistent with this study, which strongly confirms the mechanism of the effects of quinoa polysaccharide components QWP and QAP on 3T3-L1 cells, laying a foundation for the potential therapeutic utility in the prevention of obesity and associated metabolic disorders. The PI3K/AKT pathway contributes to regulating cell proliferation [27]. NOD-like receptors (NLRs) are recognized as essential regulators in disorders related to inflammation [28]. In CT vs. QAPHs, the proteasome pathway is significantly enriched, which is related to the pathogenesis of many diseases. Cell proteolysis is carried out by a complex cascade of enzymes and shows a high degree of specificity for many of its substrates. This study demonstrates that polysaccharides derived from quinoa have the ability to hinder the growth and cell division of 3T3-L1 cells by inhibiting the PI3K/AKT pathway and the NOD-like receptor signaling pathway.

3.6. Structural Analysis

At the genetic level, sequencing reads can be conducted by comparing the reference genome for expression analysis, expression difference analysis, and enrichment analysis, and a series of structural analyses can also be performed on the transcript level. We compare the spliced transcript sequence with known transcripts to obtain transcripts without annotation information. As shown in Figure S5, 91.7% belong to new potential transcripts with at least one alternative splicing site shared with known transcripts. The transcript fragment is completely in a known intron, which accounts for 2.8% and 2.1%, respectively. Exons are overlaid on the anti-strand of a known gene, and 3.3% of them belong to an unknown fragment.

3.7. Transcription Factor Analysis

Transcription Factor (TF) is a protein molecule that has the ability to selectively attach to a certain sequence located upstream of the 5′ end of a gene. It forms a complex with RNA polymerase II, which initiates the process of transcription [29]. In this study, the DEGs predicted to be transcription factors were counted. According to the family information, Figure S6 shows the number of differentially expressed transcription factors in the blank group and each sample group. The results indicated that the Zf-C2H2 transcription factor family had the largest number of enriched differential transcription factors. In the comparison of CT and QWPL treatment groups, 12 upregulated and 23 downregulated transcription factors were identified. The Zf-C2H2 transcription factor family has the most enriched differential transcription factors between CH and QWPH treatment groups, including seven upregulated and nineteen downregulated transcription factors. Zinc finger (Zf) is a compact protein structure that is defined by the binding of one or more zinc ions (Zn^{2+}) to promote folding and contains a variety of different protein structures. Despite the wide range of these proteins, the vast majority typically serve as interacting modules that bind DNA, RNA, proteins, or other useful small molecules, and changes in structure are primarily employed to modify the binding specificity of a particular protein.

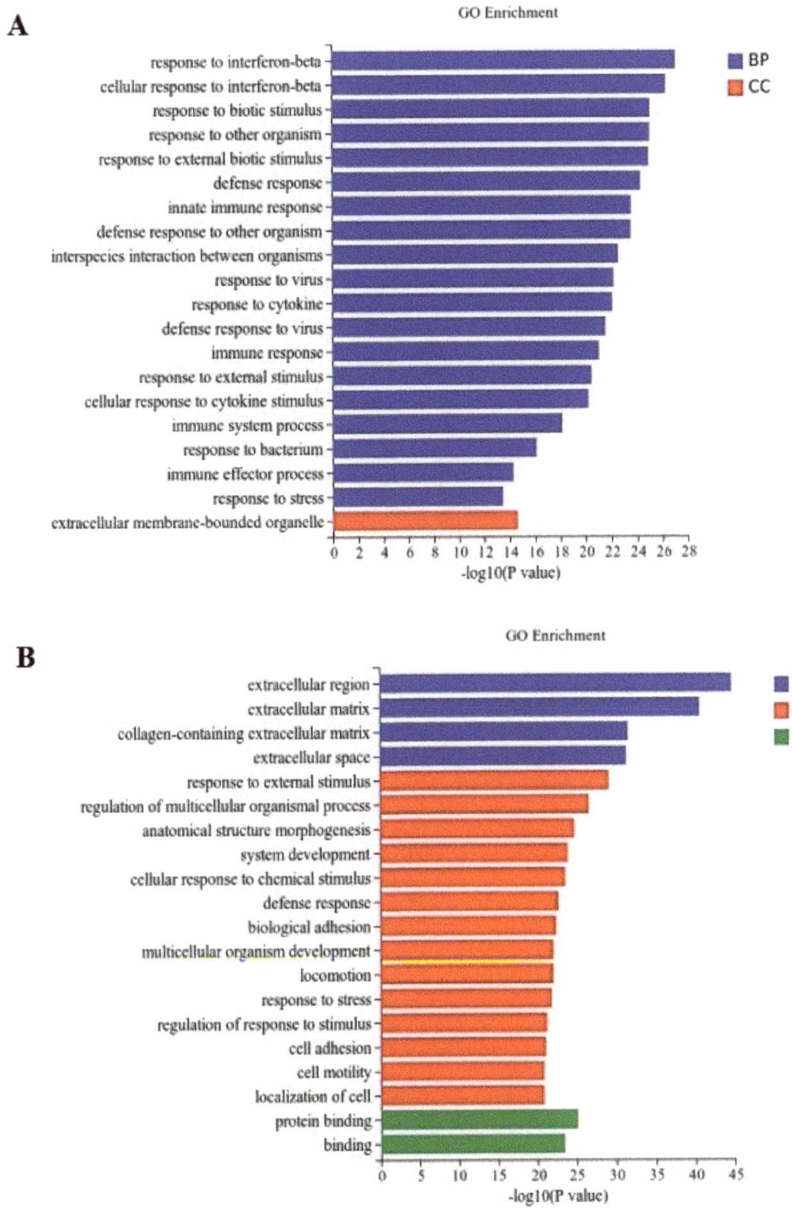

Figure 5. Histogram difference genes' GO enrichment between CT and QAP treatment. The abscissa is −log10 (*p*-value) enriched by GO Term, and the ordinate is GO Term. CC stands for cellular component, BP stands for biological process, and MF stands for molecular function. (**A**) CT vs. QAPLs; (**B**) CT vs. QAPHs. CT represents control group. QAPL represents low-concentration quinoa alkaline-extracted polysaccharide. QAPH represents high-concentration quinoa alkaline-extracted polysaccharide.

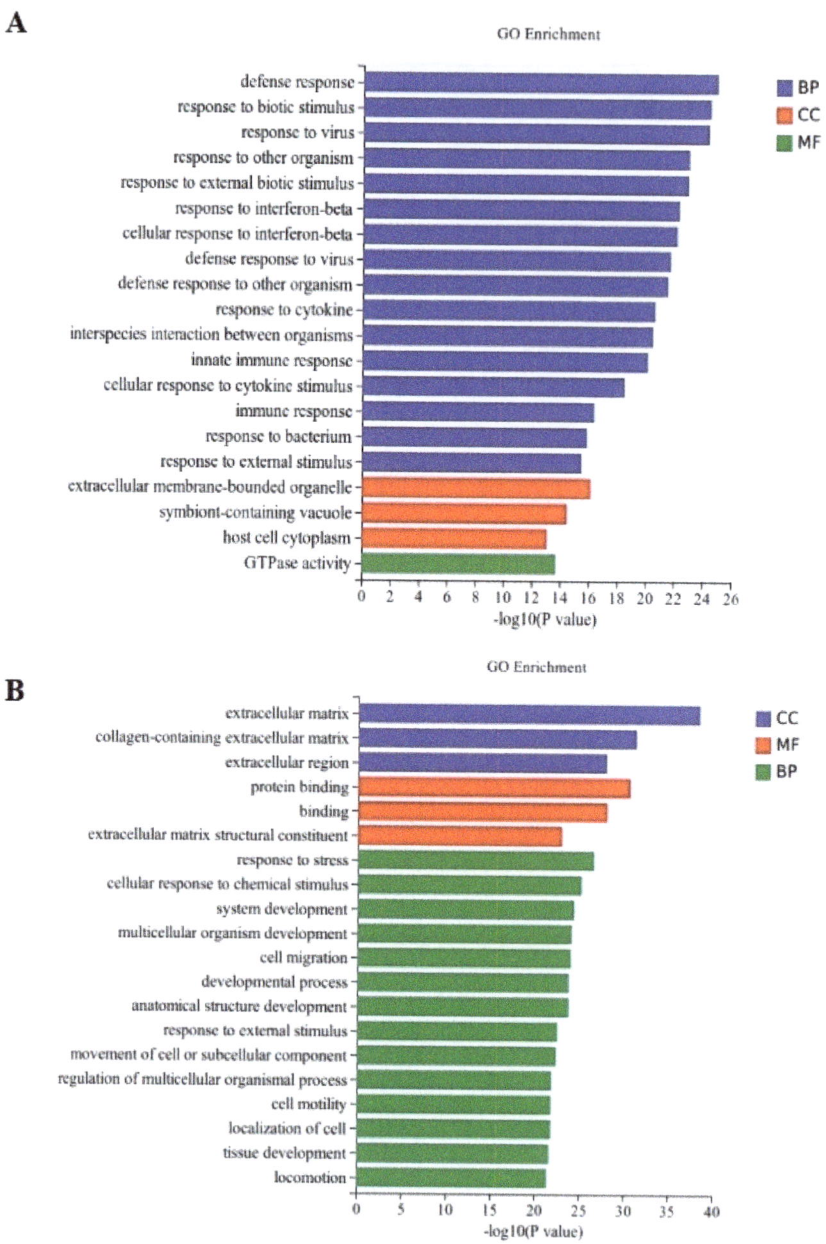

Figure 6. Histogram difference genes' GO enrichment between CT and QWP treatment. The abscissa is −log10 (p-value) enriched by GO Term, and the ordinate is GO Term. CC stands for cellular component, BP stands for biological process, and MF stands for molecular function. (**A**) CT vs. QWPLs; (**B**) CT vs. QWPHs. CT represents control group. QWPL represents low-concentration quinoa water-extracted polysaccharide. QWPH represents high-concentration quinoa water-extracted polysaccharide.

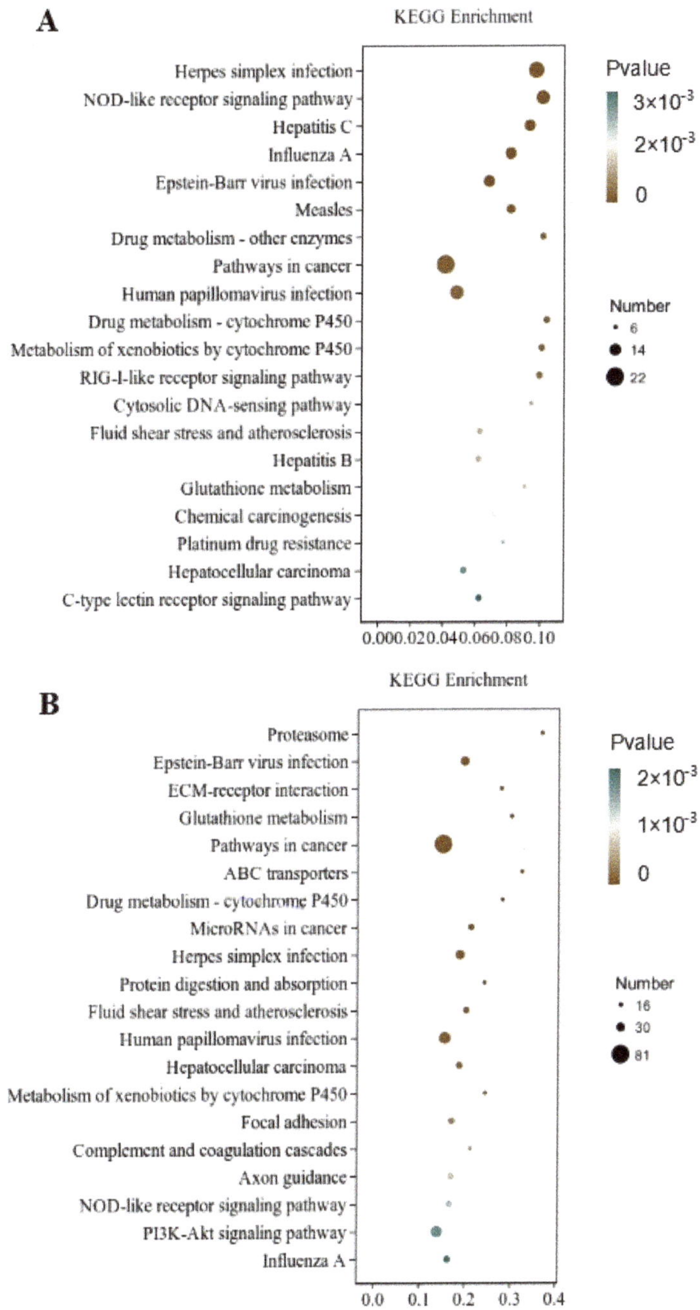

Figure 7. KEGG pathway rich enrichment. (**A**) CT vs. QAPLs; (**B**) CT vs. QAPHs. CT represents the control group. QAPL represents low-concentration quinoa alkaline-extracted polysaccharide. QAPH represents high-concentration quinoa alkaline-extracted polysaccharide.

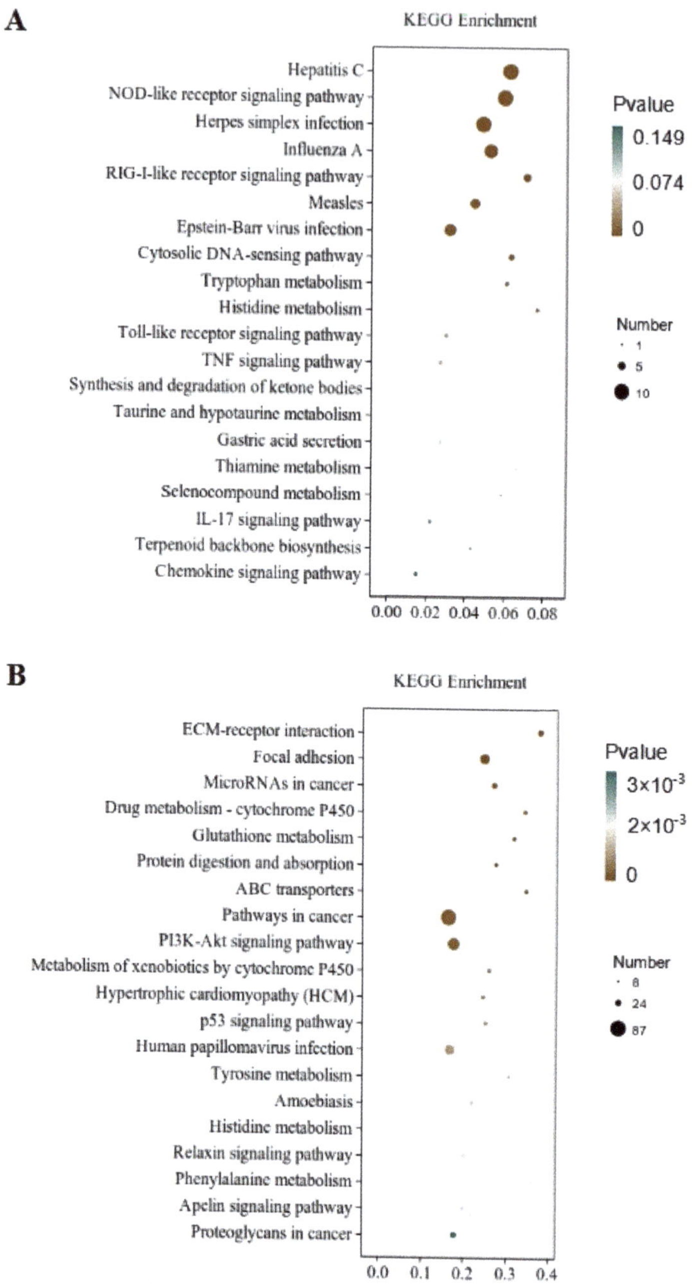

Figure 8. KEGG pathway rich enrichment. (**A**) CT vs. QWPLs; (**B**) CT vs. QWPHs. CT represents the control group. QWPL represents low-concentration quinoa water-extracted polysaccharide. QWPH represents high-concentration quinoa water-extracted polysaccharide.

This study explored the regulatory mechanisms of QWP and QAP inhibiting proliferation of 3T3-L1 preadipocytes. Metabolomics technology should be used to find the intracellular differential metabolites caused by the action of quinoa polysaccharides and to explore the relevant metabolic pathways in the process of differentiation and proliferation.

3.8. Quantitative RT-PCR (qPCR) Validation

To validate the findings of transcriptome sequencing and conduct a more detailed analysis of gene expression patterns related to anti-obesity functions, Atp13A4 and Gbgt1 were selected for quantitative RT-PCR due to their distinct expression patterns. The RT-PCR results demonstrated that the expressions of Atp13A4 were significantly higher ($p < 0.05$) in 3T3-L1 preadipocytes treated with QAPHs and QWPHs compared to the control group. Conversely, the expression of Gbgt1 was significantly reduced ($p < 0.05$, Figure 9) in the treated group.

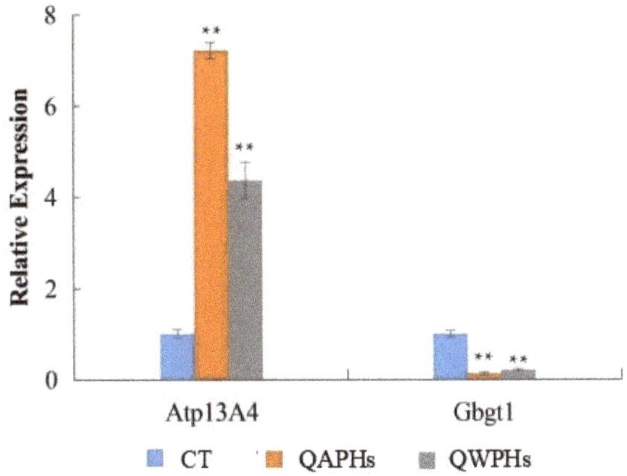

Figure 9. qPCR validation of DEGs. Mean ± SD. $n = 3$. ** $p < 0.01$ vs. CT group. CT represents the control group. QAPH represents the high-concentration quinoa alkaline-extracted polysaccharide. QWPH represents the high-concentration quinoa water-extracted polysaccharide. DEG represents differentially expressed gene.

4. Conclusions

To summarize, the quinoa polysaccharide has a notable ability to prevent the growth of 3T3-L1 preadipocytes in vitro. This impact is achieved through the interplay of many pathways and signaling molecules. Interference with many signaling pathways, such as NOD-like receptor signaling pathway, Herpes simplex infection, Hepatitis C, PI3K-Akt signaling pathway, and many factors, such as Atp13A4 and Gbgt1, affect the proliferation and apoptosis of 3T3-L1 preadipocytes. Subsequently, the differential lipid metabolites produced by quinoa polysaccharides on 3T3-L1 preadipocytes could be analyzed in conjunction with lipidomics to explore their anti-obesity effects. This study established a solid theoretical basis for investigating the anti-obesity mechanisms of QWP and QAP. In addition, quinoa polysaccharides can be further used as a kind of anti-obesity agent.

Supplementary Materials: The following supporting information can be downloaded at: https://www.mdpi.com/article/10.3390/foods13152311/s1, Figure S1: Agarose gel electrophoresis of total RNA, M: Marker, 1–3: CT, 4–6: QWPLs, 7–9: QWPHs, 10–12: QAPLs, 13–15: QAPHs; Figure S2: Distribution of basic group. The abscissa shows the base position in Reads (5′–>3′). The ordinate shows the proportion of bases. (a) CT, (b) QAPLs, (c) QAPHs, (d) QWPLs, (e) QWPHs; Figure S3: (a) The abscissa shows the log10 (FPKM) value of the gene, and the ordinate shows the gene distribution density corresponding to the expression level, (b) PCA analysis between samples. Different colors represent different groups; Figure S4: correlation of gene expression levels between samples; Figure S5: New transcript analysis. (a) New Transcript Class Code Distribution (i: the transcript fragment is completely in a known intron, j: new potential transcripts: at least one alternative splicing site shared with known transcripts, x: Exons are overlaid on the anti-strand of a known gene, u: unknown fragment, transcript in the intergenic region). (b) Sequence number of different databases; Figure S6: (a) Analysis of transcription factors; Differentially expressed transcription factor distribution, (b) CT vs. QAPLs, (c) CT vs. QAPHs, (d) CT vs. QWPLs, (e) CT vs. QWPHs; Table S1: The values of RIN and 28S/18S in all RNA samples.; Table S2: Quality of sequencing data.

Author Contributions: Conceptualization, C.T. and S.G.; methodology, C.T. and S.G.; validation, Y.L.; formal analysis, resources, G.R.; data curation, C.T.; writing—original draft preparation, C.T.; writing—review and editing, Y.L. and G.R.; supervision, G.R.; funding acquisition, G.R. All authors have read and agreed to the published version of the manuscript.

Funding: This research was funded by China Postdoctoral Science Foundation under Grant Number (2023M731402), Jiangsu Funding Program for Excellent Postdoctoral Talent (2022ZB765), and Tianfu Scholar Project (2022-17) with Chengdu University.

Institutional Review Board Statement: Not applicable.

Informed Consent Statement: Not applicable.

Data Availability Statement: The original contributions presented in the study are included in the article/Supplementary Materials, further inquiries can be directed to the corresponding authors.

Acknowledgments: We also thank Shanghai Personal Biotechnology Co., Ltd. for technical support.

Conflicts of Interest: The authors declare no conflicts of interest.

References

1. Ren, G.; Teng, C.; Fan, X.; Guo, S.; Zhao, G.; Zhang, L.; Zou, L. Nutrient composition, functional activity and industrial applications of quinoa (*Chenopodium quinoa Willd.*). *Food. Chem.* **2023**, *410*, 135290. [CrossRef] [PubMed]
2. Teng, C.; Qin, P.; Shi, Z.; Zhang, W.; Yang, X.; Yao, Y.; Ren, G. Structural characterization and antioxidant activity of alkali-extracted polysaccharides from quinoa. *Food. Hydrocoll.* **2020**, *113*, 106392. [CrossRef]
3. Tan, M.; Zhao, Q.; Zhao, B. Physicochemical properties, structural characterization and biological activities of polysaccharides from quinoa (*Chenopodium quinoa willd.*) seeds. *Int. J. Biol. Macromol.* **2021**, *193*, 1635–1644. [CrossRef] [PubMed]
4. Vilcacundo, R.; Miralles, B.; Carrillo, W.; Hernandez-Ledesma, B. In vitro chemopreventive properties of peptides released from quinoa (*Chenopodium quinoa willd.*) protein under simulated gastrointestinal digestion. *Food Res. Int.* **2018**, *105*, 403–411. [CrossRef]
5. Tang, C.; Wang, Y.; Chen, D.; Zhang, M.; Xu, J.; Xu, C.; Liu, J.; Kan, J.; Jin, C. Natural polysaccharides protect against diet-induced obesity by improving lipid metabolism and regulating the immune system. *Food Res. Int.* **2023**, *127*, 113192. [CrossRef]
6. Zhu, F. Dietary fiber polysaccharides of amaranth, buckwheat and quinoa grains: A review of chemical structure, biological functions and food uses. *Carbohydr. Polym.* **2020**, *248*, 116819. [CrossRef]
7. Verboven, K.; Wouters, K.; Gaens, K.; Hansen, D.; Bijnen, M.; Wetzels, S.; Stehouwer, C.D.; Goossens, G.H.; Schalkwijk, C.G.; Blaak, E.E. Abdominal subcutaneous and visceral adipocyte size, lipolysis and inflammation relate to insulin resistance in male obese humans. *Sci. Rep.* **2018**, *8*, 4677. [CrossRef]
8. Dodd, G.T.; Xirouchaki, C.E.; Eramo, M.; Mitchell, C.A.; Andrews, Z.B.; Henry, B.A. Intranasal targeting of hypothalamic ptp1b and tcptp reinstates leptin and insulin sensitivity and promotes weight loss in obesity. *Cell. Rep.* **2019**, *28*, 2905–2922. [CrossRef] [PubMed]
9. Singh, P.; Rai, S.N. Factors affecting obesity and its treatment. *Obes. Med.* **2019**, *16*, 100140. [CrossRef]
10. Xu, Y.; Du, J.; Zhang, P.; Zhao, X.; Zhu, L. Microrna-125a-5p mediates 3T3-L1 preadipocyte proliferation and differentiation. *Molecules* **2018**, *23*, 317. [CrossRef]
11. Hao, Y.; Qin, S.; Yang, L.; Jiang, J.; Zhu, W. Marmin from the blossoms of citrus maxima (*burm.*) merr. exerts lipid-lowering effect via inducing 3T3-L1 preadipocyte apoptosis. *J. Funct. Foods.* **2021**, *82*, 104513. [CrossRef]
12. Teng, C.; Shi, Z.; Yao, Y.; Ren, G. Structural Characterization of Quinoa Polysaccharide and Its Inhibitory Effects on 3T3-L1 Adipocyte Differentiation. *Foods* **2020**, *9*, 1511. [CrossRef]
13. Ying, J.; Wang, J.; Ji, H.; Lin, C.; Pan, R.; Zhou, L. Transcriptome analysis of phycocyanin inhibitory effects on skov-3 cell proliferation. *Gene* **2016**, *585*, 58–64. [CrossRef]
14. Tu, J.; Deng, L.; Ling, Y.; Zhu, K.; Cai, Z. Transcriptome profiling reveals multiple pathways responsible for the beneficial metabolic effects of Smilax glabra flavonoids in mouse 3T3-L1 adipocytes. *Biomed. Pharmacother.* **2020**, *125*, 110011. [CrossRef]
15. Huang, J.; Qu, Q.; Guo, Y.; Xiang, Y.; Feng, D. Tankyrases/β-catenin signaling pathway as an anti-proliferation and anti-metastatic target in hepatocarcinoma cell lines. *J. Cancer* **2020**, *11*, 432–440. [CrossRef]
16. Quick, J.; Ashton, P.; Calus, S.; Chatt, C.; Gossain, S.; Hawker, J. Rapid draft sequencing and real-time nanopore sequencing in a hospital outbreak of salmonella. *Genome Biol.* **2015**, *16*, 114. [CrossRef]
17. Trapnell, C.; Pachter, L.; Salzberg, S.L. TopHat: Discovering splice junctions with RNA-Seq. *Bioinformatics* **2009**, *25*, 1105–1111. [CrossRef]
18. Hu, Y.; Zhang, J.; Zou, L.; Fu, C. Chemical characterization, antioxidant, immune-regulating and anticancer activities of a novel bioactive polysaccharide from *Chenopodium quinoa* seeds. *Int. J. Biol. Macromol.* **2017**, *99*, 622–629. [CrossRef]

19. Wu, J.; Ye, J.; Kong, W.; Zhang, S.; Zheng, Y. Programmed cell death pathways in hearing loss: A review of apoptosis, autophagy and programmed necrosis. *Cell. Prolif.* **2020**, *53*, e12915. [CrossRef]
20. Xu, J.; Tan, Z.C.; Shen, Z.Y.; Shen, X.J.; Tang, S.M. Cordyceps cicadae polysaccharides inhibit human cervical cancer hela cells proliferation via apoptosis and cell cycle arrest. *Food Chem. Toxicol.* **2021**, *148*, 111971. [CrossRef]
21. Kanagasabapathy, G.; Chua, K.H.; Malek, S.; Vikineswary, S.; Kuppusamy, U.R. Amp-activated protein kinase mediates insulin-like and lipo-mobilising effects of β-glucan-rich polysaccharides isolated from *Pleurotus sajor-caju* (Fr.), singer mushroom, in 3T3-L1 cells. *Food Chem.* **2014**, *145*, 198–204. [CrossRef]
22. Liu, H.; Wang, J.; Liu, M.; Zhao, H.; Yaqoob, S.; Zheng, M.; Cai, D.; Liu, J. Antiobesity Effects of Ginsenoside Rg1 on 3T3-L1 Preadipocytes and High Fat Diet-Induced Obese Mice Mediated by AMPK. *Nutrients* **2018**, *10*, 830. [CrossRef]
23. Darby, A.C.; Armstrong, S.D.; Bah, G.S.; Kaur, G.; Hughes, M.A.; Kay, S.M. Analysis of gene expression from the wolbachia genome of a filarial nematode supports both metabolic and defensive roles within the symbiosis. *Genome Res.* **2012**, *22*, 2467–2477. [CrossRef]
24. Stoecker, K.; Sass, S.; Theis, F.J.; Hans, H.; Michael, W.P. Inhibition of fat cell differentiation in 3T3-L1 pre-adipocytes by all-trans retinoic acid: Integrative analysis of transcriptomic and phenotypic data. *BDQ* **2017**, *11*, 31–44. [CrossRef]
25. Elisa, B.; Eleonora, D.G.; Elga, F.B. A novel 3q29 deletion associated with autism, intellectual disability, psychiatric disorders, and obesity. *Am. J. Med. Genet. Part B* **2015**, *171*, 290–299. [CrossRef]
26. Hult, A.K.; Eoin, M.S.; Mattias, M.; Olsson, M.L. Gbgt1 is allelically diverse but dispensable in humans and naturally occurring anti-fors1 shows an abo-restricted pattern. *Transfusion* **2018**, *8*, 2036–2045. [CrossRef]
27. Osaki, M.; Oshimura, M.; Ito, H. Pi3k-akt pathway: Its functions and alterations in human cancer. *Apoptosis* **2004**, *9*, 667–676. [CrossRef]
28. Liu, P.; Lu, Z.; Liu, L.; Li, R.; Liang, Z.; Shen, M. Nod-like receptor signaling in inflammation-associated cancers: From functions to targeted therapies. *Phytomedicine* **2019**, *64*, 152925. [CrossRef]
29. Wu, A.C.K.; Patel, H.; Chia, M.; Moretto, F.; Frith, D.; Snijders, A.P. Repression of divergent noncoding transcription by a sequence-specific transcription factor. *Mol. Cell.* **2018**, *72*, 942–954. [CrossRef]

foods

MDPI

Article

The Liquid-Fermentation Formulation of *Sanghuangporus sanghuang* Optimized by Response Surface Methodology and Evaluation of Biological Activity of Extracellular Polysaccharides

Yuhan Gao [1,2], Xiaomin Li [1,3], Hui Xu [1,3], Huijuan Sun [4], Junli Zhang [4], Xiaoping Wu [1,3] and Junsheng Fu [1,3,*]

[1] College of Life Sciences, Fujian Agriculture and Forestry University, Fuzhou 350002, China; 13161198688@163.com (Y.G.); lxm2651277345@163.com (X.L.); xuhui189500@163.com (H.X.); fjwxp@126.com (X.W.)
[2] Chemical and Biomolecular Engineering Department, College of Design and Engineering, National University of Singapore, Singapore 117585, Singapore
[3] Mycological Research Center, Fujian Agriculture and Forestry University, Fuzhou 350002, China
[4] Tibet Academy of Agricultural and Animal Husbandry Sciences, Lasa 850000, China; m18898002012@163.com (H.S.); xzyyxh888@163.com (J.Z.)
* Correspondence: fujunsheng81@163.com

Abstract: *Sanghuangporus sanghuang* is a rare fungus growing on mulberry trees that has immense medicinal value. This study aimed to optimize the liquid-fermentation-media formulation and culture conditions for large-scale culturing of *S. sanghuang* by performing one-way testing and response surface methodology. The antioxidant and anticancer activities of the extracellular polysaccharides from *S. sanghuang* were also analyzed. The optimal formulation and growth conditions for *S. sanghuang* were as follows: glucose, 30.2 ± 0.37 g/L; yeast extract, 14.60 ± 0.05 g/L; dandelion powder, 1.24 ± 0.01 g/L; shaker speed, 150 r/min; and temperature, 25 °C. We obtained 13.99 ± 0.42 g/L of mycelium biomass by culturing *S. sanghuang* for 15 days with the optimized formulation. This was 2-fold higher than the mycelial mass obtained with the sub-optimal formulation. The extracellular fungal polysaccharides showed significant antioxidant activity against ABTS and DPPH free radicals, and significantly reduced the in vitro growth and survival of several cancer cell lines. The anticancer activity of the extracellular fungal polysaccharides was significantly higher in the human glioma cells than in other cancer cell lines. In summary, this study optimized the liquid media formulation and conditions for the large-scale culturing of *S. sanghuang*. Furthermore, the extracellular polysaccharides from *S. sanghuang* showed significant antioxidant and anticancer activities.

Keywords: *Sanghuangporus sanghuang*; response surface optimization; extracellular polysaccharides; biological activity

Citation: Gao, Y.; Li, X.; Xu, H.; Sun, H.; Zhang, J.; Wu, X.; Fu, J. The Liquid-Fermentation Formulation of *Sanghuangporus sanghuang* Optimized by Response Surface Methodology and Evaluation of Biological Activity of Extracellular Polysaccharides. *Foods* **2024**, *13*, 1190. https://doi.org/10.3390/foods13081190

Academic Editor: Montserrat Dueñas Paton

Received: 21 March 2024
Revised: 11 April 2024
Accepted: 12 April 2024
Published: 13 April 2024

1. Introduction

Sanghuangporus sanghuang is a medicinal fungus that has been used in traditional medicine for centuries in Southeast Asia [1]. *S. sanghuang* parasitizes on mulberry trees and is found in the north-eastern regions of Asia [2]. *S. sanghuang* is associated with anticancer activities [3] as well as antioxidant and anti-inflammatory properties [4]. The mycelium of *S. sanghuang* is rich in histones and improves sleep [5]. *Sanghuangporus* spp. are rich in polysaccharides, flavonoids, terpenes and other active substances. *Sanghuangporus* spp. polysaccharide is an important active substance of *Sanghuangporus* spp. Intracellular polysaccharides are mainly derived from fermentation broth [6–9]. The extracellular polysaccharides of *S. sanghuang* are widely-studied as bioactive compounds and show a broad range of pharmacological activities, including anticancer [10], hypoglycemic [11], antioxidant [12], and immunity-enhancing [13] properties. Liu et al. [14] found that *Phellinus baumii* polysaccharide can promote the proliferation of RAW264.7 cells, enhance their phagocytic ability, promote their secretion of IL-6 and TNF-α cytokines, and have immunomodulatory activity. Dong et al. [15] have shown that the crude extracellular polysaccharide (EPS) in *Phellinus igniarius* liquid fermentation broth can significantly inhibit the growth of mouse sarcoma S180 and mouse liver cancer H22.

As *S. sanghuang* has immense medicinal value, there is an increased demand for various natural products and health-improving nutritional food supplements containing *S. sanghuang*. Therefore, there is an increasing demand for culturing *S. sanghuang* on a large-scale. However, the yield of wild *S. sanghuang* is poor and insufficient to meet the medical and market demands. Although the cultivation of fungi such as *Sanghuangporus vaninii*, *Inonotus hispidus*, and *Sanghuangporus baumii* has been achieved successfully, the cultivation of *Sanghuangporus sanghuang* has remained difficult. Because the types of mulberry trees and habitats (temperature, humidity, light, oxygen concentration, etc.) that parasitize *Sanghuangporus sanghuang* are varied, their quality evaluation lacks scientificity and rigor. The current artificial cultivation technology is not yet mature, and the growth cycle is long, the yield is low, and it is susceptible to diseases and insect pests, so the cultivation of *Sanghuangporus sanghuang* is still difficult. Liquid medium is the preferred method for the large-scale cultivation of fungal mycelium because it enables full contact of the fungal clumps with the medium after inoculation and promotes the immediate absorption of nutrients. The liquid medium also enables rapid release of the fungal metabolites. Liquid culturing is simple and the formulation is flexible and amenable to adjustments. Therefore, liquid culturing is routinely used for large-scale fungal culturing and the extraction of fungal metabolites [16]. In this study, we used the one-way testing and response surface methodology to optimize the optimal carbon source, nitrogen source, and exogenous growth factors in the liquid-fermentation-media formulation for improving the yield of *S. sanghuang*. Furthermore, we performed chemical antioxidant and MTT assays to determine the antioxidant and anticancer activities of the extracellular polysaccharides of *S. sanghuang* to provide a theoretical basis for future industrial applications and drug development.

2. Materials and Methods

2.1. Materials

2.1.1. Strains

Three *S. sanghuang* strains (MS-1, MS-3, and MS-7) growing on the mulberry plants were obtained from the Mycological Research Center of Fujian Agriculture and Forestry University.

2.1.2. Cancer Cell Lines

HepG-2 (human liver cancer cell line), MCF-7 (human breast cancer cell line), and PC-3 (human prostate cancer cell line) cells were purchased from Saibaikang Biotechnology Co., Ltd. (Shanghai, China). HCT-116 (human colon cancer cell line) as well as T98G and U251 (human brain glioma cell lines) cells were purchased from Procell Life Technology Co., Ltd. (Wuhan, China).

2.1.3. Reagents

Potato-Dextrose-Agar (PDA)-enriched medium was prepared with an extract containing 200 g potatoes (peeled), 20 g Agar, 20 g glucose, 2% agar, 5 g peptone, 1.5 g $MgSO_4$, 2 g KH_2PO_4, 0.01 g vitamin B_1, and up to 1 L of deionized water. The medium was sterilized at 121 °C for 30 min. The PDB-enriched medium was prepared with an extract containing 200 g potatoes (peeled), 20 g glucose, 5 g peptone, 1.5 g $MgSO_4$, 2 g KH_2PO_4, and 0.01 g vitamin B_1 and made up to 1 L with deionized water. The medium was sterilized at 121 °C for 30 min.

RPMI 1640 cell culture medium, DMEM cell culture medium, MEM cell culture medium, penicillin–streptomycin solution, and fetal bovine serum were used to prepare media for culturing the human cancer cell lines and stored in the refrigerator at 4 °C

PBS buffer (pH 7.4) was prepared with NaCl (8.5 g), KCl (0.2 g), $Na_2HPO_4 \cdot 12H_2O$ (2.85 g), and KH_2PO_4 (0.27 g) in 1 L deionized water and sterilized at 121 °C for 30 min.

2.2. *Methodology*

2.2.1. Strain Activation

Strain activation was performed using the method described by Ding et al. [17]. Briefly, the test tubes of seeds of MS-1, MS-3, and MS-7 strains of *Sanghuangporus sanghuang* were inoculated into the PDA enrichment medium, wrapped with a sealing film to prevent contamination, and cultured at a constant temperature of 25 °C until the mycelium completely covered the surface of the petri dish.

2.2.2. Estimation of Mycelial Biomass

The mycelial biomass was estimated using the method described previously by Jia [18]. Briefly, five 7-mm-diameter fungal clusters were incubated individually in flasks containing 100 mL PDB growth medium. The cultures were grown in the dark for 15 days at 25 °C on a shaker set at 160 rpm/min. The liquid in the culture bottles was vacuum filtered to harvest the mycelia. The filtration was repeated three times. The harvested mycelia were washed with distilled water, dried in an oven at 65 °C, and weighed to record the mycelial biomass.

2.2.3. The Effects of Carbon Sources on the Growth Rate of *S. sanghuang*

Glucose was replaced as the carbon source in the PDB-enrichment-medium formulation by maltose, sucrose, fructose, xylose, mannitol, mannose, lactose, or galactose (20 g/L each). The control (CK) group formulation did not have any additional carbon source. The media were inoculated with 7 mm × 7 mm × 5 mm *S. sanghuang* clumps and cultured for 15 days in the dark at 25 °C on a constant shaker set at 150 r/min. The mycelia were harvested from all of the flasks and the biomasses were estimated as described above.

2.2.4. Effects of Nitrogen Sources on the Growth Rate of *S. sanghuang*

Yeast extract, peptone, ammonium tartrate, ammonium sulfate, ammonium nitrate, urea, and beef extract were used as alternative nitrogen sources in the PDB-enrichment-medium formulation. The formulation of the CK group did not have any additional nitrogen sources. The media were inoculated with 7 mm × 7 mm × 5 mm *S. sanghuang* clumps and cultivated in a shaker at 25 °C and 150 r/min for 15 days in the dark. Then, the mycelia were harvested and weighed to determine the effects of different nitrogen sources on the growth of *S. sanghuang*.

2.2.5. Effects of Different Concentrations of Carbon and Nitrogen Sources on the Growth Rate of *S. sanghuang*

Based on the results of one-way carbon and nitrogen source screening, the growth rates for different strains of *Sanghuangporus sanghuang* were analyzed using different concentrations of carbon and nitrogen. Each experiment was performed with three replicates for each group. The carbon source concentrations in the media were 10, 15, 20, 25, or 30 g/L. The nitrogen source concentrations in the media were 2.5, 5, 7.5, 10, or 15 g/L. The culture media were inoculated with 7 mm × 7 mm × 5 mm *S. sanghuang* clumps and cultured at 25 °C in a shaker set at 150 rpm/min for 15 days in the dark. The mycelial biomasses were estimated as previously described.

2.2.6. Effects of Exogenous Additives on the Growth of *S. sanghuang*

Previous experiments showed that dandelion powder significantly increased the growth of *S. sanghuang* mycelium on the plate. This suggested that dandelion powder acted as a growth factor for *S. sanghuang*. Therefore, we added 0, 0.5, 0.75, 1, 1.25, or 1.5 g/L of dandelion powder to the PDB-enriched medium, inoculated the media with 7 mm × 7 mm × 5 mm *S. sanghuang* clumps, and cultured them on a shaker at 25 °C and 150 r/min in the dark for 15 days. The mycelium biomass was estimated as previously described.

2.2.7. Response Surface Methodology for Optimizing the *S. sanghuang* Growth Medium Formulation

The response surface methodology (RSM) was used to optimize the concentrations of different carbon and nitrogen sources and other additional factors for growing *S. sanghuang*. The Box–Behnken design principle and Design Expert 11 software (ver. 11) were used to perform three levels of response surface experiments to determine the optimal concentrations of the carbon source, nitrogen source, and dandelion powder for obtaining the maximum mycelial biomass. Table 1 shows the concentrations of different factors used in the media formulations and levels of the response surface experiments.

Table 1. Factors and levels of Box-Behnken test of *Sanghuangporus sanghuang*.

Level	Factor		
	A: Carbon Source (g/L)	**B: Nitrogen Source (g/L)**	**C: Dandelion Powder (g/L)**
−1	25	12.5	1
0	30	15	1.25
1	35	17.5	1.5

Accuracy of the RSM prediction results was verified by liquid-fermentation-media culturing of the fungus with the best-predicted formulation. The mycelia were collected, dried and weighed. The weight of the mycelia was compared with that obtained under sub-optimal conditions.

2.2.8. Extraction of Extracellular Polysaccharides

The fermentation liquid was collected after growing the fungus, vacuum filtered thrice, and centrifuged at 10,000 rpm for 3 min to remove the mycelial remnants. Then, the fermentation liquid was concentrated by spin distillation to 1/5th of the original volume. Subsequently, ethanol and concentrated fermentation liquid were mixed in a 1:4 ratio by volume. Alcohol precipitation was performed overnight at 4 °C. The precipitate was harvested after centrifugation and re-solubilized with water. We added Sevag reagent (chloroform:n-butanol volume ratio is 4:1) into the crude polysaccharide solution, shook it thoroughly and centrifuged at 5000 rpm/min for 5 min, and then mixed the water phase with chloroform for separation. We added an equal volume of Sevag reagent to the water phase and repeated several times until there was no obvious protein layer. Then, the protein-free solution was incubated with ether and methanol to remove the impurities. Finally, the crude polysaccharide solution was dialyzed in running water for 48 h, freeze-dried, and stored at −20 °C.

2.2.9. Chemical Antioxidant Activity of the Extracellular Polysaccharides

An ABTS free radical scavenging assay was performed using the protocol described by Li et al. [19] with slight modifications. Different concentrations of the extracellular polysaccharide and vitamin C(VC) solutions (0.025, 0.05, 0.25, 0.5, 1, 2, and 5 mg/mL) were prepared and added into wells of a 96-well plate. Then, an equal volume of the ABTS solution (7 mM) was added into each well. In the blank group, an equal volume of distilled water was added instead of the polysaccharide solution. The reaction was performed at 25 °C. Absorbance was measured at 734 nm in an enzyme-labeling instrument. The antioxidant activity was calculated using the average value from five replicates for each concentration of the extracellular polysaccharide solution and a calibration curve of VC with the following formula:

$$\text{ABTS free radical scavenging rate (\%)} = [1 - (A_x - A_{x0})/A_0] \times 100,$$

where A_x is the absorbance value for the experimental group, A_0 is the absorbance value for the blank group, and A_{x0} is the absorbance value for the control group.

Assessment of DPPH radical scavenging activity was performed according to the method described by Luo et al. [20] with slight modifications. Different concentrations (0.025, 0.05, 0.25, 0.5, 1, 2, and 5 mg/mL) of the extracellular polysaccharide and VC solutions were prepared and aliquoted into 96-well plates. Then, an equal volume of the DPPH solution (0.2 mM of 80% ethanol solution) was added into each well. In the blank group, an equal volume of anhydrous ethanol was used instead of the polysaccharide solution. The plate was incubated for 20 min at 25 °C. Then, absorbance was measured at 517 nm using an enzyme-labeling instrument. DPPH radical scavenging activity was calculated using VC as the positive control group according to the following formula:

$$\text{DPPH radical scavenging rate (\%)} = [1 - (A_x - A_{x0})/A_0] \times 100,$$

where A_x is the absorbance value of the experimental group, A_0 is the absorbance value of the blank group, and A_{x0} is the absorbance value of the control group.

2.2.10. Ferric Reducing Antioxidant Power (FRAP) Assay

The FRAP assay was performed according to the method described by Benzie et al. [21] with slight modifications. Different concentrations (0.025, 0.05, 0.25, 0.5, 1, 2, and 5 mg/mL) of the extracellular polysaccharides were prepared. Then, 30 μL aliquots of the polysaccharide solutions were added to each well of a 96-well plate and incubated with 180 μL of FRAP solution for 20 min at 25 °C. The blank group contained 30 μL of deionized water and 180 μL of FRAP solution. A standard curve was prepared by incubating 30 μL aliquots of 0.025, 0.1, 0.2, 0.4, 0.8, and 1 mmol/L $FeSO_4$ solutions with 180 μL of FRAP solution at 25 °C for 20 min. Absorbance was measured at 593 nm in enzyme-labeling instrument. Five replicates were used for each group. The average value was calculated using the following formula:

$$\text{FRAP value} = A_x - A_0 - A_{x0},$$

where A_x is the absorbance value of the experimental group, A_0 is the absorbance value of the blank group, and A_{x0} is the absorbance value of the control group.

2.2.11. Effects of Extracellular Polysaccharides on Cancer Cell Proliferation

Based on the protocol described by Zhang et al. [22] with slight modifications, cancer cells were cultured in cell culture medium containing 10% fetal bovine serum and 1% penicillin/streptomycin mixture in a humidified incubator at 37 °C and 5% CO_2. Then, 200 μL of cancer cells (HepG-2, MCF-7, PC-3, HCT-116, U251, and T98G) in the logarithmic phase of growth were seeded in the 96-well plates and cultured for 24 h. The culture medium was removed after observing adherence of cells. In the experimental group, the cancer cells were incubated with 200 μL of cell culture medium containing different concentrations of three extracellular polysaccharides (0.025, 0.05, 0.25, 0.5, 1, 2 and 5 mg/mL). We used five replicates for each group. The blank group wells contained 200 μL of cell culture medium without any cells. In the control (CK) group, the cells were cultured with the medium for 24 h. Subsequently, after removing the medium, we added 10 μL of 5 mg/mL MTT solution and 90 μL of cell culture solution to each well and incubated for 4 h. Then, the culture solution was removed from all the wells carefully. We then added 150 μL of DMSO to each well. After mixing on a shaker for 15 min, the absorbance value (OD value) of the solution was determined at a wavelength of 490 nm in an enzyme-labeling instrument.

2.3. Statistical Analysis

SPSS 25.0 statistical software was used for statistical analysis. The data were expressed as mean ± standard deviation (x ± s). The t-test was used to determine statistically significant differences between two groups of data, whereas ANOVA was used to compare the data between multiple groups. $p < 0.05$ was considered statistically significant. $p < 0.01$ was considered highly statistically significant.

3. Results

3.1. Identification of an Optimal Carbon Source for Culturing *S. sanghuang*

Firstly, we investigated the effects of different carbon sources in the PDB-enrichment medium, including maltose, sucrose, glucose, fructose, xylose, mannitol, mannose, lactose, and galactose, on the growth of *S. sanghuang*. As shown in Figure 1, the effects of different carbon sources on the mycelium biomass (high biomass to low) of *S. sanghuang* were as follows: glucose > mannose > fructose > maltose > mannitol > lactose > sucrose > galactose > CK > xylose. The three strains of *S. sanghuang* showed similar carbon source requirements. Glucose was the most effective and xylose was the least effective carbon source. Therefore, glucose was selected as the carbon source for culturing *S. sanghuang* using the liquid fermentation medium.

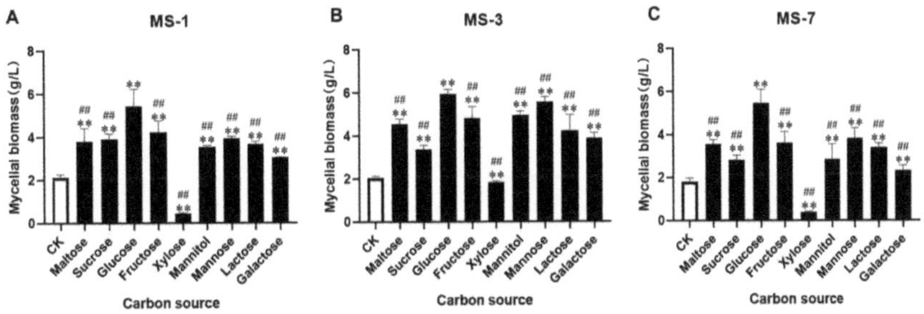

Figure 1. The effects of different carbon sources on the growth of strains of *S. sanghuang*. (**A**) MS-1, (**B**) MS-3, and (**C**) MS-7. Note: ** represents $p < 0.01$ for comparisons with the CK group; ## represents $p < 0.01$ for comparisons with the optimal carbon source.

3.2. Identification of an Optimal Nitrogen Source for Culturing *S. sanghuang*

Next, we investigated the effects of different nitrogen sources in PDB-enrichment medium on the growth of *S. sanghuang*. As shown in Figure 2, the effects of different nitrogen sources (high biomass to low) on the mycelium biomass of *S. sanghuang* were as follows: yeast extract > beef extract > peptone > CK > ammonium tartrate > ammonium sulfate > ammonium nitrate > urea. The requirement for a nitrogen source was high for all three strains of *S. sanghuang*. Yeast extract was the most effective nitrogen source, whereas ammonium tartrate, ammonium sulfate, ammonium nitrate, and urea were the least effective nitrogen sources. Therefore, yeast extract was selected as the nitrogen source for the liquid fermentation medium used to grow *S. sanghuang*.

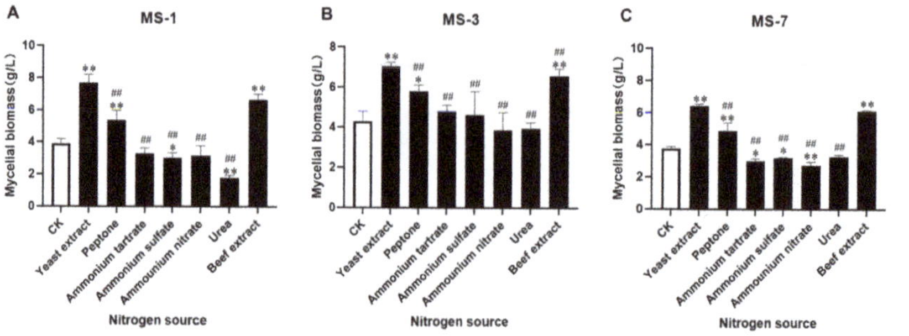

Figure 2. The effects of different nitrogen sources on the growth of *S. sanghuang*. (**A**) MS-1; (**B**) MS-3; (**C**) MS-7. Note: * represents $p < 0.05$ and ** represents $p < 0.01$ for comparisons with the CK group; ## represents $p < 0.01$ for comparisons with the optimal nitrogen source.

3.3. *Optimizing the Concentrations of the Carbon and Nitrogen Sources and the Growth Factors for Culturing S. sanghuang*

3.3.1. Optimizing the Glucose Concentration for Culturing *S. sanghuang*

As shown in Figure 3, PDB-enrichment medium with 30 g/L glucose showed the maximal and fastest growth of *S. sanghuang*. The liquid culture medium with 30 g/L glucose showed the optimal growth of *S. sanghuang*. The mycelial growth of *S. sanghuang* was sub-optimal when the glucose concentration was < 30 g/L or >30 g/L.

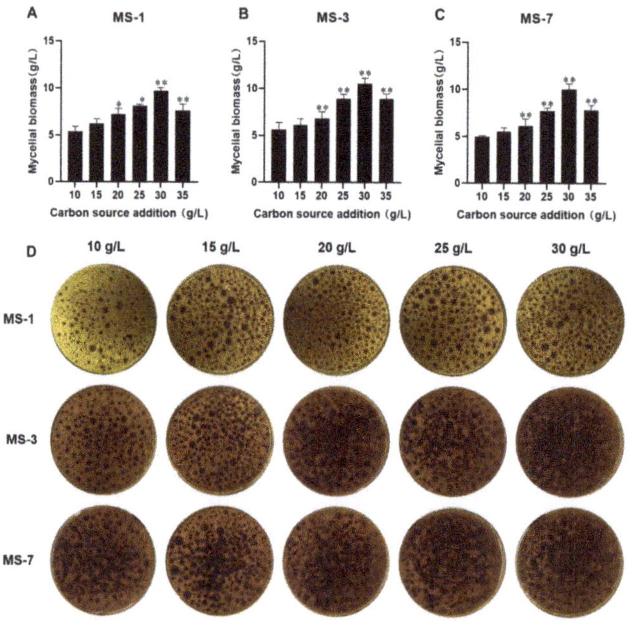

Figure 3. The effects of different concentrations of glucose on the growth of *S. sanghuang*. The mycelial biomasses of (**A**) MS-1; (**B**) MS-3; and (**C**) MS-7. (**D**) Fermentation liquid media with different glucose concentrations for growing MS-1, MS-3, and MS-7 strains of *S. sanghuang*. Note: * represents $p < 0.05$ and ** represents $p < 0.01$ when compared with the previous group.

3.3.2. Optimizing the Yeast Extract Concentration for Culturing *S. sanghuang*

As shown in Figure 4, the mycelial growth rate and biomass of *S. sanghuang* increased proportionately with increasing concentrations of yeast extract in the growth medium and they reached maximums at a concentration of 15 g/L yeast extract. The solubility of the yeast extract was also maximal at 15 g/L. Therefore, 15 g/L yeast extract was selected for the growth medium of *S. sanghuang*.

3.3.3. Optimizing the Growth Factor Concentration for Culturing *S. sanghuang*

After optimizing the concentrations of the carbon and nitrogen sources, we investigated the effects of dandelion powder as an exogenous additive to improve the growth of *S. sanghuang*. As shown in Figure 5, the optimal concentration of dandelion powder for the maximal growth of *S. sanghuang* was 1.25 g/L. The mycelial biomass for the three strains of *S. sanghuang* grown with media containing 1.25 g/L dandelion powder was 1.29~1.5 fold higher than the blank group. The mycelial growth increased gradually, peaked in media containing 1.25 g/L dandelion powder, and subsequently declined in media with higher concentrations of dandelion powder (>1.25 g/L). These results showed that higher concentrations of dandelion powder suppressed the growth of *S. sanghuang*. Therefore, the optimal concentration of dandelion powder in the growth medium for *S. sanghuang* was 1.25 g/L.

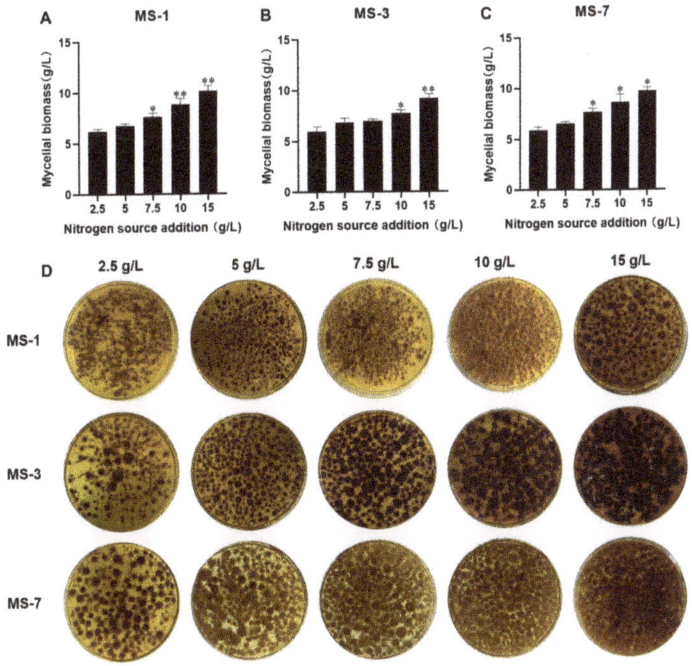

Figure 4. The effects of different yeast extract concentrations on the growth of *S. sanghuang*. The mycelial biomasses of (**A**) MS-1; (**B**) MS-3; and (**C**) MS-7 strains of *S. sanghuang*. (**D**) Fermentation liquid media with different yeast extract concentrations for growing MS-1, MS-3, and MS-7 strains of *S. sanghuang*. Note: * represents $p < 0.05$ and ** represents $p < 0.01$ compared with the previous group.

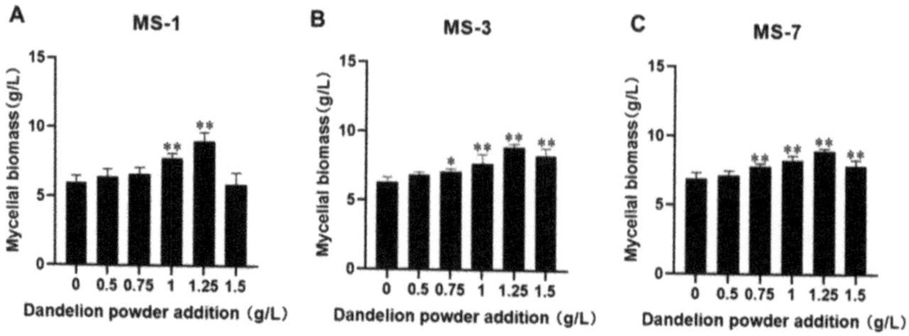

Figure 5. The effects of dandelion powder on the growth of strains of *S. sanghuang*. (**A**) MS-1; (**B**) MS-3; (**C**) MS-7. Note: * represents $p < 0.05$ and ** represents $p < 0.01$ compared with the CK group.

3.4. *Response Surface Methodology Results*

3.4.1. Box–Behnken's Central Combination Design Results

Response surface analysis tests were performed with the concentrations of glucose, yeast extract, and dandelion powder as independent variables and the mycelial biomass of *S. sanghuang* as the response value. The experimental scheme and the results are shown in Tables 2–4.

Table 2. Response surface design and test results of MS-1.

No.	A: Carbon Source (g/L)	B: Nitrogen Source (g/L)	C: Dandelion Powder (g/L)	Mycelial Biomass (g/L)
1	25	17.5	1.25	8.1
2	25	15	1	8.9
3	35	15	1.5	9.5
4	30	15	1.25	14.06
5	30	17.5	1	7.2
6	30	15	1.25	13.83
7	35	17.5	1.25	9.1
8	25	15	1.5	8
9	30	12.5	1.5	8.2
10	30	12.5	1	10.2
11	35	12.5	1.25	10.5
12	30	15	1.25	13.76
13	30	15	1.25	13.46
14	30	17.5	1.5	7.66
15	30	15	1.25	14.03
16	35	15	1	9.16
17	25	12.5	1.25	10.1

Table 3. Response surface design and test results of MS-3.

No.	A: Carbon Source (g/L)	B: Nitrogen Source (g/L)	C: Dandelion Powder (g/L)	Mycelial Biomass (g/L)
1	30	12.5	1	10.13
2	25	12.5	1.25	9.96
3	25	15	1	8.76
4	30	15	1.25	13.86
5	30	12.5	1.5	8.26
6	30	15	1.25	13.56
7	35	17.5	1.25	9.33
8	25	17.5	1.25	8.16
9	30	15	1.25	13.86
10	30	17.5	1.5	8.03
11	25	15	1.5	8.06
12	35	12.5	1.25	9.96
13	30	15	1.25	13.26
14	30	17.5	1	7.93
15	35	15	1.5	9.03
16	30	15	1.25	13.46
17	35	15	1	9.13

Table 4. Response surface design and test results of MS-7.

No.	A: Carbon Source (g/L)	B: Nitrogen Source (g/L)	C: Dandelion Powder (g/L)	Mycelial Biomass (g/L)
1	25	15	1	8.93
2	30	12.5	1.5	8.5
3	35	17.5	1.25	8.6
4	30	15	1.25	13.83
5	35	15	1.5	9.4
6	30	15	1.25	14.17
7	25	12.5	1.25	10.8
8	30	17.5	1.5	8.06
9	30	17.5	1	6.5
10	30	15	1.25	14
11	30	12.5	1	9.8
12	30	15	1.25	13.8
13	25	17.5	1.25	8.6
14	35	15	1	9.46
15	35	12.5	1.25	10.5
16	25	15	1.5	10.3
17	30	15	1.25	13.9

3.4.2. Multiple Quadratic Regression Fitting and ANOVA

Next, we performed experiments using the Box–Behnken's central combination design. Then, using Design Expert software (ver. 11), we performed regression analysis of the mycelial biomass for different growth conditions and compared the data by analysis of variance (ANOVA) (Tables 5–7).

Table 5. Variance analysis and significance tests of MS-1.

Source	Sum of Squares	Df	Mean Squares	F-Value	*p*-Value
Model	98.62	9	10.96	205.43	<0.0001
A	1.25	1	1.25	23.4	0.0019
B	6.02	1	6.02	112.87	<0.0001
C	0.5512	1	0.5512	10.33	0.0148
AB	0.09	1	0.09	1.69	0.2351
AC	0.3844	1	0.3844	7.21	0.0313
BC	1.51	1	1.51	28.36	0.0011
A2	15.22	1	15.22	285.41	<0.0001
B2	25.82	1	25.82	484.13	<0.0001
C2	38.82	1	38.82	727.83	<0.0001
Residual	0.3734	7	0.0533		
Lack of fit	0.1387	3	0.0462	0.788	0.5601
Pure error	0.2347	4	0.0587		
Cor Total	98.99	16			
R^2 = 0.9962; Adjusted R^2 = 0.9914; Predicted R^2 = 0.9739					

Table 6. Variance analysis and significance tests of MS-3.

Source	Sum of Squares	Df	Mean Squares	F-Value	*p*-Value
Model	85.4	9	9.49	168.92	<0.0001
A	0.7875	1	0.7875	14.02	0.0072
B	2.95	1	2.95	52.56	0.0002
C	0.8256	1	0.8256	14.7	0.0064
AB	0.3422	1	0.3422	6.09	0.0429
AC	0.09	1	0.09	1.6	0.2461
BC	0.9702	1	0.9702	17.27	0.0043
A2	17.61	1	17.61	313.46	<0.0001
B2	20.43	1	20.43	363.6	<0.0001
C2	33.25	1	33.25	591.84	<0.0001
Residual	0.3932	7	0.0562		
Lack offit	0.1212	3	0.0404	0.5942	0.6512
Pure error	0.272	4	0.068		
Cor Total	85.8	16			
R^2 = 0.9954; Adjusted R^2 = 0.9895; Predicted R^2 = 0.9724					

Table 7. Variance analysis and significance tests of MS-7.

Source	Sum of Squares	Df	Mean Squares	F-Value	*p*-Value
Model	97.54	9	10.84	310.37	<0.0001
A	0.0561	1	0.0561	1.61	0.2455
B	7.68	1	7.68	220.04	<0.0001
C	0.3081	1	0.3081	8.82	0.0208
AB	0.0225	1	0.0225	0.6444	0.4485
AC	0.5112	1	0.5112	14.64	0.0065
BC	2.04	1	2.04	58.56	0.0001
A2	9.52	1	9.52	272.67	<0.0001
B2	33.28	1	33.28	952.99	<0.0001
C2	35.75	1	35.75	1023.75	<0.0001
Residual	0.2444	7	0.0349		
Lack offit	0.1546	3	0.0515	2.3	0.2196
Pure error	0.0898	4	0.0224		
Cor Total	97.78	16			
R^2 = 0.9962; Adjusted R^2 = 0.9914; Predicted R^2 = 0.9739					

As shown in Tables 4–6 models with $p < 0.05$ and $R^2 > 0.98$ as a coefficient of determination were considered to be well fitted for optimal mycelial growth. If the *p*-value was

>0.05, then the model was not well fitted, and this suggested that the original model was more accurate. Therefore, the model was used to analyze the experimental results and predict the response values. The following multiple regression equations were obtained for the mycelial biomass of *Sanghuangporus sanghuang* (Y) based on independent variables such as glucose (A), yeast extract (B), and dandelion powder (C):

$$Y_{MS\text{-}1} = 13.828 + 0.395A - 0.8675B - 0.2625C + 0.15AB + 0.31AC + 0.615BC - 1.9015A^2 - 2.4765B^2 - 3.0365C^2$$

$$Y_{MS\text{-}3} = 13.6 + 0.31375A - 0.6075B - 0.32125C + 0.2925AB + 0.15AC + 0.4925BC - 2.045A^2 - 2.2025B^2 - 2.81C^2$$

$$Y_{MS\text{-}7} = 13.94 - 0.08375A - 0.98B - 0.19625C + 0.075AB - 0.3575AC + 0.715BC - 1.50375A^2 - 2.81125B^2 - 2.91375C^2$$

3.4.3. Significance Test Results

To test the validity of the multiple regression equations, the mathematical model for the estimated mycelium biomass of *S. sanghuang* was evaluated by ANOVA and the partial regression coefficients of the factors were analyzed (Tables 5–7). In the primary term, the regression coefficient B was highly significant. This suggested that the concentration of the yeast extract significantly influenced the growth of *S. sanghuang*. In the secondary term, the partial regression coefficients A, B, and C reached a level of high significance. The regression coefficient of the interaction term BC was more significant than those of the other factors. This suggested that the interaction between yeast extract and dandelion powder significantly influenced the growth of *S. sanghuang*. The out-of-fit term was not significant, and the model was appropriate. The goodness-of-fit R value was > 0.98, which was close to the real data point and suggested good fitting of the equation to the experimental set-up.

3.4.4. Response Surface Analysis of Two-by-Two Interactions of Factors Affecting the Culturing of *S. sanghuang*

Based on the regression equation, we performed response surface analysis to determine the effects of media containing different amounts of glucose, yeast extract, and dandelion powder on the growth of *S. sanghuang*.

As shown in Figure 6, the optimal concentrations of yeast extract and dandelion powder in the enrichment medium were necessary for the maximal growth of *S. sanghuang*. Therefore, according to results shown in Figure 6 and the ANOVA data, the order of effects for the growth of *S. sanghuang* were as follows: BC > AC > AB. The first-order partial derivatives of the quadratic regression equation were used for the analysis. The concentrations of different factors for obtaining the maximum response value (Y) (*S. sanghuang* biomass) were as follows: (1) the theoretical mycelial biomass for the MS-1 strain was 13.93 g/L when the medium contained 30.46 g/L of the carbon source, 14.55 g/L of the nitrogen source, and 1.24 g/L of dandelion powder; (2) the theoretical mycelial biomass for the MS-3 strain was 13.66 g/L when the medium contained 30.32 g/L of the carbon source, 14.64 g/L of the nitrogen source, and 1.23 g/L of the dandelion powder; and (3) the theoretical mycelial biomass for the MS-3 strain was 14.03 g/L when the medium contained 29.83 g/L of the carbon source, 14.57 g/L of the nitrogen source, and 1.25 g/L of the dandelion powder.

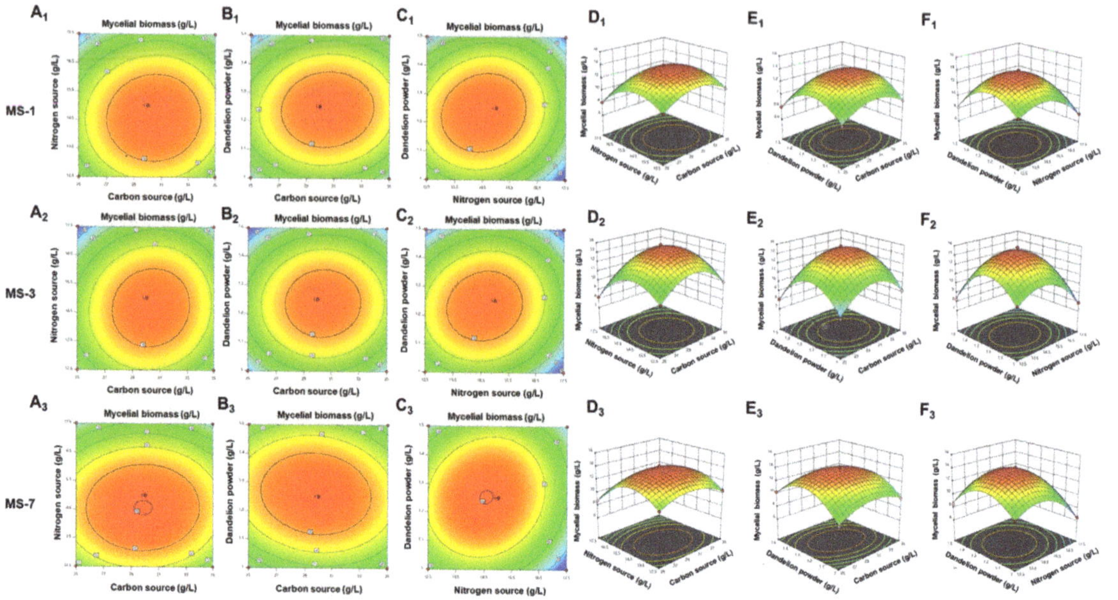

Figure 6. Response surface stereograms based on the Box–Behnken design for the optimization of liquid-fermentation-media components for growing *S. sanghuang*. ((**A1–A3**)–(**C1–C3**)) Corresponding contour values. (**D1–D3**) Interactions between different concentrations of carbon and nitrogen sources. (**E1–E3**) Interactions between different concentrations of the carbon source and the dandelion powder. (**F1–F3**) Interactions between different concentrations of the nitrogen source and the dandelion powder.

3.4.5. Validation of Fit Optimization

To verify the accuracy of the model prediction, we performed a validation test with three sets of replicates using optimized growth medium components based on the response surface test. The results of the validation test are shown in Figure 7. The mycelium yield for the MS-1 strain with the optimized formulation was 14.0 4 ± 0.61 g/L. This was 2.36 times higher than the sub-optimal formulation (5.94 ± 0.305 g/L) and slightly higher than the predicted value (13.93 g/L). The mycelial yield for the MS-3 strain with the optimized formulation was 13.82 ± 0.3 g/L. This was 2.21 fold higher than the sub-optimal formulation (6.23 ± 0.305 g/L) and slightly higher than the predicted value (13.66 g/L). The mycelial yield for the MS-7 strain with the optimized formulation was 14.12 ± 0.31 g/L. This was 2.39 times higher than the sub-optimal formulation (5.9 ± 0.3 g/L) and slightly greater than the predicted value (14.03 g/L). These results confirmed the goodness-of-fit between the model and the actual values. Therefore, the fermentation medium formulation that had been optimized according to the response surface test method provided the necessary theoretical and scientific basis for the commercial cultivation of *S. sanghuang* in the future.

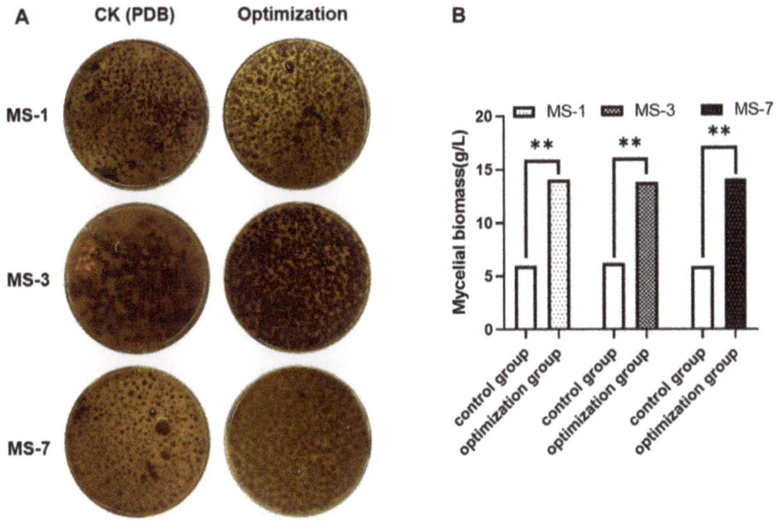

Figure 7. Validation of the optimized liquid fermentation formulation for growing *S. sanghuang*. (**A**) The CK group (PDB) and optimized-fermentation-liquid-growth formulations for *S. sanghuang*. (**B**) The mycelial biomass for the MS-1, MS-3, and MS-7 strains of *S. sanghuang* with the control and optimized formulations. Note: ** represents $p < 0.01$ compared with the CK (PDB) group.

3.5. Antioxidant Activity of Extracellular Fungal Polysaccharides

As shown in Figure 8, all three strains of *S. sanghuang* showed strong antioxidant activity. The *S. sanghuang* extracellular polysaccharides showed significantly high ABTS free radical scavenging activity that was comparable with the Vc control group, but the DPPH free radical scavenging activity of the *S. sanghuang* extracellular polysaccharides was about 50% lower than that of the Vc control group. The total antioxidant capacity of the *S. sanghuang* extracellular polysaccharides was weak at lower concentrations but increased rapidly at higher concentrations and was comparable with the Vc control group.

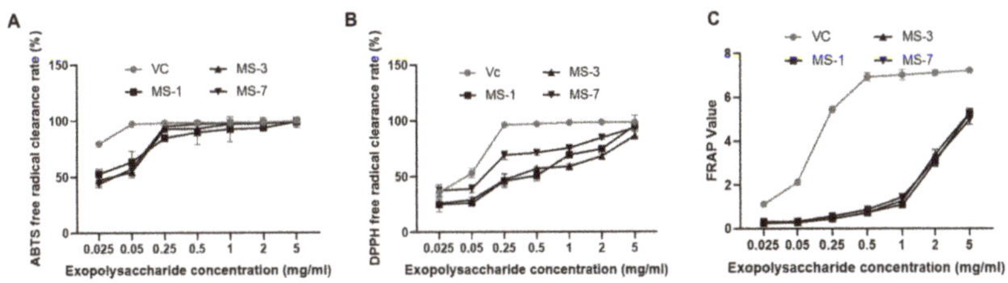

Figure 8. Antioxidant activities of the polysaccharides extracted from *S. sanghuang*. (**A**) ABTS-radical scavenging capacity of *S. sanghuang*. (**B**) DPPH radical scavenging capacity of the polysaccharides extracted from *S. sanghuang*. (**C**) FRAP assay results show the total antioxidant capacity of the polysaccharides extracted from *S. sanghuang*.

The three strains of *S. sanghuang* showed differences in their scavenging ability for the two free radicals. In the ABTS free radical scavenging experiments, the MS-1 strain showed the strongest scavenging ability (EC_{50} = 0.02 mg/mL) followed by the MS-3 strain (EC_{50} = 0.03 mg/mL) and the MS-7 strain (EC_{50} = 0.032 mg/mL). In the DPPH free radical scavenging experiment, the MS-7 strain showed the strongest scavenging ability (EC_{50} = 0.08 ± 0.00229 mg/mL) followed by the MS-1 strain (EC_{50} = 0.273 ± 0.0071 mg/mL) and the MS-3 strain (EC_{50} = 0.291 ± 0.0025 mg/mL). The results of the FRAP assay demonstrated that the total antioxidant capacity of the *S. sanghuang* extracellular polysaccharide was significantly high. The maximum FRAP values were 5.26 ± 0.11, 5.27 ± 0.19, and 4.97 ± 0.22 for the MS-1, MS-3, and MS-7 strains, respectively, when the polysaccharide concentration was 5 mg/mL.

3.6. Evaluation of the Anticancer Activity of the Extracellular Fungal Polysaccharides from S. sanghuang

Next, we analyzed the in vitro anticancer activities of the *S. sanghuang* extracellular polysaccharides on six types of cancer cells. The extracellular polysaccharides of *S. sanghuang* inhibited the growth of all six types of cancer cell lines in the following order (high to low inhibitory effect): U251 > T98G > PC-3 > HCT-116 > MCF-7 > HepG-2 (Figure 9).

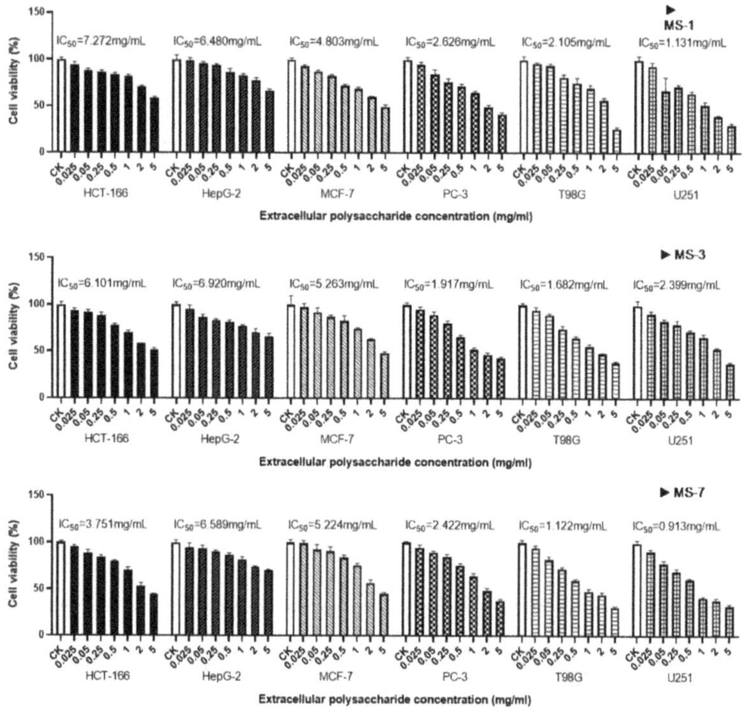

Figure 9. MTT assay results show the effects of the polysaccharides extracted from *S. sanghuang* on the growth of various cancer cells.

The extracellular polysaccharides of *S. sanghuang* showed the strongest inhibitory effects in the U251 human glioma cells. Furthermore, extracellular polysaccharides of the MS-7 strain (half inhibitory concentration of IC_{50} = 1.56 ± 0.96 mg/mL) showed higher anticancer efficacy than those from the MS-1 and MS-3 strains. The extracellular polysaccharides of *S. sanghuang* also significantly inhibited growth of human glioblastoma T98G cells. The extracellular polysaccharides of the MS-1 strain significantly inhibited the growth of T98G cells ($p < 0.01$) with an IC_{50} of 1.66 ± 0.44 mg/mL. The extracellular polysaccharides of *S. sanghuang* showed higher pro-apoptotic effects on the PC-3 prostate cancer cells. The MS-3 strain showed stronger pro-apoptotic effects than the other two strains (IC_{50} = 1.917 mg/mL). *S. sanghuang* showed weaker anticancer efficacy on the colon cancer cell line, HCT-116. Furthermore, the MS-7 strain showed higher anticancer activity on the HCT-116 cells than did the MS-1 and MS-3 strains, with an IC_{50} value of 3.751 mg/mL. The apoptosis rate was 44.33% in the HCT-116 cells after 24 h of treatment with the highest concentration of the MS-7 polysaccharides. At the highest concentrations, the anticancer activities of the MS-1 and MS-3 polysaccharides were weaker in the HCT-116 cells, with survival rates of 58.35% and 51.52%, respectively. The pro-apoptotic effects of the *S. sanghuang* polysaccharides were stronger on the breast cancer MCF-7 cells, with a survival rate ranging from 45.38% to 49.20% at polysaccharide concentrations of 5 mg/mL. The pro-apoptotic effects of *S. sanghuang* extracellular polysaccharides in the HepG-2 (hepatocellular carcinoma cells) cells were poor. At the 5 mg/mL polysaccharide concentration, the survival rate of the HepG2 cells ranged between 66.04% and 69.89%. Moreover, the IC_{50} values were comparable in the HepG-2 cells for the polysaccharides from the MS-1 (6.48 mg/mL), MS-3 (6.92 mg/mL), and MS-7 (6.589 mg/mL) strains.

4. Discussion

Sanghuangporus spp. is a rare medicinal fungus. Polysaccharide is an important active substance of its fruiting body and mycelium; it is a type of compound with hypoglycemic, hypolipidemic, antioxidant, anti-tumor, and immunomodulatory activities [22]. *Sanghuangporus* spp. is a perennial fungus. Since the growth of wild *S. sanghuang* fruiting bodies is limited by natural conditions and has a long growth cycle, its output cannot meet market demand. At present, the cultivation scale of *S. sanghuang* is small and unevenly distributed, and the planting technology is immature, making it difficult to guarantee the yield and quality of *S. sanghuang*, and it is difficult to increase the added value of *S. sanghuang* products [23]. There is a large talent gap in the resource evaluation, standard formulation, improved variety selection, and technology promotion of *S. sanghuang*, which restricts the high-quality development of a *S. sanghuang* industry [23]. Therefore, the artificial cultivation technology of *S. sanghuang* is still in the bottleneck period, and liquid fermentation culture of *S. sanghuang* has become another research hotspot. Liquid fermentation culture can quickly obtain a large amount of metabolites and can directly extract physiologically active substances for medicinal research. Compared with solid fermentation, liquid fermentation has the advantages of continuous production, low cost, simple operation, short fermentation time, high yield, and high efficiency.

Therefore, development of the optimal liquid-fermentation-media formulation and conditions were necessary for the large-scale production of *S. sanghuang*. In this study, we comprehensively analyzed the effects of different carbon and nitrogen sources, as well as exogenous growth factors, to develop the optimal liquid-fermentation-media formulation for maximal mycelial yield. We also performed the response surface method optimization test using a Box–Behnken design to analyze interactions between various factors. Furthermore, we analyzed the antioxidant and anticancer activities of the extracellular polysaccharides from three strains of *S. sanghuang*. Previous studies have shown that, although different strains of *S. sanghuang* show similar morphology, their requirements for carbon, nitrogen, and growth factors vary considerably. Our data showed that glucose was the optimal carbon source for the liquid fermentation medium used to grow *S. sanghuang*. This was comparable with the composition of the PDA medium that was used to grow *S. Sanghuang* in a previous study [24]. Jiang et al. also reported that glucose was the most significant requirement for the growth of *S. sanghuang* [25]. Furthermore, our data showed that yeast extract was the most suitable nitrogen source for growing *S. Sanghuang*. Organic nitrogen sources such as yeast extract, beef extract, and peptone were favorable for the optimal growth of *S. sanghuang*, but inorganic nitrogen sources such as ammonium tartrate, ammonium sulfate, ammonium nitrate, and urea were unfavorable for the growth of *S. sanghuang*. These results were comparable with the findings reported by Cheng et al. [11]. Since the nutritional conditions significantly alter the growth of *S. sanghuang*, a comprehensive study of nutrients and growth factors was necessary to develop an efficient, affordable, and optimal culture medium for obtaining maximal mycelial yield. Many researchers have chosen to add growth factors of natural components to promote the growth or the production of bioactive compounds of *S. sanghuang*. For example, based on the study of *Sanghuangporus baumii* medium of *Morus alba* L., Wang et al. reported that a mulberry bark extract promoted the growth of *S. sanghuang* [26]. In this study, 1.25 g/L dandelion powder significantly promoted the growth of *S. sanghuang*.

Furthermore, we analyzed the antioxidant and anticancer activities of the extracellular polysaccharides from *S. sanghuang* by in vitro experiments. The polysaccharides from *S. sanghuang* showed stronger ABTS-radical scavenging ability than DPPH radical scavenging ability. The total antioxidant capacity was also stronger at higher concentrations of the polysaccharides. When the concentration of extracellular polysaccharide is 0.5 mg/mL, the scavenging rate of ABTS free radicals by extracellular polysaccharide can reach 90% and above, which shows that the antioxidant capacity of *S. sanghuang* extracellular polysaccharides is substantial. Within a certain concentration range, increasing the polysaccharide concentration can improve the scavenging ability of ABTS and DPPH free radicals. The antioxidant effects of the exopolysaccharide of the *S. sanghuang* can provide a theoretical basis for its application in medical dressings, facial cleansers, facial masks, antioxidants, etc. *S. sanghuang* also has triterpenoids, flavonoids, and other compounds, but its biological activity needs to be further explored. Lang et al. [27] found that *Phellinus baumii* flavonoids have strong reducing power and strong scavenging ability for DPPH free radicals and hydroxyl free radicals, and the antioxidant capacity is dose-dependent on the flavonoid concentration. Li [28] reported that the flavonoid compounds of sanghuang had strong DPPH free radical scavenging ability and iron reduction ability.

MTT assay results showed that the polysaccharides from *S. sanghuang* significantly decreased the survival rate of U251 and T98G cells (human glioma cells), and PC-3 cells (human prostate cancer cells). The anticancer effects of the polysaccharides from *S. sanghuang* were poorer on the remaining three types of cancer cells. Therefore, in future studies, further investigations are necessary to determine the mechanisms by which the extracellular polysaccharides of *S. sanghuang* inhibit the growth and survival of cancer cells. The unique anticancer effects of sanghuang polysaccharides has made this a hot topic in functional food research at home and abroad. Liu et al. [14] have shown that *Phellinus baumii* polysaccharides can inhibit the proliferation of Hela and SGC-7901 cells by hindering cell division. How *S. sanghuang* extracellular polysaccharides exert anticancer effects remains to be further studied in later trials. Guo's [29] research found that *S. sanghuang* triterpenoids have inhibitory effects on PC-3, HepG-2, MDA-MB-231, HCT-116, and other cancer cells, and that they have a certain anticancer potential. In the future, we may measure the metabolic profile of *S. sanghuang* and analysis its correlation with biological activity.

5. Conclusions

The optimal formula for liquid fermentation of *S. sanghuang* obtained after optimization in this study is as follows: glucose 30.2 $\pm$ 0.33 g/L, yeast extract 14.58 $\pm$ 0.05 g/L, and dandelion powder 1.24 $\pm$ 0.01 g/L; under these conditions, the mycelium biomass can reach 13.99 $\pm$ 0.42 g/L, which is 2-fold higher than the formula used before optimization. Further research found that the extracellular polysaccharides from *S. sanghuang* showed high in vitro antioxidant activity against ABTS and DPPH radicals as well as strong total antioxidant capacity; at the same time, the extracellular polysaccharide of *S. sanghuang* has a good inhibitory effect on the tested cancer cells. Among them, U251 and T98G cells are more sensitive. This study shows that the exopolysaccharides of *S. sanghuang* have good antioxidant activity and anticancer activity.

Author Contributions: Conceptualization, Y.G. and X.L.; methodology, Y.G., X.L. and H.X.; software, Y.G.; validation, Y.G., X.L. and H.S.; formal analysis, H.S.; investigation, J.Z.; resources, J.Z.; data curation, Y.G. and H.X.; writing—original draft preparation, Y.G. and X.L.; writing—review and editing, Y.G., X.L. and J.F.; visualization, X.L. and X.W.; supervision, X.L.; project administration, J.F.; funding acquisition, J.F. All authors have read and agreed to the published version of the manuscript.

Funding: This research was funded by the Key Development Project of the Science and Technology Department of Tibet, under grant number XZ202301ZY0013N.

Institutional Review Board Statement: Not applicable.

Informed Consent Statement: Not applicable.

Data Availability Statement: The original contributions presented in the study are included in the article, further inquiries can be directed to the corresponding author.

Conflicts of Interest: The authors declare no conflicts of interest.

References

1. Wu, S.H.; Dai, Y.C.; Hattori, T.; Yu, T.W.; Wang, D.M.; Parmasto, É.K.; Chang, H.-Y.; Shih, S.-Y. Species clarification for the medicinally valuable 'sanghuang' mushroom. *Bot. Stud.* **2012**, *53*, 135–149.
2. Zhu, L.; Song, J.; Zhou, J.L.; Si, J.; Cui, B.K. Species diversity, phylogeny, divergence time, and biogeography of the genus *Sanghuangporus* (*Basidiomycota*). *Front. Microbiol.* **2019**, *10*, 812. [CrossRef] [PubMed]
3. Wang, W.; Song, J.; Lu, N.; Yan, J.; Chen, G. *Sanghuangporus sanghuang* extract inhibits the proliferation and invasion of lung cancer cells in vitro and in vivo. *Nutr. Res. Pract.* **2023**, *17*, 1070–1083. [CrossRef] [PubMed]
4. Li, I.; Lu, T.; Lin, T.; Chen, A.Y.; Chu, H.; Chen, Y.; Li, T.; Chen, C. Hispidin-enriched *Sanghuangporus sanghuang* mycelia SS-MN4 ameliorate disuse atrophy while improving muscle endurance. *J. Cachexia Sarcopenia Muscle* **2023**, *14*, 2226–2238. [CrossRef] [PubMed]
5. Li, I.; Chang, F.; Kuo, C.; Chu, H.; Li, T.; Chen, C. Pilot Study: Nutritional and preclinical safety investigation of fermented Hispidin-Enriched *Sanghuangporus sanghuang* mycelia: A promising functional food material to improve sleep. *Front. Nutr.* **2022**, *8*, 788965. [CrossRef]
6. Li, T.; Wang, Q.; Yang, Y.; Song, D. The mechanism of polysaccharide synthesis of *Sanghuangporus sanghuang* based on multi-omic analyses and feedback inhibition. *Carbohydr. Polym.* **2023**, *321*, 121288. [CrossRef] [PubMed]
7. Liu, J.; Song, J.; Chen, W.; Sun, L.; Zhao, Y.; Zong, Y.; He, Z.; Du, R. Assessment of cytotoxicity, acute, subacute toxicities and antioxidant activities (in vitro) of *Sanghuangporus vaninii* crude polysaccharide. *J. Ethnopharmacol.* **2024**, *319*, 117284. [CrossRef] [PubMed]
8. Wang, H.; Ma, J.; Wu, D.; Gao, N.; Si, J.; Cui, B. Identifying bioactive Iingredients and antioxidant activities of wild *Sanghuangporus* species of medicinal fungi. *J. Fungi* **2023**, *9*, 242. [CrossRef] [PubMed]
9. Xie, C.Y.; Li, J.M.; Guo, L.L.; Zhang, S.Y.; Zhu, W.Q.; Cai, L.S. Comparison and optimization of determination methods of extracellular polysaccharides infermentation broth of Sanghuang mushroom. *Edible Fungi* **2022**, *44*, 69–73.
10. Qu, Y.; Yang, H.; Li, S.; Li, L.; Li, Y.; Wang, D. The involvement of Th1 cell differentiation in the anti-tumor effect of purified polysaccharide from *Sanghuangporus vaninii* in colorectal cancer via multi-omics analysis. *Int. J. Biol. Macromol.* **2023**, *237*, 123927. [CrossRef]
11. Cheng, J.; Song, J.; Wei, H.; Wang, Y.; Huang, X.; Liu, Y.; Lu, N.; He, L.; Lv, G.; Ding, H.; et al. Structural characterization and hypoglycemic activity of an intracellular polysaccharide from *Sanghuangporus sanghuang* mycelia. *Int. J. Biol. Macromol.* **2020**, *164*, 3305–3314. [CrossRef] [PubMed]
12. Jiang, P.; Yuan, L.; Cai, D.; Jiao, L.; Zhang, L. Characterization and antioxidant activities of the polysaccharides from mycelium of *Phellinus pini* and culture medium. *Carbohydr. Polym.* **2015**, *117*, 600–604. [CrossRef] [PubMed]
13. Zhu, X.; Guo, R.; Su, X.; Shang, K.; Tan, C.; Ma, J.; Zhang, Y.; Lin, D.; Ma, Y.; Zhou, M.; et al. Immune-enhancing activity of polysaccharides and flavonoids derived from *Phellinus igniarius* YASH1. *Front. Pharmacol.* **2023**, *14*, 1124607. [CrossRef] [PubMed]
14. Liu, M.M.; Zeng, P.; Li, X.T.; Shi, L.G. Antitumor and immunomodulation activities of polysaccharide from *Phellinus baumii*. *Int. J. Biol. Macromol.* **2016**, *91*, 1199–1205. [CrossRef]
15. Dong, W.; Ning, L.; Lu, W.D.; Li, C.C.; Chen, R.P.; Jia, X.N.; Wang, L.; Guo, L.Z. Tumor-inhibitory and liver-protective effects of *Phellinus igniarius* extracellular polysaccharides. *World J. Microbiol. Biotechnol.* **2009**, *25*, 633–638. [CrossRef]
16. Jong, C.; Jin, M.; Ju, Y.; Li, D.; Zhu, H.; Jo, T.H.; Kim, G. Optimization of Technical Parameters on the Medium Formulation and Culture Conditions in Production of *Pleurotus eryngii* Liquid Spawn. *Int. J. Sci. Res. Sci. Technol.* **2021**, *8*, 1019–1036. [CrossRef]
17. Ding, Q.L. Comparison on mycelium activity and mixed fungi-resistant capacity of different strains of *Pleurotus eryngii*. *Fujian Agric. Sci. Technol.* **2015**, *7*, 4–6.
18. Jia, J.; Xie, X.C.; Song, Y.; Sun, Y.; Xu, Y.J.; Yao, S. Optimizing culture medium and antioxidant activity of extracellular polysaccharides in liquid fermentation of *Stropharia rugosoannulata*. *North. Hortic.* **2021**, *495*, 105–115.
19. Li, Y.H.; Yin, C.M.; Fan, X.Z.; Shi, D.F.; Yao, F.; Li, C.; Liu, Y.; Cheng, Y.Q.; Gao, H. In Vitro Anti-oxidant, hypoglycemic, and hypouricemic activities of *Sanghuangporus vaninii* extracts. *Mod. Food Sci. Technol.* **2022**, *38*, 71–80.
20. Luo, Q.; Xu, J.W. Monosaccharide composition and antioxidant activity of extracellular polysaccharides and intracellular polysaccharides in liquid ferment of *Ganoderma lucidum*. *Acta Edulis Fungi* **2021**, *28*, 135–142.
21. Benzie, I.F.; Strain, J.J. The ferric reducing ability of plasma (FRAP) as a measure of "antioxidant power": The FRAP assay. *Anal. Biochem.* **1996**, *239*, 70–76. [CrossRef]
22. Zhang, W.J. *Optimizatian of Liquid Fermentation Conditions of Phellinus igniarius and Extraction of Phellinus igniarius Polysaccharide*; Changchun University: Changchun, China, 2022.
23. Lai, Q.H.; Li, B.; Zhao, H.; Sun, C.Q.; He, X.M.; Deng, L. Development advantages, existing problems and strategies of *Sanghuangporus sanghuang* cultivation of Dazhou City. *Edible Med. Mushrooms* **2023**, *31*, 302–305.

24. Liu, C.Y.; Yuan, Y.; Hao, X.Y.; Liu, K.; Wu, X.P.; Fu, J.S. A comparative study of intracellular and extracellular polysaccharides in mycelia of *Cordyceps cicadae* against oxidative damage of hepatocytes. *Mycosystema* **2020**, *39*, 421–433.
25. Jiang, F.; Zhang, H.; Feng, J.; Yang, Y.; Feng, N.; Wu, D. Flavonoid synthesis by magnesium acetate stress in liquid fermentation of *Sanghuangporus sanghuang*. *Mycosystema* **2017**, *36*, 1141–1151.
26. Wang, S.; Liu, Z.; Wang, X.; Li, Y.; Liu, R.; Zou, L. Isolation, Culture and Bioactive Substances Elicitation of *Sanghuangporus baumii*. 2021. Available online: https://www.researchsquare.com/article/rs-1105519/v1 (accessed on 20 March 2024). [CrossRef]
27. Lang, M.Z.; Liu, M.M.; Xie, H.Q.; Li, C.H.; Chen, Y.J.; Shen, Z.F.; Li, Y.G.; Shi, L.G. Optimizing ultrasonic-assisted ethanol extracting conditions of total flavonoids from *Phellinus baumii* and its antioxidant effect in vitro. *Acta Sericologica Sin.* **2019**, *45*, 262–268.
28. Li, T. *Identification of Different Sanghuang Strains and Extraction Process and Function of Total Flavonoids from Solid Cultures*; Huazhong Agricultural University: Wuhan, China, 2023.
29. Guo, T. *Extraction, Purification and Antitumor Activity of Total Triterpenoids from Sanghuangporus Sanghuang*; Hebei University of Economics and Business: Shijiazhuang, China, 2023.

antioxidants

MDPI

Review

Ginger Bioactives: A Comprehensive Review of Health Benefits and Potential Food Applications

Muhammad Nouman Shaukat [1], Akmal Nazir [2] and Biagio Fallico [1,*]

1 Department of Agriculture, Food and Environment, University of Catania, Via S. Sofia 100, 95123 Catania, Italy; nouman.shaukat@phd.unict.it

2 Department of Food Science, College of Agriculture and Veterinary Medicine, United Arab Emirates University, Al Ain P.O. Box 15551, United Arab Emirates; akmal.nazir@uaeu.ac.ae

* Correspondence: biagio.fallico@unict.it; Tel.: +39-95-7580214

Abstract: Ginger is an herbaceous and flowering plant renowned for its rhizome, which is widely employed as both a spice and an herb. Since ancient times, ginger has been consumed in folk medicine and traditional cuisines for its favorable health effects. Different in vitro and in vivo studies have disclosed the advantageous physiological aspects of ginger, primarily due to its antioxidant, anti-inflammatory, antimicrobial, and anti-carcinogenic properties. These health-promoting features are linked to the variety of bioactive compounds that are present in ginger. Following the advancement in consumer awareness and the industrial demand for organic antioxidants and functional ingredients, the application of ginger and its derivatives has been broadly investigated in a wide range of food products. The prominent features transmitted by ginger into different food areas are antioxidant and nutraceutical values (bakery); flavor, acceptability, and techno-functional characteristics (dairy); hedonic and antimicrobial properties (beverages); oxidative stability, tenderization, and sensorial attributes (meat); and shelf life and sensorial properties (film, coating, and packaging). This review is focused on providing a comprehensive overview of the tendencies in the application of ginger and its derivatives in the food industry and concurrently briefly discusses the beneficial aspects and processing of ginger.

Keywords: ginger; bioactives; antioxidants; gingerols; health benefits; food applications; bakery; dairy; meat

Citation: Shaukat, M.N.; Nazir, A.; Fallico, B. Ginger Bioactives: A Comprehensive Review of Health Benefits and Potential Food Applications. *Antioxidants* **2023**, *12*, 2015. https://doi.org/10.3390/antiox12112015

Academic Editors: Yongliang Zhuang, Hu Hou, Zhipeng Yu and Shengbao Cai

Received: 23 October 2023
Revised: 15 November 2023
Accepted: 16 November 2023
Published: 18 November 2023

1. Introduction

Ginger (*Zingiber officinale* Roscoe) plant, producing an irregular bumpy rhizome, is extensively cultivated in the southeast region of Asia. Accurate information regarding ginger plant's origin is not available due to a very long cultivation history in this region. In ancient Greek history (40–90 AD), the Greeks were familiar to the dietary and culinary uses of ginger. In the 13th century, during his visit to China and Sumatra, Marco Polo became acquainted with ginger and imported it to Europe. In the same era, the Arabs transported the ginger from India to East Africa, while the Portuguese spread it to West Africa [1].

It is an important dietary ingredient in daily South Asian cuisine and traditional Chinese medicine. In the entire history, ginger rhizome, commonly known simply as ginger, is not merely utilized as a spice or herb to impart various food attributes like flavor, aroma, food preservation, and nutritional reform, but it has also been administered for the mitigation of various ailments such as asthma, flu, indigestion, and gastrointestinal discomfort [2].

Tropical and subtropical climates are favorable for ginger cultivation because it cannot combat very low temperatures. When ginger is utilized as a vegetable or for the production of ginger preserve, pickles, candies, and beverages, the recommended harvesting time is about 4–5 months after plantation, whereas when its intended employment is dried ginger and for the manufacturing of bleached and dehydrated ginger or functional products like ginger oil and oleoresin, the recommended harvesting period is around 8–10 months [3]. The height of a ginger plant is usually around 2 feet with yellow bloomed flowers characterized by a pungent smell. The underground stem or rhizome of ginger, attributed with a hot and strong spicy aroma flavor, used for medicinal and culinary purposes, is also one of the most abundantly used spices belonging to the *Zingiberaceae* family [3,4]. The ginger plant has a very close relevancy to other herbal plants like cardamom (*Elettaria cardamomum*) and turmeric (*Curcuma longa*) as they are members of the same family (*Zingiberaceae*). Ginger comprises carbohydrates (60–70%), water (9–12%), proteins (9%), lipids (3–6%), ash (8%), crude fiber (3–8%), and essential oil (2–3%). Apart from these macronutrients, ginger also contains various bioactive compounds and some mineral elements like sodium, potassium, calcium, magnesium, and phosphorus [5]. There is also a possibility of synthesizing new compounds when there is any alteration in the ginger's growth cycles, for example, during the storage of ginger. The time period, humidity level, and temperature are also the influencing factors of the normal composition of ginger [6]. The presence of ketones in the chemical composition bestows the unique spicy aroma to these plants [7].

Ginger has been the subject of extensive research in the past owing to its possible health advantages. Several authors have reviewed the available literature and discussed the bioactive compounds present in ginger, the techniques used for their extraction, health promoting features, the encapsulation of ginger bioactives, and some specific or brief food applications of ginger and its derivatives [6,8–13]. However, there is a lack of reviews regarding the application of ginger to different food categories. It is also imperatively important to mention that the primary objective of incorporating ginger and its derivatives into a food product is to improve the nutraceutical value and shelf life of the relevant product behind which the lead character, directly or indirectly, belongs to the antioxidant compounds and their antioxidant activities. Therefore, the present paper is designed to furnish a comprehensive review of ginger application in different food sectors, while also briefly discussing the bioactive components, the effect of processing and extraction methods, and potential health benefits.

2. Bioactive Compounds in Ginger

The nutritional and nutraceutical benefits of ginger are closely associated with the presence of various bioactive compounds. These bioactive compounds, reported in different studies, are classified into three classes: gingerols, volatile oils, and diarylheptanoids [14]. The chemical components of ginger include volatiles (geraniol, borneol, terpineol, zingiberenol, cineole, limonene, camphene, curcumene, zingiberol, geranyl acetate, linalool, α-Farnesene, α-sesquiphellandrene, α-zingiberene, β-bisabolene, β-elemene, and β-phellandrene) and non-volatiles (gingerols, shogaols, zingerone, and paradols). These non-volatile phytochemicals, having pungent odors and flavors, are the characteristic bioactive constituents of ginger. The distinctive aroma flavor of ginger is the resultant blend of these three non-volatile active compounds (gingerol, shogaol, and zingerone), which constitute one to three percent of fresh ginger by weight. There are two common components, vanillyl (4-hydroxy-3-methoxphenyl) and the ketone functional group, that are found in every pungent compound of ginger [7,15]. The chemical structures of the major bioactive compounds that are present in ginger extract and essential oil are presented in Figure 1.

Gingerol is a pungent oil, yellow in color, that is also found in crystalline solid form with a low melting point. Gingerols are classified into 4-, 6-, 8-, 10-, and 12-gingerol based on the length of the unbranched alkaline chain. 6-Gingerol is the most abundantly available type of gingerol, while considerable amounts of 8-gingerol and 10-gingerol are also found in fresh ginger [16,17]. 6-Gingerol is a moderately pungent and vital active compound of ginger with anti-inflammatory, anti-bacterial, and anti-carcinogenic properties. 6-Gingerol is reported to have different health-beneficial properties like anti-carcinogenic, antioxidant, anti-inflammatory, cardiotonic and hypotensive, anti-emetic, anti-pyretic, anti-rheumatic, anti-ulcer, anti-prostaglandin, and many other properties. 8-Gingerol is the most pungent and one of the most active phytochemicals in ginger. It has been cited with many health benefits by various researchers [18,19].

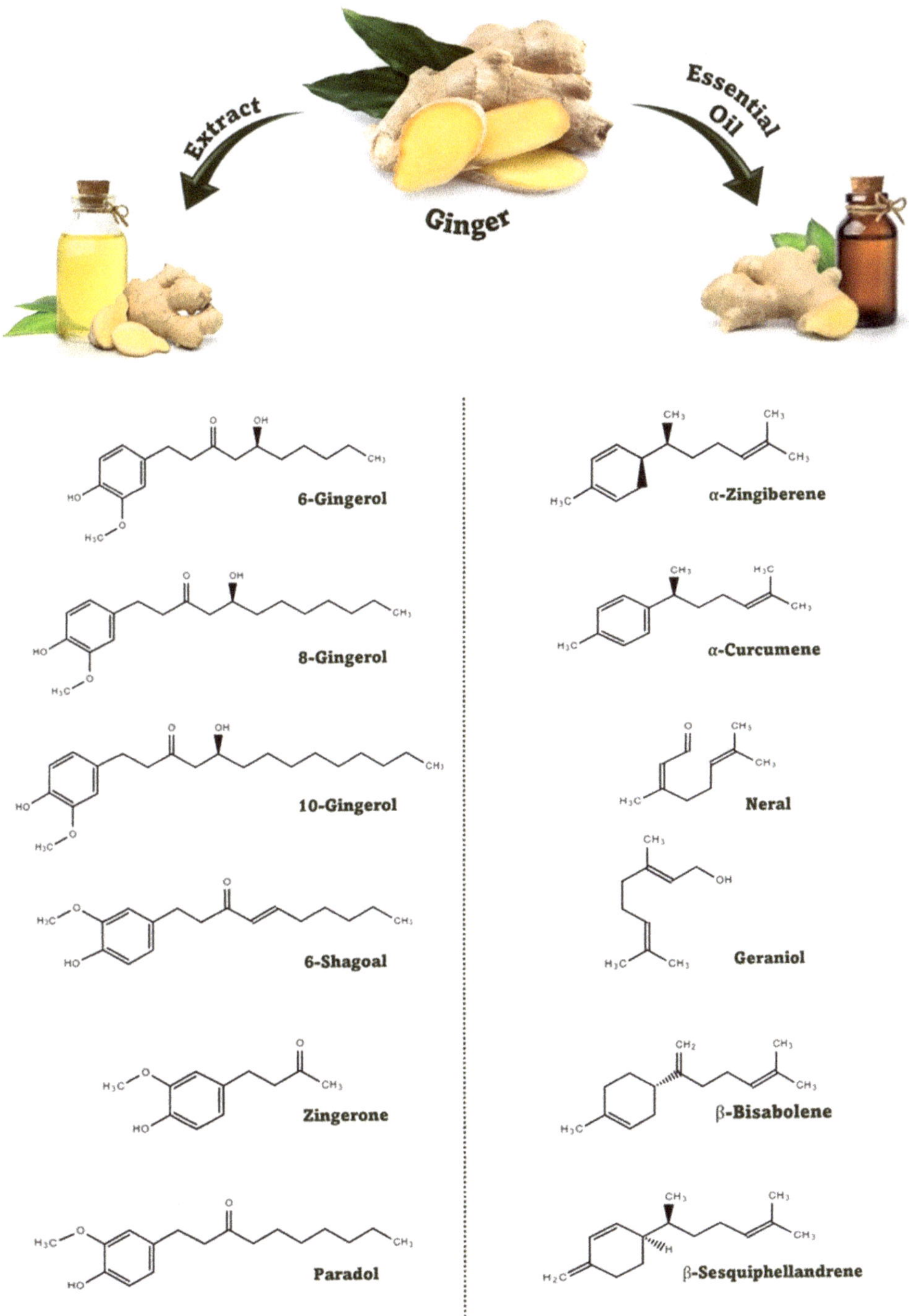

Figure 1. Chemical structures of major bioactive compounds present in ginger.

4. Beneficial Aspects of Bioactive Compounds

As mentioned above, the functional and bioactive compounds present in ginger are associated with multiple health benefits and advantageous characteristics. The exploitation and characterization of the versatile biological and functional competencies of ginger are also crucial for further research. These multifaceted bioactivities of ginger bioactives and their modes of action are discussed here in this section.

4.1. Antioxidant Activity

Any disturbance or adjustment with the cell's antioxidant defense system can lead to the excessive production of Reactive Oxygen Species (ROS), and eventually, to an increase in oxidative stress. For the recovery of normal levels of free radicals and oxidative stress, various natural antioxidants have been analyzed for the neutralization of free radicals and encounter oxidative stress for the prevention of different health disorders [35]. The maintenance of a normal ROS level in a living organism is very important for normal functioning, which can be achieved through the provision of hydrogen atoms or electrons to free radicals to hinder the ROS production and convert them into antioxidant lipid complexes or thermally stable products [36].

In the past two decades, the exploration of natural plant resources has become very attractive and beneficial, most probably due to the presence of bioactives, less toxicity, and the lower production costs [37]. Ginger bioactive compounds have significant antioxidant potential to restore or maintain normal oxidative stress levels. Ginger leaves were reported to have more antioxidative potential than ginger rhizome and ginger flowers [32,38]. The higher in vitro antioxidant activity (88.93 $\pm$ 0.03% DPPH; 88.23 $\pm$ 0.98% ABTS) was noted in 6-gingerol. The daily oral administration of 6-gingerol (50–75 mg/Kg body weight) to mice for three weeks enhanced their blood glucose levels, reduced their oxidative stress, and also improved their insulin signaling. A higher antioxidant activity of shogaols was recorded as 89.01 $\pm$ 0.6% and 90.2 $\pm$ 0.11% for ABTS and DPPH, respectively [39,40]. The highest antioxidant potential of 6-shogaol compared to other phenols of ginger is due to the existence of a special functional group (α, β-unsaturated carbonyl), which is efficiently involved in the regulation of glutathione. Glutathione is the most amply available type of thiol in animal cells, and it is engaged in the cell detoxification mechanism against detrimental substances and intracellular redox management [41].

4.2. Anti-Inflammatory Activity

Inflammation is a phase in the natural healing process of the body to combat intruders for the prevention of infection or injury. The basic aim of inflammation is to abrogate cellular damage, withdraw and absorb necrotic cells and tissues, and regain the stability and conformity of the intracellular environment [42]. When cells go through severe inflammation, a large number of mononuclear immune cells are penetrated, which stimulates the production of inflammatory interleukins, cytokines, and tumor necrosis factors including inflammation-promoting cytokines, such as tumor necrosis factor α (TNF-α), interleukin-1β (IL-1β), and interleukin-6 (IL-6), which are very significant for the inflammation process. A recent study focused on the anti-inflammatory characteristics of ginger mentioned that ginger consumption could substantially lower the TNF-α serum levels but has a minor impact on IL-6 [43].

It was also observed that the anti-neuroinflammatory capacity of gingerols enhanced with the increase in the alkyl chain length as 10-gingerol's anti-neuroinflammatory activity was higher than the other active compounds of ginger [44]. Furthermore, different in vivo studies have also supported these claims through the practical implication of ginger bioactives. Nanoparticles derived from ginger that were orally administered to mice, having acute and chronic colitis, at the dosage rate of 0.3mg/mice on a daily basis for one week efficiently halted intestinal inflammation by decreasing the pro-inflammatory cytokines like TNF-α, IL-1β, and IL-6 and increasing the levels of anti-inflammatory cytokines like IL-10 (Interleukin-10) and IL-22 (Interleukin-22). In mice suffering with dextran sulfate sodium-induced colitis, when 6-shogaol-equipped nanoparticles (equivalent to 15 mg/kg of 6-shogaol) were orally administered on a daily basis for a week, they not only diminished the colitis symptoms but also boosted the healing process [45,46].

4.3. Antimicrobial Activity

Currently, chemical and synthetic antimicrobials are being replaced with natural and organic antimicrobials derived from plants, animals, and microorganisms, both in health management and in food preservation. These biologically derived antimicrobials from plant sources are preferred due to their multi-oriental management, multi-targeted approach, drug resistance, and lower toxicity [47]. Ginger and its derivatives have been appraised for their tremendous anti-bacterial, anti-viral, and anti-fungal activities [48]. Microorganisms tend to make biofilm as their strong defense against antimicrobials. A study carried out by Chakotiya et al. [49] reported that ginger has shown an inhibitory action against *Psedomonas aeruginosa* (a multi-drug-resistant strain) by disturbing the membrane stability and preventing biofilm formation.

Ginger essential oil possesses excellent anti-bacterial activity against food-borne pathogenic bacteria such as *Escherichia coli* and *Staphylococcus aureus*. A study reported the bactericidal mechanism of ginger essential oil through the demolishment of the cell membrane activity and through intervening with the energy metabolism because of some macromolecular elements like proteins and nucleic acids [50]. Another study expressed that fresh ginger oil had a moderate antimicrobial effect on *Candida*, *Aspergillus niger*, and *P. aeruginosa* and a mild effect against *Saccharomyces cerevisiae*, and no inhibition was detected against *Bacillus subtilis*, *Trichoderma* spp., and *Pencillium* spp. Ginger oil obtained from dried ginger was observed to have higher anti-bacterial activity against *P. aeruginosa* [51]. There is a close correlation between anti-bacterial and antioxidant properties as the blocking of antioxidant activity could result in the detention of anti-bacterial activity. Processing and storage conditions could also be made changes with functional attributes of ginger resulting in difference in antioxidant and anti-bacterial activities [52].

4.4. Anti-Carcinogenic Potential

Cancer is one of the major causes of death in the world, and in 2020, approximately 19 million cancer cases were reported, out of which about 10 million cases worldwide were fatal [53]. In recent years, ginger has been extensively investigated for its anti-carcinogenic activities against different types of cancer, like cervical, breast, colorectal, and prostate cancers [54,55]. Chronic inflammation plays a key function in carcinogenesis as it has a close connection with all crucial stages of cancer formation [56]. Ginger and its active compounds have been identified with anti-carcinogenic properties against all stages of cancer development including cancer initiation, promotion, progression, and drug resistance. An in vitro study disclosed that a fraction loaded with polyphenols derived from dried ginger powder restrained the progression of gastric adenocarcinoma cells and colorectal cancer cells [57].

In vivo and in vitro studies were carried out to evaluate the action mechanism and cytotoxic effects of ginger against prostate cancer [58]. It was reported that 6-gingerol, 6-shogaol, 10-gingerol, and 10-shogaol showed considerable anti-proliferative effects on prostate cancer cells in humans through the deregulation of the protein expression of MRP1 (multi-drug resistance-associated protein 1) and GSTπ (Glutathione-S-transferase). The proliferation of PC-3 prostate cancer cells was synergistically inhibited through the bifold consolidation of ginger bioactives like 6-gingerol, 8-gingerol, 10-gingerol, and 6-shogaol [59]. 6-Gingerol stimulated the increase in ROS formation in human gastric adenocarcinoma (AGS) cells, which promote apoptosis and reduce the mitochondrial membrane potential [60]. The in vitro application of 6-shogaol (10–20 μM) to PC12 cells for 24 h improvised the phase II antioxidant compounds at the pheochromocytoma cell line in mice, which effectively inhibited the proliferation of oxidative stress in PC12 cells. On the contrary, 6-gingerol failed to protect the PC12 cells from the oxidative stress, possibly due to the absence of an α, β-unsaturated ketone, which is an indicator of the α, β-unsaturated ketone unit's importance in cytoprotection [61].

4.5. Neuroprotective Activity

There is a possibility of an onset of neurodegenerative diseases such as Alzheimer's disease and Parkinson's disease, and old people are at an especially high risk. Many recent studies have expressed that ginger possesses anti-neuroinflammatory capacity and has a significant role in memory function, which might be useful in the management and inhibition of neurodegenerative diseases [62,63]. It was also disclosed that 10-gingerol from fresh ginger exhibits powerful anti-neuroinflammatory activity, proven through a lipopolysaccharide-activated BV2 microglia culture model. 10-Gingerol restricted the articulation of pro-inflammatory genes by inhibiting NF-κB activation, which ultimately led to a decrease in the levels of IL-6, IL-1β, NO, and TNF-α. [44]. The phytochemicals derived from ginger such as gingerols, shogaols, and others, at a concentration of 20 μM, have been reported with strong neuroprotection capacities, especially against Alzheimer's disease. Ginger could lower the Aβ-species and cerebral plaques, inhibit the cerebral inflammation, amend the tau protein, and restrain Aβ-induced apoptosis through different pathways [64].

4.6. Cardiovascular Protection

Worldwide, the dominant cause of premature deaths (17.9 million people per year) is cardiovascular disease [65]. Various studies revealed that ginger can efficiently provide protection against cardiovascular diseases by reducing the levels of blood pressure and blood lipids. The level of high-density lipoprotein cholesterol (HDL-C), also known as good cholesterol due to its defensive stature against cardiovascular diseases, increased in the serum, while a decrease in body weight was observed in mice on a high-fat diet supplemented with 0.5% of lyophilized ginger extract for a period of about two weeks. Ginger extract also facilitated high-density lipoprotein (HDL) production through enhancing the levels of lecithin–cholesterol acyltransferase RNA and apoliprotein A-1 in the liver. Another study also supported the role of ginger extract in the reduction in low-density lipoprotein (LDL) and total cholesterol, and the integrated daily oral application of aqueous ginger extract (250 mg/kg of feed) for about a month and aerobic exercise increased the HDL levels in high-fat-diet-fed rats [66,67].

4.7. Anti-Obese Activity

Obesity or overweight is a great concern throughout the world, and it emerges from the disturbance in the energy metabolism. Basically, there are two types of adipose (fat) tissues classified on the basis of their arrangements and activities. The core function of white adipose tissues is to serve as energy depositories, while brown adipose tissues are responsible for metabolism improvement and energy expenditure. The transformation of white adipose tissues to brown adipose tissues is a novel approach to tackle obesity. 6-Gingerol could stimulate the cell browning via an AMPK-dependent approach in 3T3-L1 adipocytes [68,69]. Gingerenone A was discovered as a better performer than gingerol and shogaol with efficient inhibition activity against adipogenesis and fat accumulation in 3T3-L1 preadipocyte cells. In this study, the oral administration of Gingerenone A (50 mg/kg of bodyweight) on a daily basis for 15 weeks also influenced the fatty acid metabolism through the in vivo activation of AMPK, debilitating diet-induced obesity [70].

4.8. Anti-Diabetic Effect

Diabetes mellitus is a chronic metabolic disease characterized by the insufficiency of insulin and/or insulin resistance, resulting in an uncontrolled increase in the blood glucose level. Persistent hyperglycemia could promote protein glycation and the generation of advanced glycation end products (AGEs). An in vitro study revealed that 6-gingerol and 6-shogaol inhibited the proliferation of diabetic complexities and confined the methylglycoxal

(MGO), the precursor of AGEs, to hinder AGE production [71]. The oral supplementation of 6-gingerol (200 mg/kg of corn oil) on a day-to-day basis for 28 days to type 2 diabetic mice stimulated glycogen synthase 1 and improved the GLUT4 (glucose transporter type 4) cell membrane presentation, which raised the glycogen storage levels in the skeletal muscles [72].

5. Food Applications

Currently, there is a growing trend towards healthier and functional food products as consumers are more conscious about healthy lifestyles and daily intakes. Due to prospering recognition and interest in functional food products, the scope for the exploration, characterization, and potential utilization of functional food ingredients and bioactive compounds has also magnified. Ginger, being a valuable functional ingredient and marvelous source of bioactives, has also been integrated into a variety of food products. Here, the food applications of ginger in its different forms (powder, juice, extract, and essential oil) will be discussed to provide a comprehensive and broader prospective of its employment. The impact of ginger incorporation on the sensorial attributes of various food products is also discussed below (Figure 3).

5.1. Bakery Products

A variety of studies have been carried out to use ginger as a potential food additive in bakery products (Table 1). For instance, the supplementation of ginger powder at a rate of 2–4% was reported, which can promise feasible textural, rheological, and antioxidant properties along with a better mineral content in bread [73]. Balestra et al. [74] formulated a functional bread using ginger with acceptable physicochemical and sensorial characteristics. They found that up to 3% of ginger powder can be incorporated into the bread without compromising the rheological properties of the dough and bread. Another study also came up with similar findings over the functional, physicochemical, and sensory properties of bread fortified with ginger powder but with a different optimized level of ginger powder (1%) than the previous study [75]. Moreover, Almasodi [76] studied the addition of ginger powder in bread as well as in rusk. They concluded that the incorporation of up to 3% and 6% of ginger powder resulted in satisfactory hedonic and textural properties for bread and rusk, respectively. In another study, garlic and ginger powders at different concentrations were also incorporated in wheat bread. Besides functionality improvement, it also increased the shelf life of the bread from 5 days to 7 days at an ambient temperature [77]. Ginger and turmeric have inhibitory effects on the growth of *Rhizopus stolonifer* on bread, but ginger may impart unfavorable flavor characteristics to the bread [78].

The aforementioned studies mainly dealt with the addition of ginger to produce functional bread. However, in some parts of the world, distinct baking products exist wherein ginger is used as a dominant flavoring ingredient. For instance, gingerbread is a traditional bakery product in the Czech Republic, particularly in the town of Pardubice, where ginger is the prevalent flavor. It is produced in a variety of shapes, types, and textures ranging from a crispy biscuit to a soft cake [79,80]. Hence, ginger has been practiced as a flavor enhancer, solely or in a combination, for a long time, but it is largely dependent on the nature of the food product, the form of the ginger to be incorporated, their mutual compatibility, and the sensorial preferences of the consumers.

Lee et al. [81] proposed that functional cookies can be prepared with a 4% incorporation of ginger powder, as this concentration was found to have overall acceptability and a very bland effect on the quality characteristics. Kaushal et al. [82] also investigated the functional and sensorial properties of cookies with the addition of ginger powder. The functional and organoleptic properties of cookies supplemented with 12% ginger powder were very favorable compared to the controls. In another investigation, Yang et al. [83] also evaluated the effect of ginger substitution on the acrylamide content in dough and biscuit properties. Ground ginger addition at the rate of 1% was reported with favorable structural and color characteristics and acceptable sensorial characteristics with a 6.2%

decrease in the acrylamide content. The substitution of ginger at higher rates decreased the acrylamide content, but the hedonic properties were not feasible. In principle, gingerols reduced the acrylamide content through inhibition where the phenol hydroxyl group of gingerol played a vital role. Kumari and Gupta [84] also utilized ginger powder and some other functional plant ingredients in the production of value-added cookies. Acceptable organoleptic characteristics were found with 2.5% of ginger powder.

The functional, nutritional, antioxidant, and sensorial properties of millet flour and ginger powder-formulated cookies were investigated by Marak et al. [85]. This study reported that ginger flour incorporation of up to 10% was found satisfactory for the textural and sensorial properties of cookies, and it also efficiently improved the functional and nutritional values of cookies with a potential lead towards shelf life extension. When cumin and ginger were utilized for the preparation of cookies, a minimal difference was observed in the dough development as compared to the control [86]. Both potential ingredients were integrated in concentrations of up to 5% in soft wheat cookies, but there was not any substantial effect on the end product. Keeping in view the acceptability and feasibility indicators, these concentrations were also recommended for commercial scale production. The antimicrobial effect of ginger was also evaluated through the addition of ginger powder into biscuits. Aerobic bacteria were decreased proportionally by increasing the amount of ginger powder in biscuits. Functional biscuits with significant activity against bacteria and mold could be produced with around a 2.5% addition of ginger powder without any compromise on the quality characteristics [87]. Adebayo-Oyetoro et al. [88] also prepared ginger spiced cookies with around a 3% addition of ginger powder.

Table 1. Some important studies on the application of ginger and its derivatives to bakery products.

Derivative	Product	Positive Effect	Reference
Ginger Powder	Wheat Bread	↑ Antioxidant Activity; Minerals; Total Phenolics; Textural Properties; Shelf Life; Sensorial Characteristics; ↓ Microbial Growth	[73,77]
	Cookies	↑ Antioxidants; Phenolic Contents; Shelf Life; ↓ Acrylamide Content; Microbial Growth; ≈ Sensorial Characteristics	[82,83,85,86]
	Cake	↑ Antioxidant Activity; Total Phenolics, Shelf Life; Sensorial Properties; ↓ Bacterial and Fungal growth	[89]
Ginger Juice	Wheat Flour Muffin	↑ Total Phenolics; Antioxidants; Sensorial Characteristics; ≈ Textural Attributes	[90]
Ginger Extract	Baked Bars	↑ Antioxidants; Phenolic Contents; Antioxidant Retention during Storage; Flavor and Color; ≈ Overall Acceptability	[91]

↑ = Increased/Improved, ↓ = Lower/Decreased, ≈ = Negligible/Unchanged.

To a large extent, Almasodi [89] successfully incorporated ginger powder in butter cake and biscuits at the rates of 5% and 10%, respectively. Storage studies revealed that, apart from providing a nutritional boost, ginger powder also supplemented bakery products with natural antimicrobial compounds. In another recipe, the water content was substituted with ginger juice at three different concentrations from 10% to 30% to analyze the quality attributes of muffins [90]. Significant differences were observed in the color, texture, and taste with the increasing amount of ginger juice. Ginger juice at 10% of the water used in the recipe was recommended as the optimum amount for muffin production. The conventional and supercritical extracts of ginger were also analyzed through supplementation in baked bars at 3% and 0.3% concentrations, respectively [91]. The supercritical extract performed better than the conventional extract to improve and maintain the phenolic contents and antioxidant activities in storage duration. A sensorial evaluation also declared the supercritical ginger extract to be better than the control and conventional extract.

As per the foregoing studies, ginger powder is the most widely employed form of ginger in bakery products, followed by ginger juice. The pertinent amount of ginger powder and processing conditions are the key factors in the acquirement of balanced functional bakery items without any alteration in the desirable characteristics of the product. So, bakery products with admirable nutraceutical values could also be produced through the precise supplementation of ginger.

5.2. Dairy Products

There is also a considerable potential for innovation and improvement in the dairy sector, where ginger, an active ingredient, has also contributed (Table 2). The supplementation of ginger powder (1%) has been displayed as a bovine yogurt product with an enhanced nutritional value, better quality characteristics, and ease in the production process [92]. Ginger powder is also formulated with dromedary milk to prepare and characterize a functional yogurt product. The resultant product showed very similar results to the previous study (bovine milk yogurt) with the recommendation of 1% ginger powder as an optimum supplementation extent. The water holding capacity, viability of the starter culture, and rate of pH reduction was increased, while the fermentation time and flow behavior was decreased, leading towards an acceptable yogurt from dromedary milk [93]. Ethiopian cottage cheese, 'Ayib', was also treated with ginger and garlic powders to analyze their impacts on the chemical arrangement and sensorial characteristics of the product [94]. Ayib treated with garlic powder was described with better overall acceptability than ginger powder, and further investigations were suggested to explore the biochemical and microbial properties of both powders.

A novel approach was also attempted to evaluate the effects of ginger, cinnamon, and carob powders on spray-dried milk powders and their agglomerates [95]. Milk powder with added ginger powder relatively performed better in terms of solubility, hygroscopicity, flowability, and cohesiveness. The incorporation of these powders significantly reduced the solubility time of milk powders, while the water activities and moisture levels of the resultant products were also within adequate extents, feasible for the safe storage of these milk powders. In another investigation, an Indigenous sub-continental sweet dairy product, Peda, was also prepared, analyzed, and finally optimized (2%) utilizing ginger powder [96]. The inclusion of ginger powder into Peda not only increased the total solid, total sugar, and ash contents, but also improved its shelf life by reducing the total plate count.

Apart from ginger powder, the effect of ginger juice on the physicochemical and hedonic attributes of yogurt was also investigated at different concentrations. An inverse relationship was observed among the amount of ginger juice, the viability of the starter culture, and the hardness of the product. Based on the finding of this study and the sensorial characteristics of the product, the ginger juice addition at the rate of 4% (v/v) was regarded as the suitable concentration in yogurt production [97], while Allam et al. [98] reported that ginger juice (2%) incorporation into fermented milk products reduced the coagulation time and accelerated the syneresis process.

Jadhav et al. [99] formulated a functional milkshake product via the inclusion of ginger juice at four different concentrations from 2.5% to 10% (v/v). The increase in the ginger juice concentration was inversely proportional to the fat, protein, and total solid contents of the product. Considering the sensorial properties and other attributes, a 2.5% ginger juice addition in the milkshake was regarded as an optimum value. Gaur et al. [100] also successfully developed an herbal milk product by optimizing the concentrations of herbal ingredients such as ginger, turmeric, and tulsi (holy basil).

Several studies were also designed for the potential exploitation of ginger into an herbal ice cream product. Herbal ice cream could be manufactured employing 4% ginger juice (w/w) with a better hedonic response than all other treatments including the control. Ginger juice improved the product's resistance against melting but lowered the fat and total solid contents in the herbal ice cream [101]. Similar findings were also reported in another investigation [102]. The incorporation of ginger juice has a pertinent effect on the thermal properties but no substantial influence on the freezing characteristics of ice cream [103]. Mule et al. [104] efficiently employed ginger juice, aloe vera juice, and *L. acidophilus*-based probiotic culture to produce probiotic ice cream at the optimized levels of 3%, 6%, and 7% (w/w), respectively.

The fortification of ginger extract into fermented dairy products is another approach to bestow the favorable characteristics of ginger to the products. Ginger extract, at an 8% concentration of the low-fat yogurt formulation, not only boosted the phenolic contents and antioxidant activities, but also played a vital role in the retention of these properties [105]. The addition of ginger extract also lowers the syneresis process in the yogurt. During the storage period, there was not any remarkable effect on the acidity, but the viscosity of the product was decreased with the incorporation of ginger extract.

In two other investigations, herbal yogurt was developed using ginger extract and evaluated for its antioxidant capacity and sensorial properties. Ginger was regarded as a better source of antioxidant properties than beet root, while the herbal yogurt containing ginger extract (2%) and goat milk was characterized with the highest antioxidant activity than other milk sources (buffalo and cow). A sensorial analysis revealed that herbal yogurt augmented with ginger extract (1%) was superior in color and appearance, while garlic extract (0.2%) was responsible for a better taste and flavor [106,107].

Ginger extract in combination with gum arabic has also been exploited in the manufacturing of a yogurt-based traditional Iranian drink, Doogh [108]. The synbiotic dairy product with maximal overall acceptability and advanced probiotic status was developed with a blend of ginger and gum arabic at the concentrations of 0.25% and 0.5%, respectively. In another attempt to produce synbiotic labneh through the efficient utilization of ginger and probiotics, an effective inhibitory effect on some food-borne pathogens was detected along with an enhanced phenolic profile and superior overall acceptability of the end product [109].

The ethical, economical, and supply chain aspects of the calf rennet compelled dairy experts to explore other potential sources of milk coagulation. Ginger and its derivatives were also analyzed as potential coagulants in cheese processing. Ginger crude extract emerged as an auspicious coagulant in soft un-ripened cheese manufactured from camel milk [110]. Ginger extract did not perform as good as commercial camel chymosin but rendered a crucial task in the value addition of the product. Further studies were also proposed to explore the coagulation capability of pure coagulation enzymes from ginger. Peshawari cheese, a traditional cheese in the Khyber Pakhtunkhwa province of Pakistan, was prepared utilizing ginger protease compared to calf rennet [111]. In a comparison between ginger protease and calf rennet, there was not any pertinent difference observed in terms of the physicochemical and microbiological properties of the Peshawari cheese. In fact, ginger protease accelerated the proteolysis and improved the organoleptic characteristics of the cheese.

Ginger extract has been productively employed in soft cheese manufacturing as a functional ingredient. Ginger extract addition promoted the quality attributes of soft cheese by reducing the pH, oxidative rancidity, and microbial load while increasing the proteolysis, cohesiveness, and smoothness of the product. Ginger extract-fortified cheese also earned higher scores for texture, flavor, and overall acceptability than the control. Ginger extract (4%) coupled with the pickling of the soft cheese was reported with superior quality characteristics [112].

Herbal ice cream with health-promoting attributes and a reduced sugar content could also be prepared by formulating ginger extract and xylitol [113]. The physical characteristics of the product were not altered significantly by incorporating herbal ingredients, but the antioxidant activity and phenolic contents were improved, and an alluring appearance was imparted to the final product. Another study was also conducted to characterize the volatile compounds of mature ginger rhizome and its potential utilization to develop a functional ice cream product [114]. The maturity of ginger rhizome exhibited a direct and inverse relationship with zingiberene and ar-curcumin, respectively. A value-added functional ice cream with enviable sensorial characteristics and innovative ginger flavoring could be produced by incorporating the gingerols from freeze-dried ginger extract.

In another study, different forms of ginger such as ginger juice, paste, candy, and powder were formulated into ice cream to analyze its physicochemical and hedonic attributes. The addition of ginger juice and paste decreased the fat, total solid, and overrun contents, while the incorporation of ginger candy and powder increased the fat, total solid, and overrun contents, but the phenolic contents and antioxidant capacity were increased in all inclusions. The product's resistance to melting improved in all preparations of the ginger, and the optimized incorporation level of all ginger forms were 4% paste, 6% juice, 10% candy, and 1% powder [115].

A novel approach of utilizing ginger essential oil in yogurt preparation was also successfully applied to evaluate the quality characteristics of the final product. Ginger essential oil enrichment (1000 ppm) lowered the syneresis and pH, while it enhanced the shelf life of the product and improved the overall acceptance of the product [116]. Efforts were also made for the efficient employment of essential oils from ginger, thyme, and clove in soft cheese formulation [117]. The inhibitory effects of these essential oils were reported against different food-borne pathogens, and the sensorial attributes were also improved with the addition of these essential oils.

Table 2. Some important investigations on the incorporation of ginger and its derivatives in dairy products.

Derivative	Product	Positive Effect	Reference
Ginger Powder	Yogurt	↑ Antioxidant Activity; Phenolic Contents; Total Solids; Viscosity; Textural Properties, Storage Stability; ↓ Syneresis; pH; Time; Energy; ≈ Sensorial Characteristics	[92,93]
	PEDA	↑ Functionality; Shelf Life; ↓ Microbial Growth; Cost	[96]
Ginger Juice	Yogurt	↑ Total Phenolic Contents; Antioxidant Total Solids; Viscosity; ↓ Coagulation Time; pH	[97,98]
	Ice Cream	↑ Antioxidant Activity; Phenolic Content; Sensorial Characteristics; Melting Resistance	[101,102,104]
Ginger Extract	Yogurt	↑ Antioxidants; Phenolic Contents; Sensorial Properties; ↓ Syneresis; pH	[105,107]
	Labneh	↑ Antimicrobial Activity; Antioxidants; Phenolic Contents; Sensorial Characteristics; ↓ Microbial Growth; pH	[109]
	Cheese	↑ Antioxidant Activity; Total Phenols; Oxidative Stability; Storage Stability; Sensorial Characteristics; Coagulation ↓ Microbial Growth	[110,112]
	Ice Cream	↑ Antioxidants; Total Phenolic Contents; Sensorial Properties; ↓ Microbial Count; ≈ Viscosity; Overrun	[113,114]
Ginger Protease	Cheese	↑ Proteolysis; Functionality; Coagulation; Sensorial Properties	[111]
Ginger Essential Oil	Soft Cheese	↑ Antioxidants; Phenolic Contents; Sensorial Characteristics; ↓ Microbial Growth	[117]

↑ = Increased/Improved, ↓ = Lower/Decreased, ≈ = Negligible/Unchanged.

In conclusion, ginger's diversified applications in dairy products have demonstrated its vast potential as a functional ingredient. Among the different forms of ginger, ginger extract came up as an extensively employed form, followed by ginger juice, powder, and essential oil. Ginger extract exhibited its aptitude for improving the overall techno-functional characteristics of dairy products and offering persuasive coagulation properties for cheese manufacturing. While the use of all forms of ginger proved to be effective in elevating the functionality and efficacy of dairy products, at the same time, some quality and hedonic characteristics were also compromised, but the incorporation of ginger extract was identified with higher functional and physicochemical improvements and minimal quality defects.

5.3. Beverages

Like other food sectors, the incorporation of functional food ingredients including ginger has also been investigated in beverages (Table 3). In the manufacturing of a ready-to-drink ginger–cocoa beverage, ginger powder and cocoa were optimized at an 8% concentration with low-fat cocoa powder, respectively, as higher concentrations of ginger and fat interfere with two decisive quality factors, sedimentation and viscosity [118]. A successful attempt was also made to develop a ready-to-drink and ready-to-reconstitute ginger-based beverage with superior functionality and a 6-month shelf life at an ambient temperature [119].

Ginger extract in combination with lemon juice rendered an excellent job as a natural preservative in a carrot–mandarin RTS drink. During storage, a few modifications were observed in the physicochemical properties, but the product characteristics remained consistent to a large extent [120]. A ginger and Indian gooseberry-based herbal beverage was also prepared and investigated by replacing sugar with aspartame, an artificial sweetener [121]. Liu et al. [122] recommended atmospheric pressure sterilization for a ginger-flavored beverage due to its feasibility and capability to retain the important volatile compounds.

Fresh ginger extract was adequately exploited in the production of a ginger-based functional carbonated beverage. The functionality and sensorial characteristics of the drink were improved with the addition of ginger rhizome extract, but the shelf life was largely associated with the carbonation level of the product [123]. In another study, supercritical extraction was mentioned as a superior extraction approach among conventional and ultrasonic extraction techniques, while mint was detected with marginally better phenolic and antioxidant profiles than ginger. A ginger–mint drink of high nutraceutical value was prepared at 0.4% concentrations of each extract with favorable physicochemical and hedonic characteristics [124].

Ginger is also a vital ingredient in Zobo drink, a traditional and indigenous Nigerian drink prepared from *Hibiscus sabdariffa* flowers [125]. The fortification of Zobo drink with ginger and garlic extracts substantially reduced the microbial load. The degradation of sensorial features with storage was observed in the control samples, while the samples that incorporated ginger and garlic not only had improved sensorial attributes but also maintained these characteristics during storage [126].

Ginger has also been formulated in various functional beverages in the form of blends with different spices, herbs, fruits, and other food ingredients. A soybean, ginger, and mango-based functional drink was developed to effectively exploit the nutritional and functional values of these ingredients, which were consequently attributed with feasible sensorial properties and overall acceptability [127]. An optimum formulation (v/v) was furnished for a functional beverage utilizing ginger extract (15.1%), tamarind extract (9.9%), turmeric extract (5.0%), sugar solution (40%), and water (30%).

Ginger extract, Kencur extract, and their combination blend were incorporated into an instant temulawak beverage, an indigenous Indonesian drink, to evaluate their impacts on the physicochemical and sensorial attributes of the final product. The inclusion of ginger extract in a concentration of up to 15% enhanced the flavor, aroma, and overall acceptance of the Temulawak drink [128]. Another study associated ginger extract with an improvement in the sensorial properties, but there was no significant advancement in the antioxidant activity with the ginger addition to a cocoa, ginger, and hibiscus flower-based beverage product [129]. A ginger-based beverage with hyperthermic activity was also found to be efficient against peripheral skin extremities in women experiencing mild cold sensitivity [130].

The integration of two different food ingredients could ameliorate the defective quality attributes of the intended food product. The fusion of a better coloring feature from tea and a favorable flavoring attribute from ginger resulted in a superior acceptable beverage product [131]. During an antioxidant evaluation in tea-ginger extract, the powder concentration was reported as the most influential process parameter than the time and temperature [132]. The supercritical ginger extract (0.3%) was preferred over the conventional ginger extract in iced ginger tea and functional coffee substitute production due to its exceptional performance, safety, purity, and overall acceptability [133,134]. Ginger extract (10%) was formulated and optimized with non-decaffeinated Robusta coffee (40%) with an improved functional value and without any adverse effects on the sensorial and quality characteristics [135].

Table 3. A few important studies on the addition of ginger and its derivatives to beverage products.

Derivative	Product	Positive Effect	Reference
Ginger Powder	Cocoa Beverage	↑ Total Phenolic Contents; Antioxidant Activity; Viscosity	[118]
Ginger Extract	Carrot and Kinnow RTS Drink	↑ Antioxidant Activity; Preservation; ↓ Microbial Growth; ≈ Sensorial Properties	[120]
	Zobo Drink	↑ Functionality; Shelf Life; Sensorial Properties; ↓ Microbial Count	[126]
	Black Tea	↑ Antioxidant Activity; Phenolic Contents; Sensorial Properties	[131]
	Ginger Iced Tea	↑ Nutraceutical Compounds; Shelf Life; Organoleptic Attributes	[133]
	Robusta Coffee	↑ Antioxidant Activity; Phenolic Contents; ≈ Sensorial Properties	[135]

↑ = Increased/Improved, ↓ = Lower/Decreased, ≈ = Negligible/Unchanged.

The above-mentioned research findings have provided insight into the exploitation of ginger in beverage products. Even in beverages, the employment of ginger extract seems to dominate over the rest of the ginger derivatives due to its better assimilation and conducive characteristics. The incorporation of ginger into different drinks not only augmented but also maintained the shelf life, nutraceutical value, and antimicrobial activity.

5.4. Meat and Fish Products

Several investigations and experiments were also conducted to figure out the effects of ginger and its derivatives in various forms on meat and fish products (Table 4). Rabbit meat burger patties were fortified with ginger powder at two different concentrations (1% and 2%) and significantly improvised the fatty acid profile by accelerating the polyunsaturated fatty acids [136]. Ginger powder also increased the antioxidant capacity and reduced the lipid peroxidation, subsequently catering a meat product with a better shelf life and enriched nutritional status. Mancini et al. [137] also designed an experiment to enhance the quality of pork burgers through ginger powder incorporation. The physiochemical attributes, antioxidant capacity, fatty acid profile, microbial growth, and lipid oxidation of the ginger powder-enriched pork burger were evaluated, and the results were in agreement with the previous study. In another investigation, ginger was designated as a potential alternative of artificial antioxidants in pasteurized canned meat products [138]. Ginger and papain powders were also competently exploited as natural tenderizing agents in buffalo calf meat [139].

A comparative analysis was conducted to assess the effects of three different spices as potential organic preservatives in raw chicken meat products. Clove powder (0.2%), even in a very small quantity, compared to ginger (3%) and garlic (2%) pastes performed very well to improve the oxidative stability, physicochemical properties, and antimicrobial activities in raw chicken meat with refrigerated storage [140]. Ginger and pineapple juices had also been examined as promising marinating agents in raw and cooked goat meat [141]. Ginger juice not only provided oxidative stability but also developed the hedonic features by according a bright red color and strong marinating odor and reducing the gamey flavor of Saanen goat meat. In another investigation, papain juice was described as a better tenderizing agent than ginger and pineapple juices as it induced a mild disruption in the muscles of goose meat [142]. Fermented ginger paste was characterized along 25 metabolites with substantial antimicrobial activity and its application on chicken meat efficiently enhanced the shelf life through a significant reduction in the microbial load [143]. Similar findings were also reported by Abdel-Naeem and Mohamed [144] regarding the application of ginger and papain extracts in camel meat patties. Sukada et al. [145] stipulated that pineapple extract had superior hedonic characteristics than papaya leaf and ginger extracts when practiced on Bali beef through the immersion method.

It was revealed that the competence level of ginger was as good as sodium ascorbate to combat oxidative rancidity, while the textural and sensorial properties of meat products with ginger were better than synthetic antioxidants [146]. Ginger extract also performed efficiently in the case of smoked buffalo meat as the microbial, physicochemical, and sensorial attributes did not exhibit any remarkable changes during the storage period. Ginger extract not only regulated and maintained the microbial count within certain limits but also enhanced the functionality of the cooked meat products.

Smoked mackerel fish was treated with ginger extract to evaluate its influence on the shelf life and hedonic characteristics [147]. Ginger extract (5%) significantly improved the stability and functional value of the smoked fish through its antimicrobial and antioxidative features. The sensorial quality of fish with the incorporation of ginger extract was also better and more persistent than other treatments even during storage. Mattje et al. [148] disclosed that the antioxidant activity of supercritical ginger extract is greater than ginger essential oil, which, consequently, was detected in the treatment of tilapia fish burgers with ginger. Supercritical ginger extract also enhanced the shelf life of fish burgers from 6 to 8 days, but it disturbed the overall perception due its unfavorable spicy flavor.

A variety of synthetic tenderizers and proteolytic enzymes had been engaged in the tenderization of meat. Currently, natural tenderizers are in demand due to certain constraints of synthetic tenderizers such as safety issues, unpleasant tastes, and over-tenderization. Ginger extract (5%) was recommended as a potential tenderizer for Yak meat as it enhanced the cooking yield, tenderness, and lipid oxidative stability during storage [149,150].

Researchers had also adopted a distinct approach to improve the meat quality through feeding ginger, as a potential dietary supplement, to the animals. Ginger and its derivatives in various forms were highly effective in inducing favorable characteristics such as lowering the cholesterol and improving the egg functionality and carcass quality in poultry animals [151,152]. The dietary augmentation of rabbits was also carried out through ginger powder supplementation. This augmentation improved the resistance to lipid oxidation without causing any intervention to the productive functions of the rabbits [153]. The administration of red ginger as an herbal–biotic feed ingredient (1.5%) competently boosted the productive activity and carcass quality in broilers [154]. Ginger extract loaded with gingerol was effectively employed to derive its health-promoting features for the prosperity of heat-stressed broilers [155]. Healthy and good-quality meat could be produced through catering ginger and rosemary essential oils to New Zealand white (NZW) rabbits. The provision of both essential oils (0.5%) promoted the feed exploitation, growth efficiency, and meat quality in growing rabbits [156]. Meanwhile, essential oils from different spices and herbs including ginger were exploited for their potential applications as natural preservatives to improve the oxidative stability in meat and meat products [157].

Table 4. Some important studies on the addition of ginger and its derivatives to meat and fish products.

Derivative	Product	Positive Effect	Reference
Ginger Powder	Rabbit Burgers	↑ Antioxidant Activity; Fatty Acid Profile; Shelf Life; ↓ Lipid Peroxidation	[136]
	Pork Burger	↑ Antioxidant Activity; PUFA; Oxidative Stability; Sensorial Characteristics; ↓ Microbial Load	[137]
Shredded Ginger	Canned Pork	↑ Antioxidant Activity; Tenderness; Organoleptic Properties; ↓ Microbial Load	[138]
Ginger Paste	Chicken Meat	↑ Antioxidants; Antimicrobial Activity; Oxidative Stability ↓ Microbial Load	[140,143]
Ginger Juice	Goat Meat	↑ Antioxidant Activity; Tenderness; Oxidative Resistance; Storage Stability; Overall Acceptability; ↓ Gamey Odor	[141]
	Beef	↑ Antioxidant Activity; Tenderness; Sensorial Properties; ↓ Cooking Loss	[145]
Ginger Extract	Camel Meat Patties	↑ Antioxidants; Lipid Stability; Collagen Solubility; Sensorial Properties	[144]
	Smoked Buffalo Meat	↑ Antioxidant Activity; Tenderness; Sensorial Characteristics; ↓ Cooking Loss; Microbial Load	[146]
	Smoked Fish	↑ Antioxidants; Shelf Life; Organoleptic Attributes; ↓ Mold Count; Oxidative Rancidity	[147]
	Tilapia Fish Burger	↑ Antioxidant Activity; Shelf Life; ↓ Lipid Oxidation	[148]
	Yak Meat	↑ Antioxidant Activity; Tenderness; Oxidative Resistance; Cooking Yield; ↓ Microbial Load	[150]
Gingerol Rich Extract	Chicken Feed	↑ Antioxidant Activity; Bioactive Compounds; Meat Quality	[155]

↑ = Increased/Improved, ↓ = Lower/Decreased

As per the aforementioned studies, ginger has an extensive application potential in the meat industry, embellishing meat products with an array of favorable characteristics. The incorporation of ginger into meat products competently ameliorated the nutritional value, oxidative stability, lipid profile, and antimicrobial activity of the meat products. The ability of ginger to provide elegant antioxidant and tenderization activities without compromising the sensorial and textural attributes of meat products has also provided another priority lead over artificial antioxidants and tenderizing agents. The inclusion of ginger in the diets of animals as a dietary supplement has also emerged as a novel and efficient approach, which remarkably improves the fatty acid profile and carcass quality of the animals.

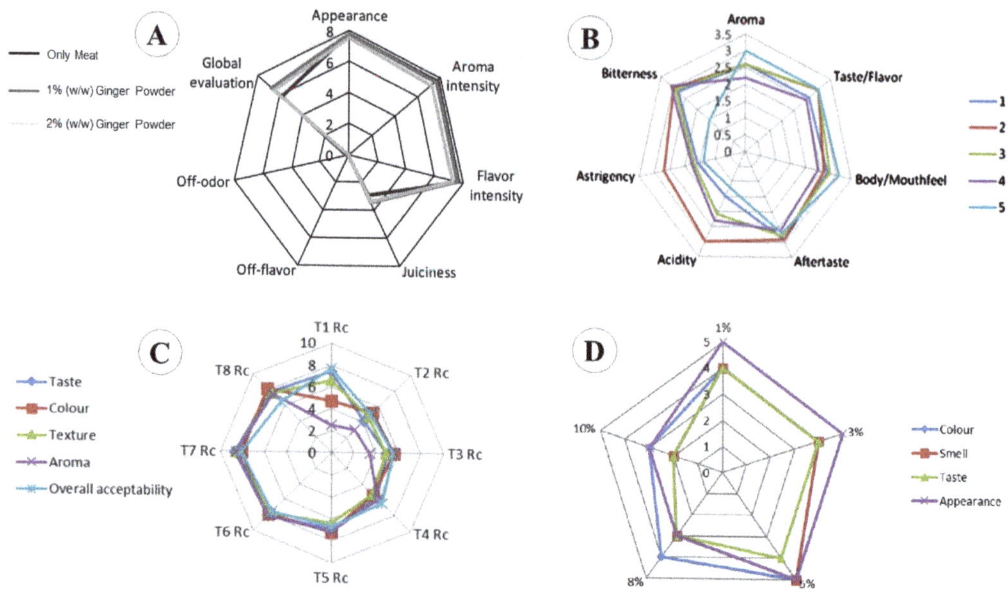

Figure 3. Sensory evaluation of different food products with incorporated ginger. (**A**) Pork burgers with added ginger powder [137]. (**B**) Comparison of ginger-based beverage with common coffee types. 1—traditional coffee; 2—ginger-based beverage; 3—instant coffee; 4—espresso coffee; 5—American/filter coffee [134]. (**C**) Different formulations of ginger-supplemented cookies. Ginger powder with concentrations of 0, 2, 4, 6, 8, 10, 12, and 14% for T1 to T8, respectively [82]. (**D**) Cottage cheese supplemented with natural ginger extract at a concentration of 1% to 10% of the total volume of the cottage cheese [158].

5.5. Film, Coating, and Packaging

Ginger and its bioactive compounds have also been examined for their potential applications as edible coatings, films, or active ingredients in packaging (Table 5). The use of glycerol as a green solvent was also practiced to extract the bioactives from ginger powder [159]. The resultant glycerol ginger extract (GGE) was incorporated in a chitosan-based edible coating for walnut kernels. This coating formulation was effective against the lipid oxidation and fungal spoilage of walnuts under storage conditions. Another study disclosed that the combination of ginger and garlic is a more efficient preservative approach than other famous preservatives such as BHT and Nisin [160]. The application of ginger–garlic extract in the vacuum packaging of Herring fish fillets increases its stability against oxidation and microbial spoilage and also imparts favorable sensorial attributes to the product.

Whey protein isolate (WPI) coating on Kashar cheese was characterized with water and fat barrier attributes, which were significantly developed with the addition of 1.5% (v/v) ginger essential oil, regarded as whey protein isolate ginger (WPIG) coating [161]. WPIG coating was also found to be more competent in the shelf life extension of Kashar cheese due to the anti-bacterial activity of ginger essential oil against *E. coli* and *S. aureus*, while Tsironi et al. [162] also expressed that WPI films enriched with 1% ginger essential oil efficiently prevented the microbial spoilage of minced lamb meat with acceptable sensorial properties until 8 days of refrigerated storage.

Sodium alginate/agar (SA) coating enhanced the shelf life and maintained the quality attributes of beef; moreover, these characteristics were substantially improved with the incorporation of ginger essential oil (SAG) into edible coating [163]. Compared to the control, these coating formulations, SA and SAG, extended the shelf life of beef at a chilling temperature by 3 to 6 days and 9 days, respectively, through their counter actions against lipid oxidation and microbial decay. Ginger essential oil- fortified (6%) sodium caseinate-based edible coating fabricated through nanoemulsion efficiently improved the shelf life of raw chicken meat [164]. This edible coating significantly enhanced the microbial resistance and embellished the sensorial characteristics of the chicken meat, but no remarkable increment in antioxidant activity was observed. Researchers also developed a peculiar coating formula based on apple peel extract and zein loaded with ginger essential oil to prolong the shelf life of raw chicken meat [165].

Fish sarcoplasmic protein and chitosan composite films enriched with ginger essential oil (up to 3%) were analyzed for their physicochemical and antioxidant features [166]. Ginger essential oil (GEO) nanoemulsion furnished the composite films with better elongation at break, antioxidant, and water barrier properties, while on the other hand, GEO was also liable to accord the inferior thermal and mechanical properties. In another investigation, a blend containing equal concentrations of ginger and cinnamon essential oils was incorporated into chitosan films for its potential utilization in pork packaging. These composite films were reported with better thickness and opacity, restricted the lipid oxidation and microbial spoilage, and enhanced the antioxidant activity in pork slices stored at a refrigeration temperature [167]. The two different studies also evaluated the effects of ginger and cinnamon essential oil (EO) supplementation on soy protein isolate (SPI) and sodium caseinate-based edible films [168,169]. The water vapor permeability was almost unchanged with both EOs, while the optical features were significantly affected by the cinnamon essential oil. In SPI films, cinnamon EO was more resistant and stretchable than ginger EO, while both EOs were incompetent to cater any improvements in the lipid oxidation stability in sodium caseinate film.

Zhang et al. [170] also formulated a Tilapia fish skin gelatin, ginger EO, and ZnO nanoparticle-based microemulsion nanofilm for the shelf life extension of fresh meat where ginger EO was responsible for better mechanical characteristics and improved functionality status with elevated antioxidant and antimicrobial properties. Researchers have also struggled to employ ginger and its bioactive compounds in active packaging to prolong the shelf life of food products. Chitosan-based bio-nanocomposite enriched with ginger EO (1%) was effectively exploited to extend the shelf life of poultry without any compromise on the sensorial properties [171].

It is evident from the foregoing literature that ginger oleoresins have promising applications in edible films, coatings, and the packaging sector, where ginger essential oil has better and extensive utilizations than ginger extract. The consolidation of ginger significantly improved the oxidative stabilities, antimicrobial activities, and sensorial properties of the intended food products. Ginger also performed better than the synthetic preservatives, improved the shelf life, and retained the quality attributes of the products.

Table 5. Some important studies on the application of ginger and its derivatives to films, coatings, and packaging.

Derivative	Product	Positive Effects	Reference
Ginger Extract	Chitosan-Based Edible Coating (Walnut)	↑ Antioxidant Activity; Shelf Life; Oxidative Stability; ↓ Fungal Spoilage	[159]
	Garlic Extract cum Vacuum Packaging (Fish Fillet)	↑ Antioxidant Activity; Shelf Life; Oxidative Stability; Sensorial Properties; ↓ Microbial Spoilage	[160]
Ginger Essential Oil	Whey Protein Isolate Coating (Cheese)	↑ Anti-bacterial Activity; Shelf Life; Water and Fat Barrier Properties	[161]
	Edible Coating (Chicken Meat)	↑ Shelf Life; Sensorial Characteristics ↓ Microbial Growth; ≈ Antioxidant Activity	[164]
	Fish Protein and Chitosan Composite Films	↑ Antioxidant Activity; Shelf Life; Water Barrier Properties	[166]
	Cinnamon EO Chitosan Films (Pork)	↑ Antioxidant Activity; Shelf Life; Film Thickness; ↓ Microbial Spoilage; Lipid Oxidation	[167]

↑ = Increased/Improved, ↓ = Lower/Decreased, ≈ = Negligible/Unchanged.

6. Conclusions and Future Prospectus

Ginger, with strong antioxidant, antimicrobial, and other functional activities, has a great potential for employment in the food industry. In this review, many studies have been presented regarding the acquirement of the bioactive compounds of ginger in different forms such as powder, extract, juice, paste, and essential oil. The application of ginger and its derivatives in a variety of food products has been extensively investigated and reported with auspicious results. Among all of the derivatives of ginger, ginger extract was found to be a more practicable, efficient, and versatile form of ginger for food applications. Ginger essential oil is another efficient variant of ginger, but its food applications are limited because of its strong aroma and immiscibility with certain food products.

These studies have also revealed another positive aspect of ginger, which is that it could be employed in a wide range of food products, covering almost all leading sectors of the food industry. The incorporation of ginger, in its various forms, thoroughly improved the functionality and bioactivity of the intended food products, but the physicochemical and sensory properties of these products were also compromised in a legitimate number of studies.

Further studies mainly focused on the efficacy, optimization, and assimilation of ginger bioactives into food products without compromising the physicochemical characteristics and sensory profiles of the products must be carried out. The probable synergistic effect of ginger bioactives with other compounds and toxicity studies of different compounds in their purest forms should also be investigated to ensure the safe use of these functional compounds.

Author Contributions: Conceptualization, M.N.S. and B.F.; writing—original draft preparation, M.N.S.; writing—review and editing, M.N.S., A.N. and B.F.; visualization, A.N.; supervision, B.F.; funding acquisition, B.F. All authors have read and agreed to the published version of the manuscript.

Funding: This research was funded by the University of Catania, Linea 2-Project, 'RiDARE'.

Institutional Review Board Statement: Not applicable.

Informed Consent Statement: Not applicable.

Data Availability Statement: Not applicable.

Conflicts of Interest: The authors declare no conflict of interest.

References

1. Elzebroek, A.T.G.; Wind, K. *Guide to Cultivated Plants*; CABI: Wallingford, CT, USA, 2008; p. 540.
2. Srinivasan, K. Ginger Rhizomes (*Zingiber officinale*): A Spice with Multiple Health Beneficial Potentials. *Pharma Nutr.* **2017**, *5*, 18–28. [CrossRef]
3. Nair, K.P. *Turmeric (Curcuma longa L.) and Ginger (Zingiber Oo Cinale Rosc.)—World's Invaluable Medicinal Spices the Agronomy and Economy of Turmeric and Ginger*; Springer: Cham, Switzerland, 2019; p. 568.
4. Baliga, M.S.; Shivashankara, A.R.; Haniadka, R.; Palatty, P.L.; Arora, R.; Fayad, R. Ginger (*Zingiber officinale* Roscoe): An Ancient Remedy and Modern Drug in Gastrointestinal Disorders. In *Bioactive Food as Dietary Interventions for Liver and Gastrointestinal Disease*, 1st ed.; Watson, R.R., Preedy, V.R., Eds.; Academic Press: Cambridge, MA, USA, 2013; pp. 187–199. [CrossRef]
5. Mahboubi, M. *Zingiber officinale* Rosc. Essential Oil, a Review on Its Composition and Bioactivity. *Clin. Phytosci.* **2019**, *5*, 12. [CrossRef]
6. Ma, R.H.; Ni, Z.J.; Zhu, Y.Y.; Thakur, K.; Zhang, F.; Zhang, Y.Y.; Hu, F.; Zhang, J.G.; Wei, Z.J. A Recent Update on the Multifaceted Health Benefits Associated with Ginger and Its Bioactive Components. *Food Funct.* **2021**, *12*, 519–542. [CrossRef]
7. Singh, R.P.; Gangadharappa, H.V.; Mruthunjaya, K. Ginger: A Potential Neutraceutical, An Updated Review. *Int. J. Pharmacogn. Phytochem. Res.* **2017**, *9*, 1227–1238. [CrossRef]
8. Mao, Q.Q.; Xu, X.Y.; Cao, S.Y.; Gan, R.Y.; Corke, H.; Beta, T.; Li, H. Bin Bioactive Compounds and Bioactivities of Ginger (*Zingiber officinale* Roscoe). *Foods* **2019**, *8*, 185. [CrossRef]
9. Kamaruddin, M.S.H.; Chong, G.H.; Mohd Daud, N.; Putra, N.R.; Md Salleh, L.; Suleiman, N. Bioactivities and Green Advanced Extraction Technologies of Ginger Oleoresin Extracts: A Review. *Food Res. Int.* **2023**, *164*, 112283. [CrossRef]
10. Dalsasso, R.R.; Valencia, G.A.; Monteiro, A.R. Impact of Drying and Extractions Processes on the Recovery of Gingerols and Shogaols, the Main Bioactive Compounds of Ginger. *Food Res. Int.* **2022**, *154I*, 111043. [CrossRef]

11. Garza-Cadena, C.; Ortega-Rivera, D.M.; Machorro-García, G.; Gonzalez-Zermeño, E.M.; Homma-Dueñas, D.; Plata-Gryl, M.; Castro-Muñoz, R. A Comprehensive Review on Ginger (*Zingiber officinale*) as a Potential Source of Nutraceuticals for Food Formulations: Towards the Polishing of Gingerol and Other Present Biomolecules. *Food Chem.* **2023**, *413*, 1–18. [CrossRef]
12. Tavares, L.; Smaoui, S.; Pinilla, C.M.B.; Ben Hlima, H.; Lopes Barros, H. Ginger: A Systematic Review of Clinical Trials and Recent Advances in Encapsulation of Its Bioactive Compounds. *Food Funct.* **2022**, *13*, 1078–1091. [CrossRef]
13. Beristain-Bauza, S.D.C.; Hernández-Carranza, P.; Cid-Pérez, T.S.; Ávila-Sosa, R.; Ruiz-López, I.I.; Ochoa-Velasco, C.E. Antimicrobial Activity of Ginger (*Zingiber officinale*) and Its Application in Food Products. *Food Rev. Int.* **2019**, *35*, 407–426. [CrossRef]
14. Liu, Y.; Liu, J.; Zhang, Y. Research Progress on Chemical Constituents of *Zingiber officinale* Roscoe. *Biomed. Res. Int.* **2019**, *2019*, 29. [CrossRef]
15. Jolad, S.D.; Lantz, R.C.; Guan, J.C.; Bates, R.B.; Timmermann, B.N. Commercially Processed Dry Ginger (*Zingiber officinale*): Composition and Effects on LPS-Stimulated PGE2 Production. *Phytochemistry* **2005**, *66*, 1614–1635. [CrossRef]
16. Sang, S.; Snook, H.D.; Tareq, F.S.; Fasina, Y. Precision Research on Ginger: The Type of Ginger Matters. *J. Agric. Food Chem.* **2020**, *68*, 8517–8523. [CrossRef]
17. Mahomoodally, M.F.; Aumeeruddy, M.Z.; Rengasamy, K.R.R.; Roshan, S.; Hammad, S.; Pandohee, J.; Hu, X.; Zengin, G. Ginger and Its Active Compounds in Cancer Therapy: From Folk Uses to Nano-Therapeutic Applications. *Semin. Cancer Biol.* **2021**, *69*, 140–149. [CrossRef]
18. Afzal, M.; Al-Hadidi, D.; Menon, M.; Pesek, J.; Dhami, M.S.I. Ginger: An Ethnomedical, Chemical and Pharmacological Review. *Drug Metabol. Drug Interact.* **2001**, *18*, 159–190. [CrossRef]
19. Vasala, P.A. Ginger. In *Handbook of Herbs and Spices*; Peter, K.V., Ed.; Woodhead Publishing: Boca Raton FL, USA, 2012; Volume 1, pp. 195–206.
20. Kuo, P.L.; Hsu, Y.L.; Huang, M.S.; Tsai, M.J.; Ko, Y.C. Ginger Suppresses Phthalate Ester-Induced Airway Remodeling. *J. Agric. Food Chem.* **2011**, *59*, 3429–3438. [CrossRef]
21. Weng, C.J.; Wu, C.F.; Huang, H.W.; Ho, C.T.; Yen, G.C. Anti-Invasion Effects of 6-Shogaol and 6-Gingerol, Two Active Components in Ginger, on Human Hepatocarcinoma Cells. *Mol. Nutr. Food Res.* **2010**, *54*, 1618–1627. [CrossRef]
22. Peng, F.; Tao, Q.; Wu, X.; Dou, H.; Spencer, S.; Mang, C.; Xu, L.; Sun, L.; Zhao, Y.; Li, H.; et al. Cytotoxic, Cytoprotective and Antioxidant Effects of Isolated Phenolic Compounds from Fresh Ginger. *Fitoterapia* **2012**, *83*, 568–585. [CrossRef]
23. Jafarzadeh, A.; Nemati, M. Therapeutic Potentials of Ginger for Treatment of Multiple Sclerosis: A Review with Emphasis on Its Immunomodulatory, Anti-Inflammatory and Anti-Oxidative Properties. *J. Neuroimmunol.* **2018**, *324*, 54–75. [CrossRef]
24. Abdullahi, A.; Ahmad, K.; Ismail, I.S.; Asib, N.; Ahmed, O.H.; Abubakar, A.I.; Siddiqui, Y.; Ismail, M.R. Potential of Using Ginger Essential Oils-Based Nanotechnology to Control Tropical Plant Diseases. *Plant Pathol. J.* **2020**, *36*, 515–535. [CrossRef]
25. Malik, A.; Shimpy; Kumar, M. Advancements in Ginger Drying Technologies. *J. Stored Prod. Res.* **2023**, *100*, 102058. [CrossRef]
26. Zagórska, J.; Czernicka-Boś, L.; Kukula-Koch, W.; Szalak, R.; Koch, W. Impact of Thermal Processing on the Composition of Secondary Metabolites of Ginger Rhizome—A Review. *Foods* **2022**, *11*, 3484. [CrossRef]
27. Ghafoor, K.; Al Juhaimi, F.; Özcan, M.M.; Uslu, N.; Babiker, E.E.; Mohamed Ahmed, I.A. Total Phenolics, Total Carotenoids, Individual Phenolics and Antioxidant Activity of Ginger (*Zingiber officinale*) Rhizome as Affected by Drying Methods. *LWT* **2020**, *126*, 109354. [CrossRef]
28. Schwertner, H.A.; Rios, D.C. High-Performance Liquid Chromatographic Analysis of 6-Gingerol, 8-Gingerol, 10-Gingerol, and 6-Shogaol in Ginger-Containing Dietary Supplements, Spices, Teas, and Beverages. *J. Chromatogr. B* **2007**, *856*, 41–47. [CrossRef]
29. Kou, X.; Wang, X.; Ji, R.; Liu, L.; Qiao, Y.; Lou, Z.; Ma, C.; Li, S.; Wang, H.; Ho, C.T. Occurrence, Biological Activity and Metabolism of 6-Shogaol. *Food Funct.* **2018**, *9*, 1310–1327. [CrossRef]
30. Ho, S.C.; Su, M.S. Optimized Heat Treatment Enhances the Anti-Inflammatory Capacity of Ginger. *Int. J. Food Prop.* **2016**, *19*, 1884–1898. [CrossRef]
31. Ozola, B.; Augspole, I.; Duma, M.; Kreicbergs, V. Bioactive Compounds in Fresh and Dried Ginger Root (*Zingiber officinale*). *Food Balt.* **2019**, *8*, 265–268.
32. Tanweer, S.; Mehmood, T.; Zainab, S.; Ahmad, Z.; Shehzad, A. Comparison and HPLC Quantification of Antioxidant Profiling of Ginger Rhizome, Leaves and Flower Extracts. *Clin. Phytosci.* **2020**, *6*, 1–12. [CrossRef]
33. Wen, L.; Zhang, Z.; Sun, D.W.; Sivagnanam, S.P.; Tiwari, B.K. Combination of Emerging Technologies for the Extraction of Bioactive Compounds. *Crit. Rev. Food Sci. Nutr.* **2020**, *60*, 1826–1841. [CrossRef]
34. Tzani, A.; Kalafateli, S.; Tatsis, G.; Bairaktari, M.; Kostopoulou, I.; Pontillo, A.R.N.; Detsi, A. Natural Deep Eutectic Solvents (NaDESs) as Alternative Green Extraction Media for Ginger (*Zingiber officinale* Roscoe). *Sustain. Chem.* **2021**, *2*, 576–599. [CrossRef]
35. Jang, T.W.; Choi, J.S.; Park, J.H. Protective and Inhibitory Effects of Acteoside from Abeliophyllum Distichum Nakai against Oxidative DNA Damage. *Mol. Med. Rep.* **2020**, *22*, 2076–2084. [CrossRef]
36. Sousa, C.M.d.M.; Silva, H.R.; Vieira, G.M., Jr.; Ayres, M.C.C.; da Costa, C.L.S.; Delton, S.A.; Cavalcante, L.C.D.; Barros, E.D.S.; Araújo, P.B.d.M.; Brandão, M.S.; et al. Total Phenolic and Antioxidant Activity of Five Medicinal Plants. *Quim. Nova* **2007**, *30*, 351–355. [CrossRef]
37. Yehya, A.H.S.; Asif, M.; Tan, Y.J.; Sasidharan, S.; Abdul Majid, A.M.S.; Oon, C.E. Broad Spectrum Targeting of Tumor Vasculature by Medicinal Plants: An Updated Review. *J. Herb. Med.* **2017**, *9*, 1–13. [CrossRef]
38. Ahmad, B.; Rehman, M.U.; Amin, I.; Arif, A.; Rasool, S.; Bhat, S.A.; Afzal, I.; Hussain, I.; Bilal, S.; Mir, M.U.R. A Review on Pharmacological Properties of Zingerone (4-(4-Hydroxy-3-Methoxyphenyl)-2-Butanone). *Sci. World J.* **2015**, *2015*, 1–6. [CrossRef]

39. Chakraborty, D.; Mukherjee, A.; Sikdar, S.; Paul, A.; Ghosh, S.; Khuda-Bukhsh, A.R. [6]-Gingerol Isolated from Ginger Attenuates Sodium Arsenite Induced Oxidative Stress and Plays a Corrective Role in Improving Insulin Signaling in Mice. *Toxicol. Lett.* **2012**, *210*, 34–43. [CrossRef]
40. Ali, A.M.A.; El-Nour, M.E.A.M.; Yagi, S.M. Total Phenolic and Flavonoid Contents and Antioxidant Activity of Ginger (*Zingiber officinale* Rosc.) Rhizome, Callus and Callus Treated with Some Elicitors. *J. Genet. Eng. Biotechnol.* **2018**, *16*, 677–682. [CrossRef]
41. Dugasani, S.; Pichika, M.R.; Nadarajah, V.D.; Balijepalli, M.K.; Tandra, S.; Korlakunta, J.N. Comparative Antioxidant and Anti-Inflammatory Effects of [6]-Gingerol, [8]-Gingerol, [10]-Gingerol and [6]-Shogaol. *J. Ethnopharmacol.* **2010**, *127*, 515–520. [CrossRef]
42. Ahmed, S.M.U.; Luo, L.; Namani, A.; Wang, X.J.; Tang, X. Nrf2 Signaling Pathway: Pivotal Roles in Inflammation. *Biochim. Biophys. Acta Mol. Basis Dis.* **2017**, *1863*, 585–597. [CrossRef]
43. Morvaridzadeh, M.; Fazelian, S.; Agah, S.; Khazdouz, M.; Rahimlou, M.; Agh, F.; Potter, E.; Heshmati, S.; Heshmati, J. Effect of Ginger (*Zingiber officinale*) on Inflammatory Markers: A Systematic Review and Meta-Analysis of Randomized Controlled Trials. *Cytokine* **2020**, *135*, 155224. [CrossRef]
44. Ho, S.C.; Chang, K.S.; Lin, C.C. Anti-Neuroinflammatory Capacity of Fresh Ginger Is Attributed Mainly to 10-Gingerol. *Food Chem.* **2013**, *141*, 3183–3191. [CrossRef]
45. Zhang, M.; Viennois, E.; Prasad, M.; Zhang, Y.; Wang, L.; Zhang, Z.; Han, M.K.; Xiao, B.; Xu, C.; Srinivasan, S.; et al. Edible Ginger-Derived Nanoparticles: A Novel Therapeutic Approach for the Prevention and Treatment of Inflammatory Bowel Disease and Colitis-Associated Cancer. *Biomaterials* **2016**, *101*, 321–340. [CrossRef]
46. Zhang, M.; Xu, C.; Liu, D.; Han, M.K.; Wang, L.; Merlina, D. Oral Delivery of Nanoparticles Loaded with Ginger Active Compound, 6-Shogaol, Attenuates Ulcerative Colitis and Promotes Wound Healing in a Murine Model of Ulcerative Colitis. *J. Crohns Colitis* **2018**, *12*, 217–229. [CrossRef]
47. Ríos, J.L.; Recio, M.C. Medicinal Plants and Antimicrobial Activity. *J. Ethnopharmacol.* **2005**, *100*, 80–84. [CrossRef]
48. Moon, Y.S.; Lee, H.S.; Lee, S.E. Inhibitory Effects of Three Monoterpenes from Ginger Essential Oil on Growth and Aflatoxin Production of Aspergillus Flavus and Their Gene Regulation in Aflatoxin Biosynthesis. *Appl. Biol. Chem.* **2018**, *61*, 243–250. [CrossRef]
49. Chakotiya, A.S.; Tanwar, A.; Narula, A.; Sharma, R.K. *Zingiber officinale*: Its Antibacterial Activity on Pseudomonas Aeruginosa and Mode of Action Evaluated by Flow Cytometry. *Microb. Pathog.* **2017**, *107*, 254–260. [CrossRef] [PubMed]
50. Wang, X.; Shen, Y.; Thakur, K.; Han, J.; Zhang, J.G.; Hu, F.; Wei, Z.J. Antibacterial Activity and Mechanism of Ginger Essential Oil against *Escherichia coli* and *Staphylococcus aureus*. *Molecules* **2020**, *25*, 3955. [CrossRef]
51. Sasidharan, I.; Menon, A.N. Comparative Chemical Composition and Antimicrobial Activity Fresh & Dry Ginger Oils (*Zingiber officinale* Roscoe). *Int. J. Curr. Pharm. Res.* **2010**, *2*, 40–43.
52. Wang, H.M.; Chen, C.Y.; Chen, H.A.; Huang, W.C.; Lin, W.R.; Chen, T.C.; Lin, C.Y.; Chien, H.J.; Lu, P.L.; Lin, C.M.; et al. *Zingiber officinale* (Ginger) Compounds Have Tetracycline-Resistance Modifying Effects against Clinical Extensively Drug-Resistant Acinetobacter Baumannii. *Phytother. Res.* **2010**, *24*, 1825–1830. [CrossRef]
53. Sarfati, D.; Gurney, J. Preventing Cancer: The Only Way Forward. *Lancet* **2022**, *400*, 540–541. [CrossRef]
54. Bray, F.; Ferlay, J.; Soerjomataram, I.; Siegel, R.L.; Torre, L.A.; Jemal, A. Global Cancer Statistics 2018: GLOBOCAN Estimates of Incidence and Mortality Worldwide for 36 Cancers in 185 Countries. *CA Cancer J. Clin.* **2018**, *68*, 394–424. [CrossRef]
55. El-Ashmawy, N.E.; Khedr, N.F.; El-Bahrawy, H.A.; Abo Mansour, H.E. Ginger Extract Adjuvant to Doxorubicin in Mammary Carcinoma: Study of Some Molecular Mechanisms. *Eur. J. Nutr.* **2018**, *57*, 981–989. [CrossRef]
56. Rajagopal, C.; Lankadasari, M.B.; Aranjani, J.M.; Harikumar, K.B. Targeting Oncogenic Transcription Factors by Polyphenols: A Novel Approach for Cancer Therapy. *Pharmacol. Res.* **2018**, *130*, 273–291. [CrossRef]
57. Sakulnarmrat, K.; Srzednicki, G.; Konczak, I. Antioxidant, Enzyme Inhibitory and Antiproliferative Activity of Polyphenolic-Rich Fraction of Commercial Dry Ginger Powder. *Int. J. Food Sci. Technol.* **2015**, *50*, 2229–2235. [CrossRef]
58. Liu, C.M.; Kao, C.L.; Tseng, Y.T.; Lo, Y.C.; Chen, C.Y. Ginger Phytochemicals Inhibit Cell Growth and Modulate Drug Resistance Factors in Docetaxel Resistant Prostate Cancer Cell. *Molecules* **2017**, *22*, 1477. [CrossRef]
59. Brahmbhatt, M.; Gundala, S.R.; Asif, G.; Shamsi, S.A.; Aneja, R. Ginger Phytochemicals Exhibit Synergy to Inhibit Prostate Cancer Cell Proliferation. *Nutr. Cancer* **2013**, *65*, 263–272. [CrossRef]
60. Mansingh, D.P.; Sunanda, O.J.; Sali, V.K.; Vasanthi, H.R. [6]-Gingerol–Induced Cell Cycle Arrest, Reactive Oxygen Species Generation, and Disruption of Mitochondrial Membrane Potential Are Associated with Apoptosis in Human Gastric Cancer (AGS) Cells. *J. Biochem. Mol. Toxicol.* **2018**, *32*, e22206. [CrossRef]
61. Peng, S.; Yao, J.; Liu, Y.; Duan, D.; Zhang, X.; Fang, J. Activation of Nrf2 Target Enzymes Conferring Protection against Oxidative Stress in PC12 Cells by Ginger Principal Constituent 6-Shogaol. *Food Funct.* **2015**, *6*, 2813–2823. [CrossRef]
62. Lim, S.; Moon, M.; Oh, H.; Kim, H.G.; Kim, S.Y.; Oh, M.S. Ginger Improves Cognitive Function via NGF-Induced ERK/CREB Activation in the Hippocampus of the Mouse. *J. Nutr. Biochem.* **2014**, *25*, 1058–1065. [CrossRef]
63. Huh, E.; Lim, S.; Kim, H.G.; Ha, S.K.; Park, H.Y.; Huh, Y.; Oh, M.S. Ginger Fermented with: Schizosaccharomyces Pombe Alleviates Memory Impairment via Protecting Hippocampal Neuronal Cells in Amyloid Beta1-42 Plaque Injected Mice. *Food Funct. R. Soc. Chem.* **2018**, *9*, 171–178. [CrossRef]
64. Iranshahy, M.; Javadi, B. Diet Therapy for the Treatment of Alzheimer's Disease in View of Traditional Persian Medicine: A Review. *Iran. J. Basic. Med. Sci.* **2019**, *22*, 1102–1117.

65. Du, H.; Li, L.; Bennett, D.; Guo, Y.; Key, T.J.; Bian, Z.; Sherliker, P.; Gao, H.; Chen, Y.; Yang, L.; et al. Fresh Fruit Consumption and Major Cardiovascular Disease in China. *N. Engl. J. Med.* **2016**, *374*, 1332–1343. [CrossRef]
66. Khosravani, M.; Azarbayjani, M.A.; Abolmaesoomi, M.; Yusof, A.; Zainal Abidin, N.; Rahimi, E.; Feizolahi, F.; Akbari, M.; Seyedjalali, S.; Dehghan, F.; et al. Ginger Extract and Aerobic Training Reduces Lipid Profile in High-Fat Fed Diet Rats. *Eur. Rev. Med. Pharmacol. Sci.* **2016**, *20*, 1617–1622.
67. Oh, S.; Lee, M.S.; Jung, S.; Kim, S.; Park, H.; Park, S.; Kim, S.Y.; Kim, C.T.; Jo, Y.H.; Kim, I.H.; et al. Ginger Extract Increases Muscle Mitochondrial Biogenesis and Serum HDL-Cholesterol Level in High-Fat Diet-Fed Rats. *J. Funct. Foods* **2017**, *29*, 193–200. [CrossRef]
68. Deng, X.; Zhang, S.; Wu, J.; Sun, X.; Shen, Z.; Dong, J.; Huang, J. Promotion of Mitochondrial Biogenesis via Activation of AMPK-PGC1α Signaling Pathway by Ginger (*Zingiber officinale* Roscoe) Extract, and Its Major Active Component 6-Gingerol. *J. Food Sci.* **2019**, *84*, 2101–2111. [CrossRef]
69. Wang, J.; Zhang, L.; Dong, L.; Hu, X.; Feng, F.; Chen, F. 6-Gingerol, a Functional Polyphenol of Ginger, Promotes Browning through an AMPK-Dependent Pathway in 3T3-L1 Adipocytes. *J. Agric. Food Chem.* **2019**, *67*, 14056–14065. [CrossRef]
70. Suk, S.; Kwon, G.T.; Lee, E.; Jang, W.J.; Yang, H.; Kim, J.H.; Thimmegowda, N.R.; Chung, M.Y.; Kwon, J.Y.; Yang, S.; et al. Gingerenone A, a Polyphenol Present in Ginger, Suppresses Obesity and Adipose Tissue Inflammation in High-Fat Diet-Fed Mice. *Mol. Nutr. Food Res.* **2017**, *61*, 1–12. [CrossRef]
71. Zhu, Y.; Zhao, Y.; Wang, P.; Ahmedna, M.; Sang, S. Bioactive Ginger Constituents Alleviate Protein Glycation by Trapping Methylglyoxal. *Chem. Res. Toxicol.* **2015**, *28*, 1842–1849. [CrossRef]
72. Samad, M.B.; Mohsin, M.N.A.B.; Razu, B.A.; Hossain, M.T.; Mahzabeen, S.; Unnoor, N.; Muna, I.A.; Akhter, F.; Kabir, A.U.; Hannan, J.M.A. [6]-Gingerol, from *Zingiber officinale*, Potentiates GLP-1 Mediated Glucose-Stimulated Insulin Secretion Pathway in Pancreatic β-Cells and Increases RAB8/RAB10-Regulated Membrane Presentation of GLUT4 Transporters in Skeletal Muscle to Improve Hyperglycemia in Leprdb/Db Type 2 Diabetic Mice. *BMC Complement. Altern. Med.* **2017**, *17*, 395. [CrossRef]
73. Amjad, A.; Sohaib, M.; Nawaz, H.; Javed, M.S.; Shah, M.; Shah, F.U.H.; Tariq, M.R.; Sajid, M.W.; Khan, A.A.; Bilal, M.; et al. Assessment of Rheological and Quality Characteristics of Bread Made by the Addition of Ginger Powder in Wheat Flour. *Food Sci. Technol. Camp.* **2022**, *42*, 47820. [CrossRef]
74. Balestra, F.; Cocci, E.; Pinnavaia, G.G.; Romani, S. Evaluation of Antioxidant, Rheological and Sensorial Properties of Wheat Flour Dough and Bread Containing Ginger Powder. *LWT Food Sci. Technol.* **2011**, *44*, 700–705. [CrossRef]
75. Özcan, M.M. The Effect of Ginger (*Zingiber officinale*) Powders at Different Concentrations on Bioactive Compounds, Antioxidant Activity, Phenolic Constituents, Nutrients and Sensory Characteristics of Wheat Bread. *Int. J. Gastron. Food Sci.* **2022**, *28*, 100532. [CrossRef]
76. Almasodi, A.G.S. Production and Evaluation of Some Bakery Products Containing Ginger Powder. *J. Food Nutr. Res.* **2018**, *6*, 205–215. [CrossRef]
77. Kaur, K.; Kaur, A. Effect of Supplementation of Garlic and Ginger Powder on Functional, Rheological and Bread Making Characteristics of Wheat Flour. *J. Res. Punjab Agric. Univ.* **2013**, *50*, 119–123.
78. Muhammad, A.; Rahman, Z.; Ayub, M.; Durrani, Y.; Ali, S.; Tabassum, A.; Shakoor, A.; Khan, M.; Khan, A. Inhibitory Effect of Ginger and Turmeric on *Rhizopus stolonifer* Growth on Bread. *J. Food Process Technol.* **2014**, *5*, 1000325. [CrossRef]
79. Cervenka, L.; Rezkova, S.; Kralovsky, J. Moisture Adsorption Characteristics of Gingerbread, a Traditional Bakery Product in Pardubice, Czech Republic. *J. Food Eng.* **2008**, *84*, 601–607. [CrossRef]
80. Ghendov-Mosanu, A.; Cristea, E.; Patras, A.; Sturza, R.; Niculaua, M. Rose Hips, a Valuable Source of Antioxidants to Improve Gingerbread Characteristics. *Molecules* **2020**, *25*, 5659. [CrossRef]
81. Lee, C.-S.; Lim, H.-S.; Cha, G.-H. Quality Characteristics of Cookies with Ginger Powder. *Korean J. Food Cook. Sci.* **2015**, *31*, 703–717. [CrossRef]
82. Kaushal, M.; Vaidya, D.; Gupta, A.; Kaushik, R.; Verma, A.K. Bioactive Compounds and Acceptance of Cookies Supplemented with Ginger Flour. *J. Pharmacogn. Phytochem.* **2019**, *8*, 185–188.
83. Yang, H.; Li, L.; Yin, Y.; Li, B.; Zhang, X.; Jiao, W.; Liang, Y. Effect of Ground Ginger on Dough and Biscuit Characteristics and Acrylamide Content. *Food Sci. Biotechnol.* **2019**, *28*, 1359–1366. [CrossRef]
84. Kumari, S.; Gupta, A. Utilization of Ashwagandha, Ginger and Shatavari Root Powder for the Preparation of Value Added "Cookies". *Int. J. Home Sci.* **2016**, *2*, 27–28.
85. Marak, N.R.; Malemnganbi, C.C.; Marak, C.R.; Mishra, L.K. Functional and Antioxidant Properties of Cookies Incorporated with Foxtail Millet and Ginger Powder. *J. Food Sci. Technol.* **2019**, *56*, 5087–5096. [CrossRef]
86. Abdel-Samie, M.A.S.; Wan, J.; Huang, W.; Chung, O.K.; Xu, B. Effects of Cumin and Ginger as Antioxidants on Dough Mixing Properties and Cookie Quality. *Cereal Chem.* **2010**, *87*, 454–460. [CrossRef]
87. Al-Timim, S.S. Effect of Ginger on Organoleptic Characteristics and Prolonging the Duration of Conservation of Laboratory Biscuit. *Iraq J. Mark. Res. Consum. Prot.* **2011**, *3*, 37–51.
88. Adebayo-Oyetoro, A.O.; Ogundipe, O.O.; Azoro, C.G.; Adeyeye, S.A.O. Production and Evaluation of Ginger Spiced Cookies from Wheat-Plantain Composite Flour. *Pac. J. Sci. Technol.* **2016**, *17*, 281–287.
89. Almasodi, A.G.S. Utilization of ginger as a Natural Antioxidant and Antimicrobial in Bakery Products. *Egypt. Arab. J. Appl. Sci. Technol.* **2021**, *1*, 1–9. [CrossRef]
90. Han, E.-J. Quality Characteristics of Muffins Containing Ginger Juice. *Korean J. Culin. Res.* **2012**, *18*, 256–266.

91. Tanweer, S.; Shehzad, A.; Butt, M.S.; Shahid, M. Phytochemical Profiling of Conventional and Supercritical Ginger Extract Based Baked Bars. *J. Food Process Technol.* **2016**, *7*, 1000596. [CrossRef]
92. Felfoul, I.; Borchani, M.; Samet-Bali, O.; Attia, H.; Ayadi, M.A. Effect of Ginger (*Zingiber officinalis*) Addition on Fermented Bovine Milk; Rheological Properties, Sensory Attributes and Antioxidanr Potential. *J. New Sci. Agric. Biotechnol.* **2017**, *44*, 2400–2409.
93. Hanou, S.; Boukhemis, M.; Benatallah, L.; Djeghri, B.; Zidoune, M.-N. Research Article Effect of Ginger Powder Addition on Fermentation Kinetics, Rheological Properties and Bacterial Viability of Dromedary Yogurt. *Adv. J. Food Sci. Technol.* **2016**, *10*, 667–673. [CrossRef]
94. Regu, M.; Yilma, Z.; Seifu, E. Effect of Garlic (*Allium sativum*) and Ginger (*Zingiber officinale*) Powder on Chemical Composition and Sensory Property of Ayib—Ethiopian Cottage Cheese. *Int. Food Res. J.* **2016**, *23*, 1226–1232.
95. Yüksel, H.; Çalışkan Koç, G.; Dirim, S.N. Physical Characterization of Spray-Dried Milk Powders and Their Agglomerates with the Addition of Carob, Cinnamon, and Ginger Powders. *Pamukkale Univ. J. Eng. Sci.* **2019**, *25*, 824–833. [CrossRef]
96. Gavhane, M.S.; Kamble, N.S.; Desale, R.J.; Ghule, B.K.; Mule, P.R. Studies on Preparation of PEDA with Ginger Powder. *Int. J. Food Agric. Vet. Sci.* **2014**, *4*, 64–68.
97. Yang, G.H.; Guan, J.J.; Wang, J.S.; Yin, H.C.; Qiao, F.D.; Jia, F. Physicochemical and Sensory Characterization of Ginger-Juice Yogurt during Fermentation. *Food Sci. Biotechnol.* **2012**, *21*, 1541–1548. [CrossRef]
98. Allam, M.G.; Gomaa, M.A.E.; Ayad, E.H.E.; Darwish, S.M. Application of Some Insensitive Probiotic Lactic Acid Bacteria and Ginger as Functional Dairy Products. *Microb. Biosyst.* **2018**, *3*, 60–73.
99. Jadhav, S.; Patil, V.N.; Naik, P.; Kadam, R.M. Studies on Chemical Quality of Ginger (*Zingiber officinale* L.) Milk Shake. *Asian J. Dairy Food Res.* **2018**, *37*, 13–17. [CrossRef]
100. Gaur, G.K.; Rani, R.; Dharaiya, C.N.; Solanki, K. Development of Herbal Milk Using Tulsi Juice, Ginger Juice and Turmeric Powder. *Int. J. Chem. Stud.* **2019**, *7*, 1150–1157.
101. Jadhav, M.S.; Nimbalkar, C.A.; Kad, V.P. Effect of Different Levels of Ginger Juice on Physico-Chemical and Sensory Characteristics of Herbal Ice Cream. *Res. J. Chem. Environ. Sci.* **2017**, *5*, 45–50.
102. David, J. Studies on Sensory Evaluation of Herbal Ice-Cream with Addition of Ginger (*Zingiber officinale*) Juice. *J. Pharma Res.* **2016**, *5*, 18–20.
103. Agrawal, A.K.; Karkhele, P.D.; Sandey, K.K.; Sahu, C.; Sinha, G. Effect of Incorporation of Ginger Juice in Various Rates on the Freezing and Thermal Properties of Ice Cream. *Asian J. Dairy Food Res.* **2015**, *34*, 92. [CrossRef]
104. Mule, S.M.; Kadam, S.S.; Dandekar, V.; Ramod, S.; Desai, B.; Narkhede, S. Manufacturing Technology and Production Cost of Ginger (*Zingiber officinale* L.) and Aloe Vera (*Aloe barbadensis*) Juice Enriched Probiotic (*L. acidophilus*) Ice Cream. *Int. J. Chem. Stud.* **2020**, *8*, 185–188. [CrossRef]
105. Molaie, S.; Latifi, Z.; Sabetgadam, M.; Khosrojerdi, M.; Razghandi, E. The Effect of Using Ginger Extract (*Zingiber officinale*) on the Antioxidant and Physicochemical Properties of Low-Fat Yogurt. *J. Food Sci. Technol.* **2022**, *18*, 265–273. [CrossRef]
106. Srivastava, P.; Prasad, S.G.M.; Ali, M.N.; Prasad, M. Analysis of Antioxidant Activity of Herbal Yoghurt Prepared from Different Milk. *Pharma Innov. J.* **2015**, *4*, 18–20. [CrossRef]
107. Simon, H.P.; Chandra, R.; Shukla, S.; Singh, S.S. Sensory Evaluation of Probiotic Herbal Yoghurt with Ginger and Garlic Extract. *Pharma Innov. J.* **2018**, *7*, 605–607. [CrossRef]
108. Haji Ghafarloo, M.; Jouki, M.; Tabari, M. Production and Characterization of Synbiotic Doogh, a Yogurt-Based Iranian Drink by Gum Arabic, Ginger Extract and B. Bifidum. *J. Food Sci. Technol.* **2020**, *57*, 1158–1166. [CrossRef]
109. Abd El-Khalek, A.B.; El-Sayed, H.S.; Ibrahim, G.A.; Sharaf, O.; Ibrahim, G.A.; El-Shafei, K.; El-Din, H.M.F.; Sharaf, O.M.; Tawfek, N.F.; Effat, B.A.; et al. Phenolic Compounds, Microbial Content and Sensory Evaluation of Synbiotic Labneh Conaining Ginger and Probiotic. *Int. J. ChemTech Res.* **2016**, *9*, 238–247.
110. Hailu, Y.; Seifu, E.; Yilma, Z. Physicochemical Properties and Consumer Acceptability of Soft Unripened Cheese Made from Camel Milk Using Crude Extract of Ginger (*Zingiber officinale*) as Coagulant. *Afr. J. Food Sci.* **2014**, *8*, 87–91. [CrossRef]
111. Hashim, M.M.; Dong, M.; Iqbal, M.F.; Li, W.; Chen, X. Ginger Protease Used as Coagulant Enhances the Proteolysis and Sensory Quality of Peshawari Cheese Compared to Calf Rennet. *Dairy Sci. Technol.* **2011**, *91*, 431–440. [CrossRef]
112. Abd El-Aziz, M.; Mohamed, S.H.S.; Seleet, F.L. Production and Evaluation of Soft Cheese Fortified with Ginger Extract as a Functional Dairy Food. *Pol. J. Food Nutr. Sci.* **2012**, *62*, 77–83. [CrossRef]
113. Chamchan, R.; Sinchaipanit, P.; Disnil, S.; Jittinandana, S.; Nitithamyong, A.; On-nom, N. Formulation of Reduced Sugar Herbal Ice Cream Using Lemongrass or Ginger Extract. *Brit. Food J.* **2017**, *119*, 2172–2182. [CrossRef]
114. Vedashree, M.; Asha, M.R.; Roopavati, C.; Naidu, M.M. Characterization of Volatile Components from Ginger Plant at Maturity and Its Value Addition to Ice Cream. *J. Food Sci. Technol.* **2020**, *57*, 3371–3380. [CrossRef]
115. Gabbi, D.K.; Bajwa, U.; Goraya, R.K. Physicochemical, Melting and Sensory Properties of Ice Cream Incorporating Processed Ginger (*Zingiber officinale*). *Int. J. Dairy Technol.* **2018**, *71*, 190–197. [CrossRef]
116. Mahalleh, A.A.; Mahalleh, A.A. Evaluation the Effect of Ginger (*Zingiber officinale*) Essential Oil on the Quality Characteristics Yogurt during Storage Time. *Sustain. Agric. Sci. Res.* **2021**, *1*, 71–87.
117. Ahmed, L.I.; Ibrahim, N.; Abdel-Salam, A.B.; Fahim, K.M. Potential Application of Ginger, Clove and Thyme Essential Oils to Improve Soft Cheese Microbial Safety and Sensory Characteristics. *Food Biosci.* **2021**, *42*, 101177. [CrossRef]
118. Faiqoh, K.E.N.; Muhammad, D.R.A.; Praseptiangga, D. Ginger-Flavoured Ready-to-Drink Cocoa Beverage Formulated with High and Low-Fat Content Powder: Consumer Preference, Properties and Stability. *Food Res.* **2021**, *5*, 7–17. [CrossRef]

119. Wadikar, D.D.; Madhura, C.V.; Premavalli, K.S. Development of a Ginger-Based Appetizer in the Form of Ready-to-Reconstitute Mix and Ready-to-Drink Beverage. *Int. J. Food Sci. Nutr.* **2011**, *62*, 404–409. [CrossRef]
120. Ullah, N.; Qazi, I.M.; Masroor, S.; Ali, I.; Khan, A.; Khan, M.; Gillani, A. Preservation of Ready to Serve Blended Carrot and Kinnow (Mandarin) Drink by Ginger Extract. *J. Food Process Technol.* **2015**, *6*, 1–6. [CrossRef]
121. Gaikwad, K.K.; Singh, S.; Shakya, B.R. Studies on the Development and Shelf Life of Low Calorie Herbal Aonla-Ginger RTS Beverage by Using Artificial Sweeteners. *J. Food Process Technol.* **2013**, *4*, 1000200. [CrossRef]
122. Liu, F.; Song, S.; Zhang, X.; Tan, C.; Karangwa, E. Effect of Sterilization Methods on Ginger Flavor Beverage Assessed by Partial Least Squares Regression of Descriptive Sensory Analysis and Gas Chromatography-Mass Spectrometry. *Eur. Food Res. Technol.* **2014**, *238*, 247–257. [CrossRef]
123. Abdulkareem, S.A.; Uthman, H.; Jimoh, A. Development and Characterization of a Carbonated Ginger Drink. *Leonardo J. Sci.* **2011**, *10*, 45–54.
124. Imran, A.; Quispe, C.; Zeeshan, A.; Imran, M.; Nadeem, M.; Gilani, S.A.; Gondal, T.A.; Tufail, T.; Aslam, F.; Rodrigues, C.F.; et al. Development and Antioxidant Characterization of Ginger-Mint Drink Prepared through Different Extraction Techniques. *J. Food Meas. Charact.* **2021**, *15*, 2576–2590. [CrossRef]
125. Adelekan, A.O.; Arisa, N.U.; Alamu, A.E.; Adebayo, Y.O.; Popoola, G.J.T. Production and Acceptability of Fruits Enhanced Zobo Drink. *Food Sci. Technol. Lett.* **2014**, *5*, 46–51.
126. Adesokan, I.A.; Abiola, O.P.; Adigun, M.O.; Anifowose, O.A. Analysis of Quality Attributes of *Hibiscus sabdariffa* (Zobo) Drinks Blended with Aqueous Extract of Ginger and Garlic. *Afr. J. Food Sci.* **2013**, *7*, 174–177. [CrossRef]
127. Cardoso, M.H.; da Costa, J.F.; Marto, R.H.; Neves, M.F.T. Soybean, Mango and Ginger Beverage: Nutritional Information, Sensory Evaluation and Consumption Intent. *Rev. Hosp. Univ. Pedro Ernesto* **2015**, *14*, 18–26. [CrossRef]
128. Septiana, A.T.; Triyanto, H.W. The Effect of Addition of Ginger Extract and Kencur Extract on the Physicochemical Properties of Instant Temulawak and the Sensory Properties of the Beverage. *Sudirman J. Nutr. Food* **2019**, *3*, 157–166. [CrossRef]
129. Awe, F.B.; Fagbemi, T.N.; Ifesan, B.O.T.; Badejo, A.A. Antioxidant Properties of Cold and Hot Water Extracts of Cocoa, Hibiscus Flower Extract, and Ginger Beverage Blends. *Food Res. Int.* **2013**, *52*, 490–495. [CrossRef]
130. Sugimoto, K.; Takeuchi, H.; Nakagawa, K.; Matsuoka, Y. Hyperthermic Effect of Ginger (*Zingiber officinale*) Extract-Containing Beverage on Peripheral Skin Surface Temperature in Women. *Evid.-Based Complement. Altern. Med.* **2018**, *2018*, 3207623. [CrossRef]
131. Makanjuola, S.A.; Enujiugha, V.N. Enhancing Sensory Perception of Plant Based Nutraceutical Drinks by Combining Plants from Different Sources: A Preliminary Study of Tea and Ginger Blend. *Prev. Nutr. Food Sci.* **2017**, *22*, 372–375. [CrossRef]
132. Makanjuola, S.A.; Enujiugha, V.N.; Omoba, O.S.; Sanni, D.M. Optimization and Prediction of Antioxidant Properties of a Tea-Ginger Extract. *Food Sci. Nutr.* **2015**, *3*, 443–452. [CrossRef]
133. Shoaib, M.; Khuram, H.; Aslam, W.; Muhmmad, S.; Shehzad, A.; Shakeel, A.; Sakandar, H.A.; Raza, H.; Omar, M.; Hafiz, A.; et al. Physico-Chemical and Sensory Profile of Conventional and Supercritical Ginger Extract Nutrified Ginger-Iced Tea. *Am. J. Food Sci. Nutr. Res.* **2015**, *2*, 154–158.
134. Švarc-Gajić, J.; Cvetanović, A.; Segura-Carretero, A.; Mašković, P.; Jakšić, A. Functional Coffee Substitute Prepared from Ginger by Subcritical Water. *J. Supercrit. Fluids* **2017**, *128*, 32–38. [CrossRef]
135. Kuswardhani, N.; Mukti, N.P.; Sari, P. Antioxidant and Sensory Properties of Ready to Drink Coffee-Ginger Made from Decaffeinated and Non-Decaffeinated Robusta Coffee Beans. *IOP Conf. Ser. Earth Environ. Sci.* **2021**, *653*, 012050. [CrossRef]
136. Mancini, S.; Preziuso, G.; Dal Bosco, A.; Roscini, V.; Parisi, G.; Paci, G. Modifications of Fatty Acids Profile, Lipid Peroxidation and Antioxidant Capacity in Raw and Cooked Rabbit Burgers Added with Ginger. *Meat Sci.* **2017**, *133*, 151–158. [CrossRef]
137. Mancini, S.; Paci, G.; Fratini, F.; Torracca, B.; Nuvoloni, R.; Dal Bosco, A.; Roscini, V.; Preziuso, G. Improving Pork Burgers Quality Using *Zingiber officinale* Roscoe Powder (Ginger). *Meat Sci.* **2017**, *129*, 161–168. [CrossRef]
138. Draszanowska, A.; Karpińska-Tymoszczyk, M.; Olszewska, M.A. The Effect of Ginger Rhizome and Refrigerated Storage Time on the Quality of Pasteurized Canned Meat. *Food Sci. Technol. Int.* **2020**, *26*, 300–310. [CrossRef]
139. Parkash, J.; Sharma, D.; Yadav, S. Application of Natural Tenderizers (Papain and Ginger) in Buffalo Calf Meat. *Pharma Innov.* **2021**, *10*, 203–205. [CrossRef]
140. Singh, P.; Sahoo, J.; Chatli, M.K.; Biswas, A.K. Shelf Life Evaluation of Raw Chicken Meat Emuslison Incorporated with Clove Powder, Ginger and Garlic Paste as Natural Preservatives at Refrigerated Storage (4+1 OC). *Int. Food Res. J.* **2014**, *21*, 1363–1373.
141. Putra, A.A.; Wattanachant, S.; Wattanachant, C. Sensory- Related Attributes of Raw and Cooked Meat of Culled Saanen Goat Marinated in Ginger and Pineapple Juices. *Trop. Anim. Sci. J.* **2019**, *42*, 59–67. [CrossRef]
142. Gao, H.Y.; Zhou, G.H.; Zeng, J.; Ma, H.J.; Pan, R.S.; Yu, X.L. Tenderization of Goose Meat by Papain, Pineapple Juice and Ginger Juice Treatment. *Adv. Mater. Res.* **2011**, *236–238*, 2349–2352. [CrossRef]
143. Muhialdin, B.J.; Kadum, H.; Fathallah, S.; Meor Hussin, A.S. Metabolomics Profiling and Antibacterial Activity of Fermented Ginger Paste Extends the Shelf Life of Chicken Meat. *LWT Food Sci. Technol.* **2020**, *132*, 109897. [CrossRef]
144. Abdel-Naeem, H.H.S.; Mohamed, H.M.H. Improving the Physico-Chemical and Sensory Characteristics of Camel Meat Burger Patties Using Ginger Extract and Papain. *Meat Sci.* **2016**, *118*, 52–60. [CrossRef]
145. Sukada, I.K.; Suberata, I.W.; Rasna, N.M.A. Immersion Effect with Extracts of Papaya Leaf, Pineapple, Ginger on Quality of Organoleptic and Bali Beef Nutrition. *Int. J. Life Sci.* **2019**, *3*, 12–24. [CrossRef]
146. Anandh, M.A.; Lakshmanan, V. Storage Stability of Smoked Buffalo Rumen Meat Product Treated with Ginger Extract. *J. Food Sci. Technol.* **2014**, *51*, 1191–1196. [CrossRef]

147. Iheagwara, M.C. Effect of Ginger Extract on Stability and Sensorial Quality of Smoked Mackerel (*Scomber scombrus*) Fish. *J. Nutr. Food Sci.* **2013**, *3*, 199. [CrossRef]
148. Mattje, L.G.B.; Tormen, L.; Bombardelli, M.C.M.; Corazza, M.L.; Bainy, E.M. Ginger Essential Oil and Supercritical Extract as Natural Antioxidants in Tilapia Fish Burger. *J. Food Process Preserv.* **2019**, *43*, e13942. [CrossRef]
149. Maiti, A.K.; Ahlawat, S.S.; Sharma, D.P.; Khanna, N. Application of Natural Tenderizers in Meat—A Review. *Agric. Rev.* **2008**, *29*, 226–230.
150. Ruitong, D.; Zhi, Y.; Yuan, L.I.; Xingmin, L.I.; Lizhen, M.A. Tenderizing and Preserving Yak Meat by Ginger Extract (*Zingiber officinale* Rose). *J. Muscle Foods* **2010**, *21*, 757–768. [CrossRef]
151. Abd El-Hack, M.E.; Alagawany, M.; Shaheen, H.; Samak, D.; Othman, S.I.; Allam, A.A.; Taha, A.E.; Khafaga, A.F.; Arif, M.; Osman, A.; et al. Ginger and Its Derivatives as Promising Alternatives to Antibiotics in Poultry Feed. *Animals* **2020**, *10*, 452. [CrossRef]
152. Khan, R.U.; Naz, S.; Nikousefat, Z.; Tufarelli, V.; Javdani, M.; Qureshi, M.S.; Laudadio, V. Potential Applications of Ginger (*Zingiber officinale*) in Poultry Diets. *Worlds Poult. Sci. J.* **2012**, *68*, 245–252. [CrossRef]
153. Mancini, S.; Secci, G.; Preziuso, G.; Parisi, G.; Paci, G. Ginger (*Zingiber officinale* Roscoe) Powder as Dietary Supplementation in Rabbit: Life Performances, Carcass Characteristics and Meat Quality. *Ital. J. Anim. Sci.* **2018**, *17*, 867–872. [CrossRef]
154. Herwati, M.; Marjuki, A. The Effect of Feeding Red Ginger (*Zingiber officinale* Rosc) as Phytobiotic on Broiler Slaughter Weight and Meat Quality. *Int. J. Poult. Sci.* **2011**, *10*, 983–986.
155. Wen, C.; Liu, Y.; Ye, Y.; Tao, Z.; Cheng, Z.; Wang, T.; Zhou, Y. Effects of Gingerols-Rich Extract of Ginger on Growth Performance, Serum Metabolites, Meat Quality and Antioxidant Activity of Heat-Stressed Broilers. *J. Therm. Biol.* **2020**, *89*, 102544. [CrossRef]
156. Elazab, M.A.; Khalifah, A.M.; Elokil, A.A.; Elkomy, A.E.; Rabie, M.M.; Mansour, A.T.; Morshedy, S.A. Effect of Dietary Rosemary and Ginger Essential Oils on the Growth Performance, Feed Utilization, Meat Nutritive Value, Blood Biochemicals, and Redox Status of Growing NZW Rabbits. *Animals* **2022**, *12*, 375. [CrossRef]
157. Pateiro, M.; Barba, F.J.; Domínguez, R.; Sant'Ana, A.S.; Mousavi Khaneghah, A.; Gavahian, M.; Gómez, B.; Lorenzo, J.M. Essential Oils as Natural Additives to Prevent Oxidation Reactions in Meat and Meat Products: A Review. *Food Res. Int.* **2018**, *113*, 156–166. [CrossRef]
158. Alibekov, R.S.; Gabrilyants, E.A.; Utebaeva, A.A.; Nurseitova, Z.T.; Konarbayeva, Z.K.; Khamitova, B.M. Cottage Cheese Fortified by Natural Additives. *Food Res.* **2021**, *5*, 152–159. [CrossRef]
159. Shaukat, M.N.; Palmeri, R.; Restuccia, C.; Parafati, L.; Fallico, B. Glycerol Ginger Extract Addition to Edible Coating Formulation for Preventing Oxidation and Fungal Spoilage of Stored Walnuts. *Food Biosci.* **2023**, *52*, 102420. [CrossRef]
160. Ahmed, E.S.; Shehata, M.G.; Abd-Rabou, H.S.; El-Menshawy, H. Extend Shelf-Life of Vacuum-Packaged Herring Fish Fillets Using Garlic and Ginger Extracts. *J. Pure Appl. Microbiol.* **2019**, *13*, 1571–1581. [CrossRef]
161. Kavas, N.; Kavas, G.; Saygili, D. Use of Ginger Essential Oil-Fortified Edible Coatings in Kashar Cheese and Its Effects on *Escherichia coli* O157:H7 and *Staphylococcus aureus*. *CyTA J. Food* **2016**, *14*, 317–323. [CrossRef]
162. Tsironi, M.; Kosma, I.S.; Badeka, A.V. The Effect of Whey Protein Films with Ginger and Rosemary Essential Oils on Microbiological Quality and Physicochemical Properties of Minced Lamb Meat. *Sustainability* **2022**, *14*, 3434. [CrossRef]
163. Zhang, B.; Liu, Y.; Wang, H.; Liu, W.; Cheong, K.; Teng, B. Effect of Sodium Alginate-Agar Coating Containing Ginger Essential Oil on the Shelf Life and Quality of Beef. *Food Control* **2021**, *130*, 108216. [CrossRef]
164. Noori, S.; Zeynali, F.; Almasi, H. Antimicrobial and Antioxidant Efficiency of Nanoemulsion-Based Edible Coating Containing Ginger (*Zingiber officinale*) Essential Oil and Its Effect on Safety and Quality Attributes of Chicken Breast Fillets. *Food Control.* **2018**, *84*, 312–320. [CrossRef]
165. Barkhordari, P.; Bazargani-Gilani, B. Effect of Apple Peel Extract and Zein Coating Enriched with Ginger Essential Oil on the Shelf Life of Chicken Thigh Meat. *J. Food Meas. Charact.* **2021**, *15*, 2727–2742. [CrossRef]
166. Cai, L.; Wang, Y. Physicochemical and Antioxidant Properties Based on Fish Sarcoplasmic Protein/Chitosan Composite Films Containing Ginger Essential Oil Nanoemulsion. *Food Bioprocess Technol.* **2021**, *14*, 151–163. [CrossRef]
167. Wang, Y.; Xia, Y.; Zhang, P.; Ye, L.; Wu, L.; He, S. Physical Characterization and Pork Packaging Application of Chitosan Films Incorporated with Combined Essential Oils of Cinnamon and Ginger. *Food Bioprocess Technol.* **2017**, *10*, 503–511. [CrossRef]
168. Atarés, L.; Bonilla, J.; Chiralt, A. Characterization of Sodium Caseinate-Based Edible Films Incorporated with Cinnamon or Ginger Essential Oils. *J. Food Eng.* **2010**, *100*, 678–687. [CrossRef]
169. Atarés, L.; De Jesús, C.; Talens, P.; Chiralt, A. Characterization of SPI-Based Edible Films Incorporated with Cinnamon or Ginger Essential Oils. *J. Food Eng.* **2010**, *99*, 384–391. [CrossRef]
170. Zhang, L.; Liu, A.; Wang, W.; Ye, R.; Liu, Y.; Xiao, J.; Wang, K. Characterisation of Microemulsion Nanofilms Based on Tilapia Fish Skin Gelatine and ZnO Nanoparticles Incorporated with Ginger Essential Oil: Meat Packaging Application. *Int. J. Food Sci. Technol.* **2017**, *52*, 1670–1679. [CrossRef]
171. Souza, V.G.L.; Pires, J.R.A.; Vieira, É.T.; Coelhoso, I.M.; Duarte, M.P.; Fernando, A.L. Shelf Life Assessment of Fresh Poultry Meat Packaged in Novel Bionanocomposite of Chitosan/Montmorillonite Incorporated with Ginger Essential Oil. *Coatings* **2018**, *8*, 177. [CrossRef]

molecules

MDPI

Article

Quality and Flavor Difference in Dry-Cured Meat Treated with Low-Sodium Salts: An Emphasis on Magnesium

Jun Xiang [1], Xuejiao Wang [1,*], Chaofan Guo [1], Liping Zang [1], Houde He [1], Xiaoyu Yin [1], Jianping Wei [2] and Jianxin Cao [1,*]

1 Faculty of Food Science and Engineering, Kunming University of Science and Technology, Kunming 650500, China; xiangjun19990903@163.com (J.X.); guochaofanfan@outlook.com (C.G.); zangliping1999@163.com (L.Z.); mocha588@outlook.com (H.H.); yinxiaoyu1234@163.com (X.Y.)

2 College of Food Science and Technology, Northwest University, Xi'an 710000, China; jianpingwei0327@nwu.edu.cn

* Correspondence: wangxuejiao173@hotmail.com (X.W.); jxcao@kust.edu.cn (J.C.)

Abstract: The present study aimed to develop low-sodium curing agents for dry-cured meat products. Four low-sodium formulations (SPMA, SPM, SP, and SM) were used for dry-curing meat. The physicochemical properties and flavor of the dry-cured meat were investigated. The presence of Mg^{2+} ions hindered the penetration of Na^{+} into the meat. The weight loss, moisture content, and pH of all low-sodium salt groups were lower than those of S. Mg^{2+} addition increased the water activity (A_W) of SPMA, SPM, and SM. Dry-curing meat with low-sodium salts promoted the production of volatile flavor compounds, with Mg^{2+} playing a more prominent role. Furthermore, low-sodium salts also promoted protein degradation and increased the content of free amino acids in dry-cured meat, especially in SM. Principal component analysis (PCA) showed that the low-sodium salts containing Mg^{2+} were conducive to improving the quality of dry-cured meat products. Therefore, low-sodium salts enriched with Mg^{2+} become a desirable low-sodium curing agent for achieving salt reduction in dry-cured meat products.

Keywords: dry-cured meat; low-sodium salt; taste; aroma; magnesium ion; salt reduction

Citation: Xiang, J.; Wang, X.; Guo, C.; Zang, L.; He, H.; Yin, X.; Wei, J.; Cao, J. Quality and Flavor Difference in Dry-Cured Meat Treated with Low-Sodium Salts: An Emphasis on Magnesium. *Molecules* **2024**, *29*, 2194. https://doi.org/10.3390/molecules29102194

Academic Editor: Carmen Cuadrado

Received: 7 March 2024
Revised: 13 April 2024
Accepted: 26 April 2024
Published: 8 May 2024

1. Introduction

Dry-meat products crafted from whole pieces, such as dry-cured loin, ham, bacon, and jerked beef, are highly valued and enjoyed globally for their distinct sensory qualities and extended shelf life [1]. Generally, a significant amount of salt is added to dry-cured meat products to reduce moisture and extend preservation time. For example, the salt content in dry-cured ham in Europe varies from 4.5% to 8% [2], while bacon ranges from 1.65% to 3.7% in Australia [3]. However, in China, the amount of salt added to dry-cured meat far exceeds that of other countries, especially in ham, where it can reach as high as 7% to 12% [4]. Excessive consumption of sodium salts increases the risk of hypertension, cardiovascular disease, stroke, liver and kidney disease, gastric cancer, and osteoporosis, etc. [5]. The World Health Organization (WHO) stated that the daily intake of salt for adults should be less than 5 g (2 g/sodium). Therefore, sodium reduction in meat products, especially dry-cured meat products, has been garnering increasing attention.

Currently, various physical processing methods are used to reduce sodium addition in dry-cured meat products, such as high pressure, ultrasonic methods, and electric field treatment [6]. There are also other methods, such as directly reducing the amount of NaCl [7], mediated pickling for salt reduction [8], and the utilization of low-sodium salts [9]. Among these methods, the partial replacement of NaCl with other metal salts is the most widely used method in traditional dry-cured meat.

Previous studies have shown that reducing sodium content can affect the quality of meat products, including sensory characteristics, flavors, bacterial colonies, and texture.

Armenteros et al. [10] and Wu et al. [11] demonstrated that using 50% KCl as a replacement for NaCl in hams resulted in a more pronounced bitter taste. Wu et al. [12] found that replacing over 40% of sodium with potassium significantly altered the distribution of volatile compounds in dry-cured meat, leading to changes in aroma and flavor. Research by Lee et al. [13] indicated that in the dry-cured ham, replacing over 50% of NaCl with KCl led to a significant decrease in tenderness, while cooking loss increased with higher concentrations of KCl. Nachtigall et al. [14] revealed that in the salted meat, under the same ionic strength, calcium chloride ($CaCl_2$) exhibited higher oxidative capacity and had a more pronounced impact on lipid profiles compared to NaCl and KCl. Li et al. [15] demonstrated that in dry-cured bacon, 40% KCl effectively inhibited the formation of biogenic amines, body alkalinity, and histamine while increasing the residual nitrate level. These findings indicated that different metal ions and substitution ratios have varying impacts on the quality of dry-cured meat products. Therefore, it is essential to explore suitable substitution levels and formulas to avoid affecting the quality of dry-cured meat.

Taking into consideration the high sodium level in dry-cured meat and the impacts linked with salt reduction on the quality and sensory properties, this study used four formulations of low-sodium salt during the processing of dry-cured meat. By measuring the physicochemical parameters, aroma, and taste of dry-cured meat products, we aim to select composite salts to enhance the quality of dry-cured meat while achieving the highest reduction in sodium content. This research will provide a theoretical basis and technical support for the development of low-sodium dry-cured meat products.

2. Results

2.1. Effect of Low-Sodium Salts on the Physicochemical Properties of Dry-Cured Meat

2.1.1. Metal Ion Content, Weight Loss, Moisture, Water Activity, and Water State of Dry-Cured Meat

Table 1 provides the measured content, theoretical content value, and permeability of each metal ion in dry-cured meat products. The contents of Na^+, K^+, and Mg^{2+} in each group of samples were proportional to the formula content, which was consistent with expectations. The amount of Na^+ added was relatively low in the groups SPMA, SPM, and SP compared to S, but they exhibited higher Na^+ permeability values. It was speculated that the introduction of K^+ would sharply promote the penetration of Na^+. However, the Na^+ permeability decreased in SM, which showed that adding Mg^{2+} to the low-sodium salt formulations with relative higher sodium contents may hinder the permeation of Na^+. Mg^{2+} can strongly combine with the polar groups of proteins and strengthen the interaction between proteins [16], thereby hindering the penetration of Na^+. Alino et al. [17] also found divalent cations decreased salt penetration into dry-cured ham. In SPMA, SPM, and SP, the K^+ permeability reaches about 100 times the theoretical value. The reason may be that the charge density of K^+ is lower than that of Na^+ and Mg^{2+}, making it easy to diffuse in the meat to produce a concentration effect and concentrate at the sampling point. The Mg^{2+} permeability was only about 20% due to the large charge density of divalent ions [18]. No significant difference was observed in the Mg^{2+} permeability of SPMA, SPM, and SM, indicating that Na^+ and K^+ had no effect on the penetration of Mg^{2+}.

As shown in Table 1, the weight loss rate of each group basically remained between 45.46% and 55.68%, though the weight loss rate of S (55.68%) was higher than that of other groups; in particular, it was significantly higher than that of SPM (47.42%) and SM (45.46%), indicating that low-sodium salt is beneficial to maintaining the moisture of dry-cured meat products. The lower weight loss of SM may be related to the permeation of ions. As shown in Table 1, SM showed the lowest permeability of Na^+ and Mg^{2+}. The moisture content of dry-cured meat products gradually decreases as the fermentation time progresses, eventually reaching a stable range, and A_w will also decrease as the moisture content decreases. A_w affects the nutrition value, color, flavor, and texture of food. In addition, A_w has an important role in food preservation. As shown in Table 1, the moisture content of each group decreased by about 65% and finally stabilized between 24% and 28%.

Although there were no significant differences between the groups, it was noteworthy that the SP had lower moisture content, which was also found in the samples salted with a mixture of NaCl and KCl [17]. As shown in Table 1, the A_w of all groups was below 0.75, among which the A_w values of SPMA (0.723) and SPM (0.729) were significantly higher than SP (0.704). In Table 1, it also can be seen that the moisture contents of SPMA, SPM, and SM were higher than that of SP. The higher moisture content was in parallel with a higher A_w value [17]. Ultimately, the moisture content and A_w of the Mg^{2+}-containing groups (SPMA, SPM, and SM) were higher; Alino, Grau, Toldra and Barat [17] also found the same results.

Table 1. Physical and chemical properties of dry-cured meat with the addition of different low-sodium salts.

Type	Index	Groups				
		S	SPMA	SPM	SP	SM
Metal ion contents and permeability (dry basis)	Na^+ (%)	5.63 ± 0.22 ^{A}	3.70 ± 0.29 ^{D}	3.48 ± 0.16 ^{D}	4.11 ± 0.17 ^{C}	4.64 ± 0.25 ^{B}
	Theoretical value	6.77	4.04	4.06	4.88	5.96
	Permeability (%)	83.25 ± 3.31 ^{B}	91.58 ± 7.19 ^{A}	85.71 ± 3.99 ^{AB}	84.22 ± 3.57 ^{AB}	77.85 ± 4.34 ^{B}
	K^+ (%)	\	3.40 ± 0.18 ^{A}	3.46 ± 0.18 ^{A}	3.27 ± 0.16 ^{A}	\
	Theoretical value	\	1.9	1.9	1.9	\
	Permeability (%)	\	1.79 ± 0.010 ^{A}	1.82 ± 0.10 ^{A}	1.72 ± 0.09 ^{A}	\
	Mg^{2+} (%)	\	0.21 ± 0.02 ^{A}	0.17 ± 0.04 ^{A}	\	0.17 ± 0.01 ^{A}
	Theoretical value	\	0.81	0.81	\	0.81
	Permeability (%)	\	25.92 ± 0.02 ^{A}	20.98 ± 0.05 ^{A}	\	20.98 ± 0.01 ^{A}
Physical and chemical properties	Weight loss (%)	55.68 ± 0.15 ^{A}	50.25 ± 1.90 ^{AB}	47.42 ± 1.34 ^{B}	51.93 ± 2.39 ^{AB}	45.46 ± 2.39 ^{B}
	Moisture (%)	27.59 ± 3.44 ^{A}	25.05 ± 1.91 ^{A}	24.56 ± 0.64 ^{A}	24.29 ± 2.41 ^{A}	24.31 ± 1.09 ^{A}
	A_w	0.707 ± 0.011 ^{BC}	0.723 ± 0.012 ^{AB}	0.729 ± 0.007 ^{A}	0.704 ± 0.024 ^{C}	0.708 ± 0.006 ^{ABC}
	pH	6.11 ± 0.10 ^{A}	5.86 ± 0.05 ^{B}	5.93 ± 0.21 ^{AB}	5.95 ± 0.09 ^{AB}	5.90 ± 0.13 ^{B}
	Shear force (N)	402.20 ± 45.90 ^{B}	338.10 ± 44.50 ^{AB}	287.10 ± 46.00 ^{B}	309.00 ± 31.90 ^{B}	269.06 ± 16.83 ^{B}
Color	L^*	49.55 ± 2.32 ^{A}	46.9 ± 1.19 ^{AB}	45.69 ± 1.67 ^{AB}	43.37 ± 4.33 ^{B}	44.61 ± 4.18 ^{AB}
	a^*	12.21 ± 1.07 ^{AB}	14.15 ± 2.41 ^{A}	11.42 ± 0.76 ^{B}	12.19 ± 0.37 ^{AB}	11.72 ± 0.63 ^{B}
	b^*	26.00 ± 2.60 ^{A}	24.05 ± 1.53 ^{AB}	23.02 ± 0.54 ^{AB}	23.67 ± 1.45 ^{AB}	21.72 ± 0.90 ^{B}

$^{A–D}$ means entries in the same column without common letters are significantly different ($p < 0.05$). S: 100% NaCl; SPMA: 59.375% NaCl+28% KCl+12% $MgCl_2$+0.625% L-arginine (Arg); SPM: 60% NaCl+28% KCl+12% $MgCl_2$; SP: 72% NaCl+28% KCl; SM: 88% NaCl+12% $MgCl_2$.

Low-field NMR can be used to analyze the distribution and migration of moisture in meat products. The relaxation time T_2 represents the strength of water mobility; that is, a shorter relaxation time represents lower mobility. As shown in Figure 1a, it can be seen that all groups contain three to four relaxation peaks, which has also been reported in the surimi-starch system [19]. The first two peaks were located between 0–1 ms and 1–10 ms, which belong to strongly bound water (T_{2b1}) and weakly bound water (T_{2b2}), respectively. The relaxation time between 30–100 ms (T_{21}) was immobilized water that interacts with the charged groups located between the fiber bundles. The relaxation time greater than 100 ms (T_{22}) was related to the free water existing outside the myofibril. As shown in Figure 1a, T_{2b2} was a major peak in dry-cured meat. The relaxation time of each group was between 3–7 ms. It can be seen from Figure 1b that S had the highest proportion of bound water, reaching 94.69%, while the immobilized water and free water were only 2.56% and 0.94%, respectively. SM enhanced the mobility of bound water, but did not increase the proportion of bound water compared to the S group. Overall, the low-sodium salt groups (SPMA, SPM, SP, and SM) increased the proportions of immobilized water and free water, and reduced the proportion of bound water, so that the dry-cured meat will not be excessively hard and thus be more acceptable to consumers.

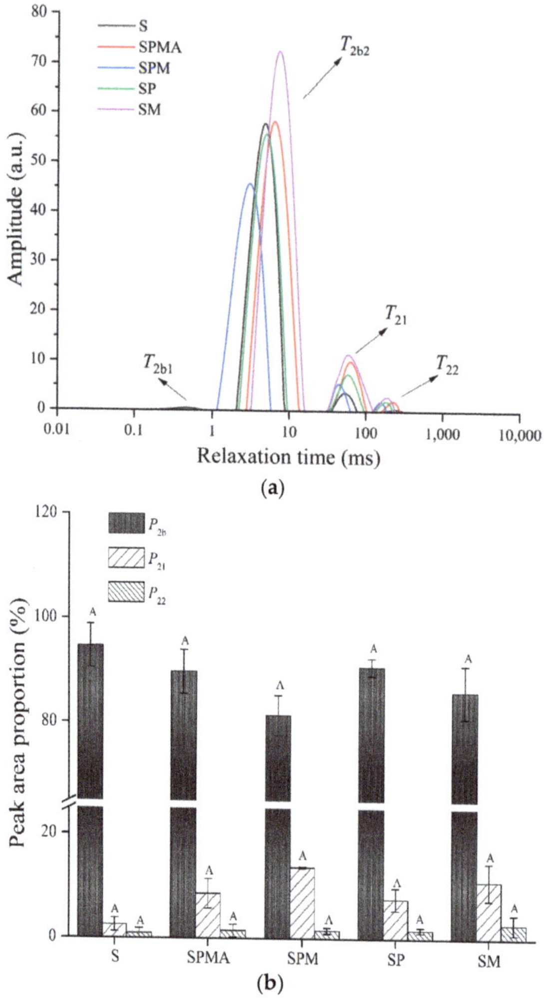

Figure 1. The distribution curves of transverse relaxation time (T_2) (**a**) of peak area proportions of T_{2b}, T_{21}, and T_{22}, (**b**) of dry-cured meat with the addition of different low-sodium salts. S: 100% NaCl; SPMA: 59.375% NaCl+28% KCl+12% $MgCl_2$+0.625% L-arginine (Arg); SPM: 60% NaCl+28% KCl+12% $MgCl_2$; SP: 72% NaCl+28% KCl; SM: 88% NaCl+12% $MgCl_2$. Different letters in the same group indicate significant differences ($p < 0.05$; Tukey test).

2.1.2. pH, Shear Force, and Color

As shown in Table 1, the pH of samples was between 5.90–6.15, which was within a good range of pH value for dry-cured meat products. Among them, the pH values of the low-sodium salt groups (SPMA, SPM, SP, and SM) were slightly lower than those of S. The pH of S (6.11) was significantly higher than SPMA (5.86) and SM (5.90), because the binding of divalent salts to proteins resulted in a decrease in the pH value of the isoelectric point [20].

Muscle is mostly composed of myofibrillar protein, and the effect of myofibrillar protein on tenderness depends not only on the cross-linking degree between protein molecules but also on the interaction between protein molecules. Regarding the shear force, the shear force values of SPMA, SPM, SP, and SM were significantly lower than that of S, which was related to their ionic strength and pH. As shown in Table 1, the pH and ionic strength of the S group were higher, which was beneficial to inhibiting the degradation of muscle fibers, resulting in greater shear force.

Color is one of the main attributes of meat and meat products which reflects freshness and influences consumer purchase intentions. In this study, when sodium was partially replaced by other cations, it had a significant impact on the color parameters, as shown in Table 1. As for L^*, S > SPMA > SPM > SM > SP, the L^* value of S was significantly higher than that of SP and slightly higher than that of SPMA, SPM and SM. Lorenzo et al. [21] also reported that the L^* value of dry-cured bacon salted with a mixture containing KCl in the formula was relatively lower. Mg^{2+} was not contained in SP, indicating that the absence of Mg^{2+} can reduce the brightness of the meat and have an adverse effect on its color. For a^*, the order of samples was SPMA > S > SP > SM > SPM. SP was significantly higher than SM and SPM; this is attributed to Arg's capacity to reduce brown metmyoglobin into bright red oxymyoglobin, improving the quality of dry-cured meat's a^* value [22], which is consistent with the findings of Zhu et al. [23] and Zheng et al. [24]. For b^*, S > SPMA > SP > SPM > SM; the b^* value of S (26.00) was significantly higher than that of SM (21.72). The presence of Mg^{2+} reduced the yellowness of dry-cured meat, which was good for consumer acceptability.

2.1.3. Effects of Low-Sodium Salt on Fat Oxidation of Dry-Cured Meat

The thiobarbituric acid reactive substances (TBARS) value refers to the result of the reaction between derivatives such as thiobarbituric acid and malondialdehyde produced by the oxidative decomposition of unsaturated fatty acids in animal fats and oils. As shown in Figure 2, the TBARS value of SM was significantly higher than that of SP. According to reports in the literature, the TBARS value increases with increasing ionic strength [25]; therefore, pork treated with higher ionic strength mixtures (SPMA and SM) had higher TBARS values, which corresponded to the ionic strength values of each. In a study regarding Brazilian low-sodium dry-fermented sausages containing blends of NaC1, KCl, and $CaC1_2$, Dos Santos et al. [26] also reported that the sample treated with a higher ionic strength of 0.7 showed a higher TBARS value.

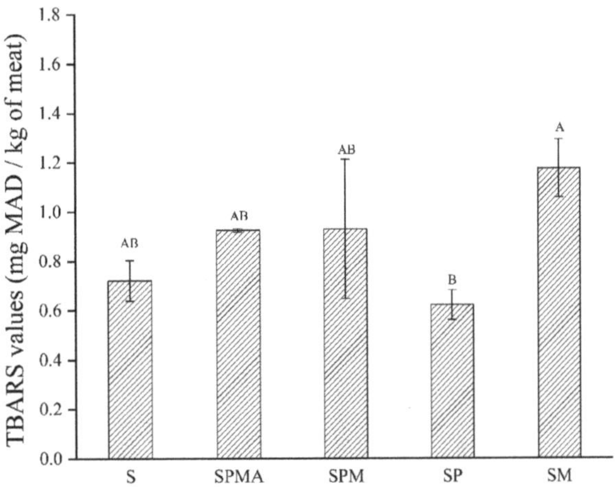

Figure 2. TBARS (Thiobarbituric acid) values of dry-cured meat with the addition of different low-sodium salts. Different letters in the same group indicate significant differences ($p < 0.05$; Tukey test) among treatments. S: 100% NaCl; SPMA: 59.375% NaCl+28% KCl+12% $MgCl_2$+0.625% L-arginine (Arg); SPM: 60% NaCl+28% KCl+12% $MgCl_2$; SP: 72% NaCl+28% KCl; SM: 88% NaCl+12% $MgCl_2$.

2.2. *Effect of Low-Sodium Salts on the Aroma of Dry-Cured Meat*

2.2.1. Volatile Compound Profile

The types of volatile organic compounds in raw pork are relatively few, but multiple types of volatile compounds are formed after dry-curing, resulting in a richer flavor. However, the number of volatile compounds in dry-cured meat was still much less than that in dry-cured ham [27]. Compared with dry-cured ham, dry-cured meat had a shorter manufacturing time (around three months versus 12 months). According to Supplementary Table S1, a total of 55 different volatile compounds from nine different chemical groups were identified in the finished dry-cured meat by HS-GC-MS, including 12 aldehydes, 12 esters, 8 alcohols, 8 ketones, 5 alkanes, 2 pyrazines, 4 aromatics, 3 olefins, and 1 furan compounds.

As shown in Figure 3, aldehydes were the most abundant volatile compounds. Aldehyde compounds in each group accounted for 42.49%–64.42% of the total volatiles, among which SP accounted for the highest proportion of 64.42%. The closest proportion to S (46.13%) was SM (42.49%). The order of aldehyde proportion content was SP > SPM > SPMA > S > SM, and there was no significant difference between each group. In addition, SM's proportions of total volatiles of esters, ketones, aromatics, and furans was also closest to that of S. The proportion of alcohols in each group was in the order of SPM > SM > SP > S > SPMA. The proportion of alcohols in SPM was significantly higher than that in S and SPMA. The proportion of esters in each group was SPMA > SM > S > SPM > SP, and the proportion of esters in SPMA was significantly higher than that in SP. The proportions of ketones and alkanes in each group were SPMA > SPM > SP > SM > S, and SPMA, SPM, and SP were significantly higher than S. The proportion of pyrazines in each group was SP > S > SPM > SM. SP was significantly higher than SM, and SPMA did not contain pyrazines. The proportion of aromatics in each group was S > SM > SP > SPM > SPMA, and S was significantly higher than SPMA. The proportion of furans in each group was SM > SPMA > SP > S > SPM, and there was no significant difference between the groups. In terms of compound types, S, SPMA, SPM, SP, and SM had 34, 39, 41, 38, and 41 volatile compounds, respectively. The findings showed that the addition of low-sodium salt increased the species of volatile compounds in dry-cured meat. Importantly, the SPMA, SPM, and SM groups containing Mg^{2+} had more volatile compounds than SP, which contains KCl.

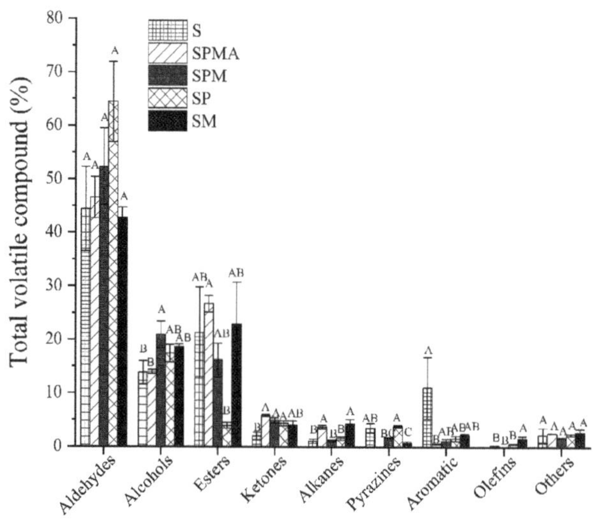

Figure 3. Percentage of various volatile compounds in the total volatile compounds. Different letters in the same group indicate significant differences ($p < 0.05$; Tukey test) among treatments. S: 100% NaCl; SPMA: 59.375% NaCl+28% KCl+12% $MgCl_2$+0.625% L-arginine (Arg); SPM: 60% NaCl+28% KCl+12% $MgCl_2$; SP: 72% NaCl+28% KCl; SM: 88% NaCl+12% $MgCl_2$.

It was noteworthy that the total concentration of aldehydes was highest in dry-cured meat compared to other compounds, which was consistent with the results in dry-cured ham [28]. It can be seen from Supplementary Table S1 that the total content of aldehydes in all groups showed the following pattern: SP > SM > SPM > SPMA > S, among which the aldehydes content in SP was significantly higher than that of SPMA and S. Among the 12 aldehydes, the OAV values of 3-Methylbutanal, 2-Methylbutyral, Pentanal, Hexanal, Heptaldehyde, Octyl aldehyde, 1-Nonanal, (*E*)-2-Heptanal, (*E*)-2-Octenal, Decyl aldehyde, and (*E*)-2-Nonenal were greater than 1 (Table 2). It is generally accepted that compounds with OAV > 1 play an important role in overall odor [29]. 3-Methylbutanal and 2-Methylbutyral originate from the degradation of amino acids [29], and these compounds contribute positively to the sensory quality of dry-cured meat products, contributing to the overall aroma by providing cheesy, nutty, and salty notes [26]. The order of their content was SM > SPM > SP > S > SPMA and SM > SP > SPM > S > SPMA, respectively. SM was significantly higher than SPMA, because SM had a higher NaCl concentration and thereby promoted the formation of 3-Methylbutanal [30]. It is worth noting that Octyl aldehyde, Hexanal, Pentanal, and 1-Nonanal (linear aldehyde) respectively represent rancid smell, nutty aroma, grassy smell, and strong grease smell. The content of these compounds was relatively high, and the OAV values were all greater than 1, indicating that fat oxidation was active during the dry-curing process of the meat. The content order of each group of these compounds was SP > SM > SPM > SPMA > S, among which SP was significantly higher than S. These compounds usually originate from fat autooxidation reactions [31,32], and adding low-sodium salt will increase the content of the aforementioned linear aldehydes because low-sodium salt promotes fat oxidation. It was speculated that the pro-oxidative activity of lower concentrations of NaCl may be further enhanced due to the presence of K^+ [11]. (*E*)-2-Heptanal and (*E*)-2-Nonenal had a green, fatty, and rancid taste, but there were no significant differences between the groups. Decyl aldehyde has a spicy waxy odor and is formed by β-oxidation of unsaturated fatty acids [33]. The Decyl aldehyde content in SP was significantly higher than that in SPMA and S, which may be explained by the fact that the presence of KCl enhances the β-oxidation reaction [33]. (*E*)-2-Octenal exhibited a sweet aroma or the smell of cucumber or melon. The (*E*)-2-Octenal in SP and SPM was significantly higher than that in SPMA, SM, and S, because SM contains more diketone compounds and amino acids (Table 3 and Table S1), while the formation of (*E*)-2-Octenal requires free amino acids and diketone compounds.

Alcohols often have aromatic, vegetal, and earthy notes and are an integral part of dry-cured cuts of meat. A total of 8 alcohols were detected in the meat, and the order of the total content of each group was as follows: SM > SP > SPM > SPMA > S, among which SM, SP, and SPM were significantly greater than SPMA and S, indicating that the low-sodium group without Arg addition significantly increased the alcohol content of dry-cured meat, because NaCl itself has antibacterial activity, whereas partial substitution will result in stronger microbial growth and higher branched-chain alcohol production [34]. Among the eight alcohols, the OAVs of 3-Methyl-1-butanol, 1-Pentanol, 1-Hexanol, 1-Octen-3-ol, 1-Heptanol, and 1-Octanol were all greater than 1. No significant difference was observed in each group for alcohols containing 3-Methyl-1-butanol, 1-Pentanol, 1-Hexanol, and 1-Octen-3-ol. 1-Heptanol has a strong aromatic odor, and the order of content in each group was SM > SP > SPM > S > SPMA, with SM being significantly higher than the other groups. 1-Octanol has a pungent odor, which may come from the oxidation of myristic acid, linoleic acid, palmitoleic acid, or oleic acid. The order of the content of each group was SP > SM > SPM > SPMA, among which SP was significantly higher than SPMA. There was no Octanol in S, probably because the high content of NaCl inhibited the growth of microorganisms and reduced the production of branched-chain alcohols.

Ester compounds are derived from the esterification reaction of carboxylic acids and alcohols. Esters containing short-chain acids have a fruity flavor, but long-chain acids produce fatty flavors [28]. The order of the total content of esters in each group was SM > SPMA > S > SPM > SP. SM was significantly higher than SP, indicating that partial replacement of magnesium increases the content of esters in dry-cured meat. Among them, the OAVs of Methyl butyrate, Methyl 2-methylbutyrate, Methyl Isovalerate, Methyl pentanoate, Methyl caproate, Ethyl caproate, γ-Caprolactone and γ-octa lactone were higher than 1, and there was no significant difference between the groups in the contents of Methyl Isovalerate, γ-Caprolactone, and γ-octa lactone. The most abundant ester in each group was methyl hexanoate, which has a pleasant smell and comes from the esterification reaction of ethanol and acid. The order of methyl hexanoate content in each group was SM > SPMA > S > SPM > SP, among which SM was significantly higher. Methyl butyrate has an apple aroma, and the content order of each group was SM > SPMA > SPM > S > SP. Methyl 2-methylbutyrate ester has a pungent, ethereal fruity smell, and the content order of each group was SM > S > SPM > SPMA > SP. Ethyl caproate has a fruity aroma like pineapple, and the content order of each group was as follows: SM > SPM > SPMA > S. The levels of Methyl butyrate, Methyl 2-methylbutyrate, and Ethyl caproate in SM were significantly higher than in the other groups.

Table 2. Volatile compounds with OAV > 1 in dry-cured meat with different low-sodium salts.

Volatile Compounds	Threshold (μg/kg)	S	SPMA	SPM	SP	SM
3-Methylbutanal	1.2	6.04	4.83	9.42	8.28	15.23
2-Methylbutyral	0.1	67.80	52.90	98.70	113.70	180.30
Pentanal	0.85	7.41	190.25	12.04	14.11	21.89
Hexanal	5	29.54	32.34	37.76	54.43	42.70
Heptaldehyde	0.26	87.15	95.81	104.73	238.46	146.08
Octyl aldehyde	0.7	36.04	37.81	46.51	123.30	56.86
1-Nonanal	1	48.00	56.39	73.83	120.54	75.31
(*E*)-2-Heptanal	2.4	ND	2.85	3.51	ND	ND
(*E*)-2-Octenal	0.25	18.56	17.08	37.72	29.36	24.60
Decyl aldehyde	0.4	14.43	14.90	29.88	46.40	24.88
(*E*)-2-Nonenal	0.05	39.80	46.80	102.80	135.20	45.60
3-Methyl-1-butanol	1.7	ND	ND	1.89	ND	ND
1-Pentanol	0.36	168.44	128.56	60.89	72.11	68.33
1-Hexanol	10	4.43	3.77	4.58	5.01	4.82
1-Octen-3-ol	1	ND	ND	39.14	38.77	54.76
1-Heptanol	4.8	1.80	1.76	2.31	3.08	6.16
1-Octanol	2.7	ND	3.83	5.30	7.08	5.52
Methyl butyrate	15.1	0.20	0.42	0.35	0.13	1.36
Methyl 2-methylbutyrate	0.00025	36,480.00	22,720.00	26,520.00	13,400.00	72,280.00
Methyl Isovalerate	2.2	5.85	3.22	3.49	ND	9.55
Methyl pentanoate	2.2	7.72	9.48	6.26	1.35	0.00
Methyl caproate	10	8.07	12.61	7.15	1.52	14.98
Ethyl caproate	1	2.06	3.35	4.68	ND	11.69
γ-Caprolactone	0.26	38.58	27.85	0.00	55.65	53.73
γ-octa lactone	0.0222	0.00	0.00	592.34	ND	ND
2-Heptanone	0.9	8.34	10.10	8.34	13.99	18.08
6-Methyl-5-hepten-2-one	0.068	24.12	34.41	43.53	ND	59.85
2,3,5-Trimethylpyrazine	0.19	52.26	ND	27.00	91.74	44.32
Tetramethylpyrazine	0.69	17.84	ND	11.57	27.93	ND
Toluene	0.527	92.45	ND	9.92	10.53	21.76
Ethylbenzene	1.2	1.61	ND	ND	1.94	2.43
Dipentene	0.21	ND	ND	ND	19.38	48.52
2-Pentyl furan	0.27	51.74	62.37	46.15	81.15	106.81

S: 100% NaCl; SPMA: 59.375% NaCl+28% KCl+12% $MgCl_2$+0.625% L-arginine (Arg); SPM: 60% NaCl+28% KCl+12% $MgCl_2$; SP: 72% NaCl+28% KCl; SM: 88% NaCl+12% $MgCl_2$.

Table 3. Contents of free amino acids in dry-cured meat with the addition of different low-sodium salts.

Content (mg/100 g)	S	SPMA	SPM	SP	SM	Threshold
Asp	35.4 ± 6.5 ^{C}	62.9 ± 7.1 ^{B}	39.2 ± 3.5 ^{C}	28.8 ± 2.9 ^{C}	96.8 ± 2.7 ^{A}	100
Glu	201.3 ± 24.1 ^{AD}	279.1 ± 6.9 ^{AB}	229.4 ± 12.2 ^{CD}	302.1 ± 5.7 ^{BC}	302.1 ± 0.6 ^{d}	30
Ser	42.9 ± 26.2 ^{A}	35.8 ± 4.1 ^{A}	29.4 ± 2.0 ^{A}	32.5 ± 5.3 ^{A}	30.7 ± 3.8 ^{A}	150
His	65.2 ± 13.0 ^{A}	82.6 ± 5.8 ^{A}	69.7 ± 3.1 ^{A}	81.2 ± 8.9 ^{A}	79.5 ± 4.5 ^{A}	20
Gly	56 ± 3.2 ^{C}	97.6 ± 3.9 ^{AB}	77.4 ± 4.0 ^{CB}	89.4 ± 5.6 ^{ABC}	118.6 ± 26.5 ^{A}	130
Thr	73.5 ± 4.5 ^{B}	100.7 ± 1.5 ^{A}	98.2 ± 5.8 ^{AB}	101.6 ± 12.5 ^{A}	116.7 ± 16.0 ^{A}	260
Arg	94.6 ± 3.3 ^{C}	137.5 ± 5.9 ^{A}	96.2 ± 0.9 ^{C}	91.4 ± 3.7 ^{C}	125.1 ± 5.2 ^{B}	50
Ala	661.8 ± 18.9 ^{B}	650.7 ± 20.3 ^{BC}	681.6 ± 28.5 ^{B}	795.1 ± 24.8 ^{A}	600.8 ± 15.0 ^{C}	60
Tyr	55.5 ± 2.6 ^{A}	40.5 ± 23.5 ^{A}	63.8 ± 4.4 ^{A}	57.8 ± 3.1 ^{A}	67.1 ± 1.7 ^{A}	-
Val	124.4 ± 3.5 ^{A}	166.3 ± 35.0 ^{A}	147.6 ± 4.4 ^{A}	159.8 ± 3.2 ^{A}	156.7 ± 6.9 ^{A}	40
Met	40.3 ± 1.7 ^{AB}	380.0 ± 8.2 ^{B}	50.7 ± 2.0 ^{A}	37.6 ± 2.7 ^{B}	46.5 ± 1.8 ^{AB}	30
Phe	54.7 ± 1.3 ^{A}	80.2 ± 29.1 ^{A}	70.2 ± 3.1 ^{A}	52.4 ± 1.4 ^{A}	65.3 ± 0.3 ^{A}	90
Ile	64.4 ± 1.0 ^{A}	82.0 ± 17.2 ^{A}	83.4 ± 2.5 ^{A}	69.7 ± 3.3 ^{A}	78.6 ± 2.1 ^{A}	90
Leu	94.8 ± 4.4 ^{D}	111.2 ± 5.5 ^{CD}	128.6 ± 6.3 ^{A}	99.3 ± 3.8 ^{CD}	117.7 ± 0.6 ^{AB}	190
Lys	175.4 ± 2.4 ^{D}	226.3 ± 9.8 ^{B}	208.7 ± 7.9 ^{BC}	205.2 ± 3.9 ^{C}	257.6 ± 8.3 ^{A}	50
Pro	94.0 ± 21.5 ^{A}	101.7 ± 27.4 ^{A}	83.2 ± 44.0 ^{A}	110.1 ± 14.2 ^{A}	127.3 ± 21.2 ^{A}	300
Cys	4.1 ± 3.0 ^{A}	5.1 ± 5.0 ^{A}	2.9 ± 1.5 ^{A}	238.6 ± 4.3 ^{A}	4.7 ± 3.8 ^{A}	
Umami FAAs	236.8 ± 30.3 ^{D}	342.0 ± 13.7 ^{B}	268.6 ± 9.4 ^{CD}	283.65 ± 0.4 ^{C}	398.8 ± 6.7 ^{A}	-
Sweet FAAs	834.1 ± 8.7 ^{B}	884.8 ± 27.2 ^{B}	886.6 ± 39.5 ^{B}	1018.6 ± 47.0 ^{A}	866.8 ± 34.1 ^{B}	-
Bitter FAAs	334.6 ± 21.5 ^{C}	396.5 ± 22.4 ^{AB}	415.6 ± 18.3 ^{A}	360.4 ± 12.4 ^{BC}	408.1 ± 7.2 ^{A}	-
Total FAAs	1938.3 ± 46.0 ^{C}	2298.3 ± 49.0 ^{AB}	2160.0 ± 86.6 ^{B}	2269.2 ± 78.1 ^{AB}	2391.7 ± 2.0 ^{A}	-

Umami amino acids include aspartic acid (Asp) and glutamic acid (Glu); sweet amino acids include alanine (Ala), glycine (Gly), serine (Ser), and threonine (Thr); bitter amino acids include histidine (His), phenylalanine (Phe), tyrosine (Tyr), and leucine (Leu). The threshold unit is mg/100g. $^{A–D}$ mean values in the same row not followed by a common letter differ significantly ($p < 0.05$). S: 100% NaCl; SPMA: 59.375% NaCl+28% KCl+12% $MgCl_2$+0.625% L-arginine (Arg); SPM: 60% NaCl+28% KCl+12% $MgCl_2$; SP: 72% NaCl+28% KCl; SM: 88% NaCl+12% $MgCl_2$.

Ketone compounds are produced when alkoxy groups are oxidized by alkyl radicals during lipid oxidation to generate ketone compounds and hydrocarbons. The order of the total content of ketones in each group was as follows: SM > SP > SPMA > SPM > S. The total content of ketones in S was significantly lower than that of other groups, indicating that the addition of low-sodium salt increased the amount of ketones in dry-cured meat. Among them, the OAVs of 2-Heptanone and 6-Methyl-5-hepten-2-one were greater than 1, and 6-Methyl-5-hepten-2-one has been considered a safe compound [35], but there was no significant difference between the contents of each group. 2-Heptanone is an oxidation product of fatty acids [30]. It has a fruity aroma similar to pear and has been detected in many dry-cured meat products, such as ham and sausages [30]. The order of the content of each group of 2-Heptanone was SM > SP > SPMA > SPM > S, where SM was significantly higher than SPM and S, among which the origin of linear 2-ketones was related to the β-oxidation activity of mold growing on the surface of dry-cured meat products [36]. The reason may be that the high ionic strength of SM enhances fat oxidation of dry-cured meat [26].

As for alkane compounds, there were very few types and contents in dry-cured meat. In addition, the threshold value was high, so their contribution to the aroma of the meat was very small. As shown in Table 2, among the alkane compounds, there was no OAV value greater than 1. Our study found that 2,3,5-Trimethylpyrazine and Tetramethylpyrazine had OAVs greater than 1. The order of the total content of pyrazine substances in each group was SP > S > SPM > SM, but SPMA does not contain pyrazines. 2,3,5-Trimethylpyrazine and Tetramethylpyrazine have the smell of corn, nuts, hazelnuts, baking, spices, and chocolate [37], and are usually formed by the condensation of two α-aminocarbonyl groups through the Strecker reaction. The order of content in each group was SP > S > SM > SPM and SP > S > SPM, respectively, among which SP was significantly higher than the other groups. It can be formed from free amino acids and reducing sugars [38]. Four aromatic compounds were also discovered in SM. The total amount in S was the highest, but there was no significant difference between the groups. In addition, three olefin compounds were found, namely 1-Octene, trans-2-Octene, and Dipentene, but there was no significant difference between the groups. Finally, a furan compound was discovered. Furan compounds provide a lot of aromas to meat products. 2-Pentylfuran is a linoleic acid oxidation product with the sensory properties of beans and green grass and a low threshold [39]. SM had the highest 2-Pentylfuran content, but there was no significant difference between the groups.

2.2.2. Aroma in Dry-Cured Meat Determined by Electronic Nose

The electronic nose radar chart intuitively shows the response intensity to various substances. Figure 4 shows the results of the response intensity of volatile substances produced by dry-cured meat under 18 types of sensors. The response substances corresponding to the 18 types of sensors are S1 (Alkanes, smoke), S2 (alcohols, aldehydes, short-chain alkanes), S3 (ozone), S4 (sulfides), S5 (organic amines), S6 (organic gases, benzophenones, aldols), S7 (short-chain alkanes, such as methane, natural gas, and biogas), S8 (short-chain alkanes, such as propane and liquefied gas), S9 (aromatic compounds, aldols), S10 (hydrogen-containing gas), S11 (alkanes, alkenes), S12 (short-chain alkanes, such as liquefied gas and methane), S13 (combustible gases, such as methane), S14 (combustible gases and smoke), S15 (alkanes, organic gases), S16 (sulfides), S17 (nitrides), S18 (ketones, alcohols). As illustrated in Figure 4, the response values of S1, S4, S5, S6, S7, S10, S11, S12, S16, S17, and S18 sensors in SM were all the highest, indicating that SM contained more volatile compounds, which was consistent with the results of GC-MS (Supplementary Table S1).

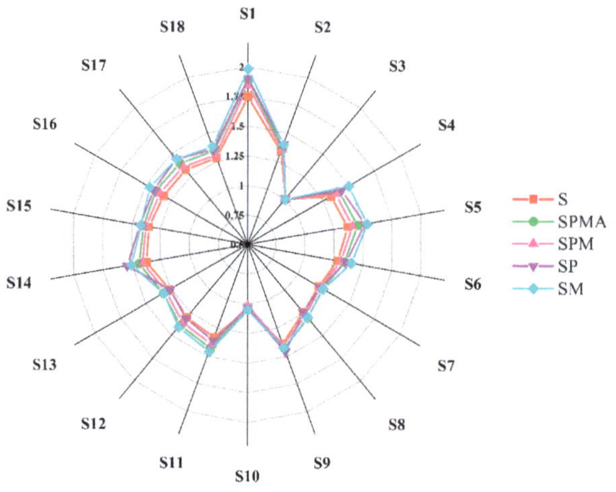

Figure 4. Radar chart of electronic nose analysis for dry-cured meat with the addition of different low-sodium salts. S: 100% NaCl; SPMA: 59.375% NaCl+28% KCl+12% $MgCl_2$+0.625% L-arginine (Arg); SPM: 60% NaCl+28% KCl+12% $MgCl_2$; SP: 72% NaCl+28% KCl; SM: 88% NaCl+12% $MgCl_2$.

2.3. *Effect of Low-Sodium Salts on the Taste of Dry-Cured Meat*

2.3.1. Molecular Weight Distribution of Peptides

The distribution of peptide molecular weight can reflect protein degradation. As shown in Figure 5, the molecular weights of the peptides in each group of dry-cured meats are mainly distributed in three ranges: molecular weights greater than 10,000 Da, 180–500 Da, and less than 180 Da. Peptides with a molecular weight greater than 10,000 Da include some macromolecular peptides or preliminarily hydrolyzed proteins [40]. The peptide molecular weight (>10,000 Da) in each group followed the order S > SPMA > SM > SPM > SP; in particular, S was significantly higher than SPM, SP, and SM, indicating that partial substitution of sodium facilitates protein degradation. The peptides with a molecular weight distribution between 180 and 500 Da may be related to polypeptides that produce flavor, salty taste, and other tastes. The order of those in each group was SP > SPM > SPMA > SM > S, among which SP and SPM were significantly higher than other groups. For molecular weights less than 180 Da, related to the content of free amino acids, the contents of SP, SM, and SPMA were greater than those of S and SPM, which implies that more free amino acids formed in SP, SM, and SPMA.

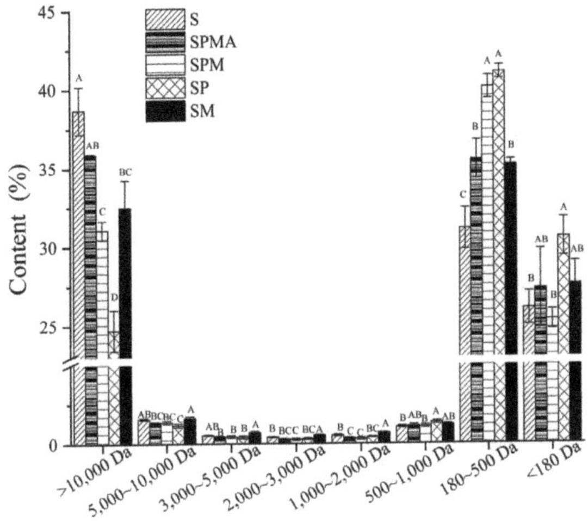

Figure 5. Molecular weight distribution in dry-cured meat with the addition of different low-sodium salts. Different letters in the same group indicate significant differences ($p < 0.05$; Tukey test) among treatments. S: 100% NaCl; SPMA: 59.375% NaCl+28% KCl+12% $MgCl_2$+0.625% L-arginine (Arg); SPM: 60% NaCl+28% KCl+12% $MgCl_2$; SP: 72% NaCl+28% KCl; SM: 88% NaCl+12% $MgCl_2$.

2.3.2. Free Amino Acid Contents in Dry-Cured Meat

The release of free amino acids (FAAs) results from protein degradation. As shown in Table 3, the promoting effect on total free amino acids (TFAAs)accumulation of KCl and $MgCl_2$ partially replacing NaCl was consistent with the results of Armenteros et al. [41] and José M et al. [42]. The TFAAs content of SM was the highest at 2391.7 mg/100g, which was significantly higher than that of S and SPM, which is related to their pH. Aksu et al. [43] reported that low pH could increase protein degradation. Table 1 also show that SM with higher TFAAs had lower pH than S and SPM.

The umami amino acids in SM were significantly higher than those in the other groups, indicating that partial replacement of NaCl with $MgCl_2$ can promote protein degradation to produce more Glu and Asp. In addition, Kęska et al. [44] found that Glu can impart salty and umami flavors to dry-cured meat and inhibit bitterness. The sweet amino acid content of SP was significantly higher than that of the other four groups, which indicates that only KCl can produce the formation of more sweet-related amino acids when it partially replaces NaCl. For bitter amino acids, the total content order was SM > SPM > SPMA > SP > S. It can be seen that the low-sodium salt groups have more bitter amino acids, but they are all within the appropriate range. In the present study, the final free amino acid content was lower than that reported by the authors of [45] in dry-cured hams, which may be due to shorter processing times.

2.3.3. Taste Analysis by Electronic Tongue

In addition to affecting the aroma, low-sodium salt also affects the taste. Therefore, an electronic tongue was used to measure the saltiness, bitterness, sourness, and richness of dry-cured meat. In terms of richness, the response values order was SM > SPMA > S > SP > SPM (Figure 6). It is worth noting that comparing SPMA and SPM revealed a small difference in Arg, but their richness was significantly different, indicating that adding a small amount of Arg could increase the richness of dry-cured meat. The response values of umami taste in each group were in the following order: SPM > S > SP > SM > SPMA. Conversely, the umami taste intensity of SPMA was significantly lower than that of SPM. These results show that adding Arg cannot improve the umami taste of dry-cured meat, but has an inhibitory effect.

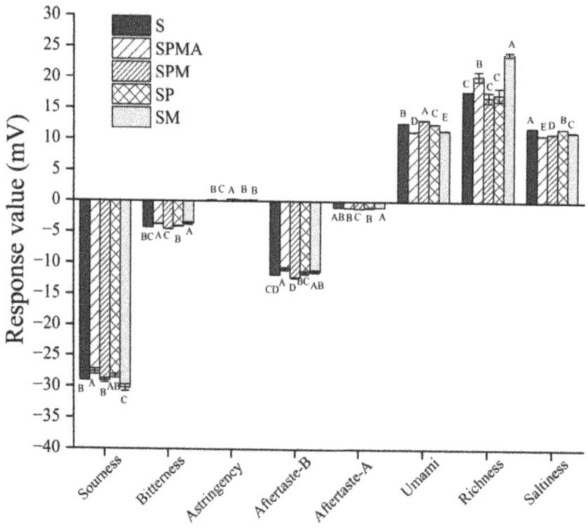

Figure 6. Electronic tongue analysis of dry-cured meat with different low-sodium salt additions. Different letters in the same group indicate significant differences ($p < 0.05$; Tukey test) among treatments. S: 100% NaCl; SPMA: 59.375% NaCl+28% KCl+12% $MgCl_2$+0.625% L-arginine (Arg); SPM: 60% NaCl+28% KCl+12% $MgCl_2$; SP: 72% NaCl+28% KCl; SM: 88% NaCl+12% $MgCl_2$.

In terms of salty attributes, the response values of each group were SM > SP > S > SPM > SPMA, which was not consistent with the sodium ion content shown in Table 1. This showed that the salty attributes were not only produced by Na^+ but can also be produced by other metal ions as well as peptides and free amino acids. As shown in Figure 5, the amount of peptides of molecular weight 180–500 Da in SM and SP was significantly higher than that in other groups, indicating that Mg^{2+} or K^+ alone can promote the production of salty peptides. As for bitterness, the response values were SM > SPMA > SP > S > SPM, among which SM and SPMA were significantly higher than SP, S, and SPM. Zhang et al. [46] also found that the presence of KCl and $MgCl_2$ enhanced the bitter taste. However, the bitterness values of each group in this study were less than the lowest response value of 4.91, which was acceptable to the senses. In terms of sourness, although there were differences in the response values between samples, the values were all less than the minimum response threshold of −13, which can be considered to be extremely weak sourness.

2.4. *Effect of Low-Sodium Salts on Sensory Scores of Dry-Cured Meat*

The dry-cured meat was evaluated in four areas: appearance, taste, texture, and overall acceptance, with scores ranging from 1 to 5. As shown in Table 4, SM received the highest scores in gloss, saltiness, umami, and fat adhesion. Its bitterness was the lowest among the low-sodium salt group. Its overall acceptance was also the highest among the low-sodium salt group, indicating that the sensory evaluators were more satisfied with SM. In terms of color, the colors of the low-sodium salt groups (SPMA, SPM, SP, and SM) were all redder than S, with SP having the highest redness, which was consistent with the color results shown in Table 1. The bitterness scores of the low-sodium salt group were higher than those of S, which was consistent with the content of bitter amino acids shown in Table 3. According to reports, substitution of 50% KCl increases the bitterness of dry-cured meat [10], but addition levels of 28% and 12% (KCl and $MgCl_2$) did not result in reports of bitterness from the assessors in our study. In terms of umami taste, SM scored the highest, which was consistent with the umami amino acid content shown in Table 3. As shown in Figure 2, SM had a higher TBARS value. It was speculated that SM had the highest degree of fat oxidation, resulting in higher adhesion. The overall acceptance scores of SM and SPMA were the highest among the low-sodium salt groups.

Table 4. Sensory evaluation scores of dry-cured meat with different low-sodium salt formulations.

	S	SPMA	SPM	SP	SM
			Score		
Redness	2.53 ± 0.99 ^A^	3.08 ± 1.14 ^A^	2.67 ± 1.23 ^A^	3.17 ± 1.10 ^A^	3.15 ± 1.32 ^A^
Homogeneity	3.11 ± 0.96 ^A^	3.33 ± 1.08 ^A^	3.17 ± 1.10 ^A^	3.00 ± 1.14 ^A^	3.22 ± 1.22 ^A^
Gloss	2.73 ± 0.91 ^A^	3.18 ± 0.87 ^A^	3.03 ± 1.12 ^A^	2.61 ± 0.78 ^A^	3.25 ± 0.97 ^A^
Saltiness	3.83 ± 1.14 ^A^	3.99 ± 0.76 ^A^	3.44 ± 0.78 ^A^	3.88 ± 0.95 ^A^	4.02 ± 0.84 ^A^
Bitterness	1.42 ± 0.77 ^A^	1.69 ± 0.89 ^A^	1.86 ± 1.16 ^A^	1.81 ± 0.93 ^A^	1.53 ± 0.81 ^A^
Umami	2.12 ± 1.09 ^A^	2.56 ± 0.99 ^A^	2.39 ± 1.05 ^A^	2.33 ± 1.08 ^A^	2.62 ± 1.10 ^A^
Sweetness	1.36 ± 0.72 ^A^	1.31 ± 0.62 ^A^	1.36 ± 0.64 ^A^	1.36 ± 0.64 ^A^	1.42 ± 0.65 ^A^
Hardness	2.93 ± 0.93 ^A^	3.00 ± 0.69 ^A^	2.96 ± 1.12 ^A^	3.32 ± 1.01 ^A^	2.99 ± 0.83 ^A^
Fat adhesion	2.33 ± 1.03 ^A^	2.44 ± 1.10 ^A^	2.22 ± 0.94 ^A^	2.28 ± 1.02 ^A^	2.72 ± 1.07 ^A^
Chewiness	3.03 ± 0.96 ^A^	3.47 ± 0.88 ^A^	3.08 ± 1.03 ^A^	3.12 ± 0.77 ^A^	3.36 ± 1.11 ^A^
Acceptance	3.31 ± 1.05 ^A^	3.08 ± 1.29 ^A^	3.04 ± 1.24 ^A^	3.06 ± 0.87 ^A^	3.08 ± 1.09 ^A^

S: 100% NaCl; SPMA: 59.375% NaCl+28% KCl+12% $MgCl_2$+0.625% L-arginine (Arg); SPM: 60% NaCl+28% KCl+12% $MgCl_2$; SP: 72% NaCl+28% KCl; SM: 88% NaCl+12% $MgCl_2$. Different letters in the same group indicate significant differences ($p < 0.05$; Tukey test).

2.5. Principal Component Analysis (PCA) of Physicochemical Properties, Aroma, Taste, and Sensory Scores

PCA can reduce the dimensionality of complex data and further reflect the overall information provided by the sample. In Figure 7, PC1 and PC2 cover 41.8% and 19.5%, respectively, with a cumulative value of 61.3%. The five treatment groups in this study were clearly separated in terms of PC1. Furthermore, SP differed from other components by PC2 through indicators such as sweet amino acids, aldehydes, pyrazines, shear force, and K^+ content. Most physicochemical properties were grouped in the negative quadrant of PC1, whereas most sensory attributes, electronic nose odor, and free amino acid content were grouped in the positive quadrant of PC1. The components detected by GC-MS, especially aldehydes, ketones, alcohols, and pyrazine, have a strong correlation with the odor detected by electronic nose. S was described as related to shear force, aromatic compound content, Na^+ content, and b^*. However, excessive shear force and pH are detrimental to meat products. SM was described as related to electronic tongue aftertaste, free amino acids, total amino acids, and electronic tongue bitterness. SM and S were the farthest apart, indicating a significant difference between them, which was conducive to the enhancement of the quality of dried-cured meat in terms of shear force, water loss, and pH. In addition, SPMA was considered to be related to A_w, ester content, TBARS, Mg^{2+} content, sensory properties, and bitter amino acids. Also, SP was described as related to electronic nose S9, sensory hardness, sweet amino acid content, aldehyde content, and K^+ content.

Importantly, SPMA, SPM, and SM containing Mg^{2+} had more volatile compounds than SP in terms of flavor (Figure 3). Sensory evaluation also found that SM had a higher overall acceptance compared to other groups. Therefore, magnesium ions contribute to improving the quality of dry-cured meat products.

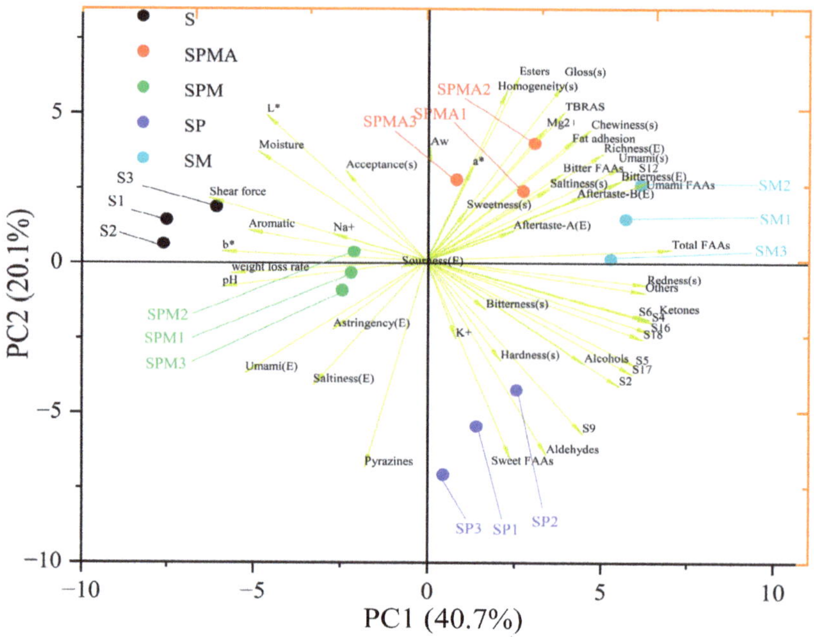

Figure 7. PCA of physicochemical properties, aroma, taste, and sensory scores. S: 100% NaCl; SPMA: 59.375% NaCl+28% KCl+12% $MgCl_2$+0.625% L-arginine (Arg); SPM: 60% NaCl+28% KCl+12% $MgCl_2$; SP: 72% NaCl+28% KCl; SM: 88% NaCl+12% $MgCl_2$.

3. Materials and Methods

3.1. Materials

Fresh pig hind legs were purchased from a local supermarket in Kunming, Yunnan Province. Food-grade magnesium chloride (NaCl), potassium chloride (KCl), magnesium chloride ($MgCl_2$), and L-arginine (Arg) were obtained from Henan Mingrui Food Additives Co., Ltd. (Zhengzhou, China). Trichloroacetic acid and thiobarbituric acid were obtained from Shanghai Titan Technology Co., Ltd. (Shanghai, China). Analytical grade KCl was sourced from Yunnan Jingrui Technology Co., Ltd. (Kunming, China). Sodium dihydrogen phosphate (NaH_2PO_4), disodium hydrogen phosphate (Na_2HPO_4), ice acetic acid (HAC), dimethyl sulfoxide (DMSO), and sodium acetate ($NaC_2H_3O_2$) were obtained from Tianjin Zhiyuan Chemical Reagent Co., Ltd. (Tianjin, China). Triton X-100 was obtained from Beijing Solarbio Science & Technology Co., Ltd. (Beijing, China). Chloroform (TCM) and concentrated nitric acid (HNO_3) were purchased from Yonghua Chemical Co., Ltd. (Suzhou, China). Dithiothreitol (DTT) was obtained from Beijing Solabao Technology Co., Ltd. Z-Phe-Arg-AMC and Z-Arg-Arg-AMC were sourced from Shanghai Zhenzhun Biotechnology Co., Ltd. Disodium EDTA was purchased from Guangzhou Saiguo Biotechnology Co., Ltd. (Guangzhou, China). Brij 35 was obtained from Shanghai Maclin Biochemical Technology Co., Ltd. (Shanghai, China). All reagents mentioned above, except for the food-grade ones, were of analytical grade.

3.2. Preparations of Dry-Cured Meat

The hind leg of the pig was uniformly cut into cubes with a weight of 500 g. Five formulas of salt were employed to cure the meat according to common low-sodium salt formulas from Finland with minor modifications [47]. The salt (5%, based on the weight of the meat) was placed on the surface of the meat and rubbed by hand to accelerate the permeation of the salt. The meat was dry-cured for 6 days at 4 °C and a relative humidity (RH) of 80%–90% before drying at 12 °C and an RH of 50% for 3 days in a constant temperature and humidity chamber (ICTHI-1100T, STIK, Stochastic Instruments Co., Ltd., Shanghai, China). For subsequent fermentation, the sample was divided into 4 stages including the early fermentation stage (20 °C, RH 60%, and 15 days), middle fermentation stage (28 °C, RH 75%, and 70 days), and late fermentation stage (20 °C, RH 65%, and 5 days). Digital images of dry-cured meat in the experiment were shown in Figure 8. Each group consisted of 3 parallel samples. The five formulas comprise one control group and four groups of low-sodium salts.

(1) S (control): 100% NaCl, ionic strength 1.160 mol/g;
(2) SPMA: 59.375% NaCl+28% KCl+12% $MgCl_2$+0.625% L-arginine (Arg), ionic strength 1.160 mol/g;
(3) SPM: 60% NaCl+28% KCl+12% $MgCl_2$, ionic strength 1.085 mol/g;
(4) SP: 72% NaCl+28% KCl, ionic strength 1.055 mol/g;
(5) SM: 88% NaCl+12% $MgCl_2$, ionic strength 1.190 mol/g.

Figure 8. The appearance of meat (**A**) during curing, both in the (**B**) early and (**C**) later stage of natural fermentation.

3.3. Sodium, Potassium, and Magnesium Content

The metal ion content in the various treated meats was determined according to national standards [48,49] with minor modification, using atomic absorption spectroscopy (Mars 6, American CEM Service Corp, Charlotte, NC, USA). The chopped sample (0.5 g) was added into 10 mL of HNO_3 for digestion using a microwave digestion system (EHD-24, Donghang Scien-Tech, Beijing, China). After cooling down, the mixture was placed on an electric heating oven to drive HNO_3 to near dryness, and adjusted to a constant volume of 50 mL with deionized water, immediately followed by the determination of sodium, potassium, and magnesium contents by atomic absorption spectrometry. The permeability was calculated based on the difference between the actual ion content and the theoretical value. The standard curves were established based on standard substances for sodium, potassium, and magnesium ions. The standard curve equations for calculating the content of sodium, potassium, and magnesium were $y = 0.3728x + 0.0298$, $R^2 = 0.9972$; $y = 02277x + 0.016$, $R^2 = 0.9944$; $y = 0.6554x + 0.4554$, $R^2 = 0.9916$, respectively. The sodium, potassium, and magnesium contents remaining in the meat and the permeability were calculated by the following equations:

$$X = \frac{C \times V \times f}{100 \times m \times (1 - w)}$$

$$P = \frac{L - X}{X} \times 100\%$$

where X is the salt content (g/100 g dry weight); C is the sodium content obtained from standard curves (mg/mL); V is the solution volume (mL); f is the dilution rate; m is the sample weight (g); and w is the moisture content. P is permeability (%); L is theoretical addition amount, that is, the amount of added metal ions.

3.4. Weight Loss

At the end of fermentation, the meat was weighed and its weight recorded. The weight loss was calculated using the following equation:

$$Weight\ loss\ (\%) = (1 - \frac{Weight\ of\ meat\ at\ the\ end\ of\ fermentation(g)}{Fresh\ meat\ weight(g)}) \times 100\%$$

3.5. Moisture Content and Water Activity (A_w)

The moisture content of dry-cured meat was assessed by the weight difference of meat before and after drying at 105 °C.

To determine the water activity (A_W), the chopped sample (2.0 g) was tiled in a water activity measuring dish so that the sample completely covered the bottom of the cassette. Then, the cassette was placed in a pre-heated water activity meter (HD-3A-4, Nanjing, China) for 20 min for water activity determination.

3.6. Low-Field ^1H NMR Measurements

A low-field nuclear magnetic resonance analyzer (LF-NMR) with a proton resonance frequency of 49.74 MHz and a permanent magnet strength of 20 ± 0.08 T was used to analyze the water state in dry-cured meat according to the method reported by Wang et al. [50] with some modifications. A quantity of meat with the same weight (4.0 g) was placed in a 30 mm diameter NMR tube, and signals were acquired using a CPMG pulse sequence. The main parameters were set as follows: TW (wait time) of 3000 ms, TE (echo time) of 1.0 ms, NECH (number of echoes) of 1000, and NS (number of scans) of 4. MultiEx Inv analysis software (NIUMAG nuclear magnetic resonance analysis application software Ver4.0) was used to fit the CPMG decay curve with a multi-exponential fitting and the SRIT algorithm was used to obtain the transverse relaxation time (T_2) curve and the corresponding NMR parameters.

3.7. pH

The meat (10 g) was put into a 150 mL centrifuge tube, containing 100 mL of 0.75% KCl solution. The mixture was homogenized for 1 min using a homogenizer (FS-2, Guoyu Instruments, Changzhou, China) and filtered with a gauze. The pH of the filtrate was measured immediately using a pH meter (PHS-3C, Yidian Scientific Instruments Co, Shanghai, China).

3.8. Shear Force

The method was slightly modified from that described by Alino, Grau, Toldra and Barat [17]. The meat was cut into a 200 mm × 200 mm × 200 mm cube and sheared on a texture analyzer (TA.XT. Plus, Stable Micro Systems, Godalming, UK) equipped with a load cell of 50 kg. The testing conditions were as follows: the pre-test probe speed was 2 mm/s, the test speed was 1 mm/s, and the post-test speed was 2 mm/s. The trigger force and compression distance were set as 5 g and 28 mm, respectively. The probe type used was HDP/BSK.

3.9. Surface Color

The color of the dry-cured meat was measured using a colorimeter (CR-400, Konica Minolta, Tokyo, Japan) with a D65 light source. The colorimeter was calibrated with a standard whiteboard (ch00) and blackboard before measurement. The probe with a lighting section diameter of Φ 8mm lightly touched the meat pieces, which were sliced to a thickness of about 3–5 mm. The observation angle was approximately 2°. The color was expressed in terms of luminosity (L^*), redness (a^*), and yellowness (b^*) values. Each measurement was performed in triplicate.

3.10. Thiobarbituric Acid (TBA) Values

Proportional amounts of lean and fat meat samples (2 g) were weighed and mixed with 17 mL 2.5% trichloroacetic acid (TCA) and 3 mL 1% thiobarbituric acid (TBA). After heating in a boiling water bath for 30 min, the mixture was removed and cooled to room temperature. Four milliliters of the supernatant were mixed with an equal volume of chloroform, shaken to mix thoroughly, and centrifuged at 3000 rpm for 10 min at room temperature. The absorbance of the supernatant was measured at 532 nm after calibration with chloroform.

3.11. Volatile Compound Measurement

The extraction of volatile compounds was carried out using the headspace solid-phase microextraction (HS-SPME) technique [51]. A chopped sample of 3.00 g was added into a 20 mL headspace vial. 1,2-dichlorobenzene (100 mg/L) of 4 μL was added as an internal standard, and the vial was sealed with a polytetrafluoroethylene (PTFE) silicone septum. The vial was then equilibrated for 10 min in a water bath at 45 °C. A divinylbenzene/carboxy/polydimethylsiloxane (DVB/CAR/PDMS) fiber was selected, exposed in the headspace of the vial, and extracted for 30 min at 45 °C. Finally, gas chromatography-mass spectrometry (GC-MS) was used for qualitative and quantitative analysis. A DB-wax (0.25 mm × 0.25 μm × 30 m) capillary column was used. The inlet temperature was 250 °C, and the SPME fiber was desorbed into the GC inlet for 5 min. The split ratio was set as 1:5. The temperature program of the oven was as follows: initial temperature of 40 °C, held for 2 min, increased to 90 °C at a rate of 3 °C /min, held for 5 min, then increased to 200 °C at a rate of 3 °C /min and to 230 °C at a rate of 15 °C /min, held for 10 min. Helium was used as the carrier gas at a flow rate of 1.5 mL/min. The interface temperature was 280 °C, and the ion source temperature was 230 °C. Electron energy was 70 eV. The scan mode was used with a mass scan range of 45~450 m/z. Compounds were identified by searching the NIST 21 mass spectral library and retention indices (RI), with a similarity > 90%. The relative content of each compound was calculated by comparing its area to that of the internal standard. Three parallels were set for each treatment, and the results were calculated on a dry basis and further analyzed by taking the average value. The OVA value was calculated according to the following equation:

$$\text{OAV} = \frac{C}{T}$$

C is compound content (μg/kg); T is the threshold value for the compound.

3.12. Electronic Tongue Detection

Peptide molecular weight was determined according to the method of Wang et al. [52]. The sample (25 g) was mixed with 100 mL of ultrapure water and homogenized using a homogenizer (T25, IKA, Staufen, Germany). The suspension was filtered using a filter paper, and then the collected filtrate was centrifuged at 4 °C and 8000 r/min for 10 min. The supernatant was filtered again using filter paper. The richness, saltiness, sourness, umami, bitterness, astringency, aftertaste-B, and aftertaste-A of the sample were analyzed using an electronic tongue (SA402B, INSENT, Kanagawa, Japan).

3.13. Peptide Molecular Weight Distribution Determination

Peptide molecular weight was determined according to the method of Zhan et al. [53], which was modified by Wang et al. [52]. An accurately weighed 3 g of chopped sample was mixed with 30 mL of ultrapure water, and homogenized at 10,000 r/min for 30 s in an ice bath. Subsequently, the mixture was centrifuged at 8000 r/min and 4 °C for

10 min. Ultimately, the supernatant was filtrated with a double-layer filter paper, and then a 0.22 μm PES filter membrane before measurement.

3.14. Free Amino Acid Measurement

A high-performance liquid chromatography (HPLC) method was used to detect the content of free amino acids in the dry-cured meat [52]. Firstly, 1 g of meat sample was accurately weighed and mixed with 25 mL of 5% TCA, followed by 30 min of ultrasonic treatment. After standing for 2 h, the suspension was centrifuged at 10,000× *g* for 10 min. Finally, the supernatant obtained after centrifugation was filtered through a 0.45 μm PES filter membrane. The chromatographic conditions were as follows: an ODS Hypersil column (250 mm × 4.6 mm × 5 μm) was used, and the column temperature was kept at 40 °C. The mobile phases A and B were 0.6 mmol/L sodium acetate and 0.15 mmol/L sodium acetate-acetonitrile-methanol (1/2/2, *v*/*v*/*v*), respectively. The flow rate was set to 1.0 mL/min.

3.15. Electronic Nose Detection

The response of volatile compounds was measured using an electronic nose (C-Nose, Baosheng Industrial Development Co., Ltd., Shanghai, China) consisting of 18 metal oxide gas sensors (S1–S18). The minced meat sample (4 g) was placed in a 20 mL headspace vial, and equilibrated at 50 °C for 50 min. The cleaned sensor array was set to zero with processed clean air as the carrier gas. Then, 200 μL of the sample was injected with a wash time of 100 s and a test time of 60 s, and the device was cleaned for 5000 s before each measurement. Three measurements were conducted for each sample.

3.16. Sensory Analysis

Selection of sensory assessors: fifty volunteers aged 20–25 years old were recruited from Kunming University of Science and Technology. According to National standard [54], a questionnaire survey was designed to understand the interests and motivations of sensory assessors, whether they dislike dry-cured meat products, whether they have taken drugs that impair sensory function or worn dentures, and whether there is sufficient time. Subsequently, the volunteers were trained in terms of identification and intensity discrimination for taste, aroma, texture, and color attributes. The training program was conducted over 4 stages in one month. Ultimately, 18 assessors (9 males and 9 females) were selected by identification and intensity tests of taste, aroma, texture, and color attributes.

According to the method described by Sugimoto et al. [55], the quality of dry-cured meat was evaluated in 5 attributes including appearance, taste, texture, and overall acceptance. Each sensory assessor evaluated the sample on a scale of 1 to 5. The quality of the meat was judged on a scale of 1 to 5. Table 5 shows the sensory evaluation criteria.

Table 5. Sensory attributes and assessment criteria for sensory evaluation.

Category	Sensory Attributes	Judgment Criteria (a Scale of 1 to 5)
Appearance	Redness	Pale pink to dark red
	Homogeneity	Non-homogeneous to homogeneous
	Glossiness	Glossy from dark to bright
Taste	Saltiness	Not distinctly salty to moderately salty
	Bitterness	Not bitter to distinctly bitter
	Umami	Not perceived to perceived
	Sweetness	Not very sweet to very sweet
Texture	Meat hardness	Loose to very hard in texture
	Fat adhesiveness	Not adhesive to appropriately adhesive
	Chewiness	Poorly chewable to very chewable
Overall acceptance	Overall acceptance	Unattractive to very attractive

Before analysis, the sample was sliced with a thickness of 2 mm, and then cooked at 90 °C in a water bath. The entire slice, including lean and fat areas, was placed on a plastic board encoded with a 3-digit number and presented to the assessors. The order of sample presentation systematically varies between evaluators to balance the impact of service order and carryover. During the evaluation period, the assessors were provided with salt-free biscuits and water for cleaning the oral cavity. All tests were conducted in a standard sensory laboratory. The sensory testing was evaluated by the ethical committee of Kunming University of Science and Technology.

3.17. Statistical Analysis

Unless otherwise stated, three parallel experiments were set up in each group. Experimental data were processed by one-way ANOVA in Minitab 21, and results were expressed as mean ± standard deviation (SD). Multiple comparisons were made by the Tukey test, with $p < 0.05$ indicating a significant level of difference.

To assess the contribution of the major physicochemical parameters, volatile compounds, electronic nose data, molecular weight distribution, free amino acids, electronic tongue data, and sensory scores in all samples, the data were evaluated by multivariate principal component analysis (PCA) using the origin 2022.

4. Conclusions

Low-sodium salt-cured meat products have good development prospects and are suitable for the needs of people with low sodium intake. In this study, four groups of low-sodium salts (SPMA, SPM, SP, and SM) and one group of pure NaCl (S) were used to process meat, and their physical and chemical properties, aroma, and taste were analyzed. It was found that Mg^{2+} can strongly combine with the polar groups of proteins to strengthen the interaction between proteins, thus hindering the penetration of salt. The Na^+ permeability of the SM was only 77.85%, which was much lower than other groups, and the salt reduction reached 24.19%. Regarding flavor, partial replacement of metal ions, except for SP, promoted the oxidation of meat and increased the types of volatile compounds. In addition, the low-sodium salt groups had lower pH values. A low pH value will increase the degradation of protein, causing more proteins in the low-sodium salt groups to degrade into small molecule peptides and amino acids, making dry-cured meat products richer in taste. The study provides a clearer direction for further exploration of marinade recipes and provides a theoretical basis for low-sodium dry-cured ham.

Supplementary Materials: The following supporting information can be downloaded at: https://www.mdpi.com/article/10.3390/molecules29102194/s1, Table S1: Content of volatile compounds in dry-cured meat with the addition of different low-sodium salts.

Author Contributions: J.X.: Methodology, Investigation, Data curation, Writing—original draft. C.G. and L.Z.: Formal analysis, Investigation, Writing—review & editing. H.H., X.Y., and J.W.: Data curation, Investigation. X.W. and J.C.: Conceptualization, Funding acquisition, Writing—review & editing, Supervision. All authors have read and agreed to the published version of the manuscript.

Funding: This research was funded by project of Yunnan Major Scientific Technological Projects [202202AG050009], Yunnan Fundamental Research Projects [202301BE070001-009], Xingdian Talent Support Plan for young talents of Yunnan Province [KKRD202246038], Yunnan Fundamental Research Projects [202301AT070144], and National Natural Science Foundation of China [32202092].

Institutional Review Board Statement: In sensory evaluation, appropriate protocols have been used to protect the rights and privacy of all participants during the implementation of the research. In this research, the sensory panel participants have given permission to participate and use their data/answers. This project had been approved by the ethics committee of Kunming University of Science and Technology, No. KMUST-MEC-221. This study does not involve any human or animal testing.

Informed Consent Statement: Informed consent was obtained from all subjects involved in the study. Written informed consent has been obtained from the patient(s) to publish this paper.

Data Availability Statement: The original contributions presented in the study are included in the article/Supplementary Material.

Conflicts of Interest: The authors declare that they have no known competing financial interests or personal relationships that could have appeared to influence the work reported in this paper.

References

1. Vidal, V.A.S.; Lorenzo, J.M.; Munekata, P.E.S.; Pollonio, M.A.R. Challenges to reduce or replace NaCl by chloride salts in meat products made from whole pieces—A review. *Crit. Rev. Food Sci. Nutr.* **2021**, *61*, 2194–2206. [CrossRef]
2. Laureati, M.; Buratti, S.; Giovanelli, G.; Corazzin, M.; Lo Fiego, D.P.; Pagliarini, E. Characterization and differentiation of Italian Parma, San Daniele and Toscano dry-cured hams: A multi-disciplinary approach. *Meat Sci.* **2014**, *96*, 288–294. [CrossRef] [PubMed]
3. WHO Collaborating Centre on Population Salt Reduction; The George Institute for Global Health. Changes in the salt levels in processed meat 2010–2017. In *Food Policy*; WHO: Geneva, Switzerland, 2017.
4. WHO. Sodium Intake for Adults and Children. Available online: https://www.who.int/en/news-room/fact-sheets/detail/salt-reduction (accessed on 25 April 2024).
5. Wang, X.; Yi, Y.; Guo, C.; Wang, X.; Yu, J.; Xia, S. Enhanced sodium release and saltiness perception of surimi gels by microwave combined with water bath heating. *Food Hydrocoll.* **2023**, *134*, 108899. [CrossRef]
6. Gong, X.; Jing, W.; Yeling, Z.; Ying, Z.; Qiujin, Z.; Linggao, L.; Dan, C.; Yanpei, H.; Sha, G.; Mingming, L. Mediated curing strategy: An overview of salt reduction for dry-cured meat products. *Food Rev. Int.* **2022**, *39*, 4565–4580.
7. Pinna, A.; Saccani, G.; Schivazappa, C.; Simoncini, N.; Virgili, R. Revision of the cold processing phases to obtain a targeted salt reduction in typical Italian dry-cured ham. *Meat Sci.* **2020**, *161*, 107994. [CrossRef] [PubMed]
8. Li, M.; Zhang, X.; Yin, Y.; Li, J.; Qu, C.; Liu, L.; Zhang, Y.; Zhu, Q.; Wang, S. Perspective of sodium reduction based on endogenous proteases via the strategy of sodium replacement in conjunction with mediated-curing. *Crit. Rev. Food Sci. Nutr.* 2023; 1–12, *Online ahead of print*.
9. Pedro, D.; Saldana, E.; Lorenzo, J.M.; Pateiro, M.; Dominguez, R.; Dos Santos, B.A.; Cichoski, A.J.; Campagnol, P.C.B. Low-sodium dry-cured rabbit leg: A novel meat product with healthier properties. *Meat Sci.* **2021**, *173*, 108372. [CrossRef]
10. Armenteros, M.; Aristoy, M.C.; Barat, J.M.; Toldra, F. Biochemical and sensory changes in dry-cured ham salted with partial replacements of NaCl by other chloride salts. *Meat Sci.* **2012**, *90*, 361–367. [CrossRef]
11. Wu, H.; Zhang, Y.; Long, M.; Tang, J.; Yu, X.; Wang, J.; Zhang, J. Proteolysis and sensory properties of dry-cured bacon as affected by the partial substitution of sodium chloride with potassium chloride. *Meat Sci.* **2014**, *96*, 1325–1331. [CrossRef]
12. Wu, H.; Zhuang, H.; Zhang, Y.; Tang, J.; Yu, X.; Long, M.; Wang, J.; Zhang, J. Influence of partial replacement of NaCl with KCl on profiles of volatile compounds in dry-cured bacon during processing. *Food Chem.* **2015**, *172*, 391–399. [CrossRef]
13. Lee, Y.S.; Zhekov, Z.G.; Owens, C.M.; Kim, M.; Meullenet, J.F. Effects of partial and complete replacement of sodium chloride with potassium chloride on the texture, flavor and water-holding capacity of marinated broiler breast fillets. *J. Texture Stud.* **2012**, *43*, 124–132. [CrossRef]
14. Nachtigall, F.M.; Vidal, V.A.S.; Pyarasani, R.D.; Dominguez, R.; Lorenzo, J.M.; Pollonio, M.A.R.; Santos, L.S. Substitution effects of NaCl by KCl and $CaCl_2$ on lipolysis of salted meat. *Foods* **2019**, *8*, 595. [CrossRef]
15. Li, F.; Zhuang, H.; Qiao, W.; Zhang, J.; Wang, Y. Effect of partial substitution of NaCl by KCl on physicochemical properties, biogenic amines and N-nitrosamines during ripening and storage of dry-cured bacon. *J. Food Sci. Technol.* **2016**, *53*, 3795–3805. [CrossRef]
16. Xiong, Y.L.; Brekke, C.J. Gelation properties of chicken myofibrils treated with calcium and magnesium chlorides. *J. Muscle Foods* **1991**, *1*, 21–36. [CrossRef]
17. Alino, M.; Grau, R.; Toldra, F.; Barat, J.M. Physicochemical changes in dry-cured hams salted with potassium, calcium and magnesium chloride as a partial replacement for sodium chloride. *Meat Sci.* **2010**, *86*, 331–336. [CrossRef]
18. Alino, M.; Grau, R.; Toldra, F.; Blesa, E.; Pagan, M.J.; Barat, J.M. Influence of sodium replacement on physicochemical properties of dry-cured loin. *Meat Sci.* **2009**, *83*, 423–430. [CrossRef]
19. Luo, H.; Guo, C.; Lin, L.; Si, Y.; Gao, X.; Xu, D.; Jia, R.; Yang, W. Combined use of rheology, LF-NMR, and MRI for characterizing the gel properties of hairtail surimi with potato starch. *Food Bioprocess Technol.* **2020**, *13*, 637–647. [CrossRef]
20. Yalınkılıç, B.; Kaban, G.; Kaya, M. Effect of sodium replacement on the quality characteristics of pastırma (a dry-cured meat product). *Food Sci. Hum. Wellness* **2023**, *12*, 266–274. [CrossRef]
21. Lorenzo, J.M.; Bermúdez, R.; Domínguez, R.; Guiotto, A.; Franco, D.; Purriños, L. Physicochemical and microbial changes during the manufacturing process of dry-cured lacón salted with potassium, calcium and magnesium chloride as a partial replacement for sodium chloride. *Food Control.* **2015**, *50*, 763–769. [CrossRef]
22. Ning, C.; Bao, P.; Zhang, D.; Li, L.; Chen, L.; Fang, H.; Tang, Y.; Zhou, C. Reduction and coordination properties of L-lysine/L-arginine/L-cysteine for the improvement of the color of cured sausage. *Food Chem.* **2020**, *312*, 126122. [CrossRef]
23. Zhu, X.; Ning, C.; Li, S.; Xu, P.; Zheng, Y.; Zhou, C. Effects of L-lysine/L-arginine on the emulsion stability, textural, rheological and microstructural characteristics of chicken sausages. *Int. J. Food Sci. Technol.* **2017**, *53*, 88–96. [CrossRef]

24. Zheng, Y.; Xu, P.; Li, S.; Zhu, X.; Chen, C.; Zhou, C. Effects of L-lysine/L-arginine on the Physicochemical properties and quality of sodium-reduced and phosphate-free pork sausage. *Int. J. Nutr. Food Sci.* **2017**, *6*, 12–18. [CrossRef]
25. Pilar, H.N.; Donkeun, P.; Rhee, K.S. Chloride salt type/ionic strength, muscle site and refrigeration effects on antioxidant enzymes and lipid oxidation in pork. *Meat Sci.* **2001**, *61*, 405–410.
26. Dos Santos, B.A.; Campagnol, P.C.B.; Fagundes, M.B.; Wagner, R.; Pollonio, M.A.R. Generation of volatile compounds in Brazilian low-sodium dry fermented sausages containing blends of NaCl, KCl, and $CaCl_2$ during processing and storage. *Food Res. Int.* **2015**, *74*, 306–314. [CrossRef] [PubMed]
27. Rosario, R.; Cava, R. Volatile profiles of dry-cured meat products from three different Iberian X duroc genotypes. *J. Agric. Food Chem.* **2007**, *55*, 1923–1931.
28. Li, F.; Feng, X.; Zhang, D.; Li, C.; Xu, X.; Zhou, G.; Liu, Y. Physical properties, compositions and volatile profiles of Chinese dry-cured hams from different regions. *J. Food Meas. Charact.* **2019**, *14*, 492–504. [CrossRef]
29. Olivares, A.; Navarro, J.L.; Flores, M. Establishment of the contribution of volatile compounds to the aroma of fermented sausages at different stages of processing and storage. *Food Chem.* **2009**, *115*, 1464–1472. [CrossRef]
30. Purriños, L.; Franco, D.; Carballo, J.; Lorenzo, J.M. Influence of the salting time on volatile compounds during the manufacture of dry-cured pork shoulder "lacón". *Meat Sci.* **2012**, *92*, 627–634. [CrossRef] [PubMed]
31. Chen, Q.; Kong, B.; Han, Q.; Xia, X.; Xu, L. The role of bacterial fermentation in lipolysis and lipid oxidation in harbin dry sausages and its flavour development. *LWT-Food Sci. Technol.* **2017**, *77*, 389–396. [CrossRef]
32. Cano-García, L.; Rivera-Jiménez, S.; Belloch, C.; Flores, M. Generation of aroma compounds in a fermented sausage meat model system by debaryomyces hansenii strains. *Food Chem.* **2014**, *151*, 364–373. [CrossRef] [PubMed]
33. Irigoyen, A.; Ortigosa, M.; Juansaras, I.; Oneca, M.; Torre, P. Influence of an adjunct culture of *Lactobacillus* on the free amino acids and volatile compounds in a Roncal-type ewe's-milk cheese. *Food Chem.* **2007**, *100*, 71–80. [CrossRef]
34. Mónica, A.; Toldrá, F.; Aristoy, M.C.; Ventanas, J.; Estévez, M. Effect of the partial replacement of sodium chloride by other salts on the formation of volatile compounds during ripening of dry-cured ham. *J. Agric. Food Chem.* **2012**, *60*, 7607–7615.
35. Api, A.M.; Belsito, D.; Botelho, D.; Bruze, M.; Burton, G.A.; Buschmann, J.; Cancellieri, M.A.; Dagli, M.L.; Date, M.; Dekant, W.; et al. RIFM fragrance ingredient safety assessment, 6-methyl-5-hepten-2-one, CAS registry number 110-93-0. *Food Chem. Toxicol.* **2021**, *156*, 112558. [CrossRef]
36. Sunesen, L.O.; Stahnke, L.H. Mould starter cultures for dry sausages—Selection, application and effects. *Meat Sci.* **2003**, *65*, 935–948. [CrossRef]
37. Laura, P.N.; de Revel, G.; Bertrand, A.; Maujean, A. Formation of flavor components by the reaction of amino acid and carbonyl compounds in mild conditions. *J. Agric. Food Chem.* **2000**, *48*, 3761–3766.
38. Koutsidis, G.; Elmore, J.S.; Oruna-Concha, M.J.; Campo, M.M.; Wood, J.D.; Mottram, D.S. Water-soluble precursors of beef flavour. Part II: Effect of post-mortem conditioning. *Meat Sci.* **2008**, *79*, 270–277. [CrossRef] [PubMed]
39. Maarse, H. *Volatile Compounds in Foods and Beverages*; Taylor & Francis: New York, NY, USA, 2017.
40. Monica, F.; M-Concepcion, A.; Spanier, A.M.; Toldra, F. Non-volatile components effects on quality of "Serrano" dry-cured ham as related to processing time. *J. Food Sci.* **1997**, *62*, 1235–1239.
41. Armenteros, M.; Aristoy, M.C.; Barat, J.M.; Toldra, F. Biochemical and sensory properties of dry-cured loins as affected by partial replacement of sodium by potassium, calcium, and magnesium. *J. Agric. Food Chem.* **2009**, *57*, 9699–9705. [CrossRef]
42. José, M.L.; Cittadini, A.; Munekata, P.E.; Domínguez, R. Physicochemical properties of foal meat as affected by cooking methods. *Meat Sci.* **2015**, *108*, 50–54.
43. Aksu, M.İ.; Erdemir, E. The effect of potassium lactate on the free amino acid composition, lipid oxidation, colour, microbiological, and sensory properties of ready-to-eat pastırma, a dry-cured and dried meat product. *J. Food Sci. Technol.* **2021**, *59*, 1288–1298. [CrossRef]
44. Kęska, P.; Stadnik, J. Taste-active peptides and amino acids of pork meat as components of dry-cured meat products: An in-silico study. *J. Sens. Stud.* **2017**, *32*, e12301. [CrossRef]
45. Pérez-Santaescolástica, C.; Carballo, J.; Fulladosa, E.; Garcia-Perez, J.V.; Benedito, J.; Lorenzo, J.M. Effect of proteolysis index level on instrumental adhesiveness, free amino acids content and volatile compounds profile of dry-cured ham. *Food Res. Int.* **2018**, *107*, 559–566. [CrossRef] [PubMed]
46. Zhang, X.; Yang, J.; Gao, H.; Zhao, Y.; Wang, J.; Wang, S. Substituting sodium by various metal ions affects the cathepsins activity and proteolysis in dry-cured pork butts. *Meat Sci.* **2020**, *166*, 108132. [CrossRef] [PubMed]
47. Blesa, E.; Alino, M.; Barat, J.M.; Grau, R.; Toldra, F.; Pagan, M.J. Microbiology and physico-chemical changes of dry-cured ham during the post-salting stage as affected by partial replacement of NaCl by other salts. *Meat Sci.* **2008**, *78*, 135–142. [CrossRef] [PubMed]
48. *GB5009.91-2017*; Determination of Potassium and Sodium in Food. National Health and Family Planning Commission of the PRC: Beijing, China, 2017.
49. *GB5009.241-2017*; Determination of Magnesium in Food. National Health and Family Planning Commission of the PRC: Beijing, China, 2017.
50. Wang, X.; Muhoza, B.; Wang, X.; Feng, T.; Xia, S.; Zhang, X. Comparison between microwave and traditional water bath cooking on saltiness perception, water distribution and microstructure of grass crap meat. *Food Res. Int.* **2019**, *125*, 108521. [CrossRef] [PubMed]

51. Chen, C.; Cheng, G.; Liu, Y.; Yi, Y.; Chen, D.; Zhang, L.; Wang, X.; Cao, J. Correlation between microorganisms and flavor of Chinese fermented sour bamboo shoot: Roles of Lactococcus and Lactobacillus in flavor formation. *Food Biosci.* **2022**, *50*, 101994. [CrossRef]
52. Wang, X.; Feng, T.; Wang, X.; Xia, S.; Yu, J.; Zhang, X. Microwave heating and conduction heating pork belly: Non-volatile compounds and their correlation with taste characteristics, heat transfer modes, and matrix microstructure. *Meat Sci.* **2022**, *192*, 108899. [CrossRef] [PubMed]
53. Zhan, H.; Hayat, K.; Cui, H.; Hussain, S.; Ho, C.-T.; Zhang, X. Characterization of flavor active non-volatile compounds in chicken broth and correlated contributing constituent compounds in muscle through sensory evaluation and partial least square regression analysis. *LWT* **2020**, *118*, 108786. [CrossRef]
54. *GBT16291.1-2012;* Sensory Analysis-General Guidance for the Selection, Training and Monitoring of Assessors-Part 1: Selected Assessors. Academia: San Francisko, CA, USA, 2012.
55. Sugimoto, M.; Obiya, S.; Kaneko, M.; Enomoto, A.; Honma, M.; Wakayama, M.; Tomita, M. Simultaneous analysis of consumer variables, acceptability and sensory characteristics of dry-cured ham. *Meat Sci.* **2016**, *121*, 210–215. [CrossRef]

foods

MDPI

Article

Synergistic Effects of High-Intensity Ultrasound Combined with L-Lysine for the Treatment of Porcine Myofibrillar Protein Regarding Solubility and Flavour Adsorption Capacity

Yongkang Xie, De Chen, Jianxin Cao, Xuejiao Wang and Xiaoyu Yin *

College of Food Science and Engineering, Kunming University of Science and Technology, Kunming 650500, China; xieyongkang1234@126.com (Y.X.); cd17852035931@163.com (D.C.); jxcao321@hotmail.com (J.C.); wangxuejiao173@hotmail.com (X.W.)
* Correspondence: yinxiaoyu@kust.edu.cn

Abstract: This study aimed to investigate the synergistic effects of high-intensity ultrasound (0, 5, 10, 15, and 20 min) in combination with L-lysine (15 mM) on improving the solubility and flavour adsorption capacity of myofibrillar proteins (MPs) in low-ion-strength media. The results revealed that the ultrasound treatment for 20 min or the addition of L-lysine (15 mM) significantly improved protein solubility ($p < 0.05$), with L-lysine (15 mM) showing a more pronounced effect ($p < 0.05$). The combination of ultrasound treatment and L-lysine further increased solubility, and the MPs treated with ultrasound at 20 min exhibited the best dispersion stability in water, which corresponded to the lowest turbidity, highest absolute zeta potential value, and thermal stability ($p < 0.05$). Based on the reactive and total sulfhydryl contents, Fourier transform infrared spectroscopy, and fluorescence spectroscopy analysis, the ultrasound treatment combined with L-lysine (15 mM) promoted the unfolding and depolymerization of MPs, resulting in a larger exposure of SH groups on the surface, aromatic amino acids in a polar environment, and a transition of protein conformation from α-helix to β-turn. Moreover, the combined treatment also increased the hydrophobic bonding sites, hydrogen-bonding sites, and electrostatic effects, thereby enhancing the adsorption capacity of MPs to bind kenone compounds. The findings from this study provide a theoretical basis for the production and flavour improvement of low-salt MP beverages and the utilisation of meat protein.

Keywords: ultrasound; low ion strength; amino acid; myofibrillar proteins; flavour; conformation

Citation: Xie, Y.; Chen, D.; Cao, J.; Wang, X.; Yin, X. Synergistic Effects of High-Intensity Ultrasound Combined with L-Lysine for the Treatment of Porcine Myofibrillar Protein Regarding Solubility and Flavour Adsorption Capacity. *Foods* **2024**, *13*, 629. https://doi.org/10.3390/foods13040629

Academic Editor: Mladen Brnčić

Received: 6 January 2024
Revised: 8 February 2024
Accepted: 16 February 2024
Published: 19 February 2024

1. Introduction

Meat and meat products are nutrient-rich and essential components of the daily human diet [1]. Meat proteins contain essential amino acids required by the human body, which are easily digested and absorbed by the body [2]. Generally, meat is ingested in its entirety as a solid mass, necessitating adequate mastication and digestive capabilities. Nevertheless, meat and meat products are inherently limited in their appeal to older individuals, dysphagic patients, and individuals with gastrointestinal and digestive disorders, which is primarily due to their impaired chewing and digestive systems, which restricts their ability to consume normal crude fibre meat [3]. Hence, there is a need to offer weakened protein in a fluid form to ensure adequate nutrient intake and satiety among this demographic [4].

Myofibrillar proteins (MPs) are the most important structural proteins in muscle, accounting for 50–60% of meat proteins. Myosin and actin are the main proteins in MPs [5]. MPs are more soluble in high-salt environments and are primarily responsible for the water holding and emulsifying ability of the meat [4]. Low-ion-strength media will significantly reduce the solubility and dispersibility of MPs and prevent the unfolding of their structures, thereby limiting their application in low-salt fluid diets [6,7]. However, excessive salt intake can bring about a series of chronic diseases, which seriously endanger human health. Therefore, it can be considered that improving the solubility and dispersion stability of MPs in low-ion-strength solutions is a prerequisite for the development of MP beverages and the expansion of meat protein utilisation [8]. Currently, many chemical (amino acid, acid or alkaline solubilization, etc.) or physical methods (ultra-high pressure, ultrasound, etc.) are available to assist in the processing of MPs to improve their solubility and dispersion stability in low-ion-strength solutions [4,9]. Moreover, researchers have paid increasing attention to the combination of two or more assistive methods (ultrasound-assisted glycation, high pressure combined with $CaCl_2$, etc.) to ameliorate the solubility of MPs [4]. Amino acids are naturally occurring compounds in meat [1]. The addition of amino acids to food is beneficial to maintain or improve the solubility of MPs [10]. Guo et al. [11] found that the introduction of arginine and L-lysine induced changes in the secondary as well as tertiary structures of the protein. The hydrophobic and sulfhydryl groups buried inside the MPs were exposed to the surface, thus increasing the solubility of the MPs. The addition of proline under low-salt conditions also resulted in an increase in MP solubility [10]. However, according to the current reports, the highest solubility can reach approximately 40% by adding amino acids, and there is still a definite gap from complete dissolution [12,13]. Moreover, the flavour of protein solutions with added amino acids is also a problem [4].

High-intensity ultrasound (HIU) is a non-ionising, non-invasive, non-polluting, and green non-thermal form of mechanical energy that has been widely applied in protein processing [14]. In recent years, many studies have demonstrated the ability of ultrasound to improve the solubility of MPs [4]. The cyclic growth and violent collapse of cavitation bubbles during ultrasonication cause high temperature, pressure, high-shear energy waves, and turbulence in the cavitation zone, which alters the structural properties of the treated protein molecules [14]. Liu et al. [8] reported that HIU with different power levels could improve the solubility as well as the dispersion stability of MPs. It was also hypothesised that physical disruption of the myosin filamentous structure may be the main reason for the increase in MP solubilization. Moreover, ultrasound has good potential for improving the flavour quality of meat products [15]. MPs can bind to flavour compounds through chemical forces such as hydrogen bonds, van der Waals forces, and hydrophobic interactions [16]. The cavitation effects of ultrasonic dissociates water molecules produces highly active hydroxyl radicals and further changes and modifies the structure and intermolecular forces of MPs, thereby affecting the flavour adsorption capacity of MPs [17]. It has been reported that different ultrasound powers can enhance the binding of MPs to furan flavour compounds [18,19]. However, the investigation of amino acids combined with ultrasound treatment for the enhancement of the solubility of MPs in low-ion-strength media has not been reported yet. Moreover, little is known about the effect on the binding ability of MPs to flavour compounds after both treatments, which are still worth exploring.

In this study, the aim was to investigate the potential of high-intensity ultrasound combined with L-lysine to improve the solubility and flavour adsorption capacity of MPs in low-ion-strength media. The synergistic effects of different ultrasound times combined with L-lysine on the pH, solubility, turbidity, particle size, and zeta potential of MPs in low-ion-strength media were investigated. The structure characterization of MPs and the binding capacity between MPs and four ketone compounds were analysed. Finally, the thermal stability of MPs treated with ultrasound and L-lysine was evaluated. This study is expected to provide a theoretical basis for the production and flavour improvement of low-salt MP beverages and the expansion of meat protein utilisation.

2. Materials and Methods

2.1. Materials and Reagents

Fresh longissimus muscle was purchased from a local Yonghui supermarket (Kunming, Yunnan Province, China), which was obtained from pig carcasses of approximately 6 months of age at 24 h post-mortem following standard industrial procedures. 2-Pentanone, 2-hexanone, 2-heptanone, and 2-octanone were purchased from Dr. Ehrenstorfer (Dr. Ehrenstorfer, Schwerin, Germany). The purity of the standard flavour compounds was between 95.50% and 99.60%. L-lysine, potassium sodium tartrate, sodium chloride, and sodium hydroxide were purchased from Aladdin (Shanghai, China). All chemical reagents used in this study were of analytical grade or higher.

2.2. Preparation of MPs

MPs were extracted from the longissimus muscle after the removal of fat and connective tissue using the method of Chen et al., with modifications [20]. Briefly, approximately 200 g of minced meat was thoroughly homogenized with 4 vol. (w/v) of 10 mM phosphate buffer (0.1 M NaCl, 2 mM $MgCl_2$, 1 mM EGTA, pH 7.0) for 60 s at 4 °C, and then the mixture was centrifuged at 2000× g for 15 min. The pellet was collected and washed twice with 4 vol. (w/v) of the same phosphate buffer and centrifugation conditions as above. After centrifugation, the MP pellet was washed three times with 4 vol. (w/v) of salt solution (0.1 M NaCl) under the same centrifugation conditions as mentioned above. The suspension underwent filtration using four layers of cheesecloth during the final wash, and then the pH was modified to 6.0 using a 0.1 M HCl solution. The final MP concentration was calculated using the Biuret method with bovine serum albumin as a standard [21]. The MPs were kept at 4 °C for use within 48 h.

2.3. Sample Treatment

In total, 50 mL of MP solution (5 mg/mL, diluted with 0.01 M phosphate buffer (0.1 M NaCl, pH 7.0) was placed in a double-walled beaker (100 mL) for every ultrasonication time [19]. An ultrasonic processor (SCIENTZ-IID, Ningbo Xinzhi Bio-technology Co., Ltd., Ningbo, China) equipped with a titanium probe with a diameter of 6.0 mm was inserted in the equivalent amounts of MP solutions with and without adding L-lysine at a depth of 1.5 cm from the bottom. The concentration of L-lysine solution for this study was 15 mM. Ultrasound treatment was conducted at a frequency of 20 kHz with different ultrasonic times (0, 5, 10, 15, and 20 min) at 300 W (pulse duration of 2 s on and 3 s off), and all the samples were treated at low temperatures (4–10 °C).

2.4. Solubility Measurements

The solubility of MP solution (20 mL) with and without adding L-lysine treated at different ultrasound times was measured as described by Shen et al. [22]. MP solution (5 mg/mL) was centrifuged at 5000× g for 20 min, the supernatant was collected, and the protein content was determined with the Biuret method. Protein solubility was expressed as the percentage of protein concentration in the supernatant after centrifugation relative to the initial protein concentration.

2.5. Turbidity Measurements

The turbidity of the MP solution (20 mL) with and without adding L-lysine treated at different ultrasound times was analysed using the method of Wang et al. [23], with modifications. The MP solution (1 mg/mL) was measured with a UV spectrophotometer (UV-1800PC, Mepda Shanghai, China) at 350 nm and 25 °C. The turbidity of MPs was expressed as their absorbance.

2.6. pH Measurements

The pH value of the MP solution (20 mL) with and without adding L-lysine treated at different ultrasound times was measured with a pH metre (FE28, Mettler Toledo, Zurich, Switzerland).

2.7. Particle Size and Zeta Potential Measurements

The particle size of the MP solution (10 mL) with and without adding L-lysine treated at different ultrasound times was measured with the Malvern MasterSizer 3000 (Malvern Instruments Ltd., Worcestershire, UK). The zeta potential was measured using the method of Chen et al. [24], with some slight modifications. The MP solution was diluted to a final concentration of 0.1 mg/mL using 0.01 M phosphate buffer (0.1 M NaCl, pH 7.0), injected into the Marvin Zeta potentiometer (Zetasizer Nano ZS90, Malvern, UK), and measured at 25 °C.

2.8. Fourier Transform Infrared Spectroscopy (FT-IR)

Infrared spectroscopy was carried out using the method of Kaewprachu et al. [25], with a slight modification. The samples were vacuum freeze-dried and then placed in a Fourier transform infrared spectrometer (TENSOR27, Bruker, Ettlingen, Germany) after KBr compression (1:50) with a scanning range of 4000–400 cm^{-1} and 64 scans with a resolution of 4 cm^{-1}, with KBr as the background single channel. The secondary structural contents of the protein samples were calculated using Peakfit 4.12 software.

2.9. Measurements of Reactive and Total Sulfhydryl Contents

The contents of reactive and total sulfhydryl groups were measured via the method of Ellman et al. [26], with some modifications. Briefly, 1 mL of sample (2 mg/mL) was thoroughly mixed with 8 mL of Tris-glycine buffer solution (86 mmol Tris, 92 mM glycine, 4 mM EDTA, 8 M urea, pH 8.0). The insoluble proteins were then removed via centrifugation at 10,000× *g* for 15 min. The supernatant was taken as 4.5 mL and reacted with 0.5 mL of Ellman's reagent (4 mg/mL DTNB). The absorbance at 412 nm was measured after 30 min using a UV spectrophotometer. The same method was used to measure reactive group content, but urea was omitted from the glycine buffer solution.

2.10. Fluorescence Measurements

The endogenous fluorescence spectra were determined with reference to the method of Guo et al. [27], with a slight modification. The MP solution (5 mL) was diluted to 0.5 mg/mL and placed in a fluorescence spectrophotometer (F-7000, Hitachi Corp., Tokyo, Japan) to determine the endogenous fluorescence spectra with an excitation wavelength of 280 nm and an emission wavelength of 300–400 nm. The slit width was 10, and the smoothing degree was 20.

2.11. Differential Scanning Calorimetry (DSC)

The thermal stability of the sample was measured with differential scanning calorimetry (DSC) (Mettler Toledo, Zurich, Switzerland). A nitrogen flow rate of 20 mL/min was set. MP samples (5–10 mg) in aluminium DSC trays were heated from 30 °C to 120 °C at a heating rate of 10 °C/min. The blank control was an empty aluminium box to enable us to obtain the DSC curve.

2.12. Solid-Phase Microextraction (SPME) and Gas Chromatography–Mass Spectrometry (GC/MS) Analysis

The four selected ketones (2-pentanone, 2-hexanone, 2-heptanone, and 2-octanone) were dissolved in methanol to form a stock solution of each ketone. Then, each ketone was added to the treated MP dispersion solution. The concentration of each ketone in the MP solution was set to 10 mg/g. The control was carried out by replacing the MP solution with an equal volume of phosphate buffer. The protein–ketone mixture was filled into 20 mL headspace vials and sealed with PTFE-faced silicone septa (Supelco, Bellefonte, PA, USA). The protein and control samples were equilibrated at room temperature (25 °C) for 16 h. Subsequently, the amount of headspace ketone in each vial was measured with SPME-GC/MS (QP2010, Shimadzu, Japan) according to the method reported by Zhou et al. [28], with modifications. The flavour compounds were extracted with an SPME fibre at 30 °C for 30 min. Subsequently, the flavour compounds absorbed were dissolved in the GC inlet at 240 °C for 5 min. The initial GC oven temperature was held at 40 °C for 3 min, increased at 3 °C to 90 °C, held for 5 min, and then increased at 25 °C to 230 °C, and finally held at 230 °C for 5 min. Results were determined by the difference in peak area of the flavour compounds between the protein and control vials and were calculated using the following equation:

$$\text{Free flavour compound (\%)} = (\text{peak area protein}/\text{peak area control}) \times 100\%$$

2.13. Statistical Analysis

Three independent batches of MPs (replicates) were prepared, and the experimental measurements were conducted in triplicate (triplicate observations) for each batch of MPs. The results were expressed as mean ± standard deviation (SD). The data were analysed using the General Linear Model procedure in the Statistix 8.1 software package (Analytical Software, St. Paul, MN, USA). The solubility, turbidity, pH, particle size, zeta potential, reactive and total sulfhydryl contents, and flavour adsorption capacity were analysed using two-way analysis of variance (ANOVA) in the factorial design procedure, which considered the ultrasound time, L-lysine (added or not), and their interactions as fixed effects and each replicate as a random effect. The determination of significant differences ($p < 0.05$) among the means was performed using Duncan's multiple comparisons.

3. Results and Discussion

3.1. Protein Solubility

As presented in Figure 1 and Table S1, ultrasound time, L-lysine (added or not), and their interactions exerted significant effects on protein solubility ($p < 0.05$). As anticipated, the untreated MPs (control) exhibited comparatively limited solubility in low-ion-strength media, at only 0.96%. The obtained results exhibited conformity with prior studies, emphasising the lack of solubility of naive MPs in low-ion-strength media [8]. The solubility of MPs only treated with ultrasound gradually increased with the increase in ultrasound time ($p > 0.05$) and reached a maximum value of 5.25% at 20 min, which was significantly higher than that of the untreated MPs ($p < 0.05$). This may be since ultrasound promotes the unfolding of myosin filaments [29]. The solubility of MPs with L-lysine alone without ultrasonic treatment (0 min) was significantly increased (22.81%). Myosin is an important component of MPs [5]. The effects of L-lysine on myosin solubility were previously reported, where a low ionic strength of L-lysine at 0.1–0.2 M had a significant effect on solubility, while a high ionic strength between 0.25 and 0.3 M had a negligible effect on solubility [12]. The presence of L-lysine was reported to inhibit myosin aggregation, possibly by interacting with acidic amino acid residues of the myosin, thereby leading to an increase in myosin solubility [30]. The solubility increased significantly ($p < 0.05$) after ultrasound combined with L-lysine treatment of MPs versus MPs with L-lysine alone without ultrasonic treatment (0 min). It was worth noting that under the same ultrasound treatment time, the solubility of MPs treated with L-lysine combined with ultrasound in low-ion-strength media was superior to that of the MPs treated with L-lysine alone without ultrasonic treatment (0 min) or with ultrasound treatment alone. The ultrasound could promote the dissolution of myosin filaments and allow them to better unfold through cavitation, such as shear force, turbulence effect, and high-velocity fluid microjet [31], as well as facilitate the interactions of protein molecules and L-lysine, leading to better solubility. This was also evidenced by the interaction of the data. As the ultrasound time increased from 0 to 20 min, the solubility of MPs with the addition of L-lysine in low-ion-strength media showed a progressive increase to a maximum of 71.42%. According to this result, the effects of MP solubility enhancement depended on ultrasound time.

3.2. Protein Turbidity

As presented in Figure 2 and Table S1, ultrasound time, L-lysine (added or not), and their interactions exerted significant effects on the protein turbidity ($p < 0.05$). The turbidity of the untreated MPs (control) was high at 0.76. The turbidity of MPs only treated with ultrasound gradually decreased with the increase in ultrasound time ($p < 0.05$), which indicated that ultrasound could make the MP solution more clear and transparent, and the

MP dispersion more uniform and consistent. The MP turbidity of L-lysine alone without ultrasonic treatment (0 min) was significantly lower than the turbidity of MP treated with ultrasound only ($p < 0.05$). Previous studies [32] have reported that the addition of L-lysine also led to a reduction in turbidity. After combining L-lysine with ultrasound treatment, the solubility of MPs was further decreased, stabilised, and reached a minimum value of 0.26 at 20 min. The turbidity of MPs treated with L-lysine and ultrasound in low-ion-strength media was lower than the turbidity of MPs treated with L-lysine alone without ultrasonic treatment (0 min) and the turbidity of MPs treated with ultrasound only at the same ultrasound treatment time. This may be due to a series of physical forces caused by the cavitation effects of ultrasound on the solution system, which breaks up protein molecules into smaller pieces. Also, ultrasonic binding of L-lysine may inhibit protein–protein covalent and noncovalent crosslinking and lower turbidity [31].

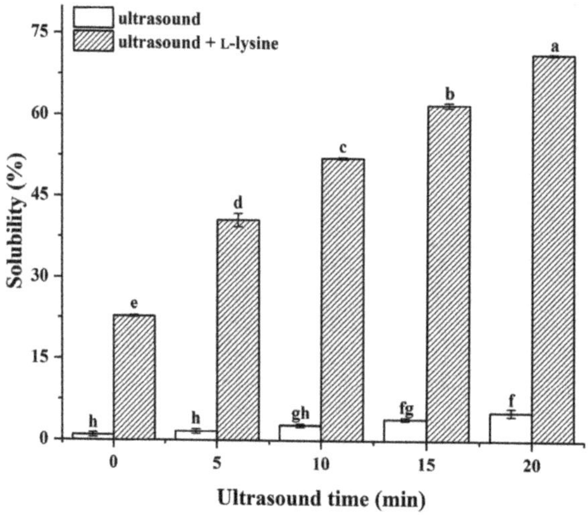

Figure 1. Effects of ultrasound combined with L-lysine on the solubility of myofibrillar proteins. The letters (a–h) above the bars indicate a significant difference ($p < 0.05$).

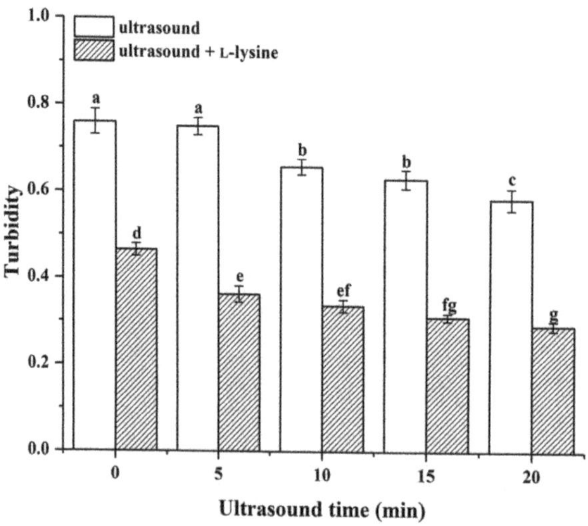

Figure 2. Effects of ultrasound combined with L-lysine on the turbidity of myofibrillar proteins. The letters (a–g) above the bars indicate a significant difference ($p < 0.05$).

3.3. pH of Protein

The pH value is an important factor influencing the solubility of proteins [33]. The effects of ultrasound combined with L-lysine on the pH of MPs is presented in Table 1. The pH of MPs only treated with ultrasound did not change significantly with the increase in ultrasound time ($p > 0.05$), suggesting that ultrasound did not affect the pH and that pH was not the main factor contributing to the improvement of the solubility of MPs. This was consistent with the findings of Liu et al. [8], who hypothesised that HIU may provide a strong physical force through cavitation, disrupting the highly ordered filamentous myocardin structure and reducing the particle sizes of MPs, leading to an increase in the solubility of MPs. L-lysine is an alkaline amino acid [34], and the pH values of MPs were increased significantly by adding L-lysine ($p < 0.05$). As the ultrasound time increased, the pH values of MPs with the addition of L-lysine in low-ion-strength media showed no significant changes ($p > 0.05$). According to Yongsawatdigul et al. [35], myosin solubility was found to vary in a "U" shape with the pH value, with the minimum solubility generally corresponding to a pH between 4.0 and 6.0. Near the isoelectric point (pH = 5.4), myosin rapidly aggregated and precipitated, resulting in smaller solubility due to the reduction of intermolecular electrostatic repulsion. However, the solubility of MPs was higher near extreme acid/base conditions. The change in pH is an important factor in the solubility of MPs.

Table 1. Effects of ultrasound combined with L-lysine on the pH value of myofibrillar protein.

Ultrasound Time (min)	pH	
	Ultrasound	Ultrasound + L-Lysine
0	6.86 ± 0.02 ^{b}	9.35 ± 0.02 ^{a}
5	6.87 ± 0.01 ^{b}	9.35 ± 0.03 ^{a}
10	6.88 ± 0.02 ^{b}	9.36 ± 0.01 ^{a}
15	6.87 ± 0.01 ^{b}	9.38 ± 0.01 ^{a}
20	6.88 ± 0.01 ^{b}	9.36 ± 0.01 ^{a}

^{a,b}—the same indexes and different letters differ significantly ($p < 0.05$).

3.4. Particle Size of Protein

The particle size of protein varies with the degree of protein aggregation [29]. As shown in Table 2, the mean particle size D_{43} of the untreated MPs (control) was relatively large, with a wide particle distribution. Consistent with the results of Li et al. [36], certain chicken breast MPs including well-organised filamentous myosin persisted in the liquid suspensions. The particle size D_{43} of MPs only treated with ultrasound gradually decreased with the increase in ultrasound time ($p < 0.05$). This may be because under the operation of the ultrasound probe, cavitation and shear forces are accompanied by microfluidics and turbulence, which disrupt the integrity of myofibrils and the dissociation of proteins [37]. The particle size D_{43} of MPs of L-lysine alone without ultrasonic treatment (0 min) was larger than that of the untreated MPs ($p < 0.05$), which showed that there were swollen myofibrils. The phenomenon of myofibril swelling in salt solutions is well-established, since it is caused by the selective binding of CL-anions to myofibrils, resulting in an increase in electrostatic repulsion between them [30]. However, one possible explanation for how L-lysine causes myofibrils to swell is that when myofibrils swell, the overall density of myofibrils decreases, which may make MPs less likely to form a precipitate during gentle centrifugation [31]. After combining L-lysine with ultrasound treatment, the D_{43} of MPs decreased. However, it is higher than that of the ultrasound treatment group alone ($p < 0.05$), which may be because the effects of L-lysine on particle size was greater than that of ultrasound on particle size. After the introduction of ultrasonic treatment, the D_{32} and D_{43} of MPs decreased by 50.26% and 59.16% as the ultrasonic time increased from 0 min to 20 min, respectively, suggesting that ultrasound contributed to the depolymerization of MPs. This is a result of the physical force that ultrasound can produce in breaking up protein particles, which reduces the size of the protein [36].

Table 2. Effects of ultrasound combined with L-lysine on the particle size of myofibrillar protein.

Group	Particle Size (μm)				
	Ultrasound Time (min)	D_{50}	D_{90}	D_{43}	D_{32}
ultrasound	0	38.39 ± 0.73 [bc]	137.63 ± 2.75 [d]	61.12 ± 2.07 [de]	22.76 ± 0.47 [b]
	5	20.96 ± 1.37 [cd]	74.41 ± 3.22 [e]	33.94 ± 2.13 [ef]	16.59 ± 1.12 [bcd]
	10	15.63 ± 1.25 [cd]	66.70 ± 0.72 [e]	29.11 ± 1.06 [f]	13.29 ± 1.16 [bc]
	15	12.03 ± 3.92 [d]	48.30 ± 2.98 [e]	22.99 ± 4.08 [f]	10.30 ± 2.69 [d]
	20	11.03 ± 4.01 [d]	28.80 ± 14.46 [e]	15.89 ± 7.30 [f]	9.79 ± 3.19 [d]
ultrasound + L-lysine	0	86.67 ± 2.75 [a]	528.48 ± 10.14 [a]	184.04 ± 4.57 [a]	39.63 ± 1.22 [a]
	5	57.50 ± 13.82 [b]	466.14 ± 28.58 [b]	141.84 ± 8.21 [b]	17.61 ± 3.75 [bcd]
	10	57.04 ± 21.73 [b]	200.95 ± 16.07 [c]	90.59 ± 11.69 [c]	19.65 ± 6.26 [bc]
	15	56.60 ± 13.66 [b]	167.65 ± 34.76 [cd]	82.22 ± 16.99 [cd]	19.61 ± 5.71 [bc]
	20	38.99 ± 4.95 [bc]	163.63 ± 12.47 [cd]	75.16 ± 6.69 [cd]	19.71 ± 1.31 [bc]

D_{90}: Particles smaller than it account for 90% of the particles. D_{50}: The particle size that 50% of the sample particles are smaller than and 50% are larger than. D_{32}: Surface-weighted mean particle size. D_{43}: Volume-weighted mean particle size. [a–f]—the same indexes and different letters differ significantly ($p < 0.05$).

3.5. Zeta potential of Protein

The zeta potential can reflect the charge that the MP surface is carrying and is crucial for maintaining the system's stability [38]. Ultrasound combined with L-lysine affected the zeta potential of MPs in Figure 3. The zeta potential of all samples was negative, indicating a greater number of positive amino acids was present on the protein surface [39]. The absolute value of the potential of MPs treated with ultrasound gradually increased with the increase in ultrasound time ($p > 0.05$), and reached its absolute maximum value at 20 min, which was significantly higher than that of the untreated protein ($p < 0.05$). The increase in the absolute value of the potential of MPs after ultrasound treatment may be due to the destruction of protein aggregates via ultrasound treatment, resulting in an increase in specific surface area and an increase in surface negative charge [40]. Relevant studies have shown that the change in protein potential may have a relationship with protein particle size [38]. Compared to the untreated MPs, the particle size of the MPs' fraction of L-lysine alone without ultrasonic treatment (0 min) increased, thereby decreasing the absolute value of the MPs' potential. After combining L-lysine with ultrasound treatment, the absolute value of the potential of MPs further increased ($p < 0.05$). It was still lower than that of the MPs treated with ultrasound treatment alone ($p < 0.05$). Protein particles can disintegrate due to the physical force of ultrasound, which reduces their size and increases their potential [41]. However, in this study, the influence of L-lysine on particle size exceeded that of ultrasound.

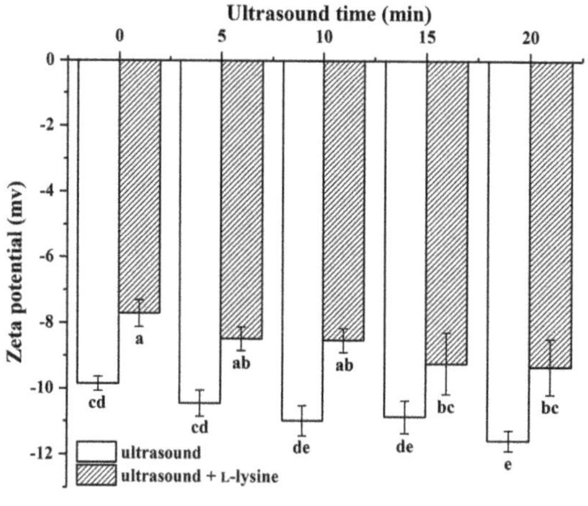

Figure 3. Effects of ultrasound combined with L-lysine on the zeta potential of myofibrillar proteins. The letters (a–e) above the bars indicate a significant difference ($p < 0.05$).

3.6. Fourier Transform Infrared Spectroscopy (FT-IR) Analysis

FT-IR spectra provide valuable information regarding the absorption energy of chemical bonds, making them effective for the structural and chemical characterization of proteins [42]. The amide I band (1600–1700 cm^{-1}) has become the most sensitive for predicting the secondary structural components of proteins in the infrared spectral region [43]. According to Table 3, the α-helix and β-turn contents of MPs treated only with ultrasound decreased slowly as the ultrasound time increased ($p > 0.05$), reaching their lowest points of 22.18% and 26.86% at 20 min, respectively. However, the β-sheet content gradually decreased ($p < 0.05$). The reason for this may be that the cavitation effect and microjet effect produced by ultrasound promoted the unfolding of protein molecules and weakened the hydrogen bonds, resulting in the disruption of the ordered structure of α-helix and β-turn, and the formation of β-sheet and a random coil. The ultrasound treatment induced conformational transitions from ordered to disordered structures [42]. The α-helix, random coil, and β-sheet of MPs were decreased by the addition of L-lysine only ($p < 0.05$). The decrease in the α-helix content indicated a change in protein conformation, thus gradually exposing the hydrophobic sites buried inside the molecule [44]. The MPs treated with L-lysine and ultrasound had no significant effect on the content of secondary structures ($p > 0.05$), which may be due to the greater effects of L-lysine on the secondary structures than ultrasound. As a result, the effects of ultrasound on the secondary structures became less significant. Guo et al. [11] found that α-helix was significantly lost after the introduction of L-lysine, regardless of ionic strength. It has also been hypothesised that the mechanism for the increase in myosin solubility is the ability of the L-lysine cation to bind to the negatively charged residues of myosin via electrostatic interaction, disrupting intra- and intermolecular ionic linkages [11]. This results in a conformational shift of the protein, loss of α-helical structure, and exposure of hydrophobic and masked SH groups to the surface. Myosin filaments depolymerize, ultimately increasing protein solubility [11].

Table 3. Effects of ultrasound combined with L-lysine on the secondary structures of myofibrillar proteins.

Group	Relative Content of Secondary Structures (%)				
	Ultrasound Time (min)	β-Sheet	Random Coil	α-Helix	β-Turn
ultrasound	0	20.58 ± 0.52 ^{c}	22.65 ± 0.91 ^{a}	24.49 ± 0.25 ^{a}	32.28 ± 0.21 ^{b}
	5	20.67 ± 0.23 ^{bc}	22. 15 ± 0.17 ^{a}	24.09 ± 0.37 ^{ab}	33.09 ± 0.25 ^{b}
	10	21.71 ± 0.59 ^{bc}	21.98 ± 0.25 ^{a}	23.74 ± 0.64 ^{ab}	32.58 ± 0.36 ^{b}
	15	22.70 ± 0.87 ^{b}	21.41 ± 0.56 ^{a}	23.49 ± 0.58 ^{ab}	32.40 ± 0.51 ^{b}
	20	28.57 ± 0.32 ^{a}	22.39 ± 0.52 ^{a}	22.18 ± 0.28 ^{bc}	26.86 ± 0.56 ^{c}
ultrasound + L-lysine	0	16.62 ± 0.29 ^{d}	18.33 ± 0.32 ^{b}	21. 10 ± 0.23 ^{c}	43.94 ± 0.45 ^{a}
	5	17.22 ± 0.43 ^{d}	18. 16 ± 0.28 ^{b}	20.79 ± 0.49 ^{c}	43.83 ± 0.39 ^{a}
	10	17.40 ± 0.52 ^{d}	18.29 ± 0.24 ^{b}	20.78 ± 0.36 ^{c}	43.53 ± 0.27 ^{a}
	15	17.41 ± 0.65 ^{d}	18.70 ± 0.31 ^{b}	21.08 ± 0.32 ^{c}	42.80 ± 0.36 ^{a}
	20	17.57 ± 0.53 ^{d}	18.99 ± 0.56 ^{b}	20.70 ± 0.21 ^{c}	42.43 ± 0.62 ^{a}

$^{a–d}$—the same indexes and different letters differ significantly ($p < 0.05$).

3.7. Total and Reactive Sulfhydryl Analysis

The sulfhydryl group is an important reactive group that reflects structural alterations in MPs [45]. The effects of ultrasound combined with L-lysine on the contents of total and reactive sulfhydryl contents in MPs are shown in Figure 4. The total sulfhydryl content of MPs treated with ultrasound alone gradually decreased with increasing ultrasound time ($p > 0.05$) and reached a minimum value of 7.62 μmol/g at 20 min, which was significantly higher than that of the untreated MPs ($p < 0.05$). These results indicated that some free sulfhydryl groups may be converted to disulfide bonds, hence enhancing the crosslinking between protein subunits [19]. The total sulfhydryl content of the MP of L-lysine alone without ultrasonic treatment (0 min) did not change significantly ($p > 0.05$). Guo et al. [11] reported similar results in the content of total sulfhydryl groups when investigating the effects of L-lysine and L-histidine on the solubility and conformational properties of porcine myosin. After combining L-lysine with ultrasound treatment, the total sulfhydryl content of MPs treated with ultrasound for 20 min was significantly lower than that of untreated MPs ($p < 0.05$). Ultrasound treatment decreased the total sulfhydryl content, probably because the physical forces generated via ultrasound disrupt the structure of the protein molecules, exposing the sulfhydryl groups originally buried inside to the surface, and the enhancement of hydrophobic interactions reduces the intermolecular sulfhydryl distances [46]. When the protein solution is exposed to oxygen, the more reactive sulfhydryl groups and shorter distances lead to the formation of disulfide bonds, which results in a decrease in total sulfhydryl content [47].

The reactive sulfhydryl content of the MPs treated with ultrasound alone exhibited a gradual increase with the extension in ultrasound duration ($p < 0.05$), which is similar to the study by Zhao et al. [19]. This observation can be attributed to the ultrasound treatment-induced boost in reactive sulfhydryl content, which facilitated the unfolding of protein structures and exposed internal functional groups such as sulfhydryl groups. An increase in reactive sulfhydryl content was observed when L-lysine was used alone without ultrasonic treatment (0 min), indicating that SH groups were exposed to the protein surface. After combining L-lysine with ultrasound treatment, the reactive sulfhydryl content of MPs further increased. It is worth noting that under the same ultrasound treatment time, the reactive sulfhydryl content of MPs treated with L-lysine combined with ultrasound in low-ion-strength media is significantly higher than that of MPs treated with L-lysine alone without ultrasonic treatment (0 min) and with ultrasound alone. This finding suggests that ultrasound treatment in combination with L-lysine further amplifies the exposure of SH groups on the protein surface. This phenomenon can be attributed to the synergistic effects of ultrasound and L-lysine, which promotes the unfolding of protein molecules and stimulates the production of hydroxyl radicals. Furthermore, this indicated that the addition of L-lysine does not disrupt the formation or cleavage of disulfide bonds but rather reveals the concealed SH groups on the protein surface [19].

The sulfhydryl group is a common site within protein molecules that is sensitive to oxidation [48]. Under conditions of oxidative stress, the sulfhydryl group is highly susceptible to attack by oxidising agents, leading to the formation of disulfide bonds [48]. The oxidative modifications of sulfhydryl groups are considered significant indicators of protein oxidation [49]. Specifically, a decrease in sulfhydryl group content often indicates an increased level of protein oxidation [50]. In this study, after combining L-lysine with ultrasound treatment, the total sulfhydryl content of ultrasound-treated MPs for 20 min was significantly lower than that of untreated MPs ($p < 0.05$), suggesting that the combination of these two factors promotes protein oxidation. While previous research has mainly focused on the negative effects of protein oxidation on the quality of meat products, recent studies have indicated that protein oxidation can also have positive influences on the quality characteristics of meat and related products to some extent. Moderate oxidative conditions can have a beneficial impact on the processing and quality characteristics of meat products. Liu et al. [51] discovered that under appropriate oxidation conditions, protein oxidation enhances protein sensitivity to enzymes, facilitating their digestion and absorption in the human body. Zhou et al. [28] used a hydroxyl radical oxidation system to perform oxidative modifications on MPs and observed that targeted oxidant concentrations significantly enhanced the binding capacity of MPs to specific flavour compounds. This enhancement contributes to the preservation of flavour in meat products throughout the processing stages. Furthermore, there have been reports indicating that HIU promotes the oxidation of proteins and lipids in meat, which is consistent with the results of our study. According to Antti et al. [52], the cavitation effects of ultrasound results in the decomposition of water molecules, leading to the generation of highly reactive hydroxyl radicals. These radicals further facilitate the formation of strong oxidising H_2O_2, causing significant oxidation of lipids and proteins in meat. Consequently, the oxidation reactions of proteins and lipids contribute to the formation of flavour compounds such as aldehydes and ketones [53]. Moreover, it has been documented that varying ultrasound powers can enhance the interaction between meat proteins and furan flavour compounds [19]. As a result, ultrasound-assisted processing techniques hold significant potential for improving the flavour quality of meat products.

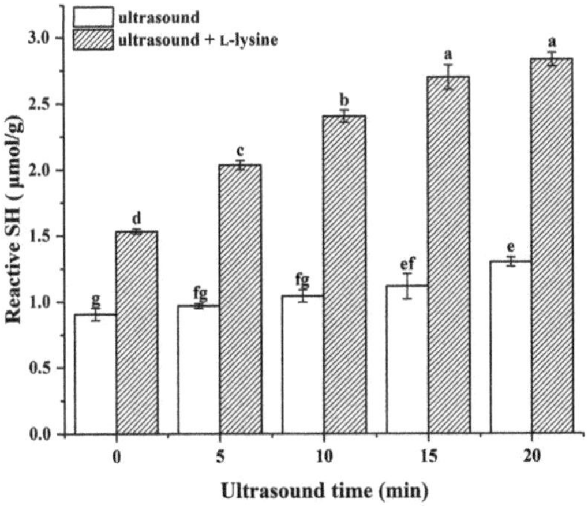

(a)

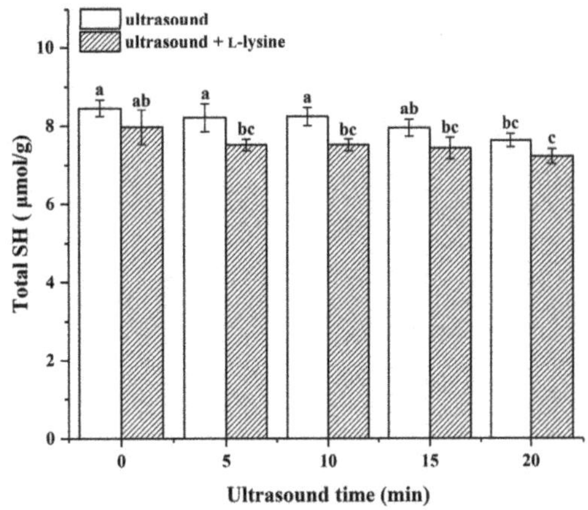

(b)

Figure 4. Effects of ultrasound combined with L-lysine on the contents of reactive sulfhydryl groups (**a**) and total sulfhydryl groups (**b**) in myofibrillar proteins. The letters (a−g) above the bars indicate a significant difference ($p < 0.05$).

3.8. Fluorescence Intensity Analysis

Fluorescence spectra have been extensively utilised for the investigation of protein conformation [54]. The intrinsic fluorescence spectra of the protein microenvironment, wherein residues such as tryptophan (Trp) act as natural chromophores, enable us to characterize the changes in residue polarity [55]. Such changes in polarity can provide insights into the alterations in the tertiary structures of proteins [55]. As depicted in Figure 5, the maximum emission wavelength (λ max) of MPs treated only with ultrasound exhibited an increase from 332.4 nm to 333 nm as the ultrasound time increased from 0 to 20 min, concomitant with a rise in fluorescence intensity. The red shift of the peak indicated the unfolding of the tertiary structure of protein, leading to a greater exposure of aromatic amino acids to the polar environment [36]. Furthermore, the increased maximum fluorescence intensity suggested that the cavitation and mechanical effects induced by ultrasound enhanced the exposure of hydrophobic groups [56]. With L-lysine alone and without ultrasonic treatment (0 min), the λ max of MPs consistently exhibited a red shift, accompanied by a continuous increase in fluorescence intensity. This observation suggested that under the current conditions, L-lysine induced a further unfolding of protein [1]. Subsequently, when L-lysine was combined with ultrasound treatment, the λ max of MPs continued to increase with increasing ultrasound time. At 20 min, the maximum value of 334.8 nm was reached. This phenomenon can be attributed to the unfolding of MP molecules due to the combined application of ultrasound and L-lysine, resulting in the exposure of chromophores such as Trp to the solvent and subsequent solvent quenching, thus influencing the fluorescence intensity. Notably, both the fluorescence intensity and the λ max of the MPs treated with the combined treatment of L-lysine and ultrasound exceeded those of the MPs treated with L-lysine alone without ultrasonic treatment (0 min) or ultrasound alone at the same ultrasonication time in low-ion-strength media. These findings indicated the maximum level of unfolding in the tertiary structures of the protein molecules [3].

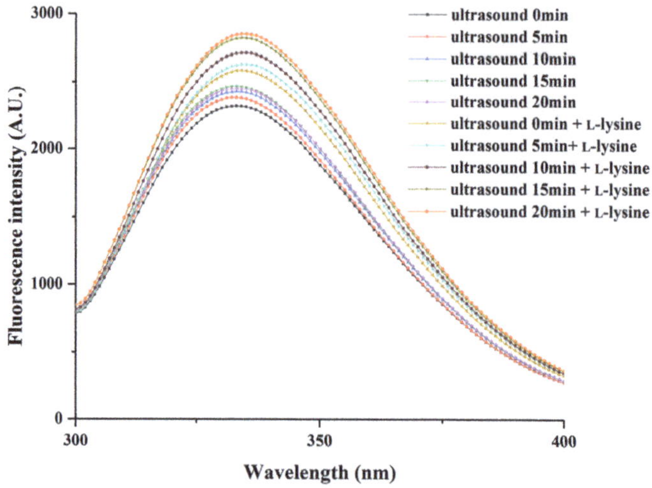

Figure 5. Effects of ultrasound combined with L-lysine on the tertiary structures of myofibrillar proteins.

3.9. DSC of Protein

Differential scanning quantities are employed to analyse energy and conformational changes during protein denaturation [57]. The higher the protein denaturation temperature, the higher the thermal stability [57]. In Table 4, upon the addition of L-lysine alone without ultrasonic treatment (0 min), the heat denaturation temperature of the MPs was elevated to 109.5 °C. The rise in thermal stability can be attributed to the basic nature of L-lysine, which increased the pH value of the solution and consequently enhanced the heat denaturation temperature [58]. Furthermore, the combination of L-lysine with ultrasound resulted in an increased stability of MPs, reaching a maximum value of 114.5 °C at 20 min, which is consistent with Bi et al., who found that ultrasound treatment enhanced the thermal stability of chickpea protein [59]. Ultrasonication disrupted the tertiary structures of MPs through mechanical and shear forces, resulting in increased exposure of SH groups to the protein surface [8]. The total sulfhydryl group content also indicated that some free sulfhydryl groups may be converted to disulfide bonds, hence enhancing the crosslinking between protein subunits [19]. Additionally, the increased thermal stability of MPs may be due to forces that reduce the interactions after HIU treatment, such as internal hydrophobic groups. This led to the unfolding of buried internal hydrophobic groups and increased intermolecular interactions, resulting in the formation of thermally stable polymers [60]. HIU assisted in preventing protein re-agglomeration during post-sonication processing by producing smaller and more stable protein particles [59]. Furthermore, this significant improvement in thermal stability may originate from the interaction between the ultrasound, L-lysine, and protein. Therefore, the combined treatment of ultrasound and L-lysine demonstrates an ability to enhance the thermal stability of MPs [4].

Table 4. Effects of thermal denaturation temperature of pork myofibrillar proteins treated with ultrasound combined with L-lysine.

Ultrasound Time (min)	Denaturation Temperature (T_d) (°C)	
	Ultrasound	Ultrasound + L-Lysine
0	98.6 ± 1.62 ^{f}	109.5 ± 0.16 ^{c}
5	100.3 ± 0.16 ^{f}	109.9 ± 0.23 ^{c}
10	104 ± 0.81 ^{e}	112.5 ± 0.24 ^{b}
15	107.3 ± 0.17 ^{d}	113.6 ± 0.26 ^{a}
20	112.1 ± 0.08 ^{b}	114.5 ± 0.43 ^{a}

$^{a–f}$—in the same indexes and different letters differ significantly ($p < 0.05$).

3.10. Binding of MPs to Ketone Compounds

The percentage of free ketone compounds in the headspace within the presence of MPs was determined as shown in Figure 6. The free flavour compounds in the vial without protein comprised 100%. If the percentage of free flavour compounds falls below 100%, it suggests the protein possesses an adsorption capacity for the flavour compounds. MPs showed an adsorption effect on four ketones [61]. In this study, the MPs showed an adsorption effect on four ketones. Generally, the molecular structure of flavour compounds plays an important role in the interaction between proteins and flavour compounds [62]. For the homologous series in ketone linearity, the flavour adsorption capacity of MPs to the four ketone compounds was in the order of 2-octanone > 2-heptanone > 2-hexanone > 2-pentanone, which is similar to some previous works [63]. This may be attributed to the enhanced hydrophobic interactions between proteins and long-chain ketone molecules, resulting in a great extent of hydrophobic binding between the proteins and the ketones; thus, the adsorption effect was more pronounced [64].

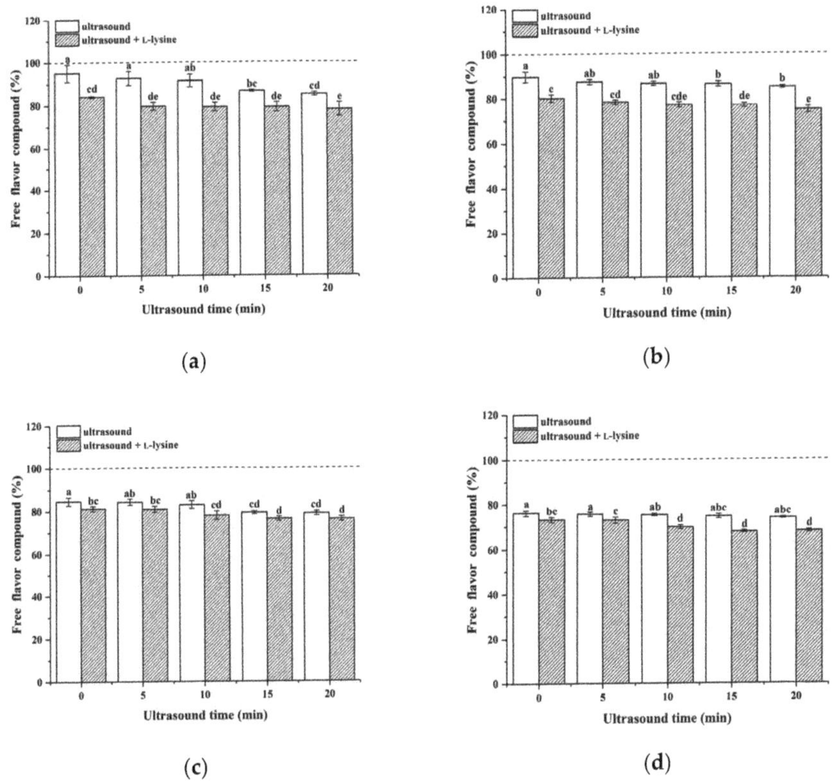

Figure 6. Effects of ultrasound combined with L-lysine treatment of pork myofibrillar proteins on the adsorption capacity of four ketones (2-pentanone, 2-hexanone, 2-heptanone, and 2-octanone). 2-pentanone (**a**), 2-hexanone (**b**), 2-heptanone (**c**), and 2-octanone (**d**). The letters (a–e) above the bars indicate a significant difference ($p < 0.05$).

Untreated MPs (control) had a higher percentage of free ketones in the headspace, which is related to steric hindrance in the ability of MPs to bind flavour compounds [17]. Except for 2-octanone, the binding capacity of MPs at 20 min of ultrasound treatment only was significantly higher than that of the untreated group. This may be due to the cavitation effects of ultrasound, which led to the unfolding of proteins, the exposure of more active sites, and the interaction of more volatiles with proteins [17]. Only the addition of L-lysine significantly increased the binding capacity of MPs for ketone compounds ($p < 0.05$). This indicated that the addition of L-lysine can increase the adsorption capacity of MPs for ketones. The adsorption capacity was further increased by combining L-lysine with ultrasound, and the maximum adsorption was reached at 20 min. The adsorption capacity of 2-pentanone was increased from 4.91% to 22.16%, that of 2-hexanone was increased from 10.30% to 24.67%, that of 2-heptanone was increased from 15.51% to 22.38%, and that of 2-octanone was increased from 24.25% to 31.67%. The adsorption capacity achieved by MPs at 20 min was better than that of MPs treated with L-lysine alone without ultrasonic treatment (0 min) and with ultrasound alone.

The results showed that the combined treatment of ultrasound and L-lysine greatly improved the flavour adsorption capacity of MPs. When the MPs were treated with ultrasound combined with L-lysine, some of the protein particles were disaggregated, leading to the exposure of their active binding sites (hydrophobic and hydrogen binding sites), which increased the interaction with flavour compounds [65]. The combination of the two also enhanced the electrostatic effect, resulting in better binding of the MPs to compounds (Figure 3). At the same time, the secondary structure of the protein unfolds and the α-helix is disrupted, leading to an increase in the hydrophobic content of the surface and thus increasing the number of hydrogen bonding sites on the protein (Table 3) [66]. As a result, hydrophobic sites that are normally buried in the molecule were exposed, which enhanced the binding of MP to the flavour compounds. In addition, ultrasound combined with L-lysine treatment of MPs resulted in the unfolding of the tertiary structure and further exposure of the active binding sites, which increased the adsorption capacity for flavour compounds (Figure 5).

4. Conclusions

This study investigated the potential of combining ultrasound treatment with L-lysine to enhance the solubility, thermal stability, and flavour compound binding ability of MPs in low-ion-strength media. The results demonstrated that a 20 min ultrasound treatment combined with L-lysine achieved optimal dispersion stability. This effect can be attributed to the unfolding and depolymerization of MPs, which exposed SH groups and aromatic amino acids and induced a conformational transition from α-helix to β-turn. Additionally, this combined treatment also enhanced the adsorption capacity of MPs for kenone compounds by facilitating the formation of hydrophobic, hydrogen, and electrostatic bonding sites. In conclusion, the synergistic use of ultrasound and L-lysine offered a promising approach for enhancing the functionality of meat-based products, resulting in improved flavours and an increased utilisation of meat proteins.

Supplementary Materials: The following supporting information can be downloaded at: https://www.mdpi.com/article/10.3390/foods13040629/s1, Figure S1: Effect of L-lysine alone on the solubility of myofibrillar proteins. The letters above the bars indicate a significant difference ($p < 0.05$); Table S1: The probability level (p-value) for experimental main factor (ultrasound time and L-lysine) and their interaction effects on MP.

Author Contributions: Data curation, investigation, and writing—original draft preparation, Y.X.; investigation, D.C.; formal analysis, data curation, and software and resources, J.C.; conceptualization and writing—reviewing and editing, X.W.; conceptualization, writing—reviewing and editing, methodology, visualization and supervision, and funding acquisition, X.Y. All authors have read and agreed to the published version of the manuscript.

Funding: This study was supported by Yunnan Fundamental Research Projects (grant NO. 202301AU070071), Xingdian Talent Support Plan for young talents of Yunnan Province (CG23166F350A) and Yunnan Major Scientific and Technological Projects (grant NO. 202202AG050009).

Data Availability Statement: The data presented in this study are available on request from the corresponding author.

Conflicts of Interest: The authors declare no conflicts of interest.

References

1. Pereira, P.; Vicente, A. Meat nutritional composition and nutritive role in the human diet. *Meat Sci.* **2013**, *93*, 586–592. [CrossRef]
2. Bohrer, B.M. Review: Nutrient density and nutritional value of meat products and non-meat foods high in protein. *Trends Food Sci. Technol.* **2017**, *65*, 103–112. [CrossRef]
3. Streimikyte, P.; Kersiene, M.; Eisinaite, V.; Jasutiene, I.; Lesauskaite, V.; Damuleviciene, G.; Knasiene, J.; Leskauskaite, D. Formulating protein-based beverages for the dysphagia diets of the elderly: Viscosity, protein quality, in vitro digestion, and consumers acceptability. *J. Sci. Food Agric.* **2020**, *100*, 3895–3901. [CrossRef] [PubMed]
4. Wang, K.; Li, Y.; Zhang, Y.; Luo, X.; Sun, J. Improving myofibrillar proteins solubility and thermostability in low-ionic strength solution: A review. *Meat Sci.* **2022**, *189*, 108822. [CrossRef] [PubMed]
5. Pérez-Juan, M.; Flores, M.; Toldrá, F. Effect of pork meat proteins on the binding of volatile compounds. *Food Chem.* **2008**, *108*, 1226–1233. [CrossRef]
6. Chen, X.; Tume, R.K.; Xu, X.L.; Zhou, G.H. Solubilization of myofibrillar proteins in water or low ionic strength media: Classical techniques, basic principles, and novel functionalities. *Crit. Rev. Food Sci.* **2017**, *57*, 3260–3280. [CrossRef] [PubMed]
7. Petit, G.; Jury, V.; de Lamballerie, M.; Duranton, F.; Pottier, L.; Martin, J.L. Salt Intake from Processed Meat Products: Benefits, Risks and Evolving Practices. *Compr. Rev. Food Sci. Food Saf.* **2019**, *18*, 1453–1473. [CrossRef] [PubMed]
8. Liu, H.; Zhang, H.; Liu, Q.; Chen, Q.; Kong, B. Solubilization and stable dispersion of myofibrillar proteins in water through the destruction and inhibition of the assembly of filaments using high-intensity ultrasound. *Ultrason. Sonochem.* **2020**, *67*, 105160. [CrossRef]
9. He, X.L.; Zhao, H.L.; Xu, Y.X.; Yi, S.M.; Li, J.R.; Li, X.P. Synergistic effects of oat β-glucan combined with ultrasound treatment on gel properties of silver carp surimi. *Ultrasonics Sonochem.* **2023**, *95*, 106406. [CrossRef]
10. Ito, Y.; Tatsumi, R.; Wakamatsu, J.-I.; Nishimura, T.; Hattori, A. The solubilization of myofibrillar proteins of vertebrate skeletal muscle in water. *Anim. Sci. J.* **2003**, *74*, 417–425. [CrossRef]
11. Guo, X.Y.; Peng, Z.Q.; Zhang, Y.W.; Liu, B.; Cui, Y.Q. The solubility and conformational characteristics of porcine myosin as affected by the presence of L-lysine and L-histidine. *Food Chem.* **2015**, *170*, 212–217. [CrossRef] [PubMed]
12. Takai, E.; Yoshizawa, S.; Ejima, D.; Arakawa, T.; Shiraki, K. Synergistic solubilization of porcine myosin in physiological salt solution by arginine. *Int. J. Biol. Macromol.* **2013**, *62*, 647–651. [CrossRef] [PubMed]
13. Li, X.; Wang, W.; Wang, S.; Shen, Y.; Pan, J.; Dong, X.; Li, S. The Solubility and Structures of Porcine Myofibrillar Proteins under Low-Salt Processing Conditions as Affected by the Presence of L-Lysine. *Foods* **2022**, *11*, 855. [CrossRef]
14. Xie, Y.; Wang, J.; Wang, Y.; Wu, D.; Liang, D.; Ye, H.; Cai, Z.; Ma, M.; Geng, F. Effects of high-intensity ultrasonic (HIU) treatment on the functional properties and assemblage structure of egg yolk. *Ultrason. Sonochem.* **2020**, *60*, 104767. [CrossRef] [PubMed]
15. Xu, J.J.; Zhang, M.; Wang, Y.C.A.; Bhandari, B. Novel Technologies for Flavor Formation in the Processing of Meat Products: A Review. *Food Rev. Int.* **2023**, *39*, 802–826. [CrossRef]
16. Liu, X.M.; Powers, J.R.; Swanson, B.G.; Hill, H.H.; Clark, S. High hydrostatic pressure affects flavor-binding properties of whey protein concentrate. *J. Food Sci.* **2005**, *70*, C581–C585. [CrossRef]
17. Sun, X.; Yu, Y.; Saleh, A.S.M.; Yang, X.; Ma, J.; Gao, Z.; Li, W.; Wang, Z.; Zhang, D. Structural changes induced by ultrasound improve the ability of the myofibrillar protein to bind flavor compounds from spices. *Ultrason. Sonochem.* **2023**, *98*, 106510. [CrossRef]
18. Khan, S.A.; Dar, A.H.; Bhat, S.A.; Fayaz, J.; Makroo, H.A.; Dwivedi, M. High Intensity Ultrasound Processing in Liquid Foods. *Food Rev. Int.* **2022**, *38*, 1123–1148. [CrossRef]
19. Zhao, X.; Qi, J.; Fan, C.; Wang, B.; Yang, C.; Liu, D. Ultrasound treatment enhanced the ability of the porcine myofibrillar protein to bind furan compounds: Investigation of underlying mechanisms. *Food Chem.* **2022**, *384*, 132472. [CrossRef]
20. Chen, Q.; Kong, B.; Han, Q.; Liu, Q.; Xu, L. The role of bacterial fermentation in the hydrolysis and oxidation of sarcoplasmic and myofibrillar proteins in Harbin dry sausages. *Meat Sci.* **2016**, *121*, 196–206. [CrossRef]
21. Gornall, A.G.; Bardawill, C.J.; David, M.M. Determination of serum proteins by means of the biuret reaction. *J. Biol. Chem.* **1949**, *177*, 751–766. [CrossRef] [PubMed]
22. Shen, H.; Zhao, M.; Sun, W. Effect of pH on the interaction of porcine myofibrillar proteins with pyrazine compounds. *Food Chem.* **2019**, *287*, 93–99. [CrossRef] [PubMed]
23. Wang, H.; Luo, Y.; Shen, H. Effect of frozen storage on thermal stability of sarcoplasmic protein and myofibrillar protein from common carp (*Cyprinus carpio*) muscle. *Int. J. Food Sci. Technol.* **2013**, *48*, 1962–1969. [CrossRef]
24. Chen, X.; Zou, Y.; Han, M.; Pan, L.; Xing, T.; Xu, X.; Zhou, G. Solubilisation of myosin in a solution of low ionic strength L-histidine: Significance of the imidazole ring. *Food Chem.* **2016**, *196*, 42–49. [CrossRef]
25. Kaewprachu, P.; Osako, K.; Benjakul, S.; Rawdkuen, S. Effect of protein concentrations on the properties of fish myofibrillar protein based film compared with PVC film. *J. Food Sci. Technol.* **2016**, *53*, 2083–2091. [CrossRef]
26. Ellman, G.L. Tissue sulfhydryl groups. *Arch. Biochem. Biophys.* **1959**, *82*, 70–77. [CrossRef] [PubMed]
27. Guo, X.; Qiu, H.; Deng, X.; Mao, X.; Guo, X.; Xu, C.; Zhang, J. Effect of Chlorogenic Acid on the Physicochemical and Functional Properties of Coregonus Peled Myofibrillar Protein through Hydroxyl Radical Oxidation. *Molecules* **2019**, *24*, 3205. [CrossRef]
28. Zhou, F.; Zhao, M.; Su, G.; Sun, W. Binding of aroma compounds with myofibrillar proteins modified by a hydroxyl-radical-induced oxidative system. *J. Agric. Food Chem.* **2014**, *62*, 9544–9552. [CrossRef]

29. Qian, S.Y.; Li, X.; Wang, H.; Mehmood, W.; Zhong, M.; Zhang, C.H.; Blecker, C. Effects of low voltage electrostatic field thawing on the changes in physicochemical properties of myofibrillar proteins of bovine (*Longissimus dorsi*) muscle. *J. Food Eng.* **2019**, *261*, 140–149. [CrossRef]
30. Li, S.; Zheng, Y.; Xu, P.; Zhu, X.; Zhou, C. l-Lysine and l-arginine inhibit myosin aggregation and interact with acidic amino acid residues of myosin: The role in increasing myosin solubility. *Food Chem.* **2018**, *242*, 22–28. [CrossRef]
31. Zhang, Y.; Ma, L.; Cai, L.; Liu, Y.; Li, J. Effect of combined ultrasonic and alkali pretreatment on enzymatic preparation of angiotensin converting enzyme (ACE) inhibitory peptides from native collagenous materials. *Ultrason. Sonochem.* **2017**, *36*, 88–94. [CrossRef] [PubMed]
32. Gao, R.C.; Shi, T.; Sun, Q.C.; Li, X.T.; McClements, D.J.; Yuan, L. Effects of L-arginine and L-histidine on heat-induced aggregation of fish myosin: Bighead carp (*Aristichthys nobilis*). *Food Chem.* **2019**, *295*, 320–326. [CrossRef]
33. Park, J.D.; Yongsawatdigul, J.; Choi, Y.J.; Park, J.W. Biochemical and conformational changes of myosin purified from Pacific sardine at various pHs. *J. Food Sci.* **2008**, *73*, C191–C197. [CrossRef] [PubMed]
34. Amiri, A.; Mousakhani-Ganjeh, A.; Shafiekhani, S.; Mandal, R.; Singh, A.P.; Kenari, R.E. Effect of high voltage electrostatic field thawing on the functional and physicochemical properties of myofibrillar proteins. *Innov. Food Sci. Emerg.* **2019**, *56*, 102191. [CrossRef]
35. Yongsawatdigul, J.; Park, J.W. Effects of alkali and acid solubilization on gelation characteristics of rockfish muscle proteins. *J. Food Sci.* **2004**, *69*, C499–C505. [CrossRef]
36. Li, K.; Fu, L.; Zhao, Y.-Y.; Xue, S.-W.; Wang, P.; Xu, X.-L.; Bai, Y.-H. Use of high-intensity ultrasound to improve emulsifying properties of chicken myofibrillar protein and enhance the rheological properties and stability of the emulsion. *Food Hydrocolloid* **2020**, *98*, 105275. [CrossRef]
37. Zhao, X.; Wu, T.; Xing, T.; Xu, X.L.; Zhou, G.H. Rheological and physical properties of O/W protein emulsions stabilized by isoelectric solubilization/precipitation isolated protein: The underlying effects of varying protein concentrations. *Food Hydrocolloid* **2019**, *95*, 580–589. [CrossRef]
38. Shigeru, H.; Shuryo, N. Contribution of Hydrophobicity, Net Charge and Sulfhydryl Groups to Thermal Properties of Ovalbumin. *Can. Inst. Food Sci. Technol. J.* **1985**, *18*, 290–295. [CrossRef]
39. Bao, P.Q.; Chen, L.; Hu, Y.; Wang, Y.; Zhou, C.L. L-Arginine and L-lysine retard aggregation and polar residue modifications of myofibrillar proteins: Their roles in solubility of myofibrillar proteins in frozen porcine (*Longissimus lumborum*). *Food Chem.* **2022**, *393*, 133347. [CrossRef]
40. Song, X.Z.; Zhou, C.J.; Fu, F.; Chen, Z.L.; Wu, Q.L. Effect of high-pressure homogenization on particle size and film properties of soy protein isolate. *Ind. Crop Prod.* **2013**, *43*, 538–544. [CrossRef]
41. Zhang, Z.Y.; Regenstein, J.M.; Zhou, P.; Yang, Y.L. Effects of high intensity ultrasound modification on physicochemical property and water in myofibrillar protein gel. *Ultrason. Sonochem.* **2017**, *34*, 960–967. [CrossRef] [PubMed]
42. Xiong, Y.; Li, Q.R.; Miao, S.; Zhang, Y.; Zheng, B.D.; Zhang, L.T. Effect of ultrasound on physicochemical properties of emulsion stabilized by fish myofibrillar protein and xanthan gum. *Innov. Food Sci. Emerg.* **2019**, *54*, 225–234. [CrossRef]
43. Wang, K.Q.; Sun, D.W.; Pu, H.B.; Wei, Q.Y. Principles and applications of spectroscopic techniques for evaluating food protein conformational changes: A review. *Trends Food Sci. Technol.* **2017**, *67*, 207–219. [CrossRef]
44. Kato, A.; Tsutsui, N.; Matsudomi, N.; Kobayashi, K.; Nakai, S. Effects of Partial Denaturation on Surface Properties of Ovalbumin and Lysozyme. *Agr. Biol. Chem.* **1981**, *45*, 2755–2760.
45. Zhang, B.; Fang, C.D.; Hao, G.J.; Zhang, Y.Y. Effect of kappa-carrageenan oligosaccharides on myofibrillar protein oxidation in peeled shrimp (*Litopenaeus vannamei*) during long-term frozen storage. *Food Chem.* **2018**, *245*, 254–261. [CrossRef] [PubMed]
46. Lin, T.M.; Park, J.W. Solubility of salmon myosin as affected by conformational changes at various ionic strengths and pH. *J. Food Sci.* **1998**, *63*, 215–218. [CrossRef]
47. Liu, H.; Wang, Z.; Badar, I.H.; Liu, Q.; Chen, Q.; Kong, B. Combination of high-intensity ultrasound and hydrogen peroxide treatment suppresses thermal aggregation behaviour of myofibrillar protein in water. *Food Chem.* **2022**, *367*, 130756. [CrossRef]
48. Lametsch, R.; Lonergan, S.; Huff-Lonergan, E. Disulfide bond within μ-calpain active site inhibits activity and autolysis. *Biochim. Biophys. Acta Proteins Proteom.* **2008**, *1784*, 1215–1221. [CrossRef]
49. Estévez, M.; Ollilainen, V.; Heinonen, M. Analysis of Protein Oxidation Markers α-Aminoadipic and γ-Glutamic Semialdehydes in Food Proteins Using Liquid Chromatography (LC)-Electrospray Ionization (ESI)-Multistage Tandem Mass Spectrometry (MS). *J. Agric. Food Chem.* **2009**, *57*, 3901–3910. [CrossRef]
50. Fu, Q.Q.; Liu, R.; Wang, H.O.; Hua, C.; Song, S.X.; Zhou, G.H.; Zhang, W.G. Effects of Oxidation (in Vitro) on Structures and Functions of Myofibrillar Protein from Beef Muscles. *J. Agric. Food Chem.* **2019**, *67*, 5866–5873. [CrossRef]
51. Liu, G.; Xiong, Y.L.L. Electrophoretic pattern, thermal denaturation, and in vitro digestibility of oxidized myosin. *J. Agric. Food Chem.* **2000**, *48*, 624–630. [CrossRef] [PubMed]
52. Grönroos, A.; Pentti, P.; Hanna, K. Ultrasonic degradation of aqueous carboxymethylcellulose: Effect of viscosity, molecular mass, and concentration. *Ultrason. Sonochem.* **2008**, *15*, 644–648. [CrossRef]
53. Feng, X.; Ahn, D.U. Volatile profile, lipid oxidation and protein oxidation of irradiated ready-to-eat cured turkey meat products. *Radiat. Phys. Chem.* **2016**, *127*, 27–33. [CrossRef]

54. Wu, M.G.; Xiong, Y.L.; Chen, J. Role of disulphide linkages between protein-coated lipid droplets and the protein matrix in the rheological properties of porcine myofibrillar protein-peanut oil emulsion composite gels. *Meat Sci.* **2011**, *88*, 384–390. [CrossRef] [PubMed]
55. Xia, W.Y.; Ma, L.; Chen, X.K.; Li, X.Y.; Zhang, Y.H. Physicochemical and structural properties of composite gels prepared with myofibrillar protein and lecithin at various ionic strengths. *Food Hydrocolloid* **2018**, *82*, 135–143. [CrossRef]
56. Keerati-u-rai, M.; Miriani, M.; Iametti, S.; Bonomi, F.; Corredig, M. Structural changes of soy proteins at the oil-water interface studied by fluorescence spectroscopy. *Colloids Surf. B* **2012**, *93*, 41–48. [CrossRef] [PubMed]
57. Xu, B.; He, W.J.; Yi, Y.; Wang, H.X.; Xu, W.; Guo, D.J. The impacting mechanism of β-sanshool in Zanthoxylum bungeanum Maxim on the structure of myofibrillar protein in duck meat. *LWT Food Sci. Technol.* **2023**, *182*, 114838. [CrossRef]
58. Xue, C.; You, J.; Zhang, H.M.; Xiong, S.B.; Yin, T.; Huang, Q.L. Capacity of myofibrillar protein to adsorb characteristic fishy-odor compounds: Effects of concentration, temperature, ionic strength, pH and yeast glucan addition. *Food Chem.* **2021**, *363*, 130304. [CrossRef]
59. Bi, C.H.; Chi, S.Y.; Zhou, T.; Zhang, J.Y.; Wang, X.Y.; Li, J.; Shi, W.T.; Tian, B.; Huang, Z.G.; Liu, Y. Effect of low-frequency high-intensity ultrasound (HIU) on the physicochemical properties of chickpea protein. *Food Res. Int.* **2022**, *159*, 111474. [CrossRef]
60. Khatkar, A.B.; Kaur, A.; Khatkar, S.K.; Bala, M.; Maan, S.; Tyagi, S.K. Valorization of ultrasound assisted restructured soy protein: Impact on the quality characteristics of instant noodles. *LWT Food Sci. Technol.* **2021**, *147*, 111599. [CrossRef]
61. Sun, X.X.; Yu, Y.M.; Saleh, A.S.M.; Yang, X.Y.; Ma, J.L.; Li, W.H.; Zhang, D.Q.; Wang, Z.Y. Understanding interactions among flavor compounds from spices and myofibrillar proteins by multi-spectroscopy and molecular docking simulation. *Int. J. Biol. Macromol.* **2023**, *229*, 188–198. [CrossRef] [PubMed]
62. Shen, H.; Huang, M.Q.; Zhao, M.M.; Sun, W.Z. Interactions of selected ketone flavours with porcine myofibrillar proteins: The role of molecular structure of flavour compounds. *Food Chem.* **2019**, *298*, 125060. [CrossRef] [PubMed]
63. Lou, X.W.; Yang, Q.L.; Sun, Y.Y.; Pan, D.D.; Cao, J.X. The effect of microwave on the interaction of flavour compounds with G-actin from grass carp (*Catenopharyngodon idella*). *J. Sci. Food Agric.* **2017**, *97*, 3917–3922. [CrossRef] [PubMed]
64. Andriot, I.; Harrison, M.; Fournier, N.; Guichard, E. Interactions between methyl ketones and beta-lactoglobulin: Sensory analysis, headspace analysis, and mathematical modeling. *J. Agric. Food Chem.* **2000**, *48*, 4246–4251. [CrossRef] [PubMed]
65. Yang, Y.; Zhang, X.; Wang, Y.; Pan, D.D.; Sun, Y.Y.; Cao, J.X. Study on the volatile compounds generated from lipid oxidation of Chinese bacon (unsmoked) during processing. *Eur. J. Lipid Sci. Technol.* **2017**, *119*, 1600512. [CrossRef]
66. Chen, X.; Xu, X.L.; Han, M.Y.; Zhou, G.H.; Chen, C.G.; Li, P.J. Conformational changes induced by high-pressure homogenization inhibit myosin filament formation in low ionic strength solutions. *Food Res. Int.* **2016**, *85*, 1–9. [CrossRef]

molecules

MDPI

Article

Fabrication of Human Milk Fat Substitute: Based on the Similarity Evaluation Model and Computer Software

Huiquan Zhu [1,2,3], Pu Zhao [1], Xiaodan Wang [1], Yunna Wang [1], Shuwen Zhang [1], Xiaoyang Pang [1,*] and Jiaping Lv [1]

1 Institute of Food Science and Technology, Chinese Academy of Agricultural Sciences, Beijing 100193, China; 13834083105@163.com (H.Z.); hscaas@163.com (P.Z.); prettyshoot@126.com (X.W.); wang_yn92@163.com (Y.W.); zswcaas@hotmail.com (S.Z.); lvjiapingcaas@126.com (J.L.)

2 Laboratory of Chemistry of Natural Molecules, Gembloux Agro-Bio Tech, University of Liege, 5030 Gembloux, Belgium

3 National Center of Technology Innovation for Dairy, Hohhot 010100, China

* Correspondence: pangxiaoyang@163.com

Abstract: We aimed to obtain the optimal formula for human milk fat substitute (HMFS) through a combination of software and an evaluation model and further verify its practicability through an animal experiment. The results showed that a total of 33 fatty acid (FA) and 63 triglyceride (TAG) molecular species were detected in vegetable oils. Palmitic acid, oleic acid, linoleic acid, 18:1/16:0/18:1, 18:2/16:0/18:2, 18:1/18:1/18:1 and 18:1/18:2/18:1, were the main molecular species among the FAs and TAGs in the vegetable oils. Based on the HMFS evaluation model, the optimal mixed vegetable oil formula was blended with 21.3% palm oil, 2.8% linseed oil, 2.6% soybean oil, 29.9% rapeseed oil and 43.4% maize oil, with the highest score of 83.146. Moreover, there was no difference in the weight, blood routine indices or calcium and magnesium concentrations in the feces of the mice between the homemade mixed vegetable oil (HMVO) group and the commercial mixed vegetable oil (CMVO) group, while nervonic acid (C24:1) and octanoic acid (C8:0) were absorbed easily in the HMVO group. Therefore, these results demonstrate that the mixing of the different vegetable oils was feasible via a combination of computer software and an evaluation model and provided a new way to produce HMFS.

Keywords: vegetable oils; evaluation model; software; in vivo experiment; intestinal flora

Citation: Zhu, H.; Zhao, P.; Wang, X.; Wang, Y.; Zhang, S.; Pang, X.; Lv, J. Fabrication of Human Milk Fat Substitute: Based on the Similarity Evaluation Model and Computer Software. *Molecules* **2024**, *29*, 2096. https://doi.org/10.3390/molecules29092096

Academic Editor: Gabriele Rocchetti

Received: 25 March 2024
Revised: 21 April 2024
Accepted: 24 April 2024
Published: 1 May 2024

1. Introduction

Human milk fat (HMF) is constructed with a core composed of triacylglycerol (TAG) and a membrane (mainly consisting of phospholipid and protein). The TAG content accounts for about 98% of the total HMF, as it is important for the growth of newborns [1]. More than 400 TAG molecular species are detected in HMF [2], in which oleic acid–palmitic acid–oleic acid (OPO) and oleic acid–palmitic acid–linoleic acid (OPL) are the main molecular species, accounting for 11.13–28.08% and 8.86–19.50% of the total TAGs, respectively [3–6]. TAGs are impacted by some factors, such as the lactation period, the region sampled, the physical condition of the voluntary mothers dietary patterns, etc. Most studies have found that the content of some TAG molecular species, such as OPO and OPL, decreases with the prolongation of lactation time [3,6]. Chen et al. (2020) reported that the content of OPL was the highest (16.55%, 19.20% and 18.67%, respectively) in breast milk samples from Zhengzhou, Wuhan and Harbin, followed by OPO (10.08, 10.22 and 12.03%, respectively) [7].

Infant formula (IF) manufacturers have been committed to meeting the nutrition needs of infants during their growth stage, so HMF substitute (HMFS) is used in IF production, which is based on the composition of HMF and selected from many plant, animal and microbial oils or fats mixed scientifically [8]. There are two main kinds of HMFS; the first arises from a variety of oil-compounding methods to adjust the fatty acid (FA) composition of baby powder, among which vegetable oil and cow milk fat are the most common sources [9]. Vegetable oils are economical and contain different FA and TAG molecular species, so they can safely mimic HMF and meet the lipid needs of infants through physical mixing. Wei et al. (2019) reported that there are 48 TAG molecular species detected in all vegetable oils, and they showed the different characteristics of TAGs among vegetable oils; for example, OPO and OOO accumulated in olive oil and linseed oil had more linolenic acid–linolenic acid–linolenic acid or linolenic acid–linoleic acid–linolenic acid [10]. The second comprises structural lipids, mainly containing OPO-structured TAGs and other structured lipids enriched with palmitic acid (C16:0) in the Sn-2 position, and they are obtained by transesterification [11]. Ghide et al. (2022) used palmitic acid–palmitic acid–palmitic acid as a substrate to synthesize OPO through rhizomucor miehei lipase, in which the percentage of Sn-2 C16:0 and Sn-1,3 oleic acid (Sn-1,3 C18:1) reached 92.93% and 57.82%, respectively [12].

Previous studies have shown that the intestinal microbial composition is affected by the FA types and TAG structure of HMFS [13]. Chen et al. (2022) carried out a 16S rRNA gene sequence analysis and metabolomics analysis, and the results showed that the abundance of bifidobacterium increased after the feeding of OPO-containing formula, which was likely due to the adhesion and proliferation of lactobacillus, and Bifidobacterium was directly affected by mono-triglyceride palmitate at the Sn-2 position [14,15]. For the influence on the intestinal flora caused by the mixed vegetable oils, Qiao et al. (2022) found that the abundances of Escherichia coli, Lactobacilli and bifidobacteria decreased after adding mixed vegetable oils to the daily diet of mice; meanwhile, the activity of protease, cellulase and amylase also decreased [16].

The model used to evaluate the similarity between HMF and HMFS was first established by Wang et al. (2010) to analyze the similarity between FAs and Sn-2 FAs in HMF and HMFS based on the reduction principle, and Hokkanen et al. (2022) reported another evaluation model based on the Bray–Curtis index [17,18]. In this study, according to the human milk data reported in our previous study [3], software based on C++ was set up, and it was used to calculate the different formulas of mixed vegetable oils according to the evaluation model. After obtaining the optimal formula, it was used for an in vivo experiment with commercial mixed vegetable oil to evaluate the intestinal flora of mice and further verify the practicability of the software. These results will provide a new approach to obtaining different formulas for HMFS and further promote the development of HMFS.

2. Results and Discussion

2.1. Fatty Acid Profile of Vegetable Oils

A total of 33 FA species were detected in seven different vegetable oils (Table 1), including 16 saturated FAs (SFAs) and 17 unsaturated FAs (UFAs). The highest number of FA species was observed in rapeseed oil and walnut oil, in both of which 28 FA species were found, followed by soybean oil (27 FA species), palm oil (26 FA species), maize oil (25 FA species), linseed oil (24 FA species) and sunflower seed oil (22 FA species). Among these FA species, C16:0 (4.670–37.093% of the total FAs), C18:1n9c (18.917–53.225% of the total FAs), linoleic acid (C18:2n6c, 10.704–61.268% of the total FAs) and linolenic acid (C18:3n6, 0.142–44.235% of the total FAs) were the main species. It was evident that C16:0, C18:1n9c, C18:2n6c and C18:3n6 showed the highest percentages in palm oil, rapeseed oil, sunflower oil and linseed oil, respectively, accounting for 37.093% of the total FAs, 53.225% of the total FAs, 61.268% of the total FAs and 44.235% of the total FAs, respectively, which are in accordance with previous results [19–23]. Moreover, some FA species showed specificity in vegetable oils; for example mecilenic acid (C14:1) was only found in palm oil, and docosanoic acid (C22:0) was only found in rapeseed oil. For polyunsaturated fatty acids (PUFAs), such as docosahexaenoic acid (DHA, C22:6n3) and nervonic acid (C24:1n9), which are important for the brain development of infants [24], small concentrations of these FAs were found in the vegetable oils. C24:1n9 was only detected in walnut oil (0.038% of the total FAs) and rapeseed oil (0.141% of the total FAs), and DHA was found in palm oil, walnut oil and maize oil, accounting for 0.078% of the total FAs, 0.113% of the total FAs and 0.026% of the total FAs, respectively.

Table 1. The content of different FAs in vegetable oils (% is shown as the average ± standard deviation).

Fatty Acid	Walnut Oil	Plam Oil	Linseed Oil	Soybean Oil	Rapeseed Oil	Sunflower Oil	Maize Oil
C4:0	0.074 ± 0.016	0.665 ± 0.021	0.072 ± 0.024	0.096 ± 0.021	0.035 ± 0.008	0.037 ± 0.003	0.069 ± 0.016
C6:0	0.202 ± 0.026	0.161 ± 0.016	0.138 ± 0.032	0.176 ± 0.023	0.195 ± 0.027	0.153 ± 0.020	0.152 ± 0.020
C8:0	2.130 ± 0.087	1.857 ± 0.423	2.558 ± 0.360	2.174 ± 0.093	2.381 ± 0.120	2.512 ± 0.121	2.509 ± 0.113
C10:0	5.805 ± 1.886	4.536 ± 0.988	4.928 ± 2.116	5.049 ± 1.639	4.666 ± 1.533	4.897 ± 1.601	5.447 ± 1.774
C11:0	0.097 ± 0.006	0.075 ± 0.015	0.023 ± 0.004	0.069 ± 0.004	0.061 ± 0.004	-	0.106 ± 0.007
C12:0	0.064 ± 0.002	0.193 ± 0.047	-	0.078 ± 0.003	0.015 ± 0.001	-	-
C13:0	0.041 ± 0.001	0.055 ± 0.014	0.096 ± 0.011	0.063 ± 0.002	0.057 ± 0.002	0.302 ± 0.009	0.031 ± 0.001
C14:0	0.091 ± 0.002	1.088 ± 0.271	0.048 ± 0.006	0.076 ± 0.002	0.066 ± 0.002	0.115 ± 0.003	0.087 ± 0.002
C14:1	0.041 ± 0.001	-	-	-	-	-	-
C15:0	0.022 ± 0.000	0.046 ± 0.014	-	0.034 ± 0.000	0.023 ± 0.000	0.055 ± 0.001	-
C16:0	9.508 ± 0.034	37.093 ± 10.59	6.117 ± 0.560	10.634 ± 0.013	4.670 ± 0.031	6.576 ± 0.028	12.923 ± 0.011
C16:1	0.112 ± 0.000	0.219 ± 0.063	0.114 ± 0.011	0.117 ± 0.000	0.175 ± 0.002	0.075 ± 0.000	0.154 ± 0.001
C17:0	0.092 ± 0.005	0.136 ± 0.027	0.112 ± 0.016	0.163 ± 0.009	0.075 ± 0.005	0.030 ± 0.002	0.132 ± 0.007
C17:1	0.118 ± 0.002	0.033 ± 0.010	0.042 ± 0.004	0.070 ± 0.001	0.055 ± 0.000	0.051 ± 0.001	0.024 ± 0.000
C18:0	3.655 ± 0.067	4.397 ± 1.356	4.100 ± 0.308	4.315 ± 0.071	2.034 ± 0.017	3.866 ± 0.043	1.915 ± 0.028
C18:1n9t	0.028 ± 0.002	0.050 ± 0.020	0.023 ± 0.001	0.025 ± 0.002	0.025 ± 0.002	0.058 ± 0.004	0.103 ± 0.007
C18:1n9c	21.285 ± 0.867	37.63 ± 15.567	19.692 ± 9.669	19.483 ± 0.740	53.225 ± 1.581	18.917 ± 0.609	26.843 ± 0.961
C18:2n6t	-	-	0.870 ± 0.113	1.377 ± 0.049	3.338 ± 0.144	-	-
C18:2n6c	49.671 ± 1.147	10.704 ± 3.366	15.736 ± 1.121	48.459 ± 0.994	20.063 ± 0.249	61.268 ± 0.913	48.049 ± 0.883
C20:0	0.155 ± 0.039	0.230 ± 0.154	0.517 ± 0.084	0.211 ± 0.052	0.465 ± 0.109	0.163 ± 0.038	0.234 ± 0.057
C18:3n6	6.041 ± 0.121	0.263 ± 0.066	44.235 ± 5.098	6.624 ± 0.147	6.307 ± 0.188	0.142 ± 0.003	0.600 ± 0.015
C20:1n9	0.205 ± 0.000	0.140 ± 0.000	0.177 ± 0.012	0.171 ± 0.004	1.147 ± 0.018	0.119 ± 0.002	0.202 ± 0.004
C18:3n3	0.061 ± 0.001	-	0.069 ± 0.000	0.026 ± 0.000	0.061 ± 0.002	-	0.084 ± 0.001
C21:0	0.017 ± 0.000	0.052 ± 0.015	-	0.047 ± 0.000	0.057 ± 0.000	0.023 ± 0.001	0.026 ± 0.000
C20:2	0.166 ± 0.000	0.035 ± 0.04	0.120 ± 0.056	0.223 ± 0.122	0.162 ± 0.087	0.359 ± 0.193	0.056 ± 0.031
C22:0	-	-	-	-	0.194 ± 0.000	-	-
C20:3n6	0.021 ± 0.001	-	-	-	-	-	0.057 ± 0.000
C22:1	-	0.077 ± 0.012	-	0.027 ± 0.002	-	-	-
C20:3n6	-	-	0.082 ± 0.001	-	0.114 ± 0.012	-	-
C20:4n6	0.063 ± 0.000	0.054 ± 0.015	0.020 ± 0.002	0.058 ± 0.001	0.027 ± 0.001	0.051 ± 0.001	0.017 ± 0.000
C24:0	0.121 ± 0.005	0.098 ± 0.020	0.112 ± 0.015	0.156 ± 0.006	0.166 ± 0.008	0.231 ± 0.010	0.155 ± 0.006
C24:1	0.038 ± 0.000	-	-	-	0.141 ± 0.002	-	-
C22:6	0.078 ± 0.002	0.113 ± 0.027	-	-	-	-	0.026 ± 0.001

2.2. *Triacylglycerol Profile of Vegetable Oils*

The stereochemical structure of TAGs is important for the properties of oil, including its absorbing and utilizable characteristics. Based on the ultraperformance liquid chromatogram–mass spectrum (UPLC-MS) method, 63 different TAG molecular species were detected in all the vegetable oils (Table 2), and 60 TAG molecular species were detected in palm oil, followed by walnut oil (56), rapeseed oil (55), linseed oil (47), sunflower seed oil (45), soybean oil (44) and maize oil (44). Among these TAG molecular species, 18:1/16:0/18:1 (OPO), 18:2/16:0/18:2 (LPL), 18:1/18:1/18:1 (OOO), 18:1/18:2/18:1 (OLO), 18:2/18:0/18:2 (LSL), 18:2/18:1/18:2 (LOL) and 18:2/18:2/18:2 (LLL) were the main TAG molecular species, constituting 2.841–21.965% of the total TAGs, 1.088–12.538% of the total TAGs, 2.690–26.833% of the total TAGs, 1.058–22.025% of the total TAGs, 0.672–11.598% of the total TAGs, 0.181–16.181% of the total TAGs and 0.070–20.694% of the total TAGs, respectively. Similar results were reported in a previous study [10]. Moreover, we could see that the OPO percentage in palm oil was the highest, which was 21.965% of the total TAGs. The most abundant contents of OOO and OLO were both observed in rapeseed oil, accounting for 26.833% and 22.025% of the total TAGs, respectively. LLL showed the highest content (20.694% of total TAGs) in the sunflower seed oil. These results have also been reported by other researchers [25,26]. The percentage of 18:1/16:0/18:2 (OPL), another important TAG molecular species in human milk, accounted for 2.709–7.778% of the total TAGs in the different vegetable oils [2,7]. Furthermore, just like FAs, the TAG molecular species also demonstrated specificity in the different vegetable oils; for example, 12:0/6:0/12:0, 12:0/8:0/12:0, 14:0/16:0/14:0, 16:0/12:0/16:0, 16:0/14:0/16:0 and 16:0/12:0/18:0 were only detected in palm oil.

Table 2. The contents of different TAGs in vegetable oil (% is shown as the average ± standard deviation).

TAG	Abbreviation	Walnut Oil	Plam Oil	Linseed Oil	Soybean Oil	Rapeseed Oil	Sunflower Oil	Maize Oil
12:0/10:0/8:0	LaCCy	0.002 ± 0.000 [a]	0.002 ± 0.001	-	-	-	-	-
12:0/6:0/12:0	LaCaLa	-	0.002 ± 0.000	-	-	-	-	-
12:0/8:0/12:0	LaCyLa	-	0.019 ± 0.012	-	-	-	-	-
12:0/6:0/14:0	LaCaM	0.001 ± 0.000	0.002 ± 0.000	-	-	-	-	-
12:0/8:0/14:0	LaCyM	0.008 ± 0.003	0.013 ± 0.003	0.006 ± 0.001	-	0.004 ± 0.001	-	-
12:0/10:0/12:0	LaCLa	0.005 ± 0.001	0.016 ± 0.003	-	-	-	-	-
12:0/12:0/12:0	LaLaLa	0.025 ± 0.005	0.203 ± 0.050	0.019 ± 0.004	0.003 ± 0.001	0.009 ± 0.003	0.002 ± 0.000	0.002 ± 0.000
12:0/10:0/14:0	LaCM	0.004 ± 0.001	0.011 ± 0.002	-	-	0.003 ± 0000	-	-
14:0/8:0/14:0	MCyM	0.005 ± 0.000	0.014 ± 0.003	0.008 ± 0.002	-	0.003 ± 0.001	-	-
12:0/14:0/12:0	LaMLa	0.018 ± 0.004	0.115 ± 0.013	0.007 ± 0.001	-	0.007 ± 0.001	0.001 ± 0.000	-
12:0/14:0/14:0	LaMM	0.01 ± 0.001	0.033 ± 0.007	0.010 ± 0.003	-	0.006 ± 0.001	-	-
12:0/16:0/12:0	LaPLa	0.008 ± 0.002	0.031 ± 0.007	0.014 ± 0.003	-	0.009 ± 0.002	-	-
14:0/16:0/12:0	MPLa	0.004 ± 0.001	0.019 ± 0.003	-	-	0.005 ± 0000	-	-
12:0/18:0/12:0	LaSLa	0.004 ± 0.000	0.009 ± 0.001		-	0.002 ± 0000	-	-
14:0/14:0/14:0	MMM	0.005 ± 0.001	0.02 ± 0.004	-	-	0.006 ± 0000	-	-
14:0/16:0/14:0	MPM	-	0.024 ± 0.002	-	-	-	-	-
16:0/12:0/16:0	PLaP	-	0.037 ± 0.004	-	-	-	-	-
16:0/14:0/16:0	PMP	-	0.096 ± 0.01	-	-	-	-	-
16:0/12:0/18:0	PLaS	-	0.007 ± 0.000	-	-	-	-	-
14:0/18:2/16:0	MLP	0.041 ± 0.005	0.346 ± 0.014	0.014 ± 0.001	0.059 ± 0.004	0.026 ± 0.002	0.039 ± 0.002	0.025 ± 0.002
16:0/18:0/16:0	PSP	0.040 ± 0.002	0.059 ± 0.001	0.113 ± 0.004	0.044 ± 0.002	0.019 ± 0.004	0.029 ± 0.001	0.033 ± 0.001
16:0/18:1/16:0	POP	1.738 ± 0.083	38.714 ± 1.722	1.737 ± 0.076	1.995 ± 0.062	0.930 ± 0.046	0.801 ± 0.024	3.073 ± 0.114
18:1/14:0/18:1	OMO	0.070 ± 0.003	0.664 ± 0.009	0.117 ± 0.006	0.078 ± 0.004	0.197 ± 0.013	0.056 ± 0.002	0.069 ± 0.004
16:0/18:2/16:0	PLP	4.051 ± 0.204	13.350 ± 1.626	1.723 ± 0.100	5.39 ± 0.214	0.87 ± 0.038	2.733 ± 0.151	6.74 ± 0.254
18:2/16:1/16:0	LPoP	0.324 ± 0.019	0.147 ± 0.006	1.884 ± 0.092	0.401 ± 0.013	0.193 ± 0.014	0.049 ± 0.002	0.116 ± 0.006
18:2/18:1/14:0	LOM	0.044 ± 0.002	0.085 ± 0.003	0.035 ± 0.002	0.046 ± 0.002	0.043 ± 0.009	0.058 ± 0.003	0.024 ± 0.001
16:0/18:3/16:0	PLnP	0.121 ± 0.005	0.063 ± 0.002	0.831 ± 0.042	0.155 ± 0.005	0.06 ± 0.008	0.003 ± 0.000	0.026 ± 0.001
18:0/16:0/18:0	SPS	0.022 ± 0.001	-	0.089 ± 0.003	0.034 ± 0.002	0.027 ± 0.011	0.019 ± 0.001	0.01 ± 0.001
18:1/12:0/18:1	OLaO	-	0.042 ± 0.002	-	-	0.040 ± 0.000	-	-
16:0/18:1/18:0	POS	0.505 ± 0.037	3.752 ± 0.268	0.917 ± 0.012	0.686 ± 0.031	0.243 ± 0.004	0.275 ± 0.018	0.312 ± 0.027
18:1/16:0/18:1	OPO	4.351 ± 0.159	21.965 ± 0.264	6.524 ± 0.273	4.226 ± 0.32	8.552 ± 0.167	2.841 ± 0.025	5.857 ± 0.316
18:2/18:0/16:0	LSP	0.764 ± 0.024	0.826 ± 0.038	0.474 ± 0.042	1.113 ± 0.059	0.145 ± 0.003	0.701 ± 0.039	0.451 ± 0.029
18:1/16:1/18:1	OPoO	5.739 ± 0.272	4.162 ± 0.348	3.383 ± 0.136	6.492 ± 0.488	3.52 ± 0.068	4.868 ± 0.048	8.432 ± 0.094
18:1/16:0/18:2	OPL	5.073 ± 0.081	3.652 ± 0.190	3.148 ± 0.605	5.802 ± 0.166	2.709 ± 0.076	4.563 ± 0.396	7.778 ± 0.358
18:2/16:0/18:2	LPL	11.273 ± 2.406	1.088 ± 0.031	3.353 ± 0.057	12.172 ± 1.369	1.309 ± 0.027	10.959 ± 0.222	12.538 ± 1.204
18:2/16:1/18:2	LPoL	1.325 ± 0.071	0.025 ± 0.003	4.647 ± 0.307	1.341 ± 0.057	0.416 ± 0.012	0.204 ± 0.012	0.263 ± 0.007
18:3/16:0/18:2	LnPL	1.372 ± 0.030	0.03 ± 0.002	5.576 ± 0.188	1.526 ± 0.015	0.398 ± 0.019	0.038 ± 0.003	0.173 ± 0.012
18:0/18:1/18:0	SOS	0.290 ± 0.010	0.402 ± 0.046	0.913 ± 0.054	0.416 ± 0.040	0.165 ± 0.012	0.163 ± 0.015	0.079 ± 0.008
20:0/18:1/16:0	AOP	0.103 ± 0.012	0.377 ± 0.059	0.124 ± 0.014	0.101 ± 0.010	0.101 ± 0.008	0.049 ± 0.004	0.112 ± 0.014
18:1/18:0/18:1	OSO	1.314 ± 0.147	1.840 ± 0.154	3.164 ± 0.273	1.460 ± 0.185	2.327 ± 0.135	1.021 ± 0.093	0.634 ± 0.023
18:0/18:2/18:0	SLS	1.250 ± 0.105	1.392 ± 0.131	2.584 ± 0.207	1.440 ± 0.176	1.709 ± 0.049	1.009 ± 0.122	0.530 ± 0.06
16:0/20:0/18:2	PAL	0.110 ± 0.005	0.081 ± 0.009	0.122 ± 0.015	0.188 ± 0.016	0.152 ± 0.012	0.071 ± 0.007	0.160 ± 0.013
18:1/18:1/18:1	OOO	4.164 ± 0.352	2.690 ± 0.451	8.485 ± 0.678	3.892 ± 0.508	26.833 ± 1.226	4.928 ± 0.205	4.257 ± 0.456
18:1/18:2/18:0	OLS	2.048 ± 0.044	0.524 ± 0.047	1.141 ± 0.074	2.466 ± 0.211	1.205 ± 0.108	1.918 ± 0.132	0.936 ± 0.104
18:1/18:2/18:1	OLO	10.805 ± 0.097	1.058 ± 0.162	13.194 ± 1.181	9.323 ± 0.898	22.025 ± 1.301	11.285 ± 0.717	12.551 ± 1.501
18:2/18:0/18:2	LSL	9.632 ± 1.027	0.672 ± 0.089	6.817 ± 0.465	9.284 ± 0.312	11.601 ± 1.000	11.598 ± 0.386	8.454 ± 0.589
18:2/18:1/18:2	LOL	10.012 ± 0.222	0.181 ± 0.013	5.649 ± 0.387	9.628 ± 0.450	4.149 ± 0.254	16.181 ± 1.053	12.001 ± 0.601
18:2/18:2/18:2	LLL	16.676 ± 0.18	0.070 ± 0.009	10.835 ± 1.571	14.753 ± 0.475	2.741 ± 0.227	20.694 ± 2.148	11.618 ± 0.099
18:2/18:3/18:2	LLnL	3.744 ± 0.303	0.010 ± 0.001	9.664 ± 1.086	3.284 ± 0.360	0.603 ± 0.020	0.078 ± 0.008	0.356 ± 0.025
22:0/18:1/16:0	DOP	0.107 ± 0.010	0.105 ± 0.019	0.202 ± 0.033	0.116 ± 0.020	0.096 ± 0.008	0.090 ± 0.014	0.070 ± 0.012
18:1/20:0/18:1	OAO	0.236 ± 0.031	0.287 ± 0.044	0.632 ± 0.068	0.251 ± 0.024	1.253 ± 0.131	0.207 ± 0.018	0.256 ± 0.037
16:0/22:0/18:2	PDL	0.155 ± 0.023	0.036 ± 0.006	0.122 ± 0.013	0.217 ± 0.024	0.133 ± 0.009	0.203 ± 0.025	0.064 ± 0.007
18:1/20:1/18:1	OGO	0.194 ± 0.022	0.085 ± 0.013	0.351 ± 0.040	0.215 ± 0.032	1.741 ± 0.185	0.141 ± 0.018	0.260 ± 0.033
18:2/18:1/20:0	LOA	0.168 ± 0.018	0.061 ± 0.008	0.117 ± 0.011	0.227 ± 0.034	0.330 ± 0.035	0.154 ± 0.018	0.206 ± 0.018
18:2/20:1/18:1	LGO	0.106 ± 0.014	0.024 ± 0.004	0.303 ± 0.031	0.133 ± 0.022	0.821 ± 0.112	0.081 ± 0.008	0.146 ± 0.013
18:2/20:0/18:2	LAL	0.285 ± 0.036	0.055 ± 0.006	0.199 ± 0.022	0.381 ± 0.061	0.645 ± 0.085	0.385 ± 0.037	0.400 ± 0.048
18:2/20:1/18:2	LGL	0.193 ± 0.023	0.010 ± 0.002	0.161 ± 0.020	0.200 ± 0.030	0.209 ± 0.020	0.195 ± 0.014	0.241 ± 0.018
24:0/16:0/18:1	TPO	0.037 ± 0.005	0.080 ± 0.017	0.148 ± 0.018	0.038 ± 0.008	0.032 ± 0.005	0.041 ± 0.007	0.051 ± 0.007
18:1/22:0/18:1	ODO	0.209 ± 0.033	0.161 ± 0.036	0.372 ± 0.052	0.220 ± 0.045	0.537 ± 0.077	0.416 ± 0.080	0.129 ± 0.017
18:2/24:0/16:0	LTP	0.069 ± 0.009	0.026 ± 0.005	0.080 ± 0.012	0.097 ± 0.020	0.035 ± 0.005	0.080 ± 0.015	0.055 ± 0.010
18:1/22:1/18:1	ODoO	0.142 ± 0.017	0.038 ± 0.008	0.305 ± 0.037	0.207 ± 0.035	0.449 ± 0.082	0.338 ± 0.064	0.065 ± 0.012
22:0/18:1/18:2	DOL	0.148 ± 0.025	-	-	0.246 ± 0.034	0.182 ± 0.030	0.415 ± 0.073	0.065 ± 0.008
18:2/22:0/18:2	LDL	0.224 ± 0.025	-	-	0.329 ± 0.056	0.163 ± 0.021	0.793 ± 0.092	0.105 ± 0.013

[a] The values represent the means ± SD. The abbreviations used are as follows: Ca, caproic acid, 6:0; Cy, caprylic acid, 8:0; C, capric acid, 10:0; La, lauric acid, 12:0; M, myristic acid, 14; P, palmitic acid, 16:0; Po, pamitoleic acid, 16:1; S, stearic acid, 18:0; O, oleic acid, 18:1; L, linoleic acid, 18:2; Ln, linolenic acid, 18:3; A, arachidic acid, 20:0; E, eicosenoic acid, 20:1; D, docosanoic acid, 22:0; Do, docosenoic acid, 22:1; T, tetracosanoic acid, 24:0.

2.3. Formula of Mixed Vegetable Oils

The software was input in Excel (Microsoft Office software 2016, Microsoft, USA) form, and the total results, including the top 1000 formulas, were also output in Excel form (Table S2). Table 3 showed the optimal mixing ratio result of the vegetable oils, which was simulated according to the evaluation model of HMF and scored with the reduction principle. The results showed that the highest score for the mixed vegetable oils was 83.146, which was mixed with 21.3% palm oil, 2.8% linseed oil, 2.6% soybean oil, 29.9% rapeseed oil and 43.4% maize oil (Table S2). There were some reasons why the highest score was not more than 90: first, some TAG molecular species, such as 18:0/16:0/18:1, 18:1/16:0/12:0, 18:1/16:0/14:0, 18:2/14:0/18:1, 18:1/16:0/16:1 and 18:1/12:0/18:2, only existed in HMF, leading to an inevitable score reduction for the mixed vegetable oils; furthermore, the deduction points for OPL and 18:0/14:0/18:1 were relatively higher. This finding might have been caused by the fact that the minimum concentrations of OPL and 18:0/14:0/18:1 in human milk were higher than those of all the vegetable oils.

Table 3. The calculation process for the optimal formula obtained by the evaluation model.

	Conditions	Ci [a]	Ei	∑Di	∑Ei	G
G 1	18:1/16:0/18:1	0.000	0.001	83.318	16.854	83.146
	18:1/16:0/18:2	0.296	2.007			
	18:1/18:1/18:1	0.000	0.000			
	18:1/18:2/18:1	0.000	0.000			
	18:2/16:0/18:2	0.000	0.000			
	16:0/18:1/16:0	0.104	0.318			
	18:2/18:1/18:2	0.000	0.000			
	18:1/14:0/18:1	0.874	2.081			
	16:0/18:2/16:0	0.135	0.279			
	18:0/16:0/18:1	1.000	1.955			
	18:1/12:0/18:1	0.977	1.444			
	18:1/18:0/18:1	0.000	0.000			
	18:1/16:0/12:0	1.000	1.404			
	18:1/16:0/14:0	1.000	1.178			
	18:2/18:2/18:2	0.000	0.000			
	18:2/14:0/18:1	1.000	0.793			
	18:1/16:0/16:1	1.000	0.682			
	18:1/12:0/18:2	1.000	0.658			
G 2	C10:0	1.067	0.412	94.468		
	C12:0	0.977	1.255			
	C14:0	0.834	0.822			
	C16:0	0.000	0.000			
	C18:0	0.245	0.341			
	C22:0	0.925	0.513			
	C16:1	0.854	0.41			
	C18:1	0.000	0.000			
	C18:2	0.000	0.000			
	C18:3	0.002	0.001			
	C20:4	0.000	0.000			
	C20:5	1.000	0.017			
	C22:6	0.753	0.067			
G 3	Sn-2	0.000	0.000	183.917		
	Sn-2	0.000	0.000			
	Sn-2	0.069	0.215			
G 4	UUU + USU	0.000	0.000	152.43		
	TAG with SFAs (carbon number < 14) at Sn-1/3 position	0.000	0.000			

[a] G is the total score for the HMFS, with a highest score of 100; G1, G2, G3 and G4 are the scores for the TAGs, FAs, Sn-2 FAs and special TAG structures, respectively, with a highest score of 50, 25, 12.5 and 12.5, respectively; ∑Ei is the sum of Ei, which is the deducted score corresponding to each evaluation index; Ci is the floating coefficient of each evaluation index; ∑Di is the sum of Di, which is the average value of each corresponding evaluation index in the HMF. TAG, triglyceride; USU, TAG connected with three unsaturated fatty acids; USU, TAG connected with two unsaturated fatty acids at the Sn-1/3 position and one saturated fatty acid at the Sn-2 position; SFA, saturated fatty acid.

2.4. Results of In Vivo Experiment

2.4.1. Influence on the Growth, Blood Routine and Serum Biochemical Indices of Sprague-Dawley Mice

An in vivo experiment was used to verify the availability of the optimal formula made by the software and evaluation model. The experimental results showed that the body weight of the Sprague-Dawley (SD) mice increased by about 10 g/day in both the group administered homemade mixed vegetable oil obtained with the software mentioned above (HMVO) and the commercial mixed vegetable oil (CMVO) group, and their food utilization rate was 38–42% (Table S3), indicating that there was no significant difference in body weight and food intake between the SD mice in the HMVO and CMVO groups. A blood routine examination, a basic examination for disease, can reflect the pathological condition of various tissues and organs in the body [16]. It was evident that the contents of red blood cells (RBCs), hemoglobin (HGB), hematocrit (HCT), mean red blood cell hemoglobin (MCH), platelets (PLTs), white blood cells (WBCs), lymphocytes (LYMs), neutrophils (NEUTs), monocytes (MONOs), eosinophils (EOs), basophils (BAEOs) and reticulocytes (RETs), the mean red blood cell volume (MCV), the mean red blood cell hemoglobin concentration (MCHC), the serum total cholesterol (CHO), the total triglyceride (TG), the low-density lipoprotein cholesterol level (LDL-c) and the high-density lipoprotein cholesterol level (HDL-c) were not impacted in the HMVO and CMVO groups, indicating that the influence caused by the HMVO and the CMVO was similar (Table S4).

2.4.2. Fatty Acid Profile of Intestinal Contents and Feces

A total of 16 FA species were detected in the small intestine contents (Table S5). Among these FAs, the percentage of C16:0 was the highest, accounting for 30.86–32.68% of the total FAs, followed by stearic acid (C18:0, 24.46–25.70% of the total FAs), C18:3n6 (11.24–14.00% of the total FAs), C20:3 (9.56–13.61% of the total FAs) and C18:1n9c (9.03–12.30% of the total FAs). Furthermore, the contents of caprylic acid (C8:0), tetracosanoic acid (C24:0) and C24:1 in the SD mice in the HMVO group were all lower than those in the CMVO group ($p < 0.05$), as a consequence of the fact that C8:0 and C24:1 in the HMVO group might be absorbed more quickly by the SD mice. C24:1 has been verified to be related to the early development of premature infants, whose brain maturation is associated with the C24:1 concentration in sphingomyelin [27]. However, for the FA profile in the SD mice's feces, we only found 11 FA species, among which butyric acid (C4:0), caproic acid (C6:0), C8:0, C18:1n9c and eicosatetraenoic acid (C20:4n6) were significantly influenced ($p < 0.05$). The percentages of C4:0, C6:0, C8:0 and C20:4n6 in the SD mice in the HMVO group were significantly lower than those in the CMVO group, while the C18:1n9c concentration in the CMVO group was higher, which indicates that the HMVO could promote the absorption of these FAs mentioned above.

2.4.3. Concentration of Calcium and Magnesium in Feces

Because calcium and magnesium ions were combined with free FAs to form calcium soap in the intestine, the concentrations of calcium and magnesium were detected in this research (Table S6). The results showed that the concentrations of calcium and magnesium were 9.72 mg/g and 0.54 mg/g in the HMVO group, respectively, and both were lower than those in the CMVO group (calcium: 10.05 mg/g; magnesium: 0.60 mg/g).

2.5. The Influence on Intestinal Flora

2.5.1. Alpha (α) and Beta Diversity Analysis

Alpha (α) diversity refers to the diversity within a particular region or ecosystem, including the Chao richness estimator index (Chao), the Sobs index (Sobs), the Shannon–Wiener diversity index (Shannon) and the Simpson diversity index (Simpson). Chao and Sobs both belong to operational taxonomic units (OUTs) [28]. In this study, the values of Chao and Sobs in the HMVO group were 326.75 and 384.58, respectively, which were significantly higher than those in the CMVO group (268.75 for Chao and 311.31 for Sobs,

$p < 0.05$), indicating that the microbial richness was obviously affected by the HMVO when compared with the CMVO. Beta diversity analysis is used for species diversity through inter-group comparative analysis and further explores the similarity or difference between different microbial communities. It was evident that the intestinal flora did not overlap in the different groups according to the principal component analysis (PCA) (Figure 1A), indicating that the intestinal flora of the HMVO and CMVO groups was different, with an Anosim value of 0.2483.

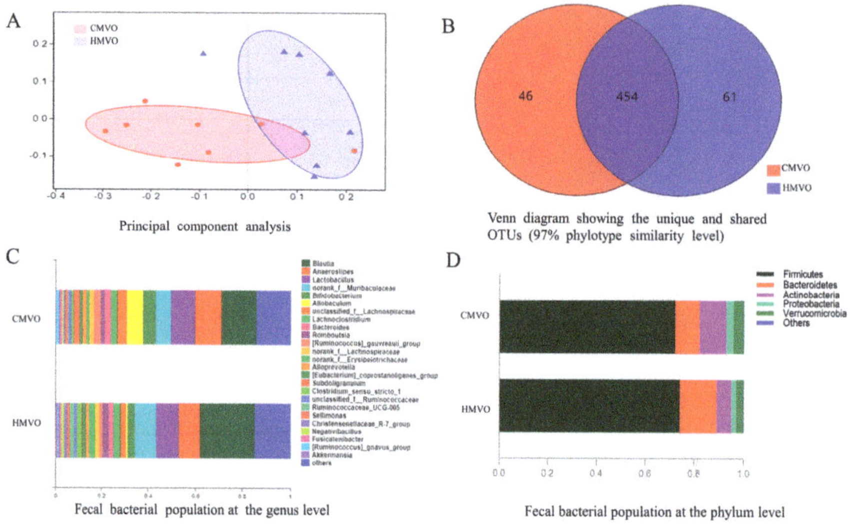

Figure 1. (**A**) shows the significant differences in the different groups through a principal component analysis; (**B**) shows the unique and shared OTUs in the different groups; (**C**) shows the fecal bacterial population at the phylum level in the different groups; and (**D**) shows the fecal bacterial population at the genus level in the different groups. HMVO = homemade mixed vegetable oil; CMVO = commercial mixed vegetable oil; OTUs: operational taxonomic units.

2.5.2. Species Composition Analysis

In order to know the similarity and overlap of the species compositions in the different samples, a Venn diagram can be used to count the number of common and unique species in multiple samples. It was evident that the number of OTUs in the HMVO and CMVO groups was 515 and 500, respectively (Figure 1B). A total of 454 OTUs in the rat fecal microflora were found in both the HMVO and CMVO groups, indicating that the microbial composition of these two groups was similar. Moreover, 46 and 61 unique OTUs were observed in the CMVO and HMVO groups, respectively. Figure 1C,D show the sample classification at the phylum and genus levels in the different groups. At the phylum classification level, firmicutes, bifidobacteria and actinobacteria were the most important microorganism species. The firmicutes and bifidobacteria were the dominant microorganisms in the HMVO and CMVO groups, whose percentages were more than 85% of the total microbial flora. Through the 16S r DNA gene method, genera were mainly detected in all the groups, and the blautia, anaerostipes, lactobacillus, muribaculaceae and bifidobacteria were the main bacteria with the highest abundances. Furthermore, the percentages of blautia and muribaculaceae in the CMVO group were lower than those in the HMVO group, while the mice fed with the HMVO had higher abundances of anaerostipes, lactobacillus and bifidobacteria.

2.5.3. Species Difference Analysis

There was no significant difference between the abundances of firmicutes, bacteroidetes, and actinobacteria in the HMVO and CMVO groups (Figure 2A,B). However, the abundances of chloroflexi and acidobacteria in the HMVO group were significantly higher compared with the CMVO group ($p < 0.05$), while the abundances of chloroflexi and acidobacteria in the HMVO group were lower compared with other microorganisms at the phylum level. At the genus level, the abundance of almost all microorganisms in the HMVO group was higher than that in the CMVO group, except allobaculum and subdoligranulum. The concentration of blautia, which can clear gas in the gut, was the most abundant. Another dominant bacteria was Ruminococcus, whose main fermentation metabolites were acetic acid and formic acid, and they mainly obtained energy by absorbing monosaccharides and degrading mucin. Therefore, based on the analyses of the different classification levels of the fecal microflora of the rats, the intestinal microflora of the rats fed the HMVO was more beneficial to the healthy development of the intestinal flora compared with the rats fed with the CMVO.

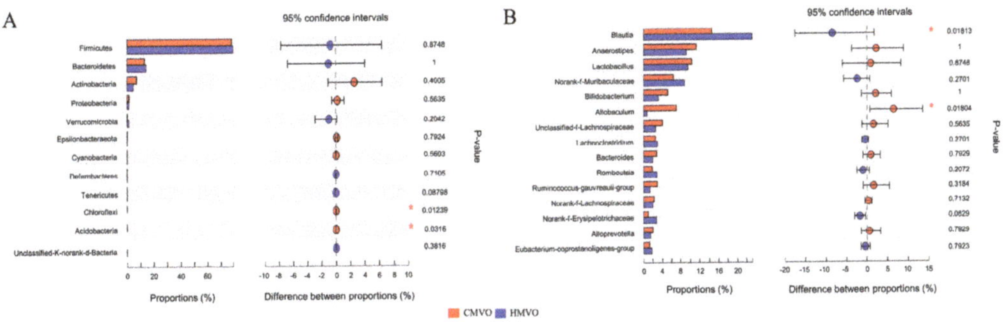

Differences in phylum classification of faecal microbial flora in rats

Differences in genus classification level of faecal microbial flora in rats

Figure 2. (**A**) shows the differences in the phylum classification of the fecal microbial flora in the different groups; (**B**) shows the differences in the genus classification level of the fecal microbial flora in the different groups. HMVO = homemade mixed vegetable oil; CMVO = commercial mixed vegetable oil; * means the significant difference between CMVO and HMVO.

3. Materials and Methods

3.1. Materials

The vegetable oils (walnut oil, palm oil, linseed oil, soybean oil, rapeseed oil, sunflower oil and maize oil) were purchased from the local market, and the internal standard (1,3 (d 5)-diheptadecanoyl-2-heptadecenoyl-glycerol (d 5-(17:0/17:1/17:0) TAGs)) was obtained from Avanti Polar Lipids (Birmingham, AL, USA). Methanol (MeOH), hydrochloric acid (HCL), ethyl alcohol, n-Hexane, acetonitrile, isopropanol (IPA), chloroform ($CHCl_3$), formic acid and ammonium formate were bought from Fisher Scientific (Pittsburgh, PA, USA). The standard mixture of 37 FA methyl esters (FAMEs) was purchased from ANPEL Laboratory Technologies Inc. (Shanghai, China).

3.2. Fatty acid Analysis

The vegetable oils were weighed to 30 mg in a glass tube, and then n-Hexane (1 mL), methanol (2 mL) and solution (methanol/HCL, 3/1, v/v) were added and then shaken for 2 min. The tube was transferred to a water bath (100 °C) for 1 h, and it was cooled to room temperature after heating. After that, distilled water (2 mL) was pipetted and vortexed for 1 min, and then the tube was separated by centrifugation (1500× g, 5 min), and the organic liquid was obtained for gas chromatography analysis (GC, Agilent 8890B, Agilent Technologies Co., Ltd., Santa Clara, CA, USA). The specific procedure for the GC column (DB-23 60 m × 0.25 mm × 0.25 μm; Sigma-Aldrich, St. Louis, MI, USA) was taken from previous studies and modified. Briefly, the initial temperature was set at 50 °C (1 min) and then increased to 180 °C at a rate of 20 °C/min; after that, the temperature still increased to 230 at a rate of 1.3 °C/min and was kept constant for 5 min [29,30].

3.3. *Triacylglycerol Analysis*

All vegetable oils (5 mg) were dissolved in solution ($CHCl_3$/methanol, 1/2, *v*/*v*) to 5 mg/mL, and then it was diluted (50 times) with methanol, which contained ammonium formate (10 mM), formic acid (0.1%) and internal standard (1,3(d5)-diheptadecanoyl-2-heptadecenoyl-glycerol (d5-(17:0/17:1/17:0) TAGs). For the setting condition of UPLC, mobile phase A was mixed with ACN and water (5:5, *v*/*v*), and mobile phase B was blended with IPA and ACN (9:1, *v*/*v*). They all contained ammonium formate (10 mM) and formic acid (0.1%). The chromatographic column was BEH C18 (1.7 μm, 2.1 mm ID × 100 mm, Waters Corporation, Milford, MA, USA), and the column temperature was 60 °C. The parameters of MS (4500QTrap, AB SCIEX, Framingham, MA, USA) are reported in detail in our previous works [3].

3.4. *Establishment of Evaluation Model*

3.4.1. Determination of the Evaluation Conditions

The principle of the similarity evaluation model concerns the optimal value distance method, which is a type of comprehensive evaluation based on distance detection. The FA and TAG composition data of the HMF were obtained from our previous works (Table S1) [3]. For the first evaluation condition group, the TAGs were selected, including 18 kinds of TAG molecular species whose average percentages were higher than 1% of the total TAGs. The second evaluation condition group was FA species, a total of 13 FA species (the proportion of these FAs was higher than 1% of the total FAs), and some PUFAs, such as DHA. Moreover, it was reported that if C14:0, C16:0 and C18:0 were connected at the Sn-2 position, these FAs would be absorbed easily by infants, and feces hardening would be reduced. Therefore, the percentages of Sn-2 C14:0, Sn-2 C16:0 and Sn-2 C18:0 were regarded as the third evaluation condition group. Finally, some special structured TAGs, whose FA species connected at the Sn-1/3 position of the TAG, were easily absorbed and digested by infants. Therefore, the fourth evaluation condition group was the specially structured TAGs, including TAGs connected with three UFAs (UUU), TAGs connected with two UFAs at the Sn-1/3 position and one SFA at the Sn-2 position (USU), and other TAG-connected SFAs with carbon numbers less than 14 at the Sn-1/3 position (Table S1).

3.4.2. Establishment of Evaluation Model

The similarity degree of the HMFS was assessed using the "deduction" principle. Due to the fact that TAGs account for a high proportion (98%) of the HMF, their highest simulation score was the highest, which was set at 50 points. The FAs were the basic components of the TAGs, so their highest simulation score was set at 25 points. The "Sn-2 FA" and the "special TAG structures" were the functional components for the development of the infants, so their highest simulation scores were set at 12.5 points and 12.5 points, respectively.

Evaluation condition:

$$G = G1 + G2 + G3 + G4 \tag{1}$$

$$G1 = 50 - \sum Ei(TAG) \tag{2}$$

$$G2 = 25 - \sum Ei(FA) \tag{3}$$

$$G3 = 12.5 - \sum Ei(Sn\text{-}2\ FA) \tag{4}$$

$$G4 = 12.5 - \sum Ei(\text{Special structure TAG}) \quad (5)$$

$$Ei = 50\ Ei = 50\ (Ci \frac{Di}{\sum Di}) \quad (6)$$

$$Ci = \frac{|Bi - Ai|}{Ai} \quad (7)$$

G is the total score of the HMFS, with a highest score of 100; G1, G2, G3 and G4 are the scores of the TAGs, FAs, Sn-2 FAs, and special TAG structures, respectively, with highest scores of 50, 25, 12.5 and 12.5, respectively; Ei is the deducted score corresponding to each evaluation index; Ci is the floating coefficient of each evaluation index; Di is the average value of the corresponding evaluation index in the HMF; and Bi is the value of each evaluation index of the HMFS, mixed with different proportions of vegetable oils. Ai is the value of the range of each evaluation index of the HMF; when Bi is within the range of breast milk, Ai is 0; when Bi is higher than the maximum value of the evaluation index, Ai takes the maximum value of the evaluation index; and when Bi is lower than the minimum value of the evaluation index, Ai takes the minimum value of the evaluation index.

3.4.3. Establishment of the Computer Software

The computer program was used to calculate the optimal formula for mixed vegetable oils, which was based on the C++ programming language. Three to five kinds of vegetable oils were selected, which is in accordance with commercial mixed vegetable oils. The minimum and maximum of each vegetable oil in each mixed formula were set at 0.001% and 100%, so the software was calculated from 0.001%, and the sum of the different vegetable oils was 100%. After obtaining a mixed vegetable oil, it was scored by the similarity evaluation model to determine the similarity between itself and the HMF.

3.5. *In Vivo Evaluation of Nutritional Function*

3.5.1. Animals, Experiment Design and Treatments

Sixteen SD mice (male, 21–28-day-old at the start of the experiment, from Beijing Huafukang Bioscience CO., LTD., Beijing, China) were housed under a 12 h/12 h light/dark cycle at 22 ± 2 °C with 55 ± 5% humidity. These mice were randomly divided into two groups, namely the HMVO (n = 8) and CMVO (n = 8) groups. To adapt to the environment, these mice were fed with an AIN-93 diet one week before the experiment, and the feeding time was 28 days, during which the mice were free to eat and drink. Their body weight and food intake were recorded every 3 days, and the food utilization rate was recorded in grams of body weight increased per 100 g feed (g weight/100 g feed). On the 28th day, all rats were sacrificed by cervical dislocation. Intestinal contents, feces and colon tissue were immediately collected in sterile freezer tubes and frozen in liquid nitrogen for further analysis. All animal-related experiments were approved by the Animal Care and Use Committee of the Institute of Food Science and Technology, Chinese Academy of Agricultural Sciences (IFST-2022-124).

3.5.2. Blood Routine and Serum Biochemical Indices

On the last day of feeding, the mice fasted for 12 h, their blood was collected and the blood routine indices were determined with a blood cell counter: RBCs, HGB, HCT, MCH, MCV, MCHC, PLTs, WBCs, LYMs, NEUTs, MONOs, EOs, BAEOs and RETs. An automatic blood biochemical analyzer was used to detect the serum biochemical indices, including CHO, TG, LDL-c and HDL-c.

3.5.3. FA Profile of Intestinal Contents and Feces

The intestinal contents or feces (50 mg) and solution (12 mL, $CHCl_3$/methanol, 2/1, v/v) were added into a tube, and they were shaken dramatically (2 min); after that, 4 mL of

$CHCl_3$ and 4 mL of distilled water were subsequently added and vortexed for 1 min. The mixing solution was centrifuged (3500× *g*, 10 min), and then organic liquid was obtained and dried by nitrogen concentration [31]. Finally, the methyl esterification of the FAs in the mice's intestinal contents or feces was carried out according to the procedure in Section 3.2.

3.5.4. Determination of Calcium in Feces

The feces were dried by freeze drying; they weighed about 10 mg and were transferred to a microwave digestion tank. Then, 5 mL of nitric acid was added, and it was left to stand for 1 h. Then, the tank was put under heated conditions (150–170 °C) for 4 h. After digestion, the sample was ultra-sounded for 5 min at 100 °C, and it was then filled to a constant volume (25 mL) for inductively coupled plasma mass spectrometry (ICP-MS). The specific parameters of the ICP-MS were as follows: The radio-frequency power was 1500 w and the flow of plasma, carrier gas, auxiliary gas and helium was 15 L/min, 0.8 L/min, 0.4 L/min and 4 mL/min, respectively. The temperature of the atomizing chamber was 2 °C, and the lifting rate was 0.3 r/s.

3.5.5. Metagenomic Detection of Intestinal Flora

A DNA extraction kit was used to extract DNA from fecal samples, and specific primers with "5 'barcode-primer-barcode 3'" were synthesized according to the specified sequencing region (V3–V4). The polymerase chain reaction products were quantified with a multifunctional enzyme marker, and then the DNA library was constructed and sequenced. The original data were filtered to obtain the optimized sequence. Afterward, the sequenced results were combined, and then they were analyzed via quantitative insights into the microbial ecology (QIIME) package, version 1.9.1, from two perspectives: an OUT cluster analysis and a taxonomic analysis. Based on the results of the OTU analysis, the diversity index and sequencing depth were detected, and a statistical analysis of the community structure was conducted at each taxonomic level. A consistency analysis was performed by combining the OTU and taxonomy analyses, and we further obtained the basic analysis results for the OTU of each group and its corresponding taxonomic lineage.

3.6. Statistical Analysis

Each vegetable oil was bought repeatedly, and they were detected three times and expressed as mean values ± standard deviation. A one-way ANOVA was performed to analyze the data, and the significant differences among groups were determined by the Duncan multiple-range tests (SPSS 24.0 software, Chicago, IL, USA). GraphPad Prism 8 (San Diego, CA, USA) was used for figure processing.

4. Conclusions

In conclusion, 33 FA species and 63 TAG molecular species were detected in seven different vegetable oils, and C16:0, C18:1n9c, C18:2n6c, C18:3n6, 18:1/16:0/18:1, 18:2/16:0/18:2, 18:1/18:1/18:1, 18:1/18:2/18:1 and 18:2/18:0/18:2 were the main species among the FAs and TAGs in the vegetable oils. Through the software and similarity evaluation model, the optimal formula for mixed vegetable oils was composed of maize oil (43.4%), rapeseed oil (29.9%), palm oil (21.3%), linseed oil (2.8%) and soybean oil (2.6%), with the highest similarity score of 83.146. The in vivo experiment showed that the weight, blood routine indices and calcium and magnesium concentrations of the mice in the HMVO and CMVO groups were similar. Moreover, the percentages of C8:0 and C24:1 in the intestinal content and feces were lower in the HMVO group compared with the CMVO group. For the microbial composition, higher abundances of anaerostipes, lactobacillus and bifidobacterium were observed in the HMVO group. These results verified the practicability of the optimal mixed vegetable oils through the integration of the similarity evaluation model and computer software. Furthermore, we should focus on the optimization of the evaluation model to accelerate the development of HMFS in the future and find more evaluation methodologies to detect similarities between HMF and HMFS.

Supplementary Materials: The following supporting information can be downloaded at https://www.mdpi.com/article/10.3390/molecules29092096/s1: Tables S1–S6: Conditions of mixed plant oil to mimic human milk; Top 1000 formulas output by the software; Food utilization and body weight of SD rats during the experiment; Results of routine blood and blood biochemistry test in each group; Fatty acids of contents in feces and small intestine of SD rats (%); The concentration of calcium and magnesium in each group (mg/g).

Author Contributions: X.P. and J.L.: conceptualization and validation. H.Z. and P.Z.: methodology. H.Z. and Y.W.: software and investigation. X.W., H.Z. and Y.W.: formal analysis. X.P. and J.L.: resources. H.Z.: data curation, writing—original draft preparation and visualization. H.Z., P.Z., X.W., Y.W., S.Z., X.P. and J.L.: writing—review and editing. X.P. and J.L.: supervision. X.P. and J.L.: project administration. X.P. and J.L.: funding acquisition. All authors have read and agreed to the published version of the manuscript.

Funding: This work was supported by the China Agriculture Research System—the National Dairy Industry and Technology System (CARS-36), the National Center of Technology Innovation for Dairy (2022-10, 2023-KFKT-21), the National Key R&D Program of China (2021YFD2100700) and the National Natural Science Foundation of China (Grant No. 32302137, No. 32172190).

Institutional Review Board Statement: The animal study protocol was approved by the Animal Care and Use Committee of the Institute of Food Science and Technology, Chinese Academy of Agricultural Sciences (IFST-2022-124).

Informed Consent Statement: Not applicable.

Data Availability Statement: The raw data supporting the conclusions of this article will be made available by the authors on request.

Conflicts of Interest: The authors declare no conflicts of interest.

References

1. Thum, C.; Wall, C.; Day, L.; Szeto, I.M.Y.; Li, F.; Yan, Y.; Barnett, M.P.G. Changes in Human Milk Fat Globule Composition Throughout Lactation: A Review. *Front. Nutr.* **2022**, *9*, 835856. [CrossRef]
2. Liu, Q.; Zhao, J.; Liu, Y.; Qiao, W.; Jiang, T.; Liu, Y.; Yu, X.; Chen, L. Advances in analysis, metabolism and mimicking of human milk lipids. *Food Chem.* **2022**, *393*, 133332. [CrossRef]
3. Zhao, P.; Zhang, S.; Liu, L.; Pang, X.; Yang, Y.; Lu, J.; Lv, J. Differences in the triacylglycerol and fatty acid compositions of human colostrum and mature milk. *J. Agric. Food Chem.* **2018**, *66*, 4571–4579. [CrossRef]
4. Kim, K.-M.; Park, T.-S.; Shim, S.-M. Optimization and validation of HRLC-MS method to identify and quantify triacylglycerol molecular species in human milk. *Anal. Methods* **2015**, *7*, 4362–4370. [CrossRef]
5. Kallio, H.; Nylund, M.; Boström, P.; Yang, B. Triacylglycerol regioisomers in human milk resolved with an algorithmic novel electrospray ionization tandem mass spectrometry method. *Food Chem.* **2017**, *233*, 351–360. [CrossRef]
6. Yuan, T.; Qi, C.; Dai, X.; Xia, Y.; Sun, C.; Sun, J.; Yu, R.; Zhou, Q.; Jin, Q.; Wei, W. Triacylglycerol composition of breast milk during different lactation stages. *J. Agric. Food Chem.* **2019**, *67*, 2272–22787. [CrossRef]
7. Chen, Y.J.; Zhou, X.H.; Han, B.; Li, S.M.; Xu, T.; Yi, H.X.; Liu, P.; Zhang, L.W.; Li, Y.Y.; Jiang, S.L.; et al. Composition analysis of fatty acids and stereo-distribution of triglycerides in human milk from three regions of China. *Food Res. Int.* **2020**, *133*, 109196. [CrossRef]
8. Karrar, E.; Albakry, Z.; Mohamed Ahmed, I.A.; Chen, C.; Yan, Y.; Brennan, C.S.; Li, J. The composition, characterisation, and health impacts of human milk lipid medium- and long-chain TAGs. *Int. J. Food Sci. Technol.* **2023**, *58*, 6192–6201. [CrossRef]
9. Delplanque, B.; Gibson, R.; Koletzko, B.; Lapillonne, A.; Strandvik, B. Lipid Quality in Infant Nutrition: Current Knowledge and Future Opportunities. *J. Pediatr. Gastroenterol. Nutr.* **2015**, *61*, 8–17. [CrossRef]
10. Wei, W.; Sun, C.; Jiang, W.; Zhang, X.; Hong, Y.; Jin, Q.; Tao, G.; Wang, X.; Yang, Z. Triacylglycerols fingerprint of edible vegetable oils by ultra-performance liquid chromatography-Q-ToF-MS. *LWT* **2019**, *112*, 108261. [CrossRef]
11. Liu, Z.; Dai, L.; Liu, D.; Du, W. Progress and perspectives of enzymatic preparation of human milk fat substitutes. *Biotechnol. Biofuels Bioprod.* **2022**, *15*, 118. [CrossRef]
12. Ghide, M.K.; Li, K.; Wang, J.; Abdulmalek, S.A.; Yan, Y. Effective Production of Human Milk Fat Substitutes Rich in 1,3-Dioleoyl-2-palmitoyl Glycerol (OPO) viaa New Strategy. *Food Biophys.* **2022**, *17*, 495–507. [CrossRef]
13. Wan, J.; Hu, S.; Jacoby, J.J.; Liu, J.; Zhang, Y.; Yu, L.L. The impact of dietary sn-2 palmitic triacylglycerols in combination with docosahexaenoic acid or arachidonic acid on lipid metabolism and host faecal microbiota composition in Sprague Dawley rats. *Food Funct.* **2017**, *8*, 1793–1802. [CrossRef]

14. Chen, Q.; Xie, Q.; Jiang, C.; Evivie, S.E.; Cao, T.; Wang, Z.; Zhao, L.; Liang, S.; Li, B.; Huo, G. Infant formula supplemented with 1,3-olein-2-palmitin regulated the immunity, gut microbiota, and metabolites of mice colonized by feces from healthy infants. *J. Dairy Sci.* **2022**, *105*, 6405–6421. [CrossRef]
15. Yaron, S.; Shachar, D.; Abramas, L.; Riskin, A.; Bader, D.; Litmanovitz, I.; Bar-Yoseph, F.; Cohen, T.; Levi, L.; Lifshitz, Y.; et al. Effect of high β-palmitate content in infant formula on the intestinal microbiota of term infants. *J. Pediatr. Gastroenterol. Nutr.* **2013**, *56*, 376–381. [CrossRef]
16. Qiao, B.; Li, X.; Zheng, T.; Tan, Z. Different Effects of Lard and Vegetable Blend Oil on Intestinal Microorganisms, Enzyme Activity and Blood Routine in Mice. *J. Oleo Sci.* **2022**, *71*, 301–310. [CrossRef]
17. Wang, Y.; Mai, Q.; Qin, X.; Yang, B.; Wang, Z.; Chen, H. Establishment of an evaluation model for human milk fat substitutes. *J. Agric. Food Chem.* **2010**, *58*, 642–649. [CrossRef]
18. Hokkanen, S.; Frey, A.D.; Yang, B.; Linderborg, K.M. Similarity Index for the Fat Fraction between Breast Milk and Infant Formulas. *J. Agric. Food Chem.* **2022**, *70*, 6191–6201. [CrossRef]
19. Guerrero-Esperanza, M.; Wrobel, K.; Wrobel, K.; Ordaz-Ortiz, J.J. Determination of fatty acids in vegetable oils by GC-MS, using multiple-ion quantification (MIQ). *J. Food Compos. Anal.* **2023**, *115*, 104963. [CrossRef]
20. Lu, C.H.; Huang, X.Y.; Shao, J.; Pei, D. A review: Role of mass spectrometry in the quality and authenticity identification of olive oil. *Int. J. Food Sci. Technol.* **2022**, *57*, 7592–7606. [CrossRef]
21. Hussain, M.; Sun, Y.; Pan, Y.; Liu, L.; Zhang, X.; Wang, Q.; Lin, S.; Qayum, A.; Hussain, K.; Li, X. Formulation, invitro digestive study, and comparative fatty acid analysis of walnut oil-based infant formula, with human milk, animal milk, and commercial infant formula. *Innov. Food Sci. Emerg. Technol.* **2023**, *84*, 103279. [CrossRef]
22. Li, H.; Han, J.; Zhao, Z.; Tian, J.; Fu, X.; Zhao, Y.; Wei, C.; Liu, W. Roasting treatments affect oil extraction rate, fatty acids, oxidative stability, antioxidant activity, and flavor of walnut oil. *Front. Nutr.* **2023**, *9*, 1077081. [CrossRef]
23. Campos, J.R.; Severino, P.; Ferreira, C.S.; Zielinska, A.; Santini, A.; Souto, S.B.; Souto, E.B. Linseed Essential Oil—Source of Lipids as Active Ingredients for Pharmaceuticals and Nutraceuticals. *Curr. Med. Chem.* **2019**, *26*, 4537–4558. [CrossRef] [PubMed]
24. Khalid, W.; Gill, P.; Arshad, M.S.; Ali, A.; Ranjha, M.M.A.N.; Mukhtar, S.; Afzal, F.; Maqbool, Z. Functional behavior of DHA and EPA in the formation of babies brain at different stages of age, and protect from different brain-related diseases. *Int. J. Food Prop.* **2022**, *25*, 1021–1044. [CrossRef]
25. Leskinen, H.; Suomela, J.P.; Kallio, H. Quantification of triacylglycerol regioisomers in oils and fat using different mass spectrometric and liquid chromatographic methods. *Rapid Commun. Mass Spectrom. RCM* **2007**, *21*, 2361–2373. [CrossRef]
26. Indelicato, S.; Bongiorno, D.; Pitonzo, R.; Di Stefano, V.; Calabrese, V.; Indelicato, S.; Avellone, G. Triacylglycerols in edible oils: Determination, characterization, quantitation, chemometric approach and evaluation of adulterations. *J. Chromatogr. A* **2017**, *1515*, 1–16. [CrossRef]
27. Li, Q.; Chen, J.; Yu, X.; Gao, J.-M. A mini review of nervonic acid: Source, production, and biological functions. *Food Chem.* **2019**, *301*, 125286. [CrossRef]
28. Guo, Q.; Hou, X.; Cui, Q.; Li, S.; Shen, G.; Luo, Q.; Wu, H.; Chen, H.; Liu, Y.; Chen, A.; et al. Pectin mediates the mechanism of host blood glucose regulation through intestinal flora. *Crit. Rev. Food Sci. Nutr.* **2023**, 1–23. [CrossRef]
29. Ahmed, T.B.; Eggesbø, M.; Criswell, R.; Uhl, O.; Demmelmair, H.; Koletzko, B. Total Fatty Acid and Polar Lipid Species Composition of Human Milk. *Nutrients* **2022**, *14*, 158. [CrossRef]
30. Dao, X.; Zhang, D.; Wang, L.; Wang, L. Analysis of human milk fatty acid composition and its correlation with diet pattern (A study in Tibetan population gathering area). *J. Food Compos. Anal.* **2023**, *116*, 105046. [CrossRef]
31. Folch, J.; Lees, M.; Stanley, G.H.S. A simple method for the isolation and purification of total lipides from animal tissues. *J. Biol. Chem.* **1957**, *226*, 497–509. [CrossRef] [PubMed]

molecules

MDPI

Article

Direct and Sensitive Electrochemical Determination of Total Antioxidant Capacity in Foods Using Nanochannel-Based Enrichment of Redox Probes

Lixia Duan [1,†], Chaoyan Zhang [2,†], Fengna Xi [2,*], Danke Su [1] and Wenhao Zhang [1,*]

1 Guangxi Medical University Cancer Hospital, Guangxi Medical University, Nanning 530021, China; duanlixia@sr.gxmu.edu.cn (L.D.); sudanke33@sina.com (D.S.)

2 School of Chemistry and Chemical Engineering, Zhejiang Sci-Tech University, Hangzhou 310018, China; 202130107346@mails.zstu.edu.cn

* Correspondence: fengnaxi@zstu.edu.cn (F.X.); zwh836@163.com (W.Z.)

† These authors have contributed equally to this work.

Abstract: Simple and sensitive determination of total antioxidant capacity (TAC) in food samples is highly desirable. In this work, an electrochemical platform was established based on a silica nanochannel film (SNF)-modified electrode, facilitating fast and highly sensitive analysis of TAC in colored food samples. SNF was grown on low-cost and readily available tin indium oxide (ITO) electrode. Fe^{3+}-phenanthroline complex-Fe(III)(phen)$_3$ was applied as the probe, and underwent chemical reduction to form Fe^{2+}-phenanthroline complex-Fe(II)(phen)$_3$ in the presence of antioxidants. Utilizing an oxidative voltage of +1 V, chronoamperometry was employed to measure the current generated by the electrochemical oxidation of Fe(II)(phen)$_3$, allowing for the assessment of antioxidants. As the negatively charged SNF displayed remarkable enrichment towards positively charged Fe(II)(phen)$_3$, the sensitivity of detection can be significantly improved. When Trolox was employed as the standard antioxidant, the electrochemical sensor demonstrated a linear detection range from 0.01 μM to 1 μM and from 1 μM to 1000 μM, with a limit of detection (LOD) of 3.9 nM. The detection performance is better that that of the conventional colorimetric method with a linear de range from 1 μM to 40 μM. Owing to the anti-interfering ability of nanochannels, direct determination of TAC in colored samples including coffee, tea, and edible oils was realized.

Keywords: total antioxidant capacity; electrochemical determination; silica nanochannel film; colored sample; food sample

Citation: Duan, L.; Zhang, C.; Xi, F.; Su, D.; Zhang, W. Direct and Sensitive Electrochemical Determination of Total Antioxidant Capacity in Foods Using Nanochannel-Based Enrichment of Redox Probes. *Molecules* **2024**, *29*, 2423. https://doi.org/10.3390/molecules29112423

Academic Editors: Liqiang Luo, Yongliang Zhuang, Hu Hou, Zhipeng Yu and Shengbao Cai

Received: 31 March 2024
Revised: 6 May 2024
Accepted: 17 May 2024
Published: 21 May 2024

1. Introduction

Oxidative stress is closely related to health. Oxidative stress refers to the accumulation of excessive free radicals or oxidizing substances in the body, exceeding the body's natural defense mechanisms, potentially causing harm to human health [1]. For instance, oxidative stress is associated with many diseases, especially cardiovascular diseases, cancer, and diabetes. Additionally, oxidative stress can lead to oxidative damage in cells and tissues and contribute to the aging process [2]. To combat oxidative stress, the human body relies on antioxidant systems, which include endogenous antioxidants and antioxidants obtained through dietary intake [3,4]. Commonly, natural foods are rich in antioxidants, such as vitamin C (VC), vitamin E, carotenoids, and polyphenols. In addition, antioxidants also serve as essential food additives. Thus, consuming foods abundant in antioxidants can help maintain the balance between oxidation and antioxidation, reducing the risk of diseases [4]. The development of selective and sensitive techniques for measuring the antioxidant capacity of foods has become increasingly important.

The total antioxidant capacity (TAC) reflects the comprehensive antioxidant ability of all antioxidants in a sample, and it is a crucial indicator for measuring the overall antioxidant content in food [5–7]. Specifically, a higher TAC value indicates that the sample has a stronger ability to resist oxidative stress and free radical damage, contributing to maintaining health and reducing the risk of diseases. Achieving accurate and rapid determination of TAC in food is of great significance. Traditional instrumental analysis methods such as high-performance liquid chromatography (HPLC), gas chromatography (GC), and surface-enhanced Raman spectrum (SERS) can accurately measure the types and concentrations of antioxidants [8–11]. However, these methods are often limited by their expensive equipment, time-consuming operations, and the need for specialized technical expertise. Colorimetric methods for TAC determination have also been developed [12]. For instance, analysis based on phenanthroline (phen), 2,2′-azino-bis(3-ethylbenzothiazoline-6-sulfonic acid (ABTS) radical cation, or 2,2-diphenyl-1-picrylhydrazyl (DPPH) has been developed [13,14]. Although colorimetric analysis is easy to operate, it cannot directly detect turbid or colored samples. Recently, electrochemical detection has gained widespread attentions. Electrochemical sensors offer advantages such as low equipment cost, high determination sensitivity, rapid response, and ease of miniaturization and integration [15–17]. Electrochemical sensors have been utilized for the assessment of TAC in different samples [18,19]. For instance, TAC in herb or beverage has been quantified utilizing a nanocomposite-graphene/poly (3,4-ethylenedioxythiophene): poly (styrenesulfonate)-modified screen-printed carbon electrode (G/PEDOT:PSS/SPCE) based on DPPH assay [18]. Multi-walled carbon nanotubes (MWCNTs)-modified electrode was also employed for the determination of TAC in berries using amperometric tyrosinase biosensor [19]. However, when detecting samples with complex matrices, electrodes are susceptible to contamination. On the one hand, biomolecules in the sample (such as proteins, DNA, RNA) or large particles tend to non-specifically adsorb to the electrode surface, thereby reducing the determination sensitivity and accuracy [20–22]. On the other hand, the presence of redox species in the sample can lead to interference in the electrochemical signals [23]. Therefore, the development of electrochemical sensor for direct analysis of TAC in food samples without complex sample pretreatment crucial.

Using a barrier film on the electrode surface is an effective means to reduce or eliminate the influence of sample matrices, thus maintaining the stability and accuracy of electrode [24–26]. In recent years, porous materials have gained significant attention due to their high surface area and size-selective capabilities [27–30]. Silica nanochannel film (SNF) is a nanoscale ultrathin film with a thickness that can be adjusted between 20 to 200 nm [31,32]. It has a vertically aligned and ordered array of nanochannels with uniform nanochannel diameter (typically ranging from 2 to 3 nm) and a high pore density (approximately 75,000 nanochannels/cm^2) [33]. As a silica thin film, SNF possesses good mechanical stability and is highly compatible with 2D planar electrode, making it suitable for modification of the electrode [34,35]. Additionally, the ultrasmall nanochannel of SNF offers selective sieving capability at the molecular level for the supporting electrode [36–38]. On the one hand, SNF-modified electrodes exhibit outstanding resistance to contamination, making them promising for direct electrochemical analysis of complex samples [39–41]. This is attributed to the electrical insulating properties of silica material and the size-selective function of its ultrasmall nanochannels, which only permit small molecules to pass through, effectively excluding larger biological molecules and solid particles [42–44]. In addition, SNF-modified electrodes possess charge-selective permeability. The nanoscale thickness and ultrahigh pore density of SNF ensure efficient diffusion and mass transport. However, unlike microscale pores, nanochannels also exhibit charge-selective capabilities at the molecular level, allowing only small molecules with matching charges to pass through [23]. This significantly reduces interference from co-existing electroactive substances and enables the enrichment of specific probe or analytes [45–47]. For instance, SNF contains numerous silanol groups with a pK_a value of approximately 2, and the negatively charged surface generated after ionization can electrostatically interact with positively charged analytes, leading to remarkable enrichment [48–50]. Therefore, SNF can preconcentrate small analytes or electrochemical probes with opposite charges and reduce interference or fouling from matrix in complex samples.

In this work, a sensitive electrochemical sensor for the determination of total antioxidant capacity (TAC) in foods was demonstrated based on SNF-modified electrode. SNF was grown on a low-cost and readily available indium-tin oxide (ITO) electrode. When 6-hydroxy-2,5,7,8-tetramethylchroman-2-carboxylic acid (Trolox) was employed as the model antioxidant, the Fe^{3+}-phenanthroline complex (Fe(III)(phen)$_3$) probe can react with Trolox to produce Fe^{2+}-phenanthroline complex (Fe(II)(phen)$_3$), leading to electrochemical signal at a constant oxidation potential. Based on this principle, the determination of the concentration of antioxidants can be achieved by measuring the oxidation current corresponding to the produced Fe(II)(phen)$_3$. The enrichment of Fe(II)(phen)$_3$ based on charge-based permeability of nanochannels and the resulting enhancement in detection sensitivity were investigated. The anti-interference capability of the SNF/ITO electrode owing to size-exclusion effect of nanochannels was studied. The detection performance of the fabricated sensor for TAC determination in colored food samples such as coffee, tea, and sesame oil was assessed.

2. Results and Discussion

2.1. Construction of An Nanochannel-Modified Electrode for Electrochemical Determination of TAC

In this work, an electrochemical platform was established for detection the total antioxidant capacity (TAC) of food samples using a SNF-modified electrode with Fe(III)(phen)$_3$ as the probe. As illustrated in Figure 1, a layer of SNF was grown on the surface of the ITO electrode using the Stöber solution growth method, which can realize the preparation of large-area SNF film with long-range order and excellent coverage [51]. Typically, the positively charged cationic surfactant, hexadecyltrimethylammonium bromide (CTAB), is adsorbed onto the negatively charged ITO and assembles into spherical micelles (SM). Subsequently, under the influence of ammonia and ethanol, siloxane precursor (TEOS) undergoes hydrolysis, forming negatively charged silicate oligomers, that further condense around the SM template. As ethanol molecules diffuse into the CTAB micelles, the alkyl interactions at the tail end of the micelles weaken, leading to an increase in micelle volume and the formation of cylindrical micelles. Then, the resulting SNF has vertically aligned nanochannels sealed by SM on the supporting ITO electrode (SM@SNF/ITO). A washing step in hydrochloric acid–ethanol solution is employed to remove the SM, resulting in electrode with open nanochannels (SNF/ITO). Due to the abundant presence of silanol groups (Si-OH) with low pK_a (~2), SNF carries a net negative charge in commonly used medium. Coupled with high surface area of SNF, SNF/ITO can significantly enrich positively charged redox probes.

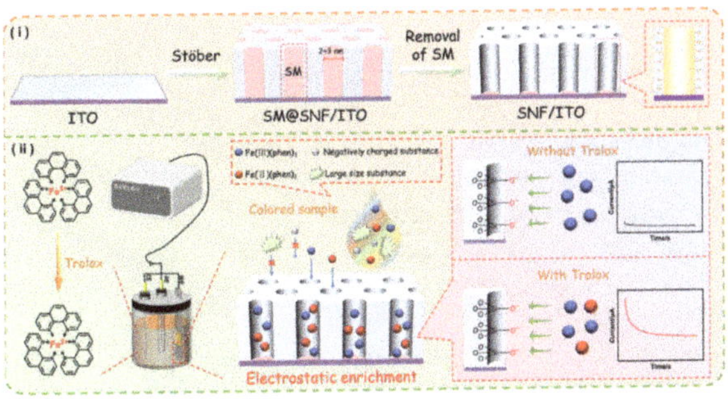

Figure 1. Schematic illustration for the preparation of the SNF-modified ITO electrode (**i**) and electrochemical determination of TAC (**ii**) using nanochannel-based enrichment of Fe(II)(phen)$_3$.

Currently, TAC can ben detected based on the colorimetric assay using phenanthroline, where the mechanism involves the reduction of Fe^{3+} to Fe^{2+} by antioxidants in the sample.

The resulting Fe^{2+} forms a stable orange-red complex with phenanthroline (Fe(II)(phen)$_3$), generating a characteristic absorption peak at 520 nm. By measuring the absorbance of Fe(II)(phen)$_3$ at 520 nm, the TAC in the sample can be calculated. Inspired by this method, the electrochemical oxidation of Fe(II)(phen)$_3$ is measured to correlate with the antioxidant content in samples, thus establishing an electrochemical method for TAC detection. Specifically, to achieve electrochemical determination of TAC, Fe(III)(phen)$_3$ was used as the probe, which can be chemically reduced to Fe(II)(phen)$_3$ by antioxidants. When the model antioxidant Trolox was present, the conversion from Fe(III)(phen)$_3$ to Fe(II)(phen)$_3$ was controlled by the redox potential of Fe^{3+}/Fe^{2+} in presence of phenanthroline ligand. Since Fe^{2+} (with a d^6 electron configuration) is a proficient cationic electron donor, phenanthroline ligand with low-lying vacant antibonding orbitals can form strong complexes with it, whereas Fe^{3+}, due to its higher charge, has inferior electron-donating properties. Consequently, phenanthroline forms a stronger and more stable complex with Fe^{2+}. Clearly, the content of Trolox is correlated with the generation of Fe(II)(phen)$_3$. In other words, the more antioxidants present, the greater the amount of Fe(II)(phen)$_3$ produced. To distinguish the two substances, chronoamperometry (current–time curve, *I*–t) was employed when different potentials were applied. Specifically, when the potential is set to the oxidation potential, the Fe(II)(phen)$_3$ is oxidized, resulting in a current signal, allowing for the quantitative determination of Fe(II)(phen)$_3$, and consequently the quantification of Trolox. In addition, the SNF/ITO electrode, which significantly enriches Fe(II)(phen)$_3$, enhances the current as well as the detection sensitivity. Due to the size-exclusion effect of the ultrasmall nanochannels in SNF, large macromolecules or particles can not interfere the determination. Combined with the advantage of electrochemical determination that is not affected by sample color, the constructed sensor can determination TAC in color or tumid samples. It also eliminates the need for long incubation and sample pretreatment processes, enabling fast determination.

2.2. *Morphological and Electrochemical Characterization of SNF/ITO*

The morphology and thickness of SNF were characterized using TEM and SEM. In the top-view TEM image of SNF (Figure 2a), it can be observed that the nanochannel size is uniformly arranged in a worm-like pattern. No defects are observed within the observed range. The pore diameter was measured to be approximately 2.6 nm, with a pore density of ~$8 \times 10^{12}/cm^2$, corresponding to a porosity of approximately 42%. The cross-sectional TEM image (Figure 2b) indicates that the film thickness is approximately 90 nm. Figure 2c,d represent the top-view and the cross-sectional view SEM images of SNF/ITO. It is evident that the SNF surface is smooth with no large particles, demonstrating that no silica nanoparticles are generated during the SNF growth process. The SEM image of cross-section of the SNF/ITO electrode reveals the three-layered structure (Figure 2d). In addition to the glass and ITO layers of the ITO electrode, the SNF layer exhibits a thickness of around 90 nm.

To confirm the integrity and charge-selective permeability of SNF, redox probes with opposite charges were chosen for cyclic voltammetry (CV) testing. The two redox probes used were the standard anionic probe $Fe(CN)_6^{3-}$ and the cationic probe $Ru(NH_3)_6^{3+}$. As shown in Figure 3a,b, when using SM@SNF/ITO as the working electrode, almost no Faradaic signal is detected in either probe solution because the supporting electrode is covered by an insulating SNF, and SM templates are present inside the nanochannels. Thus, the mass transfer of the redox probes to the underlying electrode is blocked. This indicates that the SNF grown on the ITO electrode surface is complete and has no cracks. After removing the SM template from the nanochannels, the Faradaic signal of the cationic probe $Ru(NH_3)_6^{3+}$ on the SNF/ITO electrode significantly increases and is greater than the current signal on the ITO electrode. Consequently, the Faradaic signal of the anionic probe $Fe(CN)_6^{4-}$ is noticeably suppressed and significantly smaller than that on the ITO electrode. It is evident that SNF exhibits remarkable permeation selectivity for redox probes with different charges. Specifically, the negatively charged nanochannels of

SNF electrostatically repel negatively charged small molecules and electrostatically enrich positively charged probes.

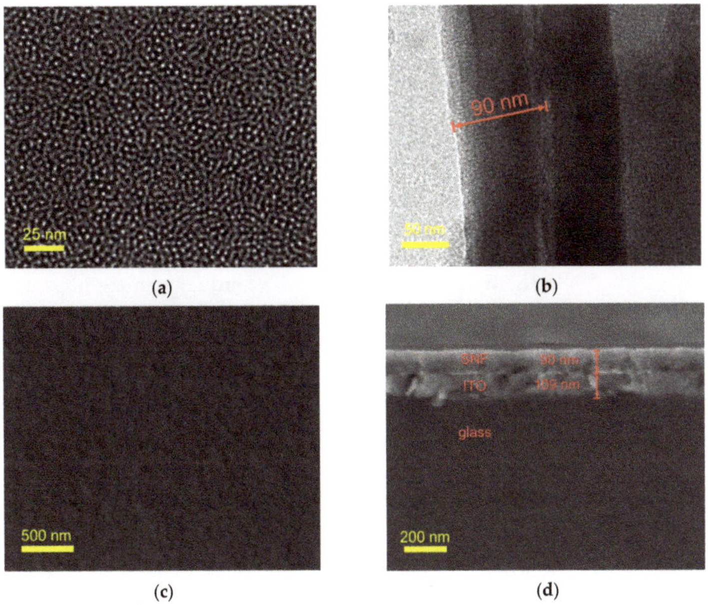

Figure 2. (**a**,**b**) Top-view and cross-sectional TEM images of SNF. (**c**,**d**) SEM image of the surface and the cross-section of SNF/ITO.

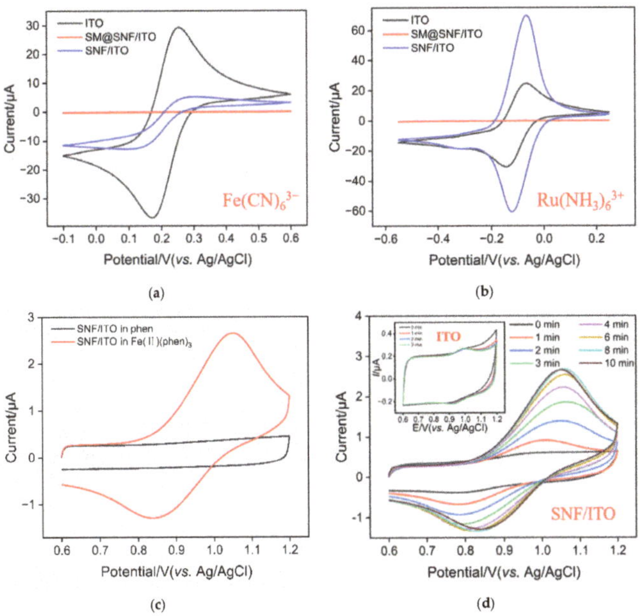

Figure 3. CV curves obtained on ITO, SNF/ITO and SM@SNF/ITO in potassium hydrogen phthalate (KHP) containing 50 μM $Fe(CN)_6^{3-}$ (**a**) or $Ru(NH_3)_6^{3+}$ (**b**). (**c**) CV curves obtained on SNF/ITO in phen (4 μM) or Fe(II)(phen)$_3$ (4 μM) solution. (**d**) CV curves obtained on ITO (inset) or the SNF/ITO electrode after stirring in Fe(II)(phen)$_3$ (1 μM) for different time.

2.3. Oxidation Potential of Fe(II)(phen)$_3$ and Its Enrichment on the SNF/ITO Electrode

To investigate the oxidation–reduction potential of Fe(II)(phen)$_3$, CV curves obtained on SNF/ITO electrodes in phenanthroline or Fe(II)(phen)$_3$ solution were investigated. As shown in Figure 3c, a significant Faradaic current is observed in the Fe(II)(phen)$_3$ solution, while there was almost no Faradaic current in the phenanthroline solution. This confirms that the redox peaks originate from the oxidation of iron atoms in the complex. From the CV curve, it can also be seen that the oxidation reaction of Fe(II)(phen)$_3$ occurs at approximately 0.7 V and reaches its peak current at around 1.05 V.

Figure 3d displayed CV curves obtained on bare ITO or SNF/ITO electrodes in Fe(II)(phen)$_3$ solution with stirring for different time. As shown, only a weak signal was observed on the ITO electrode. Even after stirring for 3 min, the signal is closely matched that obtained without stirring, indicating no enrichment in the stirring process. In contrast, the SNF/ITO electrode shows a gradual increase in peak current with increasing stirring time, reaching equilibrium at approximately 8 min of stirring. This current value is approximately 10 times that measured on the ITO electrode, indicating that SNF/ITO exhibits a significant enrichment effect on Fe(II)(phen)$_3$. Owing to the abundant silanol groups, SNF carries a negative charge and thus demonstrates a pronounced enrichment effect on the positively charged Fe(II)(phen)$_3$, resulting in a significantly increased peak current.

2.4. Feasibility Validation for the Electrochemical Sensor and Condition Optimization

When the SNF/ITO electrode was immersed in Fe(III)(phen)$_3$ solution, Trolox was added and current–time curves (*I*–t) in absence or presence of Trolox were measured using chronoamperometry when a constant oxidative potential (1 V) was applied. Results obtained on the ITO electrode were also measured as control. As shown in Figure 4a, both the bare ITO and SNF/ITO electrodes do not exhibit noticeable signals in absence of Trolox. Obviously, in a solution containing only the oxidized form of Fe(III)(phen)$_3$, the application of oxidative potential does not generate Faradaic current. However, upon the addition of Trolox to the solution, a portion of Fe(III)(phen)$_3$ is chemically reduced to Fe(II)(phen)$_3$. Consequently, the application of a continuous oxidative potential results in a current response, attributed to the electrochemical oxidation of Fe(II)(phen)$_3$ in the solution. Thus, when an antioxidant is added, Fe(III)(phen)$_3$ is chemically reduced to Fe(II)(phen)$_3$, and the oxidation current of Fe(II)(phen)$_3$ can be measured by subsequently applying an oxidative potential. This current is positively correlated with the concentration of Fe(II)(phen)$_3$ as well as the amount of added antioxidant. In addition, the signal obtained on the SNF/ITO electrode in presence of Trolox increased by 8 fold (0.1115 μA) compared to that of the ITO electrode (0.0137 μA). Thus, the nanochannel layer improve the signal owning to the enrichment of Fe(II)(phen)$_3$. Based on this, an electrochemical platform for sensitive TAC determination can be constructed.

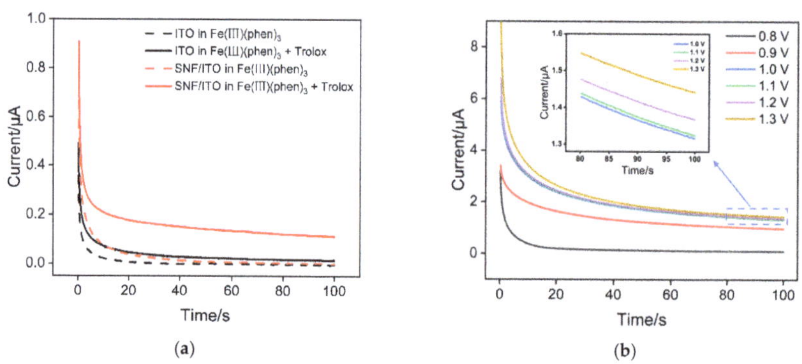

Figure 4. (**a**) *I*–t curves obtained on ITO or SNF/ITO in Fe(III)(phen)$_3$ (2 mM) with or without Trolox (1 μM). (**b**) *I*–t curves obtained on SNF/ITO by applying different potentials in Fe(III)(phen)$_3$ solution containing 100 μM Trolox.

The influence of the applied potential was investigated. As shown in Figure 4b, when a lower potential is applied, the current signal obtained increases with increasing potential. When the potential exceeds +1 V, the current signal stabilizes, indicating that at this potential, all Fe(II)(phen)3 has been oxidized. Therefore, a potential of +1 V was consistently used in subsequent experiments for measuring the oxidation current of Fe(II)(phen)3.

2.5. Electrochemical Determination of Trolox

Owing to similar structure to a form of vitamin E (α-tocopherol), Trolox is commonly applied as standard in antioxidant determination. Upon reacting Fe(III)(phen)$_3$ with various concentrations of Trolox to form Fe(II)(phen)$_3$, a +1 V oxidative potential was applied to measure the electrochemical oxidation current of Fe(II)(phen)$_3$ using chronoamperometry. The results are shown in Figure 5a. It is evident that higher concentrations of added Trolox result in higher electrochemical oxidation currents for Fe(II)(phen)$_3$. The linear regression curves in Figure 5b demonstrate that there is a good linear relationship between the current (I) and Trolox concentration (C) in the ranges of 0.01 μM to 1 μM and 1 μM to 1000 μM. At lower Trolox concentrations (0.01 μM to 1 μM), higher sensitivity is achieved (I = 0.0672C + 0.0472, R^2 = 0.999). This is due to the more pronounced enrichment effect of SNF/ITO electrodes on Fe(II)(phen)$_3$ at lower concentrations. At higher Trolox concentrations, the concentration of Fe(II)(phen)$_3$ on the electrode surface tends to saturate, resulting in slightly lower sensitivity compared to that of low concentration (I = 0.0152C + 0.0535, R^2 = 0.986). The determination limit calculated based on a three signal-to-noise (S/N = 3) is 3.9 nM. The detection of Trolox using the colorimetric method was also investigated. As shown in Figure 5c, the linear determination range was from 1 μM to 40 μM. The linear determination range and the lowest detectable concentration are not as good as that of the electrochemical method. Table S1 (in Supplementary Materials) summarizes the comparison between Trolox detection performance using different methods [12,19,52–54]. The LOD obtained using the fabricated sensor is lower than that obtained by differential pulse voltammetry (DPV) detection using a copper-neocuproine complex ([Cu(Nc)$_2$]$^{2+}$) probe [52], or a square-wave voltammetry (SWV) detection using a 2,2-diphenyl-1-picrylhydrazyl (DPPH) probe [19], or electrochemiluminescence (ECL) detection based on luminol emitter [53,54], or colorimetric detection using Fe(III)(phen)$_3$ probe [12]. Thus, the fabricated sensor also has advantages of convenient fabrication, simple detection, and high sensitivity, demonstrating great potential for TAC determination.

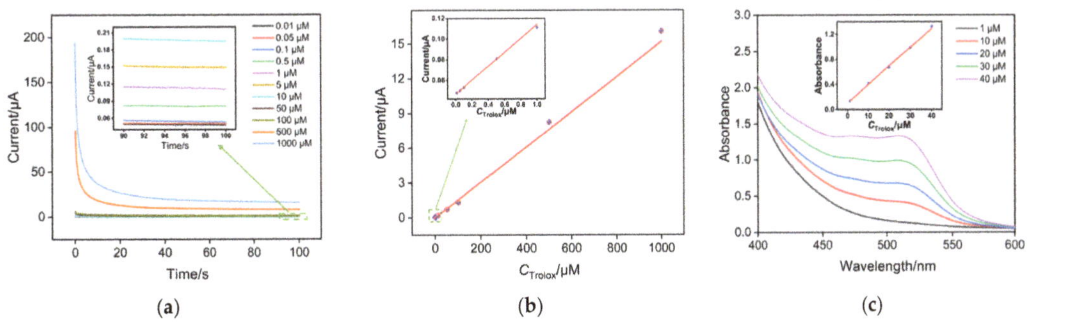

Figure 5. (**a**) I–t curves obtained on SNF/ITO electrodes in Fe(III)(phen)$_3$ (2 mM) with different concentrations of Trolox. Inset is a magnified view of the I–t curve in the low concentration region. (**b**) Calibration curves for Trolox determination. Inset shows a magnified view at low concentrations. (**c**) Spectra obtained for determination of different concentrations of Trolox using the colorimetric method. Inset is the corresponding linear regression curve. Error bars represent the standard deviations of three measurements.

2.6. Real Sample Analysis

To validate the capability of the prepared electrodes for real sample analysis, coffee, tea, and sesame oil were selected as colored samples to assess their TAC using a standard addition method. TAC was evaluated using Trolox as a reference. Specifically, each sample was directly added to the Fe(III)(phen)$_3$ detection solution. Then, different concentrations of Trolox standard solutions were added. The electrochemical signals were measured using SNF/ITO electrodes to obtain a linear regression curve between the electrochemical signal and Trolox concentration. Extrapolating this linear regression curve, the absolute value of its intersection with the x-axis (intercept in x-axis) corresponds to the equivalent concentration of Trolox of the detected solution. Figure 6a–c showed the linear relationship between the current (I) and the Trolox concentration (C) for different samples. The TAC values in the analysis solution of coffee, tea, and sesame oil samples were found to be 17.62 μM, 43 μM, and 11.18 μM, respectively. These correspond to 251.71 μmol/g, 860 μmol/g, and 22.36 μmol/g with respect to the starting coffee, tea, and sesame oil, respectively. The insets in Figure 6a–c display photos of the analyzed samples, demonstrating that the electrochemical sensor developed in this study can be directly used for the determination of colored samples. In addition, the electrochemical detection obtained on the fabricated sensor was also compared with those obtained from the phen-based colorimetric method. Considering the possible errors in detection of colored samples by the colorimetric method, two colorless samples including the diluted VC tablet solution and diluted lemon juice were detected (inset of Figure 6d). TAC of each sample was measured by the electrochemical method (I–t method) using the fabricated sensor or the colorimetric method. As shown in Figure 6d, the results obtained from both detection methods for two samples were nearly identical, demonstrating the reliability of the electrochemical detection of TAC by the sensor constructed in this study.

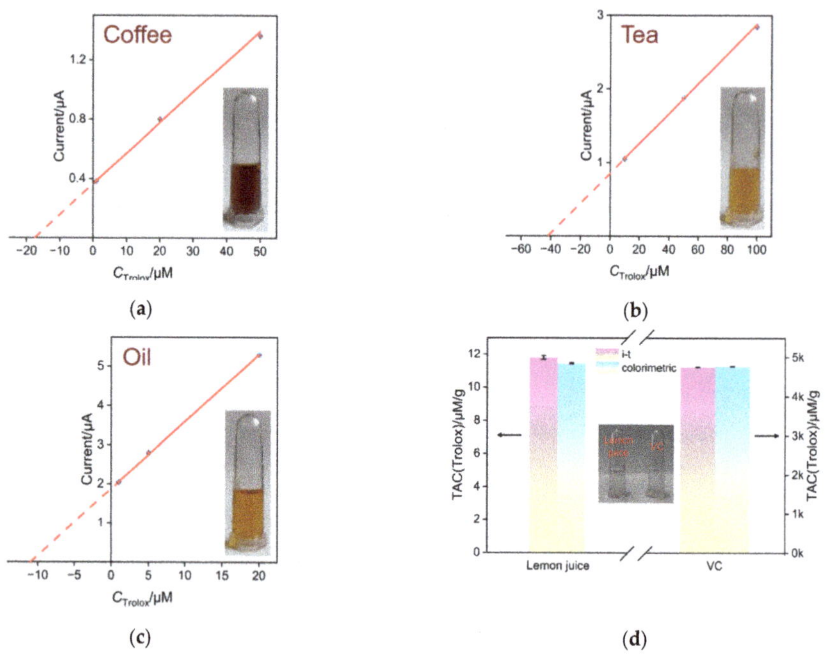

Figure 6. The linear relationship obtained for analysis coffee (**a**), tea (**b**), and sesame oil (**c**) samples. Inset in each figure displays photograph of the analyzed samples. (**d**) TAC of VC tablets or lemon detected by the I–t method using the fabricated sensor or standard colorimetric method. The inset shows the digital photo of the detected solution. TAC was calculated according to the weight of the original VC tablet or lemon.

2.7. Regeneration Performance of the Developed Electrochemical Sensor

The regeneration performance of the developed sensor was examined. After electrochemical determination in a solution containing Trolox, SNF/ITO electrodes had adsorbed a significant amount of Fe(III)$(phen)_3$ or Fe(II)$(phen)_3$ molecules in their nanochannels. Subsequently, they were immersed in an HCl–ethanol solution and stirred for 3 min to remove the adsorbed molecules. After washing the electrodes, electrochemical determination was carried out in a buffer solution or the detection solution. As shown in Figure 7a, the regenerated electrode exhibits minimal current signal in the buffer solution, demonstrating that the adsorbed substances on the electrode could be thoroughly removed through a simple washing process. When the regenerated electrode is re-used in the detection solution, the signal remains nearly unchanged. After five regenerations, the relative standard deviation (RSD) of the measured current signal is 2.9%, indicating good regeneration performance.

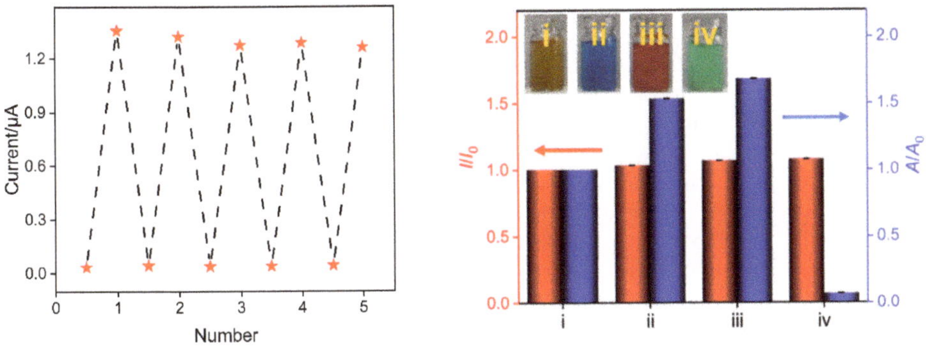

(a) (b)

Figure 7. (**a**) The regeneration performance of the developed electrochemical sensor. The signal points below in the figure represent the peak currents measured by the regenerated electrodes in the buffer solution. The signal points above represent the peak currents measured by the regenerated electrodes in the determination solution. (**b**) The relative signal ratio obtained using electrochemical or colorimetric determination. I and I_0 and A and A_0 represent the current signal or absorbance in the presence and absence of colored matrix, respectively. Inset shows photographs of colored samples. The Fe(III)$(phen)_3$ detection solution (i) was added with methylene blue (ii), phenol red (iii), or nickel chloride (iv) to simulate colored matrices.

2.8. Comparison between Electrochemical and Colorimetric Methods for Analyze Colored Samples

The determination performance obtained using the constructed electrochemical sensor or common phen-based colorimetric determination was compared for Trolox determination in colored samples. The results are shown in Figure 7b. The Fe(III)$(phen)_3$ detection solution (i) was added with methylene blue (ii), phenol red (iii), or nickel chloride (iv) to simulate colored matrices, and the determination performance of Trolox in these colored matrices was investigated. I_0 and I represent the current signals obtained in the absence and presence of colored substances, while A_0 and A represent the absorbance obtained in the absence and presence of colored matrices. It can be observed that the presence of colored substances significantly affects the determination results of the colorimetric method, while the results of electrochemical determination are hardly affected. This demonstrates that our determination method can be used for accurate determination of colored samples.

3. Materials and Methods

3.1. Chemicals and Materials

Trolox, phenoline hydrochloride (phen), ammonium ferric sulfate dodecahydrate ($NH4Fe(SO_4)_2 \bullet 12H_2O$), ferrous chloride ($FeCl_2$), tetraethoxysilane (TEOS, 98%), potassium ferricyanide ($K_3[Fe(CN)_6]$), ruthenium hexamethonium chloride ($Ru(NH_3)_6Cl_3$), and

3-aminopropyltriethoxysilane (GPTMS) were purchased from Aladdin Biochemical Technology Co., Ltd. (Shanghai, China). Cetyltrimethylammonium bromide (CTAB), acetone, ethanol (99.8%) and concentrated hydrochloric acid (HCl, 38%) were purchased from Kermel Chemical Reagent Company (Tianjin, China). All reagents were used without further purification. To synthesize Fe(III)(phen)$_3$, a solution of $NH_4Fe(SO_4)_2$ (3 mL, 0.5 M, dissolved in 1 M HCl) was combined with phen solution (15 mL, 0.4 M). To ensure complete coordination of the iron ions, an excess of the phen solution was employed. The obtained solution was diluted to 2 mM (based on the molar quantity of the iron ions) before use. The Trolox stock solution (0.1 M) was prepared in ethanol and diluted to achieve desired concentrations.

Indium tin oxide (ITO) conductive glass (square resistance <17 Ω/sq, ITO thickness: 100 ± 20 nm) was purchased from Kaiwei Optoelectronics Technology Co., Ltd. (Zhuhai, China). A simple pretreatment was carried out before use. Briefly, ITO glass was cut to a size of 2.5 cm × 5 cm, sonicated for 60 min in NaOH (1 M), followed by ultrasonic cleaning with acetone, ethanol and deionized water for 30 min in turn.

3.2. *Characteriaztions and Instrumentations*

Transmission electron microscopy (TEM) and scanning electron microscopy (SEM) were employed to examine the structural characteristics of the synthesized SNF material. TEM analysis was conducted using a Hitachi HT7700 microscope operating at an acceleration voltage of 200 kV. For TEM sample preparation, the SNF layer was meticulously scraped off the surface of an ITO substrate using a sharp blade and subsequently dispersed in ethanol via ultrasonication. The resulting dispersion was carefully deposited onto a copper grid, allowed to dry under a lamp, and then subjected to TEM imaging. SEM characterization was carried out on a Zeiss ULTRA 55 microscope using an acceleration voltage of 5 kV. UV–vis absorption spectroscopy (UV–vis) was performed with a Shimadzu UV-2450 spectrometer. Electrochemical characterization, including cyclic voltammetry (CV) or amperometry (I–t) test, was conducted using a CHI660D electrochemical workstation (Chinstrument, Shanghai, China). These electrochemical measurements were carried out with three commercial electrodes, featuring an Ag/AgCl reference electrode saturated with potassium chloride as the reference electrode, a platinum wire as the counter electrode, and ITO- or SNF-modified electrodes as the working electrode.

3.3. *Preparation of the SNF-Modified ITO Electrode*

SNF modification layer were grown through the Stöber solution growth method on the supporting ITO electrode [51,55,56]. Specifically, CTAB (0.160 g) was completely dissolved in a mixture of 70 mL of ethanol and 30 mL of water. Subsequently, 100 μL of 10% ammonia and 80 μL of TEOS were added to the solution under continuous stirring to obtain a clear and transparent precursor solution without any visible bubbles. Then, clean ITO glass (2.5 cm × 5 cm) was immersed in the precursor solution. SNF was grown at 60 °C for 24 h. After the growth process was completed, the electrode was removed, thoroughly rinsed with deionized water, dried using nitrogen gas. Then, the obtained electrode was placed in an oven at 100 °C overnight for aging, resulting in a SNF-modified electrode containing micelle (SM) template within the nanochannels. The obtained electrode was denoted as SM@SNF/ITO. The electrode was then cut, and its area was controlled to 0.5 cm × 1 cm using insulating glue. To remove the SM template, the SM@SNF/ITO electrode was immersed in a 0.1 M HCl–ethanol solution and stirred for 5 min. The resulting electrode featuring open channels was designated as SNF/ITO.

3.4. *Electrochemical Determination of Trolox*

Electrochemical determination was performed using a Fe(III)(phen)$_3$ solution (2 mM) as the determination solution. Specifically, different concentrations of Trolox were introduced into the Fe(III)(phen)$_3$ solution. Then, a fixed potential of +1 V was applied, and the oxidation current produced by the chemically reduced Fe(II)(phen)$_3$ was measured using chronoamperometry.

3.5. Real Sample Analysis

In the analysis of real samples, coffee (Shefei Brand, obtained from local market, Hangzhou, China), tea leaves (Anji Brand, obtained from local market, Hangzhou, China) and sesame oil (Golden Brand, Nanjing, China) require no complex pretreatment. Specifically, coffee was prepared by dissolving coffee powder in deionized water (1 g/100 mL). Tea leaves were soaked in distilled water (10 g/100 mL) for 30 min, and the supernatant was collected. Sesame oil was dispersed in ethanol using ultrasonication (10 g/100 mL). These samples were mixed directly with the Fe(III)(phen)$_3$ determination solution for analysis. TAC in the VC tablet or lemon was also analyzed. Specifically, the VC tablet was dissolved in ultrapure water and diluted to 1.1 mg/mL. The lemon juice squeezed out from lemon was diluted with ultrapure water to 0.1 g/mL. TAC was then detected by the standard addition method using the electrochemical method constructed in this work or the phen-based colorimetric method using characteristic absorption of Fe(II)(phen)$_3$ at 520 nm. TAC was calculated based on the weight of the initial sample and expressed as the Trolox equivalent per gram of sample (μM/g).

4. Conclusions

In summary, a rapid and sensitive electrochemical platform was established for determination of TAC in food samples based on nanochannel-modified electrodes. Employing the Stöber solution growth method, multiple nanochannel-modified electrodes (SNF/ITO) can be readily prepared on low-cost and easily obtainable ITO electrodes. The used Fe(III)(phen)$_3$ probe can be chemically reduced to Fe(II)(phen)$_3$ in the presence of antioxidants, which can be electrochemically oxidized at a high potential. Determination of antioxidants was achieved through the measurement of the oxidation current of Fe(II)(phen)$_3$. Due to the electrostatic enrichment of Fe(III)(phen)$_3$ by the negatively charged nanochannels of SNF, the SNF/ITO electrode exhibits significantly enhanced electrochemical signals as well as the detection sensitivity. Additionally, the size-exclusion effect of the ultrasmall nanochannels of SNF and the insulating nature of the SiO_2 structure provide the SNF/ITO electrode with good anti-interference capability, enabling direct detection of colored samples. Thus, the constructed electrochemical sensor exhibits good determination performance compared to traditional colorimetric methods. The fabricated sensor can be directly applied to the measurement of TAC in colored samples such as coffee, tea, and edible oils. Owing to the easy fabrication, fast detection, and high sensitivity, the developed electrochemical sensor holds great potential for TAC determination in food samples without the need for complex sample pretreatment.

Supplementary Materials: The following supporting information can be downloaded at: https://www.mdpi.com/article/10.3390/molecules29112423/s1, Table S1: Comparison between Trolox detection performance using different methods.

Author Contributions: Investigation, L.D. and C.Z.; data curation, L.D. and C.Z.; writing—original draft preparation, L.D. and D.S.; conceptualization and supervision, F.X. and W.Z.; writing—review and editing, F.X., D.S. and W.Z. All authors have read and agreed to the published version of the manuscript.

Funding: This study was funded by the Key Research and Development Program of Guangxi (AB23026079) and the Innovation Project of Guangxi Graduate Education (YCBZ2023109).

Institutional Review Board Statement: Not applicable.

Informed Consent Statement: Not applicable.

Data Availability Statement: The data presented in this study are available on request from the corresponding author.

Conflicts of Interest: The authors declare no conflict of interest.

References

1. Griendling, K.K.; Camargo, L.L.; Rios, F.J.; Alves-Lopes, R.; Montezano, A.C.; Touyz, R.M. Oxidative Stress and Hypertension. *Circ. Res.* **2021**, *128*, 993–1020. [CrossRef] [PubMed]
2. Miglioranza Scavuzzi, B.; Holoshitz, J. Endoplasmic reticulum stress, oxidative stress, and rheumatic diseases. *Antioxidants* **2022**, *11*, 1306. [CrossRef] [PubMed]
3. Złotek, U.; Jakubczyk, A.; Gawlik-Dziki, U. Antioxidant in food safety and sustainability. *Foods* **2022**, *11*, 433. [CrossRef]
4. Nedić, O.; Penezić, A.; Minić, S.; Radomirović, M.; Nikolić, M.; Ćirković Veličković, T.; Gligorijević, N. Food antioxidants and their interaction with human proteins. *Antioxidants* **2023**, *12*, 815. [CrossRef]
5. Gilnezhad, J.; Firoozbakhtian, A.; Hosseini, M.; Adel, S.; Xu, G.; Ganjali, M.R. An enzyme-free Ti(3)C(2)/Ni/Sm-LDH-based screen-printed-electrode for real-time sweat detection of glucose. *Anal. Chim. Acta* **2023**, *1250*, 340981. [CrossRef]
6. Liang, Y.; Li, R.; Sun, H.; Dan, J.; Su, Z.; Kang, Y.; Zhang, Q.; Shi, S.; Wang, J.; Zhang, W. Functionalized natural melanin nanoparticle mimics natural peroxidase for total antioxidant capacity determination. *Sens. Actuators B Chem.* **2022**, *359*, 131541–131547. [CrossRef]
7. Tolou-Ghamari, Z.; Mazdak, H. Determination of total antioxidant capacity in patients with bladder cancer and its correlation with the results of pathology. *Adv. Hum. Biol.* **2023**, *13*, 223–226. [CrossRef]
8. Cacho, J.I.; Campillo, N.; Viñas, P.; Hernández-Córdoba, M. Determination of synthetic phenolic antioxidants in edible oils using microvial insert large volume injection gas-chromatography. *Food Chem.* **2016**, *200*, 249–254. [CrossRef]
9. Tu, Q.; Lin, Z.; Liu, J.; Dai, H.; Yang, T.; Wang, J.; Decker, E.; McClements, D.J.; He, L. Multi-phase detection of antioxidants using surface-enhanced Raman spectroscopy with a gold nanoparticle-coated fiber. *Talanta* **2020**, *206*, 120197. [CrossRef]
10. Wang, L.; Luo, Y.; Wu, Y.; Xia, F.; Wu, Z. Quickly verifying the antioxidant contribution of the individual composition in natural antioxidants by HPLC-free radical scavenging detection. *LWT-Food Sci. Technol.* **2018**, *96*, 461–468. [CrossRef]
11. Zhang, X.; Liu, Q.; Chen, Z.; Zuo, X. Colorimetric sensor array for accurate detection and identification of antioxidants based on metal ions as sensor receptors. *Talanta* **2020**, *215*, 120935–120943. [CrossRef] [PubMed]
12. Bener, M.; Apak, R. Ferric-o-phenanthroline adsorbed on a Nafion membrane: A novel optical sensor for antioxidant capacity measurement of food extracts. *Sens. Actuators B Chem.* **2017**, *247*, 155–162. [CrossRef]
13. Kusznierewicz, B.; Staroszczyk, H.; Malinowska-Pańczyk, E.; Parchem, K.; Bartoszek, A. Novel ABTS-dot-blot method for the assessment of antioxidant properties of food packaging. *Food Packag. Shelf* **2020**, *24*, 100478–100484. [CrossRef]
14. Tohidi, B.; Rahimmalek, M.; Arzani, A. Essential oil composition, total phenolic, flavonoid contents, and antioxidant activity of thymus species collected from different regions of Iran. *Food Chem.* **2017**, *220*, 153–161. [CrossRef] [PubMed]
15. Munteanu, I.G.; Apetrei, C. A review on electrochemical sensors and biosensors used in assessing antioxidant activity. *Antioxidants* **2022**, *11*, 584. [CrossRef] [PubMed]
16. Yan, F.; Wang, M.; Jin, Q.; Zhou, H.; Xie, L.; Tang, H.; Liu, J. Vertically-ordered mesoporous silica films on graphene for anti-fouling electrochemical detection of tert-butylhydroquinone in cosmetics and edible oils. *J. Electroanal. Chem.* **2021**, *881*, 114969. [CrossRef]
17. Huang, C.; Zhang, S.; Ma, X.; Yan, F.; Tang, W. Highly sensitive electrochemical determination of butylated hydroxyanisole in food samples using electrochemical-pretreated three-dimensional graphene electrode modified with silica nanochannel film. *Nanomaterials* **2024**, *14*, 569. [CrossRef]
18. Frangu, A.; Ashrafi, A.M.; Sýs, M.; Arbneshi, T.; Metelka, R.; Adam, V.; Vlček, M.; Richtera, L. Determination of trolox equivalent antioxidant capacity in berries using amperometric tyrosinase biosensor based on multi-walled carbon nanotubes. *Appl. Sci.* **2020**, *10*, 2497. [CrossRef]
19. Tirawattanakoson, R.; Rattanarat, P.; Ngamrojanavanich, N.; Rodthongkum, N.; Chailapakul, O. Free radical scavenger screening of total antioxidant capacity in herb and beverage using graphene/PEDOT: PSS-modified electrochemical sensor. *J. Electroanal. Chem.* **2016**, *767*, 68–75. [CrossRef]
20. Zhou, L.; Hou, H.; Wei, H.; Yao, L.; Sun, L.; Yu, P.; Su, B.; Mao, L. In vivo monitoring of oxygen in rat brain by carbon fiber microelectrode modified with antifouling nanoporous membrane. *Anal. Chem.* **2019**, *91*, 3645–3651. [CrossRef]
21. Yan, F.; Chen, J.; Jin, Q.; Zhou, H.; Sailjoi, A.; Liu, J.; Tang, W. Fast one-step fabrication of a vertically-ordered mesoporous silica-nanochannel film on graphene for direct and sensitive detection of doxorubicin in human whole blood. *J. Mater. Chem. C* **2020**, *8*, 7113–7119. [CrossRef]
22. Wang, M.; Lin, J.; Gong, J.; Ma, M.; Tang, H.; Liu, J.; Yan, F. Rapid and sensitive determination of doxorubicin in human whole blood by vertically-ordered mesoporous silica film modified electrochemically pretreated glassy carbon electrodes. *RSC Adv.* **2021**, *11*, 9021–9028. [CrossRef]
23. Zhou, L.; Ding, H.; Yan, F.; Guo, W.; Su, B. Electrochemical detection of Alzheimer's disease related substances in biofluids by silica nanochannel membrane modified glassy carbon electrodes. *Analyst* **2018**, *143*, 4756–4763. [CrossRef]
24. Han, Q.; Zhang, T.; Wang, M.; Yan, F.; Liu, J. Disposable electrochemical sensors for highly sensitive detection of chlorpromazine in human whole blood based on the silica nanochannel array modified screen-printed carbon electrode. *Molecules* **2022**, *27*, 8200. [CrossRef] [PubMed]
25. Huang, J.; Zhang, T.; Dong, G.; Zhu, S.; Yan, F.; Liu, J. Direct and sensitive electrochemical detection of bisphenol A in complex environmental samples using a simple and convenient nanochannel-modified electrode. *Front. Chem.* **2022**, *10*, 900282. [CrossRef]
26. Zhang, C.; Zhou, X.; Yan, F.; Lin, J. N-doped graphene quantum dots confined within silica nanochannels for enhanced electrochemical detection of doxorubicin. *Molecules* **2023**, *28*, 6443. [CrossRef]

27. Zhou, H.; Ma, X.; Sailjoi, A.; Zou, Y.; Lin, X.; Yan, F.; Su, B.; Liu, J. Vertical silica nanochannels supported by nanocarbon composite for simultaneous detection of serotonin and melatonin in biological fluids. *Sens. Actuators B Chem.* **2022**, *353*, 131101. [CrossRef]
28. Gong, J.; Tang, H.; Wang, M.; Lin, X.; Wang, K.; Liu, J. Novel three-dimensional graphene nanomesh prepared by facile electro-etching for improved electroanalytical performance for small biomolecules. *Mater. Design* **2022**, *215*, 110506. [CrossRef]
29. Zhou, H.; Dong, G.; Sailjoi, A.; Liu, J. Facile pretreatment of three-dimensional graphene through electrochemical polarization for improved electrocatalytic performance and simultaneous electrochemical detection of catechol and hydroquinone. *Nanomaterials* **2022**, *12*, 65. [CrossRef]
30. Deng, X.; Lin, X.; Zhou, H.; Liu, J.; Tang, H. Equipment of vertically-ordered mesoporous silica film on electrochemically pretreated three-dimensional graphene electrodes for sensitive detection of methidazine in urine. *Nanomaterials* **2023**, *13*, 239. [CrossRef]
31. Walcarius, A. Electroinduced surfactant self-assembly driven to vertical growth of oriented mesoporous films. *Acc. Chem. Res.* **2021**, *54*, 3563–3575. [CrossRef] [PubMed]
32. Yu, R.; Zhao, Y.; Liu, J. Solid electrochemiluminescence sensor by immobilization of emitter ruthenium(ii)tris(bipyridine) in bipolar silica nanochannel film for sensitive detection of oxalate in serum and urine. *Nanomaterials* **2024**, *14*, 390. [CrossRef] [PubMed]
33. Yan, F.; Luo, T.; Jin, Q.; Zhou, H.; Sailjoi, A.; Dong, G.; Liu, J.; Tang, W. Tailoring molecular permeability of vertically-ordered mesoporous silica-nanochannel films on graphene for selectively enhanced determination of dihydroxybenzene isomers in environmental water samples. *J. Hazard. Mater.* **2021**, *410*, 124636. [CrossRef] [PubMed]
34. Walcarius, A. Mesoporous materials and electrochemistry. *Chem. Soc. Rev.* **2013**, *42*, 4098–4140. [CrossRef] [PubMed]
35. Gong, J.; Zhang, T.; Luo, T.; Luo, X.; Yan, F.; Tang, W.; Liu, J. Bipolar silica nanochannel array confined electrochemiluminescence for ultrasensitive detection of SARS-CoV-2 antibody. *Biosens. Bioelectron.* **2022**, *215*, 114563. [CrossRef] [PubMed]
36. Luo, X.; Zhang, T.; Tang, H.; Liu, J. Novel electrochemical and electrochemiluminescence dual-modality sensing platform for sensitive determination of antimicrobial peptides based on probe encapsulated liposome and nanochannel array electrode. *Front. Nutr.* **2022**, *9*, 962736. [CrossRef]
37. Yan, Z.; Zhang, S.; Liu, J.; Xing, J. Homogeneous electrochemical aptamer sensor based on two-dimensional nanocomposite probe and nanochannel modified electrode for sensitive detection of carcinoembryonic antigen. *Molecules* **2023**, *28*, 5186. [CrossRef]
38. Ma, X.; Liao, W.; Zhou, H.; Tong, Y.; Yan, F.; Tang, H.; Liu, J. Highly sensitive detection of rutin in pharmaceuticals and human serum using ITO electrodes modified with vertically-ordered mesoporous silica-graphene nanocomposite films. *J. Mater. Chem. B* **2020**, *8*, 10630–10636. [CrossRef] [PubMed]
39. Zhou, P.; Yao, L.; Chen, K.; Su, B. Silica Nanochannel Membranes for Electrochemical Analysis and Molecular Sieving: A Comprehensive Review. *Crit. Rev. Anal. Chem.* **2019**, *50*, 424–444. [CrossRef]
40. Cui, Y.; Zhang, S.; Zhou, X.; Yan, F.; Hu, W. Silica nanochannel array on co-electrodeposited graphene-carbon nanotubes 3D composite film for antifouling detection of uric acid in human serum and urine samples. *Microchem. J.* **2023**, *190*, 108632. [CrossRef]
41. Zou, Y.; Zhou, X.; Xie, L.; Tang, H.; Yan, F. Vertically-ordered mesoporous silica films grown on boron nitride-graphene composite modified electrodes for rapid and sensitive detection of carbendazim in real samples. *Front. Chem.* **2022**, *10*, 939510. [CrossRef] [PubMed]
42. Serrano, M.B.; Despas, C.; Herzog, G.; Walcarius, A. Mesoporous silica thin films for molecular sieving and electrode surface protection against biofouling. *Electrochem. Commun.* **2015**, *52*, 34–36. [CrossRef]
43. Nasir, T.; Herzog, G.; Hebrant, M.; Despas, C.; Liu, L.; Walcarius, A. Mesoporous silica thin films for improved electrochemical detection of paraquat. *ACS Sens.* **2018**, *3*, 484–493. [CrossRef]
44. Wang, K.; Yang, L.; Huang, H.; Lv, N.; Liu, J.; Liu, Y. Nanochannel array on electrochemically polarized screen printed carbon electrode for rapid and sensitive electrochemical determination of clozapine in human whole blood. *Molecules* **2022**, *27*, 2739. [CrossRef] [PubMed]
45. Huang, J.; Xu, S.; Yan, F.; Liu, J. Electrochemiluminescence enzyme biosensors for ultrasensitive determination of glucose using glucose dehydrogenase immobilized on vertical silica nanochannels. *Sens. Actuators B Chem.* **2024**, *402*, 135119. [CrossRef]
46. Huang, J.; Zhang, T.; Zheng, Y.; Liu, J. Dual-mode sensing platform for cancer antigen 15-3 determination based on a silica nanochannel array using electrochemiluminescence and electrochemistry. *Biosensors* **2023**, *13*, 317. [CrossRef]
47. Zhang, Y.; Zhang, S.; Liu, J.; Qin, D. Label-free homogeneous electrochemical aptasensor based on size exclusion/charge-selective permeability of nanochannel arrays and 2D nanorecognitive probe for sensitive detection of alpha-fetoprotein. *Molecules* **2023**, *28*, 6935. [CrossRef] [PubMed]
48. Yang, L.; Zhang, T.; Zhou, H.; Yan, F.; Liu, Y. Silica nanochannels boosting $Ru(bpy)_3^{2+}$-mediated electrochemical sensor for the detection of guanine in beer and pharmaceutical samples. *Front. Nutr.* **2022**, *9*, 987442. [CrossRef]
49. Gong, J.; Zhang, T.; Chen, P.; Yan, F.; Liu, J. Bipolar silica nanochannel array for dual-mode electrochemiluminescence and electrochemical immunosensing platform. *Sens. Actuators B Chem.* **2022**, *368*, 132086. [CrossRef]
50. Huang, J.; Fan, X.; Yan, F.; Liu, J. Vertical Silica Nanochannels and o-Phenanthroline Chelator for the Detection of Trace Fe(II). *ACS Appl. Nano Mater.* **2024**, *7*, 7743–7752. [CrossRef]

51. Teng, Z.; Zheng, G.; Dou, Y.; Li, W.; Mou, C.-Y.; Zhang, X.; Asiri, A.M.; Zhao, D. Highly ordered mesoporous silica films with perpendicular mesochannels by a simple stöber-solution growth approach. *Angew. Chem. Inter. Ed.* **2012**, *51*, 2173–2177. [CrossRef] [PubMed]
52. Trofin, A.E.; Trincă, L.C.; Ungureanu, E.; Ariton, A.M. CUPRAC voltammetric determination of antioxidant capacity in tea samples by using screen-printed microelectrodes. *J. Anal. Method Chem.* **2019**, *2019*, 8012758. [CrossRef] [PubMed]
53. Nagatani, N.; Inoue, Y.; Araki, A.; Ushijima, H.; Hattori, G.; Sakurai, Y.; Ogidou, Y.; Saito, M.; Tamiya, E. Rapid sensing of antioxidant capacity based on electrochemiluminescence induced by electrochemically generated reactive oxygen species. *Electrochim. Acta* **2016**, *222*, 580–586. [CrossRef]
54. Bushira, F.A.; Wang, P.; Hussain, A.; Li, H.; Jin, Y. Integration of high-entropy oxide with nitrogen-doped graphene for the ultrasensitive electrochemiluminescence detection of trolox and dopamine. *ACS Appl. Nano Mater.* **2023**, *6*, 4747–4753. [CrossRef]
55. Zhou, X.; Gu, X.; Zhang, S.; Zou, Y.; Yan, F. Magnetic graphene oxide and vertically-ordered mesoporous silica film for universal and sensitive homogeneous electrochemiluminescence aptasensor platform. *Microchem. J.* **2024**, *200*, 110315. [CrossRef]
56. Yan, F.; Ma, X.; Jin, Q.; Tong, Y.; Tang, H.; Lin, X.; Liu, J. Phenylboronic acid-functionalized vertically ordered mesoporous silica films for selective electrochemical determination of fluoride ion in tap water. *Microchim. Acta* **2020**, *187*, 470. [CrossRef]

foods

MDPI

Article

Antibacterial Activity and Mechanism of Self-Assembly Spermidine-Capped Carbon Dots against *Staphylococcus aureus*

Tianqi Cui [1,2], Ya Fan [1], Yaping Liu [1], Xuejing Fan [1], Yuxue Sun [2], Guiguang Cheng [1,*] and Jianjun Cheng [2,*]

1 Faculty of Food Science and Engineering, Kunming University of Science and Technology, Kunming 650550, China
2 College of Food Science, Northeast Agricultural University, Harbin 150030, China
* Correspondence: ggcheng@kmust.edu.cn (G.C.); jjcheng@neau.edu.cn (J.C.)

Abstract: This paper investigated the antibacterial mechanism of spermidine-capped carbon dots (S-PCDs) against *Staphylococcus aureus*. The results showed that there were a large number of amino groups on the surface of S-PCDs and they had a high positive charge (+47.06 mV), which could be adsorbed on the negatively charged bacterial surface through electrostatic interaction and changed the permeability of the bacterial cell membrane. The extracellular protein and nucleic acid contents of *S. aureus* treated with S-PCDs were 5.4 and 1.2 times higher than those of the control group, respectively. The surface folds and defects of the bacterial cell membrane, and the leakage of cell contents were observed using SEM and TEM. The expression of metabolic oxidation regulatory genes *dmpI*, *narJ* and *narK* was upregulated and the intracellular ROS generation was induced, causing bacterial oxidative stress and eventually bacterial death. S-PCDs can effectively inhibit biofilm formation and had low cytotoxicity. The S-PCD treatment successfully inhibited microbial reproduction when pasteurized milk was stored at 25 °C and 4 °C. These results provide important insights into the antimicrobial mechanism of S-PCDs and lay the foundation for their application in the food field as a potentially novel bacteriostatic nanomaterial.

Keywords: carbon dots; antibacterial mechanism; *S. aureus*; membrane damage; ROS generation

Citation: Cui, T.; Fan, Y.; Liu, Y.; Fan, X.; Sun, Y.; Cheng, G.; Cheng, J. Antibacterial Activity and Mechanism of Self-Assembly Spermidine-Capped Carbon Dots against *Staphylococcus aureus*. *Foods* **2024**, *13*, 67. https://doi.org/10.3390/foods13010067

Academic Editor: Isabelle Adt

Received: 25 November 2023
Revised: 19 December 2023
Accepted: 21 December 2023
Published: 23 December 2023

1. Introduction

Food safety is one of the main concerns of the global health and food industry [1]. Foodborne diseases caused by foodborne pathogens account for the highest proportion of food safety incidents [2]. *Staphylococcus aureus* is a classic gram-positive pathogen that produces seven different toxins that are common causes of foodborne illness [3]. *S. aureus* can contaminate milk and its dairy products at all stages of production, processing and storage, leading to foodborne illness and major safety concerns [4]. At present, nanoparticles (NPs) have been widely used as antimicrobial agents in the food field due to their ability to effectively inhibit pathogenic microorganisms' growth [5,6]. However, most NPs are potentially toxic [7]; therefore, the development of safe NPs with high antibacterial activity to inhibit *S. aureus* in food is essential.

Carbon dots (CDs) are a novel carbon nanomaterial with a size below 10 nm, which have been widely used in many fields due to their potential for a wide range of technological applications [8]. The size, surface charge and surface functional groups of CDs affect their antibacterial properties. The surface chemical modification of CDs by specific groups can cause strong interaction between CDs and bacterial cells, thus seriously damaging bacterial cell membranes [9]. In addition, CDs can induce elevated ROS levels which can disrupt normal bacterial metabolism by damaging DNA, proteins and enzymes in cells [10]. Wu et al. (2022) synthesized levofloxacin-based carbon spots (LCDs) using the one-pan hydrothermal method. LCDs have an effective antimicrobial property and destroy bacteria based on the dual antibacterial mode [11]. Different raw materials have great influence on the antibacterial properties of CDs. Polyethyleneimine (PEI) is a kind of compound with antibacterial activity. In addition, the characteristics of polycation, high charge and biocompatibility make spermidine suitable for surface modification of NPs [12]. Azevedo et al. (2014) used PEI to synthesize CDs with anti-biofilm and antibacterial activity against *S. aureus* and other bacteria [13]. Li et al. (2016) adopted a two-step method to synthesize spermidine-based fluorescent carbon quantum dots (SPD-CQDs), which showed strong antibacterial activity against *S. aureus* multidrug-resistant *E. coli* and methicillin-resistant *S. aureus* [14]. In addition to the advantages of simple synthesis method, low cost, good dispersion, controllable surface chemical structure, etc., CDs also have low toxicity and high biocompatibility, which makes CDs an antibacterial material widely used in the field of smart food packaging.

In this study, spermidine-capped carbon dots (S-PCDs) were synthesized using a hydrothermal reaction, and their morphologies and chemical structures were characterized by TEM, FT-IR, XPS and other technologies. The aim was to assess the antibacterial activity, antibacterial mechanism and antibiofilm formation activities of S-PCDs against *S. aureus*. The cytotoxicity of S-PCDs was also examined. Furthermore, the S-PCDs were applied to the storage of pasteurized milk at 25 °C and 4 °C, respectively, to explore their feasibility as additives in food production. The aim of this work is to provide a new effective method for the control of *S. aureus* in the food field.

2. Materials

Spermidine (99%) and kanamycin were purchased from Yuanye Biotechnology Co., Ltd. (Shanghai, China). Polyethyleneimine (PEI_{10000}) was purchased from Macrlin Biochemical Technology Co., Ltd. (Shanghai, China). NCTC L929 fibroblast cells and L929 medium, Calcein/PI Cell Viability/Cytotoxicity Assay Kit and Cell Counting Kit-8 (CCK-8) were purchased from Procell Life Science & Technology Co., Ltd. (Wuhan, China), Beyotime Biotechnology (Shanghai, China) and Labgic Technology Co., Ltd. (Beijing, China). Live/Dead BacLight Bacterial Viability Kit (L7012) was purchased from Thermo Fisher Scientific (Waltham, MA, USA).

2.1. *Staphylococcus aureus Culture*

S. aureus is a strain isolated and purified from rotten turbot in our laboratory. The *S. aureus* was cultured in LB medium using the method of Liu et al. (2022) [15]. Approximately 1×10^8 CFU/mL of bacterial suspension was obtained after 24 h culture at 37 °C for further experiments.

2.2. *Synthesis of S-PCDs*

S-PCDs were synthesized according to Cui et al. (2021) [16] and the synthesis process is shown in Figure 1a. Amounts of 2.5 g PEI and 2.5 g spermidine were added to 250 mL deionized water, and then transferred into polytetrafluoroethylene-lined hydrothermal reaction autoclave and heated at 220 °C for 3 h to form S-PCDs.

a

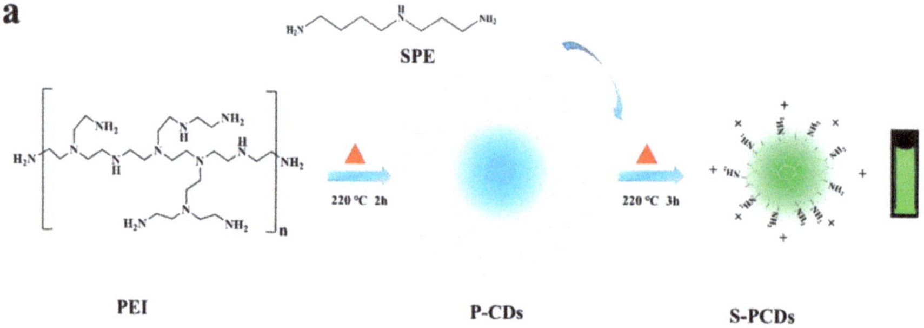

Figure 1. *Cont.*

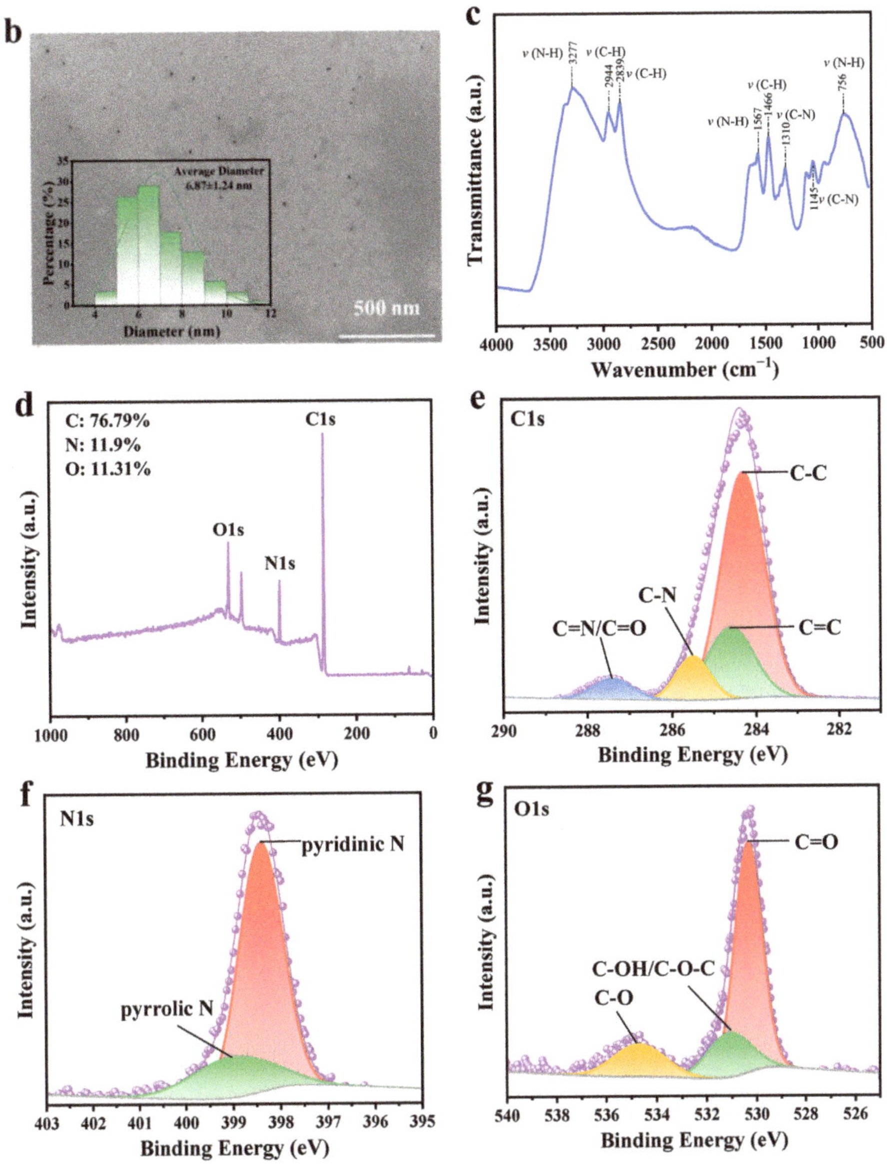

Figure 1. Structure characterization of S-PCDs. (**a**) Hydrothermal reaction synthesized spermidine-capped carbon dots (S-PCDs). (**b**) TEM images and particle size distribution, (**c**) FT-IR spectra, (**d**) XPS wide sweep and high resolution, and (**e**) C1, (**f**) N1 and (**g**) O1 spectra of S-PCDs.

2.3. Characterization of S-PCDs

According to the method of Pan et al. (2022) [10], the morphology, surface chemical elements and functional groups of S-PCDs were characterized by transmission electron microscopy (TEM), X-ray photoelectron spectra (XPS) and FT-IR.

The liquid samples of S-PCDs were dropped onto a copper mesh and dried in a vacuum drying oven under dust-free conditions, and then electron microscope operating (H-700, Hitachi, Tokyo, Japan) at 200 kV was used to obtain TEM images of S-PCDs. Nano measurer 1.2 software was used to analyze the particle size of approximately 120 monodispersed S-PCDs in TEM images.

S-PCD solution was dripped onto the cover glass and dried in a drying oven. After this was repeated several times, a K-Alpha XPS system (Thermo Fisher Scientific, USA) was used to collect the XPS data of S-PCDs.

Freeze-dried S-PCDs powders were placed on infrared crystals of a Nicolet iS10 FT-IR spectrometer (Thermo Fisher, USA) and then the chemical functional groups of S-PCDs were identified. FT-IR spectra were recorded at 4 cm^{-1} resolution in a range of 4000–500 cm^{-1} against a background of ambient air.

2.4. Antibacterial Activity of S-PCDs

2.4.1. MIC of S-PCDs against *S. aureus*

The 96-well microplate was used to determine the MIC [17]. S-PCDs diluted solution (256 μg/mL) was prepared in sterile broth. An amount of 200 μL S-PCDs was added to 1# well and LB medium (100 μL) was added to wells 2–10. Solution (100 μL) was removed from well 1 to well 2. After mixing, moved 100 μL solution from well 2 to well 3 and repeated the process to well 10. Then, 100 μL solution in well 10 was removed and discarded; the final solution was 100 μL per well. And then, each well was supplemented with 10^6 CFU/mL of bacterial suspension. The final S-PCDs concentrations in wells 1–10 were 128, 64, 32, 16, 8, 4, 2, 1, 0.5 and 0.25 μg/mL. Sterile LB broth (200 μL) was added to wells 11–12 as control group.

2.4.2. Bacteriostatic Circle and Bacteriostatic Rate

The antibacterial activity of S-PCDs against *S. aureus* was determined using Oxford cup drilling method. The bacterial suspension was mixed well with 25 mL of LB solid medium and poured into petri dish with an Oxford cup at the bottom. S-PCDs (180 μL) with 0, 1/4, 1/2, 1 and 2 MIC were added to the well formed by removing the Oxford cup. The diameters of bacteriostatic circles of 4 μg/mL PEI, spermidine and kanamycin against *S. aureus* were determined and the bacteriostasis rate analyzed as follows [18]:

$$\text{Bacteriostasis rate (\%)} = (D_{sample} - D_{control})/D_{sample} \times 100\%$$

where D_{sample} and $D_{control}$ are the diameters of the bacteriostatic circles produced with S-PCDstreatment and in the control group.

2.4.3. Spread Plate Method to Determine Antibacterial Activity

Briefly, bacterial suspension was co-cultured with 0, 1/4, 1/2, 1 and 2 MIC S-PCDs at 37 °C for 24 h. Subsequently, 100 μL of diluted co-cultures were evenly spread on the LB nutrient agar and incubated at 37 °C for 24 h. The antibacterial activity of S-PCDs and 4 μg/mL PEI, spermidine and kanamycin was calculated by counting colony-forming units (CFUs) on plates.

2.4.4. Bacteriostatic and Bactericidal Curves of *S. aureus*

The bacteriostatic and bactericidal curves of S-PCDs were determined [19]. For growth curve, the final concentration of *S. aureus* was diluted to 10^6 CFU/mL. Then, the suspension of bacteria in each test tube was treated with S-PCDs of 0, 1/16, 1/8, 1/4, 1/2 and 1 MIC. The 180 μL of co-cultures was removed to a 96-well microtitration plate every 2 h and then the OD_{600} nm was measured using a microplate reader (Bio-Rad, Hercules, CA, USA) to monitor the growth of *S. aureus*.

As for time–kill kinetics analysis, the 10^6 CFU/mL suspension of *S. aureus* was treated with S-PCDs of 1, 2, 4 and 8 MIC. After being further cultured at 37 °C for 24 h, the co-cultures (1 mL) were taken every 2 h. Then, dilution (100 μL) was evenly applied to LB agar plates and incubated; *S. aureus* suspension without S-PCD treatment was set as control.

2.5. Antibacterial Mechanism of S-PCDs

2.5.1. The Damage of Membrane Permeability

Propyl iodide (PI) uptake test was determined [20]. The suspension of *S. aureus* was centrifugated at 6000× *g* for 15 min to obtain the bacterial cells, and they were re-suspended in 10 mL PBS. The 10^6 CFU/mL of bacterial cells was treated with 0 and 1 MIC of S-PCDs for 24 h at 37 °C. The intensity of the co-culture was determined using a fluorescence spectrophotometer (Hitachi, Tokyo, Japan) after the reaction of PI in the dark at 37 °C for 10 min. The emission wavelength was set at 500–700 nm and the excitation wavelength was set at 495 nm.

2.5.2. Determination of Extracellular Nucleic Acid and Protein Content

Nucleic acid assay was performed with reference to Wang et al. (2021) [21]. The *S. aureus* suspension was treated with 1 MIC of S-PCDs and cultured at 37 °C for 12 h. The co-culture was removed every 2 h and centrifuged for 15 min at 1000× *g*. The absorbance at 260 nm of supernatant was determined by a T9s UV–vis spectrophotometer (General Instrument Co. Ltd., Beijing, China).

The effects of S-PCDs on soluble proteins of bacteria were determined using a BCA kit. The *S. aureus* suspension was treated with S-PCD of 1 MIC and cultured for 12 h. The co-culture was removed and centrifuged at 1000× *g* for 15 min every 2 h. The BCA kit was used to measure the protein content.

2.5.3. Microscope Observation of Morphology of *S. aureus*

The co-cultures of *S. aureus* suspension and 1 MIC S-PCDs were centrifuged at 500× *g* for 15 min to obtain the bacterial cells. After overnight fixation in glutaraldehyde (2.5%), the cells of *S. aureus* were washed again with PBS 3 times and dehydrated with ethanol solutions (30, 50, 70, 90 and 100%) for 15 min each. The morphology and surface characteristics of the strain were observed using SEM (Hitachi, Tokyo, Japan) after it was sputter-coated with gold.

The methods of bacterial culture, cleaning and fixation are the same as above. The bacterial suspension treated with SP-CDs was diluted to an appropriate concentration and adsorbed using a special copper mesh for TEM. The structure of fully dried *S. aureus* was observed using an H-700 TEM (Hitachi, Tokyo, Japan).

2.5.4. Confocal Laser Scanning Microscopy Detected Bacterial Viability

S. aureus was stained with a L7012 Live/Dead bacterial staining kit [10]. The co-cultures of *S. aureus* suspension and S-PCDs were centrifuged at 10,000× *g* for 10 min to obtain the bacterial cells, then washed with 0.85% NaCl 3 times and re-suspended in 2 mL 0.85% NaCl. The bacterial cells (1 mL) were treated with 3 μL SYTO 9 and PI (SYTO 9:PI = 1:1) in a dark environment for 15 min. Fluorescent images were obtained using a CLSM (Leica TCS SP5, Heidelberg, Germany).

2.5.5. Intracellular Reactive Oxygen Species (ROS)

S-PCDs of 1, 2 and 4 MIC were added into the 10^6 CFU/mL of *S. aureus* suspension and co-incubated at 37 °C for 24 h. After centrifugation at 6000× *g* for 5 min, the bacterial cells were re-suspended in 10 mL PBS. The suspension was supplemented with 2′, 7′-dichlorodihydrofluorescein diacetate (10 μL, DCFH−DA, 10 μmol/L) and incubated for 30 min, and then the fluorescence intensity was detected. Bacteria without any treatment were used as a control group. The excitation wavelength was 484 nm and the emission wavelength was 525 nm. The ROS generation capacity was expressed by the relative ROS level [22]: $ROS_{sample}/ROS_{control}$.

2.5.6. Gene Expression

The relative gene expression analysis was performed according to Zheng et al. (2021) [23]. Quantitative real-time polymerase chain reaction (qRT-PCR) was determined using the gene

9600 Quantitative PCR Instrument (Bori, Hangzhou, China). The expression of 3 indicator genes that can maintain intracellular redox balance, namely, 3 oxidation regulatory genes *dmpI, narJ* and *narK*, was detected after S-PCD treatment, and gyrB was used as the internal reference gene. Primers are shown in Table 1.

Table 1. RT-PCR primers.

Gene	Forward Primer 5′ to 3′	Reverse Primer 3′ to 5′
dmpI	TGATGCCAATCGTCAATG	CCCGTTGTTTTTCTACG
narJ	GAACGTGGGCAAATGTTAG	TTGAAGCATCAACGGTAG
narK	TATTCCCGATATTTTTCTTAAGCC	CTGATGTAACACCAACAGAG
gyrB	TCAATACAGGTTTTAGAGGGGTTA	AACCATTCAATACTTCATCGACG

2.6. Antibiofilm Activity of S-PCDs

Firstly, the antibiofilm activity of S-PCDs was determined using crystal violet staining [24]. The final concentrations of S-PCDs were 0, 1/4, 1/2 and 1 MIC in the suspensions containing 200 μL *S. aureus*, and they were cultured for 24 h, the supernatant removed, and then they were rinsed with PBS 3–5 times and stained with 0.4% crystal violet for 5 min. After rinsing with sterile water and drying for 10 min, each well was supplemented with 200 μL 95% ethanol and OD_{570} was measured using an enzyme-labeled meter. Each sample was repeated three times.

Secondly, the spread plate method was used to measure the number of bacteria in the biofilm [25]. Amounts of 2 mL LB broth and 20 μL *S. aureus* suspension were added to 24-well plates with an aseptic coverslip at the bottom, and treated with 0, 1/4 MIC, 1/2 MIC and 1 MIC of S -PCDs at 37 °C for 24 h. Then, the coverslip was cleaned with PBS 3 times and then placed in LB broth for 2 min vortex to completely release the bacterial cells. The 10-fold dilutions of bacterial suspension were inoculated in LB agar to count the total number of bacterial colonies at 37 °C for 24 h.

The biofilm formation was observed using CLSM [26]. According to above method, bacteria were allowed to grow on the coverslip at 37 °C for 48 h to form biofilm. The biofilm formation was then detected by staining in the dark for 15 min with the L7012 LIVE/DEAD bacterial staining kit. A CLSM (Leica TCS SP5, Heidelberg, Germany) was used to record the bacterial fluorescent images.

2.7. Safety Evaluation of S-PCDs

CCK-8 method was used to evaluate the effects of S-PCDs on L929 cell viability and proliferation [5]. L929 cells were passaged in L929 medium at 37 °C in 5% carbon dioxide and 95% air. S-PCDs with different gradient concentrations (0, 2, 4, 8, 16, 32 and 64 μg/mL) were prepared using sterile L929 medium. After the cells were cultured on a 96-well plate until wall attachment, the medium was discarded and different concentrations of S-PCD solutions were added. After incubating again for 24 h, the cell cultures were incubated with 10% (*v*/*v*) CCK-8 solution for 3 h and the absorbance at 450 nm wavelength was measured to calculate the cell viability.

In addition, the cells treated with S-PCDs were stained with calcitanin AM and PI according to Dead/Live kit instructions. And a fluorescence microscope (Nikon A1 MP, Tokyo, Japan) was taken to obtain the corresponding fluorescence images.

2.8. Effects of S-PCDs on the Growth of S. aureus in Pasteurized Milk

The *S. aureus* cells were diluted to approximately 1×10^6 CFU/mL with 20 mL of fresh pasteurized milk and incubated at 25 °C and 4 °C, after adding 1 MIC of S-PCDs. The control group did not contain S-PCDs. Microbial concentrations were determined using colony counting method after 1, 12, 24, 48 and 72 h of incubation. In addition, the pasteurized milk after 72 h of treatment was poured into a flat dish and left to stand for 30 min to observe the precipitation formed in the pasteurized milk [27].

2.9. Statistical Analysis

All experiments and their associated statistical analyses were conducted in triplicate and the results were expressed as mean ± standard deviation (SD). A significance level of 0.05 was used. The statistical analyses were performed using SPSS 26.0 (IBM SPSS Statistics, Chicago, IL, USA) software and the drawings were performed using Origin 2022b (Origin Lab Corporation, Northampton, MA, USA).

3. Results and Discussion

3.1. Performance Characterization of S-PCDs

S-PCDs can show a bright fluorescence emission phenomenon under a 365 nm violet lamp (Figure 1a), which indicated the successful synthesis of CDs by hydrothermal reaction. The S-PCDs were spherical particles with good dispersion without obvious aggregation (Figure 1b). The particle size of most S-PCDs ranged from 3.92 nm to 11.4 nm and the mean diameter of the S-PCDs was 6.87 nm. The diameter of carbon dots is usually less than 10 nm [28] and the diameter of the S-PCDs conforms to the typical particle size distribution range of carbon dots.

The absorption peak near 3277 cm^{-1} is related to the amino/amide N-H bond stretching vibration (Figure 1c). The characteristic absorption peaks of C−H appeared in the vicinity of 2944 and 2839 cm^{-1}, and the peak at 1567 cm^{-1} was attributed to N−H deformation vibration. There were obvious C−H characteristic absorption peaks near 1466 cm^{-1}, and the infrared absorption peaks at 1310 and 1145 cm^{-1} were related to the stretching vibration of C−O.

The surface chemical composition of S-PCDs was determined using XPS. The results of an XPS element sweep showed that the signal peaks at 284.1, 398.5 and 530.1eV of XPS binding energy were C1s, N1s and O1s respectively (Figure 1d). As the main elements of S-PCDs, the C, N and O relative contents were 76.79%, 11.31% and 11.9%, respectively. Because it is easy for the N-rich precursor to participate in the synthesis reaction, the content of the N element in the S-PCDs was high. Figure 1e shows the XPS high resolution spectrum of C1s. It can be seen that the main forms of the C element in S-PCDs were C−C/C=C (284.3–284.6 eV), C−N (285.4 eV) and C=O/C=N (287.3 eV) [29]. C−C/C=C is the main component of S-PCDs' skeleton, which is related to intramolecular dehydration. In addition, according to the C1 peak splitting results, the peak splitting area of XPS related to the C=N/C=O bond was small, indicating that the S-PCDs contained only a small amount of carbonyl groups on the surface, while the peak splitting area of XPS related to the C−N bond was large, indicating that spermidine can effectively participate in the formation of CD carbon nuclei. The N1 spectrum in Figure 1f shows the XPS signal peaks of 398.4 eV and 399.0 eV. There is no peak above 401.0 eV, indicating that the sample is without the N−N bonding structure [29]. The main peak of 398.4 eV usually belongs to the sp^2 N atom, while the signal peak at 399.0 eV was assigned to bridging N atoms in $N-(C)_3$ or N bonded with H atoms [30,31]. From the O1 spectrum (Figure 1g), it can be seen that the main forms of the O element in S-PCDs were C=O (530.8 eV), C−OH/C−O−C (531.2 eV) and C−O (534.6 eV) [31].

3.2. Antibacterial Activity of S-PCDs

The MIC of S-PCDs against *S. aureus* was 16 μg/mL. S-PCDs and kanamycin showed clear inhibition circles with diameters of 22.2 and 27.3 mm, respectively, while spermidine and polyethyleneimine showed almost no visible inhibition circles (Figure 2a). Furthermore, the diameter of the inhibition circles and the inhibition rate increased significantly ($p < 0.05$) with increasing the S-PCD concentration. When treated with 2 MIC S-PCDs, the inhibition circle diameter reached 25.8 mm and the inhibition rate of 70.11% was not significantly different from that of kanamycin (71.97%) (Figure 2b).

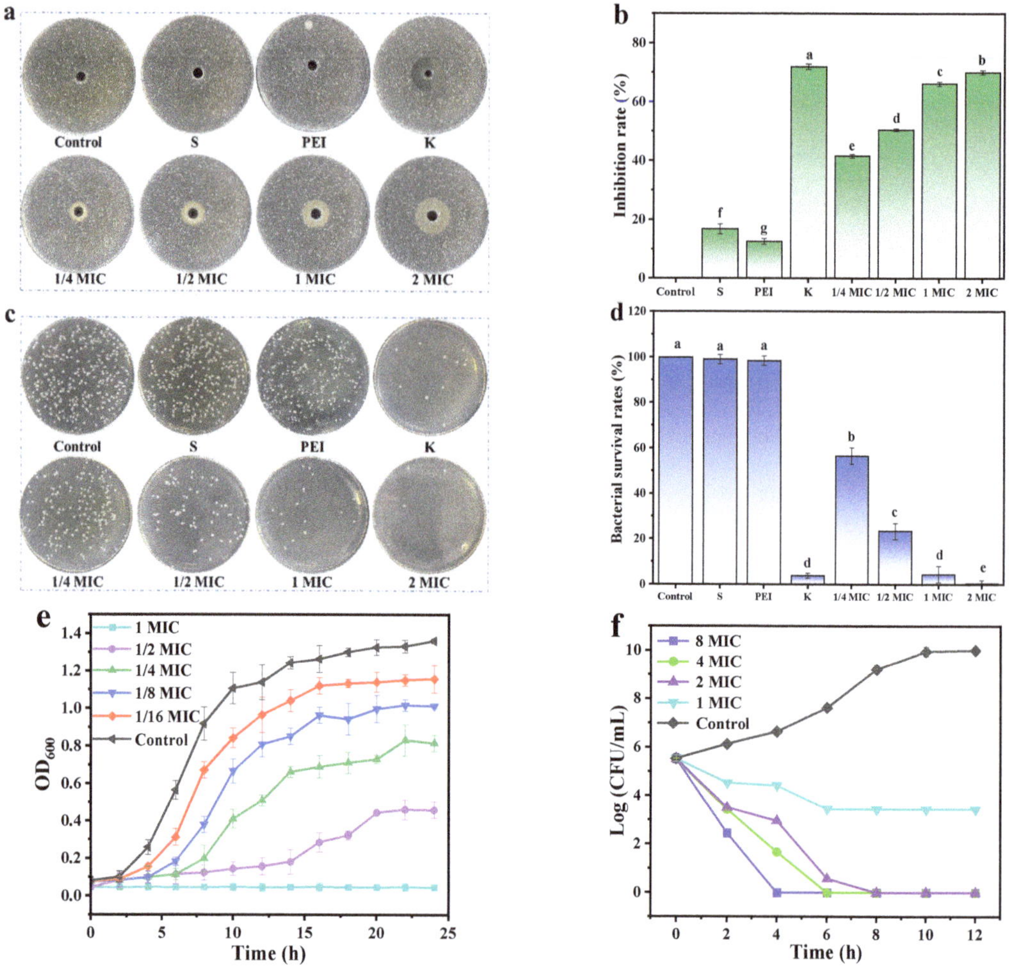

Figure 2. Antibacterial activity of S-PCDs. (**a**) Inhibition zone diameter, (**b**) inhibition rate, (**c**) antibacterial activity, and (**d**) bacterial survival rates of S-PCDs on *S. aureus*. *S. aureus* was treated with different concentrations of S-PCDs. The same concentrations of S and PEI as S-PCDs were used as negative control, and kanamycin was used as positive control. (**e**) *Bacteriostatic curve* and (**f**) bactericidal curve of different concentrations of S-PCDs against *S. aureus*. The average values of different letters in (**b**,**d**) were significantly different when $p < 0.05$.

The number of colonies of *S. aureus* did not change after spermidine and polyethyleneimine treatment (Figure 2c). The bacterial viability tended to zero when S-PCDs were treated at 1 MIC and higher concentrations, and there was a significant decrease in the total number of colonies and a significant increase in bacterial viability as the concentration of S-PCDs increased (Figure 2d).

The time-dependent bacteriostatic curve was determined by measuring the OD_{600} of the co-cultures of 1/16–1 MIC S-PCDs and *S. aureus* suspension (Figure 2e). The OD_{600} of *S. aureus* increased between 4 and 10 h, and the OD_{600} decreased significantly after treatment with S-PCDs in different concentrations. As the concentration of S-PCDs gradually increased from 1/16 MIC to 1 MIC, the inhibition curve gradually decreased with time. The concentration-dependent inhibition of S-PCDs was consistent with the results of the inhibition circle assay.

The bactericidal curve of S-PCDs on *S. aureus* showed that *S. aureus* untreated with S-PCDs grew rapidly at 4–8 h of incubation (Figure 2f). When *S. aureus* was treated with MIC and higher S-PCD concentrations for 12 h, the total number of colonies showed a continuous decreasing trend with time. In addition, as the concentration of S-PCDs increased, the total number of colonies decreased in the same treatment time. Two MIC S-PCDs could almost kill *S. aureus* within 8 h and 4 MIC S-PCDs could almost kill *S. aureus* within 6 h. The results demonstrated that S-PCDs had good bacteriostatic activity against *S. aureus* in a concentration-dependent manner.

3.3. Cell Membrane Damage of *S. aureus*

The zeta potentials of S-PCDs and *S. aureus* were +47.06 and −10.5 mV (Figure 3a). The zeta potential of *S. aureus* increased to +30.14 mV after treatment with S-PCDs. The results suggested that the high positive charge of the S-PCDs positively shifts the negative charge of *S. aureus* and electrostatic interaction may occur between the two. This electrostatic interaction may cause CDs to further destroy the surface charge and physiological metabolic activities of *S. aureus*, eventually leading to bacterial death [17].

PI can only enter the interior of the cells to bind nucleic acids to produce a fluorescence signal after the bacterial membrane is broken. The PI uptake of the S-PCD treatment group increased significantly with incubated time, and the fluorescence intensity reached 2.1 times that of the control group after culturing for 12 h (Figure 3b). The fluorescence intensity of the control group remained unchanged. The results suggested that S-PCDs can alter the permeability of cell membranes and cause bacterial lysis, allowing PI to enter the cell interior and bind to DNA.

The extracellular protein content of *S. aureus* after S-PCD treatment was significantly increased ($p < 0.001$) (Figure 3c). OD_{260} nm increased rapidly after culturing for 2–6 h and gradually stabilized after culturing for 8 h (Figure 3d). The extracellular protein and nucleic acid content were 1.41 and 3.86 mg/mL, respectively, which were 5.4 and 1.2 times those of the control group when cultured for 12 h. The results indicated than S-PCDs could disrupt the integrity of the membrane and lead to a significant leakage of cytoplasmic contents from bacterial cytoplasm. Leakage of cytoplasmic contents such as DNA, RNA and proteins is an indicator of irreversible damage to bacterial membranes [21].

The damage of S-PCDs to *S. aureus* was further evaluated using TEM and SEM. TEM images showed that the untreated *S. aureus* grew well, and its cell membrane had a complete, smooth and clear morphological structure (Figure $3e_1$). After S-PCD treatment, the cell wall and membrane of *S. aureus* became irregular, and the cells became transparent, accompanied by intracellular leakage (Figure $3e_2$). In addition, untreated cells showed complete spherical structures with clear edges and smooth cell walls (Figure $3e_3$). However, after 24 h treatment with S-PCDs, many membrane defects and leakage of cell contents were found on the bacterial cell surface (Figure $3e_4$), indicating that the bacteria were damaged. The results of SEM and TEM demonstrated that the adsorption of S-PCDs on the bacterial wall or outer membrane caused physical or mechanical damage to it, resulting in the overall physiological dysfunction of the bacteria and even the leakage of cytoplasmic components, and finally led to the death of bacteria.

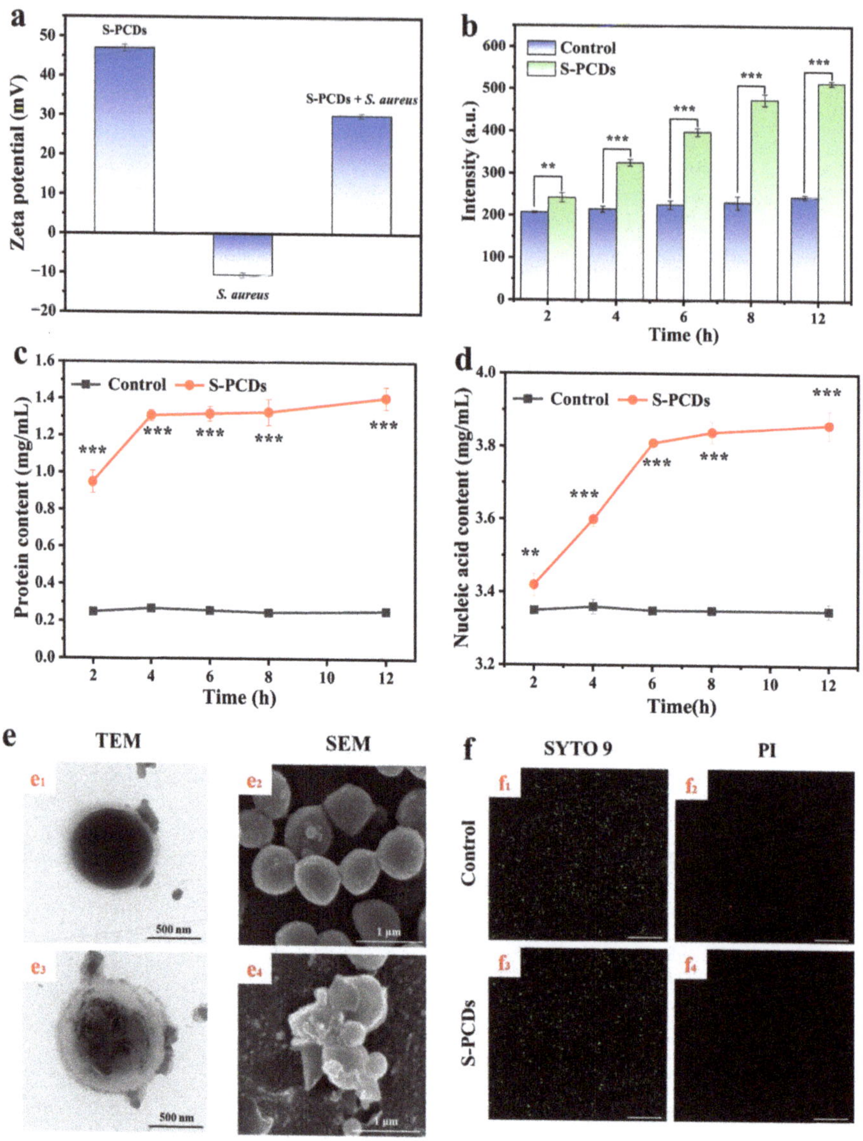

Figure 3. Bacterial cell membrane damage. (**a**) Zeta-potential of S-PCDs, *S. aureus* (with or without S-PCD treatment). (**b**) Uptake of PI in *S. aureus*. Effect of S-PCDs on (**c**) the extracellular protein content and (**d**) the extracellular nucleic acid content of *S. aureus*. (**e**) TEM and SEM images of *S. aureus*. *S. aureus* without treatment (e_1, e_2, f_1 and f_2) or with treatment of S-PCDs (e_3, e_4, f_3 and f_4). (**f**) CLSM images of *S. aureus* were stained by SYTO 9/PI dual fluorescence staining. All bacteria alive and dead were stained by SYTO 9 showing green fluorescent intensity. Only dead bacteria can be stained by PI emitting red fluorescent intensity. For SYTO 9: $\lambda ex = 488$ nm; for PI: $\lambda ex = 552$ nm. The scale bar is 75 μm. ** $p < 0.01$, *** $p < 0.001$.

The DNA of all live and dead bacteria can be stained by SYTO 9, while PI can only stain bacteria with damaged cell membranes [32]. Both the S-PCD treated group and the control group stained by SYTO 9 showed a large amount of green fluorescence. The red fluorescence of the S-PCD treatment group was significantly enhanced compared with the control group, indicating that a large number of dead bacteria were induced to emit red fluorescence after S-PCD treatment, which further confirmed the effective bactericidal ability of S-PCDs.

3.4. ROS Generation of *S. aureus*

The results of ROS production are shown in Figure 4a. The fluorescence intensity of *S. aureus* was significantly stronger than that of the control group after treatment with 1–4 MIC S-PCDs, indicating that S-PCDs stimulated ROS production. After 1 MIC concentration of S-PCDs treatment, the ROS production of *S. aureus* was 5.25 times that of the control group. And the fluorescence intensity of *S. aureus* treated with S-PCDs was significantly increased with increasing concentration of S-PCDs ($p < 0.05$). When treated with 4 MIC of S-PCDs, the ROS production reached 6.52 times that of the control group. Many nanoparticles exhibit high antibacterial activity against a variety of pathogens, including drug-resistant bacteria, and the antibacterial mechanism is through increasing physical damage to cell membranes. In addition, they can exert antibacterial activity through photoinduction or thermal induction of free radicals and ROS [33]. This study showed that, after the interaction between CDs and bacteria, ROS production can be induced through light-independent reactions, so that ROS can induce oxidative stress and destroy bacterial cell membranes to produce antibacterial effects.

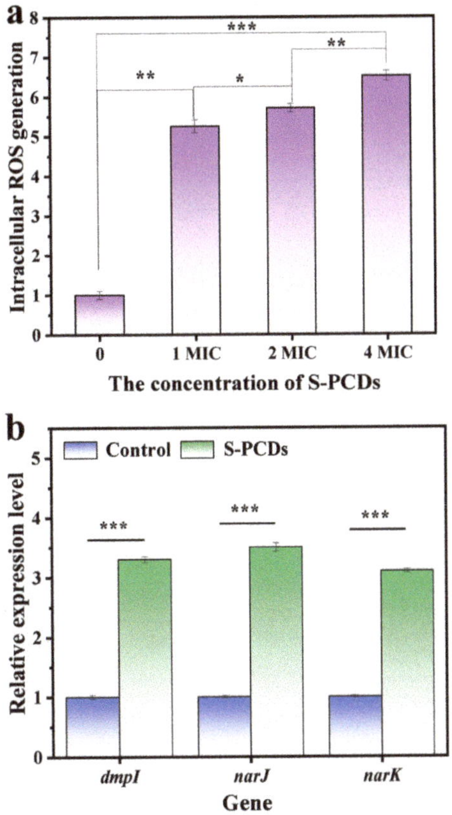

Figure 4. ROS generation in *S. aureus*. (**a**) The generation of ROS in *S. aureus* treated with S-PCDs by using the dye DCFH-DA. (**b**) Comparison on relative expression level of *dmpI*, *narJ* and *nark* in *S. aureus* after treatment with S-PCDs. * $p < 0.05$, ** $p < 0.01$. *** $p < 0.001$.

The nanomaterials can contribute to the disruption of the metabolic balance of intracellular oxidation reactions by upregulating genes encoding oxidation-stimulating enzymes (such as *dmpI*, *narJ* and *nark*) to produce large amounts of ROS [22]. The expression of three indicator genes, namely the three oxidative regulatory genes *dmpI*, *narJ* and *narK*, which can maintain intracellular redox balance, was detected after S-PCD treatment. The *dmpI* gene encodes 4-oxalocrotonate tautomerase, which participates in the oxidative catabolism of toluene, o-xylene, 3-ethyltoluene and 1,2, 4-tritoluene into corresponding intermediates in the citric acid cycle, and produces ROS as a by-product. The *narJ* and *narK* genes are important participants in respiratory nitrate reductases, which transfer electrons from NADH or NADPH to nitrate, assisting in nitrate transport and nitrate reduction [34]. The results showed that treatment with S-PCDs could induce the expression of three oxidation genes to be significantly upregulated to 3.3, 3.5 and 3.1 times ($p < 0.001$), respectively (Figure 4b), which was also consistent with the data generated by ROS. Since ROS is a by-product of intracellular redox metabolism, the difference in ROS production in the S-PCD treatment group may be due to the different redox reactions of bacteria to them [23]. Oxidative stress-induced lipid peroxidation leads to cell membrane rupture, which causes DNA degradation induced by oxidative stress after CDs enters bacterial cells, resulting in bacterial death [35].

3.5. Antibiofilm Activity of S-PCDs

The inhibition rates of S-PCDs with 1/4, 1/2 and 1 MIC were 21.34%, 53.42% and 81.34%, respectively. The compact structure of the biofilm protects the bacteria from antibacterial substances. Therefore, the viable bacteria number in the biofilm formed after S-PCD treatment was measured. S-PCDs showed effective inhibition in a dose-dependent manner. The number of living bacteria was reduced to 3.89 log CFU/cm^2 when treated with 1 MIC S-PCDs (Figure 5b). CLSM explored the effect of S-PCDs on *S. aureus* biofilm, and the results are shown in Figure 5c. The green fluorescence intensity decreased and the red fluorescence intensity increased significantly after 1/2 MIC S-PCD treatment compared with the control group. S-PCDs inhibited the biofilm of *S. aureus* by reducing the cell viability and the number of viable bacteria.

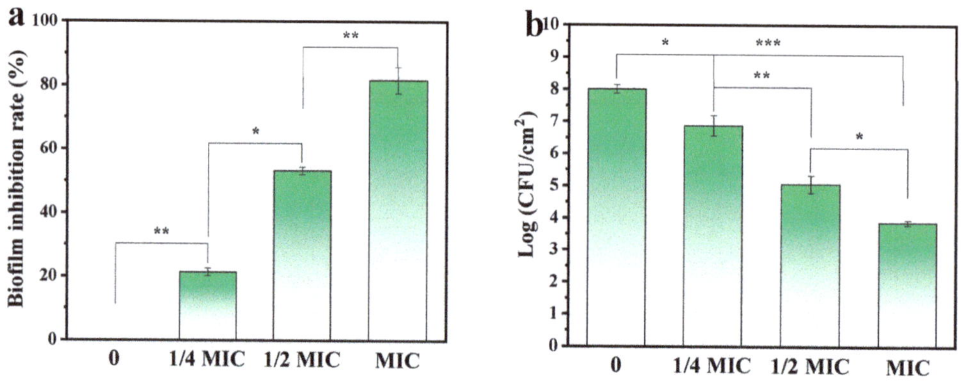

Figure 5. *Cont.*

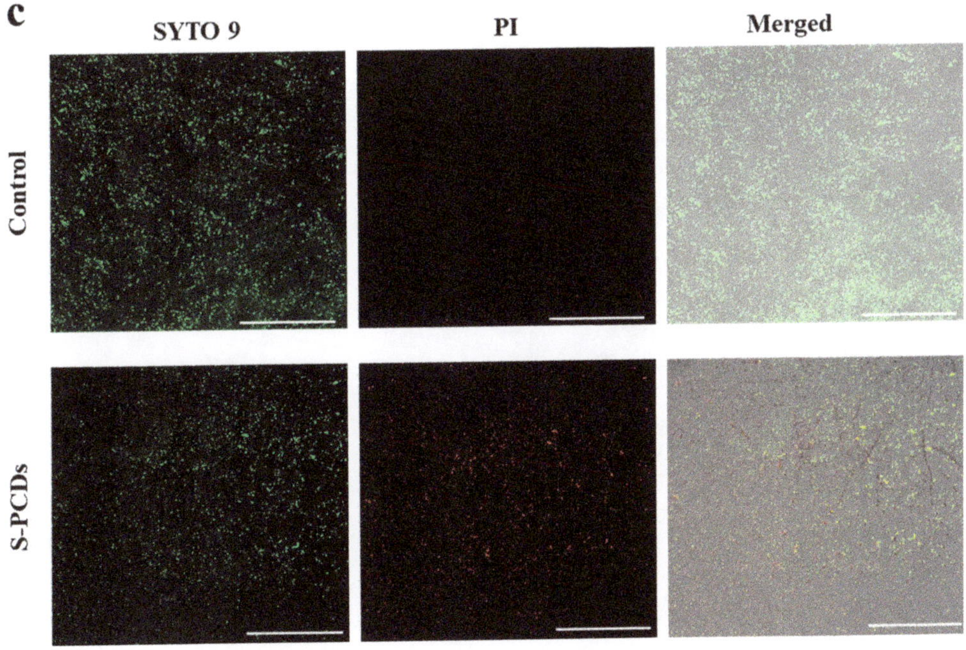

Figure 5. Antibiofilm activity of the S-PCDs. (**a**) Effects of S-PCDs on biofilm formation of *S. aureus*. (**b**) The number of viable bacteria from biofilms on coverslips surfaces. (**c**) CLSM images of *S. aureus* were stained by SYTO 9/PI dual fluorescence staining. All bacteria alive and dead were stained by SYTO 9 showing green fluorescent intensity. Only dead bacteria can be stained by PI emitting red fluorescent intensity. For SYTO 9: λex = 488 nm; for PI: λex = 552 nm. The scale bar is 250 μm. * $p < 0.05$, ** $p < 0.01$, *** $p < 0.001$.

3.6. Safety Evaluation of S-PCDs

The cytotoxicity of S-PCDs on L929 cells was determined using a CCK-8 kit. As shown in Figure 6a, when 2 μg/mL of spermidine, polyethyleneimine and S-PCDs was added to the medium and cultured for 48 h, cell viability was >80% in all groups. However, the cell viability of S-PCDs was significantly enhanced compared to that of spermidine and polyethyleneimine, thus further improving safety, which is consistent with the reduced toxicity of synthetic carbon dots reported by most investigators. The cell viability of S-PCDs at 0–128 μg/mL was >95% after 24 h of incubation (Figure 6b). In addition, L929 cells treated with 2–128 μg/mL S-PCDs proliferated normally and showed similar cell morphology to the control group, with almost no dead cells detected (Figure 6c). The results indicated that S-PCDs have excellent biocompatibility and a better safety profile.

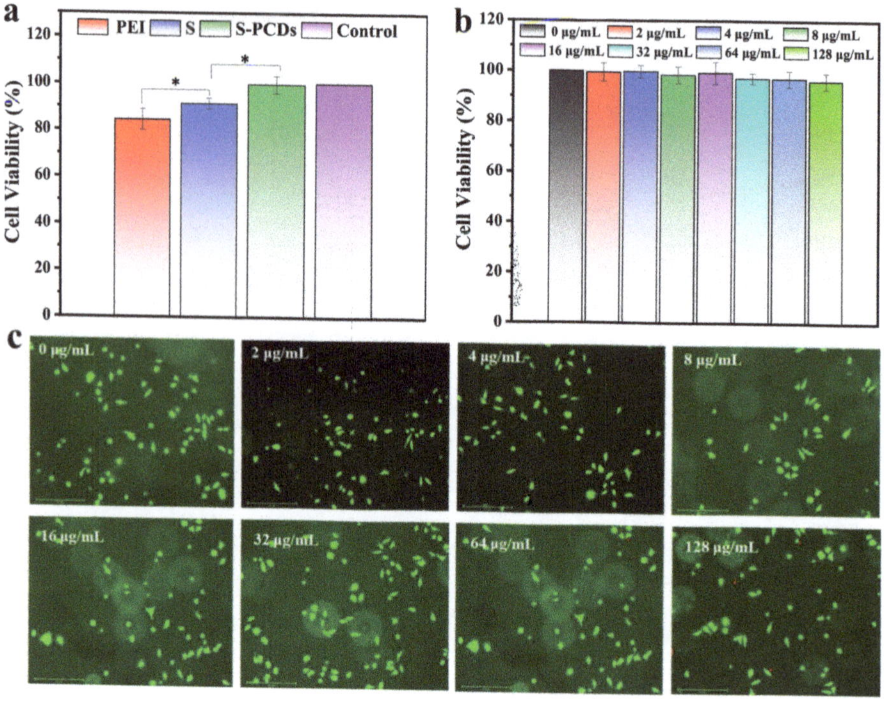

Figure 6. Safety evaluation of S-PCDs. (**a**) Cell viability of L929 fibroblasts treated with PEI, S and S-PCDs (2 μg/mL) for 24 h. (**b**) Cell viability of L929 fibroblasts treated with different concentrations of S-PCDs (0, 2, 4, 8, 16, 32, 64 and 128 μg/mL) for 24 h. (**c**) Live/Dead fluorescence staining images after 24 h incubation. * $p < 0.05$.

3.7. Effects of S-PCDs on Growth of S. aureus in Pasteurized Milk

The results of pasteurized milk treated with S-PCDs after *S. aureus* infection are shown in Figure 7. When the pasteurized milk was stored at 25 °C for different times, S-PCDs inhibited the growth of *S. aureus* (Figure 7a). After 24 h and 72 h treatment with S-PCDs, the concentration of *S. aureus* decreased from 7.48 to 5.98 $\log_{10}$ CFU/mL and 9.95 to 6.95 $\log_{10}$ CFU/mL, respectively ($p < 0.01$). The results showed that S-PCDs had a significant inhibitory effect on *S. aureus* in pasteurized milk. *S. aureus* in pasteurized milk survived and grew normally at 4 °C for 72 h (Figure 7b). The antibacterial activity of S-PCDs was significantly different after 24 h compared with the control group ($p < 0.05$). Compared with the control group, the number of microorganisms was significantly decreased after S-PCD treatment for 72 h ($p < 0.001$). This indicated that S-PCD treated milk can be stored for a long time. The morphology of the treated pasteurized milk sample was then observed by laying it on a flat plate. After growing *S. aureus* at 25 °C for 72 h, a large amount of precipitation was observed in all samples (Figure 7c), accompanied by a strong rancid smell. Since proteins and lipids in milk are agglomerated under the growth and metabolism of microorganisms, S-PCD treatment can effectively reduce the formation of precipitation, which means that its growth and metabolism are inhibited.

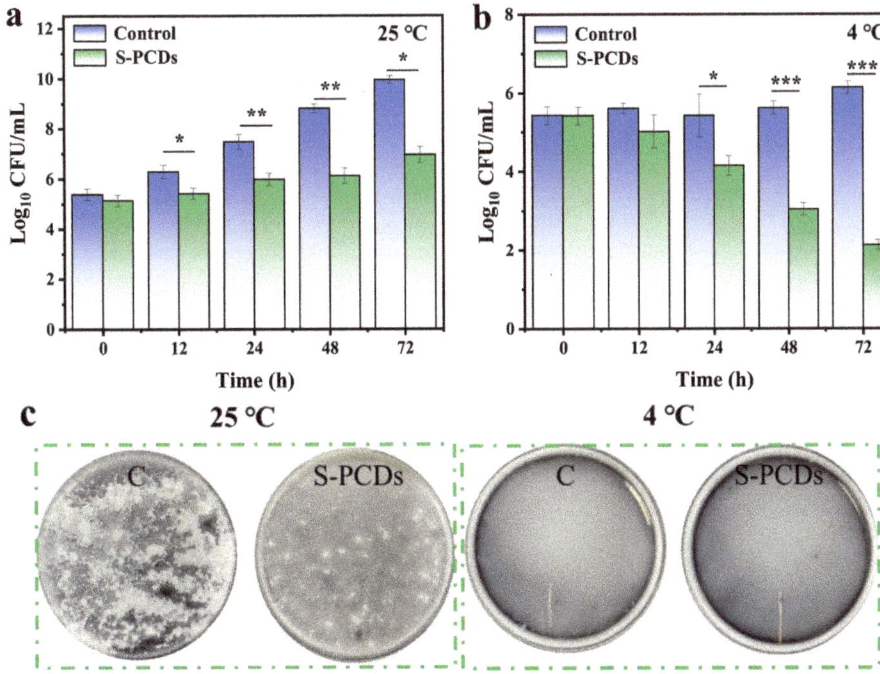

Figure 7. S-PCDs on the growth of *S. aureus* in pasteurized milk. The mode of *S. aureus* infected with pasteurized milk when treated with nisin and CAR used alone or in combination at (**a**) 25 °C and (**b**) 4 °C, and (**c**) the formation of precipitation in treated pasteurized milk. * $p < 0.05$, ** $p < 0.01$, *** $p < 0.001$.

4. Conclusions

In summary, S-PCDs were synthesized and characterized by hydrothermal synthesis. The antibacterial activity, antibacterial mechanism and antibiofilm activity of S-PCDs were systematically studied using *Staphylococcus aureus* as a model. The results showed that the antibacterial activity of S-PCDs was stronger than that of the two raw materials. The high positive charge and small size of the surface of S-PCDs enable them to inhibit *S. aureus* by damaging the membrane and producing ROS. In addition, S-PCDs can effectively inhibit the formation of biofilms. Safety evaluation results showed that S-PCDs had low cytotoxicity. S-PCDs inhibited the number of microorganisms in the pasteurized milk infected with *S. aureus* and maintained the quality of the pasteurized milk at 4 °C and 25 °C. This study clearly elucidates the interaction between S-PCDs and *S. aureus*, provides new insights into the role of carbon points in food, and further supports the in-depth application of nanomaterials.

Author Contributions: Methodology, Y.L., G.C. and J.C.; Software, Y.F., X.F. and Y.S.; Validation, X.F. and Y.S.; Investigation, Y.F.; Data curation, T.C. and Y.L.; Writing—original draft, T.C.; Writing—review & editing, G.C. and J.C.; Supervision, G.C. and J.C.; Project administration, G.C. and J.C.; Funding acquisition, Y.L., X.F. and G.C. All authors have read and agreed to the published version of the manuscript.

Funding: This research was funded by Major Science and Technology Project of Yunnan Province (202202AG050009), Yunnan High-Level Talent Innovation Program (YNWR-QNBJ-2020-226 and YNWR-QNBJ-2020-246), High-level Talent Promotion and Training Project of Kunming (2022SCP003) and Yunnan Fundamental Research Projects (202201BE070001-051).

Institutional Review Board Statement: Not applicable.

Informed Consent Statement: Not applicable.

Data Availability Statement: Data is contained within the article.

Conflicts of Interest: The authors declare no conflict of interest.

References

1. Lai, D.; Zhou, A.; Tan, B.K.; Tang, Y.; Sarah Hamzah, S.; Zhang, Z.; Lin, S.; Hu, J. Preparation and photodynamic bactericidal effects of curcumin-beta-cyclodextrin complex. *Food Chem.* **2021**, *361*, 130117. [CrossRef] [PubMed]
2. Bajpai, V.K.; Bahuguna, A.; Kumar, V.; Khan, I.; Alrokayan, S.H.; Khan, H.A.; Simal-Gandara, J.; Xiao, J.; Na, M.; Sonwal, S.; et al. Cellular antioxidant potential and inhibition of foodborne pathogens by a sesquiterpene ilimaquinone in cold storaged ground chicken and under temperature-abuse condition. *Food Chem.* **2022**, *373*, 131392. [CrossRef] [PubMed]
3. Tenderis, B.; Kilic, B.; Yalcin, H.; Simsek, A. Impact of sodium lactate, encapsulated or unencapsulated polyphosphates and their combinations on *Salmonella typhimurium*, *Escherichia coli* O157:H7 and *Staphylococcus aureus* growth in cooked ground beef. *Int. J. Food Microbiol.* **2020**, *321*, 108560. [CrossRef] [PubMed]
4. Kang, S.; Kong, F.; Shi, X.; Han, H.; Li, M.; Guan, B.; Yang, M.; Cao, X.; Tao, D.; Zheng, Y.; et al. Antibacterial activity and mechanism of lactobionic acid against *Pseudomonas fluorescens* and Methicillin-resistant *Staphylococcus aureus* and its application on whole milk. *Food Control* **2020**, *108*, 106876. [CrossRef]
5. Wu, Z.; Tang, S.; Deng, W.; Luo, J.; Wang, X. Antibacterial chitosan composite films with food-inspired carbon spheres immobilized AgNPs. *Food Chem.* **2021**, *363*, 130342. [CrossRef] [PubMed]
6. Niu, J.; Tang, G.; Tang, J.; Yang, J.; Zhou, Z.; Gao, Y.; Chen, X.; Tian, Y.; Li, Y.; Li, J.; et al. Functionalized Silver Nanocapsules with Improved Antibacterial Activity Using Silica Shells Modified with Quaternary Ammonium Polyethyleneimine as a Bacterial Cell-Targeting Agent. *J. Agric. Food Chem.* **2021**, *69*, 6485–6494. [CrossRef] [PubMed]
7. Li, H.; Huang, J.; Song, Y.; Zhang, M.; Wang, H.; Lu, F.; Huang, H.; Liu, Y.; Dai, X.; Gu, Z.; et al. Degradable Carbon Dots with Broad-Spectrum Antibacterial Activity. *ACS Appl. Mater. Interfaces* **2018**, *10*, 26936–26946. [CrossRef] [PubMed]
8. Jian, H.J.; Wu, R.S.; Lin, T.Y.; Li, Y.J.; Lin, H.J.; Harroun, S.G.; Lai, J.Y.; Huang, C.C. Super-Cationic Carbon Quantum Dots Synthesized from Spermidine as an Eye Drop Formulation for Topical Treatment of Bacterial Keratitis. *ACS Nano* **2017**, *11*, 6703–6716. [CrossRef]
9. Zheng, K.; Setyawati, M.I.; Leong, D.T.; Xie, J. Antimicrobial silver nanomaterials. *Coord. Chem. Rev.* **2018**, *357*, 1–17. [CrossRef]
10. Pan, T.; Chen, H.; Gao, X.; Wu, Z.; Ye, Y.; Shen, Y. Engineering efficient artificial nanozyme based on chitosan grafted Fe-doped-carbon dots for bacteria biofilm eradication. *J. Hazard Mater.* **2022**, *435*, 128996. [CrossRef]
11. Wu, L.-N.; Yang, Y.-J.; Huang, L.-X.; Zhong, Y.; Chen, Y.; Gao, Y.-R.; Lin, L.-Q.; Lei, Y.; Liu, A.-L. Levofloxacin-based carbon dots to enhance antibacterial activities and combat antibiotic resistance. *Carbon* **2022**, *186*, 452–464. [CrossRef]
12. Nakamura, T.; Tamura, A.; Murotani, H.; Oishi, M.; Jinji, Y.; Matsuishi, K.; Nagasaki, Y. Large payloads of gold nanoparticles into the polyamine network core of stimuli-responsive PEGylated nanogels for selective and noninvasive cancer photothermal therapy. *Nanoscale* **2010**, *2*, 739–746. [CrossRef] [PubMed]
13. Azevedo, M.M.; Ramalho, P.; Silva, A.P.; Teixeira-Santos, R.; Pina-Vaz, C.; Rodrigues, A.G. Polyethyleneimine and polyethyleneimine-based nanoparticles: Novel bacterial and yeast biofilm inhibitors. *J. Med. Microbiol.* **2014**, *63*, 1167–1173. [CrossRef] [PubMed]
14. Li, Y.J.; Harroun, S.G.; Su, Y.C.; Huang, C.F.; Unnikrishnan, B.; Lin, H.J.; Lin, C.H.; Huang, C.C. Synthesis of Self-Assembled Spermidine-Carbon Quantum Dots Effective against Multidrug-Resistant Bacteria. *Adv. Healthc. Mater.* **2016**, *5*, 2545–2554. [CrossRef] [PubMed]
15. Liu, Y.-X.; An, X.-L.; Xu, Y.-N.; Hao, Y.-J.; Piao, X.-C.; Jin, M.-Y.; Lian, M.-L. Antibacterial and antibiofilm properties of dichloromethane fraction of extracts from adventitious roots of *Eurycoma longifolia* against *Staphylococcus aureus*. *LWT* **2022**, *162*, 113438. [CrossRef]
16. Cui, F.; Sun, J.; Ji, J.; Yang, X.; Wei, K.; Xu, H.; Gu, Q.; Zhang, Y.; Sun, X. Carbon dots-releasing hydrogels with antibacterial activity, high biocompatibility, and fluorescence performance as candidate materials for wound healing. *J. Hazard Mater.* **2021**, *406*, 124330. [CrossRef] [PubMed]
17. Hao, X.; Huang, L.; Zhao, C.; Chen, S.; Lin, W.; Lin, Y.; Zhang, L.; Sun, A.; Miao, C.; Lin, X.; et al. Antibacterial activity of positively charged carbon quantum dots without detectable resistance for wound healing with mixed bacteria infection. *Mater. Sci. Eng. C Mater. Biol. Appl.* **2021**, *123*, 111971. [CrossRef]
18. Cui, T.; Bai, F.; Sun, M.; Lv, X.; Li, X.; Zhang, D.; Du, H. Lactobacillus crustorum ZHG 2-1 as novel quorum-quenching bacteria reducing virulence factors and biofilms formation of *Pseudomonas aeruginosa*. *LWT* **2020**, *117*, 108696. [CrossRef]
19. Bag, A.; Bhattacharyya, S.K.; Pal, N.K.; Chattopadhyay, R.R. In vitro antibacterial potential of *Eugenia jambolana* seed extracts against multidrug-resistant human bacterial pathogens. *Microbiol. Res.* **2012**, *167*, 352–357. [CrossRef]
20. Shi, Y.G.; Zhang, R.R.; Zhu, C.M.; Xu, M.F.; Gu, Q.; Ettelaie, R.; Lin, S.; Wang, Y.F.; Leng, X.Y. Antimicrobial mechanism of alkyl gallates against *Escherichia coli* and *Staphylococcus aureus* and its combined effect with electrospun nanofibers on Chinese Taihu icefish preservation. *Food Chem.* **2021**, *346*, 128949. [CrossRef]
21. Wang, X.; He, S.; Yuan, L.; Deng, H.; Zhang, Z. Synthesis, Structure Characterization, and Antioxidant and Antibacterial Activity Study of Iso-orientin-Zinc Complex. *J. Agric. Food Chem.* **2021**, *69*, 3952–3964. [CrossRef] [PubMed]

22. Yu, S.; Li, G.; Zhao, P.; Cheng, Q.; He, Q.; Ma, D.; Xue, W. NIR-Laser-Controlled Hydrogen-Releasing PdH Nanohydride for Synergistic Hydrogen-Photothermal Antibacterial and Wound-Healing Therapies. *Adv. Funct. Mater.* **2019**, *29*, 1905697. [CrossRef]
23. Zheng, K.; Setyawati, M.I.; Leong, D.T.; Xie, J. Overcoming bacterial physical defenses with molecule-like ultrasmall antimicrobial gold nanoclusters. *Bioact. Mater.* **2021**, *6*, 941–950. [CrossRef]
24. Damiano, S.; Forino, M.; De, A.; Vitali, L.A.; Lupidi, G.; Taglialatela-Scafati, O. Antioxidant and antibiofilm activities of secondary metabolites from *Ziziphus jujuba* leaves used for infusion preparation. *Food Chem.* **2017**, *230*, 24–29. [CrossRef]
25. Lin, W.T.; Zhang, Y.Y.; Tan, H.L.; Ao, H.Y.; Duan, Z.L.; He, G.; Tang, T.T. Inhibited Bacterial Adhesion and Biofilm Formation on Quaternized Chitosan-Loaded Titania Nanotubes with Various Diameters. *Materials* **2016**, *9*, 155. [CrossRef]
26. Huang, J.; Liu, Y.; Yang, L.; Zhou, F. Synthesis of sulfonated chitosan and its antibiofilm formation activity against *E. coli* and *S. aureus*. *Int. J. Biol. Macromol.* **2019**, *129*, 980–988. [CrossRef]
27. Li, Q.; Yu, S.; Han, J.; Wu, J.; You, L.; Shi, X.; Wang, S. Synergistic antibacterial activity and mechanism of action of nisin/carvacrol combination against *Staphylococcus aureus* and their application in the infecting pasteurized milk. *Food Chem.* **2022**, *380*, 132009. [CrossRef]
28. Zhi, B.; Gallagher, M.J.; Frank, B.P.; Lyons, T.Y.; Qiu, T.A.; Da, J.; Mensch, A.C.; Hamers, R.J.; Zeev, R.; Howard, F.D. Investigation of phosphorous doping effects on polymeric carbon dots: Fluorescence, photostability, and environmental impact. *Carbon* **2018**, *129*, 438–449. [CrossRef]
29. Zhan, Y.; Shang, B.; Chen, M.; Wu, L. One-Step Synthesis of Silica-Coated Carbon Dots with Controllable Solid-State Fluorescence for White Light-Emitting Diodes. *Small* **2019**, *15*, e1901161. [CrossRef]
30. Cui, Y.; Zhang, J.; Zhang, G.; Huang, J.; Liu, P.; Antonietti, M.; Wang, X. Synthesis of bulk and nanoporous carbon nitride polymers from ammonium thiocyanate for photocatalytic hydrogen evolution. *J. Mater. Chem.* **2011**, *21*, 13032. [CrossRef]
31. Yang, S.; Gong, Y.; Zhang, J.; Zhan, L.; Ma, L.; Fang, Z.; Vajtai, R.; Wang, X.; Ajayan, P.M. Exfoliated graphitic carbon nitride nanosheets as efficient catalysts for hydrogen evolution under visible light. *Adv. Mater.* **2013**, *25*, 2452–2456. [CrossRef] [PubMed]
32. Wang, W.-N.; Zhang, C.-Y.; Zhang, M.-F.; Pei, P.; Zhou, W.; Zha, Z.-B.; Shao, M.; Qian, H.-S. Precisely photothermal controlled releasing of antibacterial agent from Bi2S3 hollow microspheres triggered by NIR light for water sterilization. *Chem. Eng. J.* **2020**, *381*, 122630. [CrossRef]
33. Girase, B.; Depan, D.; Shah, J.S.; Xu, W.; Misra, R.D.K. Silver–clay nanohybrid structure for effective and diffusion-controlled antimicrobial activity. *Mater. Sci. Eng. C* **2011**, *31*, 1759–1766. [CrossRef]
34. Qi, X.; Shah, H.; Nawaz, A.; Xie, W.; Akram, M.Z. Antibacterial Carbon-Based Nanomaterials. *Adv. Mater.* **2018**, *31*, e1804838.
35. Bardhan, S.; Pal, K.; Roy, S.; Das, S.; Chakraborty, A.; Karmakar, P.; Basu, R.; Das, S. Nanoparticle Size-Dependent Antibacterial Activities in Natural Minerals. *J. Nanosci. Nanotechnol.* **2019**, *19*, 7112–7122. [CrossRef]

processes

MDPI

Article

The Effect of Sugar Replacement with Different Proportions of Khalas Date Powder and Molasses on the Nutritional and Sensory Properties of Kleicha

Raya S. A. Almuziree and Raghad M. Alhomaid *

Department of Food Science and Human Nutation, College of Agriculture and Veterinary Medicine, Qassim University, Buraydah 51425, Saudi Arabia; 421200156@qu.edu.sa
* Correspondence: r.alhomaid@qu.edu.sa

Abstract: Evidence from the scientific community suggests that high levels of added sugar consumption contribute to the global epidemics of glucose intolerance, diabetes, and cardiovascular disease. The study aims to develop a local traditional food product (Kleicha) with healthy ingredients and to take advantage of the most productive crops in Saudi Arabia, namely dates, where Khalas date powder and molasses will be manufactured as an alternative to sugar in the Kleicha product. Six Kleicha samples with different fillings were manufactured as follows: Kleicha filled with sugar filling (KS), Kleicha with Khalas date powder filling (KD), Kleicha filled with Khalas date molasses filling (KM), Kleicha filled with sugar and Khalas date powder (1:1) filling (KSD), Kleicha filled with sugar and Khalas date molasses (1:1) filling (KSM), and Kleicha filled with Khalas date powder and Khalas date molasses (1:1) filling (KDM). In order to evaluate the prepared Kleicha samples, the proximate chemical composition, mineral content, phytochemical content and their antioxidant activities, sugar profiles using HPLC, in vitro glycemic index, glycemic load, and sensory evaluation were investigated. The results indicated that KM had the highest moisture content, while the KS sample had the lowest content. KD and KM had significantly higher ash contents compared with the other Kleicha samples. The protein and fat contents did not differ significantly. KD, KSD, and KDM demonstrated a significantly higher dietary fiber content than the other Kleicha samples. KS had the highest value of available carbohydrates. Regarding mineral content, the KDM sample had the highest K, Na, Ca, and P contents, while KD and KM had the highest Mg, Fe, and Mn contents. The Cu content indicated that KM had the highest content, representing a 25% increase compared with KS. Similarly, the Zn content in KM and KSD was significantly higher than in the other Kleicha samples. Replacing sucrose with Khalas date or molasses significantly changed the sucrose, glucose, and fructose contents. The phenolic content in KD, KM, and KDM was higher compared with KS, KSM, and KSD. Furthermore, the oxidative activity increased associatively with the addition of Khalas date powder and molasses. The percentage of sucrose decreased in KD and KDM. There was no significant difference in the general acceptance rate between the control sample and the other samples containing Khalas date powder, Khalas date molasses, and sugar or their mixture. In conclusion, it is possible to replace the sugar in the filling of the Kleicha with Khalas date powder and molasses or their mixture as it increases the health benefits; scaling up is recommended.

Keywords: Kleicha; dates; Khalas date powder; Khalas date molasses; sugar substitution; nutrition

Citation: Almuziree, R.S.A.; Alhomaid, R.M. The Effect of Sugar Replacement with Different Proportions of Khalas Date Powder and Molasses on the Nutritional and Sensory Properties of Kleicha. *Processes* **2023**, *11*, 3077. https://doi.org/10.3390/pr11113077

Academic Editors: Elwira Sieniawska, Shengbao Cai, Yongliang Zhuang, Zhipeng Yu and Hu Hou

Received: 7 October 2023
Revised: 21 October 2023
Accepted: 24 October 2023
Published: 26 October 2023

1. Introduction

Dates are rich in certain nutrients and are widely consumed in many countries, particularly in the Arab countries. Date palm is the highest fruit that contains sugars, and these ingredients vary according to the nature of the fruit, while dates are a concentrated source of sugar [1]. It is easily absorbed by the body and quickly reaches the blood and tissues. They are considered to have a lower glycemic load than other refined sugars. Dates are a rich source of multiple vitamins, minerals, polyphenols [2], and fiber [3]. The date fruit industry in Saudi Arabia is the largest in the world.

Date fruit (*P. dactylifera*) is one of the most important crops. It has been cultivated primarily for cultural, nutritional, environmental, religious, and social development purposes [4,5]. Dates can be found in numerous varieties, which differ from each other based on shape and organoleptic properties [4,6–8]. Date sugar, a by-product of date fruit processing, has the potential to replace white sugar in some recipes. Nutritionally, 100 g of cane sugar has 387 kcal of energy, 0 g of fiber, 0.05 mg of iron, 2 mg of potassium, 0.01 mg of zinc, and 0.007 mg of copper. On the other hand, 100 g of dates contains 282 calories, 8 g of fiber, 1.02 mg of iron, 656 mg of potassium, 0.29 mg of zinc, and 0.206 mg of copper [1,9]. As an alternative to sugar cane, date sugar may be used as a sweetener and would supply more than simply calories because of its high nutrient content and possible health benefits. Date fruit mainly comprises carbohydrates, ranging from 40 to more than 80% based on dry weight, depending on the cultivated variety. The evaluated sugars in dates, such as glucose, fructose, and sucrose, were found to vary in amounts, ranging from 2 to 95%, depending on the dates' variety; the highest glucose content of 95.4% was found in the Khalas variety. Date fruit is considered to be a rich source of dietary fiber, ranging from 2 to 8%; the highest fiber content of 8% was found in the Deglet-Noor variety, mostly cultivated in Algerian and Tunisia. The well-known Moroccan date variety Medjool also contains a high amount of fiber, reaching 6.7%. In comparison, the lowest amount of 2.7% was found in the Lulu variety from the United Arab Emirates [10]. Median amounts of fiber of 4.35% were found in Sukkari dates cultivated from Saudi Arabia [11]. Dates' dietary fiber consists of good quality fiber fractions such as β-glucans, arabinoxylans, and cellulose, providing an excellent source of dietary fiber, even better than that of cereals [8]. Dates contain extremely low amounts of fat and protein [7,12,13]. Dates are considered a relatively good source of different vitamins, mainly B complex: thiamine, riboflavin, and niacin [10]. Other studies have reported a relatively high content of tocopherols based on fresh weight for α-, β-, and γ-tocopherol [14]. Some date varieties may also contain vitamin C in relative amounts but higher than that found in dried fruits such as apricots, figs, and raisins [7,13]. Minerals such as potassium, calcium, and magnesium are found in relatively high amounts in dates, ranging from 0.05–0.9%. The highest element found in dates is mostly potassium, followed by calcium, iron, phosphorus, sodium, and copper [5,12]. The high potassium and low sodium content in dates provide an excellent product for hypertensive individuals [15]. Several detected phenolic compounds were found in various date varieties [16], representing date fruits as rich sources of polyphenols. Such a high and wide variety of phenolic compounds can play a major role in elevating the antioxidant potential by inhibiting lipid peroxidation and evaluating antioxidant activities [17]. Dates also contain carotenoids and phytosterols, which are reported to have a strong antioxidant capacity [8]. Dates' bioactive components have been demonstrated to promote several health-beneficial effects. Incorporating other nutritive ingredients besides dates can provide intensive boosters to dates' functional properties [1,18].

Khalas dates are considered one of the best dates in the Al Qassim region and one of the excellent commercial varieties desired in the markets as they retain their good flavor when stored for a long time. It is a great source of carbohydrates, dietary fiber, and important minerals [19], and the high sources of antioxidants help protect the body against harm from harmful free radicals [20]. In addition, it is common in Middle Eastern cooking and can be eaten fresh or dried. Date powder is a raw product of dates that varies according to the type and quality of dates. Dates were dried and ground only and not separated from all their properties and nutritional benefits due to their soluble sugars, fiber, polyphenols, vitamins, and mineral content [1,18]. Today, it could provide a healthier substitute for white sugar and other artificial sweeteners [1]. It is also added to many products, foods, and sweets because of its benefits, as it is used as an alternative to sugar in many food industries [18,21,22]. Date powder helps lower blood cholesterol; prevent atherosclerosis, cancer of the large intestine, and hemorrhoids; reduce the formation of gallstones; and facilitate pregnancy, childbirth, and puerperium because it contains good fiber and fast-digesting sugar [23–25]. Date molasses is a dark brown drink of thick texture, tastes very sweet, and is extracted from the fruits of dates. The methods of making molasses are as follows: The extraction process either occurs warmly or without heat, producing molasses. But, for each method, a different molasses quality in color grade is obtained; the cold method produces yellowish-red molasses (golden), and the hot method results in a dark brown color [26].

Traditional foods in Saudi Arabia are integral to the country's culture and heritage. These foods have been passed down from generation to generation and reflect the region's unique history, geography, and climate. One is the traditional sweet (Kleicha), which contains flour dough with sugar and fat, and is filled with sugar. It is a local product high in calories and added sugar. According to the traditional method, a cup of wheat flour, white flour, two tablespoons of powdered milk, two tablespoons of sugar, half a cup of olive oil, and a cup of warm water re mixed to form a dough. The dough is filled with a cup of fine sugar; a tablespoon of ground black lemon; a pinch of salt; and a teaspoon of cardamom, cinnamon, and ginger. Then, the surface of the Kleicha is brushed with whipped eggs with milk mix. Khalas date powder and molasses could be used as a substitute for their great health benefits. The idea came to replace the processed sugar used in manufacturing Kleicha with natural sweeteners to protect the consumer and the national economy. The effect of the replacement process according to different proportions and its impact on nutritional and sensory specifications will be studied in this study.

2. Materials and Methods

2.1. Ingredients

Khalas date fruits, date molasses, brown wheat flour, white flour, olive oil, powdered milk, and sugar were obtained from the date market in Buriydah City, Qassim region, SA. Date fruits were dried with a multi-functional dryer (Didactaitalia, cod 947227, Torino, Italy) for 3 days at a temperature of 50 °C. Then, in the labs of the Food Science and Human Nutrition, College of Agriculture and Veterinary Medicine, Qassim University, dried dates were ground with a powerful grinder (Santos, VITA-MAX CORP-Light Industrial Food Preparing Machine Model, VM0122E, St Joseph, MO, USA) to obtain homogenous Khalas date powder.

2.2. Formulation of Different Fillings of Kleicha

The flavoring spices and condiments added to all of the samples were tested and fixed: ground black lemon, ground cinnamon, ginger, ground cardamom, and citric acids. These spices were mixed in the portion of 24.5, 24.5, 24.5, 24.5, and 2 g to create a 100 g mix, respectively. Then, 50 g was added to each 1 kg of the filling mix, and the Kleicha samples were prepared as follows: Kleicha filled with flavored sucrose (KS) with prepared flavoring spices and condiments (1 kg mixed with 50 g); Kleicha filled with flavored Khalas date (KD) at the same flavoring portion; Kleicha filled with flavored Khalas date molasses (KM); Kleicha filled with a mix of flavored sucrose and Khalas date (1:1, *w:w*) (KSD); Kleicha filled with a mix of flavored sucrose and Khalas date molasses (1:1, *w:w*) (KSM); and Kleicha filled with a mix of flavored Khalas dates and Khalas date molasses (1:1, *w:w*) (KDM).

2.3. Manufacturing of Kleicha

The Kleicha was manufactured in the College of Agriculture and Veterinary Medicine factories on Saturday, 21 January 2023. The ingredients were prepared, then the dry materials were kneaded (5 kg of white flour, 5 kg of wheat flour, 400 g of sugar, 400 g of powdered milk, 30 g of salt, 90 g of cardamoms, and 90 g of ginger) with the fatty substance (2.5 L of virgin olive oil) for 15 min, and then approximately 5 L of water was gradually added until the desired consistency and texture were reached. The kneading process was carried out in a spiral arm kneader (Gam 16—Kneading dough machine, Thermotechnika Crown Cool, Bratislava, Slovakia). The dough was left for half an hour before shaping. The Kleicha samples were standardized at 50 g dough and then filled with 10 g of different filling mixtures. Then, the Kleicha ball dough was topically stamped with the traditional shape of the Kleicha. Then, the surface of the prepared Kleicha was brushed with approximately 5 g of whipped eggs with milk mix (8:2, *w:w*). The Kleicha was left for 10 min before baking; then, it was baked in a rotating oven at a temperature of 190 °C for 30 min. Then, the Kleicha was taken out and left to cool down before packaging.

2.4. Proximate Chemical Composition and Minerals of Kleicha

The formulated Kleicha were subjected to chemical analysis (moisture, ash, dietary fibers, and available carbohydrates) according to established methods [27]. The content of minerals, including sodium and potassium, was determined using flame photometry. In contrast, the calcium, magnesium, iron, copper, manganese, and zinc contents were analyzed using atomic absorption spectroscopy using the protocol of A.O.A.C. [27]. According to the description of Borah et al. [28], phosphorus was measured using a conventional colorimetric technique.

2.5. Phytochemicals Analysis of Kleicha

The TPC in the Kleicha samples was determined using the Folin–Ciocalteu reagent, according to Nsimba et al. [29], with slight modifications. Next, a microplate reader was used to determine the absorbance at 765 nm (BioTek, Winooski, VT, USA). Gallic acid (GA) solution was used to create a standard curve against which the results could be compared. Measurements of TPC were reported in terms of milligrams of gallic acid equivalents (GAE) per 100 g (mg of GAE 100 g^{-1} dw). The TFs content of Kleicha samples using the same extract was determined as Mohdaly et al. [30] described. By reacting the methanolic extract aliquots with sodium acetate (5%), the TFL concentrations in the Kleicha samples were calculated. Optical density (OD) was measured at 440 nm after 150 min with the addition of $AlCl_3$ (2%), as reported by Kumaran and Karunakaran [31]. The TF and TFL contents were then expressed as quercetin equivalents (QE mg) per 100 g dry matter (mg QE 100 g^{-1}). The total carotenoids (TCs) were determined by performing repeated extractions of 1 g of dried sample with a combination of acetone and petroleum ether (1:1, *v:v*), as described by Yuan et al. [32]. The top layer was gathered, rinsed thoroughly with water, and then mixed with the unfiltered extracts. The volume of the solution was calibrated by adding petroleum ether. The amount of TCs was calculated as mg 100 g^{-1} dry weight after the OD was spectrophotometrically measured at 450 nm.

2.6. Antioxidant Capacity Determination

For the purpose of determining the radical scavenging activity, Khalifa et al. [33] used a spectrophotometric method based on the bleaching of DPPH radicals in a purple solution with minor modifications. After that, a microplate reader (BioTek, Winooski, USA) detected absorbance at 517 nm. Trolox was utilized to create a standard curve and draw the Trolox calibration curve using the DPPH radical scavenging activity. The antiradical activity was assessed in µmol TE 100 g^{-1} of Trolox equivalents (TE) against ABTS radicals cation. The customized Lu et al. method of [34] was used to assess the Kleicha samples' radical scavenging ability (RSA). A Trolox calibration curve was plotted as a function of the percentage of the ABTS radical cation scavenging activity. The final results were expressed as micromoles of Trolox equivalents (TE) per 100 g (µmol of TE 100 g^{-1}).

2.7. Sugar Profile by HPLC

The Kleicha samples were prepared and homogenized to obtain homogeneous Kleicha powder, and then appropriately extracted. The sucrose, glucose, and fructose contents in the Kleicha samples were tested following the modified method of AOAC 982.14 using HPLC [35]. In brief, the appropriate crushed Kleicha samples were extracted three times with aqueous ethanolic solution (50%, *v:v*); the volume reached 100 mL, was centrifuged at 5000× *g* for 15 min, and filtered using 0.45 µm filter and injected into the HPLC. An HPLC system (Agilent Technologies, Palo Alto, CA, USA) equipped with an autosampler and Refractive index detector (RID) was used. The solvent system contained acetonitrile/d.H_2O (80:20, *v:v*) as the mobile phase. Then, 20 µL of the prepared sample extract was automatically injected, the flow rate was adjusted at 2 mL min^{-1}, and separation was performed at 30 °C. The areas under the present peaks were calculated, and sugar profile data were presented as a g 100 g^{-1} Kleicha sample.

2.8. In Vitro Glycemic Index of Kleicha

The Kleicha samples were digested according to Englyst et al. [36]. For this, 100 mg of the material was weighed into 50 mL tubes with 10 glass beads (5 mm diameter). The tubes received 2 mL of HCl (0.05 M) with pepsin (5 mg mL^{-1}, The Curated Chemical Collection, Old Dutch Road Victor, NY, USA). The tubes were incubated for 30 min in a shaking water bath at 37 °C. Next, we added 4 mL of sodium acetate buffer (0.5 M, pH 5.2), 1 mL of enzyme solution with 0.104 g pancreatin (Sigma, Steinheim, Germany, P7545), and 14.45 U amyloglucosidase (260 U mL^{-1}, Sigma-Aldrich, Steinheim, Germany) to each tube. In a shaking water bath, the tubes were incubated horizontally at 37 °C. Eppendorf tubes were filled with 100 µL aliquots every 20 min at 0 to 160 min intervals and mixed with 1 mL of 100% ethanol. These solutions were centrifuged at 800× *g* for 10 min and analyzed for glucose content using Glucose liquicolor (GOD-POP) reagent (human, Wiesbaden, Germany) on a microplate reader (BioTek, Winooski, USA) at 510 nm. The hydrolysis index (HI) measured starch digestion, and the predicted glycemic index (GI) measured sample digestibility against white bread. The hydrolysis index (HI) is the ratio of the sample's hydrolysis curve area to the white bread sample's hydrolysis curve. HI was estimated as follows (Equation (1)):

$$\mathrm{HI} = \frac{\text{The area under the curve of the sample}}{\text{The area under the curve of white bread}} \tag{1}$$

The in vitro GI was determined using the following equation (Equation (2)) of Goñi et al. [37]

$$\mathrm{GI} = 39.71 + 0.549\mathrm{HI} \tag{2}$$

$$\text{Glycemic Load (GL)} = \frac{\mathrm{GI} \times \text{Available carbohydrate (g)}}{100} \tag{3}$$

2.9. Sensory Evaluation of Kleicha

A sensory assessment of the samples was carried out by appointing 14 qualified and experienced arbitrators. The sensory assessment was carried out at room temperature of 25 °C, and an evaluation was carried out on numbered dishes containing the samples in terms of appearance, external color, internal color, texture, taste, aroma, and general acceptance using the 7-Hedonic scale method, where 7 indicates strong liking, 4 indicates neutrality, and 1 indicates a low level of liking. The results of the sensory assessment traits are reported according to the method described [38].

2.10. Statistical Analysis

One-way analysis of variance (ANOVA) was used for the statistical analysis, and SPSS version 25 was used to compile the results (IBM Corp., Armonk, NY, USA, Released 2013). Tukey's test was used for multiple comparisons, and the significance threshold was set at 0.05. The data were interpreted as though they came from a fully randomized study. Numbers were shown as means and standard deviations, according to Steel et al. [39].

3. Results

3.1. Proximate Composition of Kleicha

The proximate chemical composition of the manufactured Kleicha samples with different filling mixtures is presented in Table 1. Based on the table, the KM sample had the highest moisture content, measuring 14.18 g 100 g^{-1}, while the KS sample had the lowest moisture content at 8.23 g 100 g^{-1}. The KD and KM samples had significantly higher ash contents, measuring 1.45 g 100 g^{-1} each, compared with the other Kleicha samples filled with sugar/date powder, sugar/date molasses, or a mixture of date powder and date molasses, which had ash contents of 1.3, 1.17, and 1.32 g 100 g^{-1}, respectively. These results indicate that filling Kleicha with Khalas dried dates increased the ash content, as Khalas dried dates contain ash content. Only the KS sample was significantly lower in crud fat than all of the Kleicha samples, with 17.66 g 100 g^{-1}, whereas no significant difference was found among the other samples. The Kleicha sample containing different fillings did not present a significant protein content difference. The findings showed that Kleicha filled with Khalas dried dates, Khalas date molasses, or their mixture presented slightly higher protein content. Furthermore, the KD, KSD, and KDM samples demonstrated a significantly higher dietary fiber content than the other Kleicha samples, with values of 6.22, 6.07, and 6.09 g 100 g^{-1}, respectively. Regarding the available carbohydrate content, the KS sample had the highest value of 57.72 g 100 g^{-1}.

Table 1. Proximate composition of Kleicha filled with different filling mixtures (mean ± SE), $n = 3$.

Treatment *	Chemical Composition g 100 g^{-1} fw					
	Moisture	Ash	Curd Fat	Crude Protein	Dietary Fiber	Available Carbohydrates
KS	8.23 ± 0.02^{e}	1.15 ± 0.01^{c}	17.66 ± 0.08^{b}	15.25 ± 0.23^{ab}	4.97 ± 0.12^{bc}	57.72 ± 0.28^{a}
KD	9.95 ± 0.14^{d}	1.45 ± 0.01^{a}	18.24 ± 0.14^{a}	15.65 ± 0.15^{a}	6.22 ± 0.14^{a}	54.72 ± 0.31^{b}
KM	14.18 ± 0.25^{a}	1.45 ± 0.01^{a}	18.27 ± 0.08^{a}	14.76 ± 0.23^{b}	4.87 ± 0.11^{c}	51.36 ± 0.12^{d}
KSD	9.51 ± 0.11^{d}	1.3 ± 0.02^{b}	18.65 ± 0.07^{a}	15.24 ± 0.15^{ab}	6.07 ± 0.14^{a}	55.30 ± 0.20^{b}
KSM	11.02 ± 0.06^{c}	1.17 ± 0.01^{c}	18.66 ± 0.12^{a}	15.59 ± 0.11^{a}	5.47 ± 0.13^{b}	53.56 ± 0.16^{c}
KDM	11.6325 ± 0.08^{b}	1.32 ± 0.03^{b}	18.43 ± 0.20^{a}	15.33 ± 0.16^{ab}	6.09 ± 0.14^{a}	53.29 ± 0.32^{c}

* Treatments: KS: Kleicha filled with sugar filling; KD: Kleicha with Khalas date powder filling; KM: Kleicha filled with Khalas date molasses filling; KSD: Kleicha filled with sugar and Khalas date powder (1:1) filling; KSM: Kleicha filled with sugar and Khalas date molasses (1:1) filling; KDM: Kleicha filled with Khalas date powder and Khalas date molasses (1:1) filling, a–e: No significant difference ($p > 0.05$) between any two means within the same column have the same superscripted letters.

3.2. Minerals Content in Kleicha

The mineral content of the manufactured Kleicha samples with different filling mixtures is presented in Table 2. Among the Kleicha samples, the KDM sample had the highest K, Na, Ca, and P contents, measuring 1696.44, 1136.06, 209.24, and 28.78 mg 100 g^{-1}, respectively. Regarding the Mg, Fe, and Mn content, the KD and KM samples had the highest values, measuring 450.81 mg 100 g^{-1} and 450.58 mg 100 g^{-1} for Mg, 12.94 mg 100 g^{-1} and 12.37 mg 100 g^{-1} for Fe, and 10.29 mg 100 g^{-1} and 10.16 mg 100 g^{-1} for Mn. The copper content exhibited significant differences among the samples, with the KM sample having the highest content at 131.97 mg 100 g^{-1}, representing a 25% increase compared with KS (control sample). This confirms that the copper content of Khalas molasses was higher than that of the dates and sugar. Similarly, the Zn content of the KM and KSD samples was significantly higher than that of the other samples, measuring 54.85 mg 100 g^{-1} and 51.14 mg 100 g^{-1}, respectively.

Table 2. The mineral contents of Kleicha filled with different filling mixtures (mean ± SE), $n = 3$.

Treatments *	Minerals Content mg 100 g^{-1}								
	K	Na	Ca	P	Mg	Fe	Mn	Cu	Zn
KS	1151.33 ± 7.77 ^{d}	905.04 ± 23.45 ^{b}	163.92 ± 3.27 ^{c}	25.61 ± 0.48 ^{bc}	402.87 ± 4.75 ^{c}	9.83 ± 0.22 ^{bc}	8.88 ± 0.27 ^{c}	98.26 ± 0.77 ^{c}	34.31 ± 0.68 ^{b}
KD	1578.37 ± 30.31 ^{b}	948.85 ± 8.33 ^{b}	190.08 ± 4.19 ^{ab}	24.01 ± 0.41 ^{c}	450.81 ± 2.64 ^{a}	12.94 ± 0.9 ^{a}	10.29 ± 0.08 ^{a}	67.91 ± 2.19 ^{e}	33.55 ± 1.12 ^{b}
KM	1598.37 ± 25.89 ^{b}	1014.82 ± 20.95 ^{b}	177.42 ± 5.93 ^{bc}	25.29 ± 1.14 ^{bc}	450.58 ± 8.68 ^{a}	12.37 ± 0.12 ^{a}	10.16 ± 0.1 ^{a}	131.97 ± 1.31 ^{a}	54.85 ± 0.76 ^{a}
KSD	1450.42 ± 12.27 ^{c}	963.88 ± 20.87 ^{ab}	182.43 ± 4.28 ^{bc}	26.06 ± 0.48 ^{abc}	415.64 ± 6.34 ^{bc}	11.49 ± 0.31 ^{ab}	10.04 ± 0.2 ^{ab}	114.87 ± 0.92 ^{b}	51.14 ± 2.76 ^{a}
KSM	1463.58 ± 2.21 ^{c}	1024.86 ± 15.59 ^{ab}	187.59 ± 7.45 ^{ab}	27.6 ± 0.33 ^{ab}	421 ± 4.55 ^{bc}	9.45 ± 0.26 ^{c}	9.49 ± 0.06 ^{bc}	75.22 ± 2.08 ^{d}	36.56 ± 3.52 ^{b}
KDM	1696.44 ± 6.38 ^{a}	1136.06 ± 83.3 ^{a}	209.24 ± 5.85 ^{a}	28.78 ± 1.01 ^{a}	433.47 ± 3.08 ^{ab}	10.24 ± 0.22 ^{bc}	9.85 ± 0.1 ^{ab}	42.67 ± 1.68 ^{f}	21.42 ± 2.4 ^{c}

* Treatments: KS: Kleicha filled with sugar filling; KD: Kleicha with Khalas date powder filling; KM: Kleicha filled with Khalas date molasses filling; KSD: Kleicha filled with sugar and Khalas date powder (1:1) filling; KSM: Kleicha filled with sugar and Khalas date molasses (1:1) filling; KDM: Kleicha filled with Khalas date powder and Khalas date molasses (1:1) filling, $^{a–f}$: No significant difference ($p > 0.05$) between any two means within the same column have the same superscripted letters.

3.3. Sugar Profile by HPLC

The sucrose content in the Kleicha samples varied (Figure 1), with KS having the highest content at 17.69 g 100 g^{-1}, while the KM, KD, and KDM samples had the lowest. Consequently, the samples containing Khalas date powder and molasses had a lower percentage of sucrose. Regarding the glucose content, the KS, KD, KM, KSD, KSM, and KDM samples contained 1.52, 5.67, 3.62, 3.25, 3.17, and 5.22 g 100 g^{-1}, respectively. Similarly, the fructose content in these samples was 1.7, 6.35, 3.96, 3.57, 3.56, and 5.92 g 100 g^{-1}, respectively. Kleicha filled with Khalas date powder or a mixture of Khalas date powder and molasses showed slightly higher glucose and fructose contents than the other samples. Conversely, the KS sample had the lowest glucose and fructose content.

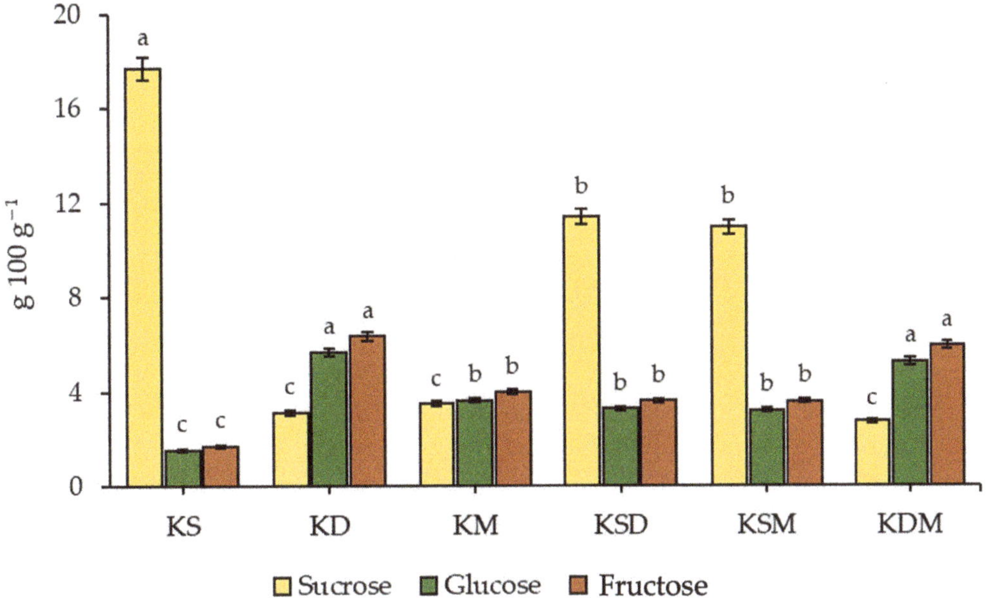

Figure 1. The sugar profile using HPLC of Kleicha filled with different filling mixtures. Treatments: KS: Kleicha filled with sugar filling; KD: Kleicha with Khalas date powder filling; KM: Kleicha filled with Khalas date molasses filling; KSD: Kleicha filled with sugar and Khalas date powder (1:1) filling; KSM: Kleicha filled with sugar and Khalas date molasses (1:1) filling; KDM: Kleicha filled with Khalas date powder and Khalas date molasses (1:1) filling (mean $\pm$ SE), $n = 3$. [a–c]: Bars of each sugar kind that do not share similar letters differ significantly ($p > 0.05$).

3.4. In Vitro Glycemic Index

The results of the in vitro glycemic index (GI) for Kleicha samples are presented in Table 3. The Kleicha sample filled with Khalas date molasses had the highest GI, significantly differing from the other samples, with a value of 40.30. On the other hand, the Kleicha samples filled with a mixture of Khalas date powder and molasses had the lowest GI at 40.15. According to the classification of foods based on GI values, the glycemic index in this study is considered low, as it is less than 55. Furthermore, the glycemic load (GL) was evaluated, and the KS sample had the highest GL at 21.21, while the KM sample had the lowest at 18.73. Foods are classified based on their glycemic load as low (GL $\leq$ 10), medium (GL 11–19), and high (GL $\geq$ 20).

Table 3. Glycemic index and glycemic load of Kleicha filled with different filling mixtures (mean ± SE), *n* = 3.

Treatment *	Glycemic Index	Glycemic Load
KS	40.21 ± 0.01 ^{bc}	21.21 ± 0.01 ^{a}
KD	40.18 ± 0.01 ^{cd}	19.49 ± 0.02 ^{c}
KM	40.30 ± 0.01 ^{a}	18.73 ± 0.01 ^{f}
KSD	40.19 ± 0.01 ^{bc}	19.78 ± 0.01 ^{b}
KSM	40.21 ± 0.02 ^{b}	19.34 ± 0.03 ^{d}
KDM	40.15 ± 0.02 ^{d}	18.95 ± 0.01 ^{e}

* Treatments: KS: Kleicha filled with sugar filling; KD: Kleicha with Khalas date powder filling; KM: Kleicha filled with Khalas date molasses filling; KSD: Kleicha filled with sugar and Khalas date powder (1:1) filling; KSM: Kleicha filled with sugar and Khalas date molasses (1:1) filling; KDM: Kleicha filled with Khalas date powder and Khalas date molasses (1:1) filling, $^{a–f}$: No significant difference ($p > 0.05$) between any two means within the same column have the same superscripted letters.

3.5. Phytochemicals Analysis of Kleicha

The results of the quantitative analysis of phytochemicals, including the total phenolic compounds (TPC), total flavonoids (TF), total flavonols (TFL), and total catechins (TC), in the Kleicha filled with different filling mixtures are presented in Table 4. Among the samples, the KD sample exhibited the highest concentration of TPC, measuring 1498.66 mg GAE 100 g^{-1}. In contrast, the Kleicha sugar-filled sample showed a significantly lower TPC content. The KD sample exhibited the most significant increase in total flavonoid (TF) content, measuring 340.67 mg QE 100 g^{-1}. Conversely, the KS sample showed a significant decrease in TF, with a 228.28 mg QE 100 g^{-1} content. Regarding the total flavanols (TFL), the KD sample demonstrated a significant increase, reaching 148.05 mg QE 100 g^{-1}, higher than the other samples, except for KDM, which had a TFL content of 134.18 mg QE 100 g^{-1}. In contrast, the KS sample had a significantly lower TFL content of 108.68 mg QE 100 g^{-1}. Similarly, the TC content of Kleicha filled with different mixtures was analyzed, and the KD and KDM samples showed the highest TC levels, measuring 425.66 mg 100 g^{-1} and 405.75 mg 100 g^{-1}, respectively. In contrast, the KS sample had the lowest TC content, with 334.46 mg 100 g^{-1}. Interestingly, there was a significant increase in TPC, TF, TFL, and TC in the KD and KDM samples compared with the Kleicha filled with sugar. Conversely, the Kleicha filled with sugar exhibited the most significant decrease in these phytochemical contents.

Table 4. The phytochemicals analysis of Kleicha filled with different filling mixtures (mean ± SE), n = 3.

Treatment *	Phytochemicals Analysis			
	TPC mg GAE 100 g^{-1}	TF mg QE 100 g^{-1}	TFL mg QE 100 g^{-1}	Carotenoids mg 100 g^{-1}
KS	701.16 ± 29.92 ^{d}	228.28 ± 9.24 $_{e}$	108.68 ± 8.16 ^{c}	334.46 ± 26.64 ^{c}
KD	1498.66 ± 22.52 ^{a}	340.67 ± 3.92 ^{a}	148.05 ± 6.15 ^{a}	425.66 ± 8.77 ^{a}
KM	1252.07 ± 33.27 ^{b}	299.19 ± 7.45 ^{bc}	115.67 ± 3.04 ^{bc}	365.31 ± 11.92 ^{bc}
KSD	1076.2 ± 32.20 ^{c}	267.74 ± 10.13 ^{cd}	120.01 ± 2.70 ^{bc}	359.24 ± 11.62 ^{bc}
KSM	971.63 ± 21.16 ^{c}	259.99 ± 11.18 ^{de}	120.94 ± 2.60 ^{bc}	366.56 ± 7.05 ^{bc}
KDM	1398.5 ± 39.71 ^{a}	310.45 ± 4.64 ^{ab}	134.18 ± 4.96 ^{ab}	405.75 ± 5.35 ^{ab}

* Treatments: KS: Kleicha filled with sugar filling; KD: Kleicha with Khalas date powder filling; KM: Kleicha filled with Khalas date molasses filling; KSD: Kleicha filled with sugar and Khalas date powder (1:1) filling; KSM: Kleicha filled with sugar and Khalas date molasses (1:1) filling; KDM: Kleicha filled with Khalas date powder and Khalas date molasses (1:1) filling, $^{a–e}$: No significant difference ($p > 0.05$) between any two means within the same column have the same superscripted letters.

3.6. Antioxidant Capacity Determination

The antioxidant activities of the manufactured Kleicha samples with different filling mixtures were evaluated using the DPPH and ABTS assays, as shown in Figure 2. The

DPPH radical scavenging activity revealed that the KS sample had the lowest antioxidant capacity, measuring 273.19 μmol 100 g^{-1}. On the other hand, the KD sample exhibited the highest antioxidant capacity, with a value of 1507.15 μmol 100 g^{-1}. Similarly, the ABTS radical scavenging activity was measured and showed values of 685.05, 1447.08, 1098.01, 921.1, 909.65, and 1345.14 for KS, KD, KM, KSD, KSM, and KDM, respectively. Notably, the KD and KDM samples demonstrated the highest antioxidant capacities compared with the other samples, while the KS sample had the lowest capacity.

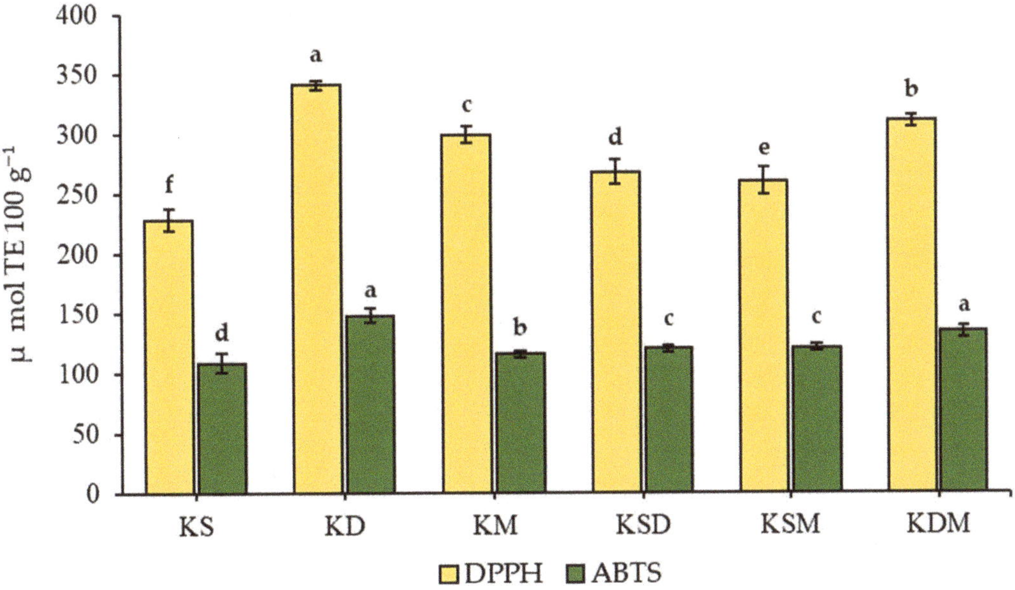

Figure 2. The DPPH and ABTS content of Kleicha filled with different filling mixtures. Treatments: KS: Kleicha filled with sugar filling; KD: Kleicha with Khalas date powder filling; KM: Kleicha filled with Khalas date molasses filling; KSD: Kleicha filled with sugar and Khalas date powder (1:1) filling; KSM: Kleicha filled with sugar and Khalas date molasses (1:1) filling; KDM: Kleicha filled with Khalas date powder and Khalas date molasses (1:1) filling (mean ± SE), $n = 3$. $^{a–f}$: Bars not sharing similar letters differed significantly ($p > 0.05$).

3.7. Sensory Evaluation

The sensory evaluation of the Kleicha samples filled with different filling mixtures was conducted, assessing six aspects, namely: appearance, external color, internal color, taste, texture, and aroma. The results of these characteristics were used to determine the general acceptance rate. The general acceptance rates for the Kleicha samples were 94.43%, 81.64%, 76.07%, 87.21%, 87.93%, and 80.29% for KS, KD, KM, KSD, KSM, and KDM, respectively (Figure 3).

Interestingly, filling the Kleicha with either Khalas date powder or Khalas date molasses alone did not indicate the highest general acceptance. However, incorporating a sugar mixture with date powder or molasses did not significantly impact the general acceptance. This suggests that incorporating Khalas date powder or molasses can improve the nutritional value of Kleicha compared with adding sugar alone.

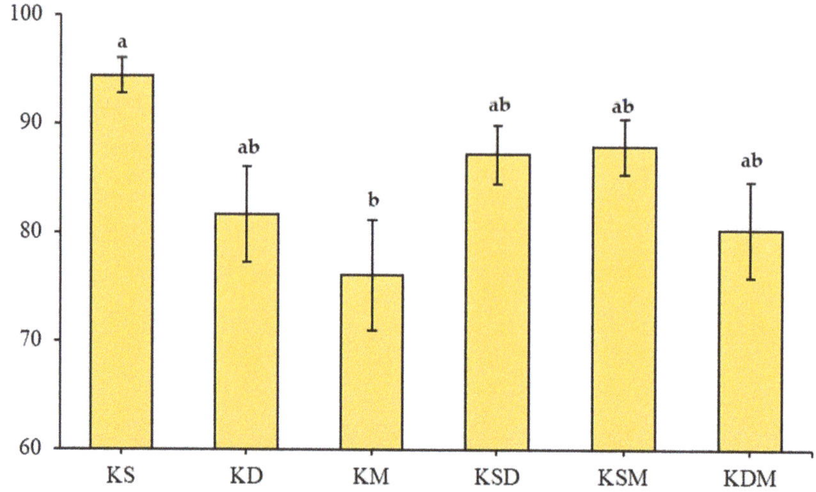

Figure 3. The general acceptance of Kleicha filled with different filling mixtures. Treatments: KS: Kleicha filled with sugar filling; KD: Kleicha with Khalas date powder filling; KM: Kleicha filled with Khalas date molasses filling; KSD: Kleicha filled with sugar and Khalas date powder (1:1) filling; KSM: Kleicha filled with sugar and Khalas date molasses (1:1) filling; KDM: Kleicha filled with Khalas date powder and Khalas date molasses (1:1) filling (mean $\pm$ SE), $n = 3$. ^{a,b}: Bars not sharing similar letters differed significantly ($p > 0.05$).

4. Discussion

The traditional cuisine of the Kingdom of Saudi Arabia plays a significant role in the history and identity of the country. These recipes have been passed down through the generations and accurately reflect the history, geography, and climate of their respective regions. Kleicha is a popular example of a traditional sweet; it is made of whole flour that has been sweetened with sucrose and flavored by a special spice mix. It is considered a sweet and calorie-dense local product. Date palm and its products, including date molasses and date powder, have been used to achieve this goal of using unhealthy ingredients like shortening and added sugars. The impact of the substitution process in varying proportions on nutritional and sensory qualities was studied. For instance, Khalas dates are highly prized as a commercial variety and are among the best in the Al Qassim area. Carbohydrates, fiber, and several essential minerals can all be found in abundant quantities [40]. It is also a high source of antioxidants that help protect the body against harm from harmful free radicals [41]. Date powder and molasses are used as an alternative to sugar in many food industries [42]. With some functional components, such as its pectin content, it aids in reducing blood cholesterol and warding off atherosclerosis. It also protects against developing colon cancer, stops hemorrhoids, lessens the chances of gallstones, and causes the nine months of pregnancy and the postpartum period to be easier [26].

In the current research, the proximate chemical composition of the manufactured Kleicha samples with different filling mixtures was determined. The increase in moisture content in the KM sample can be attributed to the formation of hydrogen bonds with water molecules, which is associated with the higher hydroxyl group content in date molasses than sugar. This results in reduced free water movement and an increase in the moisture content of prepared Kleicha. A study by Nadeem et al. [21] focused on replacing sugar with date powder in muffins showed similar results. Its samples, which had the highest percentage of date substitution, exhibited an increased moisture content. The KD and KM samples had significantly higher ash contents compared with the other Kleicha samples filled with sugar/date powder, sugar/date molasses, or a mixture of date powder and date molasses. These results indicate that filling Kleicha with Khalas dried dates increases the ash content, as Khalas dried dates have an ash content of 1.8 g 100 g^{-1} [20]. This finding aligns with the study by [21], where the sample with the highest ash content had the highest date content. Comparing the curd fat content with the Kleicha samples, the KS sample had significantly lower crude fat content compared with the other Kleicha samples. However, there were no significant differences observed among the remaining samples. It is worth noting that dates generally have a low fat content, with dried dates containing only 0.38 g of fat per 100 g [41]. Regarding the protein content, there was no clear significant difference observed among the Kleicha samples with different fillings. However, Kleicha filled with Khalas dried dates, Khalas date molasses, or a mixture of both showed a slightly higher protein content. Dates contain proteins in small amounts, with an average protein content of 2.14 g 100 g^{-1} in dried dates. The study conducted by Nadeem et al. [21] also confirmed that the sample with the highest protein content was T4 (100% date powder), which is consistent with the results of this study. Furthermore, the KD, KSD, and KDM samples demonstrated a significantly higher dietary fiber content than the other Kleicha samples. This increase in dietary fiber content can be attributed to incorporating Khalas date powder, as dried Khalas dates have a dietary fiber content of 8.0 g 100 g^{-1} [20]. This finding is consistent with the previous study by Nadeem et al. [21], which also showed increased dietary fiber content by incorporating date powder. The remarkable changes in filled Kleich with the date or its by-products might be related to the high nutritional values regarding the macro- and micro-nutrients, as previously indicated [1,9].

The mineral content in the KDM sample had the highest K, Na, Ca, and P contents. This can be attributed to the high mineral content in Khalas dates, which contain 742 mg 100 g^{-1}, 82 mg 100 g^{-1}, 85 mg 100 g^{-1}, and 74 mg 100 g^{-1} of dry weight for K, Na, Ca, and P, respectively [20,40]. Date molasses also contributes to the mineral content, with Ca and P present at a rate of 21.8 mg 100 g^{-1} and 48.2 mg 100 g^{-1}, respectively [40]. These findings align with a study by Gamal et al. [41], which demonstrated increased potassium, sodium, calcium, and phosphorus contents with the fortification of biscuits with wheat bran and date fruit. Regarding the Mg, Fe, and Mn content, the KD and KM samples had the highest values. Khalas dates contain 62 mg 100 g^{-1} of magnesium and 0.6 mg 100 g^{-1} of iron in their dry weight [20,42], while date molasses contains 52.5 mg 100 g^{-1} of magnesium and 0.4 mg 100 g^{-1} of iron [40]. These results are consistent with the study by Gamal et al. [41], which also showed an increased Mg content with the fortification of biscuits with dates. The copper content exhibited significant differences among the samples, with the KM sample having the highest content. This confirms that the Cu content of Khalas molasses is higher than that of dates and sugar. Similarly, the Zn content of the KM and KSD samples was significantly higher than that of the other samples. Date molasses contributes to the copper and zinc content, with Cu present at a rate of 0.08 mg 100 g^{-1} and Zn at 0.2 mg 100 g^{-1} [40]. These findings align with the study by Gamal et al. [41], which demonstrated an increased zinc content and the percentage of dates in biscuits. The obtained results agree with the reviewed data by Barakat and Alfheeaid [1]. Also, adding dates and their by-products enriched the minerals content as relatively expected from previous literature [5,12]. Interestingly, the newly formulated Kleisha could be a functional product for diabetes and hypertensive individuals [15].

The sucrose content in the Kleicha samples varied, with KS having the highest content while the KM, KD, and KDM samples had the lowest. This can be attributed to the low sucrose content in Khalas date powder and molasses, approximately 0.2 g per 100 g [40]. Consequently, the samples containing Khalas date powder and molasses had a lower percentage of sucrose. Kleicha filled with Khalas date powder or a mixture of Khalas date powder and molasses showed slightly higher glucose and fructose contents than the other samples. Conversely, the KS sample had the lowest glucose and fructose content. This can be attributed to the predominance of monosaccharides (glucose and fructose) in Khalas dates, with approximately 26.6 g 100 g^{-1} and 22.8 g 100 g^{-1} of glucose and fructose, respectively [43].

The Kleicha sample filled with Khalas date molasses had the highest GI, significantly differing from the other samples. On the other hand, the Kleicha samples filled with a mixture of Khalas date powder and molasses had the lowest GI. According to the classification of foods based on GI values, the glycemic index in this study is considered low, as it was less than 55. This can be attributed to Khalas dates having a low glycemic index of 35.5 [44]. A low-glycemic-index diet has been associated with protection against obesity, colon cancer, and breast cancer. A study by Sameen et al. [45] found that date molasses reduced the GI when used to replace sugar in ice cream. Furthermore, GL was evaluated, and the KS sample had the highest GL, while the KM sample had the lowest one. Foods are classified based on their GL as low (GL $\leq$ 10), medium (GL 11–19), and high (GL $\geq$ 20). Date molasses is classified as having a low glycemic load with a value of 11.29 [46], and dates, in general, are considered to have a medium glycemic load of 13.58 [47]. A high glycemic load is associated with the rapid release of triglycerides into the bloodstream [48].

The results of phytochemicals analysis in the prepared Kleicha filled with different filling mixtures indicated the KD sample exhibited the highest concentration of TPC. In contrast, the Kleicha sugar-filled sample showed a significantly lower TPC content. These findings align with a study by Mohamed et al. [49], which investigated the phytochemical composition of 16 commercial date cultivars, including Khalas, grown in Saudi Arabia. The study reported that Khalas dates contain approximately 149 mg GAE 100 g^{-1} DW of TPC. Additionally, the results of this study support the findings of a study by Manickavasagan et al. [50]. The study demonstrated that incorporating dates and molasses into IDLI led to an increase in TPC content. Specifically, the IDLI sample supplemented with chopped dates contained 75.1 mg 100 g^{-1} of TPC. Supplemented samples with date syrup contained 82.1 mg 100 g^{-1}, compared with the control sample, which had only 4.5 mg 100 g^{-1} TPC. These results are consistent with the findings of Al-Farsi and Lee [51], who emphasized the potential of dates and date molasses to enhance the phytochemical content of food products. The KD sample exhibited the most significant increase in TF content. Conversely, the KS sample showed a significant decrease in TF. Regarding TFL, the KD sample demonstrated a significant increase higher than the other samples, except for KDM, which had a valuable TFL content. The KS sample had a significantly lower TFL content. It is worth noting that Khalas dates had a TF content of 27 mg QE 100 g^{-1} DW [49]. Similarly, the TC content of the KD and KDM samples showed the highest TC levels. In contrast, the KS sample had the lowest TC content. This difference can be attributed to the high carotene content in dried Khalas dates, which contain approximately 2900 μg 100 g^{-1} [20]. Notably, the major carotenoids found in dates are lutein, β-carotene, and neoxanthin [19]. Interestingly, there was a significant increase in TPC, TF, TFL, and TC in the KD and KDM samples compared with the Kleicha filled with sugar. Conversely, the Kleicha filled with sugar exhibited the most significant decrease in these phytochemical contents. The antioxidant activities, such as DPPH and ABTS radical scavenging activity, revealed that the KS sample had the lowest antioxidant capacity. The KD sample exhibited the highest antioxidant capacity. Notably, the KD and KDM samples demonstrated the highest antioxidant capacities compared with the other samples, while the KS sample had the lowest capacity. This can be attributed to the fact that Khalas dates contain significant levels of oxidative activity, with 8.25 mg GAE of DPPH radical scavenging and 1.53 mg GAE of ABTS radical scavenging [49].

The sensory evaluation of Kleicha samples filled with different filling mixtures was conducted, assessing six aspects. The KS sample scored the highest acceptance, followed by KSM and KSD. Interestingly, filling the Kleicha with either Khalas date powder or Khalas date molasses alone did not indicate the highest acceptance score. This suggests that incorporating Khalas date powder or molasses can improve the nutritional value of Kleicha compared with adding sugar alone. This finding aligns with a study by Assirey [12], where the control sample with 100% sugar had a high general acceptance. There was no significant difference between the control and cookie samples where sugar was replaced with 25% and 50% date powder or 75% date molasses. In another study by Gamal et al. [41], the results indicated that the control sample had the highest general acceptance. There was no significant difference between the control and biscuit samples fortified with 10% and 20% date powder, while the sample fortified with 40% had the lowest acceptance score.

5. Conclusions

In conclusion, our research intended to create a less unhealthy version of the traditional food product Kleicha by substituting Khalas date powder and molasses for sugar. Six samples of Kleicha were analyzed, all of which had different fillings such as sugar, Khalas date powder, Khalas date molasses, and a few different mixes of these. Samples supplemented with Khalas date powder or molasses showed an elevated phenolic content, a marker of greater health benefits. Furthermore, the oxidative activity increased when Khalas date powder was added, whereas the sucrose content dropped when Khalas date powder and molasses were used. More importantly, there was no discernible difference in acceptance rates between the control group and the group that had fillings based on dates. Based on these results, replacing sugar with Khalas date powder, molasses, or a combination of the two in Kleicha can have positive health effects while maintaining the original flavor.

Author Contributions: Conceptualization, R.M.A. and R.S.A.A.; methodology, R.S.A.A.; investigation, R.M.A. and R.S.A.A.; data curation, R.M.A.; writing—original draft preparation, R.S.A.A.; writing—review and editing, R.M.A.; visualization, R.S.A.A.; supervision, R.M.A.; project administration, R.M.A. All authors have read and agreed to the published version of the manuscript.

Funding: The project was funded by the Deanship of Scientific Research in Qassim University (CAVM-2022-1-1-J-27842).

Data Availability Statement: Data are contained within the article.

Acknowledgments: The authors gratefully acknowledge Qassim University, represented by the Deanship of "Scientific Research, on the financial support for this research under the number (CAVM-2022-1-1-J-27842) during the academic year 1444 AH/2022 AD". The authors also thank Qassim University for its technical support.

Conflicts of Interest: The authors declare no conflict of interest.

References

1. Barakat, H.; Alfheeaid, H.A. Date Palm Fruit (*Phoenix dactylifera*) and Its Promising Potential in Developing Functional Energy Bars: Review of Chemical, Nutritional, Functional, and Sensory Attributes. *Nutrients* **2023**, *15*, 2134. [CrossRef] [PubMed]
2. Nadeem, M.; Qureshi, T.M.; Ugulu, I.; Riaz, M.N.; An, Q.; Khan, Z.I.; Ahmad, K.; Ashfaq, A.; Bashir, H.; Dogan, Y. Mineral, Vitamin and Phenolic Contents and Sugar Profiles of Some Prominent Date Palm (*Phoenix dactylifera*) Varieties of Pakistan. *Pak. J. Bot.* **2019**, *51*, 171–178. [CrossRef] [PubMed]
3. Elleuch, M.; Besbes, S.; Roiseux, O.; Blecker, C.; Deroanne, C.; Drira, N.-E.; Attia, H. Date Flesh: Chemical Composition and Characteristics of the Dietary Fibre. *Food Chem.* **2008**, *111*, 676–682. [CrossRef]
4. Ibrahim, S.A.; Ayad, A.A.; Williams, L.L.; Ayivi, R.D.; Gyawali, R.; Krastanov, A.; Aljaloud, S.O. Date Fruit: A Review of the Chemical and Nutritional Compounds, Functional Effects and Food Application in Nutrition Bars for Athletes. *Int. J. Food Sci.* **2021**, *56*, 1503–1513. [CrossRef]
5. Siddeeg, A.; Zeng, X.-A.; Ammar, A.-F.; Han, Z. Sugar Profile, Volatile Compounds, Composition and Antioxidant Activity of Sukkari Date Palm Fruit. *J. Food Sci. Technol.* **2019**, *56*, 754–762. [CrossRef] [PubMed]
6. Echegaray, N.; Pateiro, M.; Gullon, B.; Amarowicz, R.; Misihairabgwi, J.M.; Lorenzo, J.M. *Phoenix dactylifera* Products in Human Health—A Review. *Trends Food Sci. Technol.* **2020**, *105*, 238–250. [CrossRef]
7. Siddiq, M.; Greiby, I. Overview of Date Fruit Production, Postharvest Handling, Processing, and Nutrition. In *Dates: Postharvest Science, Processing Technology and Health Benefits*; John Wiley & Sons: Hoboken, NJ, USA, 2013; pp. 1–28.
8. Hussain, M.I.; Farooq, M.; Syed, Q.A. Nutritional and Biological Characteristics of the Date Palm Fruit (*Phoenix dactylifera* L.)—A Review. *Food Biosci.* **2020**, *34*, 100509. [CrossRef]
9. United States Department of Agriculture (USDA). 2020. Available online: https://fdc.nal.usda.gov/fdc-app.html (accessed on 22 May 2023).
10. Amadou, I. Date Fruits: Nutritional Composition of Dates (Balanites Aegyptiaca Delile and *Phoenix dactylifera* L.). In *Nutritional Composition of Fruit Cultivars*; Elsevier: Amsterdam, The Netherlands, 2016; pp. 215–233.

11. Aljutaily, T.; Barakat, H.; Moustafa, M.M.; Rehan, M. Incorporation of Sukkari Date in Probiotic-Enriched Fermented Camel Milk Improves the Nutritional, Physicochemical, and Organoleptical Characteristics. *Fermentation* **2021**, *8*, 5. [CrossRef]
12. Assirey, E.A.R. Nutritional Composition of Fruit of 10 Date Palm (*Phoenix dactylifera* L.) Cultivars Grown in Saudi Arabia. *J. Taibah Univ. Sci.* **2015**, *9*, 75–79. [CrossRef]
13. Ayad, A.A.; Williams, L.L.; Gad El-Rab, D.A.; Ayivi, R.; Colleran, H.L.; Aljaloud, S.; Ibrahim, S.A. A Review of the Chemical Composition, Nutritional and Health Benefits of Dates for Their Potential Use in Energy Nutrition Bars for Athletes. *Cogent Food Agric.* **2020**, *6*, 1809309. [CrossRef]
14. Hamad, I.; AbdElgawad, H.; Al Jaouni, S.; Zinta, G.; Asard, H.; Hassan, S.; Hegab, M.; Hagagy, N.; Selim, S. Metabolic Analysis of Various Date Palm Fruit (*Phoenix dactylifera* L.) Cultivars from Saudi Arabia to Assess Their Nutritional Quality. *Molecules* **2015**, *20*, 13620–13641. [CrossRef] [PubMed]
15. Ralston, R.; Lee, J.; Truby, H.; Palermo, C.; Walker, K. A Systematic Review and Meta-Analysis of Elevated Blood Pressure and Consumption of Dairy Foods. *J. Hum. Hypertens.* **2012**, *26*, 3–13. [CrossRef] [PubMed]
16. Eid, N.M.; Al-Awadi, B.; Vauzour, D.; Oruna-Concha, M.J.; Spencer, J.P. Effect of Cultivar Type and Ripening on the Polyphenol Content of Date Palm Fruit. *J. Agric. Food Chem.* **2013**, *61*, 2453–2460. [PubMed]
17. Razali, N.S.M.; Wenyin, B.; Arjunan, R.D.; Hashim, H.; Abdullah, A. Total Phenolic Content and Antioxidant Activities of Date Fruit Extracts. *Malays. Appl. Biol.* **2019**, *48*, 103–108.
18. Alfheeaid, H.A.; Barakat, H.; Althwab, S.A.; Musa, K.H.; Malkova, D. Nutritional and Physicochemical Characteristics of Innovative High Energy and Protein Fruit- and Date-Based Bars. *Foods* **2023**, *12*, 2777. [CrossRef]
19. Al-Farsi, M.A.; Lee, C.Y. Nutritional and Functional Properties of Dates: A Review. *Crit. Rev. Food Sci. Nutr.* **2008**, *48*, 877–887. [CrossRef] [PubMed]
20. Al-Farsi, M.; Alasalvar, C.; Morris, A.; Baron, M.; Shahidi, F. Comparison of Antioxidant Activity, Anthocyanins, Carotenoids, and Phenolics of Three Native Fresh and Sun-Dried Date (*Phoenix dactylifera* L.) Varieties Grown in Oman. *J. Agric. Food Chem.* **2005**, *53*, 7592–7599. [CrossRef] [PubMed]
21. Nadeem, M.; Qureshi, T.M.; Ahmad, M.M.; Riaz, M.N.; Ameer, A. Development of Free Flowing Date Powder and Its Utilization in Muffins to Enhance Nutritional Value. *J. Agric. Res.* **2017**, *55*, 03681157.
22. Abd El-Hady, E.S.A.; Alfheeaid, H.A.; Abd El-Razik, M.M.; Althwab, S.A.; Barakat, H. Nutritional, Physicochemical, and Organoleptic Properties of Camel Meat Burger Incorporating Unpollinated Barhi Date Fruit Pulp. *Int. J. Food Sci.* **2022**, *2022*, 4581821. [CrossRef]
23. Vayalil, P.K. Date Fruits (*Phoenix dactylifera* L.): An Emerging Medicinal Food. *Crit. Rev. Food Sci. Nutr.* **2012**, *52*, 249–271. [CrossRef]
24. Al-Dashti, Y.A.; Holt, R.R.; Keen, C.L.; Hackman, R.M. Date Palm Fruit (*Phoenix dactylifera*): Effects on Vascular Health and Future Research Directions. *Int. J. Mol. Sci.* **2021**, *22*, 4665. [CrossRef]
25. Ali, Z.; Li, J.; Zhang, Y.; Naeem, N.; Younas, S.; Javeed, F. Dates (*Phoenix dactylifera*) and Date Vinegar: Preventive Role against Various Diseases and Related in Vivo Mechanisms. *Food Rev. Int.* **2022**, *38*, 480–507. [CrossRef]
26. Al-Mudhaffar, A.W. *Dates and Sugar Technology*; Aldar Al-Jameaeh for Printing, Publishing and Translation; University of Kufa Iraq: Kufa, Iraq, 2019; p. 450.
27. A.O.A.C. *Official Methods of Analysis of the Association of Official Analytical Chemists*; Association of Official Analytical Chemists: Washington, DC, USA, 2000.
28. Borah, S.; Baruah, A.M.; Das, A.K.; Borah, J. Determination of Mineral Content in Commonly Consumed Leafy Vegetables. *Food Anal. Methods* **2009**, *2*, 226–230. [CrossRef]
29. Nsimba, R.Y.; Kikuzaki, H.; Konishi, Y. Antioxidant Activity of Various Extracts and Fractions of *Chenopodium Quinoa* and *Amaranthus* Spp. Seeds. *Food Chem.* **2008**, *106*, 760–766. [CrossRef]
30. Mohdaly, A.A.A.; Hassanien, M.F.R.; Mahmoud, A.; Sarhan, M.A.; Smetanska, I. Phenolics Extracted from Potato, Sugar Beet, and Sesame Processing by-Products. *Int. J. Food Prop.* **2013**, *16*, 1148–1168. [CrossRef]
31. Kumaran, A.; Karunakaran, R.J. In Vitro Antioxidant Activities of Methanol Extracts of Five Phyllanthus Species from India. *LWT Food Sci. Technol.* **2007**, *40*, 344–352. [CrossRef]
32. Yuan, G.-f.; Sun, B.; Yuan, J.; Wang, Q.-m. Effects of Different Cooking Methods on Health-Promoting Compounds of Broccoli. *J. Zhejiang Univ. Sci. B* **2009**, *10*, 580–588. [CrossRef] [PubMed]
33. Khalifa, I.; Barakat, H.; El-Mansy, H.A.; Soliman, S.A. Optimizing Bioactive Substances Extraction Procedures from Guava, Olive and Potato Processing Wastes and Evaluating Their Antioxidant Capacity. *J. Food Chem. Nanotechnol.* **2016**, *2*, 170–177. [CrossRef]
34. Lu, J.; Zhao, H.; Chen, J.; Fan, W.; Dong, J.; Kong, W.; Sun, J.; Cao, Y.; Cai, G. Evolution of Phenolic Compounds and Antioxidant Activity During Malting. *J. Agric. Food Chem.* **2007**, *55*, 10994–11001. [CrossRef]
35. Association of Official Analytical Chemists. *Aoac Official Method 982.14.: Glucose, Fructose, Sucrose, and Maltose in Presweetened Cereals—Liquid Chromatographic Method*; Association of Official Analytical Chemists: Washington, DC, USA, 2007.
36. Englyst, H.N.; Veenstra, J.; Hudson, G.J. Measurement of Rapidly Available Glucose (Rag) in Plant Foods: A Potential In Vitro Predictor of the Glycaemic Response. *Br. J. Nutr.* **1996**, *75*, 327–337. [CrossRef]
37. Goñi, I.; Garcia-Alonso, A.; Saura-Calixto, F. A Starch Hydrolysis Procedure to Estimate Glycemic Index. *Nutr. Res.* **1997**, *17*, 427–437. [CrossRef]

38. Peter Ikechukwu, A.; Okafor, D.; Kabuo, N.; Ibeabuchi, J.; Odimegwu, E.; Alagbaoso, S.; Njideka, N.; Mbah, R. Production and Evaluation of Cookies from Whole Wheat and Date Palm Fruit Pulp as Sugar Substitute. *Int. J. Eng. Technol. Manag. Appl. Sci.* **2017**, *4*, 1–31.
39. Steel, R.G. *Pinciples and Procedures of Statistics a Biometrical Approach*, 3rd ed.; McGraw-Hill: Boston, MA, USA, 1997.
40. Aleid, S.M.; Haddadin, J.S. Valorization and Chemical Constituents Assessments of Khalas Dates Fruit, Syrup and Pits. *Curr. Res. Nutr. Food Sci.* **2023**, *11*, 77–87. [CrossRef]
41. El-Sharnouby, G.A.; Aleid, S.M.; Al-Otaibi, M.M. Nutritional Quality of Biscuit Supplemented with Wheat Bran and Date Palm Fruits (*Phoenix dactylifera* L.). *Food Nutr. Sci.* **2012**, *3*, 322–328.
42. Ahmed, J.; Al-Jasass, F.M.; Siddiq, M. Date Fruit Composition and Nutrition. In *Dates*; John Wiley & Sons: Hoboken, NJ, USA, 2013; pp. 261–283.
43. Al-Saikhan, M.S.; Sallam, A. Impact of Chemical and Non-Chemical Thinning Treatments on Yield and Fruit Quality of Date Palm. *J. Food Res.* **2015**, *4*, 18–29. [CrossRef]
44. Brand-Miller, J.C. Glycemic Load and Chronic Disease. *Nutr. Rev.* **2003**, *61*, S49–S55. [CrossRef] [PubMed]
45. Sameen, A.; Manzoor, M.F.; Khan, M.I.; Sahar, A.; Saddique, A. Quality Evaluation of Ice Cream Prepared with *Phoenix dactylifera* Syrup as a Substitute of Sugar. *Pak. J. Food Sci.* **2016**, *26*, 226–233.
46. Julai, K.; Sridonpai, P.; Ngampeerapong, C.; Tongdonpo, K.; Suttisansanee, U.; Kriengsinyos, W.; On-Nom, N.; Tangsuphoom, N. Effects of Extraction and Evaporation Methods on Physico-Chemical, Functional, and Nutritional Properties of Syrups from Barhi Dates (*Phoenix dactylifera* L.). *Foods* **2023**, *12*, 1268. [CrossRef]
47. Jarrar, A.H.; Kamal-Eldin, A.; Bataineh, M.a.; Al Dhaheri, A.S. Glycemic Index (Gi) and Glycemic Load (Gl) Values for Dried Bisr and Tamr Dates. *Emir. J. Food Agric.* **2019**, *31*, 88–94. [CrossRef]
48. Liu, S.; Manson, J.E.; Stampfer, M.J.; Holmes, M.D.; Hu, F.B.; Hankinson, S.E.; Willett, W.C. Dietary Glycemic Load Assessed by Food-Frequency Questionnaire in Relation to Plasma High-Density-Lipoprotein Cholesterol and Fasting Plasma Triacylglycerols in Postmenopausal Women. *Am. J. Clin. Nutr.* **2001**, *73*, 560–566. [CrossRef]
49. Mohamed, S.A.; Awad, M.A.; El-Dengawy, E.-R.F.; Abdel-Mageed, H.M.; El-Badry, M.O.; Salah, H.A.; Abdel-Aty, A.M.; Fahmy, A.S. Total Phenolic and Flavonoid Contents and Antioxidant Activities of Sixteen Commercial Date Cultivars Grown in Saudi Arabia. *RSC Adv.* **2016**, *6*, 44814–44819. [CrossRef]
50. Manickavasagan, A.; Mathew, T.; Al-Attabi, Z.; Al-Zakwani, I. Dates as a Substitute for Added Sugar in Traditional Foods? A Case Study with Idli. *Emir. J. Food Agric.* **2013**, *25*, 899–906. [CrossRef]
51. Al-Farsi, M.A.; Lee, C.Y. 25 the Functional Values of Dates. In *Dates: Production, Processing, Food, and Medicinal Values*; CRC Press: Boca Raton, FL, USA, 2012; p. 351.

foods

MDPI

Article

Analysis of Volatile Profile and Aromatic Characteristics of Raw Pu-erh Tea during Storage Based on GC-MS and Odor Activity Value

Jie Guo [1], Zhihao Yu [1], Meiyan Liu [1], Mengdi Guan [1], Aiyun Shi [2], Yongdan Hu [1], Siyu Li [1], Lunzhao Yi [1] and Dabing Ren [1,*]

[1] Faculty of Food Science and Engineering, Kunming University of Science and Technology, Kunming 650500, China; guojie101298@163.com (J.G.); 13324915711@163.com (Z.Y.); lmy13562173653@163.com (M.L.); guan1230414@163.com (M.G.); kunlan881221@126.com (Y.H.); mnblsy@163.com (S.L.); yilunzhao@kust.edu.cn (L.Y.)

[2] Yunnan TAETEA Group Co., Ltd., Kunming 650500, China; 18787997350@163.com

* Correspondence: rendabing425@kust.edu.cn

Abstract: Volatile constituents are critical to the flavor of tea, but their changes in raw Pu-erh tea (RAPT) during storage have not been clearly understood. This work aimed to investigate the volatile composition and their changes at various storage durations. The volatile profile of RAPT was determined using headspace solid-phase microextraction in combination gas chromatography–mass spectrometry. A total of 130 volatile compounds were identified in RAPT samples, and 64 of them were shared by all samples. The aroma attributes of RAPT over a storage period ranging from 0 to 10 years were assessed through the combination of odor activity value (OAV), aroma characteristic influence(ACI) value, and multivariate statistical analysis. The results revealed that RAPT exhibited a distinct floral and fruity aroma profile after storage for approximately 3–4 years. A notable shift in aroma was observed after 3–4 years of storage, indicating a significant turning point. Furthermore, the likely notable shift after 10 years of storage may signify the second turning point. According to the odor activity value (OAV $\geq$ 100), eight key volatile compounds were identified: linalool, α-terpineol, geraniol, trans-β-ionone, α-ionone, (E,E)-2,4-heptadienal, 1-octanol, and octanal. Combining OAV ($\geq$100) and ACI ($\geq$1), five compounds, namely linalool, (E,E)-2,4-heptadienal, (Z)-3-hexen-1-ol, 2,6,10,10-tetramethyl-1-oxaspiro [4.5]dec-6-ene, and octanal, were identified as significant contributors to the aroma. The results offer a scientific foundation and valuable insights for understanding the volatile composition of RAPT and their changes during storage.

Keywords: raw Pu-erh tea; storage; volatile compounds; HS-SPME-GC-MS; aroma characteristic

Citation: Guo, J.; Yu, Z.; Liu, M.; Guan, M.; Shi, A.; Hu, Y.; Li, S.; Yi, L.; Ren, D. Analysis of Volatile Profile and Aromatic Characteristics of Raw Pu-erh Tea during Storage Based on GC-MS and Odor Activity Value. *Foods* **2023**, *12*, 3568. https://doi.org/10.3390/foods12193568

Academic Editor: Anna Arnoldi

Received: 28 August 2023
Revised: 12 September 2023
Accepted: 22 September 2023
Published: 26 September 2023

1. Introduction

Tea has a rich historical background and is highly appreciated by a diverse group of consumers owning to its unique aroma, taste, and biological properties. The principal categories of Chinese tea include green tea, yellow tea, white tea, oolong tea, black tea, dark tea or fermented tea, and Pu-erh tea [1,2]. Pu-erh tea is predominantly produced in Yunnan province of China and utilizes the fresh leaves of Yunnan's distinctive large-leaf tea (*Camellia sinensis*) cultivar as material [3,4]. Pu-erh tea consists of raw Pu-erh tea (RAPT) and ripe Pu-erh tea [5]. The manufacturing process for RAPT bears resemblance to that of traditional green tea, involving stages such as withering, de-enzyming, rolling, drying, and autoclaving–compressing. It is worth mentioning that while green tea typically has a limited shelf life of 1–2 years, RAPT tea can be stored for an extended period exceeding 5 years. The prolonged storage duration significantly contributes to the alteration and enhancement of the flavor profile of Pu-erh green tea, thus assuming a crucial role.

An abundant and distinct aroma plays a crucial role in determining the quality of tea, and the different aroma compounds contribute to the various sensory characteristics of the tea [6]. In a previous study, over 200 volatile compounds were identified in RAPT, with a predominant presence of alcohols, ketones, aldehydes, and methoxy compounds [7]. Methoxyphenolic compounds (stale/musty aroma notes) and ketones (woody or floral aroma notes) play a vital role in the special flavor of ripened Pu-erh tea. In RAPT, the floral aroma (38.62%) and fruity aroma (16.55%) were the major aroma categories [8]. Moreover, the amounts of alcohols contribute to the floral-like of tea, such as geraniol. The majority of aldehydes reported in tea are known to contribute to the citrus-like and green flavor of tea infusion. Also, some of them are responsible for bitter-almond-like, honey-like, and even fatty odor notes. Ketones constitute the second largest group of odorants that have been reported in tea to date, contributing to flowery, fruity, and woody aroma notes [9]. In summary, RAPT contains a diverse range of aroma compounds, and the composition of these compounds impacts the flavor profile of tea.

Generally, the aroma composition of tea is affected by various factors. Previous studies have investigated the impact of manufacturing process, such as de-enzyming and withering, on the aroma composition of tea. Fan et al. [10] reported that the de-enzyming and autoclaving–compressing treatments significantly changed the volatile composition of RAPT. Storage is also a crucial factor in the change in and formation of tea aroma. Qi et al. [11] found that aged white tea developed a sweet fragrance with herbal aroma, while the fresh white tea displays a fresh, delicate fragrance with mild green aroma. Additionally, the aroma of black tea undergoes significant changes during storage. Meng and co-workers reported that after five years of storage, methoxy benzene volatiles became the storage markers of Keemun congou black tea, contributing to a stale odor [12]. Then, a study revealed the aroma changes during storage of An tea (AT); the results showed that, as the storage time increased, the tea exhibited an increase in stale and woody aromas, for example, 4-ethyl-1,2-dimethoxybenzene (0–2.26 μg/L) and 1,2,3-trimethoxy-5-methyl-benzene (0–14.81 μg/L) [13]. This suggests that storage has a significant impact on the volatile composition and the formation of tea aroma. Nevertheless, it should be noted that there currently are no comprehensive investigations evaluating and comparing the flavor attributes of RAPT during storage, particularly with regard to key aroma compounds and the change in aroma.

In recent years, many fields have benefitted from the advancements in instrumental technology, including the rapid development of food flavor analysis techniques. One such technique involves the combination of gas chromatography (GC) with mass spectrometry (MS), known as GC-MS, which has been widely used for odor analysis and quality classification in food, especially in relation to tea [14–17]. Sample extraction is a crucial step in the analysis of volatile components using GC-MS, with headspace solid phase microextraction (HS-SPME) being one of the prevailing methods employed for this purpose. Nowadays, HS-SPME combined with GC-MS (HS-SPME-GC-MS) has been widely used for the determination of volatile compounds in various food products due to its high sensitivity and good reproducibility [18–20]. In a recent study, HS–SPME–GC–MS was employed to comprehensively compare and analyze the flavor-active compounds in GABA white tea processed by natural withering and combined withering. The results showed that combined withering was more favorable for the formation of floral and sour-fruity aroma compared with natural withering. It was also revealed that 2-heptanol plays an important role in contributing to the aroma of GABA white tea [21].

This study aimed to comprehensively analyze the volatile composition of RAPT and to investigate its changes during storage, considering the significant effects of storage on its aroma. To achieve this, an efficient HS-SPME–GC–MS method was used to extract and detect the volatile constituents in RAPT samples collected at different storage stages. The obtained GC–MS data were processed and interpreted using various chemometric methods, including ANOVA, PCA, HCA, and heatmap analysis. By combining OAV and ACI values, the key volatile compounds and major aroma-contributing compounds were identified, providing insights into the influence of volatile component changes on the aroma characteristics of RAPT at different storage times. These findings will contribute to a better understanding of the volatile composition and the unique aroma of RAPT during storage.

2. Materials and Methods

2.1. Chemicals and Reagents

De-ionized water was produced using a Milli-Q purification system (Milli-Q A10, Merck, Darmstadt, Germany). N-alkanes were obtained from o2si smart solutions Corporation (chromatographic pure grade, C8–C40). Sodium chloride (NaCl) was obtained from Innochem (Beijing, China). Decanoic acid ethyl ester was obtained from Sigma-Aldrich (purity of 98%).

2.2. Tea Samples

In this work, tea samples aged between 0 and 10 years were utilized. The samples denoted as T0, T1, T2, T3, T4, T5, T6, T7, T8, T9, and T10 were obtained from the products of the RAPT 7542 series (Yunnan TAETEA Group Co., Kunming, China), which exhibits a cake-like morphology. All samples were derived from the same tea variety, cultivated at the same location (Xishuangbanna, Yunnan, China), harvested during the same season (Spring tea), and processed using identical methodologies. Following production, the samples were stored in Guangdong Province. Prior to analysis, 100 g of each tea sample was passed through a 60 mesh sieve to obtain the tea powder and subsequently refrigerated at −20 °C for future use.

2.3. Extraction of Volatile Compounds by HS-SPME

The volatile compounds of RAPT were extracted by HS-SPME. Firstly, volatile compounds were extracted and enriched from RAPT samples using a 50/30 μm DVB/CAR/PDMS fiber head. In order to prevent contamination, the extracted fibers needed to be maintained at 270 °C for 1 hour in the GC-MS inlet before use.

For the HS-SPME process, a tea powder/water ratio (1:10 g/mL) was used. The extraction temperature was set at 80 °C, and the extraction time was 40 min. Precisely weighed, 0.200 g of tea powder was added into a 10 mL headspace vial, followed by the addition of 1.8 g of NaCl and 10 μL of ethyl decanoate (0.04 mg/mL) as an internal standard. Subsequently, 2 mL of boiling water was added, and the headspace vial was immediately sealed. The extracted fibers were then inserted into the vial and kept at 80 °C for 40 min. Once the extraction was complete, the SPME fiber needle was removed and inserted into the GC injection port for desorption, which lasted for 5 min at 260 °C.

2.4. Volatile Component Profiling via GC–MS

GC-MS (QP2010 Shimadzu, Kyoto, Japan) and an HP-5MS UI quartz capillary column (30 m × 0.25 mm × 0.50 μm) were employed in this experiment. The inlet temperature was 260 °C. The carrier gas was high-quality helium (>99.999%) at a flow rate of 1 mL/min; the sample was injected by split flow with a 5:1 split ratio. The column temperature was 50 °C and increased to 80 °C at a rate of 10 °C/min. Then, the temperature was increased to 90 °C at a rate of 3 °C/min, 90 °C for 3 min and heated up to 120 °C at a rate of 3 °C/min, 120 °C for 3 min, which then continued at 3 °C/min to 170 °C. Finally, the temperature was increased to 230 °C at a rate of 15 °C/min and held for 4 min.

Mass spectrometry (MS) conditions: The ion source was an EI source with an electron ionization energy of 70 eV and a scan range of m/z 30–540 amu. The temperature of the ion source was 230 °C, and the temperature of the interface was 260 °C. The solvent delay time was set to 3.2 min.

2.5. Identification and Annotation of Volatile Compounds

The data were screened from the database in the system for compounds with >80% similarity, and retention indices (*RIs*) were determined using a mixture of n-alkanes (C8–C40). The *RIs* were calculated as follows:

$$RIs = 100n + \frac{100(t_i - t_n)}{t_{n+1} - t_n} \tag{1}$$

where t_i is the retention time of the compound to be measured; n is the number of carbon atoms in the n-alkane mixture; and t_n and t_{n+1} are the retention times of the n-alkane standard mixture with n and n + 1 carbon atoms, respectively ($t_n < t_i < t_{n+1}$) [22].

After that, the volatile compounds were matched and annotated using the National Library of Standards and Technology (NIST) spectral library (https://webbook.nist.gov/chemistry/name-ser/ (accessed on 1 March 2023)).

The content of each volatile compound was determined using the internal standard method and was calculated based on the following equation. Here, decanoic acid ethyl ester (0.04 mg/mL) was used as the internal standard compound.

$$Y = \frac{ACV}{mA_{IS}} \times 100000 \tag{2}$$

where Y represents the content of each volatile compound (μg/100 g); C is the concentration of the internal standard (mg/mL); V denotes the volume of the internal standard (mL); m is the mass of the added sample (g); and A and A_{IS} represent the peak areas of the volatile compounds and the internal standard, respectively [10].

The relative content of each volatile component was calculated as follows: Relative content (%) = single component content/total content × 100% [10].

2.6. Calculation of Odor Activity Value (OAV) and Aroma Character Impact (ACI)

The odor activity value (*OAV*) represents the ratio of the dilution concentration of volatile compounds in water to its threshold value, which can better represent the degree of contribution of volatile compounds to the aroma [23,24], and the *OAV* value is calculated as follows:

$$OAV = \frac{C}{OT} \tag{3}$$

where C is the concentration of the compound in the sample (μg/mL) and OT is the odor threshold of the compound in water (μg/mL) [25].

In addition to analyzing the contribution of a single compound to the aroma in RAPT, the contribution of a mixture of compounds in RAPT to the aroma is also analyzed, i.e., the relative degree of contribution of each component in the mixture to the aroma is analyzed using the aroma characteristic influence (*ACI*) value, calculated as follows:

$$ACI(\%) = \frac{C_i T_i}{\sum_k C_k T_k} \times 100 = \frac{O_i}{\sum_k O_k} \times 100 \tag{4}$$

where C_i denotes the concentration of a given compound (i), T_i is the odor threshold of the compound in water reported in the literature, O_i is the *OAV* value of compound (i), and O_k is the *OAV* value of any compound [26].

2.7. Statistical Analysis

All experiments were carried out independently, and the results were expressed as mean plus standard deviation (n = 3). The software SPSS 25 (IBM Inc., Chicago, IL, USA) was used to determine significant differences ($p < 0.05$) for volatile compounds. The flavor characteristics of volatile compounds in RAPT were determined based on the data reported in the literature and on websites and were used for significance analysis. The used plotting software included Origin Pro version 2023 and an online plotting webpage (https://www.chiplot.online (accessed on 1 June 2023)).

3. Results and Discussions

3.1. Volatile Profile of RAPT with Various Years of Storage

In the present work, an HS-SPME-GC-MS method was optimized and developed to profile the volatile compounds in RAPT with different storage durations (1–10 years). Based on the optimized method, qualitative and quantitative analyses of the volatile compounds were performed. As illustrated in Figure S1 (total ion chromatograms (TIC)), a large number of volatile compounds were detected in RAPT, and the profile of RAPT with different years of storage varied from each other. The data analysis was performed using a commercial software (GSMSsolution, version 4.30, labSolition, Shimadzu, Kyoto, Japan). A total of 130 volatile compounds were finally identified by matching with the NIST library and comparing the retention indexes (RIs), and the results are presented in Table S1. These compounds include alcohols (22%), aldehydes (12.1%), ketones (18.2%), alkenes (14.4%), esters (8.3%), acids (8.3%), pyrroles (3.8%), and others (12.9%), as illustrated in Figure 1B. In terms of the number of detected compounds, alcohols, ketones, and alkenes are the top three classes of compounds in RAPT. Figure 1A shows the compounds identified in each year and the years in which those compounds were present. Of them, 64 compounds (approximately 49.23%) were shared by all RAPT samples regardless the years of storage. Furthermore, from Table S1 and Figure 1A, it is worth noting that a few of the compounds, such as 1,2,3-trimethoxybenzene, hexanal, and 2-hexenal, were produced with a storage time of over two years.

As illustrated in Figure 1 and Table S1, the composition and contents of volatile compounds in RAPT underwent apparent changes during storage. At the same years of storage, the difference in contents between compounds was also significant ($p < 0.05$). As shown in Figure 1C and Table S1, alcohol compounds are the most abundant component, which is independent from the years of storage. In a previous study, most of the alcohols in tea, such as benzyl alcohol, phenylethyl alcohol, and (Z)-3,7-dimethyl-2,6-octadien-1-ol, contribute to the rose-like and citrus-like aroma characteristics [9]. These alcohols are produced from the degradation of Strecker aldehydes or the hydrolysis of glycoside precursors [27]. In addition to alcohols, aldehydes, ketones, alkenes, and esters also showed relatively high contents in RAPT. The total concentration of volatiles generally fluctuated during storage (Figure 1C). Moreover, after 5 years of storage, the total volatile contents decreased, and although the overall content remains relatively stable, it is still lower compared with that in the samples with shorter storage durations. This distinction may be attributed to the differences between RAPT and other tea processing techniques.

To investigate the variation in the volatile profile of RAPT samples stored for different durations, this work conducted a principal component analysis (PCA) and a hierarchical cluster analysis (HCA) using the semi-quantitative results of 64 volatile compounds. The PCA scoring plot (Figure 1D) demonstrates a clear differentiation of the samples based on the first two principal components, which collectively account for 65.3% of the total variance. The RAPT samples can be broadly categorized into four distinct groups, indicating significant changes in the volatile profile during storage. The results of the HCA and PCA revealed a clear classification of the samples into four categories: T0, T1, T2, and T5 formed one category; T3 and T4 formed another category; T6, T7, T8, and T9 constituted a third category; and T10 represented a separate category (Figure 1D,E). Notably, the samples with a storage duration of 10 years (T10) exhibited a distinct separation from other samples, suggesting a significant change in the volatile composition following a decade of storage. In addition, the samples with a storage duration of 1–5 years were predominantly located in the left region of the scoring plot. Overall, the volatile profile of RAPT may experience a significant shift after 3–5 years of storage, representing the first turning point, followed by another notable transformation after 10 years of storage, potentially signifying the second turning point. The loading plot (Figure S2) indicated that volatile compounds underwent significant changes during the storage period, particularly in linalool, α-terpineol, 3,7-dimethyl-1,5,7-octatrien-3-ol, geraniol, cis-5-ethenyltetrahydro-α,α,5-trimethyl-2-furanmethanol, and (R)-5,6,7,7a-tetrahydro-4,4,7a-trimethyl-2(4H)-benzofuranone. These results further demonstrated the non-negligible influence of storage on the volatile composition of RAPT.

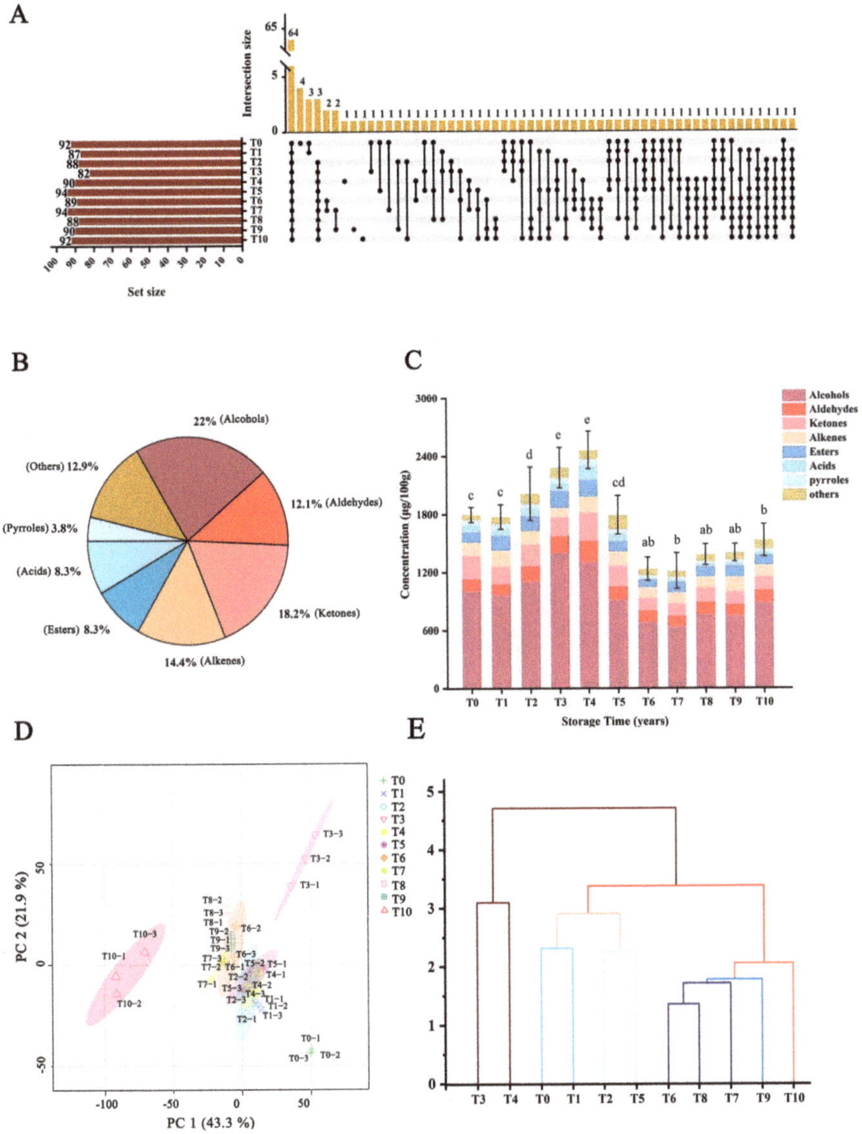

Figure 1. Annotation of volatile compounds during RAPT storage: (**A**) upset diagram; (**B**) pie diagram. The proportion of different types of volatile compounds in RAPT: (**C**) total volatile compound content and classification. (**D**) Principal component analysis (PCA). (**E**) Hierarchical cluster analysis (HCA).

3.2. Dynamic Changes in Different Compounds during Storage

As indicated by the quantitative results, the contents of many volatile compounds varied greatly during the storage of RATP. To visually illustrate the change patterns of volatile compounds in RAPT for different storage years, this study employed a heatmap analysis, as depicted in Figure 2.

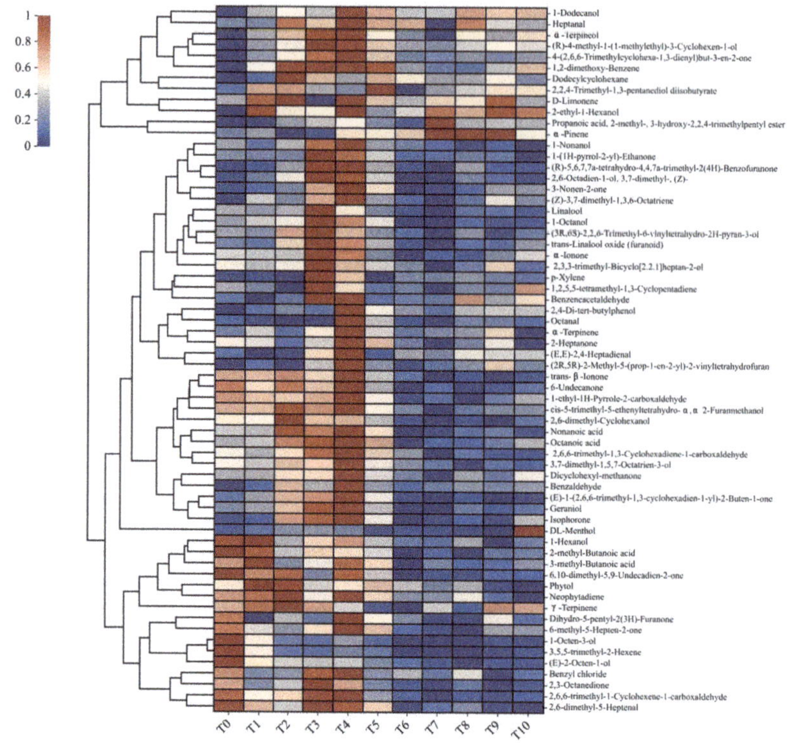

Figure 2. Heat map of differential volatile compounds in the samples of RAPT at various storage time points.

As illustrated in Figure 2, three distinct patterns of changes can be observed in the content of these compounds. The first pattern consists of a gradual decrease in content over time, the second pattern involves an increase in content with increasing storage time, and the third pattern exhibits an initial increase followed by a gradual decrease after 5 years. The first group of volatile compounds includes 1-octen-3-ol, 3,5,5-trimethy-2-hexene, (E)-2-octen-1-ol, 1-hexanol, 2-methl-butanoic acid, 3-methyl-butanoic acid, and 6,10-dimethyl-5,9-undecadien-2-one. These seven volatile compounds showed a tendency to decrease in content during storage, and their dominant aroma characteristics were described as mushroom-like, green, roasted, fatty, and cheese-like. It is worth noting that most of these compounds belong to the alcohol chemical class, which is an interesting finding. Alcohols have been found to be the main volatile compounds in RAPT and contribute significantly to its characteristic aroma [10]. Thus, it is hypothesized that tea with shorter storage times may have stronger grassy, roasted, and fatty aromas. In addition, the second group of compounds includes α-pinene, D-limonene, 2-ethyl-1-hexanol, and DL-menthol. These volatile compounds are mainly associated with a woody and mint-like aroma. This finding explains the disappearance of green and roasted aroma, and the emergence of woody and mint-like aroma in RAPT. Furthermore, most of these compounds showed a pattern towards the third group of compounds (approximately 75%). For instance, benzaldehyde increased from 14.77 μg/100 g (stored for 0 years) to 46.78 μg/100 g (stored for 5 years) and then slowly decreased. Similarly, 2,4-Di-tert-butylphenol (16.86–35.94) increased from 16.86 to 35.94 (μg/100 g) initially and then slowly decreased. Here, benzaldehyde is produced through the deamination of amino acids by aminotransferase [28]. Interestingly, a majority of the third group of volatile compounds decreased in 5 years of storage (T5), which is consistent with a previous study indicating that five years of storage is a significant turning point [29]. Moreover, most of the third group of volatile compounds contributes to the floral and fruity aroma. This result suggested that the green and roasted aromas diminish over time, while the floral and fruity aromas gradually become more prominent during the 0–5 years of storage.

Figure S3 shows a decrease in the relative contents of 12 volatile compounds during storage. These compounds include 1-hexanol, 2,6-dimethy-5-heptenal, 1-octen-3-ol, 2,3-octanedione, 2,6,6-trimethyl-1,3-cyclohexadiene-1-carboxaldehyde, 3,5,5-trimethyl-2-hexene, (E)-2-octen-1-ol, 2-methyl-butanoic acid, 1-ethyl-1H-pyrrole-2-carboxaldehyde, 6,10-dimethyl-5,9-undecadien-2-one, trans-β-ionone, and 6-undecanone. Specifically, they have predominantly green, fatty, floral, fruity, and roasted aroma. Moreover, the relative contents of some volatile compounds have been increased during storage, notably α-ionone, isophorone, 1,2-dimethoxy-benzene, and 1,2,5,5-trimethyl-1,3-cyclopentadiene. These compounds contribute to violet-like, woody, fruity, vanilla, and phenolic-like aromas. It is speculated that these changes in volatile compounds may explain the development of a woody aroma in RAPT with extended storage periods.

3.3. *Aroma Characteristics of RAPT during Storage*

The odor activity value (OAV) is a quantitative measure used to assess the odor potency of a compound or mixture and is typically determined by measuring the concentration of a volatile compound in a sample and by comparing it to a corresponding odor threshold. Based on the OAV, the contribution of each volatile compound can be evaluated, and the correlation between the quantified volatile compounds and the aroma can be determined. In general, aroma compounds with OAV greater than or equal to one are considered as active aromatic compounds and are believed to significantly contribute to the overall aroma profile of tea [6,30]. It is also important to consider the relative contribution of the components in a mixture, in addition to assessing the contribution of individual compounds. Therefore, the contribution of the mixed compounds to the aroma was further analyzed by calculating the aroma contribution index (ACI). Based on an OAV of $\geq$1, forty-three volatile compounds were identified as potential aroma compounds (Table S2). Figure 3 shows the change in these 43 volatile compounds during storage. Interestingly, most of the OVAs had their highest values at T3 and T4, and T4 had 22. In addition, according to their different aroma characteristics, the 43 volatiles were classified into nine groups: green, fruity, floral, roasted, fatty, chemical, cheese-like, stale, and woody. These aroma attributes have been previously used to describe the aroma characteristics of RAPT in other studies.

Table S2 shows that the aroma compounds in RAPT decreased with increasing storage time, including 1-octen-3-ol, 2-methyl-butanoic acid, and 1-ethyl-1H-pyrrole. These compounds are related to green and roasted aroma attributes. On the one hand, linalool (OAV = 8040.2) and 1-octanol (OAV = 620.8) exhibited the highest OAVs at T3 (3 years of storage). (E,E)-2,4-heptadienal (OAV = 5892) and octanal (OAV = 4131.4) had the highest OAVs at T4 (4 years of storage). On the other hand, DL-menthol (OAV = 1.8) reached its maximum value at T10 (10 years of storage) and was characterized by a peppermint and woody aroma. This could explain the formation of woody and mint-like aroma in RAPT after 10 years of storage. Moreover, (Z)-3,7-dimethyl-2,6-octadien-1-ol was found to contribute to the aroma profile of RAPT during storage. Its OAVs gradually increase from 2.7 at T0 to a peak value (6.2) at T3, before slowly decreasing with increasing storage time to 2.7 at T10. Furthermore, it is interesting to note that α-ionone (OAV = 562.5) and linalool (OAV = 8040.2) have violet-like, citrus-like, and flowery aroma characteristics, reaching their peak levels at T3. Meantime, geraniol (OAV = 258.3), octanal (OAV = 4131.4), and α-terpineol (OAV = 491.5) reached the highest at T4, displaying rose-like, citrus-like, and flowery aroma characteristics. This suggests that the storage of RAPT for about 3 to 4 years results in an abundance of floral and fruity aroma. Moreover, the contribution of aroma compounds in the mixture was compared using the aroma contribution index (ACI). Five volatile compounds were screened based on their high OAVs (≥100) and ACIs (≥1). These compounds include linalool (OAV: 2691.1–8040.2, ACI: 0.72–8.28), (E,E)-2,4-heptadienal (OAV: 3248.6–5892, ACI: 0.55–7.76), (Z)-3-hexen-1-ol (OAV: 0–151,243.8, ACI: 0–40.33), 2,6,10,10-tetramethyl-1-oxaspiro [4.5]dec-6-ene (OAV: 0–345,510, ACI: 0–76.48), and octanal (OAV: 1320.7–4131.4, ACI: 0.25–3.52).

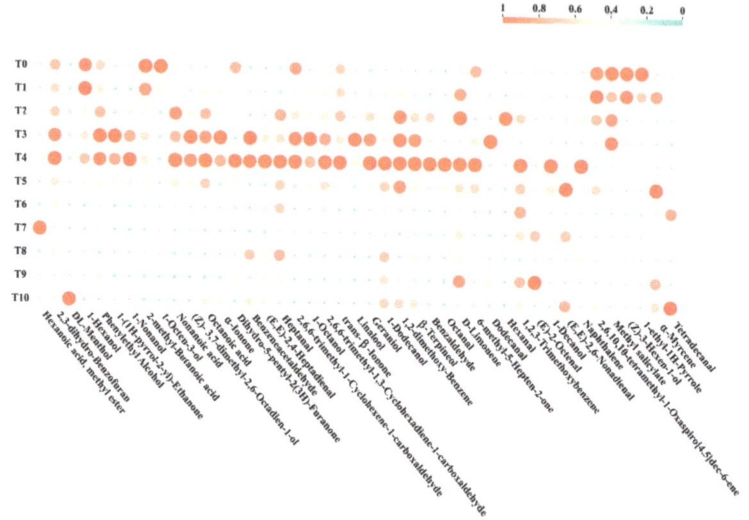

Figure 3. Based on OAV > 1 analysis of 43 volatile compounds and during RAPT storage.

3.4. *Identification of the Key Aroma Characteristics of RAPT during Storage*

Floral, fruity, and green aromas are considered favorable aroma characteristics in tea. Therefore, we analyzed the floral and fruity aromas of volatile compounds in RAPT. In Figure 4A, it is shown that most of the volatile compounds in RAPT have pleasant floral and fruity aromas during storage. Previous results have also indicated that the floral and fruity aromas increased over time during the first four years of storage. RAPT is unique compared with other teas because it undergoes autoclaving and compressing treatments. These treatments result in an increase in the content of certain aroma compounds such as phytol, linalool, and geraniol, which have a significant contribution to the floral aroma of tea [31]. Consequently, RAPT has a distinctive floral and fruity aroma, which may be dramatically changed during storage. As shown in Figure 4B, the roasted aroma, particularly that of 1-ethyl-1H-pyrrole, gradually diminished over the storage period. After ten years of storage, the green aroma decreased, while the woody aroma increased. However, it is important to note that the floral and fruity aromas remained dominant, indicating their overall persistence in the RAPT.

In addition, the intensity of the floral and fruity were gradually enhanced during storage. Hence, eight aroma compounds were screened with OAV ≥ 100, namely linalool (OAV: 2691.01–8040.2), α-terpineol (OAV: 250.8–491.5), geraniol (OAV: 99.5–258.3), trans-β-ionone (OAV: 40730.3–108493), α-ionone (OAV: 274.1–562.5), (E,E)-2,4-heptadienal (OAV: 3248.6–5892), 1-octanol (OAV: 95.2–620.8), and octanal (OAV: 1320.7–4131.4), as illustrated in Table S2. Interestingly, all of them contribute to the aroma of floral and fruity. Notably, (E,E)-2,4-heptadienal is one of the dominant volatile compounds in high-grade Jingshan green tea, and it is produced through the degradation of linolenic acid, providing flowery aroma [32]. In oolong tea (*Camellia sinensis*), geraniol and linalool are the main contributors to the floral, rose-like, and sweet aroma characteristics. In addition, Yang et al. [33] found that trans-β-ionone contributed to the aroma characteristics of Rougui tea. Therefore, these eight volatile compounds are considered to be the key contributors to the floral and fruity characteristics during RAPT storage.

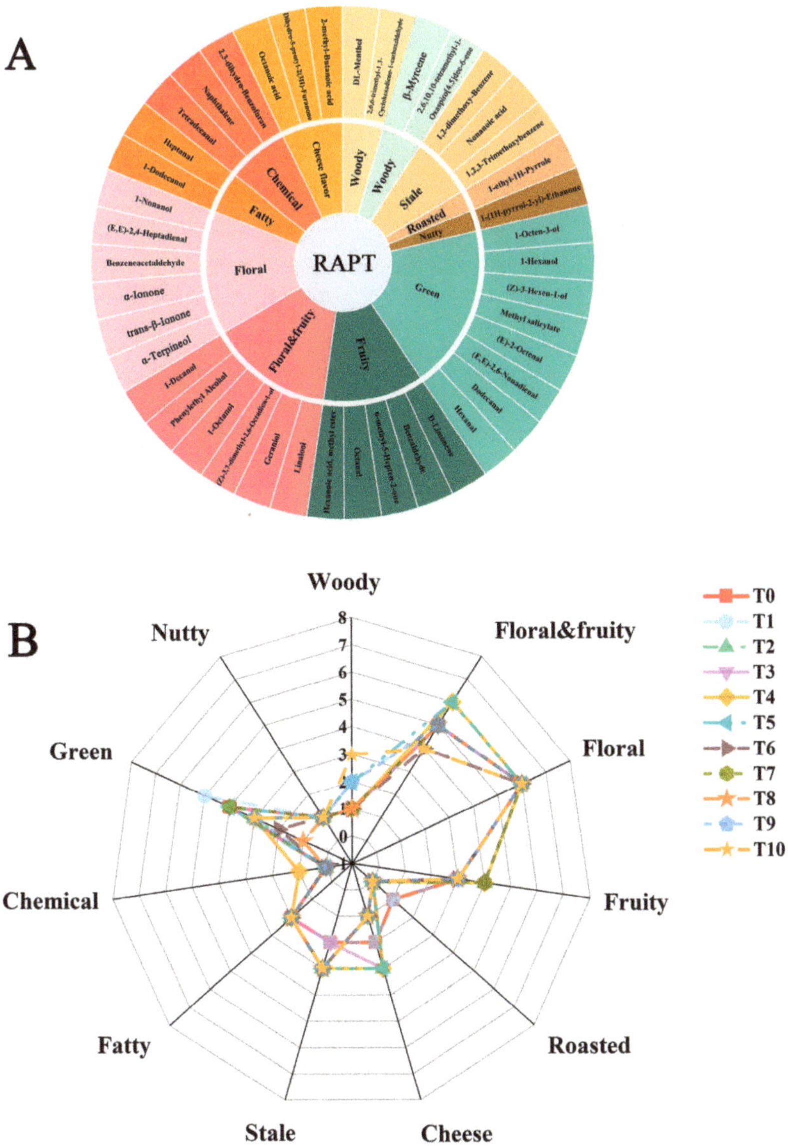

Figure 4. Based on OAV analysis of volatile compounds during RAPT storage. (**A**) Flavor wheel of 43 key volatile compounds in samples at different storage times. (OAV > 1). (**B**) Radar plots of flavor characteristics of RAPT during storage. (OAV > 1).

3.5. Metabolic Pathways of the Key Aaroma Compounds

The changes in aroma characteristics may be influenced by several factors during storage. Geraniol and linalool, which are monoterpenes, serve as the primary constituents responsible for the pleasant floral and fruity scent found in plants. These compounds are synthesized in tea through the metabolic conversion of two primary precursors, namely isopentenyl diphosphate (IPP) and dimethylallyl pyrophosphate (DMAPP). The mevalonate pathway (MVA) plays a critical role in the synthesis of secondary metabolites, including sterols, sesquiterpenes, and triterpenes within the cytoplasm and endoplasmic reticulum. The 2-C-methyl-D-erythritol-4-phosphate pathway (MEP) and carotenoid cleavage dioxygenase (CCD) are two key metabolic pathways involved in the biosynthesis of monoterpenes and oxidation of carotenoids, respectively [34]. It is hypothesized that changes in the metabolism of microorganisms during storage may result in changes in aroma characteristics. Trans-β-ionone and α-ionone are secondary oxidation products derived from carotenoids, including β-carotene, α-carotene, phytoene, lutein, and lycopene, which undergo oxidation by CCD to form primary oxidation products and then are further converted into secondary oxidation products [28]. Furthermore, the conversion of linolenic acid into (E,E)-2,4-heptadienal occurs through the involvement of leucoanthocyanidin reductase (LOX) and glutathione peroxidase (HPL). Previous work has shown that LOX continues to facilitate the synthesis of aldehydes from fatty acids throughout the fermentation stage [35]. The storage of RAPT, in addition to being a natural aging and fermentation process, is also believed to contribute to the changes in aroma. These changes may be attributed to microbial metabolism, enzymatic digestion, and natural oxidation.

4. Conclusions

In the present study, a method combining HS-SPME with GC-MS was used to profile the volatile compounds in RAPT stored for different years (0–10 years). In total, 130 volatile compounds including alcohols, aldehydes, ketones, alkenes, and esters were identified, and 64 of them were shared by all tea samples. Alcohols, ketones, and alkenes were determined as the top three classes of volatiles in RAPT. It was revealed that storage has a significant effect on the volatile composition of RAPT. A notable transformation was observed after 3–4 years of storage, signifying the first turning point. Then, the volatile profile of RAPT with a storage duration of 10 years (T10) experienced a distinct separation from other samples; the likely dramatic shift after 10 years of storage may signify the second turning point. A total of 43 volatile compounds were selected based on their OAVs and aroma profiles, and most of them contributed to the floral and fruity aromas. These floral and fruity compounds displayed a pattern of initial increase, followed by subsequent decrease throughout the storage period. In contrast, the roasted aroma, especially from 1-ethyl-1H-pyrrole, was found to diminish during the storage period. Linalool, (E,E)-2,4-heptadienal, (Z)-3-hexen-1-ol, 2,6,10,10-tetramethyl-1-oxaspiro [4.5]dec-6-ene, and octanal (OAVs of $\geq$100 and ACIs of $\geq$1) were identified as the primary contributors to the aroma profile. Ultimately, eight key compounds, namely linalool, α-terpineol, geraniol, trans-β-ionone, α-ionone, (E,E)-2,4-heptadienal, 1-octanol, and octanal, were selected based on their association with floral and fruity aromas. The results of our study enhance comprehension regarding the composition of volatile compounds in RAPT over varying storage durations. However, the changes in volatile compounds were limited by the samples and storage environment, and the effects of different storage environments need to be further studied. In addition, enzymes and microorganisms can be analyzed to explore the pathways and mechanisms of conversion between substances during storage.

Supplementary Materials: The following supporting information can be downloaded at https://www.mdpi.com/article/10.3390/foods12193568/s1. Table S1. Volatile compounds identified in RAPT samples from different storage durations. Table S2. Key volatile compounds identified in RAPT samples from different years of storage. Figure S1. Total ion chromatograms (TIC) of aroma compounds in RAPT. Figure S2. Multivariate statistical analysis of volatile compounds and during RAPT storage: (a) loadings plot; (b–g) comparison of the accurate contents of differential compounds screened by loadings plot. Figure S3. Heat map of differential volatile compounds in the samples of RAPT at various storage time points (relative content). References [36–39] are cited in Supplementary Materials.

Author Contributions: D.R., Y.H. and L.Y. conceived the original idea. J.G., M.G. and Z.Y. carried out the experiment. J.G. and M.L. performed the data analysis. J.G. wrote the manuscript with the support from S.L., Y.H. and D.R. A.S. provided the tea samples. All authors discussed the results and contributed to the final manuscript. All authors have read and agreed to the published version of the manuscript.

Funding: This research was funded by the Yunnan Major Scientific and Technological Projects, grant number 202202AG050009.

Data Availability Statement: The data are unavailable due to privacy or ethical restrictions.

Conflicts of Interest: Author Aiyun Shi was employed by the company Yunnan TAETEA Group Co., Ltd. The remaining authors declare that the research was conducted in the absence of any commercial or financial relationships that could be construed as a potential conflict of interest.

References

1. Wang, Q.; Qin, D.; Huang, G.; Jiang, X.; Fang, K.; Wang, Q.; Ni, E.; Li, B.; Pan, C.; Li, H.; et al. Identification and Characterization of the Key Volatile Flavor Compounds in Black Teas from Distinct Regions Worldwide. *J. Food Sci.* **2022**, *87*, 3433–3446. [CrossRef] [PubMed]
2. Wong, M.; Sirisena, S.; Ng, K. Phytochemical Profile of Differently Processed Tea: A Review. *J. Food Sci.* **2022**, *87*, 1925–1942. [CrossRef] [PubMed]
3. Lee, L.K.; Foo, K.Y. Recent Advances on the Beneficial Use and Health Implications of Pu-Erh Tea. *Food Res. Int.* **2013**, *53*, 619–628. [CrossRef]
4. Ding, Q.; Zheng, W.; Zhang, B.; Chen, X.; Zhang, J.; Pang, X.; Zhang, Y.; Jia, D.; Pei, S.; Dong, Y.; et al. Comparison of Hypoglycemic Effects of Ripened Pu-Erh Tea and Raw Pu-Erh Tea in Streptozotocin-Induced Diabetic Rats. *RSC Adv.* **2019**, *9*, 2967–2977. [CrossRef]
5. Karwowska, K.; Kozłowska-Tylingo, K.; Skotnicka, M.; Śmiechowska, M. Theogallin-to-Gallic-Acid Ratio as a Potential Biomarker of Pu-Erh Teas. *Foods* **2023**, *12*, 2453. [CrossRef] [PubMed]
6. Yin, X.; Huang, J.; Huang, J.; Wu, W.; Tong, T.; Liu, S.; Zhou, L.; Liu, Z.; Zhang, S. Identification of Volatile and Odor-Active Compounds in Hunan Black Tea by SPME/GC-MS and Multivariate Analysis. *LWT* **2022**, *164*, 113656. [CrossRef]
7. Gong, Z.; Watanabe, N.; Yagi, A.; Etoh, H.; Sakata, K.; Ina, K.; Liu, Q. Compositional Change of Pu-Erh Tea during Processing. *Biosci. Biotechnol. Biochem.* **1993**, *57*, 1745–1746. [CrossRef]
8. Xu, Y.-Q.; Wang, C.; Li, C.-W.; Liu, S.-H.; Zhang, C.-X.; Li, L.-W.; Jiang, D.-H. Characterization of Aroma-Active Compounds of Pu-Erh Tea by Headspace Solid-Phase Microextraction (HS-SPME) and Simultaneous Distillation-Extraction (SDE) Coupled with GC-Olfactometry and GC-MS. Food Anal. *Methods* **2016**, *9*, 1188–1198. [CrossRef]
9. Zhai, X.; Zhang, L.; Granvogl, M.; Ho, C.-T.; Wan, X. Flavor of Tea (Camellia Sinensis): A Review on Odorants and Analytical Techniques. *Compr. Rev. Food Sci. Food Saf.* **2022**, *21*, 3867–3909. [CrossRef]
10. Fan, X.; Chen, N.; Cai, F.; Ren, F.; Zhong, J.; Wang, D.; Shi, L.; Ren, D.; Yi, L. Effects of Manufacturing on the Volatile Composition of Raw Pu-Erh Tea with a Focus on de-Enzyming and Autoclaving–Compressing Treatments. *LWT* **2021**, *137*, 110461. [CrossRef]
11. Qi, D.; Miao, A.; Cao, J.; Wang, W.; Chen, W.; Pang, S.; He, X.; Ma, C. Study on the Effects of Rapid Aging Technology on the Aroma Quality of White Tea Using GC–MS Combined with Chemometrics: In Comparison with Natural Aged and Fresh White Tea. *Food Chem.* **2018**, *265*, 189–199. [CrossRef] [PubMed]
12. Meng, T.; Zhipeng, X.; Ai, H.; Jiayu, C.; Tianji, Y.; Zhengquan, L. Effect of 1–20 Years Storage on Volatiles and Aroma of Keemun Congou Black Tea by Solvent Extraction-Solid Phase Extraction-Gas Chromatography-Mass Spectrometry. *LWT* **2021**, *136*, 110278. [CrossRef]
13. Shen, S.; Wu, H.; Li, T.; Sun, H.; Wang, Y.; Ning, J. Formation of Aroma Characteristics Driven by Volatile Components during Long-Term Storage of An Tea. *Food Chem.* **2023**, *411*, 135487. [CrossRef] [PubMed]
14. Cheng, L.; Wang, Y.; Zhang, J.; Zhu, J.; Liu, P.; Xu, L.; Wei, K.; Zhou, H.; Peng, L.; Zhang, J.; et al. Dynamic Changes of Metabolic Profile and Taste Quality during the Long-Term Aging of Qingzhuan Tea: The Impact of Storage Age. *Food Chem.* **2021**, *359*, 129953. [CrossRef] [PubMed]
15. Guo, X.; Song, C.; Ho, C.-T.; Wan, X. Contribution of L-Theanine to the Formation of 2,5-Dimethylpyrazine, a Key Roasted Peanutty Flavor in Oolong Tea during Manufacturing Processes. *Food Chem.* **2018**, *263*, 18–28. [CrossRef] [PubMed]
16. Li, J.; Gong, Y.; Li, J.; Fan, L. Stir-Frying Treatment Improves the Color, Flavor, and Polyphenol Composition of Flos Sophorae Immaturus Tea. *J. Food Compos. Anal.* **2023**, *116*, 105045. [CrossRef]
17. Wang, D.; Wang, C.; Su, W.; Lin, C.-C.; Liu, W.; Liu, Y.; Ni, L.; Liu, Z. Characterization of the Key Aroma Compounds in Dong Ding Oolong Tea by Application of the Sensomics Approach. *Foods* **2023**, *12*, 3158. [CrossRef]
18. Chen, L.; Ning, F.; Zhao, L.; Ming, H.; Zhang, J.; Yu, W.; Yi, S.; Luo, L. Quality Assessment of Royal Jelly Based on Physicochemical Properties and Flavor Profiles Using HS-SPME-GC/MS Combined with Electronic Nose and Electronic Tongue Analyses. *Food Chem.* **2023**, *403*, 134392. [CrossRef] [PubMed]
19. Gong, Y.; Kerrihard, A.; Pegg, R. Characterization of the Volatile Compounds in Raw and Roasted Georgia Pecans by HS-SPME-GC-MS. *J. Food Sci.* **2018**, *83*, 2753–2760. [CrossRef]
20. Lin, Y.; Wang, Y.; Huang, Y.; Song, H.; Yang, P. Aroma Identification and Classification in 18 Kinds of Teas (Camellia Sinensis) by Sensory Evaluation, HS-SPME-GC-IMS/GC $\times$ GC-MS, and Chemometrics. *Foods* **2023**, *12*, 2433. [CrossRef]
21. Li, Y.; Wu, T.; Deng, X.; Tian, D.; Ma, C.; Wang, X.; Li, Y.; Zhou, H. Characteristic Aroma Compounds in Naturally Withered and Combined Withered γ-Aminobutyric Acid White Tea Revealed by HS-SPME-GC-MS and Relative Odor Activity Value. *LWT* **2023**, *176*, 114467. [CrossRef]

22. van Den Dool, H.; Kratz, P.D. A Generalization of the Retention Index System Including Linear Temperature Programmed Gas—Liquid Partition Chromatography. *J. Chromatogr. A* **1963**, *11*, 463–471. [CrossRef] [PubMed]
23. Liu, Y.; He, C.; Song, H. Comparison of SPME Versus SAFE Processes for the Analysis of Flavor Compounds in Watermelon Juice. *Food Anal. Methods* **2018**, *11*, 1677–1689. [CrossRef]
24. Wang, C.; Li, J.; Wu, X.; Zhang, Y.; He, Z.; Zhang, Y.; Zhang, X.; Li, Q.; Huang, J.; Liu, Z. Pu-Erh Tea Unique Aroma: Volatile Components, Evaluation Methods and Metabolic Mechanism of Key Odor-Active Compounds. *Trends Food Sci. Technol.* **2022**, *124*, 25–37. [CrossRef]
25. Wang, B.; Qu, F.; Wang, P.; Zhao, L.; Wang, Z.; Han, Y.; Zhang, X. Characterization Analysis of Flavor Compounds in Green Teas at Different Drying Temperature. *LWT* **2022**, *161*, 113394. [CrossRef]
26. Ma, L.; Gao, M.; Zhang, L.; Qiao, Y.; Li, J.; Du, L.; Zhang, H.; Wang, H. Characterization of the Key Aroma-Active Compounds in High-Grade Dianhong Tea Using GC-MS and GC-O Combined with Sensory-Directed Flavor Analysis. *Food Chem.* **2022**, *378*, 132058. [CrossRef]
27. Zhu, J.; Niu, Y.; Xiao, Z. Characterization of the Key Aroma Compounds in Laoshan Green Teas by Application of Odour Activity Value (OAV), Gas Chromatography-Mass Spectrometry-Olfactometry (GC-MS-O) and Comprehensive Two-Dimensional Gas Chromatography Mass Spectrometry (GC × GC-qMS). *Food Chem.* **2021**, *339*, 128136. [CrossRef]
28. Liu, Y.; Chen, Q.; Liu, D.; Yang, L.; Hu, W.; Kuang, L.; Huang, Y.; Teng, J.; Liu, Y. Multi-Omics and Enzyme Activity Analysis of Flavour Substances Formation: Major Metabolic Pathways Alteration during Congou Black Tea Processing. *Food Chem.* **2023**, *403*, 134263. [CrossRef]
29. Ren, Y.; Hou, Y.; Granato, D.; Zha, M.; Xu, W.; Zhang, L. Metabolomics, Sensory Evaluation, and Enzymatic Hydrolysis Reveal the Effect of Storage on the Critical Astringency-Active Components of Crude Pu-Erh Tea. *J. Food Compos. Anal.* **2022**, *107*, 104387. [CrossRef]
30. Russo, P.; Tufariello, M.; Renna, R.; Tristezza, M.; Taurino, M.; Palombi, L.; Capozzi, V.; Rizzello, C.G.; Grieco, F. New Insights into the Oenological Significance of Candida Zemplinina: Impact of Selected Autochthonous Strains on the Volatile Profile of Apulian Wines. *Microorganisms* **2020**, *8*, 628. [CrossRef]
31. Su, D.; Xu, T.; Li, Y.; Zhou, H. Flavor Evolution in Raw Pu-Erh Tea during Manufacturing Using Different Processing Types. *LWT* **2022**, *154*, 112905. [CrossRef]
32. Flaig, M.; Qi, S.; Wei, G.; Yang, X.; Schieberle, P. Characterization of the Key Odorants in a High-Grade Chinese Green Tea Beverage (Camellia Sinensis; Jingshan Cha) by Means of the Sensomics Approach and Elucidation of Odorant Changes in Tea Leaves Caused by the Tea Manufacturing Process. *J. Agric. Food Chem.* **2020**, *68*, 5168–5179. [CrossRef] [PubMed]
33. Yang, P.; Song, H.; Lin, Y.; Guo, T.; Wang, L.; Granvogl, M.; Xu, Y. Differences of Characteristic Aroma Compounds in Rougui Tea Leaves with Different Roasting Temperatures Analyzed by Switchable GC-O-MS and GC × GC-O-MS and Sensory Evaluation. *Food Funct.* **2021**, *12*, 4797–4807. [CrossRef]
34. Schwab, W.; Davidovich-Rikanati, R.; Lewinsohn, E. Biosynthesis of Plant-Derived Flavor Compounds. *Plant J.* **2008**, *54*, 712–732. [CrossRef]
35. Chen, Q.; Zhu, Y.; Liu, Y.; Liu, Y.; Dong, C.; Lin, Z.; Teng, J. Black Tea Aroma Formation during the Fermentation Period. *Food Chem.* **2022**, *374*, 131640. [CrossRef] [PubMed]
36. Guo, X.; Ho, C.-T.; Schwab, W.; Wan, X. Aroma Profiles of Green Tea Made with Fresh Tea Leaves Plucked in Summer. *Food Chem.* **2021**, *363*, 130328. [CrossRef]
37. Hao, Z.; Feng, J.; Chen, Q.; Lin, H.; Zhou, X.; Zhuang, J.; Wang, J.; Tan, Y.; Sun, Z.; Wang, Y.; et al. Comparative Volatiles Profiling in Milk-Flavored White Tea and Traditional White Tea Shoumei via HS-SPME-GC-TOFMS and OAV Analyses. *Food Chem. X* **2023**, *18*, 100710. [CrossRef]
38. Tatsu, S.; Matsuo, Y.; Nakahara, K.; Hofmann, T.; Steinhaus, M. Key Odorants in Japanese Roasted Barley Tea (Mugi-Cha)—Differences between Roasted Barley Tea Prepared from Naked Barley and Roasted Barley Tea Prepared from Hulled Barley. *J. Agric. Food Chem.* **2020**, *68*, 2728–2737. [CrossRef]
39. Zheng, Y.; Zhang, C.; Ren, D.; Bai, R.; Li, W.; Wang, J.; Shan, Z.; Dong, W.; Yi, L. Headspace Solid-Phase Microextraction Coupled with Gas Chromatography-Mass Spectrometry (HS-SPME-GC-MS) and Odor Activity Value (OAV) to Reveal the Flavor Characteristics of Ripened Pu-Erh Tea by Co-Fermentation. *Front. Nutr.* **2023**, *10*. [CrossRef]

foods

MDPI

Article

Effect of Ultrasound Pretreatment on the Moisture Migration and Quality of *Cantharellus cibarius* Following Hot Air Drying

Mianli Sun, Yuan Xu, Yangyue Ding, Ying Gu, Yongliang Zhuang and Xuejing Fan *

Faculty of Food Science and Engineering, Kunming University of Science and Technology, No. 727 South Jingming Road, Kunming 650500, China; 20212114045@stu.kust.edu.cn (M.S.); 20212214082@stu.kust.edu.cn (Y.X.); dingyangyue77@163.com (Y.D.); guying@kust.edu.cn (Y.G.); ylzhuang@kmust.edu.cn (Y.Z.)

* Correspondence: fanxuejing@kust.edu.cn; Tel.: +86-871-130-1900-6983

Abstract: The effect of different ultrasound pretreatment powers (0–500 W) before hot air drying on the moisture migration and quality of *Cantharellus cibarius* (*C. cibarius*) was investigated in this study. The results showed that the ultrasound pretreatment accelerated the drying rate. When the ultrasound power was 400 W, the drying time of *C. cibarius* was reduced by 18.90% compared with the control group. The low-field nuclear magnetic resonance (LF-NMR) and magnetic resonance imaging (MRI) results showed that the ultrasound pretreatment increased the water mobility in *C. cibarius*. The scanning electron microscopy (SEM) results revealed that the ultrasound pretreatment promoted the expansion of intercellular pores. In addition, the rehydration capacity and quality characteristics of the ultrasound-pretreated dried *C. cibarius* were better than those of the control group. Overall, this study concluded that ultrasound pretreatment is a promising pretreatment method for the hot air drying of *C. cibarius* products to reduce the total drying time significantly and improve the retention rate of the total phenolics and flavonoids of dried *C. cibarius*.

Keywords: ultrasound pretreatment; *Cantharellus cibarius*; hot air drying; moisture migration; antioxidant activity

Citation: Sun, M.; Xu, Y.; Ding, Y.; Gu, Y.; Zhuang, Y.; Fan, X. Effect of Ultrasound Pretreatment on the Moisture Migration and Quality of *Cantharellus cibarius* Following Hot Air Drying. *Foods* **2023**, *12*, 2705. https://doi.org/10.3390/foods12142705

Academic Editor: Filipa V. M. Silva

Received: 23 June 2023
Revised: 11 July 2023
Accepted: 11 July 2023
Published: 14 July 2023

1. Introduction

Cantharellus cibarius (*C. cibarius*), one of the most popular edible chanterelle mushrooms, is a highly appreciated foodstuff due to its special apricot flavor, its richness in nutrients, and its low fat content [1]. In particular, the phenolic substances (such as catechin, pyrogallol, myricetin, phenolic and protocatechuic acids) contained in *C. cibarius* make its total antioxidant content superior to that of any mushroom studied [2]. It can be used fresh, dried, or processed as pickled products. Although fresh *C. cibarius* has a more delicious and unique flavor [1], its shelf life is short and it rots easily due to its high water activity, high water content, and lack of a hard outer layer to protect it from physical and microbial attacks [3]. Moreover, the fresh *C. cibarius* harvesting season is limited and expensive. Therefore, it is necessary to take some measures to extend the shelf life of fresh *C. cibarius* and maintain its nutritional quality as much as possible. Drying is a widely used processing technology in mushroom preservation that prevents spoilage by reducing the moisture content to a level that allows safe storage. Many techniques have been described for mushroom drying, such as hot air drying, microwave drying, freeze drying, and sun drying [4].

Hot air drying technology is used to dry edible mushrooms because of its simple operation and low cost. However, a longer drying time leads to the degradation of heat-sensitive compounds and the loss of characteristic flavor compounds [5]. Moreover, improper temperature control would also cause product shrinkage, reduce rehydration capacity, and destroy nutrients [6]. Nowadays, with the increasing demand for high-quality foods, traditional hot air drying is unable to meet the needs of consumers [7]. To overcome the traditional drying process's limitations and produce high-quality products, it is necessary to explore an effective pretreatment method.

The pretreatment process before thermal drying can be divided into chemical pretreatment (liquid phase and gas phase) and physical pretreatment (thermal and non-thermal pretreatment) [8]. As a chemical pretreatment, osmotic pretreatment has the advantage of improving the drying process and maintaining product quality [8]. Pei et al. observed that osmotic dehydration accelerates moisture diffusion during drying and shortens the drying time [9]. However, in the case of a highly concentrated osmosis solution, material migration leads to the loss of minerals, vitamins, and pigments. Similarly, although blanching as a physical pretreatment can effectively inactivate various enzymes and destroy micro-organisms, thereby improving quality and promoting the drying speed, it will cause changes in product quality, soluble nutrients, pigments and flavors. Ultrasound, as a non-thermal food processing technique, can improve drying efficiency and inactivate enzymes while avoiding the detrimental effects of heat on the nutritional and sensory properties of foods compared to thermal processing [10,11]. Ultrasonic pretreatment is increasingly being used in the food industry as a non-thermal food processing technique (drying, thawing, extraction) [10,12,13].

In addition, the ultrasonic waves propagate through the material, causing a series of rapid alternating compression expansion and cavitation cycles, resulting in microchannels within the sample. Water molecules can diffuse to the surface through the microscopic channels, increasing the effective water diffusivity [14]. Ultrasound pretreatment before hot air drying has been found to significantly enhance the drying efficiency of mushrooms over untreated samples and to reduce the loss of nutritious substances [15,16]. Furthermore, ultrasonication causes the rehydration capacity of the drying products to increase [17]. Low-field nuclear magnetic resonance (LF-NMR) and magnetic resonance imaging (MRI) have become powerful tools for studying the dynamic behavior of water in food processing due to their high efficiency, low cost, and non-destructive nature. LF-NMR is used to study the mobility and distribution of water, while MRI images are used to observe the internal distribution of water and changes during the drying process [18]. However, there is limited research on the effect of different ultrasonic pretreatment powers on moisture migration and *C. cibarius* quality.

In this study, different ultrasonic powers were applied before hot air drying to evaluate the impact of ultrasound pretreatment on the overall quality of dried *C. cibarius*. Moreover, the moisture migration during drying was analyzed using the LF-NMR and MRI techniques. An effective pretreatment method is presented according to the comprehensive assessment, which is designed to improve the drying characteristics and quality of *C. cibarius*.

2. Materials and Methods

2.1. Raw Materials

Fresh *C. cibarius* mushrooms were purchased from the local wild mushroom market in Kunming, Yunnan (China), with an initial water content of 91.65 ± 0.01 g/100 g. The raw materials were wrapped in plastic and stored in a refrigerator at 4 °C to prevent water loss and ensure the quality of the materials. Folin-Ciocalteu's phenol reagent, aluminum nitrate, sodium nitrate, 2, 2-diphenyl-1-picrohydrazyl were purchased from Aladdin Biochemical Technology Co., Ltd. (Shanghai, China). All chemicals used were of analytical grade.

2.2. Ultrasound Pretreatment

The *C. cibarius* was pretreated by ultrasound in an Ultrasonic Cell Disruptor (SCIENTZ-IID, Ningbo Xinzhi Biotechnology Co., Ltd., Ningbo, China). *C. cibarius* (40 ± 1 g) and deionized water (240 mL) were placed in a beaker and maintained at 15 ± 1 °C using an ice bath [19]. The ultrasonic pretreatment power was set at 0, 100, 200, 300, 400, and 500 W for 20 min.

2.3. Hot Air Drying

The pretreated samples were weighed and dried in a hot air dryer (101 A-1 E, Shanghai Experimental Instrument Factory Co., Ltd., Shanghai, China) at 60 ± 1 °C until reaching a constant weight. To analyze the moisture content of the mushrooms during drying, the samples were weighed every 30 min.

2.3.1. Measurement of Moisture Ratio

The moisture ratio (MR) was calculated according to Equation (1) [20]:

$$MR = \frac{m_t - m_e}{m_0 - m_e} \tag{1}$$

where m_t, m_0, and m_e are the moisture content, initial moisture content, and equilibrium moisture content (g water/g dry matter) at any time, respectively. For longer drying times, the value of m_e is small compared to m_t or m_0. Therefore, Equation (1) can be simplified to Equation (2):

$$MR = \frac{m_t}{m_0} \tag{2}$$

2.3.2. Measurement of Drying Rate

The drying rate (DR) was calculated using Equation (3):

$$DR = \frac{m_t - m_{t+\Delta t}}{\Delta t} \tag{3}$$

where m_t and $m_{t+\Delta t}$ are the moisture content at t and t + Δt (g water/g dry matter), respectively; moreover, t is the time (min) and Δt is the time difference (min) [21].

2.4. LF-NMR Measurement

The transverse relaxation times (T_2) of the samples with different pretreatments were determined using an LF-NMR analyzer (Niumag, Suzhou, Jiangsu, China). The T_2 of each sample was measured via scanning with a Carr–Purcell–Meiboom–Gill (CPMG) pulse sequence The 90° and 180° pulse durations were 20.0 μs and 40.0 μs, respectively. The time waiting TW = 3000 ms, the number of echoes NECH was 7000, and the cumulative number Ns was 4 [22].

2.5. MRI Measurement

The ultrasound-pretreated C. *cibarius* samples were dried with hot air for 0, 90, and 180 min. The internal moisture distribution of the samples was examined via MRI at the above times. The MRI image analysis parameter settings were: time echo (TE): 20 ms, repetition time: 500 ms, and 256 × 256 matrix. Proton density-weighted imaging was performed using the SE sequences [22].

2.6. Scanning Electron Microscopy (SEM)

The microstructure of the pileus and stipe of the samples subjected to different ultrasound power pretreatments was observed via SEM (Tescan Mira4+ Energy Spectrum xplore 30 EDS, TESCAN Trading Co., Ltd., Kohoutovice, Czech Republic) according to Li et al., with small modifications [15]. A thin section was taken from each dried sample, sputter-coated with gold, and examined via SEM. Images of the pileus and stipes were analyzed at 5000× and 500× magnifications, respectively.

2.7. Color Determination

The color parameters of the samples were analyzed using a colorimeter [23] (NR110, Shenzhen 3NH Technology Co., Ltd., Shenzhen, China). The total chromatic aberration (ΔE) was calculated by Equation (4):

$$\Delta E = \sqrt{(L^* - L_0{}^*)^2 + (a^* - a_0{}^*)^2 + (b^* - b_0{}^*)^2} \quad (4)$$

where $L_0{}^*$, $a_0{}^*$, and $b_0{}^*$ are the color values of the fresh product and L^*, a^*, and b^* are the color values of the dry product.

2.8. Rehydration Ratio

The rehydration ratio of the dried *C. cibarius* was determined according to Zhao et al., with small modifications [24]. A quantity of dried *C. cibarius* pretreated with different ultrasound powers was selected and immersed in distilled water at a temperature maintained at 40 °C and then weighed every one hour until a constant weight was reached. The rehydration rate (RR) was obtained from Equation (5):

$$RR = \frac{m_f}{m_g} \quad (5)$$

where m_f is the weight of the rehydrated samples and m_g is the weight of the initial dried samples (g).

2.9. Determination of Bioactive Substances and Antioxidant Activities

The dried *C. cibarius* samples were crushed and passed through a 40-mesh sieve. Then, they were extracted via ethanol-assisted sonication.

2.9.1. Antioxidant Activity (DPPH)

The DPPH free radical-scavenging activity of the sample extracts was determined according to the method of Wang et al., with minor modifications [25]. The DPPH was then calculated using Equation (6):

$$\text{DPPH Radical Scavenging Activity } \% = \frac{1 - (A_i - A_j)}{A_0} \times 100 \quad (6)$$

where A_i, A_0, and A_j are the absorbance of the DPPH solution when adding the antioxidant, without the antioxidant and control, respectively.

2.9.2. Total Phenolics Content (TPC)

The phenolic content of the mushroom was estimated using the Folin–Ciocalteau assay described by Bao et al., with minor modifications [26]. The results were expressed as the mg of gallic acid equivalents (GAE) per 100 g of dry weight (DW).

2.9.3. Total Flavonoid Compounds

The total flavonoid content of the mushroom was measured using the aluminum chloride colorimetric assay described by Konieczynski et al., with slight modifications [27]. The results were expressed as the mg of rutin equivalent (RE) per 1 g dry weight (DW).

2.10. Statistical Analysis

All the analyses were repeated at least three times, and the data were expressed as the mean ± standard deviation. An analysis of variance (ANOVA, Tukey's test) was performed to determine whether the data showed significant differences within the same group. Statistic for Windows (Version 8.1, Analytical Software, St Paul, MN, USA) was used for the analysis. A significance level of 0.05 was used.

3. Results and Discussion

3.1. Drying Characteristics of C. cibarius

The drying rate and moisture ratio of *C. cibarius* pretreated with different ultrasound powers during the hot air drying process are depicted in Figure 1. As shown in Figure 1A, the moisture ratio decreased exponentially with increasing drying time. Compared with the control group, the total drying time decreased significantly with increasing ultrasound power ($p < 0.05$). When the ultrasound power reached 400 W, the moisture ratio decreased the fastest and the drying time dropped significantly from 271.67 min (control group) to only 220.33 min ($p < 0.05$). These results indicated that the ultrasound pretreatment had a positive effect on the drying of *C. cibarius*. A similar decreasing tendency of drying time with the application of ultrasonic pretreatment was reported in the convective drying of apple slices [28], heat pump drying of scallops [11] and pulsed fluidized bed microwave freeze drying of Chinese yam [29]. This could be due to the cavitation effect of the ultrasound treatment and the rapid shrinkage and expansion of the sample, which promote the formation of microchannels within the sample and facilitate the diffusion of water during the drying process [19]. However, there were no significant differences in the total drying time between the 400 W and 500 W ultrasound treatment groups. This could be because the higher ultrasound power further destroyed the internal structure of the *C. cibarius*. Ni et al. also reported that pretreatment with different ultrasound powers resulted in different effects on the drying process of goji berries [30].

As shown in Figure 1B, before drying for 30 min, the drying rate of all the groups showed a sharp upward trend, and then it showed a downward trend with the increase in drying time. The drying rate of *C. cibarius* pretreated with different ultrasound powers was higher than that of the control group in the first 90 min. Furthermore, the highest drying rate was investigated in the 400 W ultrasound pretreatment group. These results indicated that the ultrasound pretreatment resulted in rapid moisture removal in the initial phase of drying and increased the drying rate. Similarly, the acceleration of the drying rate was also due to the fact that the ultrasonic treatment changed the microstructure of the sample and promoted water loss during the drying process, thus increasing the drying rate [19]. After 120 min of drying, the drying rate of the control group was higher than that of the other groups, while the drying rate of the other groups gradually decreased. This is due to the fact that the ultrasonic treatment accelerated the removal of free water during the drying of the samples, while the remaining bound and fixed water was not sensitive to hot air drying. In contrast, the samples in the blank group had free water throughout the hot air drying process, resulting in a higher drying rate at a later stage than in the pretreatment group [31].

3.2. LF-NMR Analysis of C. cibarius

To investigate the migration rate and distribution of water molecules, and to understand the effect of ultrasound pretreatment on the water status of *C. cibarius* tissue during the drying process, the transverse relaxation time (T_2) of the samples dried for 0, 90, and 180 min was measured using LF-NMR (Figure 2). A total of three peaks were revealed in the T_2 curves of the *C. cibarius* samples: T_{21} (0.01–1 ms), T_{22} (1–100 ms), and T_{23} (100–10,000 ms). The water molecules of T_{21}, T_{22}, and T_{23} are defined as bound water, immobilized water, and free water, respectively [32]. The length of the relaxation time is closely related to the degree of freedom of water. The longer the relaxation time, the higher the degree of freedom of water and the easier its removal. Hence, the order of these three forms of moisture removed from the samples during the drying process was as follows: $T_{21} < T_{22} < T_{23}$ [33]. Compared to the untreated group, the water distribution in the ultrasonically treated samples changed. It is possible that ultrasound alters the internal structure of the chanterelles, leading to a redistribution of water within the cells. A similar situation was found in a study of water changes during the microwave vacuum drying of Lotus (*Nelumbo nucifera Gaertn.*) [34] and infrared drying of potato chips [35] after ultrasonic pretreatment.

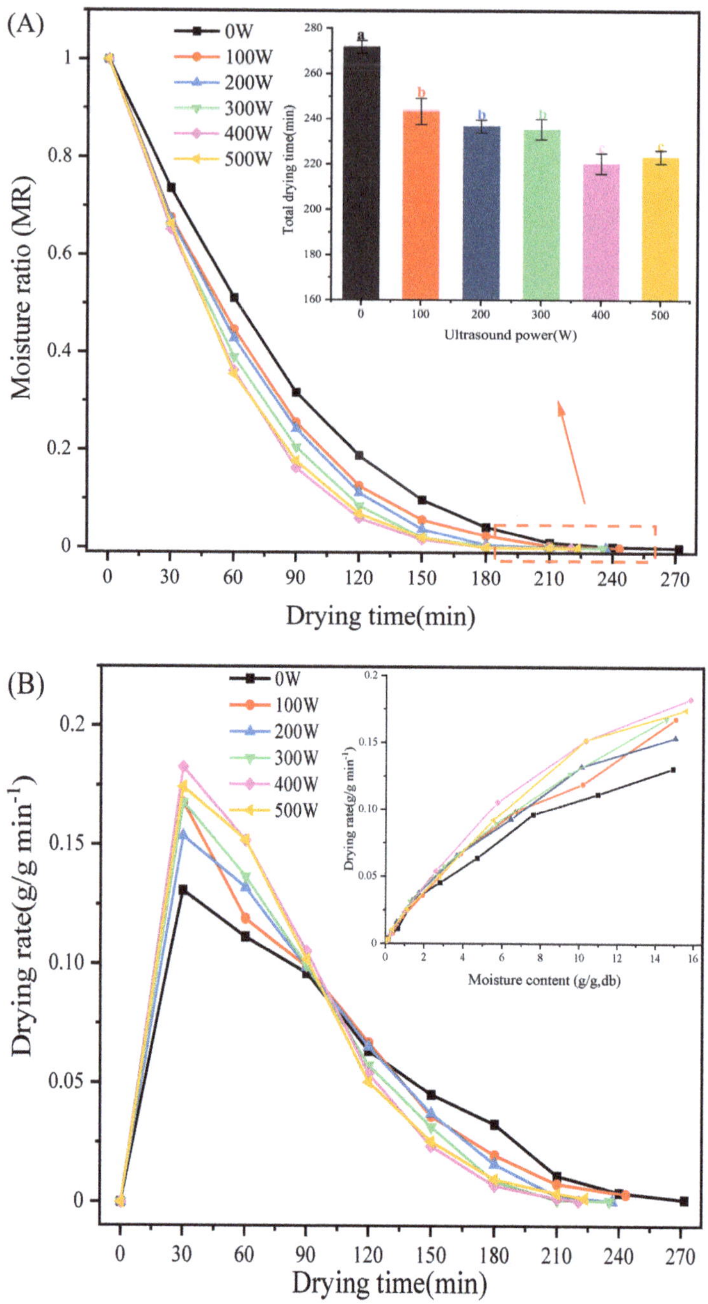

Figure 1. (**A**) Effects of the ultrasound pretreatments on the moisture ratio (MR)—time profiles and final moisture content of *C. cibarius* after hot air drying. (**B**) Drying rate (DR) versus drying time curve for mushrooms with different drying pretreatments under hot air drying (60 °C). Different letters indicate significant difference ($p < 0.05$); Same letter means no significant difference ($p > 0.05$).

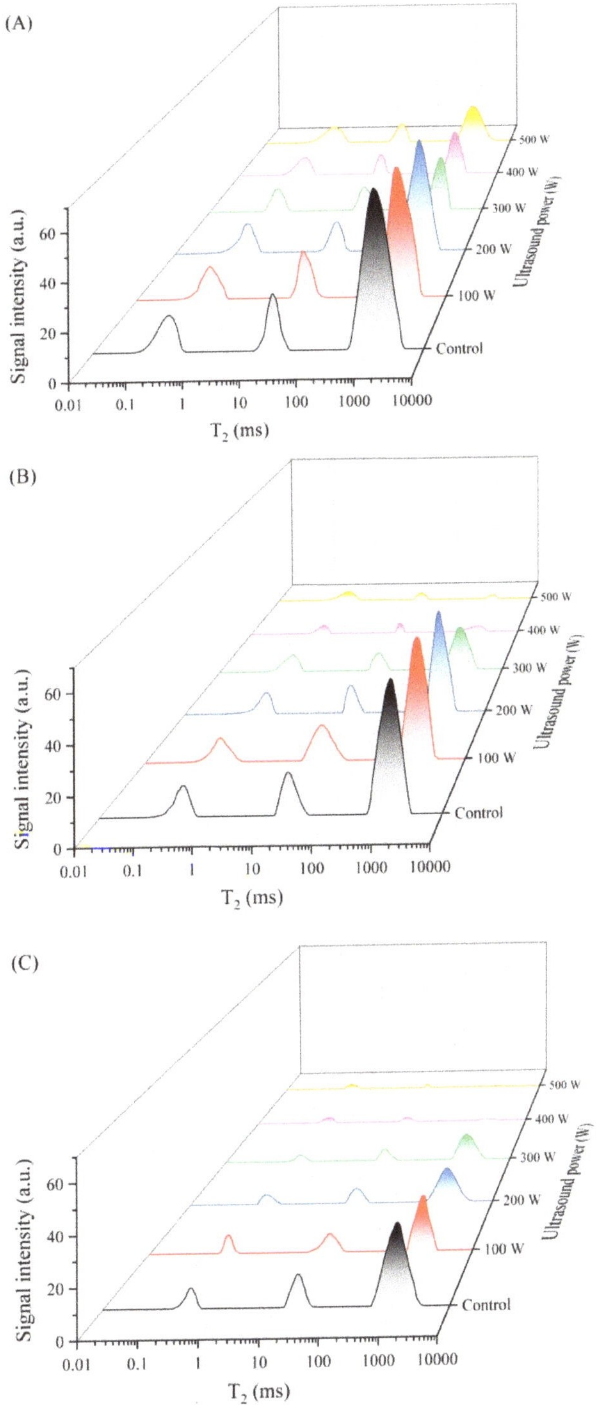

Figure 2. T_2 relaxation spectra of *C. cibarius* during drying with different ultrasonic pretreatments: (**A**) 0, (**B**) 90, and (**C**) 180 min.

As shown in Figure 2, with increasing drying time, the T_2 values of all the samples decreased significantly, which indicates that water was gradually lost. This was consistent with the moisture ratio trend of *C. cibarius* in the drying process. Compared with the control group, the ultrasound pretreatment significantly affected the water distribution in the samples. As the ultrasonic power increased, the signal strength of the sample T_2 relaxation curve decreased continuously, especially the signal strength of T_{23}. This suggests that the degree of freedom and the amount of free water decrease during drying, while higher drying temperatures also accelerate changes in water mobility and distribution [22]. These results show that the microchannels formed by the ultrasonic pretreatment reduce the resistance to water diffusion during drying, promote water flow, and accelerate water loss [36]. When the drying time reached 90 min, the signal intensity of T_{23} in the samples pretreated by ultrasound at 400 and 500 W decreased significantly. Moreover, the signal intensities of T_{21} and T_{22} were very weak. When the drying time reached 180 min, there was little signal intensity of T_{23} in the samples subjected to ultrasound pretreatment at 400 and 500 W. Only a few signal intensities of T_{21} and T_{22} were present. These results were consistent with the change in the drying rate, suggesting that the samples had dried almost to a constant weight and that there was a little water, mainly in the form of bound water [37]. Moreover, the smaller the proportion of free water in the entire drying process and in the final dry product, the better the drying effect. Zhang et al. have also found that the water loss was mainly from free water and immobile water during the drying process [18].

3.3. *MRI Analysis of Drying of C. cibarius*

Figure 3 shows MRI images of *C. cibarius* pretreated by ultrasound with different powers during the drying process. In the MRI images, the number of mobile hydrogen atoms is proportional to the color. The red color indicates high water content and high signal intensity, and the blue color indicates the opposite [38]. As shown in Figure 3, the water distribution in the sample was affected by the ultrasonic pretreatment. The moisture in the fresh samples before drying was unevenly distributed, and the water content in the stipe was higher than that in the pileus. With the prolongation of the drying time, the water content decreased from the central part toward the epidermis, which was mainly due to the dehydration of the surface of the samples and drying shrinkage [22]. Compared with the control group, the water content in the ultrasound-pretreated samples decreased significantly in the drying process. After drying for 90 min, it can be seen that the imaging signal inside the sample was weakened, which proves that the moisture content was reduced. Especially when the ultrasonic power reached 400 and 500 W, the obtained imaging signal strength was poor. When the drying time was 180 min, the proton signal of the control group was still high. However, there was almost no proton signal in the images of the samples pretreated with ultrasound (400 and 500 W). This phenomenon was caused by the fact that the ultrasound pretreatment increased the convective mass transfer and promoted the evaporation of water in the hot air drying process [39]. At the same time, in the process of hot air drying, the external water of the chanterelles was removed by the influence of hot air, while the internal water continued to migrate to the outside [32]. These results were similar to the T_2 relaxation curve of the samples during the drying process. Additionally, the decrease in the water content in the sample also led to a decrease in the drying rate in the later drying period (Section 3.1). The scanning electron microscopy results also confirmed that the microstructure of the control group hindered water diffusion (Figure 4).

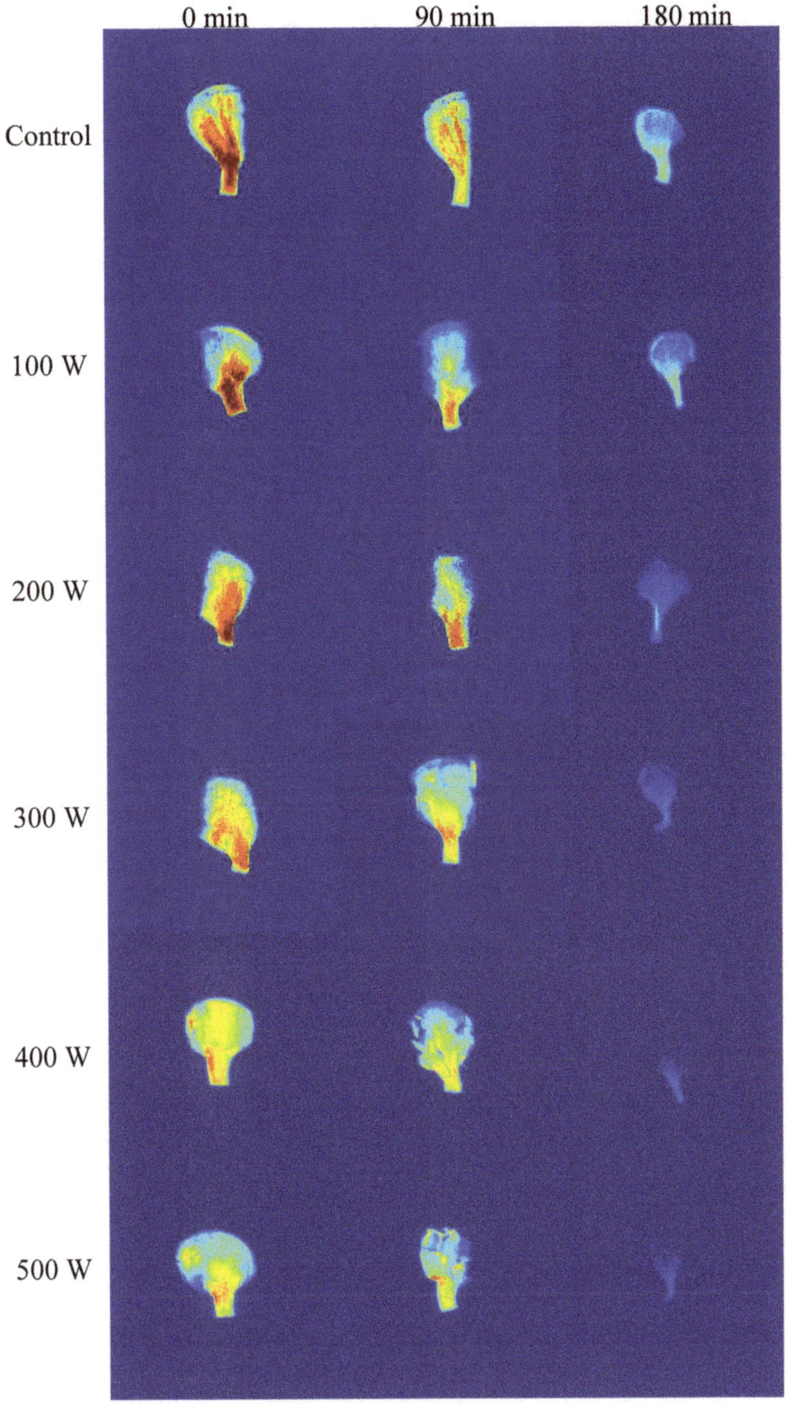

Figure 3. MRI of *C. cibarius* during drying under different ultrasonic pretreatments.

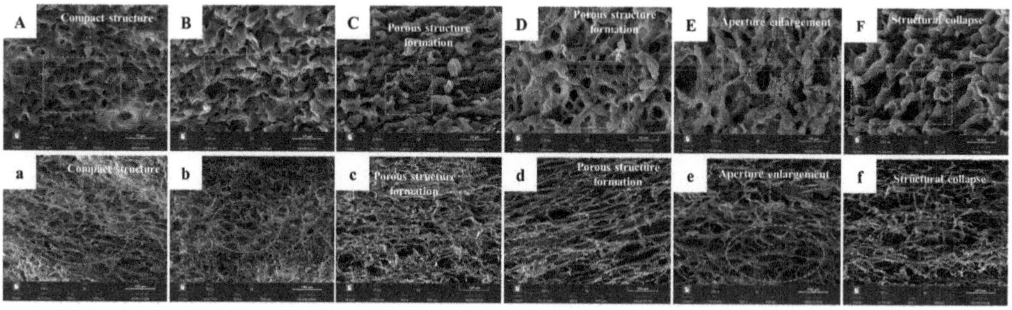

Figure 4. Microstructure of dried *C. cibarius* pileus under different ultrasound power pretreatment conditions: (**A**) control, (**B**) 100 W, (**C**) 200 W, (**D**) 300 W, (**E**) 400 W, and (**F**) 500 W. Microstructure of the longitudinal section of dried *C. cibarius* stipe under different ultrasonic power pretreatment conditions: (**a**) control, (**b**) 100 W, (**c**) 200 W, (**d**) 300 W, (**e**) 400 W, and (**f**) 500 W.

3.4. Effect of Different Ultrasonic Pretreatment Powers on Sample Microstructure

The effect of the different ultrasound pretreatment powers on the microstructure of hot air-dried *C. cibarius* is shown in Figure 4. Although a honeycomb ultra-structure and pores were observed in all the samples, significant differences in the microstructure were also observed between the untreated and treated samples. As seen in Figure 4A,a, the microstructure of the pileus and stipe of the control group was compact. After the ultrasound pretreatment with different powers, the intercellular space and pores expanded, and the intercellular space of the samples was more evident with the increase in ultrasound power until the ultrasound power reached 400 W. This was due to the formation of microchannels and the larger honeycomb structure caused by acoustic cavitation during the ultrasound pretreatment process [35]. Similarly, the formation of microchannels in Pleurotus eryngii was promoted by the high-frequency vibration and cavitation of ultrasonic waves [19]. In a study involving the ultrasonic treatment of potato chips, it was found that with the increase in the ultrasonic pretreatment power, more porous cell structures were formed inside the dry potato chips [35]. When the ultrasound power was 500 W, the internal pore structures of the pileus showed partial collapse of the cellular tissue. This might be because the frequent ultrasonic vibration with high power caused contractions and expansions, which caused the sample to repeatedly squeeze and release [24]. The study by Zhang et al. also found that high-power ultrasound destroyed the microstructure of the sample, which adversely affected the finished product [40]. At the same time, applying ultrasonic waves before drying may increase the structural damage caused by temperature rises during drying [41]. Therefore, the structure of the untreated samples could hinder the outward diffusion of moisture in the drying process. The large pore size and intercellular space in the ultrasound-pretreated samples increased the water fluidity in the tissue and the drying rate, decreasing the drying time. Similarly, Zhao et al. found that ultrasound application changed the microstructure of shiitake mushrooms and increased the water loss rate during the drying process [24]. Moreover, ultrasound pretreatment also reduced the drying time of kiwifruit, sweet potatoes, celery, and strawberries by increasing the porous structure and promoting the diffusion of internal water [42].

3.5. Effect of Different Ultrasonic Pretreatment Powers on the Color of Samples

Color is one of the critical parameters of dried products, which determines market value and consumer acceptance [28]. The L^*, a^*, b^*, and ΔE values of *C. cibarius*, including the pileus and stipe, are listed in Table 1. Compared to fresh samples, the L^* values of all the dried samples decreased significantly. The decreased L^* values of the dried samples could be attributed to the long Maillard reaction time during the drying process [43]. In the dried samples, the L^* values of the control group were lower than those of the ultrasound-pretreated group. This could be because the ultrasound pretreatment reduced the drying time, which avoided adverse effects on product quality due to prolonged drying. In a study on the color of potato chips after ultrasonic pretreatment, Wu et al. also found that the brightness of the potato chips after ultrasonic pretreatment increased and was higher than that of the fresh samples [35]. With increasing ultrasound power, the L^* value increased and then decreased. When the ultrasound power was 400 W, the L^* value was closest to that of the fresh samples. This indicates that although the hot air drying process caused enzymatic browning of the product, the 400 W treatment significantly reduced the drying time of the mushrooms and did not cause severe enzymatic browning [32]. However, the L^* value of the dried samples pretreated by ultrasound at 500 W decreased significantly. This may be because the 500 W ultrasonic treatment disrupted the cells, causing the release of the enzymes responsible for browning on the surface and making the samples look dark [44].

Table 1. Color measurement of fresh and dried *C. cibarius* pileus and stipe.

Ultrasound Power	Pileus				Stipe			
	L^*	a^*	b^*	ΔE	L^*	a^*	b^*	ΔE
Fresh	77.62 ± 0.76^{a}	6.23 ± 0.43^{e}	$73.61 \pm 0.42^{a\,b}$	——	86.58 ± 0.40^{a}	0.13 ± 0.03^{b}	10.51 ± 0.72^{d}	——
Control	66.83 ± 1.18^{d}	17.64 ± 0.91^{a}	66.18 ± 3.34^{d}	17.55 ± 2.03^{a}	71.40 ± 0.43^{d}	6.64 ± 0.55^{a}	30.75 ± 0.51^{c}	$26.14 \pm 0.31^{c,d}$
100 W	71.38 ± 0.35^{c}	$7.99 \pm 0.64d^{e}$	$71.84 \pm 0.65^{a,b,c}$	6.82 ± 0.59^{c}	71.44 ± 0.49^{d}	6.80 ± 0.11^{a}	31.85 ± 0.99^{c}	27.08 ± 0.61^{c}
200 W	71.47 ± 0.35^{c}	8.49 ± 0.93^{c}	$72.15 \pm 3.62^{a,b,c}$	8.59 ± 0.51^{c}	$72.65 \pm 0.45^{c,d}$	6.92 ± 1.74^{a}	38.93 ± 3.42^{a}	32.43 ± 2.30^{a}
300 W	71.82 ± 0.38^{c}	$9.52 \pm 0.90^{c,d}$	$68.83 \pm 0.18^{b,c,d}$	8.23 ± 1.12^{c}	73.53 ± 0.52^{c}	7.55 ± 1.15^{a}	$35.19 \pm 1.08^{a,b,c}$	$28.51 \pm 1.72^{b,c}$
400 W	75.48 ± 0.10^{b}	$6.99 \pm 0.58^{d,e}$	74.89 ± 0.87^{a}	2.65 ± 1.01^{d}	83.36 ± 0.54^{b}	1.91 ± 0.29^{b}	$33.06 \pm 2.09^{b,c}$	22.87 ± 1.77^{d}
500 W	68.16 ± 0.53^{d}	14.61 ± 1.53^{b}	$67.98 \pm 0.66^{c,d}$	13.95 ± 0.60^{b}	71.34 ± 1.00^{d}	7.95 ± 1.92^{a}	$37.58 \pm 1.57^{a,b}$	$32.09 \pm 0.62^{a\,b}$

Different letters in the same column indicated significant difference ($p < 0.05$); The same letter indicates no significant difference ($p > 0.05$).

Compared to the fresh samples, the a^* and b^* values increased, indicating that the redness and yellowness increased during the drying process due to sample pigmentation. After the ultrasound pretreatment, the a^* and b^* values were lower than those of the control group. This phenomenon could be attributed to the enzymatic browning of the samples with a long drying process. The a^* and b^* values decreased and then increased with increasing ultrasound power. When the ultrasound power was 400 W, the a^* and b^* values decreased to the lowest level. Moreover, there was no significant difference between the fresh pileus and ultrasound-pretreated pileus at 400 W. These results indicate that the ultrasound pretreatment at 400 W had the least influence on the color of the dried *C. cibarius*. Similarly, the study by Shi et al. found that the brightness of dried shiitake mushroom slices was increased by ultrasonic pretreatment, which maintained the brightness of the dried mushroom slices [45]. Çakmak et al. made a similar discovery in relation to the drying of mushrooms [17].

3.6. Effect of Different Ultrasonic Pretreatment Powers on the Rehydration Ratio of the Samples

In this study, the effect of ultrasound pretreatment with different powers on the rehydration ratio (RR) of dried *C. cibarius* was investigated (Table 2). The results showed that the RR of the ultrasound-pretreated group was significantly higher than that of the control group (5.73) ($p < 0.05$). Similarly, Zhu et al. also found that the rehydration rate of dried scallops pretreated with ultrasound was higher than that of dried scallops without pretreatment [11]. The RR first increased significantly and then decreased slightly with increasing ultrasound power. When the ultrasound power was 400 W, the RR reached a maximum (6.11) and the sample showed excellent water absorption ability. The poor rehydration ability of the dried samples in the control group could be due to the decrease in the size of the available pores in the tissue during the long drying process and the irreversible shrinkage of the tissue structure caused by drying [46]. The ultrasound pretreatment not only reduced the drying time (Section 3.1) and shrinkage but also promoted the formation of a porous microstructure that increased the moisture absorption capacity during the rehydration process [24]. In addition, the RR slightly decreased when the ultrasound power was 500 W. This could be because the high power pretreatment caused the mushroom to form a porous structure, and at the same time, part of the inner wall collapsed, reducing the rehydration capacity [44]. Jarahizadeh and Dinani used different ultrasonic powers to treat potato chips prior to convection drying and found that increasing the ultrasonic power reduced the rehydration capacity of potato chip dry products [47].

Table 2. Effect of different ultrasound power pretreatments on the moisture diffusion coefficient, rehydration ratio (g/g) and antioxidant activity (%) of *C. cibarius*.

Ultrasound Power	Rehydration Ratio (g/g)	Antioxidant Activity (%)
Control	5.73 ± 0.14^{c}	77.76 ± 0.53^{c}
100 W	$5.87 \pm 0.13^{b,c}$	78.61 ± 0.36^{c}
200 W	$5.99 \pm 0.13^{a,b}$	80.11 ± 0.24^{b}
300 W	$6.00 \pm 0.08^{a,b}$	$80.97 \pm 0.96^{a,b}$
400 W	6.11 ± 0.16^{a}	81.61 ± 0.86^{a}
500 W	$5.99 \pm 0.07^{a,b}$	$80.90 \pm 0.46^{a,b}$

Different letters in the same column indicated significant difference ($p < 0.05$); The same letter indicates no significant difference ($p > 0.05$).

3.7. Effect of Different Ultrasonic Pretreatment Powers on Nutritional Quality

The contents of total phenolics and flavonoids in dried *C. cibarius* showed a similar trend. As shown in Figure 5, the total phenolics and flavonoids contents of the samples pretreated with different ultrasound powers were higher than those of the control group. With increasing ultrasound power, the contents of phenol and flavonoids increased significantly ($p < 0.05$) and then decreased slightly. These results indicate that the ultrasound pretreatment effectively retained the total phenolics and flavonoids contents of the dried *C. cibarius*. The retention of the total phenolics content could be attributed to the inactivation of enzymes by the ultrasound pretreatment and the inhibition of the polyphenol oxidase-induced oxidation of phenolic compounds [45]. In addition, the long drying time in the control group caused prolonged exposure of the *C. cibarius* to high temperature and oxygen, resulting in strong oxidation and the loss of total phenolic substances [48]. Li et al. also reported that ultrasound pretreatment reduced the loss of phenolic compounds in dried mushroom slices [15]. However, the total phenolics and flavonoids contents of the dried samples decreased slightly under the 500 W ultrasound pretreatment. This might be because the high-power ultrasound pretreatment destroyed the sample cells, while the intracellular compounds diffused into the surrounding medium, resulting in the decreased extraction of phenolic and flavonoid compounds [35].

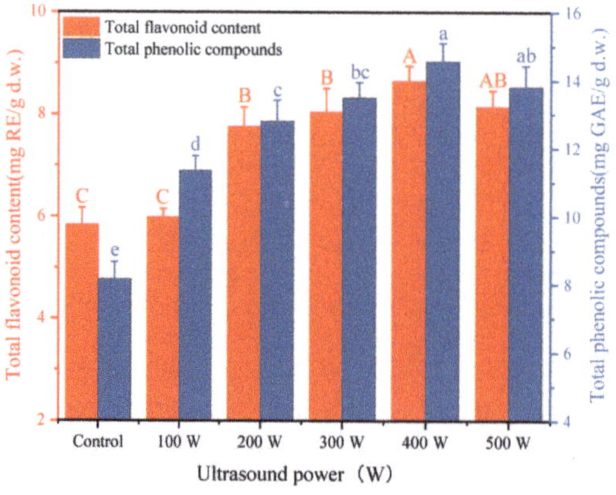

Figure 5. Total phenolics and flavonoids contents of dried *C. cibarius* pretreated with different ultrasound powers. Different letters indicate significant difference ($p < 0.05$). Capital letters indicate significance analysis for total flavonoid content and small letters indicate significance analysis for total phenolic compounds.

A similar trend in the antioxidant activity of dried *C. cibarius* pretreated by ultrasound with different powers was also observed (Table 2). Compared to the control group, the antioxidant activity of the ultrasound-pretreated samples increased significantly ($p < 0.05$). With increasing ultrasound power, the antioxidant activity increased at first and then decreased significantly ($p < 0.05$). When the ultrasound power was 400 W, the antioxidant activity reached a maximum (81.61). This could be because the ultrasound pretreatment could have shortened the drying time of the samples, protecting *C. cibarius* from elevated temperature and oxygen damage. In addition, the ultrasound pretreatment could have promoted the release of the non-phenolic antioxidant ergothioneine from the protein structure, resulting in an improvement in the antioxidant properties [49]. Similar results were also found for ultrasonically pretreated dried mushroom slices [21].

4. Conclusions

The effects of different ultrasonic pretreatment powers on the water migration of *C. cibarius* and its quality were evaluated. The experimental results of this study showed that ultrasonic pretreatment before hot air drying significantly decreased the total drying time of *C. cibarius* compared to the untreated samples. When the ultrasonic pretreatment power was 400 W, the total drying time was reduced by 18.90%. The ultrasound pretreatment played a positive role in the drying of *C. cibarius*. In particular, the pretreatment at 400 W provided a good condition for improving the retention rate of the DPPH, total phenols and total flavonoids of the dried samples and accelerating the drying rate. The ultrasonic pretreatment formed a porous structure in the mushroom tissue, which accelerated the drying rate, and the dried product had good color and rehydration ability. This study revealed that 400 W ultrasonic pretreatment could be recommended as a drying condition to obtain high-quality dried *C. cibarius*. However, improper temperature control in the ultrasonic pretreatment process will have adverse effects, so it is necessary to further optimize the pretreatment temperature control program to obtain better pretreatment products.

Author Contributions: Data curation, writing—original draft, investigation, M.S.; software and resources, Y.X.; investigation and formal analysis, Y.G. and Y.Z.; conceptualization, writing—reviewing and editing, X.F. and Y.D. All authors have read and agreed to the published version of the manuscript.

Funding: This research was funded by the Yunnan Major Scientific and Technological Projects (Grant No. 202202AG050009), Yunnan Fundamental Research Projects (Grant Nos. 202301AU070032, 202201BE070001-055, 202201AU070106 and 202101BE070001-052), Science and Technology Projects in Yunnan Province (Grant No. 202202AE090007), National Natural Science Foundation of China (Grant No. 32102073), and Yunnan Science and Technology Planning Project (Grant No. 202102AE090021).

Data Availability Statement: The data used to support the findings of this study can be made available by the corresponding author upon request.

Conflicts of Interest: The authors declare that they have no known competing financial interests or personal relationships that could have appeared to influence the work reported in this paper.

References

1. Ozturk, B.; Havsut, E.; Yildiz, K. Delaying the postharvest quality modifications of Cantharellus cibarius mushroom by applying citric acid and modified atmosphere packaging. *LWT-Food Sci. Technol.* **2021**, *138*, 110639. [CrossRef]
2. Sumic, Z.; Tepic, A.; Vidovic, S.; Vakula, A.; Vladic, J.; Pavlic, B. Process Optimization of Chanterelle (*Cantharellus Cibarius*) Mushrooms Vacuum Drying. *J. Food Process. Preserv.* **2017**, *41*, e12822. [CrossRef]
3. Oliveira, F.; Sousa-Gallagher, M.J.; Mahajan, P.V.; Teixeira, J.A. Development of shelf-life kinetic model for modified atmosphere packaging of fresh sliced mushrooms. *J. Food Eng.* **2012**, *111*, 466–473. [CrossRef]
4. Politowicz, J.; Lech, K.; Sanchez-Rodriguez, L.; Szumny, A.; Carbonell-Barrachina, A.A. Volatile composition and sensory profile of *Cantharellus cibarius* Fr. as affected by drying method. *J. Sci. Food Agric.* **2017**, *97*, 5223–5232. [CrossRef] [PubMed]
5. Zhang, L.H.; Zhang, M.; Mujumdar, A.S. Development of flavor during drying and applications of edible mushrooms: A review. *Dry. Technol.* **2021**, *39*, 1685–1703. [CrossRef]
6. Tian, Y.T.; Zhao, Y.T.; Huang, J.J.; Zeng, H.L.; Zheng, B.D. Effects of different drying methods on the product quality and volatile compounds of whole shiitake mushrooms. *Food Chem.* **2016**, *197*, 714–722. [CrossRef]

7. Niu, D.; Wang, L.H.; Zeng, X.A.; Wen, Q.H.; Brennan, C.S.; Tang, Z.S.; Wang, M.S. Effect of ethanol adaption on the inactivation of *Acetobacter* sp. by pulsed electric fields. *Innov. Food Sci. Emerg.* **2019**, *52*, 25–33. [CrossRef]
8. Deng, L.Z.; Mujumdar, A.S.; Zhang, Q.; Yang, X.H.; Wang, J.; Zheng, Z.A.; Gao, Z.J.; Xiao, H.W. Chemical and physical pretreatments of fruits and vegetables: Effects on drying characteristics and quality attributes—A comprehensive review. *Crit. Rev. Food Sci.* **2019**, *59*, 1408–1432. [CrossRef]
9. Pei, F.; Xiao, K.P.; Chen, L.F.; Yang, W.J.; Zhao, L.Y.; Fang, Y.; Ma, N.; Mariga, A.M.; Hu, Q.H. Mass transfer characteristics during ultrasound-assisted osmotic dehydration of button mushroom (*Agaricus bisporus*). *J. Food Sci. Technol.* **2019**, *56*, 2213–2223. [CrossRef]
10. Singla, M.; Sit, N. Application of ultrasound in combination with other technologies in food processing: A review. *Ultrason. Sonochem.* **2021**, *73*, 105506. [CrossRef]
11. Zhu, Z.Z.; Zhao, Y.; Zhang, Y.X.; Wu, X.T.; Liu, J.; Shi, Q.L.; Fang, Z.X. Effects of ultrasound pretreatment on the drying kinetics, water status and distribution in scallop adductors during heat pump drying. *J. Sci. Food Agric.* **2021**, *101*, 6239–6247. [CrossRef] [PubMed]
12. Rashid, M.T.; Jatoi, M.A.; Safdar, B.; Wali, A.; Aadil, R.M.; Sarpong, F.; Ma, H.L. Modeling the drying of ultrasound and glucose pretreated sweet potatoes: The impact on phytochemical and functional groups. *Ultrason. Sonochem.* **2020**, *68*, 105226. [CrossRef] [PubMed]
13. Khan, M.K.I.; Ghauri, Y.M.; Alvi, T.; Amin, U.; Khan, M.I.; Nazir, A.; Saeed, F.; Aadil, R.M.; Nadeem, M.T.; Babu, I.; et al. Microwave assisted drying and extraction technique; kinetic modelling, energy consumption and influence on antioxidant compounds of fenugreek leaves. *Food Sci. Technol.* **2022**, *42*, 1–9. [CrossRef]
14. Prithani, R.; Dash, K.K. Mass transfer modelling in ultrasound assisted osmotic dehydration of kiwi fruit. *Innov. Food Sci. Emerg.* **2020**, *64*, 102407. [CrossRef]
15. Li, X.; Li, J.; Wang, R.; Rahaman, A.; Zeng, X.A.; Brennan, C.S. Combined effects of pulsed electric field and ultrasound pretreatments on mass transfer and quality of mushrooms. *LWT-Food Sci. Technol.* **2021**, *150*, 112008. [CrossRef]
16. Wali, A.; Ma, H.L.; Aadil, R.M.; Zhou, C.S.; Rashid, M.T.; Liu, X. Effects of multifrequency ultrasound pretreatment on the enzymolysis, ACE inhibitory activity, and the structure characterization of rapeseed protein. *J. Food Process. Preserv.* **2017**, *41*, e13413. [CrossRef]
17. Cakmak, R.S.; Tekeoglu, O.; Bozkir, H.; Ergun, A.R.; Baysal, T. Effects of electrical and sonication pretreatments on the drying rate and quality of mushrooms. *LWT-Food Sci. Technol.* **2016**, *69*, 197–202. [CrossRef]
18. Zhang, B.; Li, M.; Qiao, Y.T.; Gao, P.; Li, L.Y.; Zheng, Z.J. Potential use of low-field nuclear magnetic resonance to determine the drying characteristics and quality of *Arctium lappa* L. in hot-blast air. *LWT-Food Sci. Technol.* **2020**, *132*, 109829. [CrossRef]
19. Su, D.B.; Sun, W.H.; Li, B.Z.; Yang, Y.; Wang, Y.; Lv, W.Q.; Li, D.; Wang, L.J. Influence of ultrasonic pretreatments on microwave hot-air flow rolling drying mechanism, thermal characteristics and rehydration dynamics of *Pleurotus eryngii*. *J. Sci. Food Agric.* **2022**, *102*, 2100–2109. [CrossRef]
20. Li, B.R.; Lin, J.Y.; Zheng, Z.A.; Duan, H.; Li, D.; Wu, M. Effects of different drying methods on drying kinetics and physicochemical properties of *Chrysanthemum morifolium* Ramat. *Int. J. Agric. Biol. Eng.* **2019**, *12*, 187–193. [CrossRef]
21. Mirzaei-Baktash, H.; Hamdami, N.; Torabi, P.; Fallah-Joshaqani, S.; Dalvi-Isfahan, M. Impact of different pretreatments on drying kinetics and quality of button mushroom slices dried by hot-air or electrohydrodynamic drying. *LWT-Food Sci. Technol.* **2022**, *155*, 112894. [CrossRef]
22. Cheng, S.S.; Li, R.R.; Yang, H.M.; Wang, S.Q.; Tan, M.Q. Water status and distribution in shiitake mushroom and the effects of drying on water dynamics assessed by LF-NMR and MRI. *Dry. Technol.* **2020**, *38*, 1001–1010. [CrossRef]
23. Yamchi, A.A.; Yeganeh, R.; Kouchakzadeh, A. Effect of ultrasonic pretreatment on drying kinetics and physio-mechanical characteristics of peach slices. *J. Food Process. Eng.* **2022**, *45*, 14053. [CrossRef]
24. Zhao, Y.Y.; Yi, J.Y.; Bi, J.F.; Chen, Q.Q.; Zhou, M.; Zhang, B. Improving of texture and rehydration properties by ultrasound pretreatment for infrared-dried shiitake mushroom slices. *Dry. Technol.* **2019**, *37*, 352–362. [CrossRef]
25. Wang, K.J.; Guo, J.T.; Cheng, J.X.; Zhao, X.H.; Ma, B.H.; Yang, X.B.; Shao, H.J. Ultrasound-assisted extraction of polysaccharide from spent Lentinus edodes substrate: Process optimization, precipitation, structural characterization and antioxidant activity. *Int. J. Biol. Macromol.* **2021**, *191*, 1038–1045. [CrossRef]
26. Bao, T.; Li, Y.T.; Xie, J.H.; Sun, C.D.; Li, X.; Pu, Y.F.; Chen, W. Systematic evaluation of bioactive components and antioxidant capacity of some new and common bayberry cultivars using an in vitro gastrointestinal digestion method. *Food Res. Int.* **2018**, *103*, 326–334. [CrossRef]
27. Konieczynski, P.; Lysiuk, R.; Wesolowski, M. Total flavonoid and ionic elements contents in 32 medicinal plants collected from natural habitats in Northern Ukraine. *J. Herb. Med.* **2021**, *29*, 100492. [CrossRef]
28. Fijalkowska, A.; Nowacka, M.; Wiktor, A.; Sledz, M.; Witrowa-Rajchert, D. Ultrasound as a Pretreatment Method to Improve Drying Kinetics and Sensory Properties of Dried Apple. *J. Food Process. Eng.* **2016**, *39*, 256–265. [CrossRef]
29. Li, L.L.; Zhang, M.; Wang, W.Q. Ultrasound-assisted osmotic dehydration pretreatment before pulsed fluidized bed microwave freeze-drying (PFBMFD) of Chinese yam. *Food Biosci.* **2020**, *35*, 100548. [CrossRef]
30. Ni, J.B.; Ding, C.J.; Zhang, Y.M.; Song, Z.Q.; Xu, W.Q. Influence of ultrasonic pretreatment on electrohydrodynamic drying process of goji berry. *J. Food Process. Preserv.* **2020**, *44*, e14600. [CrossRef]

31. Wang, J.; Xiao, H.W.; Ye, J.H.; Wang, J.; Raghavan, V. Ultrasound Pretreatment to Enhance Drying Kinetics of Kiwifruit (*Actinidia deliciosa*) Slices: Pros and Cons. *Food Bioprocess. Technol.* **2019**, *12*, 865–876. [CrossRef]
32. Zeng, S.Y.; Wang, B.; Lv, W.Q.; Wang, L.J.; Liao, X.J. Dynamic analysis of moisture, dielectric property and microstructure of ginger slices during microwave hot-air flow rolling drying. *Food Control* **2022**, *134*, 108717. [CrossRef]
33. Ozel, B.; Dag, D.; Kilercioglu, M.; Sumnu, S.G.; Oztop, M.H. NMR relaxometry as a tool to understand the effect of microwave heating on starch-water interactions and gelatinization behavior. *LWT-Food Sci. Technol.* **2017**, *83*, 10–17. [CrossRef]
34. Zhao, Y.T.; Wang, W.W.; Zheng, B.D.; Miao, S.; Tian, Y.T. Mathematical modeling and influence of ultrasonic pretreatment on microwave vacuum drying kinetics of lotus (*Nelumbo nucifera* Gaertn.) seeds. *Dry. Technol.* **2017**, *35*, 553–563. [CrossRef]
35. Wu, X.F.; Zhang, M.; Ye, Y.F.; Yu, D.X. Influence of ultrasonic pretreatments on drying kinetics and quality attributes of sweet potato slices in infrared freeze drying (IRFD). *LWT-Food Sci. Technol.* **2020**, *131*, 109801. [CrossRef]
36. Wu, X.F.; Zhang, M.; Mujumdar, A.S.; Yang, C.H. Effect of ultrasound-assisted osmotic dehydration pretreatment on the infrared drying of Pakchoi Stems. *Dry. Technol.* **2020**, *38*, 2015–2026. [CrossRef]
37. Zhu, D.S.; Liang, J.Y.; Liu, H.; Cao, X.H.; Ge, Y.H.; Li, J.R. Sweet cherry softening accompanied with moisture migration and loss during low-temperature storage. *J. Sci. Food Agric.* **2018**, *98*, 3651–3658. [CrossRef]
38. Tan, M.Q.; Lin, Z.Y.; Zu, Y.X.; Zhu, B.W.; Cheng, S.S. Effect of multiple freeze-thaw cycles on the quality of instant sea cucumber: Emphatically on water status of by LF-NMR and MRI. *Food Res. Int.* **2018**, *109*, 65–71. [CrossRef]
39. Jiang, N.; Liu, C.Q.; Li, D.J.; Zhang, J.; Zhang, Z.Y.; Huang, J.P.; Yu, Z.F. Effect of Thermosonic Pretreatment and Microwave Vacuum Drying on the Water State and Glass Transition Temperature in *Agaricus bisporus* Slices. *Food Bioprocess Technol.* **2018**, *11*, 172–184. [CrossRef]
40. Zhang, T.; Wang, J.; Feng, J.Q.; Liu, Y.Q.; Suo, R.; Jin, J.Y.; Wang, W.X. Ultrasonic pretreatment improves the gelation properties of low-salt Penaeus vannamei (*Litopenaeus vannamei*) surimi. *Ultrason. Sonochem* **2022**, *86*, 106031. [CrossRef]
41. Nowacka, M.; Wiktor, A.; Sledz, M.; Jurek, N.; Witrowa-Rajchert, D. Drying of ultrasound pretreated apple and its selected physical properties. *J. Food Eng.* **2012**, *113*, 427–433. [CrossRef]
42. Pandiselvam, R.; Aydar, A.Y.; Kutlu, N.; Aslam, R.; Sahni, P.; Mitharwal, S.; Gavahian, M.; Kumar, M.; Raposo, A.; Yoo, S.; et al. Individual and interactive effect of ultrasound pre-treatment on drying kinetics and biochemical qualities of food: A critical review. *Ultrason. Sonochem.* **2023**, *92*, 1–20. [CrossRef] [PubMed]
43. Rahaman, A.; Siddeeg, A.; Manzoor, M.F.; Zeng, X.A.; Ali, S.; Baloch, Z.; Li, J.; Wen, Q.H. Impact of pulsed electric field treatment on drying kinetics, mass transfer, colour parameters and microstructure of plum. *J. Food Sci. Technol.* **2019**, *56*, 2670–2678. [CrossRef] [PubMed]
44. Mothibe, K.J.; Zhang, M.; Mujumdar, A.S.; Wang, Y.C.; Cheng, X.F. Effects of Ultrasound and Microwave Pretreatments of Apple Before Spouted Bed Drying on Rate of Dehydration and Physical Properties. *Dry. Technol.* **2014**, *32*, 1848–1856. [CrossRef]
45. Shi, D.F.; Yin, C.M.; Feng, X.; Zhou, R.R.; Fan, X.Z.; Qiao, Y.; Yao, F.; Huang, W.; Liu, Y.; Gao, H.; et al. Effect of Ultrasound and Cellulase Pre-treatment on the Water Distribution, Physical Properties, and Nutritional Components of *Lentinula edodes* Chips. *Food Bioprocess Technol.* **2020**, *13*, 625–636. [CrossRef]
46. Gamboa-Santos, J.; Montilla, A.; Soria, A.C.; Carcel, J.A.; Garcia-Perez, J.V.; Villamiel, M. Impact of power ultrasound on chemical and physicochemical quality indicators of strawberries dried by convection. *Food Chem.* **2014**, *161*, 40–46. [CrossRef] [PubMed]
47. Jarahizadeh, H.; Dinani, S.T. Influence of applied time and power of ultrasonic pretreatment on convective drying of potato slices. *Food Sci. Biotechnol.* **2019**, *28*, 365–376. [CrossRef]
48. Nowacka, M.; Wiktor, A.; Anuszewska, A.; Dadan, M.; Rybak, K.; Witrowa-Rajchert, D. The application of unconventional technologies as pulsed electric field, ultrasound and microwave-vacuum drying in the production of dried cranberry snacks. *Ultrason. Sonochem.* **2019**, *56*, 1–13. [CrossRef]
49. Kalaras, M.D.; Richie, J.P.; Calcagnotto, A.; Beelman, R.B. Mushrooms: A rich source of the antioxidants ergothioneine and glutathione. *Food Chem.* **2017**, *233*, 429–433. [CrossRef]

MDPI AG
Grosspeteranlage 5
4052 Basel
Switzerland
Tel.: +41 61 683 77 34

MDPI Books Editorial Office
E-mail: books@mdpi.com
www.mdpi.com/books

www.ingramcontent.com/pod-product-compliance
Lightning Source LLC
Chambersburg PA
CBHW061544220926
51850CB00005B/244
* 9 7 8 3 7 2 5 8 2 0 4 7 4 *